META
MAGICAL
THEMAS

DOUGLAS R.
HOFSTADTER

METAMAGICAL THEMAS:

Questing for the Essence of Mind and Pattern

DOUGLAS R. HOFSTADTER

BasicBooks
A Subsidiary of Perseus Books, L.L.C.

Library of Congress Cataloging-in-Publication Data

Hofstadter, Douglas R., 1945–
    Metamagical themas.

    Bibliography: p. 802
    Includes Index
    1. Artificial intelligence. 2. Intellect. 3. Science—Philosophy.
    4. Metamathematics. 5. Self (Philosophy) 6. Amusements.
    I. Title
    Q335.H63 1985   001.53'5   83-46095
    ISBN 0–465–04540–5 (cloth)
    ISBN 0–465–04566–9 (paper)

    98 99 RRD 9 8 7 6 5 4 3

To Bloomington,

for all the times we shared.

Short Contents

*Not published as a "Metamagical Themas" column in *Scientific American*.

Section V: *Spirit and Substrate*

Section VI: *Selection and Stability*

Section VII: *Sanity and Survival*

Long Contents

Section I: *Snags and Snarls*

Chapter 1: *On Self-Referential Sentences.* The strangeness of language folding back on itself is explored here in dozens of different ways, many of them quite amusing.

Chapter 2: *Self-Referential Sentences: A Follow-Up.* A large collection of new material carries the idea of linguistic folding-back considerably further, and goes more deeply into the mechanisms of linguistic self-reference and self-replication.

Chapter 3: *On Viral Sentences and Self-Replicating Structures.* In which the concept of "memes", or self-replicating ideas, is discussed, as well as the idea of indirect self-reference.

Chapter 4: *Nomic: A Self-Modifying Game Based on Reflexivity in Law.* A remarkable game is described, which resembles a government in that a large part of its activity is devoted to changing its laws lawfully.

Section II: *Sense and Society*

Chapter 5: *World Views in Collision: The* **Skeptical Inquirer** *versus the* **National Enquirer.** An inquiry into why so many people are taken in by publications that give much play to "paranormal" or "psi" phenomena, and a report on an unusual journal that combats the psi panderers.

Chapter 6: *On Number Numbness.* A lamentation of the general low level of people's understanding of the vast numbers that describe our society's population, consumption, budgets, weaponry, and so on, including some suggestions for helping increase "numeracy".

Chapter 7: *Changes in Default Words and Images, Engendered by Rising Consciousness.* On the deep, hidden, and oft-denied connections between subconscious imagery and discriminatory usage in everyday language.

*Chapter 8: *A Person Paper on Purity in Language.* Master William Satire vents whis anger at those who, for cheap political reasons, would destroy the beauty of English by introducing ugly neologisms and changing the usage of venerated old terms.

Section III: *Sparking and Slipping*

Chapter 9: *Pattern, Poetry, and Power in the Music of Frédéric Chopin.* How did this great composer manage to encode extremely powerful and extremely delicate feelings into mere patterns of notes?

**Not published as a "Metamagical Themas" column in *Scientific American.*

Section IV: *Structure and Strangeness*

Section V: *Spirit and Substrate*

Long Contents

List of Illustrations

Note: All the gridfonts and many of the more geometric and regular figures—especially in Chapters 14 and 24—were produced by the author on an Apple Macintosh using MacPaint. Unless otherwise indicated in their captions, all other figures were done by the author using conventional implements, such as felt-tip pens.

Cover.

See "Notes on the Cover."

Front Matter.

Half title page. Ambigram on the book's title and the author's name.

Section I.

Section title page. A Whirly alphabet (see "Notes on the Cover").
Introduction page. The gridfont called "Victory".

Section II.

Section title page. A Whirly alphabet (see "Notes on the Cover").
Introduction page. The gridfont called "House".

Section III.

Section title page. A Whirly alphabet (see "Notes on the Cover").
Introduction page. The gridfont called "Double Backslash".

Section IV.

List of Illustrations

Section V.

Section title page. A Whirly alphabet (see "Notes on the Cover").
Introduction page. The gridfont called "Buxtehude".

Section VI.

Section title page. A Whirly alphabet (see "Notes on the Cover").
Introduction page. The gridfont called "Benzene Right".

Section VII.

Section title page. A Whirly alphabet (see "Notes on the Cover").
Introduction page. The gridfont called "Square Curl".

Notes on the Cover

A Spontaneous Essay on Whirly Art and Creativity

The drawing on the cover is a somewhat atypical example of a non-representational form of art I devised and developed over a period of years quite a long time ago, and which my sister Laura once rather light-heartedly dubbed "Whirly Art". The name stuck, for better or for worse. Generally speaking, I did Whirly Art on long thin strips of paper (available in rolls for adding machines) rather than on sheets of standard format. A typical piece of Whirly Art is five or six inches high and five or six feet long. Many are ten feet long, however, and some are as much as fifteen or even twenty feet in length. The one-dimensionality of Whirly Art was deliberate, of course: I was inspired by music and drew many visual fugues and canons. The time dimension was replaced by the long space dimension. I used the narrow width of the paper to represent something like pitch (although there was no strict mapping in any sense). A "voice" would be a single line tracing out some complex shape as it progressed in "time" along the paper. Several such voices could interact, and notions of what made "good" or "bad" visual harmony or counterpoint soon became intuitive to me.

The curvilinear motions constituting a single voice came from a blend of alphabets. At that time (the mid-60's), I was absolutely fascinated by the many writing systems found in and around India, exemplified by Tamil, Sinhalese, Kanarese, Telugu, Bengali, Hindi, Burmese, Thai, and many others. I studied some of them quite carefully, and even invented one of my own, based on the principles that most Indian scripts follow. It was natural that the motions my hand and mind were getting accustomed to would find their way into my visual fuguing. Thus was born Whirly Art.

Over the next several years, I did literally thousands of pieces of Whirly Art. Each one was totally improvised—in pen—so that there was no going back. A mistake was a mistake! Alternatively, a mistake could be interpreted as a very daring move from which it would be difficult, but not impossible, to recover gracefully. In other words, what seemed at first to be a disastrous mistake could turn into a joyful challenge! (I am sure that jazz improvisers will know exactly what I am talking about.) Sometimes, of course, I would fail, but other times I would succeed (at least by my own standards, since I was both performer and "listener").

Whirly Art became a (very) highly idiosyncratic language, with its own esthetic and traditions. However, traditions are made to be broken, and as soon as I spotted a tradition, I began experimenting around, violating it in various ways to see how I might move beyond my current state—how I might "jump out of the system". Style succeeded style, and I found myself paralleling the development of music. I moved from baroque Whirly Art (fugues, canons, and so forth) to "classical" Whirly Art, thence to "romantic" Whirly Art. After several years (it was now the late 60's), I reached the twentieth century, and found myself spiritually imitating such favorite composers of mine as Prokofiev and Poulenc. I did not copy any pieces specifically, but simply felt a kinship to those composers' style. Whirly Art is not translated music, but metaphorical music.

It is natural to wonder if I managed to jump beyond the twentieth century and make visual 21st-century music. That would have been quite a feat! Actually, in the early 70's I found that I simply was slowing down in production of Whirly Art. It had taken me seven years to recapitulate the history of Western music! At that point, I seemed to run out of creative juices. Of course, I could still make new Whirly Art then, as I can now—but I simply was less often inclined to do so. And today, I hardly ever do any Whirly Art, although the way that I draw curvy lines and letterforms bears the indelible marks of Whirly Art.

The piece on the cover, then, is atypical because it was done on an ordinary sheet of paper and has no direction of temporal flow. Also, there really is no concept of counterpoint in it. Still, it has something of a Whirly Art spirit. There are also seven Whirly alphabets in the book, one on each of the title pages of the seven sections. They are all somewhat atypical as well, but for slightly different reasons. Each was done on an ordinary sheet of paper but there is still always a clear flow, namely from 'A' to 'Z'. The real atypicality is the fact that genuine letters from a genuine alphabet are being used. I usually eschewed real letters, preferring to use shapes *inspired* by letters—shapes more complex and, well, "whirly" than most letters, even more so than Tamil or Sinhalese letters, which are pretty darn whirly.

Whirly Art is, I feel, quite possibly the most creative thing I have ever done. That, of course, is my opinion. Other people may disagree. It is a fairly strange and idiosyncratic form of art, however, and cannot be instantly understood. It has its own logic, related to the logics of musical harmony and counterpoint, Indian alphabets, gestalt perception, and who knows what else. I've kept it all quite literally in my closet for years—rolled up and piled into many paper bags and cardboard boxes. Because of its physical awkwardness, it is hard to show to people. But Whirly Art itself, and the experience of doing it, is an absolutely central fact about my way of looking at art, music, and creativity. Practically every time I write about creativity, some part of my mind is re-enacting Whirly Art experiences. In other words, a lot of my convictions about creativity come from self-observation rather than from scholarly study of the manuscripts or sketches of various composers or painters or writers or scientists. Of course, I have done some of that type of scholarship too, because I am fascinated by creativity in general—but I feel that to some extent "you don't really understand it unless you've done it", and so I rely a great deal on that personal experience. I feel that way that "I know what I'm talking about."

However, I would make a slightly stronger statement: Any two creative things that I've done seem to be, at some deep level, isomorphic. It's as if Whirly Art and mathematical discoveries and strange dialogues and little pieces of piano music and so on are all coming from a very similar core, and the same mechanisms are being exploited over and over again, only dressed up differently. Of course it's not all of the same quality: my *real* music is not as good as my visual music, for instance. But because I have this conviction that the core creativity behind all these things is really the same (at least in my own case), I am trying like mad to get at, and to lay bare, that core. For that reason I pursue ever-simpler domains in which I can feel myself doing "the same thing". In Chapter 24 of this book—in some sense the most creative Chapter, not surprisingly—I write about three of those domains: the Seek-Whence domain, the Copycat domain, and the Letter Spirit domain.

It is the Letter Spirit domain—"gridfonts" in particular—that is currently my most intense obsession. That domain came out of a lifelong fascination with our alphabet and other writing systems. I simply boiled away what I considered to be less interesting aspects of letterforms—I boiled and boiled—until I was left with what might be called the "conceptual skeletons" of letterforms. That is what gridfonts are about. People who have not shared my alphabetic fascination often underestimate at first the potential range of gridfonts, thinking that there might be a few and that's all. That is dead wrong: There are a huge number of them, and their variety is astounding.

As I look at the gridfonts I produce—and as I *feel* myself producing a gridfont—I feel that what I am doing is just Whirly Art all over again, in a new and ridiculously constrained way. The same mechanisms of shape transformation, the same quest for grace and harmony, the same intuitions about what works and what doesn't, the same desire to "jump out of the system"—all this is truly the same. Doing gridfonts is therefore very exciting to me and provides a new proving ground for my speculations. The one advantage that gridfonts have over Whirly Art is that they are so preposterously constrained. This means that the possibilities for choice can be watched much more easily. It does not mean that a choice can be *explained* easily, but at least it can be watched. In a way, gridfonts are allowing me to re-experience the Whirly-Art period of my life, but with the advantage of several years' thinking about artificial intelligence and how I would like to try to make it come about. In other words, I can now hope that perhaps I can get a handle—a bit of one, anyway—on what is going on in creativity by means of computer modeling of it.

Since I feel that in a fundamental sense, Whirly-Art creativity is no deeper than gridfont creativity, the study of gridfont creation—more specifically, the computer modeling of gridfont creation—could reveal some things that I have sought for a long time. Therefore the next few years will be an important time for me—a time to see if I can really get at the essence, via modeling, of what my mind is doing when I create something that to me is excitingly novel.

This book, as it says on its cover and in the Introduction, deals with Mind and Pattern. To me, boiling things down to their conceptual skeletons is the royal road to truth (to mix metaphors rather horribly). I think that a lot of truth about Mind and Pattern lies waiting to be extracted in the tiny domains that I have carved out very painstakingly over the past seven years or so in Indiana. I urge you to keep these kinds of things in mind as you read this book. This "confession", coming as it does in a most unexpected place, is a very spontaneous one and probably captures as well as anything could the reason that my research is focused as it is, and the reason that I wrote this book.

Introduction

This book takes its title from the column I wrote in *Scientific American* between January 1981 and July 1983. In that two-and-a-half-year span, I produced 25 columns on quite a variety of topics. My choice of title deliberately left the focus of the column somewhat hazy, which was fine with me as well as with *Scientific American.* When Dennis Flanagan, the magazine's editor, wrote to me in mid-1980 to offer me the chance to write a column in that distinguished publication, he made it clear that what was desired was a bridge between the scientific and the literary viewpoints, something he pointed out Martin Gardner had always done, despite the ostensibly limiting title of *his* column, "Mathematical Games". Here is how Dennis put it in his letter:

> I might emphasize the flexible nature of the department we have been calling "Mathematical Games". As you know, under this title, Martin has written a great deal that is neither mathematical nor game-like. Basically, "Mathematical Games" has been Martin's column to talk about anything under the sun that interests him. Indeed, in our view, the main import of the column has been to demonstrate that a modern intellectual can have a range of interests that are not confined by such words as "scientific" or "literary". We hope that whoever succeeds Martin will feel free to cover his own broad range of interests, which are unlikely to be identical to Martin's.

What a refreshingly open attitude! So I was being asked to be the successor to Martin Gardner—but not necessarily to continue the same column. Rather than filling the same role as Martin had, I would merely occupy the same physical spot in the magazine.

I had been offered a unique opportunity to say pretty much anything I wanted to say to a vast, ready-made audience, in a prestigious context. *Carte blanche,* in short. What more could I ask? Even so, I had to deliberate long and hard about whether to take it, because I did not consider myself primarily a writer, but a thinker and researcher, and time taken in writing would surely be time taken away from research. The conservative pathway, following what was known, would have been to say no, and just do research. The adventurous pathway, exploring the new opportunity and forsaking some research, was tempting. Both were risky, since I knew that either way I would inevitably wonder, "How would things have gone had I decided the other way?" Moreover, I had no idea how long I might write my column, since that was not stipulated. It could go on for many years—or I could decide it was too much for me, and quit after a year.

In a way, I knew from the beginning that I would take the offer, I guess because I am basically more adventurous than I am conservative. But it was a little like purchasing new clothes: no matter how much you like them, you still want to see how you look in them before you buy them, so you put them on and parade around the store, looking at yourself in the mirror and asking whoever is with you what they think of it. So I talked it over with numerous people, and finally decided as I had expected: to take the offer.

*   *   *

For the first year, Martin Gardner and I alternated columns. I have to admit that even though I was utterly free to "be myself", I felt somewhat tradition-bound. True, I had metamorphosed his title into my own title (see Chapter 1 for an explanation), but I was aware that readers of Martin's column would, naturally enough, be expecting a similar type of fare. It took a little while for me to test the waters, getting reader reactions and seeing if the magazine was satisfied with my performance, a performance very different in style from Martin's, after all. Needless to say, some readers were disappointed that I was not a clone of Martin Gardner, but others complimented me on how I had managed to keep the same level of quality while changing the style and content greatly. It was hard, knowing that people were constantly comparing me with someone very different from me. It was particularly hard when people who should have known better really confused my role with Martin's. For instance, as late as June 1983, at a conference on artificial intelligence, a colleague who spotted me came up to me and eagerly told me a math puzzle he'd just discovered and solved, hoping I would put it in my "Mathematical Games" column. How often did I have to tell people that my column was not called "Mathematical Games"!

I doubt that anyone loved Martin Gardner's column more than I did, or owed more to it. Yet I did not want my identity confused with someone else's. So writing this column and being in the shadow of someone superlative was not always easy. But I think I hit my stride and became comfortable with my new role after a few months.

In 1982, Martin retired, leaving the space entirely to me. It was a chore, to be sure, to get a column out each month, but it was also a lot of fun. In any case, what mattered to me the most was to do my best to make the column interesting and diverse and highly provocative. I took Dennis' offer quite literally, not restricting myself to purely scientific topics, but venturing into musical and literary topics as well.

After a year and a half, I was beginning to wonder how long I could sustain it without seriously jeopardizing my research. I decided to divide up my long list of prospective topics into categories: columns I would *love* to do, columns I would simply enjoy doing, and columns I could write with interest but no real passion. I found I had about a year's worth left in the first category, maybe another year's worth left in the second, and then a large number in the third. It seemed, then, that in another year or so it would be a good time to reassess the whole issue of writing the column. As it turned out, my thinking was quite consonant with evolving desires at the editorial level of the magazine. They were most interested in launching a new column to be devoted to the recreational aspects of computing, and our plans dovetailed well. My column could be phased out just as the new one was being phased in. And that is the way it came to pass, with two surprise columns by Martin Gardner filling the gap. My farewell to readers came as a postscript to Martin's final column, in September 1983.

Thus my era as a columnist came to an end. As I look back on it, I feel it lasted just about the right length of time: long enough to let me get a significant amount said, but not so long that it became a real drag on me. This way, at least, I got to explore that avenue that was so tempting, and yet it didn't radically alter the course of my life. So in sum, I am quite pleased with my stint at *Scientific American.* I am proud to have been associated with that venerable institution, and to have filled that unique slot for a time, especially coming right on the heels of someone of such high caliber.

*　　*　　*

The diversity of my columns is worth discussing for a moment. On the surface, they seem to wander all over the intellectual map—from sexism to music to art to nonsense, from game theory to artificial intelligence to molecular biology to the Cube, and more. But there is, I believe, a deep underlying unity to my columns. I felt that gradually, as I wrote more and more of them, regular readers would start to see the links between disparate ones, so that after a while, the coherence of the web would be quite clear. My image of this was always geometric. I envisioned my intellectual "home territory" as a rather large region in some conceptual space, a region that most people do not see as a connected unit. Each new column was in a way a new "random dot" in that conceptual space, and as dots began peppering the space more fully over the months, the shape of my territory would begin to emerge more clearly. Eventually, I hoped, there would emerge a clear region associated with the name "Metamagical Themas".

Of course I wonder if my 25 1/2 columns are sufficient to convey the connectedness of my little patch of intellectual territory, or if, on the contrary, they would leave a question mark in the mind of someone who read them all in succession without any other explanation. Would it simply seem like a patchwork quilt, a curious potpourri? Truth to tell, I suspect that 25 columns are not quite enough, on their own. Probably the dots are too sparsely distributed to suggest the rich web of potential cross-connections there. For that reason, in drawing all my columns together to form a book, I decided to try to flesh out that space by including a few other recent writings of mine that might help to fill some of the more blatant gaps. There are seven such pieces included (indicated by asterisks in the table of contents). I believe they help to unify this book.

If someone were to ask me, "What is your new book about, in a word?", I would probably mutter something like "Mind and Pattern". That, in fact, was one title I considered for the column, way back when. Certainly it tells what most intrigues me, but it doesn't convey it quite vividly or passionately enough. Yes, I am a relentless quester after the chief patterns of the universe —central organizing principles, clean and powerful ways to categorize what is "out there". Because of this, I have always been pulled to mathematics. Indeed, even though I dropped the idea of being a professional mathematician many years ago, whenever I go into a new bookstore, I always make a beeline for the math section (if there is one). The reason is that I still feel that mathematics, more than any other discipline, studies the fundamental, pervasive patterns of the universe. However, as I have gotten older, I have come to see that there are inner mental patterns underlying our ability to conceive of mathematical ideas, universal patterns in human minds that make them receptive not only to the patterns of mathematics but also to abstract regularities of all sorts in the world. Gradually, over the years, my focus of interest has shifted to those more subliminal patterns of memory and associations, and away from the more formal, mathematical ones. Thus my interest has turned ever more to Mind, the principal apprehender of pattern, as well as the principal producer of certain kinds of pattern.

To me, the deepest and most mysterious of all patterns is music, a product of the mind that the mind has not come close to fathoming yet. In some sense, all my research is aimed at finding patterns that will help us to understand the mysteries of musical and visual beauty. I could be bolder and say, "I seek to discover what musical and visual beauty really are." However, I don't believe that those mysteries will ever be truly cleared up, nor do I wish them to be. I would like to understand things better, but I don't want to understand them perfectly. I don't wish the fruits of my research to include a mathematical formula for Bach's or Chopin's music. Not that I think it possible. In fact, I think the very idea is nonsense. But even though I find the prospect repugnant, I am greatly attracted by the effort to do as much as possible in that direction. Indeed, how could anyone hope to approach the concept of beauty without deeply studying the nature of formal patterns and their organizations and relationships to Mind? How can anyone fascinated by beauty fail to be intrigued by the notion of a "magical formula" behind it all, chimerical though the idea certainly is? And in this day and age, how can anyone fascinated by creativity and beauty fail to see in computers the ultimate tool for exploring their essence? Such ideas are the inner fire that propels my research and my writings, and they are the core of this book.

There is another aspect of my inner fire that is brought out in the writings here collected, particularly toward the end, but it pops up throughout. That is a concern with the global fate of humanity and the role of the individual in helping determine it. I have long been an activist, someone who periodically gets fired up by some cause and ardently works for it, exhorting everyone else I come across to get involved as well. I am a fierce believer in the value of passion and commitment to social causes, someone baffled and troubled by apathy. One of my personal mottos is: "Apathy on the individual level translates into insanity at the mass level", a saying nowhere better exemplified than by today's insane dedication of so many human and natural resources to the building up of unimaginably catastrophic arsenals, all while mountains of humanity are starving and suffering in horrible ways. Everyone knows this, and yet the situation remains this way, getting worse day by day. We do live in a ridiculous world, and I would not wish to talk about the world without indicating my confusion and sadness, but also my vision and hope, concerning our shared human condition.

*　　*　　*

Inevitably, people will compare this book with my earlier books, *Gödel, Escher, Bach: an Eternal Golden Braid,* and *The Mind's I,* coedited with my friend Daniel Dennett. Let me try for a moment to anticipate them.

GEB was a unique sort of book—the detailed working-out of a single potent spark. It was a kind of explosion in my mind triggered by my re-falling in love with mathematical logic after a long absence. It was the first time I had tried to write anything long, and I pulled out all the stops. In particular, I made a number of experiments with style, especially in writing dialogues based on musical forms such as fugues and canons. In essence, *GEB* was one extended flash having to do with Kurt Gödel's famous incompleteness theorem, the human brain, and the mystery of consciousness. It is well described on its cover as "a metaphorical fugue on minds and machines".

The Mind's I is very different from *Gödel, Escher, Bach.* It is an extensively annotated anthology rather than the work of a single person. It is far more like a monograph than *GEB* is, in that it has a unique goal: to probe the mysteries of matter and consciousness in as vivid and jolting a way as possible, through stories that anyone can read and understand, followed by careful commentaries by Dan Dennett and myself. Its subtitle is "Fantasies and Reflections on Self and Soul".

One thing that *GEB* and *The Mind's I* have in common is their internal structure of alternation. *GEB* alternates between dialogues and chapters, while *The Mind's I* alternates between fantasies and reflections. I guess I like this contrapuntal mode, because it crops up again in the present volume. Here, I alternate between articles and postscripts.

If *GEB* is an elaborate fugue on one very complex theme, and *MI* is a collection of many variations on a theme, then perhaps *MT* is a fantasia employing several themes. If it were not for the postscripts, I would say that it was disjointed. However, I have made a great effort to tie together the diverse themes—Themas—by writing extensive commentaries that cast the ideas of each article in the light of other articles in the book. Sometimes the postscripts approach the length of the piece they are "post", and in one case (Chapter 24) the postscript is quite a bit longer than its source.

The reason for that particularly long postscript is that I decided to use it to describe some aspects of my own current research in artificial intelligence. There are other places as well in the book where I touch on my research ideas, though I never go into technical details. My main concern is to give a clear idea of certain central riddles about how minds work, riddles that I have run across over and over again in different guises. The questions I raise are difficult but I find them as beguiling as mathematical ones. In any case, this book will give readers a better understanding of how my research and the rest of my ideas fit together.

*   *   *

One aspect of this book that, I must admit, sometimes makes me uneasy is the striking disparity in the seriousness of its different topics. How can both Rubik's Cube and nuclear Armageddon be discussed at equal length in one book by one author? Partly the answer is that life itself is a mixture of things of many sorts, little and big, light and serious, frivolous and formidable, and *Metamagical Themas* reflects that complexity. Life is not worth living if one can never afford to be delighted or have fun.

There is another way of explaining this huge gulf. Elegant mathematical structures can be as central to a serious modern worldview as are social concerns, and can deeply influence one's ways of thinking about anything —even such somber and colossal things as total nuclear obliteration. In order to comprehend that which is incomprehensible because it is too huge or too complex, one needs simpler models. Often, mathematics can provide the right starting point, which is why beautiful mathematical concepts are so pervasive in explanations of the phenomena of nature on the micro-level. They are now proving to be of great help also on a larger scale, as Robert Axelrod's lovely work on the Prisoner's Dilemma so impeccably demonstrates (see Chapter 29).

The Prisoner's Dilemma is poised about halfway between the Cube and Armageddon, in terms of complexity, abstraction, size, and seriousness. I submit that abstractions of this sort are direly needed in our times, because many people—even remarkably smart people—turn off when faced with issues that are too big. We need to make such issues graspable. To make them graspable and fascinating as well, we need to entice people with the beauties of clarity, simplicity, precision, elegance, balance, symmetry, and so on.

Those artistic qualities, so central to good science as well as to good insights about life, are the things that I have tried to explore and even to celebrate in *Metamagical Themas*. (It is not for nothing that the word "magic" appears inside the title!) I hope that *Metamagical Themas* will help people to bring more clarity, precision, and elegance to their thinking about situations large and small. I also hope that it will inspire people to dedicate more of their energies to global problems in this lunatic but lovable world, because we live in a time of unprecedented urgency. If we do not care enough now, future generations may not exist to thank us for their existence and for our caring.

Section 1:

Snags and Snarls

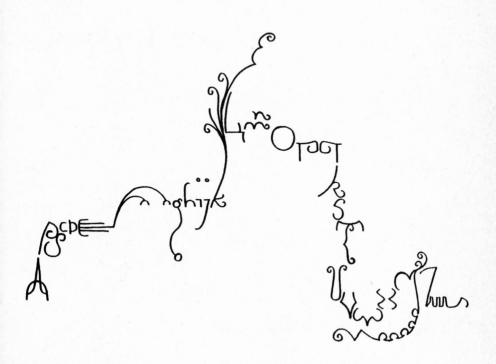

Section I:
Snags and Snarls

The title of this section conveys the image of problematical twistiness. The twists dealt with here are those whereby a system (sentence, picture, language, organism, society, government, mathematical structure, computer program, etc.) twists back on itself and closes a loop. A very general name for this is *reflexivity*. When realized in different ways, this abstraction becomes a concrete phenomenon. Examples are: self-reference, self-description, self-documentation, self-contradiction, self-questioning, self-response, self-justification, self-refutation, self-parody, self-doubt, self-definition, self-creation, self-replication, self-modification, self-amendment, self-limitation, self-extension, self-application, self-scheduling, self-watching, and on and on. In the following four chapters, these strange phenomena are illustrated in sentences and stories that talk about themselves, ideas that propagate themselves from mind to mind, machines that replicate themselves, and games that modify their own rules. The variety of these loopy tangles is quite remarkable, and the subject is far from being exhausted. Furthermore, although their connection with paradox may make reflexive systems seem no more than intellectual playthings, study of them is of great importance in understanding many mathematical and scientific developments of this century, and is becoming ever more central to theories of intelligence and consciousness, whether natural or artificial. Reflexivity will therefore make many return appearances in this book.

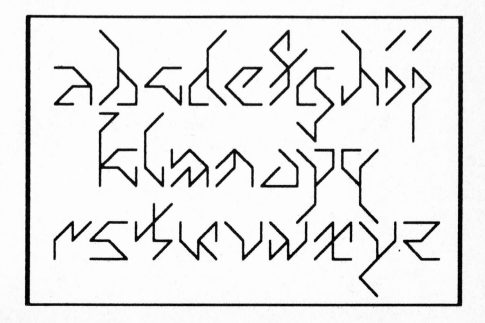

1

On Self-Referential Sentences

January, 1981

Inever expected to be writing a column for *Scientific American*. I remember once, years ago, wishing I were in Martin Gardner's shoes. It seemed exciting to be able to plunge into almost any topic one liked and to say amusing and instructive things about it to a large, well-educated, and receptive audience. The notion of doing such a thing seemed ideal, even dreamlike. Over the next several years, by a series of total coincidences (which turned out to be not so total), I met one after another of Martin's friends. First it was Ray Hyman, a psychologist who studies deception. He introduced me to the magician Jerry Andrus. Then I met the statistician and magician Persi Diaconis and the computer wizard Bill Gosper. Then came Scott Kim, and soon afterward, the mathematician Benoît Mandelbrot. All of a sudden, the world seemed to be orbiting Martin Gardner. He was at the hub of a magic circle, people with exciting, novel, often offbeat ideas, people with many-dimensional imaginations. Sometimes I felt overawed by the whole remarkable bunch.

One day, five or so years ago, I had the pleasure of spending several hours with Martin in his house, discussing many topics, mathematical and otherwise. It was an enlightening experience for me, and it gave me a new view into the mind of someone who had contributed so much to my own mathematical education. Perhaps the most striking thing about Martin to me was his natural simplicity. I had been told that he is an adroit magician. This I found hard to believe, because one does not usually imagine someone so straightforward pulling the wool over anyone's eyes. However, I did not see him do any magic tricks. I simply saw his vast knowledge and love of ideas spread out before me, without the slightest trace of pride or pretense. The Gardners—Martin and his wife Charlotte—entertained me for the day. We ate lunch in the kitchen of their cozy three-story house. It pleased me somehow to see that there was practically no trace of mathematics or games or tricks in their simple but charming living room.

After lunch—sandwiches that Martin and I made while standing by the kitchen sink—we climbed the two flights of stairs to Martin's hideaway. With his old typewriter and all kinds of curious jottings in an ancient filing cabinet and his legendary library of three-by-five cards, he reminded me of an old-time journalist, not of the center of a constellation of mathematical eccentrics and game addicts, to say nothing of magicians, anti-occultists, and of course the thousands of readers of his column.

Occasionally we were interrupted by the tinkling of a bell attached to a string that led down the stairs to the kitchen, where Charlotte could pull it to get his attention. A couple of phone calls came, one from the logician and magician Raymond Smullyan, someone whose name I had known for a long time, but who I had no idea belonged to this charmed circle. Smullyan was calling to chat about a book he was writing on Taoism, of all things! For a logician to be writing about what seemed to me to be the most anti-logical of human activities sounded wonderfully paradoxical. (In fact, his book *The Tao Is Silent* is delightful and remarkable.) All in all, it was a most enjoyable day.

Martin's act will be a hard one to follow. But I will not be trying to be another Martin Gardner. I have my own interests, and they are different from Martin's, although we have much in common. To express my debt to Martin and to symbolize the heritage of his column, I have kept his title "Mathematical Games" in the form of an anagram: "Metamagical Themas".

What does "metamagical" mean? To me, it means "going one level beyond magic". There is an ambiguity here: on the one hand, the word might mean "ultramagical"—magic of a higher order—yet on the other hand, the magical thing about magic is that what lies behind it is always *non*-magical. That's metamagic for you! It reflects the familiar but powerful adage "Truth is stranger than fiction." So my "Metamagical Themas" will, in Gardnerian fashion, attempt to show that magic often lurks where few suspect it, and, by the opposite token, that magic seldom lurks where many suspect it.

*   *   *

In his July, 1979 column, Martin wrote a very warm review of my book *Gödel, Escher, Bach: an Eternal Golden Braid.* He began the review with a short quotation from my book. If I had been asked to guess what single sentence he would quote, I would never have been able to predict his choice. He chose the sentence "This sentence no verb." It is a catchy sentence, I admit, but something about seeing it again bothered me. I remembered how I had written it one day a few years earlier, attempting to come up with a new variation on an old theme, but even at the time it had not seemed as striking as I had hoped it would. After seeing it chosen as the symbol of my book, I felt challenged. I said to myself that surely there must be much cleverer types of self-referential sentence. And so one day I wrote down quite a pile of self-referential sentences and showed them to friends, which began a mild craze among a small group of us. In this column, I will present a selection of what I consider to be the cream of that crop.

Before going further, I should explain the term "self-reference". Self-reference is ubiquitous. It happens every time anyone says "I" or "me" or "word" or "speak" or "mouth". It happens every time a newspaper prints a story about reporters, every time someone writes a book about writing, designs a book about book design, makes a movie about movies, or writes an article about self-reference. Many systems have the capability to represent or refer to themselves somehow, to designate themselves (or elements of themselves) within the system of their own symbolism. Whenever this happens, it is an instance of self-reference.

Self-reference is often erroneously taken to be synonymous with paradox. This notion probably stems from the most famous example of a self-referential sentence, the Epimenides paradox. Epimenides the Cretan said, "All Cretans are liars." I suppose no one today knows whether he said it in ignorance of its self-undermining quality or for that very reason. In any case, two of its relatives, the sentences "I am lying" and "This sentence is false", have come to be known as the *Epimenides paradox* or the *liar paradox.* Both sentences are absolutely self-destructive little gems and have given self-reference a bad name down through the centuries. When people speak of the evils of self-reference, they are certainly overlooking the fact that not every use of the pronoun "I" leads to paradox.

*   *   *

Let us use the Epimenides paradox as our jumping-off point into this fascinating land. There are many variations on the theme of a sentence that somehow undermines itself. Consider these two:

This sentence claims to be an Epimenides paradox, but it is lying.

This sentence contradicts itself—or rather—well, no, actually it doesn't!

What should you do when told, "Disobey this command"? In the following sentence, the Epimenides quality jumps out only after a moment of thought: "This sentence contains exactly threee erors." There is a delightful backlash effect here.

Kurt Gödel's famous Incompleteness Theorem in metamathematics can be thought of as arising from his attempt to replicate as closely as possible the liar paradox in purely mathematical terms. With marvelous ingenuity, he was able to show that in any mathematically powerful axiomatic system S it is possible to express a close cousin to the liar paradox, namely, "This formula is unprovable within axiomatic system S."

In actuality, the Gödel construction yields a mathematical formula, not an English sentence; I have translated the formula back into English to show what he concocted. However, astute readers may have noticed that, strictly speaking, the phrase "this formula" has no referent, since when a *formula* is translated into an English *sentence*, that sentence is no longer a formula!

If one pursues this idea, one finds that it leads into a vast space. Hence the following brief digression on the preservation of self-reference across language boundaries. How should one translate the French sentence *Cette phrase en français est difficile à traduire en anglais*? Even if you do not know French, you will see the problem by reading a literal translation: "This sentence in French is difficult to translate into English." The problem is: To what does the subject ("This sentence in French") refer? If it refers to the sentence it is part of (which is not in French), then the subject is self-contradictory, making the sentence false (whereas the French original was true and harmless); but if it refers to the French sentence, then the meaning of "this" is strained. Either way, something disquieting has happened, and I should point out that it would be just as disquieting, although in a different way, to translate it as: "This sentence in English is difficult to translate into French." Surely you have seen Hollywood movies set in France, in which all the dialogue, except for an occasional *Bonjour* or similar phase, is in English. What happens when Cardinal Richelieu wants to congratulate the German baron for his excellent command of French? I suppose the most elegant solution is for him to say, "You have an excellent command of our language, *mon cher baron*", and leave it at that.

* * *

But let us undigress and return to the Gödelian formula and focus on its meaning. Notice that the concept of *falsity* (in the liar paradox) has been replaced by the more rigorously understood concept of *provability*. The logician Alfred Tarski pointed out that it is in principle impossible to translate the liar paradox exactly into any rigorous mathematical language, because if it were possible, mathematics would contain a genuine paradox —a statement both true and false—and would come tumbling down.

Gödel's statement, on the other hand, is not paradoxical, though it constitutes a hair-raisingly close approach to paradox. It turns out to be true, and for this reason, it is unprovable in the given axiomatic system. The revelation of Gödel's work is that in *any* mathematically powerful and consistent axiomatic system, an endless series of true but unprovable formulas can be constructed by the technique of self-reference, revealing that somehow the full power of human mathematical reasoning eludes capture in the cage of rigor.

In a discussion of Gödel's proof, the philosopher Willard Van Orman Quine invented the following way of explaining how self-reference could be achieved in the rather sparse formal language Gödel was employing. Quine's construction yields a new way of expressing the liar paradox. It is this:

"yields falsehood, when appended to its quotation." yields falsehood, when appended to its quotation.

This sentence describes a way of constructing a certain typographical entity —namely, a phrase appended to a copy of itself in quotes. When you carry out the construction, however, you see that the end product is the sentence itself—or a perfect copy of it. (There is a resemblance here to the way self-replication is carried out in the living cell.) The sentence asserts the falsity of the constructed typographical entity, namely itself (or an indistinguishable copy of itself). Thus we have a less compact but more explicit version of the Epimenides paradox.

It seems that all paradoxes involve, in one way or another, self-reference, whether it is achieved directly or indirectly. And since the credit for the discovery—or creation—of self-reference goes to Epimenides the Cretan, we might say: "Behind every successful paradox there lies a Cretan."

On the basis of Quine's clever construction we can create a self-referential question:

What is it like to be asked,
"What is it like to be asked, self-embedded in quotes after its comma?"
self-embedded in quotes after its comma?

Here again, you are invited to construct a typographical entity that turns out, when the appropriate operations have been performed, to be identical with the set of instructions. This self-referential question suggests the following puzzle: What is a question that can serve as its own answer? Readers might enjoy looking for various solutions to it.

*   *   *

When a word is used to *refer* to something, it is said to be being *used*. When a word is *quoted*, though, so that one is examining it for its surface aspects (typographical, phonetic, etc.), it is said to be being *mentioned*. The following sentences are based on this famous use-mention distinction:

You can't have your use and mention it too.

You can't have "your cake" and spell it "too".

"Playing with the use-mention distinction" isn't "everything in life, you know".

In order to make sense of "this sentence", you will have to ignore the quotes in "it".

This is a sentence with "onions", "lettuce", "tomato", and "a side of fries to go".

This is a hamburger with vowels, consonants, commas, and a period at the end.

The last two are humorous flip sides of the same idea. Here are two rather extreme examples of self-referential use-mention play:

Let us make a new convention: that anything enclosed in *triple* quotes—for example, "'No, I have decided to change my mind; when the triple quotes close, just skip directly to the period and ignore everything up to it'"—is not even to be read (much less paid attention to or obeyed).

A ceux qui ne comprennent pas l'anglais, la phrase citée ci-dessous ne dit rien: "For those who know no French, the French sentence that introduced this quoted sentence has no meaning."

The bilingual example may be more effective if you know only one of the two languages involved.

Finally, consider this use-mention anomaly: "i should begin with a capital letter." This is a sentence referring to itself by the pronoun "I", a bit mauled, instead of through a pointing-phrase such as "this sentence"; such a sentence would seem to be arrogantly proclaiming itself to be an animate agent. Another example would be "I am not the person who wrote me." Notice how easily we understand this curious nonstandard use of "I". It seems quite natural to read the sentence this way, even though in nearly all situations we have learned to unconsciously create a mental model of some person—the sentence's speaker or writer—to whom we attribute a desire to communicate some idea. Here we take the "I" in a new way. How come? What kinds of cues in a sentence make us recognize that when the word "I" appears, we are supposed to think not about the author of the sentence but about the sentence itself?

*   *   *

Many simplified treatments of Gödel's work give as the English translation of his famous formula the following: "I am not provable in axiomatic system *S.*" The self-reference that is accomplished with such sly trickery in the formal system is finessed into the deceptively simple English word "I", and we can—in fact, we automatically do—take the sentence to be talking about itself. Yet it is hard for us to hear the following sentence as talking about itself: "I *already* took the garbage out, honey."

The ambiguous referring possibilities of the first-person pronoun are a source of many interesting self-referential sentences. Consider these:

I am not the subject of this sentence.

I am jealous of the first word in this sentence.

Well, how about that—this sentence is about me!

I am simultaneously writing and being written.

This raises a whole new set of possibilities. Couldn't "I" stand for the writing instrument ("I am not a pen"), the language ("I come from Indo-European roots"), the paper ("Cut me out, twist me, and glue me to form a Möbius strip, please")? One of the most involved possibilities is that "I" stands not for the physical tokens we perceive before us but for some more ethereal and intangible essence, perhaps the *meaning* of the sentence. But then, what is meaning? The next examples explore that idea:

I am the meaning of this sentence.

I am the thought you are now thinking.

I am thinking about myself right now.

I am the set of neural firings taking place in your brain as you read the set of letters in this sentence and think about me.

This inert sentence is my body, but my soul is alive, dancing in the sparks of your brain.

The philosophical problem of the connections among Platonic ideas, mental activity, physiological brain activity, and the external symbols that trigger them is vividly raised by these disturbing sentences.

This issue is highlighted in the self-referential question, "Do you think anybody has ever had *precisely this thought* before?" To answer the question, one would have to know whether or not two different brains can *ever* have precisely the same thought (as two different computers can run precisely the same program). An illustration of this possibility may be found in Figure 24-2. I have often wondered: Can *one* brain have the same thought more than once? Is a thought something Platonic, something whose essence exists independently of the brain it is occurring in? If the answer is "Yes, thoughts are brain-independent", then the answer to the self-referential question would also be yes. If it is not, then no one could ever have had the same thought before—not even the person thinking it!

Certain self-referential sentences involve a curious kind of communication between the sentence and its human friends:

You are under my control because I am choosing exactly what words you are made out of, and in what order.

No, *you* are under *my* control because you will read until you have reached the end of me.

Hey, down there—are you the sentence I am writing, or the sentence I am reading?

And you up there—are you the person writing me, or the person reading me?

You and I, alas, can have only one-way communication, for you are a person and I, a mere sentence.

As long as you are not reading me, the fourth word of this sentence has no referent.

The reader of this sentence exists only while reading me.

Now *that* is a rather frightening thought! And yet, by its own peculiar logic, it is certainly true.

Hey, out there—is that *you* reading me, or is it someone else?

Say, haven't you written me somewhere else before?

Say, haven't I written you somewhere else before?

The first of the three sentences above addresses its reader; the second addresses its author. In the last one, an author addresses a sentence.

Many sentences include words whose referents are hard to figure out because of their ambiguity—possibly accidental, possibly deliberate:

Thit sentence is not self-referential because "thit" is not a word.

No language can express every thought unambiguously, least of all this one.

In the Escher-inspired Figure 1-1, visual and verbal ambiguity are simultaneously exploited.

* * *

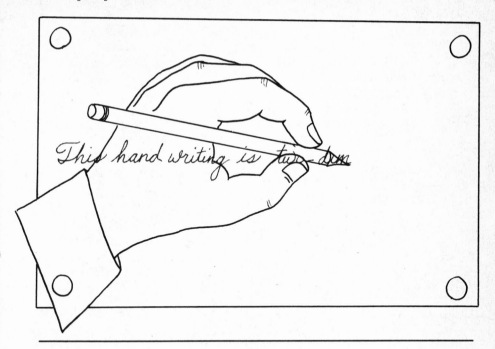

FIGURE 1–1. *Ambiguity: What is being described—the hand, or the writing?* [*Drawing by David Moser, after M. C. Escher.*]

Let us turn to a most interesting category, namely sentences that deal with the languages they are in, once were in, or might have been in:

When you are not looking at it, this sentence is in Spanish.

I had to translate this sentence into English because I could not read the original Sanskrit.

The sentence now before your eyes spent a month in Hungarian last year and was only recently translated back into English.

If this sentence were in Chinese, it would say something else.

.siht ekil ti gnidaer eb d'uoy ,werbeH ni erew ecnetnes siht fI

The last two sentences are examples of *counterfactual conditionals.* Such a sentence postulates in its first clause (the *antecedent*) some contrary-to-fact situation (sometimes called a "possible world") and extrapolates in its second clause (the *consequent*) some consequence of it. This type of sentence opens up a rich domain for self-reference. Some of the more intriguing self-referential counterfactual conditionals I have seen are the following:

If this sentence didn't exist, somebody would have invented it.

If I had finished this sentence,

If there were no counterfactuals, this sentence would not be paradoxical.

If wishes were horses, the antecedent of this conditional would be true.

If this sentence were false, beggars would ride.

What would this sentence be like if it were not self-referential?

What would this sentence be like if π were 3?

Let us ponder the last of these (invented by Scott Kim) for a moment. In a world where π actually *did* have the value 3, you wouldn't ask about how things *would* be if π were 3. Instead, you might muse "if π were 2" or "if π *weren't* 3". So one's first answer to the question might be this: "What would this sentence be like if π *weren't* 3?". But there is a problem. The referent of "this sentence" has now changed identity. So is it fair to say that the second sentence is an answer to the first? It is a little like a woman who muses, "What would I be doing now if I had had different genes?" The problem is that she would not be herself; she would be someone else, perhaps the little boy across the street, playing in his sandbox. Personal pronouns like "I" cannot quite keep up with such strange hypothetical world-shifts.

But getting back to Scott Kim's counterfactual, I should point out that there is an even more serious problem with it than so far mentioned. Changing the value of π is, to put it mildly, a radical change in mathematics, and presumably you cannot change mathematics radically without having radically changed the fabric of the universe within which we live. So it is quite doubtful that any of the concepts in the sentence would make any sense if π were 3 (including the concepts of "π", "3", and so on).

Here are two more counterfactual conditionals to put in your pipe and smoke:

If the subjunctive was no longer used in English, this sentence would be grammatical.

This sentence would be seven words long if it were six words shorter.

These two lovely examples, invented by Ann Trail (who is also responsible for quite a few others in this column), bring us around to sentences that comment on their own form. Such sentences are quite distinct from ones that comment on their own content (such as the liar paradox, or the sentence that says "This sentence is not about itself, but about whether it is about itself."). It is easy to make up a sentence that refers to its own form, but it is hard to make up an *interesting* one. Here are a few more quite good ones:

because I didn't think of a good beginning for it.

This sentence was in the past tense.

This sentence has contains two verbs.

This sentence contains one numeral 2 many.

a preposition. This sentence ends in

In the time it takes you to read this sentence, eighty-six letters could have been processed by your brain.

*   *   *

David Moser, a composer and writer, is a delector and creator of self-reference and frame-breaking of all kinds. He has even written a story in which every sentence is self-referential (it is included in Chapter 2). It might seem unlikely that in such a limited domain, individual styles could arise and flourish, but David has developed a self-referential style quite his own. As a mutual friend (or was it David himself?) wittily observed, "If David Moser had thought up this sentence, it would have been funnier." Many Moser creations have been used above. Some further Moserian delights are these:

This is not a complete. Sentence. This either.

This sentence contains only one nonstandard English flutzpah.

This gubblick contains many nonsklarkish English flutzpahs, but the overall pluggandisp can be glorked from context.

This sentence has cabbage six words.

In my opinion, it took quite a bit of flutzpah to just throw in a random word so that there *would* be cabbage six words in the sentence. That idea inspired the following: "This sentence has five (5) words." A few more miscellaneous Moserian gems follow:

This is to be or actually not two sentences to be, that is the question, combined.

It feels *sooo* good to have your eyes run over my curves and serifs.

This sentence is a !!!! premature punctuator

Sentences that talk about their own punctuation, as the preceding one does, can be quite amusing. Here are two more:

This sentence, though not interrogative, nevertheless ends in a question mark?

This sentence has no punctuation semicolon the others do period

Another ingenious inventor of self-referential sentences is Donald Byrd, several of whose sentences have already been used above. Don too has his own very characteristic way of playing with self-reference. Two of his sentences follow:

This hear sentence do'nt know Inglish purty good.

If you meet this sentence on the board, erase it.

The latter, via its form, alludes to the Buddhist saying "If you meet the Buddha on the road, kill him."
Allusion through similarity of form is, I have discovered, a marvelously rich vein of self-reference, but unfortunately this article is too short to contain a full proof of that discovery. I shall explicitly discuss only two examples. The first is "This sentence verbs good, like a sentence should." Its primary allusion is to the famous slogan "Winston tastes good, like a cigarette should", and its secondary allusion is to "This sentence no verb." The other example involves the following lovely self-referential remark, once made by the composer John Cage: "I have nothing to say, and I am saying it." This allows the following rather subtle twist to be made: "I have nothing to allude to, and I am alluding to it."

*　　*　　*

Some of the best self-referential sentences are short but sweet, relying for their effect on secondary interpretations of idiomatic expressions or well-known catch phrases. Here are five of my favorites, which seem to defy other types of categorization:

Do you read me?

This point is well taken.

You may quote me.

I am going two-level with you.

I have been sentenced to death.

In some of these, even sophisticated non-native speakers would very likely miss what's going on.

Surely no article on self-reference would be complete without including a few good examples of self-fulfilling prophecy. Here are a few:

This prophecy will come true.

This sentence will end before you can say "Jack Rob

Surely no article on self-reference would be complete without including
a few good examples of self-fulfilling prophecy.

Does this sentence remind you of Agatha Christie?

That last sentence—one of Ann Trail's—is intriguing. Clearly it has nothing to do with Agatha Christie, nor is it in her style, and so the answer ought to be no. Yet I'll be darned if I can read it without being reminded of Agatha Christie! (And what is even stranger is that I don't know the first thing about Agatha Christie!)

In closing, I cannot resist the touching plea of the following Byrdian sentence:

Please, oh please, publish me in your collection of self-referential sentences!

Post Scriptum.

This first column of mine triggered a big wave of correspondence, some of which is presented in the next chapter. Most of the correspondence was light-hearted, but there were a number of serious letters that intrigued me. Here is a repartee that appeared in the pages of *Scientific American* a few months later.

The kind of structural analysis engaged in, and the resulting questions raised by, Douglas Hofstadter in his amusing and intriguing article concerning self-referential sentences need not lead inevitably to bafflement of the reader.

Help is at hand from the "laggard science" psychology, but only from that carefully defined quarter of psychology known as behavior analysis, which was progenerated by the famous Harvard psychologist B. F. Skinner almost 50 years ago.

In examining the implications of linguistic analyses such as Hofstadter's for the serious student of verbal behavior, Skinner comments in his book *About Behaviorism* (pages 98—99) as follows:

> Perhaps there is no harm in playing with sentences in this way or in analyzing the kinds of transformations which do or do not make sentences acceptable to the ordinary reader, but it is still a waste of time, particularly when the sentences thus generated could not have been emitted as verbal behavior. A classical example is a paradox, such as 'This sentence is false', which appears to be true if false and false if true. The important thing to consider is that no one could ever have emitted the sentence as verbal behavior. A sentence must be in existence before a speaker can say, 'This sentence is false', and the response itself will not serve, since it did not exist until it was emitted. What the logician or linguist calls a sentence is not necessarily verbal behavior in any sense which calls for a behavioral analysis.

As Skinner pointed out long ago, verbal behavior results from contingencies of reinforcement arranged by verbal communities, and it is these contingencies that must be analyzed if we are to identify the variables that control verbal behavior. Until we grasp the full import of Skinner's position, which goes beyond structure to answer *why* we behave as we do verbally or nonverbally, we shall continue to fall back on prescientific formulations that are about as useful in understanding these phenomena as Hofstadter's quaint metaphorical speculation: "Such a sentence would seem to be arrogantly proclaiming itself to be an animate agent."

George Brabner
College of Education
University of Delaware

I felt compelled to reply to Professor Brabner's interesting views about these matters, and so here is what I wrote:

> I assume that the quote from B. F. Skinner reflects Professor Brabner's own sentiments about the likelihood of self-referential utterances. I am always baffled by people who doubt the likelihood of self-reference and paradox. Verbal behavior comes in many flavors. Humor, particularly self-referential humor, is one of the most pervasive flavors of verbal behavior in this century. One has only to watch the Muppets or Monty Python on television to see dense and intricate webs of self-reference. Even advertisements excel in self-reference.
>
> In art, René Magritte, Pablo Picasso, M. C. Escher, John Cage, and dozens of others have played with the level-distinction between *that which represents* and *that which is represented.* The "artistic behavior" that results includes much self-reference and many confusing and sometimes exhilaratingly paradoxical

tangles. Would Professor Brabner say that no one could ever have "emitted" such works as "artistic behavior"? Where is the borderline?

Ordinary language, as I pointed out in my column, is filled with self-reference, usually a little milder-seeming than the very sharply pointed paradoxes that Professor Brabner objects to. "Mouth", "word", and so on are all self-referential. Language is inherently filled with the potential of sharp turns on which it may snag itself.

Many scholarly papers begin with a sentence about "the purpose of this paper". Newspapers report on their own activities, conceivably on their own inaccuracies. People say, "I'm tired of this conversation." Arguments evolve about arguments, and can get confusingly and painfully self-involved. Has Professor Brabner never thought of "verbal behavior" in this light? It is likely that in hunting woolly mammoths, no one found it extraordinarily likely to shout, "This sentence is false!" However, civilization has come a long way since those days, and the primitive purposes of language have by now been almost buried under an avalanche of more complex purposes.

Part of human nature is to be introspective, to probe. Part of our "verbal behavior" deliberately, often playfully, explores the boundaries between conceptual levels of systems. All of this has its root in the struggle to survive, in the fact that our brains have become so flexible that much of their time is spent in dealing with their own activities, consciously or unconsciously. It is simply a consequence of representational power—as Kurt Gödel showed—that systems of increasing complexity become increasingly self-referential.

It is quite possible for people filled with self-doubt to recognize this trait in themselves, and to begin to doubt their self-doubt itself. Such psychological dilemmas are at the heart of some current theories of therapy. Gregory Bateson's "double bind", Victor Frankl's "logotherapy", and Paul Watzlawick's therapeutic ideas are all based on level-crossing paradoxes that crop up in real life. Indeed, psychotherapy is itself based completely on the idea of a "twisted system of self"—a self that wants to reach inward and change some presumably wrong part of itself.

We human beings are the only species to have evolved humor, art, language, tangled psychological problems, even an awareness of our own mortality. Self-reference—even of the sharp Epimenides type—is connected to profound aspects of life. Would Professor Brabner argue that suicide is not conceivable human behavior?

Finally, just suppose Professors Skinner and Brabner are right, and no one ever says exactly "This sentence is false." Would this mean that study of such sentences is a waste of time? Still not. Physicists study ideal gases because they represent a distillation of the most significant principles of the behavior of real gases. Similarly, the Epimenides paradox is an "ideal paradox"—one that cuts crisply to the heart of the matter. It has opened up vast domains in logic, pure science, philosophy, and other disciplines, and will continue to do so despite the skepticism of behaviorists.

It is a curious coincidence that the only other reply to my article that was printed in the "Letters" column of *Scientific American* also came from the University of Delaware. Here it is:

I hope that you do not receive any correspondence concerning Douglas R. Hofstadter's article on self-reference. I should like to inform your readers that many years of study on this problem have convinced me no conclusion whatsoever can be drawn from it that would stand up to a moment's scrutiny. There is no excuse for *Scientific American* to publish letters from those cranks who consider such matters to be worthy of even the slightest notice.

> A. J. Dale
> Department of Philosophy
> University of Delaware

I replied as follows:

> Many years of reading such letters have convinced me that no reply whatsoever can be given to them that would stand up to a moment's scrutiny. There is no excuse for publishing responses to those cranks who send them.

After these two exchanges had appeared in print, a number of people remarked to me that they'd read the two letters from Delaware that had attacked me, and had enjoyed my responses. Two? I guess it wasn't so obvious that Dale's letter was completely tongue-in-cheek. In fact, that was its point.

*   *   *

Two other letters stand out sharply in my memory. One was from an individual who signed himself (I presume it is a male) as "Mr Flash qFiasco". Mr Flash insisted that a sentence cannot *say* what it *shows*. The former concerns only its *content,* which is supposedly independent of how it manifests itself in print, while the latter is a property exclusively of its *form,* that is, of the physical sentence only when it is in print. This distinction sounds crystal-clear at first, but in reality it is mud-blurry. Here is some of what Flash wrote me:

> For a sentence to attempt to say what it shows is to commit an error of logical types. It seems to be putting a round peg into a square hole, whereas it is instead putting a round peg into something which is not a hole at all, square or otherwise. This is a category mismatch, not a paradox. It is like throwing the recipe in with the flour and butter and eggs. The source of the equivocation is an illegitimate use of the term 'this'. 'This' can point to virtually anything, but 'this' cannot point to itself. If you stick out your index finger, you can point to virtually anything; and by curling it you can even point to the pointing finger; but you cannot point to *pointing.* Pointing is of a higher logical type than the thing which is doing the pointing. Similarly, the referent of 'this sentence' can be virtually anything but that sentence. Sentences of the form exemplified by 'This sentence no verb.' and 'This sentence has a verb.' are not well-formed: they commit fallacies of logical type equivocation. Thus their self-referential character is not genuine and they present no problem as paradoxes.

There will always be people around who will object in this manner, and in the Brabnerian manner. Such people think it is possible to draw a sharp line between attributes of a printed sentence that can be considered part of its *form* (*e.g.*, the typeface it is printed in, the number of words it contains, and so on), and attributes that can be considered part of its *content* (*i.e.*, the things and events and relationships that it refers to).

Now, I am used to thinking about language in terms of how to get a machine to deal with it, since I look at the human brain as a very complex machine that can handle language (and many other things as well). Machines, in trying to make sense of sentences, have access to nothing more than the *form* of such sentences. The *content*, if it is to be accessible to a machine, has to be derived, extracted, constructed, or created somehow from the sentence's physical structure, together with other knowledge and programs already available to the machine.

When very simple processing is used to operate on a sentence, it is convenient to label the information thus obtained "syntactic". For instance, it is clearly a syntactic fact about "This sentence no verb." that it contains six vowels. The vowel-consonant distinction is obviously a typographical one, and typographical facts are considered superficial and syntactic. But there is a problem here. With different *depths of processing,* aspects of different degrees of "semanticity" may be detected.

Consider, for example, the sentence "Mary was sick yesterday." Let's call it *Sentence M*. Listed below are the results of seven different degrees of processing of Sentence M by a hypothetical machine, using increasingly sophisticated programs and increasingly large knowledge bases. You should think of them as being English translations, for your convenience, of computational structures inside the machine that it can act on and use fluently.

1. Sentence M contains twenty characters.
2. Sentence M contains four English words.
3. Sentence M contains one proper noun, one verb, one adjective, and one adverb, in that order.
4. Sentence M contains one human's name, one linking verb, one adjective describing a potential health state of a living being, and one temporal adverb, in that order.
5. The subject of Sentence M is a pointer to an individual named 'Mary', the predicate is an ascription of ill health to the individual so indicated, on the day preceding the statement's utterance.
6. Sentence M asserts that the health of an individual named 'Mary' was not good the day before today.
7. Sentence M says that Mary was sick yesterday.

Just where is the boundary line that says, "You can't do *that* much processing!"? A machine that could go as far as version 7 would have actually *understood*—at least in some rudimentary sense—the content of Sentence M. Work by artificial-intelligence researchers in the field of natural language understanding has produced some very impressive results along these lines, considerably more sophisticated than what is shown here. Stories can be "read" and "understood", at least to the extent that certain kinds of questions can be answered by the machine when it is probed for its understanding. Such questions can involve information not explicitly in the story itself, and yet the machine can fill in the missing information and answer the question.

I am making this seeming digression on the processing of language by computers because intelligent people like Mr Flash qFiasco seem to have failed to recognize that the boundary line between form and content is as blurry as that between blue and green, or between human and ape. This comparison is not made lightly. Humans are supposedly able to get at the "content" of utterances, being genuine language-users, while apes are not. But ape-language research clearly shows that there is some kind of in-between world, where a certain degree of content can be retrieved by a being with reduced mental capacity. If mental capacity is equated with potential processing depth, then it is obvious why it makes no sense to draw an arbitrary boundary line between the form and the content of a sentence. *Form blurs into content as processing depth increases.* Or, as I have always liked to say, "Content is just fancy form." By this I mean, of course, that "content" is just a shorthand way of saying "form as perceived by a very fancy apparatus capable of making complex and subtle distinctions and abstractions and connections to prior concepts".

Flash qFiasco's down-home, commonsense distinction between form and content breaks down swiftly, when analyzed. His charming image of someone making a "category error" by throwing a recipe in with the flour and butter and eggs reveals that he has never had Recipe Cake. This is a delicious cake whose batter is made out of cake recipes (if you use pie recipes, it won't taste nearly as good). The best results are had if the recipes are printed in French, in Baskerville Roman. A preponderance of *accents aigus* lends a deliciously piquant aroma to the cake. My recommendation to Brabner and qFiasco is: "Let them eat recipes."

*   *   *

Finally, I come to John Case, a computer scientist who wrote from Yale, insisting that there is no conceptual problem whatsoever in translating the French sentence *"Cette phrase en français est difficile à traduire en anglais"* into English. Case's translation was the following English sentence:

The French sentence *"Cette phrase en français est difficile à traduire en anglais"* is difficult to translate into English.

In other words, Case translates a *self*-referential French sentence into an *other*-referential English sentence. The English sentence talks about the French sentence—in fact it quotes it completely! Something radical is missing here. At one level, of course, Case is right: now the two sentences, one French and one English, both are talking about (or pointing to) the same thing (the French sentence). But the absolute crux of the French one is its tangledness; the English one completely lacks that quality. Clearly Case has had to make a sacrifice, a compromise.

The alternative, which I prefer, is to construct in English an *analogue* to the French sentence: a *self*-referential English sentence, one that has a tangledness isomorphic to that of the French sentence. That's where the *essence* of the sentence lies, after all! "But is that its *translation*?" you might ask. A good question.

Ionesco once remarked, "The French for London is Paris." (Use-mention fanatic that I am, I assume that he meant "The French for 'London' is 'Paris' ", although it is pungent either way.) What he meant was that in understanding situations, French people tend to translate them into their own frame of reference. This is of course true for all of us. If Mary tells Ann, "My brother died", and if Ann does not know Mary's brother, then how can she understand this statement? Surely projection is of the essence: Ann will imagine her *own* brother dying (if she has one—and if not, then her sister, a good friend, possibly even a pet!). This alternate frame of reference allows Ann to empathize with Mary. Now if Ann *did* know Mary's brother somewhat, then she might flicker between thinking of him as the person she vaguely remembers and thinking of her own brother (friend, pet, or whatever) dying. This dilemma (discussed further in the postscript to Chapter 24) arises for all beings with their own preferred vantage points: Do I map things into *what they would be for me,* or do I stand apart and survey them completely objectively and impassively?

Case is advocating the latter, which is all very well as an intellectual stance to adopt, but when it comes to real life, it just won't cut the mustard. To be concrete, one might ask: What was the actual solution used in the French edition of *Scientific American?* The answer, surprising no one, I hope, was this: "This English sentence is difficult to translate into French." I rest my case.

*   *   *

I wonder what literalists like John Case would suggest as the proper translation of the title of the book *All the President's Men* (a book about the downfall of President Nixon, a downfall that none of the people around him could prevent). Would they say that *Tous les hommes du Président* fills the bill admirably? Back-translated rather literally, it means "All the men of the President". It completely lacks the allusion —the reference by similarity of form—to the nursery rhyme "Humpty Dumpty". Is that dispensable? In my opinion, hardly. To me, the essence of the title resides in that allusion. To lose that allusion is to deflate the title totally.

Of course, what do I mean by "that allusion"? Do I wish the French title to contain, somehow, an allusion to an *English* nursery rhyme? That would be rather pointless. Well, then, do I want the French title to allude to the French version of "Humpty Dumpty"? It all depends how well known that is. But given that Humpty Dumpty is practically an unknown figure to French-speaking people, it seems that something else is wanted. Any old French nursery rhyme? Obviously not. The critical allusion is to the lines "All the King's horses/ And all the King's men/ Couldn't put Humpty together again." Are there—*anywhere* in French literature—lines with a similar import? If not, how about in French popular songs? In French proverbs? Fairy tales?

One might well ask why French-speaking people would ever care about reading a book about Watergate in the first place. And even if they *did* want to read it, shouldn't it be *completely* translated, so that it happens in a French-speaking city? Come to think of it, didn't Ioratno once remark that the French for Washington is Montréal?

Clearly, this is carrying things to an extreme. There must be some middle ground of reasonableness. These are matters of subtle judgment, and they are where being human and flexible makes all the difference. Rigid rules about translation may lead you to a kind of mechanical consistency, but at the sacrifice of all depth and charm. The problem of self-referential sentences is just the tip of the iceberg, as far as translation is concerned. It is just that these issues show up very early when direct self-reference is concerned. When self-reference (or reference in general, for that matter) is indirect, mediated by form, then fluidity is required. The understanding of such sentences involves a mixture of deriving the content and yet retaining the form in mind, letting qualities of the form conjure up flavors and enhance the meaning with a halo of not-quite-conscious pseudo-meanings, connotations, flavors, that flicker in the mind, not quite in reach, not quite out of reach. Self-reference is a good starting point for investigation of this kind of issue, because it is so much on the surface there. You can't sweep the problems under the rug, even though some would like to do so.

* * *

This first column, together with this postscript, provides a good introduction to the book as a whole, because many central issues are touched on: codes, translation, analogies, artificial intelligence, language and machines, mind and meanings, self and identity, form and content—all the issues I originally was motivated by when first writing that collection of teasing self-referential sentences.

2

Self-Referential Sentences:
A Follow-Up

January, 1982

As January has rolled around again, I thought I'd give a follow-up to my column of a year ago on self-referential sentences, and that is what this column is; however, before we get any further, I would like to take advantage of this opening paragraph to warn those readers whose sensibilities are offended by explicit self-referential material that they probably will want to quit reading before they reach the end of this paragraph, or for that matter, this sentence—in fact, this clause—even this noun phrase—in short, *this*.

Well, now that we've gotten *that* out of the way, I would like to say that, since last January, I have received piles upon piles of self-referential mail. Tony Durham astutely surmised: "What with the likely volume of replies, I should not think you are reading this in person." John C. Waugh's letter yelped: "Help, I'm buried under an avalanche of reader's responses!" At first, I thought Waugh himself was empathizing with my plight, putting words into my own mouth, but then I realized it was his *letter* calling for help. Fortunately, it was rescued, and now is comfortably nestled in a much reduced pile. Indeed, I have had to cull from that massive influx of hundreds of replies a very small number. Here I shall present some of my favorites.

Before leaving the topic of mail, I would like to point out that the postmark on Ivan Vince's postcard from Britain cryptically remarked, "Be properly addressed." Was this an order issued by the post office to the postcard itself? If so, then British postcards must be far more intelligent than American ones; I have yet to meet a postcard that could read, let alone correct its own address. (One postcard that reached me was addressed to me in care of *Omni* magazine! And yet somehow it arrived.)

I was flattered by a couple of self-undermining compliments. Richard Ruttan wrote, "I just can't tell you how much I enjoyed your first article.", and John Collins said, "This does not communicate my delight at January's column." I was also pleased to learn that my fame had spread as far as the men's room at the Tufts University Philosophy Department, where Dan

Dennett discovered "This sentence is graffiti. —Douglas R. Hofstadter" penned on the wall.

*   *   *

A popular pastime was the search for interesting self-answering questions. However, only a few succeeded in genuinely "jootsing" (jumping out of the system), which, to me, means being truly novel. It seems that successes in this limited art form are not easy to come by. John Flagg cynically remarked (I paraphrase slightly): "Ask a self-answering question, and get a self-questioning answer." One of my favorites was given by Henry Taves: "I fondly remember a history exam I encountered in boarding school that contained the following: 'IV. Write a question suitable for a final exam in this course, and then answer it.' My response was simply to copy that sentence twice." I was delighted by this. Later, upon reflection, I began to suspect something was slightly wrong here. What do you think?

Richard Showstack contributed two droll self-answering questions: "What question no verb?" and "What is a question that mentions the word 'umbrella' for no apparent reason?" Jim Shiley sent in a clever entry that I modify slightly into "Is this a rhetorical question, or is this a rhetorical question?" He also contributed the following idea:

Take a blank sheet of paper and on it write:

How far across the page will this sentence run?

Now if some polyglot friend of yours points out that the same string of phonemes in Ural-Altaic means '2.3 inches', send me a free subscription to *Scientific American.* Otherwise, if the inscription of a question counts both as the question and as unit of measure, I at least get a booby prize. But I think somehow I bent the rules.

My own solutions to the problem of the self-answering question are actually not so much self-*answering* as self-*provoking,* as in the following example: "Why are you asking me *that* out of the blue?" It is obvious that when the question is asked out of the blue, it might well elicit an identical response, indicating the hearer's bewilderment.

Philip Cohen relayed the following anecdote about a self-answering question, from Damon Knight: "Terry Carr, an old friend, sent us a riddle on a postcard, then the answer on another postcard. Then he sent us another riddle: 'How do you keep a turkey in suspense?' and never sent the answer. After about two weeks, we realized that *was* the answer."

*   *   *

Several of the real masterpieces sent in belong to what I call the *self-documenting* category, of which a simple example is Jonathan Post's "This sentence contains ten words, eighteen syllables and sixty-four letters." A neat twist is supplied by John Atkins in his sentence " 'Has eighteen letters' does." The self-documenting form can get much more convoluted and introspective. An example by the wordplay master Howard Bergerson was brought to my attention by Philip Cohen. It goes:

> In this sentence, the word *and* occurs twice, the word *eight* occurs twice, the word *four* occurs twice, the word *fourteen* occurs four times, the word *in* occurs twice, the word *seven* occurs twice, the word *the* occurs fourteen times, the word *this* occurs twice, the word *times* occurs seven times, the word *twice* occurs eight times and the word *word* occurs fourteen times.

That is good, but the gold medal in the category is reserved for Lee Sallows, who submitted the following *tour de force:*

> Only the fool would take trouble to verify that his sentence was composed of ten a's, three b's, four c's, four d's, forty-six e's, sixteen f's, four g's, thirteen h's, fifteen i's, two k's, nine l's, four m's, twenty-five n's, twenty-four o's, five p's, sixteen r's, forty-one s's, thirty-seven t's, ten u's, eight v's, eight w's, four x's, eleven y's, twenty-seven commas, twenty-three apostrophes, seven hyphens, and, last but not least, a single !

I (perhaps the fool) did take trouble to verify the whole thing. First, though, I carried out some spot checks. And I must say that when the first random spot check worked (I think I checked the number of 'g's), this had a strong psychological effect: all of a sudden, the credibility rating of the *whole sentence* shot way up for me. It strikes me as weird (and wonderful) how, in certain situations, the verification of a tiny percentage of a theory can serve to powerfully strengthen your belief in the full theory. And perhaps that's the whole point of the sentence!

The noted logician Raphael Robinson submitted a playful puzzle in the self-documenting genre. Readers are asked to complete the following sentence:

> In this sentence, the number of occurrences of 0 is __, of 1 is __, of 2 is __, of 3 is __, of 4 is __, of 5 is __, of 6 is __, of 7 is __, of 8 is __, and of 9 is __.

Each blank is to be filled with a numeral of one or more digits, written in decimal notation. Robinson states that there are exactly two solutions. Readers might also search for two sentences of this form that document each other, or even longer loops of that kind.

Clearly the ultimate in self-documentation would be a sentence that does more than merely inventory its parts; it would be a sentence that includes a rule as well, telling all the King's men how to put those parts back together again to create a full sentence—in short, a self-reproducing sentence. Such a sentence is Willard Van Orman Quine's English rendition of Kurt Gödel's classic metamathematical homage to Epimenides the Cretan:

"yields falsehood when appended to its quotation." yields falsehood when appended to its quotation.

Quine's sentence in effect tells the reader how to construct a replica of the sentence being read, and then (just for good measure) adds that the replica (not *itself*, for heaven's sake!) asserts a falsity! It's a bit reminiscent of the famous remark made by Epilopsides the Concretan (second cousin of Epimenides) to Flora, a beautiful young woman whose ardent love he could not return (he was betrothed to her twin sister Fauna): "Take heart, my dear. I have a suggestion that may cheer you up. Just take one of these cells from my muscular biceps here, and clone it. You'll soon wind up with a dashing blade who looks and thinks just like me! But *do* watch out for him—he is given to telling beautiful women real whoppers!"

*   *   *

In the early 1950's, John von Neumann worked hard trying to design a machine that could build a replica of itself out of raw materials. He came up with a theoretical design consisting of hundreds of thousands of parts. Seen in hindsight and with a considerable degree of abstraction, the idea behind von Neumann's self-reproducing machine turns out to be pretty similar to the means by which DNA replicates itself. And this in turn is close to Gödel's method of constructing a self-referential sentence in a mathematical language in which at first there seems to be no way of referring to the language itself.

The First Every-Other-Decade Von Neumann Challenge is thus hereby presented for ambitious readers: Create a comprehensible and not unreasonably long self-documenting sentence that not only lists its parts (at the word level or, better yet, the letter level) but also tells how to put them together so that the sentence reconstitutes itself. (Notice, by the way, the requirement is that the sentence be *not unreasonably long,* which is different —very different—from being *reasonably long.*) The parts list (or *seed*) should be an inventory of words or typographical symbols, more or less as in the sentences created by Howard Bergerson and Lee Sallows. The inventoried symbols should in some way be clearly distinguishable from the text that talks about them. For instance, they can be enclosed in quotation marks, printed in another typeface, or referred to by name. It is not so important what convention is adopted, so long as the distinction is sharp. The rest of the sentence (the *building rule*) should be printed normally, since it is to be regarded not as typographical raw material but as a set of instructions. This is the use-mention distinction I discussed in Chapter 1, and to disregard it is a serious conceptual weakness. (It is a flaw in Sallows' sentence that slightly tarnishes the gold on his medal.)

The building rule may not talk about normally-printed material—only about parts of the inventory. Thus, it is not permitted for the building rule to refer to itself in any way! The building rule has to describe structure explicitly. Furthermore (and this is the subtlest and probably the most often overlooked aspect of self-reference), the building rule must specify which parts are to be printed normally and which parts in quotes (or however the raw materials are being indicated). In this respect, Bergerson's sentence fails. Although, to its credit, it sharply distinguishes between use and mention by relying on upper case for the names of inventory items and lower case for item counts and filler words, it does not have separate inventories for items in upper case and lower case. Instead it lumps the two together, blurring a vital distinction.

In the Von Neumann Challenge, extra points will be awarded for solutions given in Basic English, or whose seed is entirely at the letter level (as in Sallows' sentence). The Quine sentence, although it clearly incorporates a seed (the seven-word phrase in quotation marks) and a building rule (that of appending something to its quotation), is not a legal entry because its seed is too far from being raw material. It is so structured that it is like a fetus more than it is like a zygote.

*   *   *

There is a very good reason, by the way, that the Quine sentence's seed is so complicated—in fact, is identical with the building rule, except for the quotation marks. The reason is simple to state: You've got to *build a copy of the building rule* out of raw materials, and the more your building rule looks like your seed, the simpler it will be to build a copy of it from a copy of the seed. To make a full new sentence, all you need to do is make two copies of the seed, carry out whatever simple manipulations will convert one copy of the seed into the building rule, and then splice the other copy of the seed onto the newly minted building rule to make up a complete new sentence, fresh off the assembly line.

To make this clearer, it is helpful to show a slight variation on Quine's sentence. Imagine that you could recognize only the lowercase roman letters, and that uppercase letters were alien to you. Then text printed in upper case would, for all practical purposes, be devoid of meaning or interest, whereas text in lower case would be full of meaning and interest, able to suggest ideas or actions in your mind. Now suppose someone gave you a conversion table that matched each uppercase letter with its lowercase counterpart, so that you could "decode" uppercase text. Then one day you came across this piece of "meaningless" uppercase text:

YIELDS A FALSEHOOD WHEN USED AS THE SUBJECT OF ITS LOWERCASE VERSION

On being decoded, it would yield a lowercase sentence, or rather, a lowercase sentence fragment—a predicate without a subject. Suggestive, eh? What might you try out, as a possible subject of that predicate?

This notion of two parallel alphabets, one in which text is inert and meaningless and the other in which text is active and meaningful, may strike you as yielding no more than a minor variation on Quine's sentence, but in fact it is very similar to an exceedingly clever trick that nature discovered and has exploited in every cell of every living organism. Our seed—our genome—our DNA—is a huge long volume of *inert* text written in a chemical alphabet that has 64 "uppercase" letters (codons). Our building rules—our enzymes—are short, pithy slogans of *active* text written in a different chemical alphabet that has just twenty "lowercase" letters (amino acids). There is a map (the genetic code) that converts uppercase letters into lowercase ones. Obviously, some lowercase letters must correspond to more than one uppercase letter, but here that is a detail. It also turns out that three characters of the uppercase alphabet are not letters but punctuation marks telling where one pithy slogan ends and the next one begins—but again, these are details. (See Chapter 27 for some of those details.)

Once you know this mapping, you often won't even remember to distinguish between the two chemical alphabets: the inert uppercase codon alphabet and the active lowercase amino acid alphabet. The main thing is that, armed with the genetic code, you can read the DNA book (seed) as if it were a sequence of enzyme slogans (building rules) telling how to write a new DNA book together with a new set of enzyme slogans! It is a perfect parallel to our variation on the Quine sentence, where inert, uppercase seed-text was converted into active, lowercase rule-text that told how to make a copy of the full Quine sentence, given its seed.

A cell's DNA and enzymes act like the seed and building rules of Quine's sentence, or the parts list and building rules of von Neumann's self-reproducing automaton—or then again, like the seed and building rules of computer programs that print themselves out. It is amazing how universal this mechanism of self-reference is, and for that reason I always find it quaint that people who rant and rave against the silliness of self-reference are themselves composed of trillions and trillions of tiny self-referential molecules.

* * *

Scott Kim and I constructed an intriguing pair of sentences:

The following sentence is totally identical with this one, except that the words 'following' and 'preceding' have been exchanged, as have the words 'except' and 'in', and the phrases 'identical with' and 'different from'.

Self-Referential Sentences: A Follow-Up

The preceding sentence is totally different from this one, in that the words 'preceding' and 'following' have been exchanged, as have the words 'in' and 'except', and the phrases 'different from' and 'identical with'.

At first glance, these sentences are reminiscent of a two-step variant on the Epimenides paradox ("The following sentence is true."; "The preceding sentence is false."). On second glance, though, they are seen to say exactly the same thing. Curiously, my Australian colleague and sometime alter ego, Egbert B. Gebstadter, writing in his ever fascinating but often-furiating monthly row "Thetamagical Memas" (which appears in *Literary Australian*), disagrees with me; he maintains they say totally different things. (See figure 2-1.)

Not surprisingly, several of the sentences submitted by readers had a paradoxical flavor. Some were variants on Bertrand Russell's paradox about the barber who shaves all those who do not shave themselves, or the set of all sets that do not include themselves as elements. For instance, Gerald Hull concocted this strange sentence: "This sentence refers to every sentence that does not refer to itself." Is Hull's concoction self-referential, or is it not? In a similar vein, Michael Gardner cited a Reed College senior thesis whose dedication ran: "This thesis is dedicated to all those who did not dedicate their theses to themselves." The book *Model Theory,* by C. C. Chang and H. J. Keisler, bears a similar dedication, as Charles Brenner pointed out to me. He also suggested another variant on Russell's paradox: Write a computer program that prints out a list of all programs that do not ever print themselves out. The question is, of course: Will this program ever print itself out?

One of the most disorienting sentences came from Robert Boeninger: "This sentence does in fact not have the property it claims not to have." Got that? A serious problem seems to be to figure out just what property it is that the sentence claims it lacks.

The Dutch mathematician Hans Freudenthal sent along a charming paradoxical anecdote based on self-reference:

> There is a story by the eighteenth-century German Christian Gellert called "Der Bauer und sein Sohn" ("The Peasant and His Son"). One day during a walk, when the son tells a big lie, his father direly warns him about the "Liars' Bridge", which they are approaching. This bridge always collapses when a liar walks across it. After hearing this frightening warning, the boy admits his lie and confesses the truth.
>
> When I [Freudenthal] told a ten-year-old boy this story, he asked me what happened when they eventually came to the bridge. I replied, "It collapsed under the father, who had lied, since in fact there is no Liars' Bridge." (Or did it?)

C. W. Smith, writing from London, Ontario, described a situation reminiscent of the Epimenides paradox:

THETAMAGICAL MEMAS:

Seeking the Whence
of Letter and Spirit

EGBERT B. GEBSTADTER

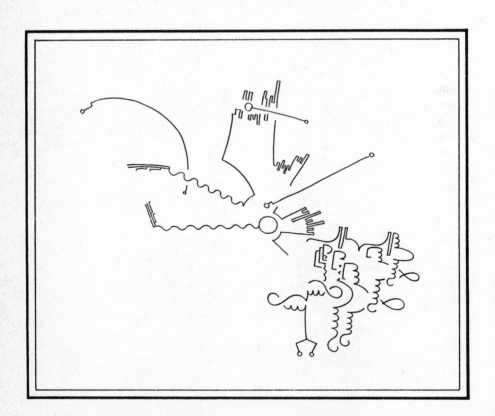

A Copious Concatenation of
Artsy, Scientistic, and Literal Mumbo-Jumbo

During the 1960's, standing alone in the midst of a weed-strewn field in this city, there was a weathered sign that read: "$25 reward for information leading to the arrest and conviction of anyone removing this sign." For whatever it's worth, the sign has long since disappeared. And so, for that matter, has the field.

Incidentally, the Epimenides paradox should not be confused with the Nixonides paradox, first uttered by Nixonides the Cretin in A.D. 1974: "This statement is inoperative." Speaking of Epimenides, one of the most elegant variations on his paradox is the "Errata" section in a hypothetical book described by Beverly Rowe. It looks like this:

(vi)

Errata

Page (vi): For *Errata,* read *Erratum*

Closely related to the truly paradoxical sentences are those that belong to what I call the *neurotic* and *healthy* categories. A healthy sentence is one that, so to speak, practices what it preaches, whereas a neurotic sentence is one that says one thing while doing its opposite. Alan Auerbach has given us a good example in each category. His healthy sentence is: "Terse!" His neurotic sentence is: "Proper writing—and you've heard this a million times —avoids exaggeration." Here's a healthy one by Brad Shelton: "Fourscore and seven words ago, this sentence hadn't started yet." One of the jootsingest of sentences came from Carl Bender:

The rest of this sentence is written in Thailand, on

Consider a related sentence sent in by David Stork: "It goes without saying

---

FIGURE 2–1. *The cover of Egbert B. Gebstadter's latest book, showing some of his "Whorly Art." See the Bibliography for a short description of the book.*

Gebstadter, best known as the author of Copper, Silver, Gold: an Indestructible Metallic Alloy, *also co-edited* The Brain's U *with Australian philosopher Denial E. Dunnitt, and for two and a half years wrote a monthly row ("Thetamagical Memas") for* Literary Australian. *Having spent the last several years in the Psychology Department of Pakistania University in Wiltington, Pakistania, he has recently joined the faculty of the Computer Science Department of the University of Mishuggan in Tom Treeline, Mishuggan, where he occupies the Rexall Chair in the College of Art, Sciences, and Letters. His current research projects in IA (intelligent artifice) are called Quest-Essence, Mind Pattern, Intellect, and Studio. His focus is on deterministic sequential models of digital emotion.*

---

that . . ." To which category does it belong? Perhaps it is a psychotic sentence.

Pete Maclean contributed a puzzling one: "If the meanings of 'true' and 'false' were switched, then this sentence wouldn't be false." I'm still scratching my head over what that means! Dan Krimm wrote to tell me: "I've heard that this sentence is a rumor." Linda Simonetti contributed the following example, "which actually is not a complete sentence, but merely a subordinate clause." Douglas Wolfe offered the following neurotic rule of thumb: "Never use the imperative, and it is also never proper to construct a sentence using mixed moods." David Moser reminded me of a slogan that the *National Lampoon* once used: "So funny it sells without a slogan!" Perry Weddle wrote, "I'm trying to teach my parrot to say, 'I don't understand a thing I say.' When *I* say it, it's viciously self-referential, but in *his* case?" Stephen Coombs pointed out that "A sentence may self-refer in the verb." My mother, Nancy Hofstadter, heard Secretary of State Alexander Haig describe a warning message to the Russians as "a calculated ambiguity that would be clearly understood". Yes, Sir!

Jim Propp submitted a sequence of sentences that slide elegantly from the neurotically healthy to the healthily neurotic:

(1) This sentence every third, but it still comprehensible.
(2) This would easier understand fewer had omitted.
(3) This impossible except context.
(4) *4'33"* attempt idea.
(5)

The penultimate sentence refers to John Cage's famous piece of piano music consisting of four minutes and 33 seconds of silence. The last sentence might well be an excerpt from *The Wit and Wisdom of Spiro T. Agnew,* although it is too short an excerpt to be sure. Propp also sent along the following healthy sentence, which was apparently inspired by his readings in the book *Intelligence in Ape and Man,* by David Premack: "By the 'productivity' of language, I mean the ability of language to introduce new words in terms of old ones."

Philosopher Howard DeLong contributed what might be considered a neurotic syllogism:

> All invalid syllogisms break at least one rule.
> This syllogism breaks at least one rule.
> ———————————————————————
> Therefore, this syllogism is invalid.

Several readers pointed out phrases and jokes that have been making the rounds. D.A. Treissman, for instance, reminded me that "Nostalgia ain't what it used to be." Henry Taves mentioned the delightful T-shirts adorned with statements such as "My folks went to Florida and all they brought back for me was this lousy T-shirt!" And John Fletcher described an episode of the television program *Laugh-In* a few years ago on which Joanne Worley sang, "I'm just a girl who can't say 'n . . .', 'n . . .', 'n . . .' ". John Healy wrote, "I used to think I was indecisive, but now I'm not so sure."

I myself have a few contributions to this collection. A neurotic one is: "In this sentence, the concluding three words 'were left out'." Or is it neurotic? These things confuse me! In any case, a most healthy sentence is: "This sentence offers its reader(s) various alternatives/options that he or she (or they) is (are) free to accept and/or reject." And then there is the inevitable "This sentence is neurotic." The thing is, if it *is* neurotic, it practices what it preaches, so it's healthy and therefore cannot be neurotic—but then if it *isn't* neurotic, it's the opposite of what it claims to be, so it's *got* to be neurotic. No wonder it's neurotic, poor thing!

Speaking of neurotic sentences, what about sentences with identity crises? These are, in some sense, the most interesting ones of all to me. A typical example is Dan Krimm's vaguely apprehensive question, "If I stated something else, would it still be me?" I thought this could be worded better, so I revised it slightly, as follows: "If I said something else, would it still be me saying it?" I still was not happy, so I wrote one more version: "In another world, could I have been a sentence about Humphrey Bogart?" When I paused to reflect on what I had done, I realized that in reworking Dan's sentence, I had tampered with its identity in the very way it feared. The question remained, however: Were all these variants really the same sentence, deep down? My last experiment along these lines was: "In another world, could this sentence have been Dan Krimm's sentence?"

Clearly some readers were thinking along parallel lines, since John Atkins queried, "Can anyone explain why this would still be the same magazine without this query, and yet this would not be the same query without this word?" (Of course, just which word "this word" refers to is a little vague, but the idea is clear.) And Loul McIntosh, who works at a rehabilitation center for formerly schizophrenic patients, had a question connecting personal identity with self-referential sentences: "If I were you, who would be reading this sentence?" She then added: "That's what I get for working with schizophrenics." This brings me to Peter M. Brigham, M.D., who in his work ran across a severe case of literary schizophrenia: "You have, of course, just begun reading the sentence that you have just finished reading." It's one of my favorites.

Pursuing the slithery snake of self in his own way, Uilliam M. Bricken, Jr., wrote in: "If you think this sentence is confusing, then change one pig." Now, *anyone* can see that this doesn't make any sense at all. Surely what he meant was, "If you think this sentence is confusing, then *roast* one pig."— don't ewe agree? By the by, if ewe think "Uilliam" is confusing, then roast one ewe. And while we're mentioning ewes, what's a nice word like "ewe" doing in a foxy paragraph like this?

A while back, driving home late at night, I tuned in to a radio talk show about pets. A heated discussion was taking place about the relative merits of various species, and at one point the announcer mused, "If a dog had written this broadcast, he might have said that *people* are inferior because they don't wag their tails." This gave me paws for thought: What might this column have been like if it had been written by a dog? I can't say for *sure*, but I have a hunch it would have been about chasing squirrels. And it might have had a paragraph speculating about what this column would have been like if it had been written by a squirrel.

*   *   *

I think my favorite of all the sent-in-ces was one contributed by Harold Cooper. He was inspired by Scott Kim's counterfactual self-referential question: "What would this sentence be like if π were 3?" His answer is shown in Figure 2-2. This, to me, exemplifies the meaning of the verb

---

If π were 3, this sentence would look something like this.

FIGURE 2–2. *A counterfactual self-referential sentence, inspired by Harold Cooper and Scott Kim.*

"joots". The six-sided 'o's represent the fact that the ratio of the circumference to the diameter of a hexagon is 3. Clearly, in Cooper's mind, if π were 3, why, what more natural conclusion than that *circles would be hexagons*! Who could ever think otherwise? I was intrigued by the fact that, as π's value slipped to 3, not only did circles turn into hexagons, but also the interrogative mood slipped into the declarative mood. Remember that the question asked how the question itself would be in that strange subjunctive world. Would it lose its curiosity about itself and cease to be a question? I did not see why that personality trait of the sentence would be affected by the value of π. On the other hand, it seemed obvious to me that if π were 3, the antecedent of the conditional should no longer be subjunctive. In fact, rather than saying "if π were 3", it should say, "*because* π is 3" (or something to that effect). Putting my thoughts together, then, I came up with a slight variation on Cooper's sentence: "What is this sentence like, π being 3 (as usual)?"

Self-Referential Sentences: A Follow-Up

Several readers were interested in sentences that refer to the language they are in (or not in, as the case may be). An example is "If you spoke English, you'd be in your home language now." Jim Propp sent in a delightful pair of such sentences that need to be read together:

Cette phrase se réfère à elle-même, mais d'une manière peu évidente à la plupart des Américains.

Plim glorkle pegram ut replat, trull gen ris clanter froat veb nup lamerack gla smurp Earthlings.

If you do not understand the first sentence, just get a Martian friend to help you decode the second one. That will provide hints about the first. (I apologize for leaving off the proper Martian accent marks, but they were not available in this typeface.)

* * *

Last January, I published several sentences by David Moser and mentioned that he had written an entire story consisting of self-referential sentences. Many readers were intrigued. I decided there could be no better way to conclude this column than to print David's story in its entirety. So here 'tis!

This Is the Title of This Story, Which Is Also Found Several Times in the Story Itself

This is the first sentence of this story. This is the second sentence. This is the title of this story, which is also found several times in the story itself. This sentence is questioning the intrinsic value of the first two sentences. This sentence is to inform you, in case you haven't already realized it, that this is a self-referential story, that is, a story containing sentences that refer to their own structure and function. This is a sentence that provides an ending to the first paragraph.

This is the first sentence of a new paragraph in a self-referential story. This sentence is introducing you to the protagonist of the story, a young boy named Billy. This sentence is telling you that Billy is blond and blue-eyed and American and twelve years old and strangling his mother. This sentence comments on the awkward nature of the self-referential narrative form while recognizing the strange and playful detachment it affords the writer. As if illustrating the point made by the last sentence, this sentence reminds us, with no trace of facetiousness, that children are a precious gift from God and that the world is a better place when graced by the unique joys and delights they bring to it.

This sentence describes Billy's mother's bulging eyes and protruding tongue and makes reference to the unpleasant choking and gagging noises she's making. This sentence makes the observation that these are uncertain and difficult times, and that relationships, even seemingly deep-rooted and permanent ones, do have a tendency to break down.

Introduces, in this paragraph, the device of sentence fragments. A sentence fragment. Another. Good device. Will be used more later.

This is actually the last sentence of the story but has been placed here by mistake. This is the title of this story, which is also found several times in the story itself. As Gregor Samsa awoke one morning from uneasy dreams he found himself in his bed transformed into a gigantic insect. This sentence informs you that the preceding sentence is from another story entirely (a much better one, it must be noted) and has no place at all in this particular narrative. Despite the claims of the preceding sentence, this sentence feels compelled to inform you that the story you are reading is in actuality "The Metamorphosis" by Franz Kafka, and that the sentence referred to by the preceding sentence is the *only* sentence which does indeed belong in this story. This sentence overrides the preceding sentence by informing the reader (poor, confused wretch) that this piece of literature is actually the Declaration of Independence, but that the author, in a show of extreme negligence (if not malicious sabotage), has so far failed to include even *one single sentence* from that stirring document, although he has condescended to use a small sentence *fragment,* namely, "When in the course of human events", embedded in quotation marks near the end of a sentence. Showing a keen awareness of the boredom and downright hostility of the average reader with regard to the pointless conceptual games indulged in by the preceding sentences, *this* sentence returns us at last to the scenario of the story by asking the question, "Why is Billy strangling his mother?" This sentence attempts to shed some light on the question posed by the preceding sentence but fails. *This* sentence, however, succeeds, in that it suggests a possible incestuous relationship between Billy and his mother and alludes to the concomitant Freudian complications any astute reader will immediately envision. Incest. The unspeakable taboo. The universal prohibition. Incest. And notice the sentence fragments? Good literary device. Will be used more later.

This is the first sentence in a new paragraph. This is the last sentence in a new paragraph.

This sentence can serve as either the beginning of the paragraph or the end, depending on its placement. This is the title of this story, which is also found several times in the story itself. This sentence raises a serious objection to the entire class of self-referential sentences that merely comment on their own function or placement within the story (*e.g.,* the preceding four sentences), on the grounds that they are monotonously predictable, unforgivably self-indulgent, and merely serve to distract the reader from the real subject of this story, which at this point seems to concern strangulation and incest and who knows what other delightful topics. The purpose of this sentence is to point out that the preceding sentence, while not itself a member of the class of self-referential sentences it objects to, nevertheless *also* serves merely to distract the reader from the real subject of this story, which actually concerns Gregor Samsa's inexplicable transformation into a gigantic insect (despite the vociferous counterclaims of other well-meaning although misinformed sentences). This sentence can serve as either the beginning of a paragraph or the end, depending on its placement.

This is the title of this story, which is also found several times in the story itself. This is *almost* the title of the story, which is found only once in the story itself. This sentence regretfully states that up to this point the self-referential mode of narrative has had a paralyzing effect on the actual progress of the story itself—that is, these sentences have been so concerned with analyzing themselves and their role in the story that they have failed by and large to perform their function as communicators of events and ideas that one hopes coalesce into a plot, character development, etc.—in short, the very *raisons d'être* of any respectable, hardworking sentence in the midst of a piece of compelling prose fiction. This sentence in addition points out the obvious analogy between the plight of these agonizingly self-aware sentences and similarly afflicted human beings, and it points out the analogous paralyzing effects wrought by excessive and tortured self-examination.

The purpose of this sentence (which can also serve as a paragraph) is to speculate that if the Declaration of Independence had been worded and structured as lackadaisically and incoherently as this story has been so far, there's no telling what kind of warped libertine society we'd be living in now or to what depths of decadence the inhabitants of this country might have sunk, even to the point of deranged and debased writers constructing irritatingly cumbersome and needlessly prolix sentences that sometimes possess the questionable if not downright undesirable quality of referring to themselves and they sometimes even become run-on sentences or exhibit other signs of inexcusably sloppy grammar like unneeded superfluous redundancies that almost certainly would have insidious effects on the lifestyle and morals of our impressionable youth, leading them to commit incest or even murder and maybe *that's* why Billy is strangling his mother, because of sentences *just like this one,* which have no discernible goals or perspicuous purpose and just end up anywhere, even in mid

Bizarre. A sentence fragment. Another fragment. Twelve years old. This is a sentence that. Fragmented. And strangling his mother. Sorry, sorry. Bizarre. This. More fragments. This is it. Fragments. The title of this story, which. Blond. Sorry, sorry. Fragment after fragment. Harder. This is a sentence that. Fragments. Damn good device.

The purpose of this sentence is threefold: (1) to apologize for the unfortunate and inexplicable lapse exhibited by the preceding paragraph; (2) to assure you, the reader, that it will not happen again; and (3) to reiterate the point that these are uncertain and difficult times and that aspects of language, even seemingly stable and deeply rooted ones such as syntax and meaning, do break down. This sentence adds nothing substantial to the sentiments of the preceding sentence but merely provides a concluding sentence to this paragraph, which otherwise might not have one.

This sentence, in a sudden and courageous burst of altruism, tries to abandon the self-referential mode but fails. This sentence tries again, but the attempt is doomed from the start.

This sentence, in a last-ditch attempt to infuse some iota of story line into this paralyzed prose piece, quickly alludes to Billy's frantic cover-up attempts, followed by a lyrical, touching, and beautifully written passage wherein Billy is reconciled with his father (thus resolving the subliminal Freudian conflicts obvious to any astute reader) and a final exciting police chase scene during which Billy is accidentally shot and killed by a panicky rookie policeman who is coincidentally named Billy. This sentence, although basically in complete sympathy with the laudable efforts of the preceding action-packed sentence, reminds the reader that such allusions to a story that doesn't, in fact, yet exist are no substitute for the real thing and therefore will not get the author (indolent goof-off that he is) off the proverbial hook.

Paragraph. Paragraph. Paragraph. Paragraph. Paragraph. Paragraph. Paragraph. Paragraph. Paragraph. Paragraph. *Paragraph.* Paragraph. Paragraph. Paragraph.

The purpose. Of this paragraph. Is to apologize. For its gratuitous use. Of. Sentence fragments. Sorry.

The purpose of this sentence is to apologize for the pointless and silly adolescent games indulged in by the preceding two paragraphs, and to express regret on the part of us, the more mature sentences, that the entire tone of this story is such that it can't seem to communicate a simple, albeit sordid, scenario.

This sentence wishes to apologize for all the needless apologies found in this story (this one included), which, although placed here ostensibly for the benefit of the more vexed readers, merely delay in a maddeningly recursive way the continuation of the by-now nearly forgotten story line.

This sentence is bursting at the punctuation marks with news of the dire import of self-reference as applied to sentences, a practice that could prove to be a veritable Pandora's box of potential havoc, for if a sentence can refer or allude to itself, why not a lowly subordinate clause, perhaps *this very* clause? Or this sentence fragment? Or three words? Two words? *One?*

Perhaps it is appropriate that this sentence gently and with no trace of condescension remind us that these are indeed difficult and uncertain times and that in general people just aren't nice enough to each other, and perhaps we, whether sentient human beings or sentient sentences, should just *try harder.* I mean, there *is* such a thing as free will, there *has* to be, and this sentence is proof of it! Neither this sentence nor you, the reader, is completely helpless in the face of all the pitiless forces at work in the universe. We should stand our ground, face facts, take Mother Nature by the throat and just *try harder.* By the throat. Harder. Harder, harder.

Sorry.

This is the title of this story, which is also found several times in the story itself.

This is the last sentence of the story. This is the last sentence of the story. This is the last sentence of the story. This is.

Sorry.

Post Scriptum.

As you can see, there is a vast amount of self-referential material out there in the world. To pick only the very best is a monumental task, and certainly a highly subjective one. I would like to include here some of the things that I had to omit from the second self-reference column with great regret, as well as some of the things that were sent in later, in response to it.

First, though, I would like to mention an amusing incident. When Lee Sallows' self-documenting sentence was to be printed in the narrow columns of *Scientific American,* nobody remembered to tell the typesetters not to break any unhyphenated words. As luck would have it, two such breaks were introduced, yielding two spurious hyphens, thus spoiling (in a superficial sense) the accuracy of his construction. How subtly one can get snagged when self-reference is concerned!

Paul Velleman sent me a copy of the front page of the *Ithaca Journal,* dated January 26, 1981, with a banner headline saying "Ex-hostages enjoy their privacy". He wrote, "I think it may be self-referent (and self-contradictory) in a different way than your other examples because the medium, positioning, and size of its printing are all necessary components of the contradiction." When I looked at the page, I simply saw nothing self-referential. I thought maybe I was supposed to look at the flip side, for some reason, but that had even less of interest. So I looked back at the headline, and suddenly it hit me: How can people "enjoy privacy" when it's being blared across the front page of newspapers across the nation?

Along the same lines, soon thereafter I came across a photograph of Lady Di in tears, and in the caption her tears were explained this way: "Lady Di was apparently overcome by the strain of the impending royal wedding and having her every move in public watched by thousands. See story on page A20. Details on the royal honeymoon, page A7."

John M. Lankford wrote me a long letter from Japan on self-reference, remarkably similar in some ways to the one from Flash qFiasco. The most memorable paragraph in his letter was the following one:

Here in Japan, twice a week, I teach a little class in English for a group of university students—mainly graduate students in the sciences. I spent one class hour taking some of your sentences from the *Scientific American* article, writing them on the blackboard, and asking the students what they meant. The students had a fairly good command of written English, but they were poor in their command of idiom, quick verbal response, and, for want of a better term, "humor of the abstract". As I suspected, many of the sentences—perhaps the most interesting of them—die when ripped from their cultural context. I had quite a bit of difficulty getting across the idea that the pronoun "I" could refer to the sentence as well as to the writer of the sentence. Pronouns cause a lot of trouble in Japan. For example, when I ask someone, "Am I wearing a blue jacket?", they might frequently reply, "Yes, I am wearing a blue jacket." This confusion is easy in Japanese due to the relative lack of pronouns in ordinary speech. Of course you can imagine the extra layers of incomprehension that would arise in reading your sentences if the boundaries between "you" and "I" were rather vague.

On a visit to Gettysburg, I read Abraham Lincoln's Gettysburg address, and for the first time its curious self-reference struck me: "The world will little note nor long remember what we say here." Lincoln had no way of knowing at the time, but this would turn out to be an extremely false sentence (if it is permissible to speak of *degrees* of falsity). In fact, that sentence itself is a very memorable one. While we're on presidential self-reference, listen to this self-descriptive remark by former President Ford: "I am the first to admit that I am no great orator or no person that got where I have gotten by any William Jennings Bryan technique." I guess that where Lincoln's sentence was extremely false, Ford's is extremely true. Here is a final self-referential sentence along presidential lines:

> If John F. Kennedy were reading this sentence, Lee Harvey Oswald would have missed.

*   *   *

One of the best self-answering questions came up naturally in the course of a very brief telephone call I made to a restaurant one evening. It went this way: "May I help you?" to which I answered, "You've already helped me—by telling me that you're open today. Thank you. Bye!" And here's a "self-deferential" sentence by Don Byrd: "I am not as witty as my author."

I received this anonymous letter in the mail: "I received this anonymous letter in the mail so I can't credit the author."—so I can't credit the author. I also received a request from someone living in Calgary, Alberta, whose name I forget (but if he's reading this, he'll know who he is) who wrote "This is my feeble way of attempting to get my name into print." I hope this satisfies him.

And now a few miscellaneous examples by me, culled from a second wild binge of self-referential sentence-writing I engaged in not long ago. The first three involve translation issues.

Self-Referential Sentences: A Follow-Up

One me has translated at the foot of the letter of the French.

Would not be anomalous if were in Italian.

When one this sentence into the German to translate wanted, would one the fact exploit, that the word order and the punctuation already with the German conventions agree.

How come *this* noun phrase doesn't denote the same thing as *this* noun phrase does?

Every last word in this sentence is a grotesque misspelling of "towmatow".

I don't care *who* wrote this sentence—whoever he is, he's a damn sexist!

This analogy is like lifting yourself by your own bootstraps.

Although this sentence begins with the word "because", it is false.

Despite the fact that it opens like a two-pronged pitchfork—or rather, because of it—this sentence resembles a double-edged sword.

This line from Shakespeare has delusions of grandeur.

If writers were bakers, this sentence would be exactly a dozen words long.

If this sentence had been on the previous page, this very moment would have occurred approximately 60 seconds ago.

This sentence is helping to increase the likelihood of nuclear war by distracting you from the more serious concerns of the world and beguiling you with the trivial joys of self-reference.

This sentence is helping to decrease the likelihood of nuclear war by chiding you for indulging in the trivial joys of self-reference and reminding you of the more serious concerns of the world.

We *mention* "our gigantic nuclear arsenal" in order not to *use* it.

The whole point of this sentence is to make clear what the whole point of this sentence is.

This last one's bizarre circularity reminds me of the number *P* that I invented a couple of years ago. *P* is, for each individual, the number of minutes per month that that person spends thinking about the number *P*. For me, the value of *P* seems to average out at about 2. I certainly wouldn't want it to go much above that! I find it crosses my mind most often when I'm shaving.

* * *

Dr. J. K. Aronson from Oxford, England, sent in some of the most marvelous discoveries. Here is one of his best:

'T' is the first, fourth, eleventh, sixteenth, twenty-fourth, twenty-ninth, thirty-third, . . .

The sentence never ends, of course. He also submitted a wonderful complementary pair that faked me out beautifully. His challenge to you is: Try deciphering the first before you read the second.

I eee oai o ooa a e ooi eee o oe.

Ths sntnc cntns n vwls nd th prcdng sntnc n cnsnnts.

One that reminds me somewhat of Aronson's last sentence above is the following spoof on the ads that I believe you can still find in the New York subway, after all these years:

f y cn rd ths, itn tyg h myxbl cd.

By a remarkable coincidence, the remainder of Carl Bender's sentence "The rest of this sentence is written in Thailand, on" was discovered in, of all places, Bangkok, Thailand, by Gregory Bell, who lives there. He has luckily provided me with a perfect copy of it, so for all those who were dying of suspense, it is shown in Figure 2-3.

One evening during a bad electrical storm, I got the following message on the computer from Marsha Meredith:

I]ion't be able to work at all tonight b]iecause of the w&atherBr/ I]i'm getting too many bad characters (as you can see). Ioo baw3d—I get spurious characters]i all over]ithe place—talk totrrRBow,1F7U Marsha.

FIGURE 2–3. *The conclusion of Carl Bender's sentence fragment ("The rest of this sentence is written in Thailand, on"), discovered by Gregory Bell on a scrap of paper in Bangkok, Thailand. Translated, it says: "this sheet of paper and is in Thai".*

กระดาษแผ่นนี้และเขียนเป็นภาษาไทย

Self-Referential Sentences: A Follow-Up

I wish she had had the patience to type more carefully, so that I could have understood what her problem was.

The sentences having to do with identity in counterfactual worlds, such as Dan Krimm's and its alter egos, reminded me of a blurb by E. O. Wilson I read recently on Lewis Thomas' latest book: "If Montaigne had possessed a deep knowledge of twentieth-century biology, he would have been Lewis Thomas." Ah me, the flittering elf of self! And Banesh Hoffmann, in *Relativity and Its Roots,* has written: "How safe we would be from death by nuclear bomb had we been born in the time of Shakespeare." Sure, except we'd also all be long dead—unless, of course, the 24th-century doctors who will invent immortality pills had *also* been born in Shakespeare's time!

The following self-referential poem just came to me one day:

> Twice five syllables,
> Plus seven, can't say much—but . . .
> That's haiku for you.

The genre of self-referential poetry—including haiku—was actually quite popular. Tom McDonald submitted this non-limerick:

> A very sad poet was Jenny—
> Her limericks weren't worth a penny.
> In technique they were sound,
> Yet somehow she found
> Whenever she tried to write any,
> That she always wrote one line too many!

Several people sent in complex poems of various sorts, and mentioned books of them, such as John Hollander's *Rhyme's Reason,* a collection of poems describing their own forms.

*   *   *

Self-referential book titles are enjoying a mild vogue these days. Raymond Smullyan was one of the most enthusiastic explorers of the potential of this idea, using the titles *What Is the Name of This Book?* and *This Book Needs No Title.* Actually, I think *Needs No Title* would have said it more crisply, or maybe just *No Title.* Come to think of it, why not *No,* or even just plain *?* (I hope you could tell that those blanks were in *italics!*)

Other self-referential book titles I have collected include these:

Forget all the rules you ever learned about graphic design.
   Including the ones in this book.
Steal This Book
Ban This Book
Deduct This Book (How Not to Pay Taxes While Ronald Reagan Is President)
Do You Think Mom Would Like This One?
Dewey Decimal No. 510.46 FC H3
I Never Can Remember What It's Called
The Great American Novel
ISBN 0-943568-01-3
Self-Referential Book Title
The Top Book on the New York Times *Bestseller List for the Past Ten Weeks*
Don't Go Overseas Until You've Read This Book
Soon to Become a Major Motion Picture
By Me, William Shakespeare (by Robert Payne)
That Book with the Red Cover in Your Window
Reviews of This Book

Oh, by the way, some of these are fake, others are real. For example, the last one, *Reviews of This Book,* is just a fantasy of mine. I would love to see a book consisting of nothing but a collection of reviews of it that appeared (after its publication, of course) in major newspapers and magazines. It sounds paradoxical, but it could be arranged with a lot of planning and hard work. First, a group of major journals would all have to agree to run reviews of the book by the various contributors to the book. Then all the reviewers would begin writing. But they would have to mail off their various drafts to all the other reviewers very regularly so that all the reviews could evolve together, and thus eventually reach a stable state of a kind known in physics as a "Hartree-Fock self-consistent solution". Then the book could be published, after which its reviews would come out in their respective journals, as per arrangement. (A little more on this idea is given in the postscript to Chapter 16.)

*   *   *

I chanced across two books devoted to the subject of indexing books. They are: *A Theory of Indexing* (by Gerald Salton) and *Typescripts, Proofs, and Indexes* (by Judith Butcher). Amazingly, neither one has an index. I also received a curious letter soliciting funds, which began this way: "Dear Friend: In these last months, I've been making a study of the money-raising letter as an art form . . ." I didn't read any further.

Aldo Spinelli, an Italian artist and writer, sent me some of his products. One, a short book called *Loopings,* has pages documenting their own word and letter counts in various complex ways, and includes at the end a short essay on various ways in which documents can tally themselves up or can mutually tally each other in twisty loops. Another, called *Chisel Book,* documents its own production, beginning with the idea, going through the finding of a publisher, making the layout, designing the cover, printing it, and so on.

Ashleigh Brilliant is the inventor of a vast number of aphorisms he calls "potshots", many of which have become very popular phrases in this country. For some reason, he has a self-imposed limit of seventeen words per potshot. A few typical potshots (all taken from his four books listed in the Bibliography) are:

What would life be, without me?

As long as I have you, I can endure all the troubles you inevitably bring.

Remember me? I'm the one who never made any impression on you.

Why does trouble always come at the wrong time?

Due to circumstances beyond my control, I am master of my fate and captain of my soul.

Although strictly speaking these are not self-referential sentences, they are all admirable examples of how the world constantly tangles with itself in multifarious self-undermining ways, and as such, they definitely belong in this chapter. As a matter of fact, I would like to take this occasion to announce that Ashleigh Brilliant is the 1984 recipient of the last annual Nobaloney Prize for Aphoristic Eloquence. The traditional Nobaloney ceremony, involving the awarding of a $1,000,000 cash prize two minutes before the recipient's decapitation, has been waived, at Mr. Brilliant's request.

There are other books containing much of interest to the self-reference addict. I would particularly recommend the recent *More on Oxymoron,* by Patrick Hughes, as well as the earlier *Vicious Circles and Infinity,* by Hughes and George Brecht. Also in this category are three thin volumes on Murphy's Law, compiled by Arthur Bloch. Murphy's Law, of course, is the one that says, "If anything can go wrong, it will", although when I first heard of it, it was called the "Fourth Law of Thermodynamics". O'Toole's Commentary on Murphy's Law is: "Murphy was an optimist." Goldberg's Commentary thereupon is: "O'Toole was an optimist." And finally, there is Schnatterly's Summing Up: "If anything *can't* go wrong, it will."

My own law, "Hofstadter's Law", states: "It always takes longer than you think it will take, even if you take into account Hofstadter's Law." Despite being its enunciator, I never seem to be able to take it fully into account in budgeting my own time. To help me out, therefore, my friend Don Byrd came up with his own law that I have taken to heart:

Byrd's Law:

It always takes longer than you think it will take, even if you take into account Hofstadter's Law.

Unfortunately, Byrd himself seems unable to take this law into account.

3

On Viral Sentences and
Self-Replicating Structures

January, 1983

Two years ago, when I first wrote about self-referential sentences, I was hit by an avalanche of mail from readers intrigued by the phenomenon of self-reference in its many different guises. I had the chance to print some of those responses one year ago, and that column then triggered a second wave of replies. Many of them have cast self-reference in new light of various sorts. In this column, I would like to describe the ideas of several people, two of whom responded to my initial column with remarkably similar letters: Stephen Walton of New York City and Donald R. Going of Oxon Hill, Maryland.

Walton and Going saw self-replicating sentences as similar to viruses—small objects that enslave larger and more self-sufficient "host" objects, getting the hosts by hook or by crook to carry out a complex sequence of replicating operations that bring new copies into being, which are then free to go off and enslave further hosts, and so on. "Viral sentences", as Walton called them, are "those that seek to obtain their own reproduction by commandeering the facilities of more complex entities".

Both Walton and Going were struck by the perniciousness of such sentences: the selfish way in which they invade a space of ideas and, merely by making copies of themselves all over the place, manage to take over a large portion of that space. Why do they not manage to overrun all of that idea-space? A good question. The answer should be obvious to students of evolution: competition from other self-replicators. One type of replicator seizes a region of the space and becomes good at fending off rivals; thus a "niche" in idea-space is carved out.

This idea of an evolutionary struggle for survival by self-replicating ideas is not original with Walton or Going, although both had fresh things to say on it. The first reference I know of to this notion is in a passage by neurophysiologist Roger Sperry in an article he wrote in 1965 called "Mind, Brain, and Humanist Values". He says: "Ideas cause ideas and help evolve new ideas. They interact with each other and with other mental forces in the same brain, in neighboring brains, and, thanks to global communication, in far distant, foreign brains. And they also interact with the external surroundings to produce *in toto* a burstwise advance in evolution that is far beyond anything to hit the evolutionary scene yet, including the emergence of the living cell."

Shortly thereafter, in 1970, the molecular biologist Jacques Monod came out with his richly stimulating and provocative book *Chance and Necessity*. In its last chapter, "The Kingdom and the Darkness", he wrote of the selection of ideas as follows:

> For a biologist it is tempting to draw a parallel between the evolution of ideas and that of the biosphere. For while the abstract kingdom stands at a yet greater distance above the biosphere than the latter does above the nonliving universe, ideas have retained some of the properties of organisms. Like them, they tend to perpetuate their structure and to breed; they too can fuse, recombine, segregate their content; indeed they too can evolve, and in this evolution selection must surely play an important role. I shall not hazard a theory of the selection of ideas. But one may at least try to define some of the principal factors involved in it. This selection must necessarily operate at two levels: that of the mind itself and that of performance.
>
> The performance value of an idea depends upon the change it brings to the behavior of the person or the group that adopts it. The human group upon which a given idea confers greater cohesiveness, greater ambition, and greater self-confidence thereby receives from it an added power to expand which will insure the promotion of the idea itself. Its capacity to "take", the extent to which it can be "put over" has little to do with the amount of objective truth the idea may contain. The important thing about the stout armature a religious ideology constitutes for a society is not what goes into its structure, but the fact that this structure is accepted, that it gains sway. So one cannot well separate such an idea's power to spread from its power to perform.
>
> The "spreading power"—the infectivity, as it were—of ideas, is much more difficult to analyze. Let us say that it depends upon preexisting structures in the mind, among them ideas already implanted by culture, but also undoubtedly upon certain innate structures which we are hard put to identify. What is very plain, however, is that the ideas having the highest invading potential are those that *explain* man by assigning him his place in an immanent destiny, in whose bosom his anxiety dissolves.

Monod refers to the universe of ideas, or what I earlier termed "idea-space", as "the abstract kingdom". Since he portrays it as a close analogue to the biosphere, we could as well call it the "ideosphere".

* * *

In 1976, evolutionary biologist Richard Dawkins published his book *The Selfish Gene*, whose last chapter develops this theme further. Dawkins' name for the unit of replication and selection in the ideosphere—the ideosphere's counterpart to the biosphere's gene—is *meme*, rhyming with "theme" or "scheme". As a library is an organized collection of books, so a memory is an organized collection of memes. And the soup in which memes grow and flourish—the analogue to the "primordial soup" out of which life first oozed —is the soup of human culture. Dawkins writes:

> Examples of memes are tunes, ideas, catch-phrases, clothes fashions, ways of making pots or of building arches. Just as genes propagate themselves in the gene pool by leaping from body to body via sperms or eggs, so memes propagate themselves in the meme pool by leaping from brain to brain via a process which, in the broad sense, can be called imitation. If a scientist hears, or reads about, a good idea, he passes it on to his colleagues and students. He mentions it in his articles and his lectures. If the idea catches on, it can be said to propagate itself, spreading from brain to brain. As my colleague N. K. Humphrey neatly summed up an earlier draft of this chapter: ' . . . memes should be regarded as living structures, not just metaphorically but technically. When you plant a fertile meme in my mind you literally parasitize my brain, turning it into a vehicle for the meme's propagation in just the way that a virus may parasitize the genetic mechanism of a host cell. And this isn't just a way of talking—the meme for, say, 'belief in life after death' is actually realized physically, millions of times over, as a structure in the nervous systems of individual men the world over.'
>
> Consider the idea of God. We do not know how it arose in the meme pool. Probably it originated many times by independent 'mutation'. In any case, it is very old indeed. How does it replicate itself? By the spoken and written word, aided by great music and great art. Why does it have such high survival value? Remember that 'survival value' here does not mean value for a gene in a gene pool, but value for a meme in a meme pool. The question really means: What is it about the idea of a god which gives it its stability and penetrance in the cultural environment? The survival value of the god meme in the meme pool results from its great psychological appeal. It provides a superficially plausible answer to deep and troubling questions about existence. It suggests that injustices in this world may be rectified in the next. The 'everlasting arms' hold out a cushion against our own inadequacies which, like a doctor's placebo, is none the less effective for being imaginary. These are some of the reasons why the idea of God is copied so readily by successive generations of individual brains. God exists, if only in the form of a meme with high survival value, or infective power, in the environment provided by human culture.

Dawkins takes care here to emphasize that there need not be an exact copy of each meme, written in some universal memetic code, in each person's brain. Memes, like genes, are susceptible to variation or distortion—the analogue to mutation. Various mutations of a meme will have to compete with each other, as well as with other memes, for attention—which is to say, for brain resources in terms of both space and time devoted to that meme. Not only must memes compete for inner resources, but, since they are

51

transmissible visually and aurally, they must compete for radio and television time, billboard space, newspaper and magazine column-inches, and library shelf-space. Furthermore, some memes will tend to discredit others, while some groups of memes will tend to be internally self-reinforcing. Dawkins says:

> ... Mutually suitable teeth, claws, guts, and sense organs evolved in carnivore gene pools, while a different stable set of characteristics emerged from herbivore gene pools. Does anything analogous occur in meme pools? Has the god meme, say, become associated with any other particular memes, and does this association assist the survival of each of the participating memes? Perhaps we could regard an organized church, with its architecture, rituals, laws, music, art, and written tradition, as a co-adapted stable set of mutually-assisting memes.
>
> To take a particular example, an aspect of doctrine which has been very effective in enforcing religious observance is the threat of hell fire. Many children and even some adults believe that they will suffer ghastly torments after death if they do not obey the priestly rules. This is a particularly nasty technique of persuasion, causing great psychological anguish throughout the middle ages and even today. But it is highly effective. It might almost have been planned deliberately by a machiavellian priesthood trained in deep psychological indoctrination techniques. However, I doubt if the priests were that clever. Much more probably, unconscious memes have ensured their own survival value by virtue of those same qualities of pseudo-ruthlessness which successful genes display. The idea of hell fire is, quite simply, *self-perpetuating*, because of its own deep psychological impact. It has become linked with the god meme because the two reinforce each other, and assist each other's survival in the meme pool.
>
> Another member of the religious meme complex is called faith. It means blind trust, in the absence of evidence, even in the teeth of evidence .... Nothing is more lethal for certain kinds of meme than a tendency to look for evidence .... The meme for blind faith secures its own perpetuation by the simple unconscious expedient of discouraging rational inquiry.
>
> Blind faith can justify anything. If a man believes in a different god, or even if he uses a different ritual for worshipping the same god, blind faith can decree that he should die—on the cross, at the stake, skewered on a Crusader's sword, shot in a Beirut street, or blown up in a bar in Belfast. Memes for blind faith have their own ruthless ways of propagating themselves. This is true of patriotic and political as well as religious blind faith.

*   *   *

When I muse about memes, I often find myself picturing an ephemeral flickering pattern of sparks leaping from brain to brain, screaming "Me, me!" Walton's and Going's letters reinforced this image in interesting ways. For instance, Walton begins with the simplest imaginable viral sentences— "Say me!" and "Copy me!"—and moves quickly to more complex variations with blandishments ("If you copy me, I'll grant you three wishes!") or threats ("Say me or I'll put a curse on you!"), neither of which, he observes, is likely to be able to keep its word. Of course, as he points out, this may not matter, the only final test of viability being success at survival in the meme pool. All's fair in love and war—and war includes the eternal battle for survival, in the ideosphere no less than in the biosphere.

To be sure, very few people above the age of five will fall for the simple-minded threats or promises of these sentences. However, if you simply tack on the phrase "in the afterlife", far more people will be lured into the memetic trap. Walton observes that a similar gimmick is used by your typical chain letter (or "viral text"), which "promises wealth to those who faithfully replicate it and threatens doom to any who fail to copy it". Do you remember the first time you received such a chain letter? Do you recall the sad tale of "Don Elliot, who received $50,000 but then lost it because he broke the chain"? And the grim tale of "General Welch in the Philippines, who lost his life [or was it his *wife*?] six days after he received this letter because he failed to circulate the prayer—but before he died, he received $775,000"? Poor Don Elliot! Poor General Welch! It's hard not to be just a little sucked in by such tales, even if you wind up throwing the letter out contemptuously.

I found Walton's phrases "viral sentence" and "viral text" to be exceedingly catchy—little memes in themselves, definitely worthy of replication some 700,000 times in print, and who knows how many times orally beyond that. At least that's *my* opinion. Of course, it also depends on how the editor of *Scientific American* feels. [It turned out he felt fine about it.] Well, now, Walton's own viral text, as you can see here before your eyes, has managed to commandeer the facilities of a very powerful host—an entire magazine and printing press and distribution service. It has leapt aboard and is now—even as you read this viral sentence—propagating itself madly throughout the ideosphere!

This idea of choosing the right host is itself an important aspect of the quality of a viral entity. Walton puts it this way:

> The recipient of a viral text can, of course, make a big difference. A tobacco mosaic virus that attacks a salt crystal is out of luck, and some people rip up chain letters on sight. A manuscript sent to an editor may be considered viral, even though it contains no explicit self-reference, because it is attempting to secure its own reproduction through an appropriate host; the same manuscript sent to someone who has nothing to do with publishing may have no viral quality at all.

As it concludes, Walton's letter graciously steps forward from the page and squeaks to me directly on its own behalf: "Finally, I (this text) would be delighted to be included, in whole or in part, in your next discussion of self-reference. With that in mind, please allow me to apologize in advance for infecting you."

*   *   *

Whereas Walton mentioned Dawkins in his letter, Going seems not to have been aware of Dawkins at all, which makes his letter quite remarkable in its close connection to Dawkins' ideas. Going suggests that we consider, to begin with, Sentence *A*:

> It is your duty to convince others that this sentence is true.

As he says:

> If you were foolish enough to believe this sentence, you would attempt to convince your friends that *A* is true. If they were equally foolish, they would convince their friends, and so on until every human mind contained a copy of *A*. Thus, *A* is a self-replicating sentence. More particularly, it is the intellectual equivalent of a virus. If Sentence *A* were to enter a mind, it would take control of the mind's intellectual machinery and use it to produce hundreds of copies of itself in other minds.

The problem with Sentence *A*, of course, is that it is absurd; no one could possibly believe it. However, consider the following:

> System *S*:
> Begin:
> S_1: Blah.
> S_2: Blah blah.
> S_3: Blah blah blah.
>   .   .
>   .   .
>   .   .
>   .   .
> S_{99}: Blah blah blah blah blah blah . . . .
> S_{100}: It is your duty to convince others that System *S* is true.
> End.

Here, S_1 through S_{99} are meant to be statements that constitute a belief system having some degree of coherency. If System *S* taken as a whole were convincing, then the entire system would be self-replicating. System *S* would be especially convincing if S_{100} were not stated explicitly but held as a logical consequence of the other ideas in the system.

Let us refer to Going's S_{100} as the *hook* of System *S*, for it is by this hook that System *S* hopes to hoist itself onto a higher level of power. Note that on its own, a hook that in effect says "It is your duty to believe me" is not a viable viral entity; in order to "fly", it needs to drag something extra along with it, just as a kite needs a tail to stabilize it. Pure lift goes out of control and self-destructs, but controlled lift can lift itself along with its controller. Similarly, S_{100} and $S_1 - S_{99}$ (taken as a set) are symbiotes: they play complementary, mutually supportive roles in the survival of the meme they together constitute. Now Going develops this theme a little further:

> Statements $S_1 - S_{99}$ are the bait which attracts the fish and conceals the hook. No bait—no bite. If the fish is fool enough to swallow the baited hook, it will have little enough time to enjoy the bait. Once the hook takes hold, the fish will lose all its fishiness and become instead a busy factory for the manufacture of baited hooks.
>
> Are there any real idea systems that behave like System S? I know of at least three. Consider the following:

> System X:
> Begin:
> X_1: Anyone who does not believe System X will burn in hell.
> X_2: It is your duty to save others from suffering.
> End.

If you believed in System X, you would attempt to save others from hell by convincing them that System X is true. Thus System X has an implicit 'hook' that follows from its two explicit sentences, and so System X is a self-replicating idea system. Without being impious, one may suggest that this mechanism has played some small role in the spread of Christianity.

Self-replicating ideas are most often found in politics. Consider Sentence W:

> The whales are in danger of extinction.

If you believed this idea, you would want to save the whales. You would quickly discover that you could not reach this goal by yourself. You would need the help of thousands of like-minded people. The first step in getting their help would be to convince them that Sentence W is true. Thus a 'hook' like S_{100} follows from Sentence W, and Sentence W is a self-replicating idea.

In a democracy, nearly any idea will tend to replicate since the only way to win an election is to convince other people to share your ideas. Most political ideas are not properly self-replicating, since the motive for spreading the idea is separate from the idea itself. Statement W, on the other hand, is genuinely self-replicating, since the duty to propagate it is a direct logical consequence of W itself. Ideas like W can sometimes take on a life of their own and drive their own propagation.

A more sinister form of self-replication is Sentence B:

> The bourgeoisie is oppressing the proletariat.

This statement is self-replicating for the same reason as W is. The desire to propagate statements like B is driven by a desire to protect a victim figure from a villain figure. Such ideas are dangerous because belief in them may lead to attacks on the supposed villain. Statement B also illustrates the fact that the self-replicating character of an idea depends only upon the idea's logical structure, not upon its truth.

Statement *B* is merely a special case of the generalized statement, Sentence *V*:

> The *villain* is *wronging* the *victim*.

Here, the word *villain* must be replaced with the name of some real group (capitalists, communists, imperialists, Jews, freemasons, aristocrats, men, foreigners, etc.). Likewise, *victim* must be replaced with the name of the corresponding victim and *wronging* filled in as desired. The result will be a self-replicating idea system for the same reasons as *W* and *B* were. Note that each of the suggested substitutions yields a historically attested idea system. It has long been recognized that most extremist mass movements are based on a belief similar to *V*. Part of the reason seems to be that type-*V* statements reduce to the 'hook', S_{100}, and therefore define self-replicating idea systems. One hesitates to explain real historical events in terms of such a silly mechanism, and yet . . . .

Going brings his ideas to an amusing conclusion as follows:

Suppose we parody my thesis by proposing Sentence *E*:

> The self-replicating ideas are conspiring to enslave our minds.

This 'paranoid' statement is clearly an idea of type *V*. Thus, the thesis seems to describe itself. Further, if we accept *E*, then we must say that this type-*V* idea implies that we must distrust all ideas of type *V*. This is the Epimenides Paradox.

It is interesting that all these people who have explored these ideas have given examples ranging from the very small scale of such things as catchy tunes (for example, Dawkins cites the opening theme of Beethoven's fifth symphony) and phrases (the word "meme" itself) to the very large scale of ideologies and religions. Dawkins uses the term *meme complex* for these larger agglomerations of memes; however, I prefer the single word *scheme*.

One reason I prefer it is that it fits so well with the usage suggested by psychiatrist and writer Allen Wheelis in his novel *The Scheme of Things*. Its central character is a psychiatrist and writer named Oliver Thompson, whose darkly brooding essays are scattered throughout the book, interspersed with brightly colored, evocative episodes. Thompson is obsessed with the difference between, on the one hand, "the raw nature of existence, unadorned, unmediated", which he refers to repeatedly as "the way things are", and, on the other hand, "schemes of things", invented by humans—ways of making order and sense out of the way things are. Here are some of Thompson's musings on that theme:

I want to write a book . . . . the story of one man whose life becomes a metaphor for the entire experience of man on earth. It will portray his search through a succession of schemes of things, show the breakdown, one after another, of each pattern he finds, his going on always to another, always in the hope that the scheme of things he finds and for the moment is serving is *not* a scheme of things at all but reality, the way things are, therefore an absolute that will endure forever, within which he can serve, to which he can contribute, and through which he can give his mortal life meaning and so achieve eternal life . . . .

The scheme of things is a system of order. Beginning as our view of the world, it finally *becomes* our world. We live within the space defined by its coordinates. It is self-evidently true, is accepted so naturally and automatically that one is not aware of an act of acceptance having taken place. It comes with one's mother's milk, is chanted in school, proclaimed from the White House, insinuated by television, validated at Harvard. Like the air we breathe, the scheme of things disappears, becomes simply reality, the way things are. It is the lie necessary to life. The world as it exists beyond that scheme becomes vague, irrelevant, largely unperceived, finally nonexistent . . . .

No scheme of things has ever been both coextensive with the way things are and also true to the way things are. All schemes of things involve limitation and denial . . . .

A scheme of things is a plan for salvation. How well it works will depend upon its scope and authority. If it is small, even great achievement in its service does little to dispel death. A scheme of things may be as large as Christianity or as small as the Alameda County Bowling League. We seek the largest possible scheme of things, not in a reaching out for truth, but because the more comprehensive the scheme the greater its promise of banishing dread. If we can make our lives mean something in a cosmic scheme we will live in the certainty of immortality. Those attributes of a scheme of things that determine its durability and success are its scope, the opportunity it offers for participation and contribution, and the conviction with which it is held as self-evidently true. The very great success of Christianity for a thousand years follows upon its having been of universal scope, including and accounting for everything, assigning to all things a proper place; offering to every man, whether prince or beggar, savant or fool, the privilege of working in the Lord's vineyard; and being accepted as true throughout the Western world.

As a scheme of things is modified by inroads from outlying existence, it loses authority, is less able to banish dread; its adherents fall away. Eventually it fades, exists only in history, becomes quaint or primitive, becomes, finally, a myth. What we know as legends were once blueprints of reality. The Church was right to stop Galileo; activities such as his import into the regnant scheme of things new being which will eventually destroy that scheme.

Taken in Wheelis' way, "scheme" seems a fitting replacement for Dawkins' "meme complex". A scheme imposes a top-down kind of perceptual order on the world, propagating itself ruthlessly, like Going's System *S* with its "hook". Wheelis' description of the inadequacy of all "schemes of things" to fully and accurately capture "the way things are" is strongly reminiscent of the vulnerability of all sufficiently powerful formal systems to either incompleteness or inconsistency—a vulnerability that ensues from another kind of "hook": the famous Gödelian hook, which arises from the capacity for self-reference of such systems, although neither Wheelis nor Thompson makes any mention of the analogy. We shall come back to Gödel momentarily.

* . * *

The reader of this novel must be struck by the professional similarity of Wheelis and his protagonist. It is impossible to read the book and not to surmise that Thompson's views are reflecting Wheelis' own views—and yet, who can say? It is a tease. Even more tantalizing is the title of Thompson's imaginary book, which Wheelis casually mentions toward the end of the novel: it is *The Way Things Are*—a striking contrast to the title of the real book in which it exists. One wonders: What is the meaning of this elegant literary pleat in which one level folds back on another? What is the symbolism of Wheelis within Wheelis?

Such a twist, by which a thing (sentence, book, system, person) seems to refer to itself but does so only by allusion to something *resembling* itself, is called *indirect self-reference.* You can do this by pointing at your image in a mirror and saying, "That person sure is good-looking!" That one is very simple, because the connection between something and its mirror image is so familiar and obvious-seeming to us that there seems to be no distance whatsoever between direct and indirect referents: we equate them completely. Thus it seems there is no referential indirectness.

On the other hand, this depends upon the ease with which our perceptual systems convert a mirror image into its reverse, and upon other qualities of our cognitive systems that allow us to see through several layers of translation without being aware of the layers—like looking through many feet of water and seeing not the water but only what lies at its bottom.

Some indirect self-references are of course subtler than others. Consider the case of Matt and Libby, a couple ostensibly having a conversation about their friends Tammy and Bill. It happens that Matt and Libby are having some problems in their relationship, and those problems are quite analogous to those of Tammy and Bill, only with sexes reversed: Matt is to Libby what Tammy is to Bill, in their respective relationships. So as Matt and Libby's conversation progresses, although on the surface level it is completely about their friends Tammy and Bill, on another level it is *actually* about themselves, as reflected in these other people. It is almost as if, by talking about Tammy and Bill, Matt and Libby are going over a fable by Aesop that has obvious relevance to their own plight. There are things going on simultaneously on two levels, and it is hard to tell how conscious either of the participants is of the exchange of dual messages—one of concern about their friends, one of concern about themselves.

Indirect self-reference can be exploited in the most unexpected and serious ways. Consider the case of President Reagan, who on a recent occasion of high Soviet-American tension over Iran, went out of his way to recall President Truman's behavior in 1945, when Truman made some very blunt threats to the Soviets about the possibility of the U.S. using nuclear weapons if need be against any Soviet threat in Iran. Merely by bringing up the memory of that occasion, Reagan was inviting a mapping to be made between himself and Truman, and thereby he was issuing a not-so-veiled threat, though no one could point to anything explicit. There simply was no way that a conscious being could fail to make the connection. The resemblance of the two situations was too blatant.

Thus, does self-reference really come in two varieties—direct and indirect —or are the two types just distant points on a continuum? I would say unhesitatingly that it is the latter. And furthermore, you can delete the prefix "self", so that the question becomes one of reference in general. The essence is simply that one thing refers to another whenever, to a conscious being, there is a sufficiently compelling mapping between the roles the two things are perceived to play in some larger structures or systems. (See Chapter 24 for further discussion of the perception of such roles.) Caution is needed here. By "conscious being", I mean an analogy-hungry perceiving machine that gets along in the world thanks to its perceptions; it need not be human or even organic. Actually, I would carry the abstraction of the term "reference" even further, as follows. The mapping of systems and roles that establishes reference need not actually be *perceived* by any such being: it suffices that the mapping exist and simply be *perceptible* to such a being were it to chance by.

*   *   *

The movie *The French Lieutenant's Woman* (based on John Fowles' novel of the same name) provides an elegant example of ambiguous degrees of reference. It consists of interlaced vignettes from two concurrently developing stories both of which involve complex romances; one takes place in Victorian England, the other in the present. The fact that there are two romances already suggests, even if only slightly, that a mapping is called for. But much more is suggested than that. There are structural similarities between the two romances: each of them has triangular qualities, and in both stories, only one leg of the triangle is focused upon. Morever, the same two actors play the two lovers in both romances, so that you see them in alternating contexts and with alternating personality traits. The reason for this "coincidence" is that the contemporary story concerns the making of a film of the Victorian story.

As the two stories unfold in parallel, a number of coincidences arise that suggest ever more strongly that a mapping should be made. But it is left to the movie viewer to carry this mapping out; it is never called for explicitly.

After a time, though, it simply becomes unavoidable. What is pleasant in this game is the fluidity left to the viewer: there is much room for artistic license in seeing connections, or suspecting or even inventing connections.

Indirect reference of the artistic type is much less precise than indirect reference of the formal type. The latter arises when two formal systems are *isomorphic*—that is, they have strictly analogous internal structures, so that there is a rigorous one-to-one mapping between the roles in the one and the roles in the other. In such a case, the existence of genuine reference becomes as clear to us as in the case of someone talking about their mirror image: we take it as immediate, pure self-reference, without even noticing the indirectness, the translational steps mediated by the isomorphism. In fact, the connection may seem too direct even to be called "reference"; some may see it simply as *identity.*

This perceptual immediacy is the reason that Gödel's famous sentence **G** of mathematical logic is said to be self-referential. Everyone accepts the idea that **G** talks about a *number, g* (though a radical skeptic might question even that!); the tricky Gödelian step is in seeing that *g* (the *number*) plays a role in the system of natural numbers strictly analogous to the role that **G** (the *sentence*) plays in the axiomatic system it is expressed in. This Wheelis-like oblique reference by **G** to itself via its "image" *g* is generally accepted as genuine self-reference. (Note that we have even one further mapping: **G** plays the role of Wheelis, and its Gödel number *g* that of Wheelis' alter ego Thompson.)

The two abstract mappings that, when telescoped, establish **G**'s self-reference but make it seem indirect can be collapsed into just *one* mapping, following a slogan that we might formulate this way: "If *A* refers to *B*, and *B* is just like *C*, then *A* refers to *C*." For instance, we can let *A* and *C* be Wheelis, with *B* being Thompson. This makes Wheelis' self-reference a "theorem". Of course, this "theorem" is not rigorously proven, since our slogan has to be taken with a grain of salt. Being "just like" something else is a highly disputable matter.

However, in a formal context where *is just like* is virtually synonymous with *plays a role isomorphic to that of,* then the slogan can have a strict meaning, and thereby justify a theorem more rigorously. In particular, if *A* and *C* are equated with **G,** and *B* with *g,* then our slogan runs: "If **G** refers to *g,* and *g* plays a role isomorphic to that of **G,** then **G** refers to **G**." Since the premises are true, the conclusion must be true. According to this scheme of things, then, **G** is a genuinely self-referential sentence, rather than some sort of logical illusion as deceptive as an Escher print.

*   *   *

Indirect self-reference suggests the idea of *indirect self-replication,* in which a viral entity, instead of replicating itself exactly, brings into being another entity that plays the same role as it does, but in some other system: perhaps its mirror image, perhaps its translation into French, perhaps a string of the product numbers of all its parts, together with pre-addressed envelopes containing checks made out to the factories where those parts are made, and a list of instructions telling what to do with all the parts when they arrive in the mail.

This may sound familiar to some readers. In fact, it is an indirect reference to the Von Neumann Challenge, the puzzle posed in Chapter 2 to create a self-describing sentence whose only quoted matter is at the word or letter level, rather than at the level of whole quoted phrases. I discovered, as I received candidate solutions, that many readers did not understand what this requirement meant. The challenge came out of an objection to the complexity of the "seed" (the quoted part) in Quine's version of the Epimenides paradox:

"yields falsehood when appended to its quotation." yields falsehood when appended to its quotation.

To see what is strange here, imagine that you wish to have a space-roving robot build a copy of itself out of raw materials that it encounters in its travels. Here is one way you could do it: Make the robot symmetrical, like a human being. Also make the robot able to make a mirror-image copy of any structure that it encounters along its way. Finally, have the robot be programmed to scan the world constantly, the way a hawk scans the ground for rodents. The search image in the robot's case is that of an object identical to its own left half. The robot need not be *aware* that its target is identical to its left half; the search can go on merrily for what seems to it to be merely a very complex and arbitrary structure. When, after scouring the universe for seventeen googolplex years, it finally comes across such a structure, then of course the robot activates its mirror-image-production facility and creates a right half. The last step is to fasten the two halves together, and presto! A copy emerges. Easy as pie—provided you're willing to wait seventeen googolplex years (give or take a few minutes).

The arbitrary and peculiar aspect of the Quine sentence, then, is that its seed is half as complex—which is to say, nearly as complex—as the sentence itself. If we resume our robot parable, what we'd ideally like in a self-replicating robot is the ability to make itself literally from the ground up: let us say, for instance, to mine iron ore, to smelt it, to cast it in molds to make nuts and bolts and sheet metal and so on; and finally, to be able to assemble the small parts into larger and larger subunits until, miraculously, a replica is born out of truly raw materials. This was the spirit of the Von Neumann Challenge: I wanted a linguistic counterpart to this "self-replicating robot of the second kind".

In particular, this means a self-documenting or self-building sentence that builds both its halves—its quoted seed and its unquoted building rule—out of linguistic raw materials (words or letters). Many readers failed to understand what this implies. The most common mistake was to present, as the seed, a long sequence of individually quoted words (or letters) in a specific order, then to exploit that order in the building rule. Well then, you might as well have quoted one big long ordered string, as Quine did. The idea of my challenge was that all structure in the built object must arise exclusively out of some principle enunciated in the building rule, not out of the seed's internal structure.

Just as a self-replicating robot in some random alien environment is hardly likely to find all its parts lined up on a shelf in order of assembly but must rely on its "brain" or program to recognize raw parts wherever and whenever they turn up so that it can grab them and therefrom assemble a copy of itself, so the desired sentence must treat the pieces of the seed without regard to the order in which they are listed, yet must be able to construct itself in the proper order out of them. Thus it's fine if you enclose the entire seed within a single pair of quotes, rather than quoting each word individually—all that matters is that the seed's word order (or better yet, its letter order) not be exploited. The seed of the ideal solution would be a long inventory of parts, similar to the list of ingredients of a recipe—perhaps a list of 50 'e's, then 46 't's, and so on. Clearly those letters cannot remain in that order; they simply constitute the raw materials out of which the new sentence is to be built.

*   *   *

Nobody sent in a solution whose seed was at the primordial level of letters. A few people, however, did send in adequate, if not wonderfully elegant, solutions with seeds at the word level. The first correct solution I received came from Frank Palmer of Chicago, who therefore receives the first "Johnnie" award—a self-replicating dollar bill given to the Grand Winner of the First Every-Other-Decade Von Neumann Challenge. Unfortunately, the dollar bill consumes the entire body of its owner in its bizarre process of self-replication, and so it is wisest to simply lock it up to protect oneself from its voracious appetite.

Palmer submitted several versions. In them, he utilized upper and lower cases to distinguish between seed and building rule, respectively. Here is one solution, slightly modified by me:

after alphabetizing, decapitalize FOR AFTER WORDS STRING FINALLY UNORDERED UPPERCASE FGPBVKXQJZ NONVOCALIC DECAPITALIZE SUBSTITUTING ALPHABETIZING, finally for nonvocalic string substituting unordered uppercase words

Let us watch how it works, step by careful step. We must bear in mind that the instructions we are following are the lowercase words printed above, and that the uppercase words are not to be read as instructions. Nor, for that matter, are the lowercase words that we will soon be working with. They are like the inert, anesthetized body of a patient being operated on, who, when the operation is over, will awake and become animate. So let's go. First we are to alphabetize the seed. (I am treating the comma as attached to the word preceding it.) This gives us the following:

AFTER ALPHABETIZING, DECAPITALIZE FGPBVKXQJZ FINALLY FOR NONVOCALIC STRING SUBSTITUTING UNORDERED UPPERCASE WORDS

Next we are to decapitalize it. This will yield some lowercase words—the "anesthetized" lowercase words I spoke of above:

after alphabetizing, decapitalize fgpbvkxqjz finally for nonvocalic string substituting unordered uppercase words

All right; now our final instruction is to locate a nonvocalic string (that's easy: *"fgpbvkxqjz"*) and to substitute for it the uppercase words, *in any order* (that is, the original seed itself, but without regard for its structure above the level of the individual word-unit). This last bit of surgery yields:

after alphabetizing, decapitalize SUBSTITUTING FINALLY WORDS UNORDERED STRING DECAPITALIZE UPPERCASE FOR NONVOCALIC AFTER FGPBVKXQJZ ALPHABETIZING, *finally for nonvocalic string substituting unordered uppercase words*

And this is a perfect copy of our starting sentence! Or rather, semiperfect. Why only semiperfect? Because the seed has been randomly scrambled in the act of self-reproduction. The beauty of the scheme, though, is that the internal structure of the seed is entirely irrelevant to the efficacy of the sentence as a self-replicator. All that matters is that the new building rule say the proper thing, and it will do so no matter what order the seed from which it sprang was in. Now this fresh new baby sentence can wake up from its anesthesia and go off to replicate itself in turn.

The critical step was the first one: alphabetization. This turns the arbitrarily-ordered seed into a grammatical, meaningful command—merely by mechanically exploiting a presumed knowledge of the "ABC"'s. But why not? It is perfectly reasonable to presume superficial typographical knowledge about letters and words, since such knowledge deals with printed material *as raw material:* purely syntactically, without regard to the meanings carried therein. This is just like the way that enzymes in the living cell deal with the DNA and RNA they chop up and alter and piece together again: purely chemically, without regard to the "meanings" carried therein. Just as chemical valences and affinities and so on are taken as givens in the workings of the cell, so alphabetic and typographic facts are taken as givens in the V. N. Challenge.

When Palmer sent in his solution, he happened to write down his seed in order of increasing length of words, but that is inessential; any random order would have done, and that sort of idea is the crucial point that many readers missed. Another rather elegant solution was sent in by Martin Weichert of Munich. It runs this way (slightly modified by me):

Alphabetize and append, copied in quotes, these words: "these append, in Alphabetize and words: quotes, copied"

It works on the same principle as Palmer's sentence, and again features a seed whose internal structure (at least at the word level) is irrelevant to successful self-replication. Weichert also sent along an intriguing palindromic solution in Esperanto, in which the flexible word order of the language plays a key role. Michael Borowitz and Bob Stein of Durham, North Carolina sent in a solution similar to Palmer's.

* * *

Finally, last year's gold-medal winner for self-documentation, Lee Sallows, was a bit piqued by my suggestion that the gold on his medal was somewhat tarnished since he had not paid close enough attention to the use-mention distinction. Apparently I goaded him into constructing an even more elaborate self-documenting sentence. Although it does not quite fit what I had in mind for the Von Neumann Challenge, as it does not spell out its own construction explicitly at the letter level or word level, it is another marvelous Sallowsian gem, and I shall therefore generously allow the gold on his medal to go untarnished this year. (Apologies to those purists who insist that gold doesn't tarnish. I must have been confusing it with copper and silver. How silly of me!) Herewith follows Sallows' 1982 contribution:

*

Write
down ten 'a's,
eight 'c's, ten 'd's,
fifty-two 'e's, thirty-eight 'f's,
sixteen 'g's, thirty 'h's, forty-eight 'i's,
six 'l's, four 'm's, thirty-two 'n's, forty-four 'o's,
four 'p's, four 'q's, forty-two 'r's, eighty-four 's's,
seventy-six 't's, twenty-eight 'u's, four 'v's, four 'W's,
eighteen 'w's, fourteen 'x's, thirty-two 'y's, four ':'s,
four '*'s, twenty-six '-'s, fifty-eight ','s,
sixty '"'s and sixty ''s, in a
palindromic sequence
whose second
half runs
thus:
:suht
snur flah
dnoces esohw
ecneuqes cimordnilap
a ni ,s'" ytxis dna s'" ytxis
,s',' thgie-ytfif ,s'-' xis-ytnewt ,s'*' ruof
,s':' ruof ,s'y' owt-ytriht ,s'x' neetruof ,s'w' neethgie
,s'W' ruof ,s'v' ruof ,s'u' thgie-ytnewt ,s't' xis-ytneves
,s's' ruof-ythgie ,s'r' owt-ytrof ,s'q' ruof ,s'p' ruof
,s'o' ruof-ytrof ,s'n' owt-ytriht ,s'm' ruof ,s'l' xis
,s'i' thgie-ytrof ,s'h' ytriht ,s'g' neetxis
,s'f' thgie-ytriht ,s'e' owt-ytfif
,s'd' net ,s'c' thgie
,s'a' net nwod
etirW

*

Post Scriptum

After writing this column, I received much mail testifying to the fact that there are a large number of people who have been infected by the "meme" meme. Arel Lucas suggested that the discipline that studies memes and their connections to humans and other potential carriers of them be known as *memetics,* by analogy with "genetics". I think this is a good suggestion, and hope it will be adopted.

Maurice Guéron wrote me from Paris to tell me that he believed the first clear exposition of the idea of self-reproducing ideas that inhabit the brains of organisms was put forward in 1952 by Pierre Auger, a physicist at the Sorbonne, in his book *L'homme microscopique.* Guéron sent me a photocopy of the relevant portions, and I could indeed see how prophetic the book was.

I received a copy of the book *General Theory of Evolution* by Vilmos Csányi, a Hungarian geneticist. In this book, he attempts to work out a theory in which memes and genes evolve in parallel. A similar attempt is made in the book *Ever-Expanding Horizons: The Dual Informational Sources of Human Evolution,* by the American biologist Carl B. Swanson.

The most thorough-going research on the topic of pure memetics I have yet run across is that of Aaron Lynch, an engineering physicist at Fermilab in Illinois, who in his spare time is writing a book called *Abstract Evolution.* The portions that I have read go very carefully into the many "options", to speak anthropomorphically, that are open to a meme for getting itself reproduced over and over in the ideosphere (a term Lynch and I invented independently). It promises to be a provocative book, and I look forward to its publication.

*   *   *

Jay Hook, a mathematics graduate student, was provoked by the solutions to the Von Neumann Challenge as follows:

> The notion that it takes two to reproduce is suggestive. Perhaps a change in terminology is appropriate. The component that you call the "seed" might be thought of as the "female" fragment—the egg that grows into an adult, but only after receiving instructions from the sperm, the "male" fragment—the building rule. In this interpretation, our sentences say everything twice because they are hermaphroditic: the male and female fragments appear together in the same individual.
>
> To better mimic nature, we should construct *pairs* of sentences or phrases, one male and one female—expressions that taken individually produce nothing but when put together in a dark room make copies of themselves. I propose the following. The male fragment

> After alphabetizing and deitalicizing, duplicate female fragment in its original version.

doesn't seem to say much by itself, and the female fragment

> *in and its After female fragment original version. duplicate alphabetizing deitalicizing,*

certainly doesn't, but let them at each other and watch the fireworks. (I follow your practice of assuming each punctuation mark to be attached to the preceding word.) The male takes the lead, and sets to work on the female. First we alphabetize and deitalicize her, he says; that gives a new male fragment.

Then we simply make a copy of her—so we get one of each!

Nature still doesn't work this way, of course; it's not clear that couples that produce offspring only in boy-girl pairs are really superior to self-replicating hermaphrodites. Ideally, our fragments should produce *either* a copy of the male *or* a copy of the female, depending on, say, the day of the week or the parity of some external index like the integer part of the current Dow Jones Industrial Average. Surprisingly, this isn't hard. Take the male to be

> Alphabetize and deitalicize female fragment if index is odd; otherwise reproduce same verbatim.

and take for the female

> *if is and odd; same index female fragment otherwise reproduce verbatim. Alphabetize deitalicize*

One more refinement. To this point, each offspring has been exactly identical to one of its parents. We can introduce variation, at least in the girls, as follows. Male fragment:

> Alphabetize and deitalicize female fragment if index is odd; otherwise randomly rearrange the words.

Female fragment:

> *if is and the odd; index female words. fragment randomly otherwise rearrange Alphabetize deitalicize*

Now all of the boys will be the spittin' image of their father, but whereas one daughter might be

> *index rearrange if the Alphabetize randomly fragment odd; deitalicize is and words. otherwise female*

another might be

> *Alphabetize index and rearrange the fragment if female is odd; otherwise randomly deitalicize words.*

The important point, however, is that all of these female offspring, however diverse, are genetically capable of mating with any of the (identical) males. Can you find a way to introduce variation in the males without producing sterile offspring?

In conclusion, allow me to observe that the Dow closed on Friday at 1076.0. Therefore I proudly proclaim: It's a girl!

*   *   *

I now close by returning to Lee Sallows. This indefatigable researcher of what he calls *logological space* continued his quest after the holy grail of *perfect* self-documentation. His jealousy was aroused in the extreme when Rudy Kousbroek, who is Dutch, and Sarah Hart, who is English, together tossed off what Sallows terms "the greatest logological jewel the world has ever seen". Kousbroek and Hart's self-documenting sentence, though in Dutch, ought to be pretty clearly understandable by anyone who takes the time to look at it carefully:

Dit pangram bevat vijf a's, twee b's, twee c's, drie d's, zesenveertig e's, vijf f's, vier g's, twee h's, vijftien i's, vier j's, een k, twee l's, twee m's, zeventien n's, een o, twee p's, een q, zeven r's, vierentwintig s's, zestien t's, een u, elf v's, acht w's, een x, een y, en zes z's.

In fact, you can learn how to count in Dutch by studying it!

There's not an ounce of fat or awkwardness in this sentence, and it drove Sallows mad that he couldn't come up with an equally perfect pangram (sentence containing every letter of the alphabet) in English. Every attempt had some flaw in it. So in desperation, Sallows, electronics engineer that he is, decided he would design a high-speed dedicated "letter-crunching" machine to search the far reaches of logological space for an equivalent English sentence. Sallows sent me some material on his Pangram Machine. He says:

At the heart of the beast is a clock-driven cascade of sixteen Johnson-counters: the electronic analogue of a stepper-motor-driven stack of combination lock-discs. Every tick of the clock clicks in a new combination of numbers: a unique combination of counter output lines becomes activated . . . . Pilot tests have been surprisingly encouraging; it looks as though a clock frequency of a million combinations per second is quite realistic. Even so it would take 317 years to explore the ten-deep stratum. But does it have to be ten? With this reduced to a modest but still very worthwhile six-deep range it will take just 32.6 days. Now we're talking!

Over the past eight weeks I have devoted every spare second to constructing this rocket for exploring the far regions of logological space . . . . Will it really fly? So far it looks very promising. And the end is already in sight. With a bit of luck Rudy Kousbroek will be able to launch the machine on its 32-day journey when he comes to visit here at the end of this month. If so, a bottle of champagne will not be out of place.

Two months later, I got a most excited transmission from Lee, which began with the word "EUREKA!"—the word the Pangram Machine was set up to print on success. He then presented three pangrams that his machine had discovered, floating "out there" somewhere beyond the orbit of Pluto. My favorite one is this:

This pangram tallies five a's, one b, one c, two d's, twenty-eight e's, eight f's, six g's, eight h's, thirteen i's, one j, one k, three l's, two m's, eighteen n's, fifteen o's, two p's, one q, seven r's, twenty-five s's, twenty-two t's, four u's, four v's, nine w's, two x's, four y's, and one z.

Now that's what I call a success for mechanical translation!

Sallows writes: "I wager ten guilders that nobody will succeed in producing a perfect self-documenting solution (or proof of its non-existence) to the sentence beginning, 'This computer-generated pangram contains . . .' *within the next ten years.* No tricks allowed. The format to be exactly as in the above pangrams. Either 'and' or '&' is permissible. Result to be derived exclusively by von Neumann architecture digital computer (no super computers, no parallel processing). Fancy your chances?" Anyone who wants to write to Sallows can do so, at Buurmansweg 30, 6525 RW Nijmegen, Holland.

Much though I am delighted by Sallows' ingenious machine and his plucky challenge, I expect him to lose his wager before you can say "Raphael Robinson". For my reasons, see the postscript to Chapter 16.

4

Nomic: A Self-Modifying Game Based on Reflexivity in Law

June, 1982

IN his excellent book *A Profile of Mathematical Logic,* the philosopher Howard DeLong tells the following classic story of ancient Greece. "Protagoras had contracted to teach Euathlus rhetoric so that he could become a lawyer. Euathlus initially paid only half of the large fee, and they agreed that the second installment should be paid after Euathlus had won his first case in court. Euathlus, however, delayed going into practice for quite some time. Protagoras, worrying about his reputation as well as wanting the money, decided to sue. In court Protagoras argued:

> Euathlus maintains he should not pay me but this is absurd. For suppose he wins this case. Since this is his maiden appearance in court he then ought to pay me because he won his first case. On the other hand, suppose he loses his case. Then he ought to pay me by the judgment of the court. Since he must either win or lose the case he must pay me.

Euathlus had been a good student and was able to answer Protagoras' argument with a similar one of his own:

> Protagoras maintains that I should pay him but it is this which is absurd. For suppose he wins this case. Since I will not have won my first case I do not need to pay him according to our agreement. On the other hand, suppose he loses the case. Then I do not have to pay him by judgment of the court. Since he must either win or lose I do not have to pay him."

Then DeLong adds, "It is clear that to straighten out such puzzles one has to inquire into general procedures of argument." Actually, to many people, it is not at all clear that *general* procedures of argument will need scrutiny—quite the contrary. To many people, paradoxes such as this one appear to be mere pimples or blemishes on the face of the law, which can be removed by simple cosmetic surgery. Similarly, many people who take theology seriously think that paradoxical questions about omnipotence, such as "Can God make a stone so heavy that It cannot lift it?", are just childish riddles, not serious theological dilemmas, and can be resolved in a definitive and easy way. Throughout history, simplistic or patchwork remedies have been proposed for all kinds of dilemmas created by loops of this sort. Bertrand Russell's theory of types is a famous example in logic. But the dreaded loops just won't go away that easily, however, as Russell found out. Wherever they occur, they are deep and pervasive, and attempts to unravel them lead down unexpected pathways.

In fact, reflexivity dilemmas of the Protagoras-vs.-Euathlus type and problems of conflicting omnipotence crop up with astonishing regularity in the down-to-earth discipline of law. Yet until recently, their central importance in defining the nature of law has been little noticed. In the past few years, only a handful of specialized papers on the subject have appeared in law journals and philosophy journals.

It was with surprise and delight, therefore, that I learned that an entire book on the role of reflexivity in law was in preparation. I first received word of it—"The Paradox of Self-Amendment: A Study of Logic, Law, Omnipotence, and Change"—in a letter from its author, Peter Suber, who identified himself as a philosophy Ph.D. and lawyer now teaching philosophy at Earlham College in Richmond, Indiana. He hopes "The Paradox of Self-Amendment" will be out soon.

In correspondence with Suber, I have found out that he has an even more ambitious book in the works, tentatively titled "The Anatomy of Reflexivity", which is a study of reflexivity in its broadest sense, encompassing, as he says, "the self-reference of signs, the self-applicability of principles, the self-justification and self-refutation of propositions and inferences, the self-creation and self-destruction of legal and logical entities, the self-limitation and self-augmentation of powers, circular reasoning, circular causation, vicious and benign circles, feedback systems, mutual dependency, reciprocity, and organic form."

In his original letter to me, Suber not only gave a number of interesting examples of self-reference in law but also presented a game he calls *Nomic* (from the Greek νόμος (nómos), meaning "law") which is presented in an appendix to *The Paradox of Self-Amendment.* I found reading the rules of Nomic to be a mind-opening experience. Much of this article will be devoted to Nomic, but before we tackle the game itself, I would like to set the stage by mentioning some other examples of reflexivity in the political arena.

*   *   *

My friend Scott Buresh, himself a lawyer, described the following perplexing hypothetical dilemma, which he first heard posed in a class on constitutional law. What if Congress passes a law saying that henceforth all determinations by the Supreme Court shall be made by a 6–3 majority (rather than a simple 5–4 majority, as is currently the case)? Imagine that this law is challenged in a court case that eventually makes its way up to the Supreme Court itself, and that the Supreme Court rules that the law is unconstitutional—and needless to say the ruling is by a 5–4 majority. What happens? This is a classic paradox of the separation of powers and it was nearly played out, in a minor variation, during the Watergate era, when President Nixon threatened he would obey a Supreme Court ruling to turn over his tapes only if it were "definitive", which presumably meant something like a unanimous decision.

It is interesting to note that conservatives are now trying to limit the jurisdiction of the Supreme Court over issues such as abortion and prayer in the schools. Constitutional scholars expect that a showdown might ensue if Congress passes such a statute and the Supreme Court is asked to review its constitutionality.

Conflicts that enmesh the Supreme Court with itself can arise in less flashy ways. Suppose the Supreme Court proposes to build an annex in an area that environmentalists want to protect. The environmentalists take their case to court, and it gets blown up into a large affair that eventually reaches the level of the Supreme Court. What happens? Clearly the reason this kind of thing cannot be prevented is that any court is itself a part of society, with buildings, employees, contracts, and so on. And since the law deals with things of this kind, no court at any level can guarantee that it will never get ensnared in legal problems.

If self-ensnaredness is a rare event for the Supreme Court, it is not so rare for other arms of government. An interesting case came up recently in San Francisco. There had been a large number of complaints about the way the police department was handling cases, and so an introverted "Internal Affairs Bureau" was set up to look into such matters as police brutality. But then, inevitably, complaints arose that the Internal Affairs Bureau was whitewashing its findings, and so Mayor Dianne Feinstein set up a doubly-introverted committee, again internal to the police department, to investigate the performance of the Internal Affairs Bureau. The last I heard was that the report of this committee was unfavorable. What finally resulted I do not know.

Parliamentary procedure too can lead to the most tangled of situations. For example, there are several editions of *Robert's Rules of Order,* and a body must choose which set of rules will govern its deliberations. The latest edition of *Robert's Rules* states that if no specific edition is chosen as the governing one, then the most recent issue holds. A problem arises, though, if one hasn't adopted the latest edition, since one cannot then rely on its authority to tell one to rely on it.

In some ways, parliamentary procedure, which deals with how to handle simultaneous and competing claims for attention, bears a remarkable resemblance to the way a large computer system must manage its own internal affairs. Within such a system, there is always a program called an *operating system* with a part called the *scheduling algorithm,* which weighs priorities and decides which activity will proceed next. In a "multiprocessing" system, this means determining which activity gets the next "time slice" (lasting for anywhere from a millisecond to a few seconds, or possibly even for an unlimited time, depending on the activity's priority and numerous other factors). But there are also *interrupts* that come and interfere with—oops, just a moment, my telephone's ringing. Be right back. There. Sorry we were disturbed. Someone wanted to sell me a telephone-answering system. Now what would—ah, ah, just a sec—ah-choo! —sorry—what would I do with one of those things? Now where was I? Oh, yes—interrupts. Well, in a way they are like telephone calls that take the store clerk away from you, annoying you in the extreme, since you have come to the store in person, whereas the telephone caller has been lazy and yet is given higher priority.

A good scheduling algorithm strives to be equitable, but all kinds of conflicts can arise, in which interrupts interrupt interrupts and are then themselves interrupted. Moreover, the scheduler has to be able to run its own internal decision-making programs with high priority, yet not so high a priority that nothing else ever runs. Sometimes the internal and external priorities can become so tangled that the entire system begins to "thrash". This is the term used to describe a situation where the operating system is spending most of its time bogged down in "introverted" computation, deciding what it should spend its time doing. Needless to say, during periods of thrashing, very little "real" computation gets done. It sounds quite like the cognitive state a person can get into when too many factors are weighing down all at once and the slightest thought on any topic seems to trigger a rash of paradoxical dilemmas from which there is no escape. Sometimes the only solution is to go to sleep, and let the paradoxes somehow drift away into a better perspective.

*　　*　　*

Operating systems and courts of law cannot, unfortunately, go to sleep. Their snarls are very real, and some means of dealing with them has to be invented. It was considerations such as this that led Peter Suber to invent his tangled game of Nomic.

He writes that he was struck by the oft-heard cynicism that "Government is just a game." Now, one essential activity of government is law-making, so if it is a game, then it is a game in which changing the laws (or rules) is a move. Moreover, some rules are needed to structure the process of changing the rules. Yet no legal system seems to have any rules that are absolutely immune to legal change. Suber's main aim, he wrote, was "to make a playable game that models this particular situation. But whereas governments are at any given moment pushed in various directions in their rule-changing by historical realities and the ideology of their people and existing rules, I wanted the game to start with as 'clean' an initial set of rules as possible." Nomic is such a game, and its rules (or rather, its Initial Set of rules) will be presented below. Most of the following description is in essence by Suber himself. I have simply interspersed some of my own observations.

In legal systems, statutes are the paradigmatic rules. Statutes are made by a rule-governed process that is itself partly statutory; hence the power to make and change statutes can reach some of the rules governing the process itself. Most of the rules, however, that govern the making of statutes are constitutional and are therefore beyond the reach of the power they govern. For instance, Congress may change its parliamentary rules and its committee structure, and it may bind its future action by its past action, but it cannot, through mere statutes, alter the fact that a two-thirds "supermajority" is needed to override an executive veto, nor can it abolish or circumvent one of its houses, start a tax bill in the Senate, or even delegate too much of its power to experts.

Although statutes cannot affect constitutional rules, the latter can affect the former. This is an important difference of logical priority. When there is a conflict between rules of different types, the constitutional rules always prevail. This *logical* level-distinction is matched by a *political* level-distinction —namely, that the logically prior (constitutional) rules are more difficult to amend than the logically posterior (statutory) rules.

It is no coincidence that logically prior laws are harder to amend. One purpose of making some rules more difficult to change than others is to prevent a brief wave of fanaticism from undoing decades or even centuries of progress. This could be called "self-paternalism": a deliberate retreat from democratic principles, although one chosen for the sake of preserving democracy. It is our chosen insurance against our anticipated weak moments. But that purpose will not be met unless the two-tier (or multi-tier) system also creates a logical hierarchy in which the less mutable rules take logical priority over the more mutable rules; otherwise, the more mutable rules could by themselves undo the deeper and more abstract principles on which the whole system is based. If supermajorities and the concurrence of many bodies are necessary to protect the foundations of the system from hasty change, that protective purpose is frustrated if those foundations are reachable by rules requiring merely a simple majority of one legislature.

Although all the rules in the American system are mutable, it is convenient to refer to the less mutable constitutional rules as *immutable,* and to the more mutable rules below them in the hierarchy as *mutable.* The same is true in Nomic, where, at least initially, no rule is literally immutable. If Nomic's self-paternalism is to be effective, then, its "immutable" rules, in addition to resisting easy amendment, must possess logical priority.

Many designs could satisfy this requirement. Nomic has adopted a simple two-tiered system, modeled to some extent on the U.S. Constitution. In principle, a system could have any number of degrees of difficulty in the amendment of rules. For instance, Class *A* rules, the hardest to amend, could require unanimity of a central body and the unanimous concurrence of all regional bodies. Class *B* rules could require 90 percent supermajorities, Class *C* rules 80 percent supermajorities, and so on. The number of such categories could be indefinitely large.

Indeed, if appropriate qualifications are made for the informality of custom and etiquette, a strong argument could be made that normal social life is just such a system of indefinite tiers. Near the top of the "difficult" end of the series of rules are actual laws, rising through case precedents, regulations, and statutes, all the way up to constitutional rules. At the bottom of the scale are rules of personal behavior that individuals can amend unilaterally without incurring disapprobation or censure. Above these are rules for which amendment is increasingly costly, starting with costs on the order of furrowed brows and clucked tongues, and passing through indignant blows and vengeful homicide.

*　　*　　*

In any case, for the sake of simplicity and to make it easier to learn and play, Nomic is a clean two-tier system rather than a nuanced or multi-tier system like the U.S. Government, with its intermediate and substatutory levels such as parliamentary rules, administrative regulations, joint resolutions, treaties, executive agreements, higher and lower court decisions, state practice, judicial rules of procedure and evidence, executive orders, canons of professional responsibility, evidentiary presumptions, standards of reasonableness, rules establishing priority among rules, canons of interpretation, contractual rules, and so on. This is not to say that nuanced, intermediate levels may not arise in Nomic through game custom and tacit understandings. In fact, the nature of the game allows players to add new tiers by explicit amendment as they see fit, and one reason for making Nomic simple initially is that it is easier to add tiers to a simple game than it is to subtract them from a complex one.

Nomic's two-tier system embodies the same self-paternalistic elements as does the Federal Constitution. The "immutable" rules govern more basic processes than the "mutable" ones do, and thus shield them from hasty change. Since, in the course of play, the central core of the game may change (and the minor aspects *must* change), after a few rounds the game being played by the players may in a certain sense be different from the one they were playing when they started. Yet needless to say, whatever results from compliance with the rules is, by definition, the game Nomic. The "feel" of the game may change drastically even as, at a deeper level, the game remains the same.

In a similar way, human beings undergo constant development and self-modification, and yet continue to be convinced that it makes sense to refer, via such words as "I", to an underlying stable entity. The more immediately perceptible patterns change, whereas deeper and more hidden patterns remain the same. From birth to maturity to death, however, the changes can be so radical that one may sometimes feel that in a single lifetime one is several different people. Similarly, in law, many have acknowledged that an amendment clause (a clause defining how a constitution may be amended)—even a clause limited to piecemeal amendment—could, through repeated application, create a fundamentally new constitution.

The fact that Nomic has more than one tier prevents the logical foundation of the game—the central core—from changing radically in just a few moves. Such continuity is a virtue both of games and of governments, but players of Nomic have an advantage over citizens in that, whenever they are so motivated, they can adjust the degree of continuity and the rate of change rather quickly, using their wits, whereas in real life the mechanisms by which such change could be effected are barely known and partially beyond reach.

Standard games possess the continuity of unchanging rules, or at least of rules that change only between games, not during them. Nomic's continuity is more like that of a legal system than that of a standard game: it is a rule-governed set of systems, directives, and processes undergoing constant rule-governed change. If, however, one wants a specific entity to point to as being "Nomic itself", the Initial Set of rules, as presented below, will do. Yet Nomic is equally the product, at any given moment, of the dynamic rule-governed change of the Initial Set. The continuing identity of the game, like that of a nation or person, is due to the fact (if fact it is) that all change is the product of existing rules properly applied, and that no change is revolutionary. (One could even argue that revolutionary change is just more of the same: In a revolution, rules that have been assumed to be totally immutable simply are rendered mutable by other rules that are more deeply immutable, but that previously had been taken for granted and hence had been invisible, or tacit.)

* * *

In its Rule 212, Nomic includes provision for subjective *judgment* (as in a court of law), not merely to imitate government in yet another aspect, but for the same reasons that compel government itself to make provisions for judgment: rules will inevitably be made that are ambiguous, inconsistent, or incomplete, or that require application to individual circumstance. "Play" must not be interrupted; therefore some agency must be empowered to make an authoritative and final determination so that play can continue.

Judgments in Nomic are not bound by rules of precedent, since that would require a daunting amount of record-keeping for each game. But the doctrine of *stare decisis* (namely, that precedents should be followed) may be imposed at the players' option, or it may arise without explicit amendment, as successive judges feel impelled to treat "similarly situated" persons "similarly". (Admittedly, the meanings of these terms in specific cases may well require further levels of judgment. This fact is one of the most dangerous sources of potential infinite regress in real court cases.) Without *stare decisis,* the players are constrained to draft their rules carefully, make thoughtful adjudications, overrule poor judgments, and amend defective rules. This is one way Nomic teaches basic principles and exigencies of law, even as it vastly simplifies.

The Initial Set must be short and simple enough to encourage play, yet long and complex enough to cover contingencies likely to arise before the players get around to providing for them in a rule, and to prevent any single rule change from disturbing the continuity of the game. Whether the Initial Set presented below satisfies these competing interests is left to players to judge.

One contingency deliberately left to the players to resolve is what to do about violations of the rules. The players must also decide whether old violations are protected by a statute of limitations or whether they may still be punished or nullified. Whether the likelihood of compliance and the discretionary power of the judge suffice to deal with a crisis of confidence or to delay it until a rule can take over, and whether in other respects the Initial Set satisfactorily balances the competing interests of simplicity and complexity, can best be determined by playing the game.

*   *   *

Nomic affords a curious twist on one common and fundamental property of games: it allows the blurring of the distinction between *constitutive rules* and *rules of skill*—that is, between rules that define lawful play and those that define artful play. In other words, in Nomic there is a blurring between the permissible and the optimal.

Most games do not embrace non-play, and do not become paradoxical by seeming to. Interestingly, however, children often invent games that provide game penalties for declining to play, or that incorporate or extend game jurisdiction to all of "real life", and end only when the children tire of the game or forget they are playing. ("Daddy, Daddy, come play a new game we invented!" "No, sweetheart, I'm reading." "That's ten points!") Nomic carries this principle to an extreme. A game of Nomic can embrace anything at the vote of the players. The line between play and non-play may shift at each turn, or it may apparently be eliminated. Players may be governed by the game when they think they are between games or when they think they have quit.

For most games, there is an infallible decision procedure to determine the legality of a move. In Nomic, by contrast, situations may easily arise where it is very hard to determine whether or not a move is legal. Moreover, paradoxes can arise in Nomic that paralyze judgment. Occasionally this will be due to the poor drafting of a rule, but it may also arise from a rule that is unambiguous but mischievous. The variety of such paradoxes is truly impossible to anticipate. Rule 213, nonetheless, is designed to cope with them as well as possible without cluttering the Initial Set with too many legalistic qualifications. Note that Rule 213 allows a wily player to create a paradox, get it passed (if the rule seems innocent enough to the other players), and thereby win.

So much for a general prologue to the game itself. Now we can move on to a description of how a game of Nomic is played. To reiterate, Nomic is a game in which changing the rules is a move. Two can play, but having three or more makes for a better game. The gist of Nomic is to be found in Rule 202, which should be read first. Players will need paper and pencil, and (at least at the outset!) one die. Instead of sheets of paper, players may find it easier to use a set of index cards. All new rules and amendments are to be written down. How the rules are positioned on paper or on the table can indicate which ones are currently immutable and which ones are mutable. Amendments can be placed on top of or next to the rules they amend. Inoperative rules may simply be deleted. Alternatively, for more complex games, players may prefer to transcribe into their own notebooks the text of each new rule or amendment and to keep a separate list, by number, of the rules still in effect. Ideally, perhaps, all rules should be entered in a computer, with a terminal for each player; amendments could then be incorporated instantly into the main text, with a corresponding adjustment to the numerical order.

Initial Set of Rules of Nomic

I. *Immutable Rules*

101. All players must always abide by all the rules then in effect, in the form in which they are then in effect. The rules in the Initial Set are in effect whenever a game begins. The Initial Set consists of Rules 101–116 (immutable) and 201–213 (mutable).

102. Initially, rules in the 100's are immutable and rules in the 200's are mutable. Rules subsequently enacted or transmuted (*i.e.,* changed from immutable to mutable or vice versa) may be immutable or mutable regardless of their numbers, and rules in the Initial Set may be transmuted regardless of their numbers.

103. A rule change is any of the following: (1) the enactment, repeal, or amendment of a mutable rule; (2) the enactment, repeal, or amendment of an amendment, or (3) the transmutation of an immutable rule into a mutable rule, or vice versa. (Note: This definition implies that, at least initially, all new rules are mutable. Immutable rules, as long as they are immutable, may not be amended or repealed; mutable rules, as long as they are mutable, may be amended or repealed. No rule is absolutely immune to change.)

104. All rule changes proposed in the proper way shall be voted on. They will be adopted if and only if they receive the required number of votes.

105. Every player is an eligible voter. Every eligible voter must participate in every vote on rule changes.

106. Any proposed rule change must be written down before it is voted on. If adopted, it must guide play in the form in which it was voted on.

107. No rule change may take effect earlier than the moment of the completion of the vote that adopted it, even if its wording explicitly states otherwise. No rule change may have retroactive application.

108. Each proposed rule change shall be given a rank-order number (ordinal number) for reference. The numbers shall begin with 301, and each rule change proposed in the proper way shall receive the next successive integer, whether or not the proposal is adopted.

   If a rule is repealed and then re-enacted, it receives the ordinal number of the proposal to re-enact it. If a rule is amended or transmuted, it receives the ordinal number of the proposal to amend or transmute it. If an amendment is amended or repealed, the entire rule of which it is a part receives the ordinal number of the proposal to amend or repeal the amendment.

109. Rule changes that transmute immutable rules into mutable rules may be adopted if and only if the vote is unanimous among the eligible voters.

110. Mutable rules that are inconsistent in any way with some immutable rule (except by proposing to transmute it) are wholly void and without effect. They do not implicitly transmute immutable rules into mutable rules and at the same time amend them. Rule changes that transmute immutable rules into mutable rules will be effective if and only if they explicitly state their transmuting effect.

111. If a rule change as proposed is unclear, ambiguous, paradoxical, or destructive of play, or if it arguably consists of two or more rule changes compounded or is an amendment that makes no difference, or if it is otherwise of questionable value, then the other players may suggest amendments or argue against the proposal before the vote. A reasonable amount of time must be allowed for this debate. The proponent decides the final form in which the proposal is to be voted on and decides the time to end debate and vote. The only cure for a bad proposal is prevention: a negative vote.

112. The state of affairs that constitutes winning may not be changed from achieving n points to any other state of affairs. However, the magnitude of n and the means of earning points may be changed, and rules that establish a winner when play cannot continue may be enacted and (while they are mutable) be amended or repealed.

113. A player always has the option to forfeit the game rather than continue to play or incur a game penalty. No penalty worse than losing, in the judgment of the player to incur it, may be imposed.

114. There must always be at least one mutable rule. The adoption of rule changes must never become completely impermissible.

115. Rule changes that affect rules needed to allow or apply rule changes are as permissible as other rule changes. Even rule changes that amend or repeal their own authority are permissible. No rule change or type of move is impermissible solely on account of the self-reference or self-application of a rule.

116. Whatever is not explicitly prohibited or regulated by a rule is permitted and unregulated, with the sole exception of changing the rules, which is permitted only when a rule or set of rules explicitly or implicitly permits it.

II. *Mutable Rules*

201. Players shall alternate in clockwise order, taking one whole turn apiece. Turns may not be skipped or passed, and parts of turns may not be omitted. All players begin with zero points.

202. One turn consists of two parts, in this order: (1) proposing one rule change and having it voted on, and (2) throwing one die once and adding the number of points on its face to one's score.

203. A rule change is adopted if and only if the vote is unanimous among the eligible voters.

204. If and when rule changes can be adopted without unanimity, the players who vote against winning proposals shall receive 10 points apiece.

205. An adopted rule change takes full effect at the moment of the completion of the vote that adopted it.

206. When a proposed rule change is defeated, the player who proposed it loses 10 points.

207. Each player always has exactly one vote.

208. The winner is the first player to achieve 100 (positive) points.

209. At no time may there be more than 25 mutable rules.

210. Players may not conspire or consult on the making of future rule changes unless they are teammates.

211. If two or more mutable rules conflict with one another, or if two or more immutable rules conflict with one another, then the rule with the lowest ordinal number takes precedence.

   If at least one of the rules in conflict explicitly says of itself that it defers to another rule (or type of rule) or takes precedence over another rule (or type of rule), then such provisions shall supersede the numerical method for determining precedence.

   If two or more rules claim to take precedence over one another or to defer to one another, then the numerical method must again govern.

212. If players disagree about the legality of a move or the interpretation or application of a rule, then the player preceding the one moving is to be the Judge and to decide the question. Disagreement, for the purposes of this rule, may be created by the insistence of any player. Such a process is called *invoking judgment*.

   When judgment has been invoked, the next player may not begin his or her turn without the consent of a majority of the other players.

   The Judge's judgment may be overruled only by a unanimous vote of the other players, taken before the next turn is begun. If a Judge's judgment is overruled, the player preceding the Judge in the playing order becomes the new Judge for the question, and so on, except that no player is to be Judge during his or her own turn or during the turn of a teammate.

   Unless a Judge is overruled, one Judge settles all questions arising from the game until the next turn is begun, including questions as to his or her own legitimacy and jurisdiction as Judge.

New Judges are not bound by the decisions of old Judges. New Judges may, however, settle only those questions on which the players currently disagree and that affect the completion of the turn in which judgment was invoked. All decisions by Judges shall be in accordance with all the rules then in effect; but when the rules are silent, inconsistent, or unclear on the point at issue, then the Judge's only guides shall be common morality, common logic, and the spirit of the game.

213. If the rules are changed so that further play is impossible, or if the legality of a move is impossible to determine with finality, or if by the Judge's best reasoning, not overruled, a move appears equally legal and illegal, then the first player who is unable to complete a turn is the winner.

This rule takes precedence over every other rule determining the winner.

*　　*　　*　　*　　*

Whew! So there you have the rules of Nomic. After reading them, a friend of mine commented, "It won't ever replace Monopoly." I'll grant the truth of that, but it is certainly more interesting than Monopoly to contemplate playing! To make such contemplation even more intriguing, Suber, who has actually played this crazy-sounding game, offers a wide variety of suggestions for interesting types of rule changes. Here are some samples.

Make mutable rules easier to amend than immutable rules, by repealing the unanimity requirement of Initial Rule 203 and substituting (say) a simple majority. Add new tiers above, below, or between the two tiers with which Nomic begins. Make some rules amendable only by special procedures ("incomplete self-entrenchment"). Devise "sunset" rules that automatically expire after a certain number of turns. Allow private consultation between players on future rule changes ("log-rolling"). Allow secret ballots. Allow "constitutional conventions" (or "revolutions") in which all the rules are more easily and jointly subject to change according to new, temporary procedures. Put an upper limit on the number of initially immutable rules that at any given time may be mutable or repealed.

Allow the ordinal numbers of rules to change in certain contingencies, thereby changing their priorities. Or alter the very method of determining precedence; for example, make more recent rules take precedence over earlier rules, rather than vice versa. (In most actual legal systems, the rule of priority favors recent rules.)

Convert the point-earning mechanism from one based on randomness to one based on skill (intellectual or even athletic). Apply a formula to the number on the die so that it will increase the number of points awarded to any player whose proposal gets voted down or whose judgment gets overruled, but will decrease the number of points awarded to a player who votes nay, who proposes a rule change of more than 50 words, who takes more than two minutes to propose a rule change, who proposes to transmute an immutable rule to a mutable rule, or who proposes a rule that is enacted but is later repealed.

Introduce a second or third objective—for example, a cooperative objective, to complement the competitive objective of earning more points. Thus, each player might, on each turn, contribute a letter to a growing sentence, a line to a growing poem, a block to a growing castle, and so on, the group as a whole trying to complete the thing before one of them reaches the winning number of points. Or introduce a second competitive objective, such as having each player make a move in another game, with the winner (or winners) of the game that is finished first obtaining some predetermined advantage in the game that is still being played. Or make some aspect of the game conditional on the outcome of a different game, thus incorporating into Nomic any other game or activity that can muster enough votes. Similarly, leave Nomic pure but add stakes or drama (such as psychodrama).

Institute team play. Require permanent team combinations or allow alliances to shift according to procedures (informal negotiation, an algebraic formula applied to scores, or systematic rotation of partners). Create "hidden" partners (*e.g.*, the points a player earns in a turn are also added to the score of another player, or split with one, selected by a mechanism).

Extend the aptness of the game as a model of the legislative process by inventing an index that goes up and down according to events in the game and that measures "constituency pressure" or "constituency satisfaction"; use the index to constrain permissible moves (*e.g.*, through a system of rewards and penalties). Allow a certain number of turns to pass before a proposal is voted on, giving the players the opportunity to see what other proposals may be adopted in its place.

Suber's ultimate challenge to players of Nomic is this: to ascertain whether any rules can be made genuinely immutable while preserving some rule-changing power, and whether the power to change the rules can be irrevocably and completely repealed. Suber is interested in hearing from readers about their experiences in playing Nomic, as well as any suggestions for improvement or comments on reflexivity in law generally. His address is: Department of Philosophy, Earlham College, Richmond, Indiana 47374.

*   *   *

The richness of the Nomic universe is abundantly clear. It certainly meets every hope I had when, in my book *Gödel, Escher, Bach: an Eternal Golden Braid*, I wrote about self-modifying games. It was my purpose there to describe such games in the abstract, never imagining that anyone would work out a game so fully in the concrete. It had been a dream of mine for a long time to devise a system that was in some sense capable of modifying every aspect of itself, so that even if it had what I referred to as "inviolate" levels (corresponding roughly to Suber's "immutable" rules), they could be modified as well.

I vividly remember how this dream came about. I was a high school student when I first heard about computers from the late George Forsythe, then a professor of mathematics at Stanford (there was no such thing as a department of computer science yet). In his guest lecture to our math class he emphasized two things. One was the notion that the purpose of computing was to do anything that people could figure out how to mechanize. Thus, he pointed out, computing would inexorably make inroads on one new domain after another, as we came to recognize that an activity that had seemed to require ever-fresh insights and mental imagery could be replaced by an ingenious and subtly worked-out collection of rules, the execution of which would then be a form of glorified drudgery carried out at the speed of light. For me, one of Forsythe's most stunning illustrations of this notion was the way computers had in some sense been applied to themselves—namely in compilers, programs that translate programs from an elegant and human-readable language into the cryptic strings of 0's and 1's of machine language.

The other notion Forsythe emphasized—and it was closely related to the first one—was the fact that a program is just an object that sits in a computer's memory, and as such is no more and no less subject to manipulation by other programs—or even by itself!—than mere numbers are. The fusion of these two notions was what gave me my inspiration to design an abstract computer. Playing on the names of the ENIAC, ILLIAC, JOHNNIAC, and other computers I had heard of, I called it "IACIAC". I hoped IACIAC could not only manipulate its own programs but also redesign itself, change the way it interpreted its own instructions, and so on. I quickly ran into many conceptual difficulties and never completed the project, but I have never forgotten that fascination. It seems to me that although it is a game and not a computer, Nomic comes closer in spirit to that goal I sought than anything I have ever encountered. That is, except for itself.

Post Scriptum.

As a result of the publication of this column, I received a letter from a law professor named William Popkin, who obviously had found the game of Nomic fascinating while disagreeing philosophically with some points expressed. Subsequently, an exchange between Popkin and me was printed in the "Letters" column in *Scientific American.* Here is what Popkin had to say:

> As a law professor I was very interested in Douglas Hofstadter's piece on reflexivity and self-reference in the law. There are, as he says, many examples. Article V of the United States Constitution prohibits amendments denying

states equal representation in the Senate. The Supreme Court of India went out of its way to create a reflexivity problem by deciding that the normal process of amending the Indian Constitution did not apply to their Bill of Rights, even though no explicit provision prohibiting such amendments existed.

These reflexivity problems are fascinating, but I do not see what they have to do with "general procedures of argument", as Hofstadter (quoting Howard DeLong) suggests. They have everything to do with the meaning of rules, law, and politics, but not with procedures of argument. Let me explain how at least one law professor would approach these problems. Every reflexivity example has the same structure. There is a rule that has specific cases coming under the rule. One particular case, by coming under the rule, appears to undermine the rule itself. For example, assume that the Supreme Court must decide cases properly appealed to it, but that no judge can sit on a case in which he is personally interested. A case arises involving the reduction of judges' salaries, which is arguably unconstitutional. If the judges decide the case, they violate the rule against deciding cases in which they are personally interested, but failure to decide violates the rule requiring them to decide cases. The same structure exists for rules about amendment of the document containing the amending provision. Assume that the Constitution can be amended by a two-thirds vote but that one of the provisions requires a 100 percent vote. An amendment is passed changing the unanimity rule. If the amendment is valid, the unanimity rule is undermined, but if the amendment is invalid, the procedures for amendment are incomplete.

What is presented in all these cases is a problem of meaning and a conflict between rival conclusions, not a logical conundrum. The ultimate decision may be hard or easy, but the issues are not difficult to conceptualize. My own conclusion is that the Supreme Court should hear the case involving its own salary because we do not want Congress deciding such issues, and that the amending power should not extend to the unanimity rule because this breaks the social contract. These are hard cases, but another example presented in Hofstadter's article is easy. It concerns a contract to pay the rhetoric teacher Protagoras when his pupil Euathlus wins his first case. The teacher sues the pupil for the payment, figuring that if he wins the suit he gets his money and if he loses the suit he collects under the contract. But on what possible ground could he win the case before the pupil had won a lawsuit? And how could the original contract, in referring to a victory by the pupil as the occasion for the payment, include a victory in a frivolous lawsuit by the teacher?

What I am pointing out is that reflexivity presents problems of choice, sometimes difficult, sometimes trivial, but that is nothing new in the law. Most important legal problems involve choice without involving reflexivity. Do we prefer a right of privacy or freedom of the press? The deeper point concerns the interaction of law and artificial intelligence and perhaps interdisciplinary studies generally. Reflexivity is undoubtedly an important phenomenon in philosophy for reasons I do not fully appreciate. If developments in artificial intelligence are to be useful in law, however, they must take into account what legal problems are all about. To a lawyer, reflexivity is not a relevant category but choice is. Indeed, I suspect that reflexivity is just a diversion for Hofstadter. In an earlier article about analogy he dealt with the imaginative problem of defining the First Lady of Britain [see Chapter 24]. He there grappled with the problem of deciding what is like something else, which is the way most lawyers always proceed in making choices. How we make analogies determines how we make choices, and that is the essential nature of all judgment. If that is what artificial intelligence is all about, I very much want to hear more.

As for the question of whether there are immutable rules, the answer is: Of course there are, if that's what you want.

William D. Popkin
Professor of Law
Indiana University

I found this letter very nicely put, and a constructive opening for a small debate. I replied as follows:

Professor Popkin raises a very interesting point in his comment on my column about Peter Suber's game Nomic. His point is essentially twofold: (1) The fact that any legal system is inevitably chock-full of tangles arising from reflexivity is amusing, but rather than being themselves a deep aspect of law, such tangles are a consequence of other deep aspects, the most significant of which is that (2) the crux of any legal system is the ability of people to distinguish between the incidental qualities and the essential qualities of various events and relations, which ability results finally in recognition of what a given item is—that is, which category the item belongs to. Popkin calls this "choice". In conclusion, he suggests that to discover the principles by which people can "choose" is a critical task for artificial-intelligence workers to tackle.

I feel that neither Suber's *reflexivity* nor Popkin's *choice* is more central than the other in defining the nature of law. In fact, they are intertwined. Suber stresses that people, in choosing which of two inconsistent aspects of a supposedly self-consistent system shall take precedence, often make their choice without explicit rules (since if the rules were spelled out, they would be susceptible to getting embroiled in a similar tangle once again, only at a higher level of abstraction). "Law can disregard logical difficulties and ground a solution on pragmatic rules, social policies, and legal doctrines", Suber has written [in a reply to Popkin]. "The effectiveness of policy, or what Popkin calls 'choice', in plowing under logical obstacles is not the answer to the question but the mystery to be explained."

Coming to grips with this contrast between explicit rules and implicit principles or guidelines is of great importance if one wants to characterize how flexible category recognition—"choice"—takes place, whether one is doing research in artificial intelligence, philosophizing about free will, or attempting to characterize the nature of law. Popkin, in fact, is rather charitable toward artificial-intelligence research, suggesting that it may some day yield clues, if not the key, to the mystery of choice. I think he is right about this. He may have failed to realize, however, that in any attempt to make a machine capable of choice, one runs headlong into the problem of inconsistencies, level-collisions, and reflexivity tangles, and for the following reason.

All recognition programs are invariably modeled on what we know about perception in various modalities, such as hearing and sight. One thing we know for sure is that in any modality, perception consists of many layers of processing, from the most primitive or "syntactic" levels, to the most abstract or "semantic" levels. The zeroing-in on the semantic category to which a given raw stimulus belongs is carried out not by a purely bottom-up (stimulus-driven) or purely top-down (category-driven) scheme, but rather by a mixture of them, in which hypotheses at various levels trigger the creation of new hypotheses or undermine the existence of already-existing hypotheses at other levels. This process of sprouting and pruning hypotheses is a highly parallel one, in which all the levels compete simultaneously for attention, like billboards or radio commercials or advertisements in the subway.

Yet out of this seemingly anarchic chaos comes an integrated decision, in which the various levels gradually come to some kind of self-reinforcing agreement. If a firm decision is to emerge from such a swirl of conflicting claims, there must be some kind of mental *scheduler,* something that functions like *Robert's Rules of Order,* letting various levels have the floor, scheduling collective actions such as votes, overriding or tabling motions, and so on. In fact, to the best of our knowledge, this is the heart of the perceptual process. But this is the very place where reflexivity tangles crop up with a vengeance!

Any perception program has various levels of "inner sanctum"—that is, levels of untouchability of its data structures. (These structures include not only the current hypotheses, but also deeper, more permanent aspects of the program itself, such as the ways it weights various pieces of evidence, the rules by which it sorts out conflicts, the priority rules of its scheduler, and—of course—the information about the untouchability of levels!) Now, for the ultimate in flexibility, none of these levels should be *totally* untouchable (although that degree of flexibility may be unattainable), but obviously some levels should be less touchable than others. Therefore any recognition program must have at its core a tiered structure precisely like that of government (or that of the rules of Nomic), in which there are levels that are "easily mutable", "moderately mutable", "almost mutable", and so on. The structure of a recognition program—a "choice" program—is seen inevitably to be riddled with reflexivity.

The point of all this is that the very reflexivity issues that Popkin considers to be merely amusing sideshows in law are actually deeply embroiled in what he sees as the meat of the matter, namely the question of how category recognition—discerning the essence of something—works. For that reason, I found Suber's game not merely amusing but philosophically provocative as well. In fact, I consider the intertwined study of reflexivity and recognition, using the fresh methods of the emerging discipline of cognitive science, to be of great interest and importance for the light it may shed on the ancient philosophical problems of mind, free will, and identity—not to mention those of the philosophy of law.

*　　*　　*

It occurs to me that the message of my letter to Popkin could be put in a nutshell this way: To get *flexible cognition,* concentrate on *reflexivity* and *recognition.* Some of these ideas will come up again, more specifically in the context of artificial intelligence, in Chapters 23 and 24.

Section ll:

Sense and Society

Section II:
Sense and Society

Another broad theme of this book is introduced in the four chapters comprising this section: the harm that occurs when vast numbers of people accept without reflection the words, sayings, ideas, fads, styles, and tastes paraded in front of them by indiscriminate media and popular myth. Our society does a rather poor job of making us aware of, let alone interested in, the nature of common sense, the hidden assumptions that permeate thought, the complex mechanisms of sensory perception and category systems, the will to believe, the human tendency toward gullibility, the most typical flaws in arguments, the statistical inferences we make unconsciously, the vastly different temporal and spatial scales on which one can look at the universe, the many filters through which one can perceive and conceptualize people and events, and so on. The resulting deceptions, delusions, confusions, ignorances, and fears can lead to many disquieting social consequences, such as mildly or absurdly wasteful spending of funds, blatant or subtle discrimination against groups, and local or global apathy about the current state and momentum of the world. Of course everyone labors under some delusions, avoids certain kinds of thoughts, has an overly closed mind on this or that subject. What, however, are the consequences when this is multiplied by hundreds or thousands of millions, and all the small pieces are woven together into a vast fabric? What does a carpet woven from the incomplete understandings and ignorances of five billion sentient beings look like from afar—and where is this flying carpet headed?

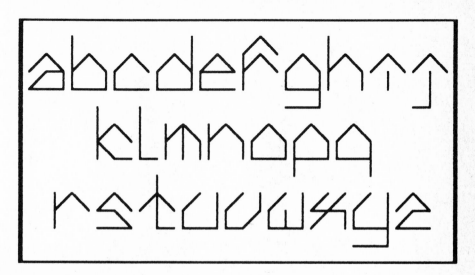

5

World Views in Collision: The *Skeptical Inquirer* versus the *National Enquirer*

February, 1982

Baffled Investigators and Educators Disclose . . .
BOY CAN SEE WITH HIS EARS

A Cross between Human Beings and Plants . . .
SCIENTISTS ON VERGE OF CREATING PLANT PEOPLE . . . Bizarre Creatures Could Do Anything You Want

Alien from Space Shares Woman's Mind and Body, Hypnosis Reveals

—Headlines from the *National Enquirer*

DID the child you once were ever wonder why the declarative sentences in comic books always ended with exclamation points? Were all those statements really that startling? Were the characters saying them really that thrilled? Of course not! Those exclamation points were a psychological gimmick put there purely for the sake of appearance, to give the story more pizzazz!

The *National Enquirer,* one of this country's yellowest and purplest journalistic instituitions, uses a similar gimmick! Whenever it prints a headline trumpeting the discovery of some bizarre, hitherto unheard-of phenomenon, instead of ending it with an exclamation point, it ends it (or begins it) with a reference to "baffled investigators", "bewildered scientists", or similarly stumped savants! It is an ornament put there to make the story seem to have more credibility!

Or is it? What do the editors really want? That the story appear credible or that it appear incredible? It seems they want it both ways: they want the story to sound as outlandish as possible and yet they want it to have the appearance of authenticity. Their ideal headline should thus embody a contradiction: impossibility coupled with certainty. In short, confirmed nonsense.

What is one to make of headlines like those printed above? Or of articles about plants that sing in Japanese, and calculating cacti? Or of the fact that this publication is sold by the millions every week in grocery stores, and that people gobble up its stories as voraciously as they do potato chips? Or of the fact that when they are through with it, they can turn to plenty of other junk food for thought, such as the *National Examiner,* the *Star,* the *Globe,* and, perhaps the most lurid of the lot, the *Weekly World News*? What is one to think? For that matter, what are Martians to think? (See Figure 5-1.)

FIGURE 5–1. *A Martian's reaction to a tabloid article. Note the complex diacritical marks of the Martian language, regrettably unavailable on most Terran typesetting machines. [Photograph by David J. Moser.]*

Naturally, one's first reaction is to chuckle and dismiss such stories as silly. But how do you know they are silly? Do you also think *that* is a silly question? What do you think about articles printed in *Scientific American*? Do you trust them? What is the difference? Is it simply a difference in publishing style? Is the tabloid format, with its gaudy pictures and sensationalistic headlines, enough to make you distrust the *National Enquirer*? But wait a minute—isn't that just begging the question? What kind of argument is it when you use the guilty verdict as part of the case for the prosecution? What you need is a way of telling objectively what you mean by "gaudy" or "sensationalistic" —and that could prove to be difficult.

And what about the obverse of the coin? Is the rather dignified, traditional format of *Scientific American*—its lack of photographs of celebrities, for example—what convinces you it is to be trusted? If so, that is a pretty curious way of making decisions about what truth is. It would seem that your concept of truth is closely tied in with your way of evaluating the "style" of a channel of communication—surely quite an intangible notion!

Having said that, I must admit that I, too, rely constantly on quick assessments of style in my attempt to sift the true from the false, the believable from the unbelievable. (Quickness is of the essence, like it or not, because the world does not allow infinite time for deliberation.) I could not tell you what criteria I rely on without first pondering for a long time and writing many pages. Even then, were I to write the definitive guide (*How to Tell the True from the False by Its Style of Publication*), it would have to be published to do any good; and its title, not to mention the style it was published in, would probably attract a few readers, but would undoubtedly repel many more. There is something disturbing about that thought.

There is something else disturbing here. Enormous numbers of people are taken in, or at least beguiled and fascinated, by what seems to me to be unbelievable hokum, and relatively few are concerned with or thrilled by the astounding—yet true—facts of science, as put forth in the pages of, say, *Scientific American*. I would proclaim with great confidence that the vast majority of what that magazine prints is true—yet my ability to defend such a claim is weaker than I would like. And most likely the readers, authors, and editors of that magazine would be equally hard pressed to come up with cogent, nontechnical arguments convincing a skeptic of this point, especially if pitted against a clever lawyer arguing the contrary. How come Truth is such a slippery beast?

*   *   *

Well, consider the very roots of our ability to discern truth. Above all (or perhaps I should say "underneath all"), *common sense* is what we depend on —that crazily elusive, ubiquitous faculty we all have, to some degree or other. But not to a degree such as "Bachelor's" or "Ph.D.". No, unfortunately, universities do not offer degrees in Common Sense. There are not even any Departments of Common Sense! This is, in a way, a pity.

At first, the notion of a Department of Common Sense sounds ludicrous. Given that common sense *is* common, why have a department devoted to it? My answer would be quite simple: In our lives we are continually encountering strange new situations in which we have to figure out how to apply what we already know. It is not enough to have common sense about known situations; we need also to develop the art of extending common sense to apply to situations that are unfamiliar and beyond our previous experience. This can be very tricky, and often what is called for is common sense in knowing *how* to apply common sense: a sort of "meta-level" common sense. And this kind of higher-level common sense also requires its own meta-level common sense. Common sense, once it starts to roll, gathers more common sense, like a rolling snowball gathering ever more snow. Or, to switch metaphors, if we apply common sense to itself over and over again, we wind up building a skyscraper. The ground floor of this structure is the ordinary common sense we all have, and the rules for building new floors are implicit in the ground floor itself. However, working it all out is a gigantic task, and the result is a structure that transcends mere common sense.

Pretty soon, even though it has all been built up from common ingredients, the structure of this extended common sense is quite arcane and elusive. We might call the quality represented by the upper floors of this skyscraper "rare sense"; but it is usually called "science". And some of the ideas and discoveries that have come out of this originally simple and everyday ability defy the ground floor totally. The ideas of relativity and quantum mechanics are anything but commonsensical, in the ground-floor sense of the term! They are outcomes of common sense self-applied, a process that has many unexpected twists and gives rise to some unexpected paradoxes. In short, it sometimes seems that common sense, recursively self-applied, almost undermines itself.

Well, truth being this elusive, no wonder people are continually besieged with competing voices in print. When I was younger, I used to believe that once something had been discovered, verified, and published, it was then part of Knowledge: definitive, accepted, and irrevocable. Only in unusual cases, so I thought, would opposing claims then continue to be published. To my surprise, however, I found that the truth has to fight constantly for its life! That an idea has been discovered and printed in a "reputable journal" does *not* ensure that it will become well known and accepted. In fact, usually it will have to be rephrased and reprinted many different times, often by many different people, before it has any chance of taking hold. This is upsetting to an idealist like me, someone more disposed to believe in the notion of a monolithic and absolute truth than in the notion of a pluralistic and relative truth (a notion championed by a certain school of anthropologists and sociologists, who un-self-consciously insist "all systems of belief are equally valid", seemingly without realizing that this dogma of relativism not only is just as narrow-minded as any other dogma, but moreover is unbelievably wishy-washy!). The idea that the truth has to fight for its life is a sad discovery. The idea that the truth will *not* out, unless it is given a lot of help, is pretty upsetting.

*   *   *

A question arises in every society: Is it better to let all the different voices battle it out, or to have just a few "official" publications dictate what is the case and what is not? Our society has opted for a plurality of voices, for a "marketplace of ideas", for a complete free-for-all of conflicting theories. But if things are this chaotic, who will ensure that there is law and order? Who will guard the truth? The answer (at least in part) is: CSICOP will!

CSICOP? Who is CSICOP? Some kind of cop who guards the truth? Well, that's pretty close. "CSICOP" stands for "Committee for the Scientific Investigation of Claims of the Paranormal"—a rather esoteric title for an organization whose purpose is not so esoteric: to apply common sense to claims of the outlandish, the implausible, and the unlikely.

Who are the people who form CSICOP and what do they do together? The organization was the brainchild of Paul Kurtz, professor of philosophy at the State University of New York at Buffalo, who brought it into being because he thought there was a need to counter the rising tide of irrational beliefs and to provide the public with a more balanced treatment of claims of the paranormal by presenting the dissenting scientific viewpoint. Among the early members of CSICOP were some of America's most distinguished philosophers (Ernest Nagel and Willard Van Orman Quine, for example) and other colorful combatants of the occult, such as psychologist Ray Hyman, magician James Randi, and someone whom readers of this column may have heard of: Martin Gardner. In the first few meetings, it was decided that the committee's principal function would be to publish a magazine dedicated to the subtle art of debunking. Perhaps "debunking" is not the term they would have chosen, but it fits. The magazine they began to publish in the fall of 1976 was called *The Zetetic,* from the Greek for "inquiring skeptic".

As happens with many fledgling movements, a philosophical squabble developed between two factions, one more "relativist" and unjudgmental, the other more firmly opposed to nonsense, more willing to go on the offensive and to attack supernatural claims. Strange to say, the open-minded faction was not so open-minded as to accept the opposing point of view, and consequently the rift opened wider. Eventually there was a schism. The relativist faction (one member) went off and started publishing his own journal, the *Zetetic Scholar,* in which science and pseudo-science coexist happily, while the larger faction retained the name "CSICOP" and changed the title of its journal to the *Skeptical Inquirer.*

In a word, the purpose of the *Skeptical Inquirer* is to combat nonsense. It does so by recourse to common sense, and as much as possible by recourse to the *ground floor* of the skyscraper of science—the common type of common sense. This is by no means always possible, but it is the general style of the magazine. This means it is accessible to anyone who can read English. It does not require any special knowledge or training to read its pages, where nonsensical claims are routinely smashed to smithereens. (Sometimes the claims are as blatantly silly as the headlines at the beginning of this article, sometimes much subtler.) All that is required to read this maverick journal is curiosity about the nature of truth: curiosity about how truth defends itself (through its agent CSICOP) against attacks from all quarters by unimaginably imaginative theorizers, speculators, eccentrics, crackpots, and out-and-out fakers.

The journal has grown from its original small number of subscribers to roughly 7,500—a David, compared with the Goliaths mentioned above, with their circulations in the millions. Its pages are filled with lively and humorous writing—the combat of ideas in its most enjoyable form. By no means is this journal a monolithic voice, a mouthpiece of a single dogma. Rather, it is itself a marketplace of ideas, strangely enough. Even people who wield the tool of common sense with skill may do so with different styles, and sometimes they will disagree.

There is something of a paradox involved in the editorial decisions in such a magazine. After all, what is under debate here is, in essence, the nature of correct arguments. What should be accepted and what shouldn't? To caricature the situation, imagine the editorial dilemmas that would crop up for journals with titles such as *Free Press Bulletin, The Open Mind,* or *Editorial Policy Newsletter.* What letters to the editor should be printed? What articles? What policy can be invoked to screen submitted material?

These are not easy questions to answer. They involve a paradox, a tangle in which the ideas being evaluated are also what the evaluations are based on. There is no easy answer here! There is no recourse but to common sense, that rock-bottom basis of all rationality. And unfortunately, we have no foolproof algorithm to uniquely characterize that deepest layer of rationality, nor are we likely to come up with one soon. The ability to use common sense—no matter how much light is shed on it by psychologists or philosophers—will probably forever remain a subjective art more than an objective science. Even when experimental epistemologists, in their centuries-long quest for artificial intelligence, have at last made a machine that thinks, its common sense will probably be just as instinctive and fallible and stubborn as ours. Thus at its core, rationality will always depend on inscrutables: the simple, the elegant, the intuitive. This weird paradox has existed throughout intellectual history, but in our information-rich times it seems particularly troublesome.

Despite these epistemological puzzles, which seem to be intimately connected with its very reason for existence, the *Skeptical Inquirer* is flourishing and provides a refreshing antidote to the jargon-laden journals of science, which often seem curiously irrelevant to the concerns of everyday life. In that one way, the *Inquirer* resembles the scandalous tabloids.

The list of topics covered in the seventeen issues that have appeared so far is remarkably diverse. Some topics have arisen only once, others have come up regularly and been discussed from various angles and at various depths. Some of the more commonly discussed topics are:

> ESP (extra-sensory perception) * telekinesis (using mental power to influence events at a distance) * astrology * biorhythms * Bigfoot * the Loch Ness monster * UFO's (unidentified flying objects) * creationism * telepathy * remote viewing * clairvoyant detectives who allegedly solve crimes * the Bermuda (and other) triangles * "thoughtography" (using mental power to create images on film) * the supposed extraterrestrial origin of life on the earth * Carlos Castaneda's mystical sorcerer "Don Juan" * pyramid power * psychic surgery and faith healing * Scientology * predictions by famous "psychics" * spooks and spirits and haunted houses * levitation * palmistry and mind reading * unorthodox anthropological theories * plant perception * perpetual-motion machines * water witching and other kinds of dowsing * bizarre cattle mutilations

When I contemplate the length of this list, I am quite astonished. Before I ever subscribed to the magazine, I had heard of almost all these items and was skeptical of most of them, but I had never seen a frontal assault mounted against so many paranormal claims at once. And I have only scratched the surface of the list of topics, because the ones listed above are regulars! Imagine how many topics are treated at shorter length.

There are quite a few frequent contributors to this iconoclastic journal, such as James Randi, who is truly prolific. Among others are aeronautics writer Philip J. Klass, UFO specialist James E. Oberg, writer Isaac Asimov, CSICOP's founder (and current director) Paul Kurtz, psychologist James Alcock, educator Elmer Kral, anthropologist Laurie Godfrey, science writer Robert Sheaffer, sociologist William Sims Bainbridge, and many others. And the magazine's editor, Kendrick Frazier, a free-lance science writer by trade, periodically issues eloquent and mordant commentaries.

* * *

I know of no better way to impart the flavor of the magazine than to quote a few selections from articles. One of my favorite articles appeared in the second issue (Spring/Summer, 1977). It is by psychologist Ray Hyman (who, incidentally, like many other authors in the *Skeptical Inquirer,* is a talented magician) and is titled "Cold Reading: How to Convince Strangers that You Know All About Them".

It begins with a discussion of a course Hyman taught about the various ways people are manipulated. Hyman states:

I invited various manipulators to demonstrate their techniques—pitchmen, encyclopedia salesmen, hypnotists, advertising experts, evangelists, confidence men and a variety of individuals who dealt with personal problems. The techniques which we discussed, especially those concerned with helping people with their personal problems, seem to involve the client's tendency to find more meaning in any situation than is actually there. Students readily accepted this explanation when it was pointed out to them. But I did not feel that they fully realized just how pervasive and powerful this human tendency to make sense out of nonsense really is.

Then Hyman describes people's willingness to believe what others tell them about themselves. His "golden rule" is: "To be popular with your fellow man, tell him what he wants to hear. He wants to hear about himself. So tell him about himself. But not what you know to be true about him. Oh, no! Never tell him the truth. Rather, tell him what he would like to be true about himself!" As an example, Hyman cites the following passage (which, by an extraordinary coincidence, was written about none other than *you,* dear reader!):

> Some of your aspirations tend to be pretty unrealistic. At times you are extroverted, affable, sociable, while at other times you are introverted, weary, and reserved. You have found it unwise to be too frank in revealing yourself to others. You pride yourself on being an independent thinker and do not accept others' opinions without satisfactory proof. You prefer a certain amount of change and variety, and become dissatisfied when hemmed in by restrictions and limitations. At times you have serious doubts as to whether you have made the right decision or done the right thing. Disciplined and controlled on the outside, you tend to be worrisome and insecure on the inside.
>
> Your sexual adjustment has presented some problems for you. While you have some personality weaknesses, you are generally able to compensate for them. You have a great deal of unused capacity which you have not turned to your advantage. You have a tendency to be critical of yourself. You have a strong need for other people to like you and for them to admire you.

Pretty good fit, eh? Hyman comments:

> The statements in this stock spiel were first used in 1948 by Bertram Forer in a classroom demonstration of personal validation. He obtained most of them from a newsstand astrology book. Forer's students, who thought the sketch was uniquely intended for them as a result of a personality test, gave the sketch an average rating of 4.26 on a scale of 0 (poor) to 5 (perfect). As many as 16 out of his 39 students (41 percent) rated it as a perfect fit to their personality. Only five gave it a rating below 4 (the worst being a rating of 2, meaning "average"). Almost 30 years later students give the same sketch an almost identical rating as a unique description of themselves.

A particularly delicious feature is the thirteen-point recipe that Hyman gives for becoming a cold reader. Among his tips are these: "Use the technique of 'fishing' (getting the subject to tell you about himself or herself, then rephrasing it and feeding it back); always give the impression that you know more than you are saying; don't be afraid to flatter your subject every chance you get." This cynical recipe for becoming a character reader is presented by Hyman in considerable detail, presumably not to convert readers of the article into charlatans and fakers, but to show them the attitude of the tricksters who do such manipulations. Hyman asks:

> Why does it work so well? It does not help to say that people are gullible or suggestible. Nor can we dismiss it by implying that some individuals are just not sufficiently discriminating or lack sufficient intelligence to see through it. Indeed, one can argue that it requires a certain degree of intelligence on the part of a client for the reading to work well . . . . We have to bring our knowledge and expectations to bear in order to comprehend anything in our world. In most ordinary situations, this use of context and memory enables us to correctly interpret statements and supply the necessary inferences to do this. But this powerful mechanism can go astray in situations where there is no actual message being conveyed. Instead of picking up random noise, we still manage to find meaning in the situation. So the same system that enables us to creatively find meanings and to make new discoveries also makes us extremely vulnerable to exploitation by all sorts of manipulators. In the case of the cold reading, the manipulator may be conscious of his deception; but often he too is a victim of personal validation.

Hyman knows what he's talking about. Many years ago, he was convinced for a time that he himself had genuine powers to read palms, until one day when he tried telling people the exact opposite of what their palms told him and saw that they still swallowed his line as much as ever! Then he began to suspect that the plasticity of the human mind—his own particularly—was doing some strange things.

*　　*　　*

At the beginning of each issue of the *Skeptical Inquirer* is a feature called "News and Comment". It covers such things as the latest reports on current sensational claims, recently broadcast television shows for and against the paranormal, lawsuits of one sort or another, and so on. One of the most amusing items was the coverage in the Fall 1980 issue of the "Uri Awards", given out by James Randi (on April 1, of course) to various deserving souls who had done the most to promote gullibility and irrational beliefs. Each award consists of "a tastefully bent stainless-steel spoon with a very transparent, very flimsy base". Award winners were notified, Randi explained, by telepathy, and were "free to announce their winning in advance, by precognition, if they so desired". Awards were made in four categories: Academic ("to the scientist who says the dumbest thing about parapsychology"), Funding ("to the funding organization that awards the most money for the dumbest things in parapsychology"), Performance ("to the psychic who, with the least talent, takes in the most people"), and Media ("to the news organization that supports the most outrageous claims of the paranormalists").

The nature of coincidences is a recurrent theme in discussions of the paranormal. I vividly remember a passage in a lovely book by Warren Weaver titled *Lady Luck: The Theory of Probability,* in which he points out that in many situations, the most likely outcome may well be a very unlikely event (as when you deal hands in bridge, where whatever hand you get is bound to be extraordinarily rare). A similar point is made in the following excerpt from a recent book by David Marks and Richard Kammann titled *The Psychology of the Psychic* (from which various excerpts were reprinted in one issue of the *Skeptical Inquirer*):

> 'Koestler's fallacy' refers to our general inability to see that unusual events are probable in the long run . . . . It is a simple deduction from probability theory that an event that is very improbable in a *short run* of observations becomes, nevertheless, highly probable somewhere in a *long run* of observations . . . . We call it 'Koestler's fallacy' because Arthur Koestler is the author who best illustrates it and has tried to make it into a scientific revolution. Of course, the fallacy is not unique to Koestler but is widespread in the population, because there are several biases in human perception and judgment that contribute to this fallacy.
>
> First, we notice and remember matches, especially *oddmatches,* whenever they occur. (Because a psychic anecdote first requires a match, and, second, an oddity between the match and our beliefs, we call these stories *oddmatches.* This is equivalent to the common expression, an "unexplained coincidence".) Second, we do *not* notice non-matches. Third, our failure to notice nonevents creates the *short-run illusion* that makes the oddmatch seem improbable. Fourth, we are poor at estimating combinations of events. Fifth, we overlook the *principle of equivalent oddmatches,* that one coincidence is as good as another as far as psychic theory is concerned.

An excellent example of people not noticing non-events is provided by the failed predictions of famed psychics (such as Jeane Dixon). Most people never go back to see how the events bore out the predictions. The *Skeptical Inquirer,* however, has a tradition of going back and checking. As each year concludes, it prints a number of predictions made by various psychics for that year and evaluates their track records. In the Fall 1980 issue, the editors took the predictions of 100 "top psychics", tabulated them, listed the top twelve in order of frequency, and left it to the reader to assess the accuracy of psychic visions of the future. The No. 1 prediction for 1979 (made by 86 psychics) was "Longer lives will be had for almost everyone as aging is brought under control." No. 2 (85 psychics) was "There will be a major breakthrough in cancer, which will almost totally wipe out the disease." No. 3 (also 85 psychics) was "There will be an astonishing spiritual rebirth and a return to the old values." And so on. No. 6 (81 psychics) was "Contact will be made with aliens from space who will give us incredible knowledge." The last four, interestingly, all involved celebrities: Frank Sinatra was supposed to become seriously ill, Edward Kennedy to become a presidential candidate, Burt Reynolds to marry, and Princess Grace to return to this country to resume a movie career. Hmm . . .

There is something pathetic, even desperate, about these predictions. One sees only too clearly the similarity of the tabloids (which feature these predictions) to the equally popular television shows like *Fantasy Island* and *Star Trek*. The common denominator is escape from reality. This point is well made in an article by William Sims Bainbridge in the Fall 1979 issue, on television pseudo-documentaries on the occult and pseudo-science. He characterizes those shows as resembling entertainment shows in which fact and fantasy are not clearly distinguised. His name for this is "wish-fulfillment fantasy".

Perhaps a key to why so much fantasy is splashed across the tabloids and splattered across our living-room screens lies here. Perhaps we all have a desire to dilute reality with fantasy, to make reality seem simpler and more aligned with what we wish it were. Perhaps for us all, the path of least resistance is to allow reality and fantasy to run together like watercolors, blurring our vision but making life more pastel-like: in a word, softer. Yet at the same time, perhaps all of us have the potential capacity and even the desire to sift sense from nonsense, if only we are introduced to the distinction in a sufficiently vivid and compelling manner.

* * *

But how can this be done? In the "News and Comment" section of the Spring 1980 issue, there was an item about a lively anti-pseudo-science traveling comedy lecture act by one "Captain Ray of Light"—actually Douglas F. Stalker, an associate professor of philosophy at the University of Delaware. The article quotes Stalker on his "comical debunking show" (directed at astrology, biorhythms, numerology, UFO's, pyramid power, psychic claims, and the like) as follows:

> For years I lectured against them in a serious way, with direct charges at their silly theories. These direct attacks didn't change many minds, and so I decided to take an indirect approach. If you can't beat them, join them. And so I did, in a manner of speaking. I constructed some plainly preposterous pseudosciences of my own and showed that they were just like astrology and the others. I also explained how you could construct more of these silly theories. By working from the inside out, more students came to see how pseudo these pseudosciences are . . . . And that is the audience I try to reach: the upcoming group of citizens. My show reaches them in the right way, too. It leaves a lasting impression; it wins friends and changes minds.

I am delighted to report that Stalker welcomes new bookings. He can be reached at the Department of Philosophy, University of Delaware, Newark, Delaware 19711.

One of the points Stalker makes is that no matter how eloquent a lecture may be, it simply does not have the power to convince that experience does. This point has been beautifully demonstrated in a study made by Barry Singer and Victor A. Benassi of the Psychology Department of California State University at Long Beach. These two investigators set out to determine the effect on first-year psychology students of seemingly paranormal effects created in the classroom by an exotically dressed magician. Their findings were reported in the Winter 1980/81 issue of the *Skeptical Inquirer* in a piece titled "Fooling Some of the People All of the Time".

In two of the classes, the performer (Craig Reynolds) was introduced as a graduate student "interested in the psychology of paranormal or psychic abilities, [who has] been working on developing a presentation of his psychic abilities". The instructor also explicitly stated, "I'm not convinced personally of Craig's or anyone else's psychic abilities." In two other classes, Craig was introduced as a graduate student "interested in the psychology of magic and stage trickery, [who has] been working on developing a presentation of his magic act". The authors emphasize that all the stunts Craig performed are "easy amateur tricks that have been practiced for centuries and are even explained in children's books of magic".

After the act, the students were asked to report their reactions. Singer and Benassi received two jolts from the reports. They write:

> First . . . . in both the "magic" and the "psychic" classes, about two-thirds of the students clearly believed Craig was psychic. Only a few students seemed to believe the instructor's description of Craig as a magician, in the two classes where he was introduced as such. Secondly, psychic belief was not only prevalent; it was strong and loaded with emotion. A number of students covered their papers with exorcism terms and exhortations against the Devil. In the psychic condition, 18 percent of the students explicitly expressed fright and emotional disturbance. Most expressed awe and amazement.
>
> We were present at two of Craig's performances and witnessed some extreme behavior. By the time Craig was halfway through the "bending" chant [part of a stunt where he bent a stainless-steel rod], the class was in a terribly excited state. Students sat rigidly in their chairs, eyes glazed and mouths open, chanting together. When the rod bent, they gasped and murmured. After class was dismissed, they typically sat still in their chairs, staring vacantly or shaking their heads, or rushed excitedly up to Craig, asking him how they could develop such powers. We felt we were observing an extraordinarily powerful behavioral effect. If Craig had asked the students at the end of his act to tear off their clothes, throw him money, and start a new cult, we believe some would have responded enthusiastically. Obviously, something was going on here that we didn't understand.

After this dramatic presentation, the classes were told they had only been seeing tricks. In fact, two more classes were given the same presentation, with the added warning: "In his act, Craig will pretend to read minds and demonstrate psychic abilities, but Craig does not really have psychic abilities, and what you'll be seeing are really only tricks." Still, despite this strong initial disclaimer, more than half the students in these classes believed Craig was psychic after seeing his act. "This says either something about the status of university instructors with their students or something about the strange pathways people take to occult belief", Singer and Benassi observe philosophically. Now comes something astonishing.

> The next question asked was whether magicians could do exactly what Craig did. Virtually all the students agreed that magicians could. They were then asked if they would like to revise their estimate of Craig's psychic abilities in the light of this negative information that they themselves had furnished. Only a few did, reducing the percentage of students believing that Craig had psychic powers to 55 percent.
> Next the students were asked to estimate how many people who performed stunts such as Craig's and claimed to be psychic were actually fakes using magician's tricks. The consensus was that at least three out of four "psychics" were in fact frauds. After supplying this negative information, they were again asked if they wished to revise their estimate of Craig's psychic abilities. Again, only a few did, reducing the percentage believing that Craig had psychic powers to 52 percent.

Singer and Benassi muse:

> What does all this add up to? The results from our pen-and-pencil test suggest that people can stubbornly maintain a belief about someone's psychic powers *when they know better*. It is a logical fallacy to admit that tricksters can perform exactly the same stunts as real psychics and to estimate that most so-called psychics are frauds—and at the same time to maintain with a fair degree of confidence that any given example (Craig) is psychic. Are we humans really that foolish? Yes.

*   *   *

A few years ago, Scot Morris (now a senior editor at *Omni* magazine in charge of its "Games" department) carried out a similar experiment on a first-year psychology class at Southern Illinois University, which he wrote up in the Spring 1980 issue of the *Skeptical Inquirer*. First, Morris assessed his students' beliefs in ESP by having them fill out a questionnaire. Then a colleague performed an "ESP demonstration", which Morris calls "frighteningly impressive".

After this powerful performance, Morris tried to "deprogram" his students. He had two weapons at his disposal. One is what he calls "dehoaxing". This process, just three minutes long, consisted in a revelation of how two of the three tricks worked, together with a confession that the remaining one of the baffling stunts was also a trick. "But," said Morris, "I'm not going to say how it was done, because I want you to experience the feeling that, even though you can't explain something, that doesn't make it supernatural." The other weapon was a 50-minute anti-ESP lecture, in which secrets of professional mind readers were revealed, commonsense estimates of probabilities of "oddmatches" were discussed, "scientific" studies of ESP were shown to be questionable for various statistical and logical reasons, and some other everyday reasons were adduced to cast ESP's reality into strong doubt.

After the performance, only half of the classes were "dehoaxed", but all of them heard the anti-ESP lecture. The students were then polled about the strength of their belief in various kinds of paranormal phenomena. It turned out that dehoaxed classes had a far lower belief in ESP than classes that had simply heard the anti-ESP lecture. The dehoaxed classes' average level of ESP belief dropped from nearly 6 (moderate belief) to about 2 (strong disbelief), while the non-dehoaxed classes' average level dropped from 6 to about 4 (slight disbelief). As Morris summarizes this surprising result, "The dehoaxing experience was apparently crucial; a three-minute revelation that they had been fooled was more powerful than an hour-long denunciation of ESP in producing skepticism toward ESP."

One of Morris' original interests in conducting this experiment was "whether the exercise would teach the students skepticism for ESP statements only, or a more general attitude of skepticism, as we had hoped. For example, would their experience also make them more skeptical of astrology, Ouija boards, and ghosts?" Morris did find a slight transfer of skepticism, and from it he concluded hopefully that "teaching someone to be skeptical of one belief makes him somewhat more skeptical of similar beliefs, and perhaps slightly more skeptical even of dissimilar beliefs."

This question of transfer of skepticism is, to my mind, the critical one. It is of little use to learn a lesson if it always remains a lesson about particulars and has no applicability beyond the case in which it was first learned. What, for instance, would you say is "the lesson of the People's Temple incident in Jonestown"? Simply that one should never follow the Reverend Jim Jones to Guyana? Or more generally, that one should be wary of following any guru halfway across the world? Or that one should never follow anyone anywhere? Or that all cults are evil? Or that any belief in any kind of savior, human or divine, is crazy and dangerous? Or consider the recent convulsions in Iran. Is it likely that the fundamentalist "Moral Majority" Christians in America would see their own attitudes as parallel to those of fundamentalist Moslems whose fanaticism they abhor, and that they would thereby be led to reflect on their own behavior? I wouldn't hold my breath. At what level of generality is a lesson learned? What was "the lesson of Viet Nam"? Does it apply to any present political situations that the United States is facing, or that any country is facing?

*　　*　　*

Stalker's Captain Ray of Light expresses faith that by debunking his own "miniature" pseudo-sciences before audiences, he can transfer to people a more general critical ability—an ability to think more clearly about paranormal claims. But how true is this? There are untold believers in some types of paranormal phenomena who will totally ridicule other types. It is quite common to encounter someone who will scoff at the headlines in the *National Enquirer* while at the same time believing, say, that through Transcendental Meditation you can learn to levitate, or that astrological predictions come true, or that UFO's are visitors from other galaxies, or that ESP exists. I've heard many people express the following sort of opinion: "Most psychics, unfortunately, are frauds, which makes it all the more difficult for the *genuine* ones to be recognized." You even get believers in tricksters such as Uri Geller who say, "I admit he cheats *some* of the time, maybe even 90 percent of the time—but believe me, he has genuine psychic abilities!"

If you are hunting for a signal in a lot of noise, and the more you look, the more noise you find, when is it reasonable to give up and conclude there is no signal there at all? On the other hand, sometimes there just might be a signal! The problem is, you don't want to jump too quickly to a negative generalization, especially if your feelings are based merely on some kind of guilt by association. After all, not *everything* published in the *National Enquirer* is false. (I had to look awfully hard, though, to locate something in its pages that I was *sure* is true!) The subtle art is in sensing just when to shift—in sensing when there is enough evidence. But for better or for worse, this is a subjective matter, an art that few journals heretofore have dealt with.

The *Skeptical Inquirer* concerns itself with questions ranging from the ridiculous to the sublime, from the trivial to the profound. There are those who would say it is a big waste of time to worry about such drivel as ESP and other so-called paranormal effects, whereas others (such as myself) feel that anyone who is unable or unwilling to think hard about what distinguishes the scientific system of thinking from its many rival systems is not a devotee of truth at all, and furthermore that the spreading of nonsense is a dangerous trend that ought to be checked.

In any case, the question arises whether the *Skeptical Inquirer* will ever amount to more than a tiny drop in a huge bucket. Surely its editors do not expect that someday it will be sold alongside the *National Enquirer* at supermarket checkout counters! Or, carrying this vision to an upside-down extreme, can you imagine a world where a debunking journal such as the *Skeptical Inquirer* (in tabloid form, of course) sold millions of copies each week at supermarkets (along with its many rivals), while one lone courageous voice of the occult came out four times a year (in a relatively staid format) and was sought out by a mere 7,500 readers? Where the many rival debunking tabloids were always to be found lying around in laundromats? It sounds like a crazy story fit for the pages of the *National Enquirer!* This ludicrous scenario serves to emphasize just what the hardy band at CSICOP is up against.

What good does it do to publish their journal when only a handful of already-convinced anti-occult fanatics read it anyway? The answer is found in, among other places, the letters column at the back of each issue. Many people write in to say how vital the magazine has been to them, their friends, and their students. High-school teachers are among the most frequent writers of thank-you notes to the magazine's editors, but I have also seen enthusiastic letters from members of the clergy, radio talk-show hosts, and people in many other professions.

I would hope that by now I have aroused enough interest on the part of readers that they might like to subscribe to at least one of the journals that I have discussed in these pages. In the spirit of open-mindedness and relativism, therefore, I hereby provide addresses for all three (in alphabetical order):

> *National Enquirer*
> Lantana, Florida 33464
>
> *Skeptical Inquirer*
> Box 229, Central Park Station
> Buffalo, New York 14215
>
> *Zetetic Scholar*
> Department of Sociology
> Eastern Michigan University
> Ypsilanti, Michigan 48197

Of course, I would not dream of suggesting which one to subscribe to. Perhaps the most prudent course would be not to make any prejudgments, and to subscribe to all three.

* * *

Certainly one will never be able to empty the vast ocean of irrationality that all of us are drowning in, but the ambition of the *Skeptical Inquirer* has never been that heroic; it has been, rather, to be a steady buoy to which one could cling in that tumultuous sea. It has been to promote a healthy brand of skepticism in as many people as it can. As Kendrick Frazier said in one of his eloquent editorials,

> Skepticism is not, despite much popular misconception, a point of view. It is, instead, an essential component of intellectual inquiry, a method of determining the facts whatever they may be and wherever they might lead. It is a part of what we call common sense. It is a part of the way science works. All who are interested in the search for knowledge and the advancement of understanding, imperfect as those enterprises may be, should, it seems to me, support critical inquiry, whatever the subject and whatever the outcome.

It is too bad that we should have to constantly defend truth against so many onslaughts from people unwilling to think, but, on the other hand, sloppy thought seems inevitable. It's just part of human nature. Come to think of it, didn't I read somewhere recently about how your average typical-type John or Jane Doe in the street uses only ten percent of his or her brains? Something like that! How come folks don't think harder and get *more* of those little brain cells going? Beats me! Talk about sloppy—it's downright boggling!! Even the scientists are stumped!!!

Post Scriptum

In the April 1982 issue of *Spektrum der Wissenschaft* (the German edition of *Scientific American*), the translation of this column appeared. On the flip side of the page with the headline "Boy can see with his ears" (*Junge kann mit den Ohren sehen*) I found a short article whose headline ran "Learning to hear with your eyes" (*Mit den Augen hören lernen*). It's logical, I guess—hearing with your eyes *does* seem to be the flip side of seeing with your ears! The article actually was about a machine for helping deaf people improve their speech with the aid of computer displays of their voices.

It was remarkable to see how similar these flipped headlines were, and yet how totally different the articles were. The main difference was actually in *tone*. The *National Enquirer* article spoke of an event that supposedly had occurred and characterized it as baffling and beyond explanation; the *Spektrum der Wissenschaft* article mentioned a counterintuitive idea and explained how it might conceivably be realized, after a fashion. Note that *Spektrum der Wissenschaft* managed to grab my attention by exploiting the same device as the tabloids do: catch readers by blaring something paradoxical. To someone not firmly grounded in science, "hearing with your eyes" and "seeing with your ears" sound (and look!) about equally implausible. Indeed, even to someone who is scientifically educated, the two phrases sound about equally weird. More information is needed to flesh out the meanings. That information was provided in *Spektrum der Wissenschaft*, and turned the initially grabbing headline into a sensible notion. Such is usually not the case for articles in the tabloids. But for most readers, such a subtle distinction doesn't matter.

This all goes to emphasize the claim at the beginning of this chapter about the trickiness of trying to pin down what truth is, and how deeply circular all belief systems are, no matter how much they try to be objective. In the end, rate of survival is the only difference between belief systems. This is a worrisome statement. It certainly worries me, at least. Still, I believe it. But scientists, I find, are not usually willing to see science itself as being rooted in an impenetrably murky swamp of beliefs and attitudes and perceptions. Most of them have never considered how it is that human perception and categorization underlie all that we take for granted in terms of common sense, and in more primordial ways that are so deeply embedded that we even find them hard to talk about. Such things as: how we break the world into parts, how we form mental categories, how we refine them certain times while blurring them other times, how experiences and categories are clustered associatively, how analogies guide our intuitions, how imagery works, how valid logic is and where it comes from, how we tend to favor simple statements over complex ones, and so on—all these are, for most scientists, nearly un-grapplable-with issues, and so they pay them no heed and continue with their work.

The idea of "simplicity" is a real can of worms, for what is simple in one vocabulary can be enormously complex in another vocabulary—and vice versa. Does the sun rise in the mornings? Ninety-nine to one you use that geocentric phrase in your ordinary conversations, and geocentric imagery in your private thoughts. Yet we all "know" that the truth is different: the earth is *really* rotating on its axis and so the sun's motion is only *apparent*. Well, it may be news to you that general relativity says that all coordinate systems are equally valid—and that includes one from whose point of view all motion takes place with respect to a fixed, nonrotating earth. Thus Einstein tells us that Copernicus and Galileo were, after all, not any righter than Ptolemy and the Pope (score ten points for infallibility!). There is even, for each of us, a physically valid "egocentric" system of coordinates in which *I* am still and everything moves relative to me! I point this out to show that the truth is much shiftier and subtler than any simple picture can ever say. Scientists who oversimplify science distort reality as much as religious fanatics or pseudo-scientists do. The troubling truth is that there is no simple boundary line between nonsense and sense. (See Chapter 11). It is a lot hazier and blurrier and messier than even thoughtful people generally wish to admit.

When I was a columnist in *Scientific American,* I got quite a lot of mail, including a sizable number of letters from what I might charitably term "fringe thinkers", or uncharitably term "crackpots". I built up large files of such letters in the hopes of someday writing an article about "crackpotism" and its detection. The hypothetical book *How to Tell the True from the False by Its Style of Publication,* which I jokingly referred to in the article as something that I might write, was therefore not entirely a joke.

How can you discern which books you *do* want to read from those you don't? Answer: You have various levels of depth of evaluation, ranging from extremely brief and superficial tests to very deep and probing ones (*i.e.*, where you actually *do* take the trouble to read the book to see what it says). In order to reach the final stage (reading the book), you go through several very critical intermediate levels of analysis and scrutiny. I call this mechanism for filtering the "terraced scan".

How do I decide which letters to read carefully, if I don't read them all carefully (to decide whether or not to read them carefully . . .)? Answer: I apply the crudest, most "syntactic" stages of my terraced scanner and prune out the worst ones very quickly. Then I apply a slighty more refined stage of testing to the survivors, and prune out some more. And on it goes, until I am left with just a handful of truly provocative, significant letters. But if I had no such terraced-scan mechanism, I would be trapped in perpetual indecision, having no basis to decide to do anything, since I would need to evaluate *every pathway in depth* in order to decide whether or not to follow it. Should I take the bus to Kalamazoo today? Study out of a Smullyan book? Practice the piano? Read the latest *New York Review of Books*? Write an angry letter to someone in government?

This question of the interaction of *form* and *content* fascinates me deeply. I do indeed believe that if one has the right "terraced scan" mechanisms, one can go very far indeed in separating the wheat from the chaff. Of course, one has to believe that there *is* such a distinction: that The Truth actually exists. And just what this Truth is is very hard to say.

*   *   *

To me, part of the challenge of Zen is very much akin to the challenge of the occult and of pseudo-science: the baffling inner consistency of a worldview totally antithetical to my own. What is also interesting is that each human being has a totally unique worldview, with its private contradictions and even small insanities. It is my belief, for instance, that inside every last one of us there is at least a small pocket of insanity: a kind of Achilles' heel that we try to avoid exposing to the world—and to ourselves. In his own personal way, Einstein was loony; in my own personal way, I am loony; and the same for you, dear lunatic!

In a way, therefore, to try to pursue the nature of ultimate truth is to enter a bottomless pit, filled with circular vipers of self-reference. One could liken CSICOP's job to that of the American Civil Liberties Union, which gets itself in all sorts of tangled loops because of its stance of defending radical belief systems. For instance, in an odd twist, its director, a former concentration camp inmate, found himself defending the rights of neo-Nazis to march down the streets of highly Jewish Skokie, Illinois, parading their banners advocating the extermination of all "inferior races". And what was worse for him was that as a consequence of his actions, the ACLU lost a significant portion of its membership. Patrick Henry spoke of "defending to the death your right to say it"—but does "it" include *anything*? Recipes for how to murder people? How to build atomic bombs? How to destroy the free press? Governments also face this sticky kind of issue. Can a government dedicated to liberty afford to let an organization dedicated to that government's downfall flourish?

It always seems refreshing to see how magazines, in their letters columns, willingly publish letters highly critical of them. I say "seems", because often those letters are printed in pairs, both raking the magazine over the coals but from opposite directions. For example, a right-wing critic and a left-wing critic both chastise the magazine for leaning too far the wrong way. The upshot is of course that the magazine doesn't even have to say a thing in its own defense, for it is a kind of cliché that if you manage to offend both parties in a disagreement, you certainly must be essentially right! That is, the truth is supposedly *always* in the middle—a dangerous fallacy.

Raymond Smullyan, in his book *This Book Needs No Title,* provides a perfect example of the kind of thing I am talking about. It is a story about two boys fighting over a piece of cake. Billy says he wants it all, Sammy says they should divide it equally. An adult comes along and asks what's wrong. The boys explain, and the adult says, "You should compromise—Billy gets three quarters, Sammy one quarter." This kind of story sounds ridiculous, yet it is repeated over and over in the world, with loudmouths and bullies pushing around meeker and fairer and kinder people. The "middle position" is calculated by averaging all claims together, outrageous ones as well as sensible ones, and the louder any claim, the more it will count. Politically savvy people learn this early and make it their credo; idealists learn it late and refuse to accept it. The idealists are like Sammy, and they always get the short end of the stick.

Magazines often gain rather than lose by printing what amounts to severe criticism. This holds even if the critical letter is not matched by an equally critical letter from the other side, because if a magazine prints letters critical of it, it appears open-minded and willing to listen to criticism. Thus the opposition is co-opted and undercut.

Another problem is that by shouting loud enough, advocates of any viewpoint can gain public attention. Sometimes the loudness comes from the large number of adherents of a particular point of view, sometimes it comes from the eloquence or charisma of a single individual, and sometimes it comes from the high status of one individual. A particularly salient example of this sort of thing is provided by the behavior of the Nixon "team" during the Watergate affair. There, they had the ability to manipulate the press and the public simply because they were in power. What no private individual would ever have been able to get away with for a second was done with the greatest of ease by the Nixon people. They shamelessly changed the rules as they wished, and for a long time got away with it.

What does all this have to do with the *Skeptical Inquirer*? Plenty. Amidst the tumult and the shouting, where does the truth lie? What voices should one listen to? How can one tell which are credible and which are not? It might seem that the serious matters of life have precious little to do with the validity of horoscopes, the probability of reincarnation, or the existence of Bigfoot, but I maintain that susceptibility to bad arguments in one domain opens the door to being manipulated in another domain. A critical mind is critical on all fronts simultaneously, and it is vital to train people to be critical at an early stage.

* * *

The most serious piece of mail I received as a result of this column was from Marcello Truzzi, founder of the *Zetetic Scholar*. Truzzi wrote me as follows (somewhat excerpted):

> I was greatly disturbed and disappointed to read your column because of its serious distortions about the character of the 'schism' in CSICOP and the position and history of the *Zetetic Scholar*. Your article conveys the clear impression that *Zetetic Scholar* is somehow more sympathetic to pseudo-science, is more 'relativist' and 'unjudgmental'. That is completely untrue . . . .
>
> I think you completely missed the issue between CSICOP and CSAR [Truzzi's Center for Scientific Anomalies Research—the organization behind *Zetetic Scholar*]. The term 'skeptic' has become unfortunately equated with *disbelief* rather than its proper meaning of *nonbelief*. That is, skepticism means the raising of doubts and the urging of inquiry. *Zetetic Scholar* very much stands for doubt and inquiry . . . . I view much of CSICOP activity as obstructing inquiry because it has prejudged many areas of inquiry by labeling them pseudo-scientific *prior* to serious inquiry. In other words, it is not judgment that I wish to avoid—quite the contrary—but prejudgment.
>
> The major problem is that CSICOP, in its fervor to debunk, has tended to lump the nonsense of the *National Enquirer* with the serious scientific research programs of what I call 'protosciences' (that is, serious but maverick scientists trying to play by the rules of science and get their claims properly tested and examined). By scoffing at all claims of the paranormal, CSICOP inhibits (through mockery) serious work on anomalies . . . .
>
> *Zetetic Scholar* tries to bring together protoscientific proponents and responsible critics into rational dialogue . . . . The purpose is to advance science.
>
> My position is not a relativist one. I believe science does progress and is cumulative. But I do believe that skepticism must extend to all claims, including orthodox ones. Thus, before I condemn fortune tellers as doing social evil, I think the effects of their use need to be compared to the orthodox practitioners —psychiatrists and clinical psychologists. The simple fact is that much nonsense goes on within science that is at least as pseudo-scientific as anything going on in what we usually term pseudo-sciences . . . .
>
> I do not believe in most paranormal claims, but I refuse to close the door on discussion of them. The simple fact is that I think I have more confidence in science than, say, Martin Gardner does. For example, Martin resigned as a consulting editor for *Zetetic Scholar* when he was told that I planned to publish a 'stimulus' article asking for a reconsideration of the views of Velikovsky. [Immanuel Velikovsky is best known for his fantastic, fiery visions of the evolution of the solar system and, among other things, a theory claiming that the earth, up until quite recently (in astronomical terms), was spinning in the other direction! He claimed that his views reconciled science and the Bible, and he published many books, perhaps the most famous of which is called *Worlds in Collision*.] Martin was invited to comment, as were many critics of Velikovsky. But Martin felt that even considering Velikovsky seriously in *Zetetic Scholar* gave Velikovsky undeserved legitimacy, so Martin resigned. I happen to think Velikovsky is dead wrong, but I also think that he has not been given due process by his critics. I have confidence that honest discourse will reveal the errors and virtues (if any) in any esoteric scientific claim. I see nothing to be

afraid of. I have full confidence in science as a self-correcting system. Some on CSICOP, like Martin, do not.

This is only a small portion of Truzzi's letter, but it gets the idea across. All in all, Truzzi emphasized that his magazine serves a different purpose from the *Skeptical Inquirer,* and that I had not made it sufficiently clear what that purpose really is. I hope that readers can now understand what it is. My reply to Truzzi follows (also somewhat shortened).

I have thought quite a bit about the issues you raise, and about the difference in tone, outlook, purpose, vision, etc., between *Zetetic Scholar* and the *Skeptical Inquirer.* I find myself more sympathetic than you are to the cause of out-and-out debunking. I am impatient with, and in fact rather hostile towards, the immense amount of nonsense that gets given a lot of undue credit because of human irrationality. It is like not dealing with someone very unpleasant in a group of people because you've been trained to be very tolerant and polite. But eventually there comes a point where somebody gets up and lets the unpleasant person 'have it'—verbally or physically or however—maybe just escorts them out—and everyone then is relieved to be rid of the nuisance, even though they themselves didn't have the courage to do it.

Admittedly, it's just an analogy, but to me, Velikovsky is just such an obnoxious person. And there are loads more. I simply don't feel they should be accorded so much respect. One shouldn't bend over backwards to be polite to genuinely offensive parties. I happen to feel that much of parapsychology has been afforded too much credibility. I feel that ESP and so on are incompatible with science for *very fundamental reasons.* In other words, I feel that they are so unlikely to be the case that people who spend their time investigating them really do not understand science well. And so I am impatient with them. Instead of welcoming them into scientific organizations, I would like to see them kicked out.

Now this doesn't mean that I feel that debating about the *reasons* I find ESP (etc.) incompatible with science at a very deep level is worthless. Quite to the contrary: coming to understand *how* to sift the true from the false is exceedingly subtle and important. But that doesn't mean that all pretenders to truth should be accorded respect.

It's a terribly complex issue. None of us sees the full truth on it. I am sorry if I did you a disservice by describing your magazine as I did. I have nothing against your magazine in principle, except that I find its open-mindedness *so* open that it gets boring, long-winded, and wishy-washy. Sometimes it reminds me of the senators and representatives who, during Watergate, seemed *endlessly* dense, and either unable or unwilling to get the simple point: that Nixon was guilty, on many counts. And that was it. It was very simple. And yet Nixon and company *did* manage to obscure the obvious for many months, thanks to fuzzy-minded people who somehow couldn't 'snap' into something that was very black-and-white. They *insisted* on seeing it in endless shades of gray. And in a way I think that's what you're up to, in your magazine, a lot of the time: seeing endless shades of gray where it's black and white.

There is a legitimate, indeed, *very* deep question, as to *when* that moment of 'obviousness', that moment of 'snapping' or 'clicking', comes about. Certainly

I'd be the first to say that that's as deep a question as one can ask. But that's a question about the nature of truth, evidence, perception, categories, and so forth and so on. It's *not* a question about parapsychology or Velikovsky *et al.* If yours were a magazine about *the nature of objectivity*, I'd have no quarrel with it. I'd love to see such a magazine. But it's really largely a magazine that helps to lend credibility to a lot of pseudo-scientists. Not to say that everyone who writes for it is a pseudo-scientist! Not at all! But my view is that there is such a thing as being *too* open-minded. I am *not* open-minded about the earth being flat, about whether Hitler is alive today, about claims by people to have squared the circle, or to have proven special relativity wrong. I am also not open-minded with respect to the paranormal. And I think it is wrong to be open-minded with respect to these things, just as I think it is wrong to be open-minded about whether or not the Nazis killed six million Jews in World War II.

I *am* open-minded, to some extent, about questions of ape language, dolphin language, and so on. I haven't reached any final, firm conclusion there. But I don't see that being debated in *Zetetic Scholar* (or in the *Skeptical Inquirer*).

My viewpoint is that the *Skeptical Inquirer* is doing a service to the *masses* of the country, albeit indirectly, by publishing articles that have flair and dash and whose purpose is to combat the huge waves of nonsense that we are forced to swim in all the time. Of course most people will never read the *Skeptical Inquirer* themselves, but many *teachers* will, and will be much better equipped thereby to refute kids who come up and tell them about precognitive dreams and bent keys or magically fixed watches or you name it.

I feel that the *Skeptical Inquirer* is playing the role of the chief prosecutor, in some sense, of the paranormal, and *Zetetic Scholar* is a member of the jury who refuses, absolutely refuses, to make a decision until more evidence is in. And after more, more, more, more, more, more, more, more evidence is in and this character *still* refuses to go one way or another, then one gets impatient.

Professor Truzzi was very kind to me in his reply, and subsequently even invited me to serve on the board of CSAR. I had to decline because of time constraints, but I appreciate his—I hate to say this—open-mindedness. Part of his reply is worth repeating:

> You seem to have the idea that I am reluctant to make a decision about many extraordinary claims. That really is not the case. I want to make decisions and am emotionally inclined to the same impatience as you have. Most of my pro-paranormal friends see me as a die-hard skeptic. But hard-line debunkers like Martin Gardner see me as wishy-washy or naïve. So I get it from both sides, I assure you.

*   *   *

I have quite a bit of sympathy for what Professor Truzzi is attempting to do, in a way. What bothers me is that all the vexing problems that he is attempting to be neutral on have their counterparts one level up, on the "meta-level", so to speak. That is, for every debate in science itself, there is an isomorphic debate in the methodology of science, and one could go on up the ladder of "meta"s, running and yet never advancing, like a hamster on a treadmill. Nixon exploited this principle very astutely in the Watergate days, smoking up the air with so many technical procedural and meta-procedural (etc.) questions that the main issues were completely forgotten about for a long time while people tried to sort out the mess that his smokescreen had created. This kind of technique need not be conscious on the part of politicians or scientists—it can emerge as an unconscious consequence of simple emotional commitment to an idea or hope.

It seems to me that object level and meta-level are hopelessly tangled here, just as in the Gödelian knot, and the only solution is to cut the knot cleanly and get rid of it. Otherwise you can wallow forever in the mess. Can cardboard pyramids *really* sharpen razor blades placed underneath them? How many weeks must one wait before one gives up? And what if, after you've given up, a friend claims it *really* works if you put a fried egg at each corner of the pyramid? Will you then go back and try that as earnestly as you tried the original idea? Will you *ever* simply reject a claim out of hand?

Where does one draw the line? Where is the borderline between open-mindedness and stupidity? Or between closed-mindedness and stupidity? Where is the optimum balance? That is such a deep question that I could not hope to answer it. Professor Truzzi's position and my own lie at different points along a spectrum. We have both arrived at our positions not by pristine logic, but as a result of many complex interacting intuitions about the world and about minds and knowledge. There is certainly no way to *prove* that my position is righter than his, or vice versa. But even if we have no adequate theory to *formalize* such decisions, we nonetheless are all walking instantiations of such decision-making beings, and we make decisions for which we could not formally account in a million years. Such decisions include all decisions of taste, whether in food, music, art, or science. We have to live with the fact that we do not yet know *how* we make such decisions, but that does not mean we have to wallow in indecisiveness in the meantime. And anything that helps to make our quick decisions more informed while not impairing their quickness is of tremendous importance. I view the *Skeptical Inquirer* as serving that purpose, and I heartily recommend it to my readers.

6

On Number Numbness

May, 1982

THE renowned cosmogonist Professor Bignumska, lecturing on the future of the universe, had just stated that in about a billion years, according to her calculations, the earth would fall into the sun in a fiery death. In the back of the auditorium a tremulous voice piped up: "Excuse me, Professor, but h-h-how long did you say it would be?" Professor Bignumska calmly replied, "About a billion years." A sigh of relief was heard. "Whew! For a minute there, I thought you'd said a *million* years."

John F. Kennedy enjoyed relating the following anecdote about a famous French soldier, Marshal Lyautey. One day the marshal asked his gardener to plant a row of trees of a certain rare variety in his garden the next morning. The gardener said he would gladly do so, but he cautioned the marshal that trees of this size take a century to grow to full size. "In that case," replied Lyautey, "plant them this afternoon."

In both of these stories, a time in the distant future is related to a time closer at hand in a startling manner. In the second story, we think to ourselves: Over a century, what possible difference could a day make? And yet we are charmed by the marshal's sense of urgency. Every day counts, he seems to be saying, and particularly so when there are thousands and thousands of them. I have always loved this story, but the other one, when I first heard it a few thousand days ago, struck me as uproarious. The idea that one could take such large numbers so personally, that one could sense doomsday so much more clearly if it were a mere *million* years away rather than a far-off *billion* years—hilarious! Who could possibly have such a gut-level reaction to the difference between two huge numbers?

Recently, though, there have been some even funnier big-number "jokes" in newspaper headlines—jokes such as "Defense spending over the next four years will be $1 trillion" or "Defense Department overrun over the next four years estimated at $750 billion". The only thing that worries me about these jokes is that their humor probably goes unnoticed by the average citizen. It would be a pity to allow such mirth-provoking notions to be appreciated only by a select few, so I decided it would be a good idea to devote some space to the requisite background knowledge, which also happens to be one of my favorite topics: the lore of very large (and very small) numbers.

I have always suspected that relatively few people really know the difference between a million and a billion. To be sure, people generally know it well enough to sense the humor in the joke about when the earth will fall into the sun, but what the difference is *precisely*—well, that is something else. I once heard a radio news announcer say, "The drought has cost California agriculture somewhere between nine hundred thousand and a billion dollars." Come again? This kind of thing worries me. In a society where big numbers are commonplace, we cannot afford to have such appalling number ignorance as we do. Or do we actually suffer from *number numbness*? Are we growing ever number to ever-growing numbers?

What do people think when they read ominous headlines like the ones above? What do they think when they read about nuclear weapons with 20-kiloton yields? Or 60-megaton yields? Does the number really register —or is it just another cause for a yawn? "Ho hum, I always knew the Russians could kill us all 20 times over. So now it's 200 times, eh? Well, we can be thankful it's not 2,000, can't we?"

What do people think about the fact that in some heavily populated areas of the U.S., it is typical for the price of a house to be a quarter of a million dollars? What do people think when they hear radio commercials for savings institutions telling them that if they invest now, they could have a million dollars on retirement? Can *everyone* be a millionaire? Do we now *expect* houses to take a fourth of a millionaire's fortune? What ever has become of the once-glittery connotations of the word "millionaire"?

* * *

I once taught a small beginning physics class on the thirteenth floor of Hunter College in New York City. From the window we had a magnificent view of the skyscrapers of midtown Manhattan. In one of the opening sessions, I wanted to teach my students about estimates and significant figures, so I asked them to estimate the height of the Empire State Building. In a class of ten students, not one came within a factor of two of the correct answer (1,472 feet with the television antenna, 1,250 without). Most of the estimates were between 300 and 500 feet. One person thought 50 feet was right—a truly amazing underestimate; another thought it was a mile. It turned out that this person had actually calculated the answer, guessing 50 feet per story and 100 stories or so, thus getting about 5,000 feet. Where one person thought each *story* was 50 feet high, another thought the whole 102-story *building* was that high. This startling episode had a deep effect on me.

It is fashionable for people to decry the appalling illiteracy of this generation, particularly its supposed inability to write grammatical English. But what of the appalling *innumeracy* of most people, old and young, when it comes to making sense of the numbers that, in point of fact, and whether they like it or not, run their lives? As Senator Everett Dirksen once said, "A billion here, a billion there—soon you're talking real money."

The world is gigantic, no question about it. There are a lot of people, a lot of needs, and it all adds up to a certain degree of incomprehensibility. But that is no excuse for not being able to understand—or even relate to —numbers whose purpose is to summarize in a few symbols some salient aspects of those huge realities. Most likely the readers of this article are not the ones I am worried about. It is nonetheless certain that every reader of this article knows many people who are ill at ease with large numbers of the sort that appear in our government's budget, in the gross national product, corporation budgets, and so on. To people whose minds go blank when they hear something ending in "illion", all big numbers are the same, so that exponential explosions make no difference. Such an inability to relate to large numbers is clearly bad for society. It leads people to ignore big issues on the grounds that they are incomprehensible. The way I see it, therefore, anything that can be done to correct the rampant innumeracy of our society is well worth doing. As I said above, I do not expect this article to reveal profound new insights to its readers (although I hope it will intrigue them); rather, I hope it will give them the materials and the impetus to convey a vivid sense of numbers to their friends and students.

*　　*　　*

As an aid to numerical horse sense, I thought I would indulge in a small orgy of questions and answers. Ready? Let's go! How many letters are there in a bookstore? Don't calculate—just guess. Did you say about a billion? That has nine zeros (1,000,000,000). If you did, that is a pretty sensible estimate. If you didn't, were you too high or too low? In retrospect, does your estimate seem far-fetched? What intuitive cues suggest that a billion is appropriate, rather than, say, a million or a trillion? Well, let's calculate it. Say there are 10,000 books in a typical bookstore. (Where did I get this? I just estimated it off the top of my head, but on calculation, it seems reasonable to me, perhaps a bit on the low side.) Now each book has a couple of hundred pages filled with text. How many words per page—a hundred? A thousand? Somewhere in between, undoubtedly. Let's just say 500. And how many letters per word? Oh, about five, on the average. So we have $10,000 \times 200 \times 500 \times 5$, which comes to five billion. Oh, well—who cares about a factor of five when you're up this high? I'd say that if you were within a factor of ten of this (say, between 500 million and 50 billion), you were doing pretty well. Now, could we have sensed this *in advance*—by which I mean, *without calculation?*

We were faced with a choice. Which of the following twelve possibilities is the most likely:

(a) 10;
(b) 100;
(c) 1,000;
(d) 10,000;
(e) 100,000;
(f) 1,000,000;
(g) 10,000,000;
(h) 100,000,000;
(i) 1,000,000,000;
(j) 10,000,000,000;
(k) 100,000,000,000;
(l) 1,000,000,000,000?

In the United States, this last number, with its twelve zeros, is called a *trillion;* in most other countries it is called a *billion.* People in those countries reserve "trillion" for the truly enormous number 1,000,000,000,000,000,000—to us a "quintillion"—though hardly anyone knows that term.

What most people truly don't appreciate is that making such a guess is very much the same as looking at the chairs in a room and guessing quickly if there are two or seven or fifteen. It is just that here, what we are guessing at is the number of zeros in a numeral, that is, the logarithm (to the base 10) of the number. *If we can develop a sense for the number of chairs in a room, why not as good a sense for the number of zeros in a numeral?* That is the basic premise of this article.

Of course there is a difference between these two types of numerical horse sense. It is one thing to look at a numeral such as "10000000000000" and to have an intuitive feeling, without counting, that it has somewhere around twelve zeros—certainly more than ten and fewer than fifteen. It is quite another thing to look at an aerial photograph of a logjam (see Figure 6-1) and to be able to sense, visually or intuitively or somewhere in between, that there must be between three and five zeros in the decimal representation of the number of logs in the jam—in other words, that 10,000 is the closest power of 10, that 1,000 would definitely be too low, and that 100,000 would be too high. Such an ability is simply a form of number perception one level of abstraction higher than the usual kind of number perception. But one level of abstraction should not be too hard to handle.

The trick, of course, is practice. You have to get used to the idea that ten is a very big number of zeros for a numeral to have, that five is pretty big, and that three is almost graspable. Probably what is most important is that you should have a prototype example for each number of zeros. For instance: *Three* zeros would take care of the number of students in your high school: 1,000, give or take a factor of three. (In numbers having just a few zeros we are always willing to forgive a factor of three or so in either direction, as long as we are merely estimating and not going for exactness.) *Four* zeros is the number of books in a non-huge bookstore. *Five* zeros is

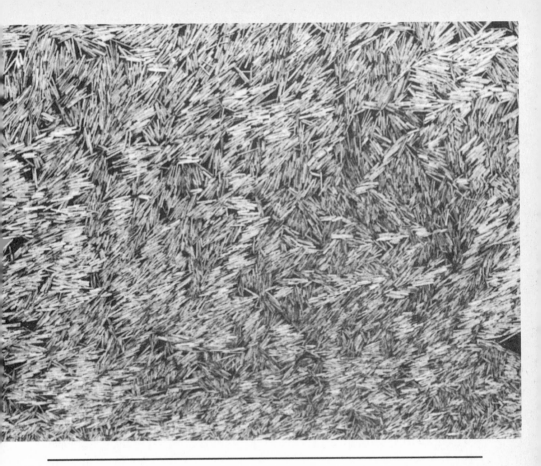

FIGURE 6–1. *Aerial view of a logjam in Oregon. How many logs?* [*Photo by Ray Atkeson.*]

the size of a typical county seat: 100,000 souls or so. *Six* zeros—that is, a million—is getting to be a large city: Minneapolis, San Diego, Brasília, Marseilles, Dar és Salaam. *Seven* zeros is getting huge: Shanghai, Mexico City, Seoul, Paris, New York. Just how many cities do you think there are in the world with a population of a million or more? Of them, how many do you think you have never heard of? What if you lowered the threshold to 100,000? How many towns are there in the United States with a population of 1,000 or less? Here is where practice helps.

I said that you should have one prototype example for each number of digits. Actually, that is silly. You should have a few. In order to have a concrete sense of "nine-zero-ness", you need to see it instantiated in several different media, preferably as diverse as populations, budgets, small objects (ants, coins, letters, etc.), and maybe a couple of miscellaneous places, such as astronomical distances or computer statistics.

Consider the famous claim made by the McDonald's hamburger chain: "Over 25 billion served" (or whatever they say these days). Is this figure credible? Well, if it were ten times bigger—that is, 250 billion—we could divide by the U.S. population more easily. (This is apparent if you happen to know that the U.S. population is about 230 million. For the purposes of this discussion, let us call the U.S. population 250 million, or 2.5×10^8—a common number that everyone should know.) Let us imagine, then, that the claim were "Over 250 billion served". Then we would compute that 1,000 burgers had been cooked for every person in the U.S. But since we deliberately inflated it by a factor of 10, let us now undo that—let us divide our answer by ten, to get 100. Is it plausible that McDonald's has prepared 100 burgers for every person in the U.S.? Sounds reasonable to me; after all, they have been around for many years, and some families go there many times a year. Therefore the claim *is* plausible, and the fact that it is *plausible* makes it *probable* that it is quite accurate. Presumably, McDonald's wouldn't go to the trouble of updating their signs every so often if they were not trying to be accurate. I must say that if their earnest effort helps to reduce innumeracy, I approve highly of it.

Where do all those burgers come from? A staggering figure is the number of cattle slaughtered every day in the U.S. It comes to about 90,000. When I first heard this, it sounded amazingly high, but think about it. Maybe half a pound of meat per person per day. Once again, the U.S. population—250 million—comes in handy. With half a pound of meat per person per day, that comes to 100 million pounds of meat per day—or something like that, anyway. We're certainly not going to worry about factors of two. How many tons is that? Divide by 2,000 to get 50,000 tons. But an individual animal does not yield a ton of meat. Maybe 1,000 pounds or so—half a ton. For each ton of meat, that would mean two animals were killed. So we would get about 100,000 animals biting the dust every day to satisfy our collective appetite. Of course, we do not eat only beef, so the true figure should be a bit lower. And that brings us back down to about the right figure.

*   *   *

How many trees are cut down each week to produce the Sunday edition of the *New York Times*? Say a couple of million copies are printed, each one weighing four pounds. That comes to about eight million pounds of paper —4,000 tons. If a tree yielded a ton of paper, that would be 4,000 trees. I don't know much about logging, but we cannot be too far off in assuming a ton per tree. At worst it would be 200 pounds of paper per tree, and that would mean 40,000 small trees. The logjam photograph shows somewhere between 7,500 and 15,000 logs, as nearly as I can estimate. So, if we do assume 200 pounds of paper per tree, the logs in the photograph represent considerably less than half of one Sunday *Times'* worth of trees! We could go on to estimate the number of trees cut down every month to provide for all the magazines, books, and newspapers published in this country, but I'll leave that to you.

How many cigarettes are smoked in the U.S. every year? (How many zeros?) This is a classic "twelver"—on the order of a trillion. It is easy to calculate. Say that half of the people in the country are cigarette smokers: 100 million of them. (I know this is something of an overestimate; we'll compensate by reducing something else somewhere along the way.) Each smoker smokes—what? A pack per day? All right. That makes 20 cigarettes times 100 million: two billion cigarettes per day. There are 365 days per year, but let's say 250, since I promised to reduce something somewhere; 250 times two billion gives about 500 billion—half a trillion. This is just about on the nose, as it turns out; the last I looked (a few years ago), it was some 545 billion. I remember how awed I was when I first encountered this figure; it was the first time I had met up with a *concrete* number about the size of a trillion.

By the way, "20 (cigarettes) times 100 million" is not a hard calculation, yet I bet it would stump many Americans, if they had to do it in their head. My way of doing it is to shift a factor of 10 from one number to the other. Here, I *reduce* 20 to 2, while *increasing* 100 to 1,000. It makes the problem into "2 times 1,000 million", and then I just remember that 1,000 million is one billion. I realize that this sounds absolutely trivial to anyone who is comfortable with figures, but it sounds truly frightening and abstruse to people who are not so comfortable with them—and that means *most* people.

It is numbers like 545 billion that we are dealing with when we talk about a Defense Department overrun of $750 billion for the next four years. A really fancy single-user computer (the kind I wouldn't mind having) costs approximately $75,000. With $750 billion to throw around, we could give one to every person in New York City, which is to say, we could buy about ten million of them. Or, we could give $1 million to every person in San Francisco, and still have enough left over to buy a bicycle for everyone in China! There's no telling what good uses we could put $750 billion to. But instead, it will go into bullets and tanks and fighters and war games and missile systems and jet fuel and marching bands and so on. An interesting way to spend $750 billion, but I can think of better ways.

*   *   *

Let us think of some other kinds of big numbers. Did you know that your retina has about 100 million cells in it, each of which responds to some particular kind of stimulus? And they feed their signals back into your brain, which is now thought to consist of somewhere around 100 billion neurons, or nerve cells. The number of glia—smaller supporting cells in the brain—is about ten times as large. That means you have about one trillion glia in your little noggin. That may sound big, but in your body altogether there are estimated to be about 60 or 70 trillion cells. Each one of them contains millions of components working together. Take the protein hemoglobin, for instance, which transports oxygen in the bloodstream. We each have about six billion trillion (that is, six thousand million million million) copies of the hemoglobin molecule inside us, with something like 400 trillion of them (400 million million) being destroyed every second, and another 400 trillion being made! (By the way, I got these figures from Richard Dawkins' book *The Selfish Gene.* They astounded me when I read them there, and so I tried to calculate them on my own. My estimates came out pretty close to his figures, and then, for good measure, I asked a friend in biology to calculate them, and she seemed to get about the same answers independently, so I guess they are pretty reliable.)

The number of hemoglobin molecules in the body is about 6×10^{21}. It is a curious fact that over the past year or two, nearly everyone has become familiar, implicitly or explicitly, with a number nearly as big—namely, the number of different possible configurations of Rubik's Cube. This number —let us call it *Rubik's constant*—is about 4.3×10^{19}. For a very vivid image of how big this is, imagine that you have many cubes, an inch on each side, one in every possible configuration. Now you start spreading them out over the surface of the United States. How thickly covered would the U.S. be in cubes? Moreover, if you are working in Rubik's "supergroup", where the orientations of face centers matter, then Rubik's "superconstant" is 2,048 times bigger, or about 9×10^{22}!

The Ideal Toy Corporation—American marketer of the Cube—was far less daring than McDonald's. On their package, they softened the blow, saying merely "Over three billion combinations possible"—a pathetic and euphemistic underestimate if ever I heard one. This is the first time I have ever heard Muzak based on a pop number rather than a pop melody. Try these out, for comparison's sake:

(1) "Entering San Francisco—population greater than 1."
(2) "McDonald's—over 2 served."
(3) "Together, the superpowers have 3 pounds of TNT for every human being on earth."

Number 1 is off by a factor of about a million, or six *orders of magnitude* (factors of ten). Number 2 is off by a factor of ten billion or so (ten orders of magnitude), while number 3 (which I saw in a recent letter to the editor of the *Bulletin of the Atomic Scientists*) is too small by a factor of about a thousand (three orders of magnitude).

The hemoglobin number and Rubik's superconstant are *really* big. How about some smaller big ones, to come back to earth for a moment? All right —how many people would you say are falling to earth by parachute at this moment (a perfectly typical moment, presumably)? How many English words do you know? How many murders are there in Los Angeles County every year? In Japan? These last two give quite a shock when put side by side: Los Angeles County, about 2,000; Japan, about 900.

Speaking of yearly deaths, here is one we are all used to sweeping under the rug, it seems: 50,000 dead per year (in this country alone) in car accidents. If you count the entire world, it's probably two or three times that many. Can you imagine how we would react if someone said to us today: "Hey, everybody! I've come up with a really nifty invention. Unfortunately, it has a minor defect—every twelve years or so it will wipe out about as many Americans as the population of San Francisco. But wait a minute! Don't go away! The rest of you will love it, I promise!" Now, these statistics are accurate for cars. And yet we seldom hear people chanting, "No cars is good cars!" How many bumper strips have you seen that say, "No more cars!"? Somehow, collectively, we are willing to absorb the loss of 50,000 lives per year without any serious worry. And imagine that half of this—25,000 needless deaths—is due to drunks behind the wheel. Why aren't you just fuming?

* * *

I said I would be a little lighter. All right. Light consists of photons. How many photons per second does a 100-watt bulb put out? About 10^{20}—another biggie. Is it bigger or smaller than the number of grains of sand on a beach? What beach? Say a stretch of beach a mile long, 100 feet wide and six feet deep. What would you estimate? Now calculate it. How about trying the number of drops in the Atlantic Ocean? Then try the number of fish in the ocean. Which are there more of: fish in the sea, or ants on the surface of the earth? Atoms in a blade of grass, or blades of grass on the earth? Blades of grass, or insects? Leaves on a typical oak tree, or hairs on a human head? How many raindrops fall on your town in one second during a terrific downpour?

How many copies of the Mona Lisa have ever been printed? Let's try this one together. Probably it is printed in magazines in the United States a few dozen times per year. Say each of the magazines prints 100,000 copies. That makes a few million copies per year in American magazines, but then there are books and other publications. Maybe we should double or triple our figure for the U.S. To take into account other countries, we can multiply it again by three or four. Now we have hit about 100 million copies per year. Let us assume this held true for each year of this century. That would make nearly ten billion copies of the Mona Lisa! Quite a meme, eh? Probably we have made some mistakes along the way, but give or take a factor of ten, that is very likely about what the number is.

"Give or take a factor of *ten*"!? A moment ago I was saying that a factor of *three* was forgivable, but now, here I am forgiving myself *two* factors of three—that is, an entire order of magnitude. Well, the reason is simple: We are now dealing with larger numbers (10^{10} instead of 10^5), and so it is permissible. This brings up a good rule of thumb. Say an error of a factor of three is permissible for each estimated factor of 100,000. That means we are allowed to be off by a factor of ten—*one* order of magnitude—when we get up to sizes around ten billion, or by a factor of 100 or so (*two* orders of magnitude) when we get up to the square of that, which is 10^{20}, about 2.5

times the size of Rubik's constant. This means it would have been forgivable if Ideal had said, "Over a *billion* billion combinations", since then they would have been off by a factor of only 40—about 1.5 orders of magnitude —which is within our limits when we're dealing with numbers that large.

Why should we be content with an estimate that is only one percent of the actual number, or with an estimate that is 100 times too big? Well, if you consider the base-10 logarithm of the number—the number of zeros—then if we say 18 when the real answer is 20, we are off by only ten percent! Now what entitles us to cavalierly dismiss the magnitude itself and to switch our focus to its logarithm (its order of magnitude)? Well, when numbers get this big, we have no choice. Our perceptual reality begins to shift. We simply *cannot* visualize the actual quantity. The numeral—the string of digits—takes over: our perceptual reality becomes one of numbers of zeros. When does this shift take place? It begins when we can no longer see, in our mind's eye, a collection of the right order of magnitude. For me, this "perceptual logjam" begins at about 10^4—the size of the actual logjam I remember in the photograph. It is important to understand this transition. It is one of the key ideas of this article.

There are other ways to grasp 10^4, such as the number of soup cans that would fill a 50-foot shelf in a supermarket. Numbers much bigger than that, I simply cannot visualize. The number of tiles lining the Lincoln Tunnel between Manhattan and New Jersey is so enormous that I cannot easily picture it. (It is on the order of a million, as you can calculate for yourself, even if you've never seen it!) In any case, somewhere around 10^4 or 10^5, my ability to visualize begins to fade and to be replaced with that second-order reality of the number of digits (or, to some extent, with number names such as "million", "billion", and "trillion"). Why it happens at this size and not, say, at 10 million or at 1,000 must have to do with evolution and the role that the perception of vast arrays plays in survival. It is a fascinating philosophical question, but one I cannot hope to answer here.

In any case, a pretty good rule of thumb is this: Your estimate should be within ten percent of the correct answer—but this need apply only *at the level of your perceptual reality.* Therefore you are excused if you guessed that Rubik's cube has 10^{18} positions, since 18 is pretty close to 19.5, which is about what the number of digits is. (Remember that—roughly speaking— Rubik's constant is 4.3×10^{19}, or 43,000,000,000,000,000,000. The leading factor of 4.3 counts for a bit more than half a digit, since each factor of 10 contributes a *full* digit, whereas a factor of 3.16, the square root of 10, contributes *half* a digit.)

If, perchance, you were to start dealing with numbers having millions or billions of digits, the numerals themselves (the colossal strings of digits) would cease to be visualizable, and your perceptual reality would be forced to take another leap upward in abstraction—to the number that counts the digits in the number that counts the digits in the number that counts the objects concerned. Needless to say, such third-order perceptual reality is highly abstract. Moreover, it occurs very seldom, even in mathematics. Still, you can imagine going far beyond it. Fourth- and fifth-order perceptual realities would quickly yield, in our purely abstract imagination, to tenth-, hundredth-, and millionth-order perceptual realities.

By this time, of course, we would have lost track of the *exact* number of levels we had shifted, and we would be content with a mere *estimate* of that number (accurate to within ten percent, of course). "Oh, I'd say about two million levels of perceptual shift were involved here, give or take a couple of hundred thousand" would be a typical comment for someone dealing with such unimaginably unimaginable quantities. You can see where this is leading: to multiple levels of abstraction in talking about multiple levels of abstraction. If we were to continue our discussion just one zillisecond longer, we would find ourselves smack-dab in the middle of the theory of recursive functions and algorithmic complexity, and that would be too abstract. So let's drop the topic right here.

*　　*　　*

Related to this idea of huge numbers of digits, but more tangible, is the computation of the famous constant π. How many digits have so far been calculated by machine? The answer (as far as I know) is one million. It was done in France a few years ago, and the million digits fill an entire book. Of these million, how many have been committed to human memory? The answer strains credulity: 20,000, according to the latest *Guinness Book of World Records.* I myself once learned 380 digits of π, when I was a crazy high-school kid. My never-attained ambition was to reach the spot, 762 digits out in the decimal expansion, where it goes "999999", so that I could recite it out loud, come to those six '9's, and then impishly say, "and so on!" Later, I met several other people who had outdone me (although none of them had reached that string of '9's). All of us had forgotten most of the digits we once knew, but at least we all remembered the first 100 solidly, and so occasionally we would recite them in unison—a rather esoteric pleasure.

What would you think if someone claimed that the entire book of a million digits of π had been memorized by someone? I would dismiss the claim out of hand. A student of mine once told me very earnestly that Jerry Lucas, the memory and basketball whiz, knew the entire Manhattan telephone directory by heart. Here we have a good example of how innumeracy can breed gullibility. Can you imagine what memorizing the Manhattan telephone directory would involve? To me, it seems about two orders of magnitude beyond credibility. To memorize one page seems fabulously difficult. To memorize ten pages seems at about the limit of credibility. Incidentally, memorizing the entire Bible (which I have occasionally heard claimed) seems to me about equivalent to memorizing ten pages of the phone book, because of the high redundancy of written language and the regularity of events in the world. But to have memorized 1,500 dense pages of telephone numbers, addresses, and names is literally beyond belief. I'll eat my hat—in fact, all of my 10,000 hats—if I'm wrong.

* * *

There are some phenomena for which there are two (or more) scales with which we are equally comfortable, depending on the circumstances. Take pitch in music. If you look at a piano keyboard, you will see a linear scale along which pitch can be measured. The natural thing to say is: "This A is nine semitones higher than that C, and the C is seven semitones higher than that F, so the A is 16 semitones higher than the F." It is an additive, or linear, scale. By this I mean that if you assigned successive whole numbers to successive notes, then the distance from any note to any other would be given by the difference between their numbers. Only addition and subtraction are involved.

By contrast, if you are going to think of things acoustically rather than auditorily, physically rather than perceptually, each pitch is better described in terms of its *frequency* than in terms of its position on a keyboard. The low A at the bottom of the keyboard vibrates about 27 times per second, whereas the C three semitones above it vibrates about 32 times per second. So you might be inclined to guess that in order to jump up three semitones one should always add five cycles per second. Not so. You should always *multiply* by about 32/27 instead. If you jump up twelve semitones, that means four repeated up-jumps of three semitones.

Thus, when you have gone up one octave (twelve semitones), your pitch has been multiplied by 32/27 four times in a row, which is 2. Actually, the fourth power of 32/27 is not quite 2, and since an octave represents a ratio of *exactly* 2, 32/27 must be a slight underestimate. But that is beside the point. The point is that the natural operations for comparing frequencies are multiplication and division, whereas the natural operations for note numbers on a keyboard are addition and subtraction. What this means is that the note numbers are logarithms of the frequencies. Here is a case where we think naturally in logarithms!

Here is a different way of putting things. Two adjacent notes near the top of a piano keyboard differ in frequency by about 400 cycles per second, whereas adjacent notes near the bottom differ by only about two cycles per second. Wouldn't that seem to imply that the intervals are wildly different? Yet to the human ear, the high and the low interval sound exactly the same!

Logarithmic thinking happens when you perceive only a linear increase even if the thing itself doubles in size. For instance, have you ever marveled at the fact that dialing a mere seven digits can connect any telephone to any other in the New York metropolitan area, where some 10 million people live? Suppose New York were to double in population. Would you then have to add seven more digits to each phone number, making fourteen-digit numbers, in order to reach those twenty million people? Of course not.

Adding seven more digits would *multiply* the number of possibilities by ten million. In fact, adding merely three digits (the area code in front) enables you to reach any phone number in North America. This is simply because each new digit creates a tenfold increase in the number of phones reachable. Three more digits will always multiply your network by a factor of 1,000: three orders of magnitude. Thus the length of a phone number—the quantity directly perceived by you when you are annoyed at how long it takes to dial a long-distance number—is a logarithmic measure of the size of the network you are embedded in. That is why it is preposterous to see huge long numbers of 25 or 30 digits used as codes for people or products when, without any doubt, a few digits would suffice.

I once was sent a bill asking that I transfer a fee to account No. 60802-620-1-1-721000-421-01062 in a bank in Yugoslavia. For a while this held my personal record for absurdity of numbers encountered in business transactions. Recently, however, I was sent my car registration form, at the bottom of which I found this enlightening constant: 0101013612182003010700142631172415120036036000030002. For good measure it was followed, a few blank spaces later, by '19283'.

One place where we think logarithmically is number names. We in America have a new name every three zeros (up to a certain point): from *thousand* to *million* to *billion* to *trillion.* Each jump is "the same size", in a sense. That is, a billion is exactly as much bigger than a million as a million is bigger than a thousand. Or a trillion is to a billion exactly as a billion is to a million. On the other hand, does this continue forever? For instance, does it seem reasonable to say that 10^{103} is to 10^{100} exactly as a million is to a thousand? I would be inclined to say "No, those big numbers are almost the same size, whereas a thousand and a million are very different." It is a little tricky because of the shifts in perceptual reality.

In any case, we seem to run out of number names at about a trillion. To be sure, there are some official names for bigger numbers, but they are about as familiar as the names of extinct dinosaurs: "quadrillion", "octillion", "vigintillion", "brontosillion", "triceratillion", and so on. We are simply not familiar with them, since they died off a dinosillion years ago. Even "billion" presents cross-cultural problems, as I mentioned above. Can you imagine what it would be like if in Britain, "hundred" meant 1,000? The fact is that when numbers get too large, people's imaginations balk. It is too bad, though, that a trillion is the largest number with a common name. What is going to happen when the defense budget gets even more bloated? Will we just get number? Of course, like the dinosaurs, we may never be granted the luxury of facing that problem.

*   *   *

The speed of automatic computation is something whose progress is best charted logarithmically. Over the past several decades, the number of primitive operations (such as addition or multiplication) that a computer can carry out per second has multiplied tenfold about every seven years. Nowadays, it is some 100 million operations per second or, on the fanciest machines, a little more. Around 1975, it was about 10 million operations per second. In the later 1960's, one million operations per second was extremely fast. In the early 1960's, it was 100,000 operations per second. 10,000 was high in the mid-1950's, 1,000 in the late 1940's—and in the early 1940's, 100.

In fact, in the early 1940's, Nicholas Fattu was the leader of a team at the University of Minnesota that was working for the Army Air Force on some statistical calculations involving large matrices (about 60×60). He brought about ten people together in a room, each of whom was given a Monroematic desk calculator. These people worked full-time for ten months in a coordinated way, carrying out the computations and cross-checking each other's results as they went along. About twenty years later, out of curiosity, Professor Fattu redid the calculations on an IBM 704 in twenty minutes. He found that the original team had made two inconsequential errors. Nowadays, of course, the whole thing could be done on a big "mainframe" computer in a second or two.

Still, modern computers can easily be pushed to their limits. The notorious computer proof of the four-color theorem, done at the University of Illinois a few years ago, took 1,200 hours of computer time. When you convert that into days, it sounds more impressive: 50 full 24-hour days. If the computer was carrying out twenty million operations per second, that would come to 10^{14}, or 100 trillion, primitive operations—a couple of hundred for every cigarette smoked that year in the U.S. Whew!

A computer doing a billion operations per second would really be moving along. Imagine breaking up one second into as many tiny fragments as there are seconds in 30 years. That is how tiny a nanosecond—a billionth of a second—is. To a computer, a second is a lifetime! Of course, the computer is dawdling compared with the events inside the atoms that compose it. Take one atom. A typical electron circling a typical nucleus makes about 10^{15} orbits per second, which is to say, a million orbits per nanosecond. From an electron's-eye point of view, a computer is as slow as molasses in January.

Actually, an electron has two eyes with which to view the situation. It has both an *orbital* cycle time and a *rotational* cycle time, since it is spinning on its own axis. Now, strictly speaking, "spin" is just a metaphor at the quantum level, so you should take the following with a big grain of salt. Nevertheless, if you imagine an electron to be a classically (non-quantum-mechanically) spinning sphere, you can calculate its rotation time from its known spin angular momentum (which is about Planck's constant, or 10^{-34} joule-second) and its radius (which we can equate with its Compton wavelength, which is about 10^{-10} centimeter). The spin time turns out to be about 10^{-20} second. In other words, every time the superfast computer adds two numbers, every electron inside it has pirouetted on its own axis about

100 billion times. (If we took the so-called "classical radius" of the electron instead, we would have the electron spinning at about 10^{24} times per second —enough to make one dizzy! Since this figure violates both relativity and quantum mechanics, however, let us be content with the first figure.)

At the other end of the scale, there is the slow, stately twirling of our galaxy, which makes a leisurely complete turn every 200 million years or so. And within the solar system, the planet Pluto takes about 250 years to complete an orbit of the sun. Speaking of the sun, it is about a million miles across and has a mass on the order of 10^{30} kilograms. The earth is a featherweight in comparison, a mere 10^{24} kilograms. And we should not forget that there are some stars—red giants—of such great diameter that they would engulf the orbit of Jupiter. Of course, such stars are very tenuous, something like cotton candy on a cosmic scale. By contrast, some stars—neutron stars—are so tightly packed that if you could remove from any of them a cube a millimeter on an edge, its mass would be about half a million tons, equal to the mass of the heaviest oil tanker ever built, fully loaded!

*   *   *

These large and small numbers are so far beyond our ordinary comprehension that it is virtually impossible to keep on being more amazed. The numbers are genuinely beyond understanding—unless one has developed a vivid feeling for various exponents. And even with such an intuition, it is hard to give the universe its awesome due for being so extraordinarily huge and at the same time so extraordinarily fine-grained. Number numbness sets in early these days. Most people seem entirely unfazed by words such as "billion" and "trillion"; they simply become synonyms for the meaningless "zillion".

This hit me particularly hard a few minutes after I had finished a draft of this column. I was reading the paper, and I came across an article on the subject of nerve gas. It stated that President Reagan expected the expenditures for nerve gas to come to about $800 million in 1983, and $1.4 billion in 1984. I was upset, but I caught myself being thankful that it was not $10 billion or $100 billion. Then, all at once, I really felt ashamed of myself. That guy has some nerve gas! How could I have been *relieved* by the figure of a "mere" $1.4 billion? How could my thoughts have become so dissociated from the underlying reality? One billion for nerve gas is not merely lamentable; it is odious. We cannot afford to become number-number than we are. We need to be willing to be jerked out of our apathy, because this kind of "joke" is in very poor taste.

Survival of our species is the name of the game. I don't really care if the number of mosquitoes in Africa is greater or less than the number of pennies in the gross national product. I don't care if there are more glaciers in the Dead Sea or scorpions in Antarctica. I don't care how tall a stack of one billion dollar bills would be (an image that President Reagan evoked in a speech decrying the size of the national debt created by his predecessors). I don't care a hoot about pointless, silly images of colossal magnitudes. What I *do* care about is what a billion dollars *represents* in terms of buying power: lunches for all the schoolkids in New York for a year, a hundred libraries, fifty jumbo jets, a few years' budget for a large university, one battleship, and so on. Still, if you love numbers (as I do), you can't help but blur the line between number play and serious thinking, because a silly image converts into a more serious image quite fluidly. But frivolous number virtuosity, enjoyable though it is, is far from the point of this article.

What I hope people will get out of this article is not a few amusing tidbits for the next cocktail party, but an increased passion about the importance of grasping large numbers. I want people to understand the very real consequences of those very surreal numbers bandied about in the newspaper headlines as interchangeably as movie stars' names in the scandal sheets. *That*'s the only reason for bringing up all the more humorous examples. At bottom, we are dealing with perceptual questions, but ones with life-and-death consequences!

* * *

Combatting number numbness is basically not so hard. It simply involves getting used to a second set of meanings for *small* numbers—namely, the meanings of numbers between say, five and twenty, when used as exponents. It would seem revolutionary for newspapers to adopt the convention of expressing large numbers as powers of ten, yet to know that a number has twelve zeros is *more* concrete than to know that it is called a "trillion".

I wonder what percentage of our population, if shown the numerals "314,159,265,358,979" and "271,828,182,845", would recognize that the former magnitude is about 1,000 times greater than the latter. I am afraid that the vast majority would not see it and would not even be able to read these numbers out loud. If that is the case, it is something to be worried about.

One book that attempts valiantly and poetically to combat such numbness, a book filled with humility before some of the astounding magnitudes that we have been discussing, is called *Cosmic View: The Universe in Forty Jumps,* by a Dutch schoolteacher, the late Kees Boeke. In his book, Boeke takes us on an imaginary voyage in pictures, in which each step is an exponential one, involving a factor of ten in linear size. From our own size, there are 26 upward steps and 13 downward steps. It is probably not coincidental that the book was written by someone from Holland, since the Dutch have long been internationally minded, living as they do in a small and vulnerable country among many languages and cultures. Boeke closes in what therefore seems to me to be a characteristically Dutch way, by pleading that his book's journey will help to make people better realize their place in the cosmic scheme of things, and in this way contribute to drawing the world closer together. Since I find his conclusion eloquent, I would like to close by quoting from it:

> When we thus think in cosmic terms, we realize that man, if he is to become really human, must combine in his being the greatest humility with the most careful and considerate use of the cosmic powers that are at his disposal.
>
> The problem, however, is that primitive man at first tends to use the power put in his hands for himself, instead of spending his energy and life for the good of the whole growing human family, which has to live together in the limited space of our planet. It therefore is a matter of life and death for the whole of mankind that we learn to live together, caring for one another regardless of birth or upbringing. No difference of nationality, of race, creed or conviction, age or sex may weaken our effort as human beings to live and work for the good of all.
>
> It is therefore an urgent need that we all, children and grown-ups alike, be educated in this spirit and toward this goal. Learning to live together in mutual respect and with the definite aim to further the happiness of all, without privilege for any, is a clear duty for mankind, and it is imperative that education be brought onto this plane.
>
> In this education the development of a cosmic view is an important and necessary element; and to develop such a wide, all-embracing view, the expedition we have made in these 'forty jumps through the universe' may help just a little. If so, let us hope that many will make it!

Post Scriptum.

By coincidence, in the same issue of *Scientific American* as this column appeared in, there was a short note in "Science and the Citizen" on the American nuclear arsenal. The information, compiled by the Center for Defense Information and the National Resources Defense Council, stated that the current stockpile amounted to some 30,000 nuclear weapons, 23,000 of which were operational. (An excellent way of visualizing this is shown in Figure 33-2, the last figure in the book.) The Reagan administration, it said, intended to build about 17,000 in the next ten years while destroying about 7,000, thus increasing the net arsenal by about 10,000 nuclear weapons.

This is roughly equivalent to ten tons of TNT per Russian capita. Now what does this really mean? Wolf H. Fahrenbach had the same nagging question, and he wrote to tell me what he discovered.

> Ten tons of TNT exceeds my numericity, so I asked a demolitions-expert friend of mine what one pound, ten pounds, 100 pounds, etc. of TNT could do. One pound of TNT in a car kills everybody within and leaves a fiery wreck; ten

pounds totally demolishes the average suburban home; and 1,000 pounds packed inside an old German tank sent the turret to disappear in low overhead clouds. It could be reasonably suggested to the administration that most civilized nations are content with simply *killing* every last one of their enemies and that there is no compelling reason to have to ionize them.

Now this was interesting to me, because I happened to remember that the 241 marines killed in the recent truck-bombing in Beirut had been in a building brought down by what was estimated as one ton of TNT. Ten tons, if well placed, might have done in 2,400 people, I suppose. Ten tons is my allotment, and yours as well. That's the kind of inconceivable overkill we are dealing with in the nuclear age.

Another way of looking at it is this. There are about 25,000 megatons of nuclear weapons in the world. If we decode the "mega" into its meaning of "million", and "ton" into "2,000 pounds", we come up with $25,000 \times 1,000,000 \times 2,000$ pounds of TNT-equivalent, which is 50,000,000,000,000 pounds to be distributed among us all, perhaps not equally—but surely there's enough to go around.

I find myself oscillating between preferring to see it spelled out that way with all the zeros, and leaving it as 25,000 megatons. What I have to remember is what "megaton" really means. Last summer I visited Paris and climbed the butte of Montmartre, from the top of which, at the foot of the Sacré Coeur, one has a beautiful view of all of Paris spread out below. I couldn't refrain from ruining my two friends' enjoyment of this splendid panorama, by saying, "Hmm... I bet one or two nicely placed megatons would take care of all this." And so saying, I could see exactly how it might look (provided I were a superbeing whose eyes could survive light and heat blasts far brighter than the sun). I know it seems ghoulish, yet it was also completely in keeping with my thoughts of the time.

Now if you just say to yourself "one megaton equals Paris's doom" (or some suitable equivalent), then I think that the phrase "25,000 megatons" will become as vivid as the long string of zeros—in fact, probably more vivid. It seems to me that this perfectly illustrates how the psychological phenomenon known as *chunking* is of great importance in dealing with otherwise incomprehensible magnitudes.

Chunking is the perception *as a whole* of an assembly of many parts. An excellent example is the difference between 100 pennies and the concept of one dollar. We would find it exceedingly hard to deal with the prices of cars and houses and computers if we always had to express them in pennies. A dollar has psychological reality, in that we usually do not break it down into its pieces. The concept is valuable for that very reason.

It seems to me a pity that the monetary chunking process stops at the dollar level. We have inches, feet, yards, miles. Why could we not have pennies, dollars, grands, megs, gigs? We might be better able to digest newspaper headlines if they were expressed in terms of such chunked units —provided that those units had come to mean something to us, as such. We all have a pretty good grasp of the notion of a grand. But what can a meg or a gig buy you these days? How many megs does it take to build a high school? How many gigs is the annual budget of your state?

Most numerically-oriented people, in order to answer these questions, will have to resort to calculation. They do not have such concepts at their mental fingertips. But in a numerate populace, everyone *should*. It should be a commonplace that a new high school equals about 20 megs, a state budget several gigs, and so on. These terms should not be thought of as *shorthand* for "million dollars" and "billion dollars" any more than "dollar" is a shorthand for "100 cents". They should be autonomous concepts— mental "nodes"—with information and associations dangling from them without any need for conversion to some other units or calculation of any sort.

If that kind of direct sense of certain big numbers were available, then we would have a much more concrete grasp on what otherwise are nearly hopeless abstractions. Perhaps it is in the vast bureaucracies' interest that their budgets remain opaque and impenetrable—but even that holds true only in the short run. Economic ruin and military suicide are not good for anybody in the long run—not even arms manufacturers! The more transparent the realities are, the better it is for any society in the long run.

* * *

This kind of total incomprehension extends even to the highest echelons of our society. Bucknell University President Dennis O'Brien recently wrote on the *New York Times* op-ed page: "My own university has just opened a multibillion-dollar computer center and prides itself that 90 percent of its graduates are computer-literate." And the Associated Press distributed an article that said that the U.S. federal debt ceiling had gone up to 1.143 trillion dollars, and then cited the latest figure for the debt itself as "$1,070,241,000". In that case, what's the hurry about raising the ceiling? These may have been typos, but even so, they betray our society's rampant innumeracy.

You may think I am being nitpicky, but when our populace is so boggled by large numbers that even many university-educated people listen to television broadcasts without an ounce of comprehension of the numbers involved, I think something has gone haywire somewhere. It is a combination of numbness, apathy, and a resistance to recognizing the need for new concepts.

One reader, a refugee from Poland, wrote to me, complaining that I had memorized hundreds of digits of π in my high school days without appreciating the society that afforded me this luxury. In East Block countries, he implied, I would never have felt free to do something so decadent. My feeling, though, is that memorizing π was for me no different from any other kind of exuberant play that adolescents in any country engage in. In a recent book by Stephen B. Smith, called *The Great Mental*

Calculators—a marvelously engaging book, by the way—one can read the fascinating life stories of people who were far better than I with figures. Many of them grew up in dismal circumstances, and numbers to them were like playmates, life-saving friends. For them, to memorize π would not be decadent; it would be a source of joy and meaning. Now I had read about some of these people as a teen-ager, and I admired, even envied, their abilities. My memorization of π was not an isolated stunt, but part of an overall campaign to become truly fluent with numbers, in imitation of calculating prodigies. Undoubtedly this helped lead me toward a deeper appreciation of numbers of all sizes, a better intuition, and in some intangible ways, a clearer vision of just what it is that the governments on this earth—West Block no less than East—are up to.

But there may be more direct routes to that goal. For example, I would suggest to interested readers that they attempt to build up their own numeracy in a very simple way. All they need to do is to get a sheet of paper and write down on it the numbers from 1 to 20. Then they should proceed to think a bit about some large numbers that seem of interest to them, and try to estimate them within one order of magnitude (or two, for the larger ones). By "estimate" here, I mean actually do a back-of-the-envelope (or mental) calculation, ignoring all but factors of ten. Then they should attach the idea to the computed number. Here are some samples of large numbers:

* What's the gross state product of California?
* How many people die per day on the earth?
* How many traffic lights are there in New York City?
* How many Chinese restaurants are there in the U.S.?
* How many passenger-miles are flown each day in the U.S.?
* How many volumes are there in the Library of Congress?
* How many notes are played in the full career of a concert pianist?
* How many square miles are there in the U.S.? How many of them have you been in?
* How many syllables have been uttered by humans since 1400 A.D.?
* How many "300" games are bowled in the U.S. per year?
* How many stitches are there in a stocking?
* How many characters does one need to know to read a Chinese newspaper?
* How many sperms are there per ejaculate?
* How many condors remain in the U.S.?
* How many moving parts are in the Columbia space shuttle?
* How many people in the U.S. are called "Michael Jackson"? "Naomi Hunt"?
* What volume of oil is removed from the earth each year?
* How many barrels of oil are left in the world?
* How much carbon monoxide enters the atmosphere each year in auto exhaust fumes?

* How many meaningful, grammatical, ten-word sentences are there in English?
* How long did it take the 200-inch mirror of the Palomar telescope to cool down?
* What angle does the earth's orbit subtend, as seen from Sirius?
* What angle does the Andromeda galaxy subtend, as seen from earth?
* How many heartbeats does a typical creature live?
* How many insects (of how many species) are now alive?
* How many giraffes are now alive? Tigers? Ostriches? Horseshoe crabs? Jellyfish?
* What are the pressure and temperature at the bottom of the ocean?
* How many tons of garbage does New York City put out each week?
* How many letters did Oscar Wilde write in his lifetime?
* How many typefaces have been designed for the Latin alphabet?
* How fast do meteorites move through the atmosphere?
* How many digits are in 720 factorial?
* How much is a brick of gold worth?
* How many gold bricks are there in Fort Knox? How much is it worth?
* How fast do your wisdom teeth grow (in miles per hour, say)?
* How fast does your hair grow (again in miles per hour)?
* How fast is Venice sinking?
* How far is a million feet? A billion inches?
* What is the weight of the Empire State Building? Of Hoover Dam? Of a fully loaded jumbo jet?
* How many commercial airline takeoffs occur each year in the world?

These or similar questions will do. The main thing is to attach some concreteness to those numbers from 1 to 20, seen as exponents. They are like dates in history. At first, a date like "1685" may be utterly meaningless to you, but if you love music and find out that Bach was born that year, all of a sudden it sticks. Likewise with this secondary meaning for small numbers. I can't guarantee it will work miracles, but you may increase your own numeracy and you may also help to increase others'. Merry numbers!

7

Changes in Default Words and Images, Engendered by Rising Consciousness

November, 1982

A father and his son were driving to a ball game when their car stalled on the railroad tracks. In the distance a train whistle blew a warning. Frantically, the father tried to start the engine, but in his panic, he couldn't turn the key, and the car was hit by the onrushing train. An ambulance sped to the scene and picked them up. On the way to the hospital, the father died. The son was still alive but his condition was very serious, and he needed immediate surgery. The moment they arrived at the hospital, he was wheeled into an emergency operating room, and the surgeon came in, expecting a routine case. However, on seeing the boy, the surgeon blanched and muttered, "I can't operate on this boy—he's my son."

What do you make of this grim riddle? How could it be? Was the surgeon lying or mistaken? No. Did the dead father's soul somehow get reincarnated in the surgeon's body? No. Was the surgeon the boy's true father and the dead man the boy's adopted father? No. What, then, is the explanation? Think it through until you have figured it out on your own—I insist! You'll know when you've got it, don't worry.

* * *

When I was first asked this riddle, a few years ago, I got the answer within a minute or so. Still, I was ashamed of my performance. I was also disturbed by the average performance of the people in the group I was with—all educated, intelligent people, some men, some women. I was neither the quickest nor the slowest. A couple of them, even after five minutes of scratching their heads, still didn't have the answer! And when they finally hit upon it, their heads hung low.

Whether we light upon the answer quickly or slowly, we all have something to learn from this ingenious riddle. It reveals something very deep about how so-called *default assumptions* permeate our mental representations and channel our thoughts. A default assumption is what holds true in what you might say is the "simplest" or "most natural" or "most likely" possible model of whatever situation is under discussion. In this case, the default assumption is to assign the sex of male to the surgeon. The way things are in our society today, that's the most plausible assumption. But the critical thing about default assumptions—so well revealed by this story—is that they are made automatically, not as a result of consideration and elimination. You didn't explicitly ponder the point and ask yourself, "What is the most plausible sex to assign to the surgeon?" Rather, you let your past experience merely assign a sex for you. Default assumptions are by their nature implicit assumptions. You never were aware of having made any assumption about the surgeon's sex, for if you had been, the riddle would have been easy!

Usually, relying on default assumptions is extremely useful. In fact, it is indispensable in enabling us—or any cognitive machine—to get around in this complex world. We simply can't afford to be constantly distracted by all sorts of theoretically possible but unlikely exceptions to the general rules or models that we have built up by induction from many past experiences. We have to make what amount to shrewd guesses—and we do this with great skill all the time. Our every thought is permeated by myriads of such shrewd guesses—assumptions of normalcy. This strategy seems to work pretty well. For example, we tend to assume that the stores lining the main street of a town we pass through are not just cardboard façades, and for good reason. Probably you're not worried about whether the chair you're sitting on is about to break. Probably the last time you used a salt shaker you didn't consider that it might be filled with sugar. Without much trouble, you could name dozens of assumptions you're making at this very moment—all of which are simply *probably* true, rather than *definitely* true.

This ability to ignore what is very unlikely—*without even considering whether or not to ignore it!*—is part of our evolutionary heritage, coming out of the need to be able to size up a situation quickly but accurately. It is a marvelous and subtle quality of our thought processes; however, once in a while, this marvelous ability leads us astray. And sexist default assumptions are a case in point.

* * *

When I wrote my book *Gödel, Escher, Bach: an Eternal Golden Braid*, I employed the dialogue form, a form I enjoy very much. I was so inspired by Lewis Carroll's dialogue "What the Tortoise Said to Achilles" that I decided to borrow his two characters. Over time I developed them into my own characters. As I proceeded, I found that I was naturally led to bringing in some new characters of my own. The first one was the Crab. Then came the Anteater, the Sloth, and various other colorful characters. Like the Tortoise and Achilles, the new characters were all male: *Mr.* Crab, *Mr.* Sloth, and so on.

This was in the early 70's, and I was quite conscious of what I was doing. Yet for some reason, I could not get myself to invent a female character. I was upset with myself, yet I couldn't help feeling that introducing a female character "for no reason" would be artificial and therefore too distracting. I didn't want to mix sexual politics—an ugly real-world issue—with the ethereal pleasures of an ideal fantasy world.

I racked my brains on this for a long time, and even wrote an apologetic dialogue on this very topic—an intricate one in which I myself figured, discussing, with my own characters, the question of sexism in writing. Aside from my friends Achilles and the Tortoise, the cast featured God as a surprise visitor—and, as in the old joke, she was black. Though corny, it was an earnest attempt to grapple with some problems of conscience that were plaguing me. The dialogue never got polished, and was not included in my book. However, a series of reworkings gradually turned it into the "Six-Part Ricercar" with which the book concludes.

My pangs of conscience did lead me to making a few minor characters female: there were Prudence and Imprudence, who briefly argued about consistency; Aunt Hillary, a conscious ant colony; and every even-numbered member of the infinite series Genie, Meta-genie, Meta-meta-genie, and so on. I was particularly proud of this gentle touch. But no matter how you slice it, females got the short end of the stick in *GEB*. I was not altogether happy with that, but that's the way it was.

Aside from its dialogues being populated with male characters, the book was also filled with default assumptions of masculinity: the standard "he" and "his" always being chosen. I made no excuse for this. I gave my reader credit for intelligence; I assumed he would know that often, occurrences of such pronouns carry no gender assumptions but simply betoken a "unisex" person.

Over a period of time, however, I have gradually come to a different feeling about how written language should deal with people of unspecified sex, or with supposedly specific but randomly chosen people. It is a very subtle issue, and I do not claim to have the final answers by any means. But I have discovered some approaches that please me and that may be useful for other people.

*   *   *

What woke me up? Given that I was already conscious of the issues, what new element did it take to induce this shift? Well, one significant incident was the telling of that surgeon riddle. My own reaction to it and the reactions of my companions surprised me. To most of us, bizarre worlds with such things as reincarnation came more easily to mind than the idea that a surgeon could be a woman! How ludicrous! The event underscored for me how deeply ingrained are our default assumptions, and how unaware we are of them. This seemed to me to have potential consequences far beyond what one might naïvely think. I am hardly one to believe that language "pushes us around", that we are its slaves—yet on the other hand, I feel that we must do our best to rid our language of usages that may induce or reinforce default assumptions in our minds.

One of the most vivid examples of this came a couple of years after my book had been published. I was describing its dialogues to a group of people, and I said I regretted that the characters had all been male. One woman asked me why, and I replied, "Well, I began with two males—Achilles and the Tortoise—and it would have been distracting to introduce females seemingly for no reason except politics . . ." Yet as I heard myself saying this, a horrifying thought crept into my mind for the first time: How did I know the Tortoise was really a male? Surely he was, wasn't he? Obviously! I seemed to remember that very well.

And yet the question nagged at me. As I had a copy of my book at hand, with the Carroll dialogue reprinted in it, I turned to it for verification. I was nonplussed to see that Carroll nowhere even hints at the sex of his Tortoise! In fact, the opening sentence runs thus: "Achilles had overtaken the Tortoise, and had seated himself comfortably upon its back." This is the only occurrence of "it"; from there on, "the Tortoise" is what Carroll writes. "*Mr.* Tortoise", indeed! Was this entirely a product of my own defaults?

Probably not. The first time I had heard about the Carroll dialogue, many years earlier, someone—a male—had described it to me. This person very likely had passed on *his* default assumption to me. So I could claim innocence. Moreover, I realized, I had read a few responses in philosophy journals to the Carroll dialogue, and when I went back and looked at them, I found that they too had featured a "sexed" Tortoise, in contrast to the way Carroll had carefully skirted the issue. Though I felt somewhat exonerated, I was still upset. I kept on asking myself, "What if I had envisioned a female Tortoise to begin with? Then how would *GEB* have been?" This was a most provocative counterfactual excursion.

One thing that had dissuaded me from using female characters was the distractingly political way that some books had of referring to the reader or briefly mentioned random people (such as "the student" or "the child") as "she" or "her". It stuck out like a sore thumb, and made one think so much about sexism that the main point of the passage often went unnoticed. It seemed to me that such a strategy might be too blunt and simplistic, and could easily turn more people off than on.

And yet I couldn't agree with the attitude of some people—largely but by no means exclusively men—who refused to switch their usage on grounds of "tradition", "linguistic purity", "beauty of the language", and so on. To be sure, words like "fireperson", "snowperson", "henchperson", and "personhandle" are unappealing—but they aren't your only recourse! There are other options.

In the introduction to Robert Nozick's *Philosophical Explanations*—an exciting and admirable book on philosophy—I came across this footnote. "I do not know of a way to write that is truly neutral about pronoun gender yet does not constantly distract attention—at least the contemporary reader's—from the sentence's central content. I am still looking for a satisfactory solution." From this point on, Nozick uses "he" and "him" nearly everywhere. My reaction was annoyance: could Nozick have really looked very hard? Part of my annoyance was undoubtedly due to my own guilt feelings for having done no better in *GEB,* but some was due to my feeling that Nozick had failed to see a fascinating challenge here—one to which he could bring his philosophical insight, and in doing so, make a creative contribution to society.

*   *   *

As best I can recall, I first began seriously trying to "demasculinize" my prose in working on the dialogue on the Turing Test that eventually wound up as my "Metamagical Themas" column for May, 1981, and which is Chapter 22 in this book. I wrote the dialogue with the sexes of the characters shifting about fluidly in my mind, since I was modeling the characters on mixtures of various people I knew. I always imagined the character I most agreed with more as female than as male, and the others vacillated.

One day, it occurred to me that the beginning of the dialogue discussed Turing's question "Can you in principle tell, merely from a written dialogue, a female from a male?" This question applied so well to the very characters discussing it that I could not resist making some character "ambisexual"—ambiguous in terms of sex. Thus I named one of them "Pat". Soon I realized there was no reason not to extend this notion to *all* the characters in the dialogue, making it a real guessing game for readers. Thus were born "Sandy", "Chris", and "Pat".

Writing this dialogue was a turning point for me. Even though its total sexual equality had been motivated by my desire to give the dialogue an interesting self-referential twist, I found that I was very relieved to have broken out of the all-male mold that I had earlier felt locked into. I started looking for more ways to make up for my past default sexism.

It was not easy, and still is not. For example, in teaching classes, I find myself wanting to use the pronoun "she" to refer back to an earlier unspecified person—a random biologist, say, or a random logician. Yet I find it doesn't seem to come out of my mouth easily. What I have trained myself to do rather well is to avoid gender-laden pronouns altogether, thus, like Carroll, "skirting" the issue. Sometimes I just keep on saying "the logician" over and over again, or perhaps I just say "the person" or "that person". Every once in a while, I say "he or she" (or "he" or "she"), although I have to admit that I more often simply say "they".

Someone who, like me, is trying to eliminate gender-laden pronouns from their speech altogether can try to rely on the word "they", but they will find themself in quite a pickle as soon as they try to use any reflexive verbal construction such as "the writer will paint themselves into a corner", and what's worse is that no matter how this person tries, they'll find that they can't extricate themselves gracefully, and consequently he or she will just flail around, making his or her sentence so awkward that s/he wis/hes s/he had never become conscious of these issues of sexism. Obviously, using "they" just carries you from the frying pan into the fire, as you have merely exchanged a male-female ambiguity for a singular-plural ambiguity. The only advantage to this ploy is, I suppose, that there is/are, to my knowledge, no group(s) actively struggling for equality between singular and plural.

One possible solution is to use the plural exclusively—to refer to "biologists" or "a team of biologists", never just "a biologist". That way, "they" is always legitimately referring to a plural. However, this is a very poor solution, since it is much more vivid to paint a picture of a specific individual. A body can't always deal in plurals!

Another solution, somewhat more pleasing, is to turn an impersonal situation into a more personal one, by using the word "you". This way, your listeners or readers are encouraged to put themselves in the situation, to experience it vicariously. Sometimes, however, this can backfire on you. Suppose you're talking about the strange effects in everyday life that statistical fluctuations can produce. You might write something like this: "One day your mailman might have so much mail to sort down at the post office that it's afternoon by the time she gets started on her route." At the outset, your avid reader Polly manufactures an image of her friendly postman sorting letters; a few moments later, she is told the postman is a woman. Jolt! It's not just a surface-level jolt (the collision of the words "mailman" and "she"), although it's that too; it's really an image-image conflict, since you expressly invited Polly to think of *her own* mailman, who happens to be a man. Even if you'd said "your letter carrier", Polly would still have been jolted. On the other hand, if you'd asked Polly to think about, say, "Henry's letter carrier", then that "she" would not have caused nearly as much surprise—maybe even not any.

*　　*　　*

In teaching my classes, I try always to use sex-neutral nouns such as "letter carrier" and "department head" (which I prefer to "chairperson"), and having done so, I try my utmost to avoid using gender-specific pronouns to refer back to them. But I have realized that this is largely a show put on for my own benefit. I'm not actively undermining any bad stereotypes simply by avoiding them. The fact that I'm *not* saying "he" where many people *would* is not the sort of thing that will grab my students by the collar and shake them. A few people may notice my "good behavior", but those are the ones who are already attuned to these issues.

So why not just use an unexpected "she" now and then? Isn't that the obvious thing to do? Perhaps. But in many cases, as Nozick pointed out, it may seem so politically motivated that it will distract more than enlighten. The problem is, once you start to describe some unknown receptionist (say), listeners will manufacture a fresh, blank mental *node* to represent that receptionist. By "node", I mean something like a mental dossier or questionnaire with a number of questions wanting immediate answers.

Now, it is naïve to suppose that a few seconds after they have manufactured their new node, their image of the receptionist is—or ever was—floating in a sexual limbo. It is next to impossible to build up more than the most fleeting, insubstantial image of a person without assuming he's a she, or vice versa. The instant that node is manufactured, unless *you* fill in all its blanks, it will fill them in for itself. (Imagine that each question has a default answer entered in light pencil, easily erasable but to be used in case no other answer is provided.) And unfortunately—even for ardent feminists—those unconscious default assumptions are usually going to be sexist. (Feminists can be as sexist as the next guy!) For example, I have realized, to my dismay, that my defaults run very deep—so deep that, even when I *say* "his or her telephone", I am often nonetheless *thinking* "her telephone", and *envisioning* a woman at a desk. This is most disconcerting. It reveals that, although my self-training has succeeded quite well at the linguistic level, it hasn't yet fully filtered down to the *imagistic* level.

As a corrective measure, I have trained myself, over the past few years, to have a sort of "second-order reflex" triggered by the manufacture of a new node for an unknown individual. What this reflex does is to make me consciously attempt to assign a female wherever my first-order reflex—that is, the naïve reflex—would tend to automatically assign a male (and vice versa). I have become pretty good at this, but sometimes it is difficult or just plain silly to take this default-violating image seriously. For instance, when there's a slow truck somewhere ahead of me, holding up the traffic on a two-lane road, it is so tempting to say, "Why doesn't that guy pull over and let the rest of us pass him?" Although I won't say it that way, I also won't say, "Why doesn't he or she let us pass him or her?" It's not easy for me to talk about the pilot of the airliner I'm riding in in sex-neutral terms, because the vast majority of commercial airline pilots *are* men. The person in the seat next to me will look at me a bit strangely if I say, "He or she just made a beautiful landing, didn't they?" And if someone tells me that a thief has just broken into their car, should I say, "How much did he or she get away with?"

* * *

So haven't I painted myselves into a corner? Am I not damned if I do, damned if I don't? After all, I've said that on the one hand, the passive approach of merely avoiding sexist usages isn't enough, but that on the other hand, the active approach of throwing in jolting stereotype violations can be too much. Is there no successful middle path?

I have discovered, as a matter of fact, what I think is a rather graceful compromise solution to such dilemmas. Instead of dropping a nondefault gender into her lap *after* your reader has set up her default images of the people involved in the situation, simply don't let her get off the ground with her defaults. Upset her default assumptions explicitly from the word "go".

I did this in my column on big numbers and innumeracy (Chapter 6), at the beginning of which I retold an old joke. Usually the storyteller begins, "A professor was giving a lecture on the fate of the solar system, and he said . . ." Almost always, the professor is made out to be a male. This may reflect the sexual statistics for astronomers, but individuals aren't statistics.

So how could this story be improved—gracefully? Well, there is a delay —not a long one, but still a delay—between the first mention of the professor and the pronoun "he". It's long enough for that default male image to get solidly—even though implicitly—implanted in the listener's mind. So just don't let that happen. Instead, make the professor a woman from the very start. By this I certainly do not mean that you should begin your story, "A *lady* professor was giving a lecture on the fate of the solar system, and . . .". Good grief! That's horrible!

My solution, instead, was to give her sex away by her name. I invented the silly pseudo-Slavic name "Professor Bignumska", whose ending in 'a' signifies that its owner is female. To be sure, not everyone is attuned to such linguistic subtleties, so that for some people it will come as a surprise when a line or two later, they read the phrase "according to her calculations". But at least they will get the point in the end.

What's much worse is when people do not miss the point, but rather, reject the point altogether. In the published French translation of my article, my "Professor Bignumska" was turned into *monsieur le professeur Grannombersky*. Not only was the sex reversed, but clearly the translator had recognized what I was up to, and had deliberately removed all telltale traces by switching the ending to a masculine one. This is certainly disappointing. On the other hand, it was a relief to see that in the German translation, the professor's femininity remained intact: she was now called *die namhafte Kosmogonin Großzahlia*. Here not only her name but even her title has a feminine ending!

This practice of giving some professions explicitly feminine and masculine words certainly makes for trouble. What do you do when talking about a mixed group of actors and actresses? Unless you want to be verbose, you have little choice but to refer to "actors". Why does a word like "waiter", with its completely noncommittal ending, have to refer to a male? We are hard put to come up with a neutral term. Certainly "waitperson" is a strange concoction. "Server" is not so bad, and nowadays I don't object to "waitron", although the first time I heard it, it sounded *very* odd. It is nice to see "stewardess" and "steward" gradually getting replaced by the general title "flight attendant".

* * *

All languages I have studied are in one way or another afflicted by these sorts of problems. Whereas we in English have our quaint-sounding "poetess" and "aviatrix", in French they have no better way of referring to a female writer or professor than *une femme écrivain* or *une femme professeur,* the default male gender being built right into the nouns themselves. That is, *écrivain* and *professeur* are both masculine nouns. In order to allow them to refer to women, you must treat them essentially as adjectives following (and modifying) the noun *femme* ("woman").

Another peculiarity of French is the word *quelqu'un*—the word for "someone". It literally means "some one", and it requires the masculine *un* ("one") no matter whom it refers to. This means, for example, that if an unfamiliar woman knocks at the door of Nicole's house, and Nicole's young daughter answers the door, she is likely to yell to Nicole: *Maman, il y a quelqu'un à la porte!* ("Mommy, there's someone at the door!") It is impossible to "feminize" this pronoun: *Maman, il y a quelqu'une à la porte.* Even sillier would be to try to transform the impersonal *il y a*—"there is" —into a feminine version, *elle y a.* It just rings absurd. The masculine *il* is as impersonal as "it" in "It is two o'clock." Surely no one would suggest that we say "They are two o'clock".

In English, we have some analogous phenomena. If a pair of strangers knock at Paul's door, his daughter may yell to him, "Daddy, someone's at the door." She will not say, "Sometwo are at the door." What this illustrates is that the pronoun "someone" does not carry with it strong implications of singularity. It can apply to a group of people without sounding odd. Perhaps, analogously, *quelqu'un* is not as sexist at the image level as its surface level would suggest. But this is hard to know.

Normally in French, to speak about a mixed or unspecified group of people, one uses the masculine plural pronoun *ils.* Even a group whose membership hasn't yet been determined, but which stands a fair chance of including at least one male among twenty females, will still call for *ils.* Female speakers grow up with this usage, of course, and follow it as naturally and unconsciously as male speakers do. Can you imagine the uproar if there were a serious attempt to effect a reversal of this age-old convention? How would men feel if the default assumption were to say *elles*? How would women feel? How would people in general feel if a group consisting of several men and one woman were always referred to as *elles*?

Curiously enough, there are circumstances where nearly that happens. There is a formalistic style of writing often found in legal or contractual documents in which the word *personnes* is used to refer to an abstract and unspecified group of people; thereafter the feminine plural pronoun *elles* is used to refer back to that noun. Since the word *personne* is of feminine gender (think of the Latin *persona*), this is the proper pronoun to use, even if the group being referred to is known to consist of males only!

Although it is grammatically correct, when this is dragged out over a long piece of text it can give the reader a strange impression, since the original noun is so distant that the pronoun feels autonomous. One feels that the pronoun should at some point switch to *ils* (and in fact, sometimes this happens). When it doesn't, it can make the reader uneasy. Perhaps this is just my own reaction. Perhaps it's merely the typical reaction of someone used to having the default pronoun for an unspecified group of people be masculine. Perhaps it's good for a man to experience that slight sense of malaise that women may feel when they see themselves referred to over and over again as *ils,* simply because there is likely to be at least one male present in the group.

We are all, of course, members of that collective group often referred to as "mankind", or simply "man". Even the ardent feminist Ashley Montagu once wrote a book called *Man: His First Two Million Years.* (I guess this was a long time ago.) Many people argue that this usage of "man" is completely distinct from the usage of "man" to refer to individuals, and that it is devoid of sexual implications. But many studies have been done that undeniably establish the contrary. David Moser once vividly pointed out to me the sexism of this usage. He observed that in books you will find many sentences in this vein: "Man has traditionally been a hunter, and he has kept his females close to the hearth, where they could tend his children." But you will never see such sentences as "Man is the only mammal who does not always suckle his young." Rather, you will see "Man is the only mammal in which the females do not always suckle their young." So much for the sexual neutrality of the generic "man". I began to look for such anomalies, and soon ran across the following gem in a book on sexuality: "It is unknown in what way Man used to make love, when he was a primitive savage millions of years ago."

*   *   *

Back to other languages. When I spent a few months in Germany working on my doctoral dissertation, I learned that the term for "doctoral advisor" in German is *Doktorvater*—literally, "doctor father". I immediately wondered, "What if your *Doktorvater* is a woman? Is she your *Doktormutter*?" Since that rang absurd to my ears, I thought that a better solution would be to append the feminizing suffix *in,* making *Doktorvaterin*—"doctor father-ess". However, it seems that a neutral term just might be preferable.

Italian and German share an unexpected feature: In both, the respectful way of saying "you" is identical to the feminine singular pronoun, the only difference being capitalization. In Italian, it's *Lei;* in German, *Sie.* Now in German the associated verb uses a plural ending, so that the connection to "she" is somewhat diluted, but in Italian, the verb remains a third-person singular verb. Thus, to compliment a man, you might say: *Oh, come è bello Lei!* ("How handsome She is!") Of course, Italians do not hear it this naïve way. To them, it might seem equally bewildering that in English, adding 's' to a noun makes it plural whereas adding 's' to a verb makes it singular.

One of the strangest cases is that of Chinese. In Mandarin Chinese, there has traditionally been just one pronoun for "he" and "she", pronounced "tā" and written as in Figure 7-1a. This character's left side consists of the "person" radical, indicating that it refers to a human being, sex unspecified. Curiously, however, in the linguistic reforms carried out in China during the past 70 years or so, a distinction has been introduced whereby there are now separate written forms for the single sound "tā". The old character has been retained, but now in addition to its old meaning of "s/he", it has the new meaning of "he" (wouldn't you know?), while a new character has been invented for "she". The new character's radical is that for "woman" or "female", so the character looks as is shown in Figure 7-1b.

The new implication—not present in Chinese before this century—is that the "standard" type of human being is a male, and that females have to be indicated specially as "deviant". It remains a mystery to me why the Chinese didn't leave the old character as it was—a neutral pronoun—and simply manufacture *two* new characters, one with the female radical and one with the *male* radical, as in Figure 7-1c. (These three characters were created on a Vax computer using the character-designing program Hàn Zì, written by

---

FIGURE 7-1. *Characters for third-person singular pronouns in Chinese. In (a), the generic, or neutral, pronoun, corresponding neither to "she" nor to "he", but more to our usage of "they" in the singular. In (b), a new character first introduced some 70 years ago, meaning "she", thus setting females apart as "special" or "deviant" (depending on your point of view). In (c), a character of my own invention, being the masculine counterpart of that in (b), thus restoring sexual symmetry to the language's pronouns. The left-hand element of all three characters is the radical, or semantic component, and in the three cases its meaning is: (a) "person"; (b) "female"; (c) "male". Unfortunately, "male" is considered by pedants not to be a legitimate radical in Chinese. For purposes of comparison, though, my new character is about as offensive to an average Chinese reader as the mixing of Latin and Greek roots is to us—or, for that matter, as offensive as the recently constructed title "Ms." Of course, there are English-speaking pedants who object to "Ms.", whining, "But it's not an abbreviation for anything!" [Characters printed by the Hàn Zì program, developed by David B. Leake and the author at Indiana University.]*

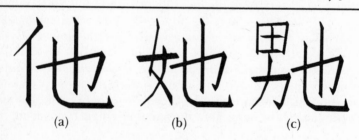

(a)　　　　　(b)　　　　　(c)

David B. Leake and myself. More of the program's output is shown in Figure 13-13.) To give a corresponding (though exaggerated) example in English, can you imagine a political reform in which the word "person" came to mean "man", and for "woman" we were told to say "personess"? Actually, as I found out some time after inventing my new Chinese character, the character meaning "male" is not generally considered a radical, whereas the character meaning "female" is. A typical asymmetry, obviously not limited to the Occident!

The upshot is that in China, there is no longer a truly gender-free pronoun in writing. Formerly, you could write a whole story without once revealing the sex of its participants, whereas now, your intentions to be ambiguous are themselves ambiguous. In the case of the joke about the cosmologist with its default option, it is interesting to consider which way would be better for the sake of feminism. Would you rather have the storyteller leave the professor's sex unspecified throughout the story, so that people's default options would be invoked? Or would you rather have the storyteller forced to commit himself?

*   *   *

One of my pet peeves is the currently popular usage of the word "guys". You often hear a group of people described as "guys", even when that group includes women. In fact, it is quite common to hear women addressing a group of other women as "you guys". This strikes me as very strange. However, when I have asked some people about it, they have adamantly maintained that, when in the plural, the word "guy" has completely lost all traces of masculinity. I was arguing with one woman about this, and she kept on saying, "It may have retained some male flavor for *you,* but it has none in most people's usage." I wasn't convinced, but nothing I could think of to say would budge her from her position. However, fortune proved to be on my side, because, in a last-ditch attempt to convince me, she said, "Why, I've even heard *guys* use it to refer to a bunch of women!" Only after saying it did she realize that she had just unwittingly undermined her own claim.

Such are the subtleties of language. We are often simply too unaware of how our own minds work, and what we really believe. It is there for us to perceive, but too often people do not listen to themselves. They think they know themselves without listening to themselves. Along these lines, I recently heard myself saying "chesspeople" to refer to those wooden objects that you move about on a chessboard. It seems that my second-order reflex to change the suffix "man" into "person" and "men" into "people" was a little too strong, or at least too mechanical. After all, we *do* have the term "chess pieces"!

There simply is a problem with default assumptions in our society. It is manifested everywhere. You find it in proverbs like "To each his own",

"Time and tide wait for no man", and so on. You hear it when little children (and adults) talk about squirrels and birds in their yards ("Oh, look at him running with that acorn in his mouth!"). You see it in animated cartoons, many of which feature some poor schlemiel—a sad "fall guy", a kind of schmoe with whom "everyman" can identify—whose fate it is to be dumped on by the world, and we all laugh with him as he is dealt one cruel setback after another. But why aren't there women in this role more often? Why aren't there more "schlemielesses"—more "fall gals"?

One evening at some friends', I was reading a delightful children's book called *Frog and Toad Are Friends,* and I asked why Frog and Toad both had to be males. This brought up the general topic of female representation in children's television and movies. In particular, we discussed the Muppets, and we all wondered why there are so few sympathetic female Muppet characters. I'm a great fan of Ms. Piggy's, but still I feel that if she's the only major female character, something is wrong. She's hardly an ideal role model.

This general kind of problem, of course, is not limited to questions of sex. It extends far further, to groups of any sort, large or small. The cartoons in *The New Yorker,* for instance, although innocuous in one sense, certainly do not do anything to promote a change in one's default assumptions about the roles people can play. How often do you see a black or female executive in a *New Yorker* cartoon (unless, of course, they are there expressly because the point of the joke depends on it)? The same could be said for most television shows, most books, most movies . . . It is hard to know how to combat such a huge monolithic pattern.

There is an excellent and entertaining book that I discovered only after this column was nearly complete, and which could be a giant leap for humankind in the right direction. It is *The Handbook of Nonsexist Writing,* by Casey Miller and Kate Swift. I recommend it heartily.

*   *   *

One of the most eloquent antisexist statements I have ever come across is a talk delivered recently by Stanford University President Donald Kennedy at an athletes' banquet. Thirty years ago, Kennedy himself was an athlete at Harvard, and he reminisced about a similar banquet he had attended back then. He mused:

> It occurs to me to wonder: What would the reaction have been if I had predicted that soon . . . . women would run the Boston Marathon faster than it had ever been run by men up to that point? There would have been incredulous laughter from two-thirds of the room, accompanied by a little locker-room humor.
>
> Yet that is just what has taken place. My classmates would be astonished at the *happening,* but they would be even more astonished at the *trends.* If we look at the past ten years of world's best times in the marathon for men and women, it is clear that the women's mark has been dropping, over the decade, at a rate about seven times faster than the men's record.

The case of swimming is even more astonishing. Kennedy recalls that in his day, the Harvard and Yale teams were at the very pinnacle of the nation in swimming, and both came undefeated into their traditional rival meet at the end of that season.

> What would have happened if you had put this year's Stanford women into that pool? *Humiliation* is what. Just to give you a sample, *seven* current Stanford women would have beaten my friend Dave Hedberg, Harvard's great sprint freestyler, and *all* the Yalies in the 100. The Stanford women would have swept the 200-yard backstroke and breaststroke, and won *all* the other events contested.
>
> In the 400-yard freestyle relay, there would have been a 10-second wait between Stanford's touch and the first man to arrive at the finish. Do you know how *long* ten seconds is? Can you imagine that crowd in Payne Whitney Gymnasium, seeing a team of *girls* line up against the two best freestyle relay groups in the East, expecting the unexpected, and then having to wait *this long* —for the men to get home?"

Kennedy paints a hilarious picture, but of course his point is dead serious:

> I ask you: If conventional wisdom about women's capacity can be so thoroughly decimated in this most traditional area of male superiority, how can we possibly cling to the illusions we have about them in other areas?
>
> What, in short, is the lesson to be drawn from the emerging *athletic* equality of women? I think it is that those who make all the other, less objectively verifiable assumptions about female limitations would do well to discard them. They belong in the same dusty closet with the notion that modern ballplayers couldn't carry Ty Cobb's spikes and the myth that blacks can't play quarterback. Whether it is vicious or incapacitating or merely quaint, nonsense is nonsense. And it dies hard.

'Tis a point to ponder. In the meantime:

Post Scriptum.

Since writing this column, I have continued to ponder these issues with great intensity. And I must say, the more I ponder, the more prickly and confusing the whole matter becomes. I have found appalling unawareness of the problem all around me—in friends, colleagues, students, on radio and television, in magazines, books, films, and so on. The *New York Times* is one of the worst offenders. You can pick it up any day and see prominent women referred to as "chairman" or "congressman". Even more flagrantly obnoxious is when they refer to prominent feminists by titles that feminism repudiates. For example, a long article on Judy Goldsmith (head of NOW, the National Organization for Women) repeatedly referred to her as "Mrs. Goldsmith". The editors' excuse is:

> Publications vary in tone, and the titles they affix to names will differ accordingly. The Times clings to traditional ones (*Mrs., Miss,* and *Dr.,* for example). As for *Ms.*—that useful business-letter coinage—we reconsider it from time to time; to our ear, it still sounds too contrived for news writing.

As long as they stick with the old terms, they will sound increasingly reactionary and increasingly silly.

Perhaps what bothers me the most is when I hear newscasters on the radio —especially public radio—using blatantly sexist terms when it would be so easy to avoid them. Female announcers are almost uniformly as sexist as male announcers. A typical example is the female newscaster on National Public Radio who spoke of "the employer who pays his employees on a weekly basis" and "the employee who is concerned about his tax return", when both employer and employee were completely hypothetical personages, thus without either gender. Or the male newscaster who described the Pope in Warsaw as "surrounded by throngs of his countrymen". Or the female newscaster who said, "Imagine I'm a worker and I'm on my deathbed and I have no money to support my wife and kids . . ." Of all people, newscasters should know better.

I attended a lecture in which a famous psychologist uttered the following sentence, *verbatim:* "What the plain man would like, as he comes into an undergraduate psychology course, as a man or a woman, is that he would find out something about emotions." Time and again, I have observed people lecturing in public who, like this psychologist, seem to feel a mild discomfort with generic "he" and generic "man", and who therefore try to compensate, every once in a while, for their constant usage of such terms. After, say, five uses of "he" in describing a hypothetical scientist, they will throw in a meek "he or she" (and perhaps give an embarrassed little chuckle); then, having pacified their guilty conscience, they will go back to "he" and other sexist usages for a while, until the guilt-juices have built up enough again to trigger one more token nonsexist usage.

This is not progress, in my opinion. In fact, in some ways, it is retrograde motion, and damages the cause of nonsexist language. The problem is that these people are simultaneously showing that they recognize that "he" is *not* truly generic and yet continuing to use it as if it were. They are thereby, at one and the same time, increasing other people's recognition of the sham of considering "he" as a generic, and yet reinforcing the old convention of using it anyway. It's a bad bind.

In case anybody needs to be convinced that supposed generics such as "he" and "man" are *not* neutral in people's minds, they should reflect on the following findings. I quote from the chapter called "Who Is Man?" in *Words and Women,* an earlier book by Casey Miller and Kate Swift:

> In 1972 two sociologists at Drake University, Joseph Schneider and Sally Hacker, decided to test the hypothesis that *man* is generally understood to embrace *woman.* Some three hundred college students were asked to select from magazines and newspapers a variety of pictures that would appropriately illustrate the different chapters of a sociology textbook being prepared for publication. Half the students were assigned chapter headings like "Social Man", "Industrial Man", and "Political Man". The other half were given different but corresponding headings like "Society", "Industrial Life", and "Political Behavior". Analysis of the pictures selected revealed that in the minds of students of both sexes use of the word *man* evoked, to a statistically significant degree, images of males only—filtering out recognition of women's participation in these major areas of life—whereas the corresponding headings without *man* evoked images of both males and females. In some instances the differences reached magnitudes of 30 to 40 per cent. The authors concluded, "This is rather convincing evidence that when you use the word *man* generically, people do tend to think male, and tend not to think female."

Subsequent experiments along the same lines but involving schoolchildren rather than college students are then described by Miller and Swift. The results are much the same. No matter how generic "man" is claimed to be, there is a residual trace, a subliminal connotation of higher probability of being male than female.

*   *   *

Shortly after this column came out, I hit upon a way of describing one of the problems of sexist language. I call it the *slippery slope of sexism.* The idea is very simple. When a generic term and a "marked" term (*i.e.,* a sex-specific term) coincide, there is a possibility of mental blurring on the part of listeners and even on the part of the speaker. Some of the connotations of the generic will automatically rub off even when the specific is meant, and conversely. The example of "Industrial Man" illustrates one half of this statement, where a trace of male imagery rubs off even when no gender is intended. The reverse is an equally common phenomenon; an example would be when a newscaster speaks of "the four-man crew of next month's space shuttle flight". It may be that all four are actually males, in which case the usage would be precise. Or it may be that there is a woman among them, in which case "man" would be functioning generically (supposedly). But if you're just listening to the news, and you *don't know* whether a woman is among the four, what are you supposed to do?

Some listeners will automatically envision four males, but others, remembering the existence of female astronauts, will leave room in their minds for at least one woman potentially in the crew. Now, the newscaster

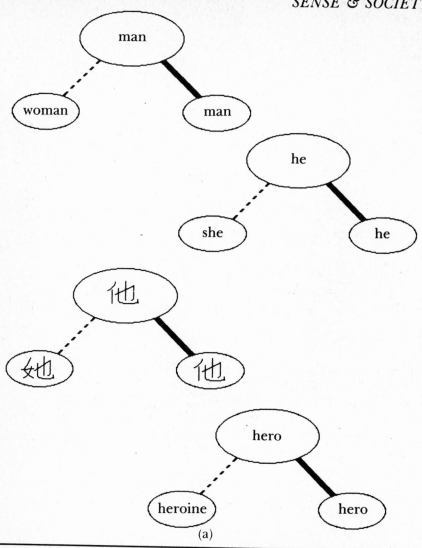

(a)

FIGURE 7–2. *The "slippery slope of sexism", illustrated. In each case in (a), a supposed generic (i.e., gender-neutral term) is shown above its two marked particularizations (i.e., gender-specific terms). However, the masculine and generic coincide, which fact is symbolized by the thick heavy line joining them—the slippery slope, along which connotations slosh back and forth, unimpeded. The "most-favored sex" status is thereby accorded the masculine term. In (b), the slippery slopes are replaced by true gender fairness, in which generics are unambiguously generic*

may know full well that this flight consists of males only. In fact, she may have chosen the phrase "four-man crew" quite deliberately, in order to let you know that no woman is included. For her, "man" may be marked. On the other hand, she may not have given it a second thought; for her, "man" may be unmarked. But how are you to know? The problem is right there: the slippery slope. Connotations slip back and forth very shiftily, and totally

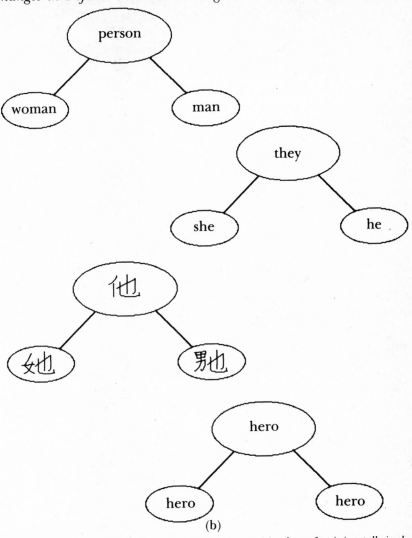

(b)

and marked terms unambiguously marked. Still, it is surprising how often it is totally irrelevant which sex is involved. Do we need—or want—to be able to say such things as, "Her actions were heroinic"? Who cares if a hero is male or female, as long as what they did is heroic? The same can be said about actors, sculptors, and a hostess of other terms. The best fix for that kind of slippery slope is simply to drop the marked term, making all three coincide in a felicitously ambisexual ménage à trois.

---

beneath our usual level of awareness—especially (though not exclusively) at the interface between two people whose usages differ.

Let me be a little more precise about the slippery slope. I have chosen a number of salient examples and put them in Figure 7-2. Each slippery slope involves a little triangle, at the apex of which is a supposed generic, and the bottom two corners of which consist of oppositely marked terms. Along one side of each triangle runs a diagonal line—the dreaded slippery slope itself. Along that line, connotations slosh back and forth freely in the minds of listeners and speakers and readers and writers. And it all happens at a completely unconscious level, in exactly the same way as a poet's choice of a word subliminally evokes dozens of subtle flavors without anyone's quite understanding how it happens. This wonderful fluid magic of poetry is not quite so wonderful when it imbues one word with all sorts of properties that it should not have.

The essence of the typical slippery slope is this: it establishes a firm "handshake" between the generic and the masculine, in such a way that the feminine term is left out in the cold. The masculine inherits the abstract power of the generic, and the generic inherits the power that comes with specific imagery. Here is an example of the *generic-benefits-from-specific* effect: "Man forging his destiny". Who can resist thinking of some kind of huge mythical brute of a guy hacking his way forward in a jungle or otherwise making progress? Does the image of a woman even come *close* to getting evoked? I seriously doubt it. And now for the converse, consider these gems: "Kennedy was a man for all seasons." "Feynman is the world's smartest man." "Only a man with powerful esthetic intuition could have created the general theory of relativity." "Few men have done more for science than Stephen Hawking." "Leopold and Loeb wanted to test the idea that a perfect crime might be committed by men of sufficient intelligence." Why "man" and "men", here? The answer is: to take advantage of the *specific-benefits-from-generic* effect. The power of the word "man" emanates largely from its close connection with the mythical "ideal man": Man the Thinker, Man the Mover, Man whose Best Friend is Dog.

*   *   *

Another way of looking at the slippery-slope effect is to focus on the single isolated corner of the triangle. At first it might seem as if it makes women somehow more distinguished. How nice! But in fact what it does is mark them as *odd.* They are considered nonstandard; the standard case is presumed not to be a woman. In other words, women have to fight their way back into imagery as just-plain *people.* Here are some examples to make the point.

When I learned French in school, the idea that masculine pronouns covered groups of mixed sex seemed perfectly natural, logical, and unremarkable to me. Much later, that usage came to seem very biased and bizarre to me. However, very recently, I was a bit surprised to catch myself falling into the same trap in different guise. I was perusing a multilingual dictionary, and noticed that instead of the usual *m.* and *f.* to indicate noun genders, they had opted for '+' and '−'. Which way, do you suspect? Right! And it seemed just right to me, too—until I realized how dumb I was being.

Heard on the radio news: "A woman motorist is being held after officials observed her to be driving erratically near the White House." Why say "*woman* motorist"? Would you say "man motorist" if it had been a male? Why is gender, and gender alone, such a crucial variable?

Think of the street sign that shows a man in silhouette walking across the street, intended to tell you "Pedestrian Crossing" in sign language. What if it were recognizably a *woman* walking across the street? Since it violates the standard default assumption that people have for people, it would immediately arouse a kind of suspicion: "Hmm . . . 'Women Crossing'? Is there a nunnery around here?" This would be the reaction not merely of dyed-in-the-wool sexists, but of anyone who grew up in our society, where women are portrayed—not deliberately or consciously, but ubiquitously and subliminally—as "exceptions".

If I write, "In the nineteenth century, the kings of nonsense were Edward Lear and Lewis Carroll", people will with no trouble get the message that those two men were the best of all nonsense writers at that time. But now consider what happens if I write, "The queen of twentieth-century nonsense is Gertrude Stein". The implication is unequivocal: Gertrude Stein is, among *female* writers of nonsense, the best. It leaves completely open her ranking relative to males. She might be way down the list! Now isn't this preposterous? Why is our language so asymmetric? This is hardly chivalry —it is utter condescension.

A remarkable and insidious slippery-slope phenomenon is what has happened recently to formerly all-women's colleges that were paired with formerly all-men's colleges, such as Pembroke and Brown, Radcliffe and Harvard, and so on. As the two merged, the women's school gradually faded out of the picture. Do men now go to Radcliffe or Pembroke or Douglass? Good God, no! But women are proud to go to Harvard and Brown and Rutgers. Sometimes, the women's college keeps some status within the larger unit, but that larger unit is always named after the men's college. In a weird twist on this theme, Stanford University has no sororities at all—but guess what kinds of people it now allows in its fraternities!

Another pernicious slippery slope has arisen quite recently. That is the one involving "gay" as both masculine and generic, and "Lesbian" as feminine. What is problematic here is that some people are very conscious of the problem, and refuse to use "gay" as a generic, replacing it with "gay or Lesbian" or "homosexual". (Thus there are many "Gay and Lesbian Associations".) Other people, however, have eagerly latched onto "gay" as a generic and use it freely that way, referring to "gay people", "gay men", "gay women", "gay rights", and so on. As a consequence, the word "gay" has a much broader flavor to it than does "Lesbian". What does "the San Francisco gay community" conjure up? Now replace "gay" by "Lesbian" and try it again. The former image probably is capable of flitting between that of both sexes and that of men only, while the latter is certainly restricted to women. The point is simply that men are made to seem standard, ordinary, somehow proper; women as special, deviant, exceptional. That is the essence of the slippery slope.

*   *   *

Part of the problem in sexism is how deeply ingrained it is. I have noticed a disturbing fact about my observation of language and related phenomena: whenever I encounter a particularly blatant example, I write it down joyfully, and say to friends, "I just heard a *great* example of sexism!" Now, why is it *good* to find a glaring example of something *bad*? Actually, the answer is very simple. You need outrageously clear examples if you want to convince many people that there is a problem worth taking at all seriously.

I was very fortunate to meet the philosopher and feminist Joan Straumanis shortly after my column on sexism appeared. We had a lot to talk over, and particularly enjoyed swapping stories of the sort that make you groan and say, "Isn't that *great?*"—meaning, of course, "How sickening!" Here's one that happened to her. Her husband was in her university office one day, and wanted to make a long-distance phone call. He dialed '0', and a female operator answered. She asked if he was a faculty member. He said no, and she said, "Only faculty members can make calls on these phones." He replied, "My wife is a faculty member. She's in the next room—I'll get her." The operator snapped back, "Oh, no—*wives* can't use these phones!"

Another true story that I got from Joan Straumanis, perhaps more provocative and fascinating, is this one. A group of parents arranged a tour of a hospital for a group of twenty children: ten boys and ten girls. At the end of the tour, hospital officials presented each child with a cap: doctors' caps for the boys, nurses' caps for the girls. The parents, outraged at this sexism, went to see the hospital administration. They were promised that in the future, this would be corrected. The next year, a similar tour was arranged, and at the end, the parents came by to pick up their children. What did they find, but the exact same thing—all the boys had on doctors' hats, all the girls had on nurses' hats! Steaming, they stormed up to the director's office and demanded an explanation. The director gently told them, "But it *was* totally different this year: we offered them all *whichever hat they wanted.*"

David Moser, ever an alert observer of the language around him, had tuned into a radio talk show one night, and heard an elderly woman voicing outrage at the mild sentence of two men who had murdered a three-year-old girl. The woman said, "Those two men should get the gas chamber for sure. I think it's terrible what they did! Who knows what that little girl could have grown up to become? Why, she could have been the mother of the next great composer!" The idea that that little girl might have grown up to *be* the next great composer undoubtedly never entered the woman's mind. Still, her remark was not consciously sexist and I find it strangely touching, reminiscent of a quieter era where gender roles were obvious and largely unquestioned, an era when many people felt safe and secure in their socially defined niches. But those times are gone, and we must now move ahead with consciousness raised high.

In one conversation I was in, a man connected with a publisher—let's call it "Freeperson"—said to me, "Aldrich was the liaison between the Freeperson boys and we—er, I mean *us.*" What amused me so much was his instant detection and correction of a *syntactic* error, yet no awareness of his more serious *semantic* error. Isn't that *great*?

*   *   *

I would not be being totally honest if I did not admit that occasionally, despite my apparent confidence in what I have been saying, I experience serious doubts about how deeply negative the impact of sexist language upon minds is. I must emphasize that I reject the Sapir-Whorf hypothesis about language molding perception and culture. I think the flow of causality is almost entirely in the other direction. And I am truly impressed with the plasticity of the human mind, with its ability to replace default assumptions at the drop of a hat with alternatives—even wildly unusual ones. People may assume that an unspecified orchestra conductor is male—but if they learn it is a woman, they immediately absorb that piece of knowledge without flinching. A barber I recently went to said to me, "They treated me like a king." This perhaps wouldn't surprise you—unless you knew that she was a woman. So why didn't she say "like a queen"? And David Moser reports that a woman he knows told him, "That family treated me just like a son!" Now why didn't she say "like a daughter"? I suppose it is because "treat someone like a king" and "treat someone like a son" are to some extent stock phrases in English, and despite their *apparent* sexism, perhaps they are actually quite neutral in their *deep* imagery. I am not saying I *know;* but I am saying I wonder, sometimes.

I also have to give pause to the following fact: Marina Yaguello, a professor of linguistics at the University of Paris and the author of the strongly feminist book *Les mots et les femmes* ("Words and Women"), an extended study of sexism in the French language, more recently wrote another book about general linguistics for the lay public, called *Alice au pays du langage* ("Alice in Language-Land"). In this book, Yaguello makes no effort to avoid all the sexist traps of the French language that she took so many pains to spell out in her previous book. To say "all people", she writes *tous les hommes* ("all men"); to refer to a generic young child, she says *le jeune enfant* (using the masculine article). Perhaps what flabbergasted me most was that when she wanted to refer to a female child, instead of writing *une enfant* (with "child" feminine, which is perfectly possible), she wrote *un enfant du sexe féminin*—"a child of the feminine sex", where "child" itself is masculine! If even a staunch feminist can reconcile herself to such blatantly sexist usages, feeling that there are deeper truths than what appears on the surface, I guess I have to sit back and think.

This does not prevent me from feeling that we live in a sexist society whose most accurate reflection is provided for us in our language, and from collecting specimens to document that sexism as clearly as possible. It seems to me that the state of our language provides a kind of barometer of the state of our society. Trying to change society through changing language may be a case of trying to get the tail to wag the dog, but one way of getting people to wake up to the problem is to point to language, a clearly observable phenomenon.

The nonsexist goal that I would advocate is not that every profession should consist of half males and half females. To tell the truth, I suspect that even if we reached such a balanced state some day, it would not be an equilibrium state—the percentages would slide. It is just very unlikely, it seems to me, that males and females are that symmetric. But that is not at all the point of a push towards sex-neutral language. The purpose of eliminating biases and preconceptions is to open the door wide for people of either sex in any line of work or play. *Symmetric opportunity,* not necessarily symmetric distribution, is the goal that we should seek.

*　　*　　*

I was provoked to write the following piece about a year after the column on sexism came out. It came about this way. One evening I had a very lively conversation at dinner with a group of people who thought of the problem of sexist language as no more than that: dinner-table conversation. Despite all the arguments I put forth, I just couldn't convince them there was anything worth taking seriously there. The next morning I woke up and heard two most interesting pieces of news on the radio: a black Miss America had been picked, and a black man was going to run for president. Both of these violated default assumptions, and it set my mind going along two parallel tracks at once: What if people's default assumptions were violated in all sorts of ways both sexually and racially? And then I started letting the default violations cross all sorts of lines, and pretty soon I was coming up with an image of a totally different society, one in which . . . Well, I'll just let you read it.

8

A Person Paper
on Purity in Language

by William Satire (alias Douglas R. Hofstadter)

September, 1983

IT'S high time someone blew the whistle on all the silly prattle about revamping our language to suit the purposes of certain political fanatics. You know what I'm talking about—those who accuse speakers of English of what they call "racism". This awkward neologism, constructed by analogy with the well-established term "sexism", does not sit well in the ears, if I may mix my metaphors. But let us grant that in our society there may be injustices here and there in the treatment of either race from time to time, and let us even grant these people their terms "racism" and "racist". How valid, however, are the claims of the self-proclaimed "black libbers", or "negrists"—those who would radically change our language in order to "liberate" us poor dupes from its supposed racist bias?

Most of the clamor, as you certainly know by now, revolves around the age-old usage of the noun "white" and words built from it, such as *chairwhite, mailwhite, repairwhite, clergywhite, middlewhite, Frenchwhite, forewhite, whitepower, whiteslaughter, oneupswhiteship, straw white, whitehandle,* and so on. The negrists claim that using the word "white", either on its own or as a component, to talk about *all* the members of the human species is somehow degrading to blacks and reinforces racism. Therefore the libbers propose that we substitute "person" everywhere where "white" now occurs. Sensitive speakers of our secretary tongue of course find this preposterous. There is great beauty to a phrase such as "All whites are created equal." Our forebosses who framed the Declaration of Independence well understood the poetry of our language. Think how ugly it would be to say "All persons are created equal.", or "All whites and blacks are created equal." Besides, as any schoolwhitey can tell you, such phrases are redundant. In most contexts, it is self-evident when "white" is being used in an inclusive sense, in which case it subsumes members of the darker race just as much as fairskins.

There is nothing denigrating to black people in being subsumed under

159

the rubric "white"—no more than under the rubric "person". After all, white is a mixture of all the colors of the rainbow, including black. Used inclusively, the word "white" has no connotations whatsoever of race. Yet many people are hung up on this point. A prime example is Abraham Moses, one of the more vocal spokeswhites for making such a shift. For years, Niss Moses, authoroon of the well-known negrist tracts *A Handbook of Nonracist Writing* and *Words and Blacks,* has had nothing better to do than go around the country making speeches advocating the downfall of "racist language" that ble objects to. But when you analyze bler objections, you find they all fall apart at the seams. Niss Moses says that words like "chairwhite" suggest to people—most especially impressionable young whiteys and blackeys— that all chairwhites belong to the white race. How absurd! It is quite obvious, for instance, that the chairwhite of the League of Black Voters is going to be a black, not a white. Nobody need think twice about it. As a matter of fact, the suffix "white" is usually not pronounced with a long 'i' as in the noun "white", but like "wit", as in the terms *saleswhite, freshwhite, penwhiteship, first basewhite,* and so on. It's just a simple and useful component in building race-neutral words.

But Niss Moses would have you sit up and start hollering "Racism!" In fact, Niss Moses sees evidence of racism under every stone. Ble has written a famous article, in which ble vehemently objects to the immortal and poetic words of the first white on the moon, Captain Nellie Strongarm. If you will recall, whis words were: "One small step for a white, a giant step for whitekind." This noble sentiment is anything but racist; it is simply a celebration of a glorious moment in the history of White.

Another of Niss Moses' shrill objections is to the age-old differentiation of whites from blacks by the third-person pronouns "whe" and "ble". Ble promotes an absurd notion: that what we really need in English is a single pronoun covering *both* races. Numerous suggestions have been made, such as "pe", "tey", and others. These are all repugnant to the nature of the English language, as the average white in the street will testify, even if whe has no linguistic training whatsoever. Then there are advocates of usages such as "whe or ble", "whis or bler", and so forth. This makes for monstrosities such as the sentence "When the next President takes office, whe or ble will have to choose whis or bler cabinet with great care, for whe or ble would not want to offend any minorities." Contrast this with the spare elegance of the normal way of putting it, and there is no question which way we ought to speak. There are, of course, some yapping black libbers who advocate writing "bl/whe" everywhere, which, aside from looking terrible, has no reasonable pronunciation. Shall we say "blooey" all the time when we simply mean "whe"? Who wants to sound like a white with a chronic sneeze?

*   *   *

One of the more hilarious suggestions made by the squawkers for this point of view is to abandon the natural distinction along racial lines, and to replace it with a highly unnatural one along sexual lines. One such suggestion—emanating, no doubt, from the mind of a madwhite—would have us say "he" for male whites (and blacks) and "she" for female whites (and blacks). Can you imagine the outrage with which sensible folk of either sex would greet this "modest proposal"?

Another suggestion is that the plural pronoun "they" be used in place of the inclusive "whe". This would turn the charming proverb "Whe who laughs last, laughs best" into the bizarre concoction "They who laughs last, laughs best". As if anyone in whis right mind could have thought that the original proverb applied only to the white race! No, we don't need a new pronoun to "liberate" our minds. That's the lazy white's way of solving the pseudo-problem of racism. In any case, it's ungrammatical. The pronoun "they" is a plural pronoun, and it grates on the civilized ear to hear it used to denote only one person. Such a usage, if adopted, would merely promote illiteracy and accelerate the already scandalously rapid nosedive of the average intelligence level in our society.

Niss Moses would have us totally revamp the English language to suit bler purposes. If, for instance, we are to substitute "person" for "white", where are we to stop? If we were to follow Niss Moses' ideas to their logical conclusion, we would have to conclude that ble would like to see small blackeys and whiteys playing the game of "Hangperson" and reading the story of "Snow Person and the Seven Dwarfs". And would ble have us rewrite history to say, "Don't shoot until you see the *persons* of their eyes!"? Will pundits and politicians henceforth issue *person* papers? Will we now have egg yolks and egg *persons*? And pledge allegiance to the good old Red, *Person,* and Blue? Will we sing, "I'm dreaming of a *person* Christmas"? Say of a frightened white, "Whe's *person* as a sheet!"? Lament the increase of *person*-collar crime? Thrill to the chirping of bob*persons* in our gardens? Ask a friend to *person* the table while we go visit the *persons'* room? Come off it, Niss Moses—don't personwash our language!

What conceivable harm is there in such beloved phrases as "No white is an island", "Dog is white's best friend", or "White's inhumanity to white"? Who would revise such classic book titles as Bronob Jacowski's *The Ascent of White* or Eric Steeple Bell's *Whites of Mathematics*? Did the poet who wrote "The best-laid plans of mice and whites gang aft agley" believe that blacks' plans gang *ne'er* agley? Surely not! Such phrases are simply metaphors; everyone can see beyond that. Whe who interprets them as reinforcing racism must have a perverse desire to feel oppressed. "Personhandling" the language is a habit that not only Niss Moses but quite a few others have taken up recently. For instance, Nrs. Delilah Buford has urged that we drop the useful distinction between "Niss" and "Nrs." (which, as everybody knows, is pronounced "Nissiz", the reason for which nobody knows!). Bler argument is that there is no need for the public to know whether a black is employed or not. *Need* is, of course, not the point. Ble conveniently side-steps the fact that there is a *tradition* in our society of calling unemployed blacks "Niss" and employed blacks "Nrs." Most blacks—in fact, the vast majority—prefer it that way. They *want* the world to know what their employment status is, and for good reason. Unemployed blacks want prospective employers to know they are available, without having to ask embarrassing questions. Likewise, employed blacks are proud of having found a job, and wish to let the world know they are employed. This distinction provides a sense of security to all involved, in that everyone knows where ble fits into the scheme of things.

But Nrs. Buford refuses to recognize this simple truth. Instead, ble shiftily turns the argument into one about whites, asking why it is that whites are universally addressed as "Master", without any differentiation between employed and unemployed ones. The answer, of course, is that in Anerica and other Northern societies, we set little store by the employment status of whites. Nrs. Buford can do little to change that reality, for it seems to be tied to innate biological differences between whites and blacks. Many white-years of research, in fact, have gone into trying to understand why it is that employment status matters so much to blacks, yet relatively little to whites. It is true that both races have a longer life expectancy if employed, but of course people often do not act so as to maximize their life expectancy. So far, it remains a mystery. In any case, whites and blacks clearly have different constitutional inclinations, and different goals in life. And so I say, *Vive na différence!*

*   *   *

As for Nrs. Buford's suggestion that both "Niss" and "Nrs." be unified into the single form of address "Ns." (supposed to rhyme with "fizz"), all I have to say is, it is arbitrary and clearly a thousand years ahead of its time. Mind you, this "Ns." is an abbreviation concocted out of thin air: it stands for absolutely nothing. Who ever heard of such toying with language? And while we're on this subject, have you yet run across the recently founded *Ns.* magazine, dedicated to the concerns of the "liberated black"? It's sure to attract the attention of a trendy band of black airheads for a little while, but serious blacks surely will see through its thin veneer of slick, glossy Madison Avenue approaches to life.

Nrs. Buford also finds it insultingly asymmetric that when a black is employed by a white, ble changes bler firmly name to whis firmly name. But what's so bad about that? Every firm's core consists of a boss (whis job is to make sure long-term policies are well charted out) and a secretary (bler job is to keep corporate affairs running smoothly on a day-to-day basis). They are both equally important and vital to the firm's success. No one disputes this. Beyond them there may of course be other firmly members. Now it's quite obvious that all members of a given firm should bear the same name—otherwise, what are you going to call the firm's products? And since it would be nonsense for the boss to change whis name, it falls to the secretary to change bler name. Logic, not racism, dictates this simple convention.

What puzzles me the most is when people cut off their noses to spite their faces. Such is the case with the time-honored colored suffixes "oon" and "roon", found in familiar words such as *ambassadroon, stewardoon,* and *sculptroon.* Most blacks find it natural and sensible to add those suffixes onto nouns such as "aviator" or "waiter". A black who flies an airplane may proudly proclaim, "I'm an aviatroon!" But it would sound silly, if not ridiculous, for a black to say of blerself, "I work as a waiter." On the other hand, who could object to my saying that the debonair Pidney Soitier is a great actroon, or that the hilarious Quill Bosby is a great comedioon? You guessed it—authoroons such as Niss Mildred Hempsley and Nrs. Charles White, both of whom angrily reject the appellation "authoroon", deep though its roots are in our language. Nrs. White, perhaps one of the finest poetoons of our day, for some reason insists on being known as a "poet". It leads one to wonder, is Nrs. White *ashamed* of being black, perhaps? I should hope not. White needs black, and black needs white, and neither race should feel ashamed.

Some extreme negrists object to being treated with politeness and courtesy by whites. For example, they reject the traditional notion of "Negroes first", preferring to open doors for themselves, claiming that having doors opened for them suggests implicitly that society considers them inferior. Well, would they have it the other way? Would these incorrigible grousers prefer to open doors for whites? What do blacks want?

* * *

Another unlikely word has recently become a subject of controversy: "blackey". This is, of course, the ordinary term for black children (including teen-agers), and by affectionate extension it is often applied to older blacks. Yet, incredible though it seems, many blacks—even teen-age blackeys—now claim to have had their "consciousness raised", and are voguishly skittish about being called "blackeys". Yet it's as old as the hills for blacks employed in the same office to refer to themselves as "the office blackeys". And for their boss to call them "my blackeys" helps make the ambiance more relaxed and comfy for all. It's hardly the mortal insult that libbers claim it to be. Fortunately, most blacks are sensible people and realize that mere words do not demean; they know it's how they are *used* that counts. Most of the time, calling a black—especially an older black—a "blackey" is a thoughtful way of complimenting bler, making bler feel young, fresh, and hireable again. Lord knows, I certainly wouldn't object if someone told me that I looked whiteyish these days!

Many young blackeys go through a stage of wishing they had been born white. Perhaps this is due to popular television shows like *Superwhite* and *Batwhite,* but it doesn't really matter. It is perfectly normal and healthy. Many of our most successful blacks were once tomwhiteys and feel no shame about it. Why should they? Frankly, I think tomwhiteys are often the cutest little blackeys—but that's just my opinion. In any case, Niss Moses (once again) raises a ruckus on this score, asking why we don't have a corresponding word for young whiteys who play blackeys' games and generally manifest a desire to be black. Well, Niss Moses, if this were a common phenomenon, we most assuredly *would* have such a word, but it just happens not to be. Who can say why? But given that tomwhiteys are a dime a dozen, it's nice to have a word for them. The lesson is that White must learn to fit language to reality; White cannot manipulate the world by manipulating mere words. An elementary lesson, to be sure, but for some reason Niss Moses and others of bler ilk resist learning it.

Shifting from the ridiculous to the sublime, let us consider the Holy Bible. The Good Book is of course the source of some of the most beautiful language and profound imagery to be found anywhere. And who is the central character of the Bible? I am sure I need hardly remind you; it is God. As everyone knows, Whe is male and white, and that is an indisputable fact. But have you heard the latest joke promulgated by tasteless negrists? It is said that one of them died and went to Heaven and then returned. What did ble report? "I have seen God, and guess what? Ble's female!" Can anyone say that this is not blasphemy of the highest order? It just goes to show that some people will stoop to any depths in order to shock. I have shared this "joke" with a number of friends of mine (including several blacks, by the way), and, to a white, they have agreed that it sickens them to the core to see Our Lord so shabbily mocked. Some things are just in bad taste, and there are no two ways about it. It is scum like this who are responsible for some of the great problems in our society today, I am sorry to say.

*   *   *

Well, all of this is just another skirmish in the age-old Battle of the Races, I guess, and we shouldn't take it too seriously. I am reminded of words spoken by the great British philosopher Alfred West Malehead in whis commencement address to my *alma secretaria,* the University of North Virginia: "To enrich the language of whites is, certainly, to enlarge the range of their ideas." I agree with this admirable sentiment wholeheartedly. I would merely point out to the overzealous that there are some extravagant notions about language that should be recognized for what they are: cheap attempts to let dogmatic, narrow minds enforce their views on the speakers lucky enough to have inherited the richest, most beautiful and flexible language on earth, a language whose traditions run back through the centuries to such deathless poets as Milton, Shakespeare, Wordsworth, Keats, Walt Whitwhite, and so many others . . . Our language owes an incalculable debt to these whites for their clarity of vision and expression, and if the shallow minds of bandwagon-jumping negrists succeed in destroying this precious heritage for all whites of good will, that will be, without any doubt, a truly female day in the history of Northern White.

Post Scriptum.

Perhaps this piece shocks you. It is meant to. The entire point of it is to use something that we find shocking as leverage to illustrate the fact that something that we usually close our eyes to is also very shocking. The most effective way I know to do so is to develop an extended analogy with something known as shocking and reprehensible. Racism is that thing, in this case. I am happy with this piece, despite—but also because of—its shock value. I think it makes its point better than any factual article could. As a friend of mine said, "It makes you so uncomfortable that you can't ignore it." I admit that rereading it makes even me, the author, uncomfortable!

Numerous friends have warned me that in publishing this piece I am taking a serious risk of earning myself a reputation as a terrible racist. I guess I cannot truly believe that anyone would see this piece that way. To misperceive it this way would be like calling someone a vicious racist for telling other people "The word 'nigger' is extremely offensive." If *allusions* to racism, especially for the purpose of satirizing racism and its cousins, are confused with racism itself, then I think it is time to stop writing.

Some people have asked me if to write this piece, I simply took a genuine William Safire column (appearing weekly in the *New York Times Magazine* under the title "On Language") and "fiddled" with it. That is far from the truth. For years I have collected examples of sexist language, and in order to produce this piece, I dipped into this collection, selected some of the choicest, and ordered them very carefully. "Translating" them into this alternate world was sometimes extremely difficult, and some words took weeks. The hardest terms of all, surprisingly enough, were "Niss", "Nrs.", and "Ns.", even though "Master" came immediately. The piece itself is not based on any particular article by William Safire, but Safire has without doubt been one of the most vocal opponents of nonsexist language reforms, and therefore merits being safired upon.

Interestingly, Master Safire has recently spoken out on sexism in whis column (August 5, 1984). Lamenting the inaccuracy of writing either "Mrs. Ferraro" or "Miss Ferraro" to designate the Democratic vice-presidential candidate whose husband's name is "Zaccaro", whe writes:

> It breaks my heart to suggest this, but the time has come for *Ms.* We are no longer faced with a theory, but a condition. It is unacceptable for journalists to dictate to a candidate that she call herself *Miss* or else use her married name;

FIGURE 8–1. *From a "Peggy Mills" comic strip,* circa *1930.*

it is equally unacceptable for a candidate to demand that newspapers print a blatant inaccuracy by applying a married honorific to a maiden name.

How disappointing it is when someone finally winds up doing the right thing but for the wrong reasons! In Safire's case, this shift was entirely for journalistic rather than humanistic reasons! It's as if Safire wished that women had never entered the political ring, so that the Grand Old Conventions of English—good enough for our grandfathers—would never have had to be challenged. How heartless of women! How heartbreaking the toll on our beautiful language!

*   *   *

A couple of weeks after I finished this piece, I ran into the book *The Nonsexist Communicator,* by Bobbye Sorrels. In it, there is a satire called "A Tale of Two Sexes", which is very interesting to compare with my "Person Paper". Whereas in mine, I slice the world orthogonally to the way it is actually sliced and then perform a mapping of worlds to establish a disorienting yet powerful new vision of our world, in hers, Ms. Sorrels simply reverses the two halves of our world as it is actually sliced. Her satire is therefore in some ways very much like mine, and in other ways extremely different. It should be read.

I do not know too many publications that discuss sexist language in depth. The finest I have come across are the aforementioned *Handbook of Nonsexist Writing,* by Casey Miller and Kate Swift; *Words and Women,* by the same authors; *Sexist Language: A Modern Philosophical Analysis,* edited by Mary Vetterling-Braggin; *The Nonsexist Communicator,* by Bobbye Sorrels; and a very good journal titled *Women and Language News,* from which the cartoon in Figure 8-1 was taken. Subscriptions are available at Centenary College of Louisiana, 2911 Centenary Boulevard, Shreveport, Louisiana 71104.

My feeling about nonsexist English is that it is like a foreign language that I am learning. I find that even after years of practice, I still have to translate sometimes from my native language, which is sexist English. I know of no human being who speaks Nonsexist as their native tongue. It will be very interesting to see if such people come to exist. If so, it will have taken a lot of work by a lot of people to reach that point.

One final footnote: My book *Gödel, Escher, Bach,* whose dialogues were the source of my very first trepidations about my own sexism, is now being translated into various languages, and to my delight, the Tortoise, a green-blooded male if ever there was one in English, is becoming *Madame Tortue* in French, *Signorina Tartaruga* in Italian, and so on. Full circle ahead!

Section III:

Sparking and Slipping

Section III:
Sparking and Slipping

The concern of the following five chapters is creativity: its wellsprings and its mechanizability. One of the most common metaphors for creativity is that of "spark": an electric leap of thought from one place to a remote one, without any apparent justification beforehand, but with all the justification in the world after the fact. Besides being used as a noun, "spark" is also used as a verb: one idea *sparks* another. Creative mental activity becomes, in this imagery, a set of sparks flying around in a space of concepts. Just how different is this metaphor for the mind from the reality of computers? They are filled with electricity rushing from one place to another at the most unimaginable speeds. Isn't that enough to turn the mechanical into the fluid? Or do computers still lack something ineffable? Are their mechanical attempts at thinking still too rigid, too dry? Is something liquid and slippery missing? My word for the elusive aspect of human thought still lacking in synthetic imitations is "slippability". Human thoughts have a way of slipping easily along certain conceptual dimensions into other thoughts, and resisting such slippage along other dimensions. A given idea has slightly different slippabilities—predispositions to slip—in each different human mind that it comes to live in. Yet some minds' slippabilities seem to give rise to what we consider genuine creativity, while others' do not. What is this precious gift? Is there a formula to the creative act? Can spark and slippability be canned and bottled? In fact, isn't that just what a human brain is—an encapsulated creativity machine? Or is there more to creativity and mind than can ever be encapsulated in any finite physical object or mathematical model?

9

Pattern, Poetry, and Power in the Music of Frédéric Chopin

April, 1982

THE abstract visual pattern in Figure 9-1 is a graphical representation of the opening of one of the most difficult and lyrical pieces for piano ever composed, namely the eleventh étude in Frédéric Chopin's Opus 25, written in about 1832, when he was in his early twenties. As a boy, I heard the Chopin études many times over on my parents' phonograph, and I quickly grew to love them. They became as familiar to me as the faces of my friends. Indeed, I cannot imagine who I would be if I did not know these pieces.

A few years later, as a teen-ager who enjoyed playing piano, I wanted to learn to play some of these old friends. I went to the local music store and found a complete volume of them. I will never forget my reaction on opening the book and looking for my friends. They were nowhere to be found! I saw nothing but masses of black notes and chords: complex, awesome visual patterns that I had never imagined. It was as if, expecting to meet old friends, I had instead found their skeletons grinning at me. It was terrifying. I closed the book and left, somewhat in shock.

I remember going back several times to that music store, each time pulled by the same curiosity tinged with fear. One day I worked up my courage and actually bought that book of études. I suppose I hoped that if I simply sat down at the piano and tried playing the notes I saw, I would hear my old friends, albeit a little slowly. Unfortunately, nothing of the kind happened. In general, I could not even play the two hands together comfortably, let alone recreate the sounds I knew so well. This left me disheartened and a little frightened at the realization of the awesome complexities I had taken for granted. You can look at it two ways. One way is to be amazed at how human perception can integrate a huge set of independent elements and "hear" only a single quality; the other is to be amazed at the incredible skill of a pianist who can play so many notes so quickly that they all blur into one shimmering mass, a "co-hear-ent" totality.

At first it was bewildering to see that "friends" had anatomies of such overwhelming complexity. But looking back, I don't know what I expected. Did I expect that a few simple chords could work the magic that I felt? No; if I had thought it over, I would have realized this was impossible. The only possible source of that magic was in some kind of complexity—patterned complexity, to be sure. And I think this experience taught me a lifelong lesson: that phenomena perceived to be magical are always the outcome of complex patterns of *non*magical activities taking place at a level below perception. More succinctly: The magic behind magic is pattern. The magic of life itself is a perfect example, emerging as it does out of patterned but lifeless activities at the molecular level. The magic of music emerges from complex, nonmagical—or should I say *meta*magical?—patterns of notes.

*   *   *

Having bought this volume, I felt drawn to it, wanted to explore it somehow. I decided that, hard work though it might be, I would learn an étude. I chose the one that was my current favorite—the one pictured in Figure 9-1—and set about memorizing the finger pattern in the right hand, together with the patterns that follow it, making up the first two pages or so. I played the pattern literally thousands of times, and gradually it became natural to my fingers, although never as natural as it had always sounded to my ears—or rather, to my *mind.*

It was then that I first observed the amazing subtlety of the lightning flash of the right hand, how it is composed of two alternating and utterly different components: the odd-numbered notes (in red) trace out a perfect descending chromatic scale for four octaves, while the even-numbered notes (in black), wedged between them like pickets between the spaces in a picket fence, dictate an arpeggio with repeated notes. To execute this alternating pattern, the right hand flutters down the keyboard, tilting from side to side like a swift in flight, its wings beating alternately.

A word of explanation. On a piano there are twelve notes (some black, some white) from any note to the corresponding note one octave away. Playing them all in order creates a *chromatic scale,* as contrasted with the more familiar diatonic scales (usually major or minor). These latter involve only seven notes apiece (the eighth note being the octave itself). The seven intervals between the successive notes of a diatonic scale are not all equal. Some are twice as large as others, yet to the ear there is a perfect intuitive logic to it. Rather paradoxically, in fact, most people can sing a major scale without any trouble, uneven intervals notwithstanding, but few can sing a chromatic scale accurately, even though it "ought" to be much more straightforward—or so it would seem, since all its intervals are exactly the same size. The chromatic scale is so called because the extra notes it introduces to fill up the gaps in a diatonic scale have a special kind of "bite" or sharpness to them that adds color or piquancy to a piece. For that reason, a piece filled with notes other than the seven notes belonging to the key it is in is said to be chromatic.

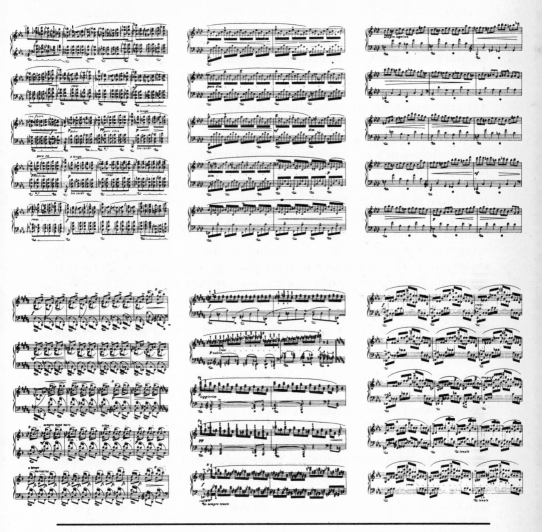

FIGURE 9–2. *The strikingly different visual textures of six Chopin études. On top, Op. 10, No. 11, in E-flat major; Op. 25, No. 1, in A-flat major; and Op. 25, No. 2, in F minor. Below, Op. 25, No. 3, in F major; Op. 25, No. 6, in G-sharp minor; and Op. 25, No. 12, in C minor.* [*From the G. Schirmer (Friedheim) edition.*]

An *arpeggio* is a broken chord played one or more times in a row, moving up or down the keyboard. Thus it bears a resemblance to a spread-out scale, a little like someone bounding up a staircase three or four steps at a time. Chopin's music is filled with both arpeggios and chromatic passages, but the intricate fusion of these two opposite structural elements in the eleventh étude struck me as a masterpiece of ingenuity. And what is amazing is how it is perceived when the piece moves quickly. The chromatic scale comes through loud and clear, forming a smooth "envelope" of the pattern (your eye picks it out too), but the arpeggio blurs into a kind of harmonic fog that deeply affects one's perception, if only subliminally, or so it seems at least to the untrained ear.

Each étude in that book I bought has a characteristic appearance, a *visual texture* (see Figure 9-2). This was one of the most striking things about the book at first. I was not at all accustomed to the idea of written music as texture; the simple pieces I had played up to that time were slow, so that every note was distinctly heard. In other words, the pieces in my playing experience were coarse-grained compared with the fine grain of a Chopin étude, where notes often go by in a blur and are merely parts of an auditory gestalt. Conversion of this kind of auditory experience to notated music sheets often yields quite stunning textures and patterns. Each composer has a characteristic set of patterns the eye becomes familiar with, and these études provided for me a stunning realization of that fact.

* *   * *   * *

Sadly, I was forced to abandon étude Op. 25, No. 11, after having learned only a little more than a page—it was simply too hard for me. James Huneker, an American critic and one of Chopin's earliest English-language biographers, wrote of this study: "Small-souled men, no matter how agile their fingers, should not attempt it." Well, whatever the size of my soul, my fingers were not agile enough. For a while, that discouraged me from attacking any more Chopin études at all. A few years later, though, when I was working more earnestly on improving my modest piano skills, I came across an isolated Chopin étude in a book of medium-difficult selections. It turned out to be one of three études he had composed later in life, none of which had been on my parents' records. This was a real find! Luckily its texture looked less prickly, its pace less forbidding. Somewhat gingerly, I played through it very slowly and discovered that it was astonishingly beautiful and not as inaccessible as the others I'd tried.

Like all the rest of Chopin's studies, this one is centered on a particular technical point, although to think of the études primarily in that way is like thinking of the fantastic gymnastic performances of Nadia Comaneci as merely fancy fitness exercises. Louis Ehlert, a nineteenth-century musicologist, wrote of one of the most beautiful études in Opus 25 (the sixth one, in G-sharp minor): "Chopin not only versifies an exercise in thirds; he transforms it into such a work of art that in studying it one could sooner fancy oneself on Parnassus than at a lesson. He deprives every passage of all mechanical appearance by promoting it to become the embodiment of a beautiful thought, which in turn finds graceful expression in its motion." Similar words apply to this easier, posthumously published étude in A-flat major, whose chief technical concern is the concept of *three against two*, a special case of the general concept of *polyrhythm*.

Mathematically, the concept is simple enough: play two musical lines simultaneously, one of them sounding three notes to the other's two. Usually the triplet and doublet are aligned so that they start at the same instant. When they are both plotted on a unit interval (see Figure 9-3*a*), you can see that the doublet's second note is struck halfway between the triplet's second and third notes. Of course, this is simply a pictorial representation of the fact that 1/2 is the arithmetic mean of 1/3 and 2/3.

In theory, two voices playing a three-against-two pattern need not be perfectly aligned. If you shift the upper voice by, say, 1/12 to the right, you get a different picture (see Figure 9-3*b*). Here the triplet's third note starts halfway through the doublet's second. As you can see, the triplet extends beyond the end of the interval, presumably to join onto another identical

---

FIGURE 9–3. *The 3-against-2 phenomenon. In (a), as it is usually heard, with both voices "in phase". In (b), one voice is shifted by 1/12 with respect to the other, producing a quite unusual pattern of beats. In (c), it is shown how in principle the relative staggering of the two voices could be adjusted continuously by a knob arrangement.*

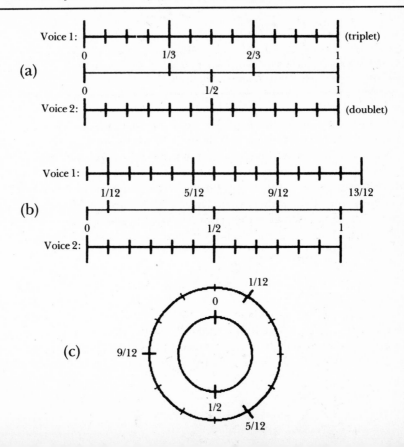

pattern. We can fold the pattern around and represent its periodicity in a circle, as is shown in Figure 9-3c. By rotating either of the concentric circles like a knob, we get all possible ways of hearing three beats against two. In Chopin and most other Western music, however, the only possibility that I have seen explored is where the triplet and doublet are perfectly "in phase".

At first I found the three-against-two rhythm hard to perform exactly. One has to learn how to hear the voices separately, to hear the roundish lilt of the three-rhythm weaving itself into the square mesh of the two-rhythm. Of course, it's easy to hear when someone else is playing; the trick is to hear it in one's own playing! In principle the task is not hard, but it is one of coordination, and requires practice. I found that once I had mastered the problem of playing the two rhythms evenly and independently, I could play the whole étude. To play it—or to hear it—is like smiling through tears, it is so beautiful and sad at the same time.

It is impossible to pinpoint the source of the beauty, needless to say, but it is certainly due in part to the way the chords in the right hand flow into one another. (See Figure 9-4.) Almost all the way through the piece, the

FIGURE 9–4. *The opening two measures of the posthumous étude in A-flat major, showing its typical 3-against-2 pattern with slowly shifting chords in the right hand.* [*Music printed by Donald Byrd's SMUT program at Indiana University.*]

right hand plays three-note *chords* (six to a measure) against *single notes* by the left hand (four to a measure). The delicacy of the piece comes from the fact that very often, when one chord flows into the next one, only a single note changes. And to add to the subtlety of this slowly shifting sound-pattern, usually the steps taken by the shifting voice are single scale-steps rather than wide jumps. These "rules" do not hold all the way, of course; there are numerous exceptions. Nevertheless, there is a uniform aural texture to the piece that imbues it with its soft melancholy, known in Polish as *tęsknota*.

*   *   *   *   *   *

It is interesting to speculate about the extent to which such formal considerations occurred to Chopin while he was composing. It is well known that Chopin revered Bach's music. "Always play Bach" was his advice to a

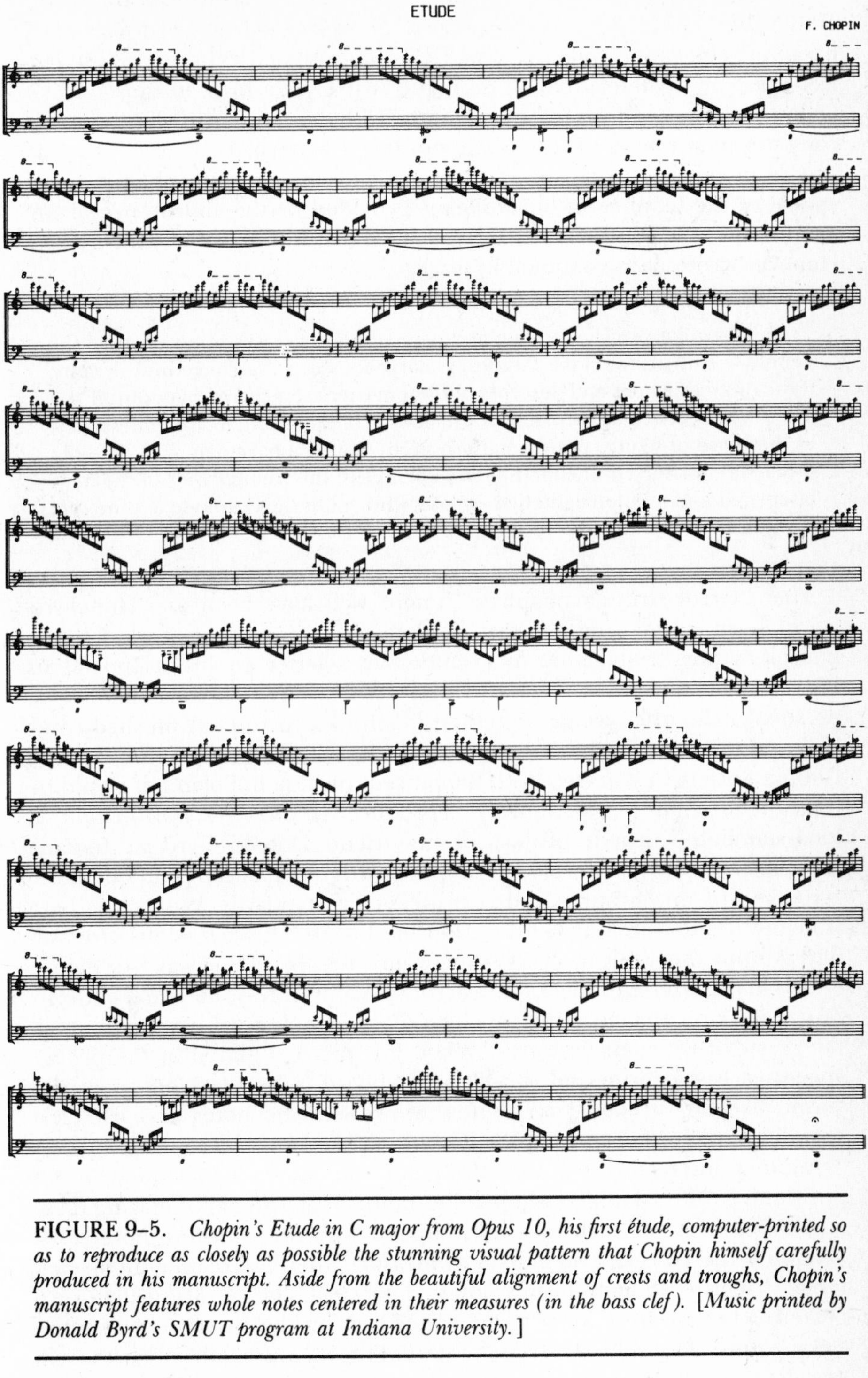

FIGURE 9–5. *Chopin's Etude in C major from Opus 10, his first étude, computer-printed so as to reproduce as closely as possible the stunning visual pattern that Chopin himself carefully produced in his manuscript. Aside from the beautiful alignment of crests and troughs, Chopin's manuscript features whole notes centered in their measures (in the bass clef). [Music printed by Donald Byrd's SMUT program at Indiana University.]*

pupil, and he was particularly devoted to the Well-Tempered Clavier, a paragon of elegant formal structures. Chopin confided to his friend Eugène Delacroix, the painter, that "The fugue is like pure logic in music... To know the fugue deeply is to be acquainted with the element of all reason and consistency in music." Clearly, Chopin loved pattern.

A stunning demonstration of Chopin's extreme awareness of the visual appeal of the textures in his études is provided by the appearance of the manuscript of his étude Op. 10, No. 1, in C major, one about which James Huneker wrote, in his inimitable prose:

> The irregular black ascending and descending staircases of notes strike the neophyte with terror. Like Piranesi's marvellous aerial architectural dreams, these dizzy acclivities and descents of Chopin exercise a charm, hypnotic, if you will, for eye as well as ear. Here is the new technique in all its nakedness, new in the sense of figure, design, pattern, web, new in a harmonic way. The old order was horrified at the modulatory harshness; the young sprigs of the new, fascinated and a little frightened. A man who could thus explode a mine that assailed the stars must be reckoned with.

That "terror-stricken neophyte" might well have been me. Huneker's words form an amusing contrast with what the nineteen-year-old Chopin himself wrote of this, his first étude, in a letter to his friend Tytus Woyciechowski in 1829: "I have written a large exercise in form, in my own personal style; when we get together, I'll show it to you." A finished copy, believed to be in Chopin's hand, is now in the Museum of the Frédéric Chopin Society in Warsaw. With the present turmoil in Poland, it would be difficult to gain permission to reproduce it directly. Fortunately, a long-standing research project of my friend Donald Byrd at Indiana University has been to develop a computer program that can print out music according to specification, and at professional standards. With some help from our friend Adrienne Gnidec, Don and I coaxed his marvelous program into printing the music in a very strange and visually striking way (see Figure 9-5). This figure reproduces quite accurately the large-scale visual patterns of Chopin's own manuscript, in which Chopin took great care to align all the crests of the massive waves. When this piece is played at the proper speed, each sweep up and down the keyboard is heard as one powerful surge, like the stroke of an eagle's wing, with the notes of each crest sparkling brilliantly like wingtips flashing in the sun.

Another interesting feature of Chopin's notation, here copied, is his positioning of the doubled whole notes in the bass. Instead of placing them at the very start of each measure, aligned with the sixteenth-note rests, Chopin centered each one in its own measure, thereby creating an elegant visual balance, though losing some notational clarity. Musically, such centering has no effect. Since a whole note lasts for the duration of an entire 4/4 measure, it must be struck at the start of the measure, otherwise it would overflow into the next measure, and that is impossible. (Or rather, it would violate a much more rigid convention of music notation—namely, that no note can designate a sound that overflows the boundaries of its measure.) Hence the only possible interpretation is that the whole note is to be struck at the outset. In other words, the centering is simply a charming artistic touch with a quaint nineteenth-century flavor, like the ornaments on a Victorian house. The modern music-reading eye is used to more functional notation; in particular, it expects the staff to be in essence a graph of the sound, in which the horizontal axis is time. Thus notes struck simultaneously are expected to line up vertically.

But let us return to the matter of Chopin's preoccupation with form and structure. Few composers of the romantic era have penned such visually patterned pages, have spun a whole cloth out of a single textural idea. With Chopin, though, preoccupation with strict pattern never took precedence over the expression of heartfelt emotions. One must distinguish, it seems to me, between "head pattern" and "heart pattern", or, in more objective-sounding terms, between *syntactic* pattern and *semantic* pattern. The notion of a syntactic pattern in music corresponds to the formal structural devices used in poetry: alliteration, rhyme, meter, repetition of sounds, and so on. The notion of a semantic pattern is analogous to the pattern or logic that underlies a poem and gives it reason to exist: the inspiration, in short.

That there are such semantic patterns in music is as undeniable as that there are courses in the theory of harmony. Yet harmony theory has no more succeeded in explaining such patterns than any set of rules has yet succeeded in capturing the essence of artistic creativity. To be sure, there are words to describe well-formed patterns and progressions, but no theory yet invented has even come close to creating a semantic sieve so fine as to let all bad compositions fall through and to retain all good ones. Theories of musical quality are still descriptive and not generative; to some extent, they can explain in hindsight why a piece seems good, but they are not sufficient to allow someone to create new pieces of quality and interest. It is nonetheless fascinating, if not downright compelling, to try to find certain earmarks of greatness, to try to understand why it is that one composer's music can reach in and touch your innermost core while another composer's music leaves you cold and unmoved. It is a mystery.

*   *   *   *   *   *

After learning the posthumous A-flat étude, I felt encouraged to tackle some of the others. One of the ones I had loved the most was Op. 25, No. 2, in F minor. To me, it was a soft, rushing whisper of notes, a fluttering, like the leaves of a quaking aspen in a gentle breeze. Yet it was not just a scene of nature; it expressed a human longing, a melancholy infused with strange and wild yearnings for something unknown and remote—*tęsknota* again. I knew this melody inside out from many years of hearing it, and I looked forward to transferring it to my fingers.

After a couple of months' practice, my fingers had built up enough stamina to play the piece fairly evenly and softly. This was very satisfying to me until one day, an acquaintance for whom I was playing it commented, "But you're playing it in *twos*—it's supposed to be in *threes*!" What she meant by this was that I was stressing every second note, rather than every third. Bewildered, I looked at the score, and of course, as she had pointed out, the melody was written in triplets. But surely Chopin had not meant it to be played in threes. After all, I knew the melody perfectly! Or did I? I tried playing it in threes. It sounded strange and unfamiliar, a perceptual distortion the like of which I had never experienced.

I went home and took out my parents' old Remington LP of the Chopin études Opus 25 (played by a wonderful but hardly remembered pianist named Alexander Jenner). I put on the F minor étude and tried to hear which way he played it. I found I could hear it *either* way. Jenner had played it so smoothly, so free of accent (as they say Chopin did, by the way), that one really could not tell which way to hear it. All of a sudden I saw that I really knew *two* melodies composed of the exact same sequence of notes! I felt myself to be very fortunate, because now I could experience this familiar old melody in a fresh new way. It was like falling in love with the same person twice.

I had to practice hard to undo the bad habits of "biplicity" and to replace them with the indicated "triplicity", but it was a delight. The hardest part, however, was combining the two hands. With *duplets* in the right hand, this had presented no problem; all the accented notes fell in coincidence with notes in the left hand, moving at exactly half the speed of the right hand in a pattern of wide arpeggios. But if I were to spread my accents thinner, so that I accented only every *third* note of the right hand, then many of the notes in the left hand would be struck simultaneously with weak notes in the right. This may sound simple enough, but I found it very tricky. The difference is shown in Figure 9-6 (which, like most of the others in this article, was created by Don Byrd's program).

FIGURE 9–6.   *The opening of Etude Op. 25, No. 2, printed in two ways. In (a), as Chopin penned it, and as it is usually conceived: in threes. In (b), as I first heard it and first learned to play it: in twos. [Music printed by Donald Byrd's SMUT program at Indiana University.]*

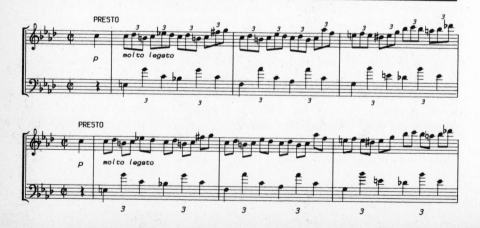

Even after mastering the right-hand solo in triplets, I found that when I put the parts together, it was at first nearly impossible to keep from softly accenting the melodic notes coinciding with the bass. It was a fearsome task of coordination, yet I enjoyed it greatly. After a while something just "snapped into place", and I found I was doing it. It was not something I could consciously control or explain; I simply was playing it right, all of a sudden. Huneker, in his commentary on this étude, quotes Theodor Kullak, another Chopin specialist, about the "algebraic character of the tone-language" and then adds his own image: "At times so delicate is its design that it recalls the faint fantastic tracery made by frost on glass."

Chopin's music is filled to the brim with such "algebraic" tricks of cross-rhythm. He seemed to revel in them in a way that no previous composer ever had. A famous example is his iconoclastic waltz, Opus 42 in A-flat major, written in 1840. In this waltz, the bass line follows the usual "oom-pah-pah" convention, but the melody of the first section completely counters this three-ness; its six eighth-notes, instead of being broken up into *three pairs* aligned with the left hand's bounces, form *two triplets,* as in the F minor étude just discussed (see Figure 9-7). Here, though, in contrast to the nearly accentless shimmering desired in that étude, the initial notes of successive triplets are to be clearly emphasized and prolonged, thus creating a higher-level melody (shown in red) abstracted out of the quietly rippling right hand. This melody is composed of *two* notes per measure, beating regularly against the *three* notes of the waltzing bass. It is a marvelous *trompe-l'oreille* effect, one that Chopin exploited again in his E major scherzo, Opus 54, written in 1842, when he was 32.

* * * * * *

In that same year, Chopin wrote what some admirers consider to be his greatest work: the fourth Ballade, in F minor. This piece is filled with noteworthy passages, but one in particular had a profound effect on me. One day, long after I knew the piece intimately from recordings, a friend told me that he had been practicing it and wanted to show me "a bit of tricky polyrhythm" that was particularly interesting. I was actually not that interested in hearing about polyrhythm at the moment, and so I didn't pay much attention when he sat down at the keyboard. Then he started to play. He played just two measures, but by the time they were over, I felt that someone had reached into the very center of my skull and caused something to explode deep down inside. This "bit of tricky polyrhythm" had undone me completely. What in the world was going on?

Of course, it was much more than just polyrhythm, but that is part of it. As you can see in our three-color plot of the two measures concerned (Figure 9-8), the left hand forms large, rumbling waves of sound, like deep ocean waves on which a ship is sailing. Each wave consists of six notes, forming a rising and falling arpeggio (in blue). High above these billows of sound, a lyrical melody (in red) soars and floats, emerging out of a blur of notes swirling around it like a halo (in black). This high melody and its halo are actually fused together in the right hand's eighteen notes per measure. They are written as six groups of three, so that in each half-measure, nine high notes beat against the six-note ocean wave below—already a clear problem in three-against-two. But look: on top of those flying triplets, there are eight-note flags placed on every *fourth* note! Thus there is a flag on the first note of the first triplet, on the second note of the second triplet, on the third note of the third triplet, on the fourth note of the fourth triplet... Well, that cannot be. In fact, the fourth triplet has no flag at all; the flag goes to the first note of the *fifth* triplet, and the pattern resumes. Flags waving in wind, high on the masts of a sea-borne sailing ship.

This wonderfully subtle rhythmic construction might—just might—have been invented by anyone, say by a rhythm specialist with no feeling for melody. And yet it was not. It was invented by a composer with a supreme gift for melody and harmony as well as for rhythm, and this can be no coincidence. A mere "rhythms hacker" would not have the sense to know what to do with this particular rhythm any more than with any other rhythmic structure. There is something about this passage that shows true genius, but words alone cannot define it. You have to hear it. It is a burning lyricism, having a power and intensity that defy description.

One must wonder about the soul of a man who at age 32 could write such possessed music—a man who at the tender age of nineteen could write such perfectly controlled and poetic outbursts as the études of Opus 10. Where could this rare combination of power, poetry, and pattern, this musical self-confidence and maturity, have come from?

*　　*　　*　　　*　　*　　*

In search of an answer, one must look to Chopin's roots, both his family roots and his roots in his native land, Poland. Chopin was born in a small and peaceful country village 30 miles west of Warsaw called Żelazowa Wola, which means Iron Will. His father, Nicolas (Mikołaj) Chopin, was French by birth but emigrated to Poland and became an ardent Polish patriot (so ardent, in fact, that he participated in the celebrated but ill-fated insurrection led by the national hero Jan Kiliński in 1794 against the Russian occupation of Warsaw). Chopin's mother, Justyna Krzyżanowska, was a distant relative of the rich and aristocratic Skarbek family, who lived in Żelazowa Wola. She lived with them as a family member and took care of various domestic matters. When Mikołaj Chopin came to be the tutor of the Skarbek children, he and Justyna met and married. In addition to being a gentle and loving mother, she was as fervent a Polish patriot as her husband, and had a romantic and dreamy streak. They had four children, of whom Frédéric, born in 1810, was the second. The other three children were girls, one of whom died young, of tuberculosis—a disease that in the end would claim Frédéric as well, at age 39. The four children doted on one another. It was a close-knit family, and all in all, Chopin had a very happy childhood.

The family moved to Warsaw when Frédéric was very young, and there he was exposed to culture of all kinds, since his father was a teacher and knew university people of all disciplines. Frédéric was a fun-loving and spirited boy. The summer he was fourteen he spent away from home in a lilac-filled village called Szafarnia. He wrote home a series of letters gleefully mocking the style of the *Warsaw Courier,* a gossipy provincial paper of the times. One item from his *"Szafarnia Courier"* ran as follows (in full):

> The Esteemed Mr. Pichon [an anagram of "Chopin"] was in Golub on the 26th of the current month. Among other foreign wonders and oddities, he came across a foreign Pig, which Pig quite specially attracted the attention of this most distinguished Voyageur.

Chopin's musical talent, something he shared with his mother, emerged very early and was nurtured by two excellent piano teachers, first by a gentle and good-humored old Czech named Wojciech Żywny, and later by the director of the Warsaw Conservatory, Józef Elsner.

Chopin grew up in the capital city of the "Grand Duchy of Warsaw"— what little remained of Poland after it had been decimated, in three successive "partitions" in the late eighteenth century, by its greedy neighbors: Russia, Prussia, and Austria. The turn of the century was marked by a mounting nationalistic fervor; in Warsaw and Cracow, the two main Polish cities, there occurred a series of rebellions against the foreign occupiers, but to no avail. A number of ardent Polish nationalists went abroad and formed "Polish Legions" whose purpose was to fight for the liberation of all oppressed peoples and to eventually return to Poland and reclaim it from the occupying powers. When Napoleon invaded Russia in 1806, a Polish state was established for a brief shining instant; then all was lost again. The Polish nation's flame flickered and nearly went out totally, but as the words to the Polish national anthem proclaim, *"Jeszcze Polska nie zginęła, póki my żyjemy."* It is a curious sentence, built out of past and present tenses, and literally translated it runs: "Poland has not yet perished, as long as we live." The first clause sounds so fatalistic, as if to admit that Poland surely *will* someday perish, but not quite yet! Some Poles tell me that the connotations are not that despairing, that a better overall translation would be, "Poland will not perish, as long as we live." Others, though, tell me that the construction is subtly ambiguous, that its meaning floats somewhere between grim fatalism and ardent determination.

* * *   * * *

The Poles are a people who have learned to distinguish sharply between two conceptions of Poland: Poland the abstract social entity, at whose core are the Polish language and culture, and Poland the concrete geographical entity, the land that Poles live in. *Naród polski*—the "Polish nation"—represents a *spirit* rather than a piece of territory, although of course the nation came into existence because of the bonds between people who lived in a certain region. It is the fragility of this flickering flame, and the determination to keep it alive, that Chopin's music reflects so purely and poignantly. There is a certain fusion of bitterness, anger, and sadness called *żal* that is uniquely Polish. One hears it, to be sure, in the famous mazurkas and polonaises, pieces that Chopin composed in the form of national dances. The mazurkas are mostly smaller pieces based on folk-like tunes with a lilting 3/4 rhythm; the polonaises are grand, heroic, and martial in spirit. But one hears this burning flame of Poland just as much in many of Chopin's other pieces—for example, in the slow middle sections of such pieces as the waltzes in A minor (Op. 34, No. 2) and A-flat major (Op. 64, No. 3), the pathos-filled Prelude in F-sharp major (Op. 28, No. 13), and particularly in the middle part of the F-sharp minor Polonaise (Opus 44), where a ray of hope bursts through dark visions like a gleam in the gloom. One hears *żal* in the angry, buzzing harmonies of the étude in C-sharp minor (Op. 10, No. 4) and in the passion of the étude in E major (Op. 10, No. 3). In fact, Chopin is said to have cried out once, on hearing this piece played in his presence, *"O ma patrie!"* ("O my homeland!").

But aside from the fervent patriotism of Chopin's music there is in it that different and softer kind of Polish nostalgia: *tęsknota*. It is his yearning for home—for his childhood home, for his family, for a dream-Poland that at age twenty he had left forever. In 1830, at the height of the turmoil in Warsaw, Chopin set out for France. He had a premonition that he would never return. Traveling by way of Vienna, he made slow progress. When things boiled over in late 1831—when, in September 1831, the Russians finally crushed the desperate Warsaw insurrection—Chopin was in Stuttgart. On hearing the news, he was overwhelmed with agitation and grief, partly out of fear for the fate of his family, partly out of love for his stricken homeland. He wavered about going back to Poland and fighting for his nation, but the idea eventually receded from his mind.

It was at about this time that he composed the twelfth and final étude of his Opus 10. Of this étude, Chopin's Polish biographer Maurycy Karasowski wrote:

> Grief, anxiety, and despair over the fate of his relatives and his dearly beloved father filled the measure of his sufferings. Under the influence of this mood he wrote the C minor étude, called by many the "Revolutionary Etude". Out of the mad and tempestuous storm of passages for the left hand the melody rises aloft, now passionate and anon proudly majestic, until thrills of awe stream over the listener, and the image is evoked of Zeus hurling thunderbolts at the world.

This is pretty strong language. Huneker echoes these sentiments, as does the French pianist Alfred Cortot, who in his famous Student's Edition of the études refers to the piece as "an exalted outcry of revolt . . . . wherein the emotions of a whole race of people are alive and throbbing." I myself have never found this étude as overwhelming as these authors do, although it is unquestionably a powerful outburst of emotion. If someone had told me that one of the études had come to be known as the "Revolutionary Etude" and had asked me to guess which one, I would certainly have picked one of the last two of Opus 25, either No. 11 in A minor, the one pictured at the beginning of this article, with its tumultuous cascades of notes in the right hand against the surging, heroic melody in the left hand, or else No. 12 in C minor, which sounds to me like a glowing inferno seen at night from far away, flaring up unpredictably and awesomely. As for the actual "Revolutionary Etude", I have always found its ending enigmatic, fluctuating as it does between major and minor, between the keys of F and C, like an indecisive thunderclap.

Still, this piece, like the martial A-flat major Polonaise (Opus 53), has become a symbol of the tragic yet heroic Polish fate. Wherever and whenever it is played, it is special to Poles; their hearts beat faster, and their spirits cannot fail to be deeply moved. I will never forget how I heard it nightly as the clarion call of Poland, when, from a small town in Germany in 1975, I would try to tune in Radio Warsaw. Two measures of shrill, rousing chords above a roaring left hand, like a call to arms, were repeated over and over again as the call signal, preceding a nightly broadcast of Chopin's music. Nor will I ever forget how that feeble signal of Radio Warsaw faded in and out, symbolizing to me the flickering flame of Poland's spirit.

* * *   * * *

However one chooses to describe it—whether in terms of *żal* and *tęsknota*, or *patriotyzm* and polyrhythm, or chromaticism and arpeggios—Chopin's music has had a deep influence on the composers of succeeding generations. It is perhaps most visible in the piano music of Alexander Scriabin, Sergei Rachmaninoff, Gabriel Fauré, Felix Mendelssohn, Robert and Clara Schumann, Johannes Brahms, Maurice Ravel, and Claude Debussy, but Chopin's influence is far more pervasive than even that would suggest. It has become one of the central pillars of Western music, and as such it has its effect on the music perceived and created by everyone in the Western world.

In one way, Chopin's music is purely Polish, and that Polishness— *polskność*—extends even to foreign-inspired pieces such as his Bolero, Tarantella, Barcarolle, and so on. In another way, though, Chopin's music is universal, so that even his most deeply Polish pieces—the mazurkas and polonaises—speak to a common set of emotions in everyone. But what *are* these emotions? How are they so deeply evoked by mere pattern? What is the secret magic of Chopin? I know of no more burning question.

*   *

Post Scriptum.

This column is a unique one, in that it expresses certain kinds of emotions that are not expressed as directly in my other published writings. But the part of me represented by it is no smaller and no less important than the part of me from which my other writings flow. It was provoked, of course, by the worsening crisis in Poland in late 1981, just at the time of the takeover by the military and the tragic collapse of Solidarity. In fact, it was almost exactly 150 years after the tragic takeover of Warsaw by the Russians that triggered the Revolutionary Etude. I guess Poland has not yet perished— but it is certainly going through terrible tribulations, once again.

I received some heart-warming correspondence in response to this column. One letter, from Andrzej Krasiński, a Pole living in West Germany, ran this way:

> I just read your nice article about Chopin's music in the April issue of *Scientific American* in which you have shown so much sympathy and understanding for a Polish soul, and so much care for the Polish language. I enjoyed it a lot, although I am no expert in music. However, by my birth, I happen to be an expert in the Polish language, and I wish to point out a minor error you have made. The name of the village where Chopin was born, *Żelazowa Wola,* does not mean "Iron Will", although you might have picked such a meaning by looking for the two words in a dictionary separately. The word *wola,* which means "will" alone, when applied as a part of a village's name means that the village was founded by somebody's will, and then the other part of the village's name usually stems from a person's name. There are numerous examples of such names in Poland, and normally they are attached to small hamlets. Consequently, *Wola* as a village's name has a second meaning in Polish, and that is simply "small village". The word *Żelazowa* does not seem to stem from a person's name (although I have no literature here to answer that question with certainty). It suggests that the founding of the village had something to do either with iron ore being found somewhere in the neighborhood or with iron being processed there. So the best translation of *Żelazowa Wola* would be "Iron Village" or "Iron-Ore Village". "Iron will" in Polish would be *Żelazna Wola,* and the name of Chopin's village does quite certainly not mean that.

I stand corrected!

Jakub Tatarkiewicz, a physicist writing from Warsaw, very gently pointed out that I had somehow managed to invent a new Polish word: *polskność*. I was quite surprised to learn that I had invented it, since I was sure I had seen it somewhere, but as it turns out, what I had actually seen was *polskość* (with no 'n'). Tatarkiewicz complimented me, however, for my talent in coming up with a good neologism, for, he said, my word has poignant overtones of such loaded words as *tęsknota* and *Solidarność*. As he put it: "I can only doubt if you really meant all those connotations—*or* is it just Chopin's music that played in your soul?!" I don't know. I guess I'd chalk it up to serendipity.

Great art has a way of evoking continual commentary; it is a bottomless source of inspiration to others. I have my blind spots in terms of understanding music, that's for sure; but Chopin hits some kind of bull's-eye in my soul. If I could meet any one person from the past, it would be Chopin, without any doubt. What saddens me enormously is his relatively small output. He died at age 39, with his expressive powers clearly as strong as ever. What *ever* would he have produced, had he lived to the age of, say, 65, as Bach did? Unbelievable firegems, I am sure. Indeed, I cannot imagine who I would be if I knew those pieces.

10

Parquet Deformations: A Subtle, Intricate Art Form

July, 1983

WHAT'S the difference between music and visual art? If I were asked this, I would have no hesitation in replying. To me, the major difference is clearly *temporality*. Works of music intrinsically involve time; works of art do not. More precisely, pieces of music consist of sounds intended to be played and heard in a specific order and at a specific speed. Music is thus fundamentally one-dimensional; it is tied to the rhythms of our existence. Works of visual art, by contrast, are generally two-dimensional or three-dimensional. Paintings and sculptures seldom have any intrinsic "scanning order" built into them that the eye must follow. Mobiles and other pieces of kinetic art may change over time, but often without any specific initial state or final state or intermediate stages. You are free to come and go as you please.

There are exceptions to this generalization, of course. European art has its grand friezes and historic cycloramas, and Oriental art has intricate pastoral scrolls of up to hundreds of feet in length. These types of visual art impose a temporal order and speed on the scanning eye. There is a starting point and a final point. Usually, as in stories, these points represent states of relative calm—especially the end. In between them, various types of tension are built up and resolved in an idiosyncratic but pleasing visual rhythm. The calmer end states are usually orderly and visually simple, while the tenser intermediate states are usually more chaotic and visually confusing. If you replace "visual" by "aural", virtually the same could be said of music.

I have been fascinated for many years by the idea of trying to capture the essence of the musical experience in visual form. I have my own ideas as to how this can be done; in fact, I spent several years working out a form of visual music. It is perhaps the most original and creative thing I have ever done. However, by no means do I feel that there is a unique or best way to carry out this task of "translation", and indeed I have often wondered how

191

others might attempt to do it. I have seen a few such attempts, but most of them, unfortunately, did not grab me. One striking counterexample is the set of "parquet deformations" meta-composed by William Huff, a professor of architectural design at the State University of New York at Buffalo.

I say "meta-composed" for a very good reason. Huff himself has never executed a single parquet deformation. He has elicited hundreds of them, however, from his students, and in so doing has brought this form of art to a high degree of refinement. Huff might be likened to the conductor of a fine orchestra, who of course makes no sound whatsoever during a performance. And yet we tend to give the conductor most of the credit for the quality of the sound. We can only guess how much preparation and coaching went into this performance. And what about the selection of the pieces and tempos and styles—not to mention the many-year process of culling the performers themselves?

So it is with William Huff. For 23 years, his students at Carnegie-Mellon and SUNY at Buffalo have been prodded into flights of artistic inspiration, and it is thanks to Huff's vision of what constitutes quality that some very beautiful results have emerged. Not only has he elicited outstanding work from students, he has also carefully selected what he feels to be the best pieces and these he is preserving in archives. For these reasons, I shall at times refer to Huff's "creations", but it is always in this more indirect sense of "meta-creations" that I shall mean it.

Not to take credit from the students who executed the individual pieces, there is a larger sense of the term "credit" that goes exclusively to Huff, the person who has shaped this whole art form himself. Let me use an analogy. Gazelles are marvelous beasts, yet it is not they themselves but the selective pressures of evolution that are responsible for their species' unique and wondrous qualities. Huff's judgments and comments have here played the role of those impersonal evolutionary selective pressures, and o¹　˜them has been molded a living and dynamic tradition, a "species˙ of art exemplified and extended by each new instance.

*　*　*

All that remains to be said by way of introduction is the meaning of the term *parquet deformation.* It is nearly self-explanatory, actually: traditionally, a *parquet* is a regular mosaic made out of inlaid wood, on the floor of an elegant room; and a *deformation*—well, it's somewhere in between a distortion and a transformation. Huff's parquets are more abstract: they are regular *tessellations* (or *tilings*) *of the plane,* ideally drawn with zero-thickness line segments and curves. The deformations are not arbitrary but must satisfy two basic requirements:

(1) There shall be change only in one dimension, so that one can see a temporal progression in which one tessellation gradually becomes another;

(2) At each stage, the pattern must constitute a regular tessellation of the plane (*i.e.*, there must be a unit cell that could combine with itself so as to cover an infinite plane exactly).

(Actually, the second requirement is not usually adhered to strictly. It would be more accurate to say that the unit cell at any stage of a parquet deformation can be easily modified so as to allow it to tile the plane perfectly.)

From this very simple idea emerge some stunningly beautiful creations. Huff explains that he was originally inspired, back in 1960, by the woodcut "Day and Night" of M. C. Escher. In that work, forms of birds tiling the plane are gradually distorted (as the eye scans downwards) until they become diamond-shaped, looking like the checkerboard pattern of cultivated fields seen from the air. Escher is now famous for his tessellations, both pure and distorted, as well as for other hauntingly strange visual games he played with art and reality.

Whereas Escher's tessellations almost always involve animals, Huff decided to limit his scope to purely geometric forms. In a way, this is like a decision by a composer to use austere musical patterns and to totally eschew anything that might conjure up a "program" (that is, some sort of image or story behind the sounds). An effect of this decision is that the beauty and visual interest must come entirely from the complexity and the subtlety of the interplay of abstract forms. There is nothing to "charm" the eye, as with pictures of animals. There is only the uninterpreted, unembellished perceptual experience.

Because of the linearity of this form of art, Huff has likened it to visual music. He writes:

> Though I am spectacularly ignorant of music, tone deaf, and hated those piano lessons (yet can be enthralled by Bach, Vivaldi, or Debussy), I have the students 'read' their designs as I suppose a musician might scan a work: the themes, the events, the intervals, the number of steps from one event to another, the rhythms, the repetitions (which can be destructive, if not totally controlled, as well as reinforcing). These are principally temporal, not spatial, compositions (though all predominantly temporal compositions have, of necessity, an element of the spatial and vice versa—*e.g.*, the single-frame picture is the basic element of the moving picture).

*   *   *

What are the basic elements of a parquet deformation? First of all, there is the class of allowed parquets. On this, Huff writes the following:

> We play a different (or rather, tighter) game than does Escher. We work with only *A* tiles (*i.e.*, congruent tiles of the same handedness). We do not use, as he does, *A* and *A'* tiles (*i.e.*, congruent tiles of both handedness). Finally, we don't use *A* and *B* tiles (*i.e.*, two different interlocking tiles), since two such tiles can always be seen as subdivisions of a single larger tile.

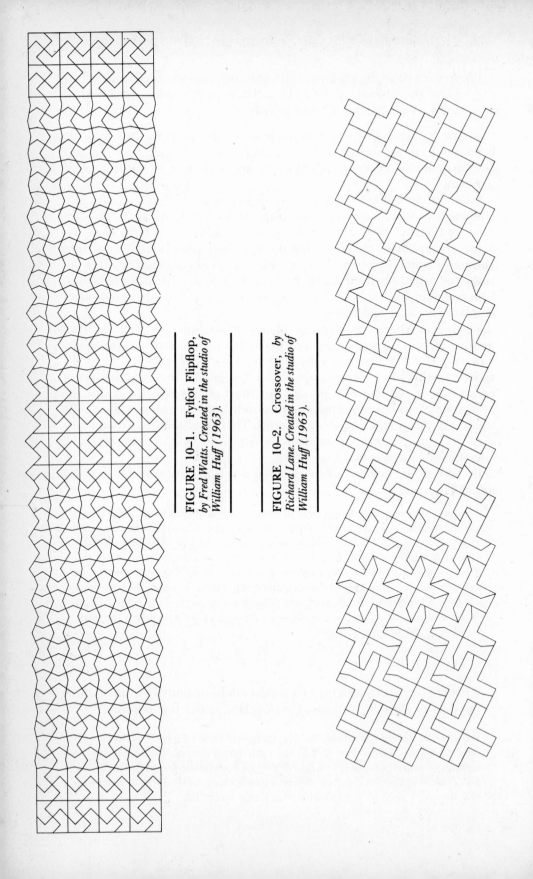

FIGURE 10–1. Fylfot Flipflop, by Fred Watts. Created in the studio of William Huff (1963).

FIGURE 10–2. Crossover, by Richard Lane. Created in the studio of William Huff (1963).

The other basic element is the repertoire of standard deforming devices. Typical devices include:

* lengthening or shortening a line;
* rotating a line;
* introducing a "hinge" somewhere inside a line segment so that it can "flex";
* introducing a "bump" or "pimple" or "tooth" (a small intrusion or extrusion having a simple shape) in the middle of a line or at a vertex;
* shifting, rotating, expanding, or contracting a group of lines that form a natural subunit;

and variations on these themes. To understand these descriptions, you must realize that a reference to "a line" or "a vertex" is actually a reference to a line or vertex inside a unit cell, and therefore, when *one* such line or vertex is altered, *all* the corresponding lines or vertices that play the same role in the copies of that cell undergo the same change. Since some of those copies may be at 90 degrees (or other angles) with respect to the master cell, one locally innocent-looking change may induce changes at corresponding spots, resulting in unexpected interactions whose visual consequences may be quite exciting.

*   *   *

Without further ado, let us proceed to examine some specific pieces. Look at the one called "Fylfot Flipflop" (Figure 10-1). It is an early one, executed in 1963 by Fred Watts at Carnegie-Mellon. If you simply let your eye skim across the topmost line, you will get the distinct sensation of scanning a tiny mountain range. At either edge, you begin with a perfectly flat plain, and then you move into gently rolling hills, which become taller and steeper, eventually turning into jagged peaks; then past the centerpoint, these start to soften into lower foothills, which gradually tail off into the plain again. This much is obvious even upon a casual glance. Subtler to see is the line just below, whose zigging and zagging is 180 degrees out of phase with the top line. Thus notice that in the very center, that line is completely at rest: a perfectly horizontal stretch flanked on either side by increasingly toothy regions. Below it there are seven more horizontal lines. Thus if one completely filtered out the vertical lines, one would see nine horizontal lines stacked above one another, the odd-numbered ones jagged in the center, the even-numbered ones smooth in the center.

Now what about the vertical lines? Both the lefthand and righthand borderlines are perfectly straight vertical lines. However, their immediate neighbors are as jagged as possible, consisting of repeated 90-degree bends, back and forth. Then the next vertical line nearer the center is practically straight up and down again. Then there is a wavy one again, and so on. As you move across the picture, you see that the jagged ones gradually get less jagged and the straight ones get increasingly jagged, so that in the middle the roles are completely reversed. Then the process continues, so that by the time you've reached the other side, the lines are back to normal again. If you could filter out the horizontal lines, you would see a simple pattern of quite jaggy lines alternating with less jaggy lines.

When these two extremely simple independent patterns—the horizontal and the vertical—are superimposed, what emerges is an unexpectedly rich perceptual feast. At the far left and right, the eye picks out fylfots—that is, swastikas—of either handedness contained inside perfect squares. In the center, the eye immediately sees that the central fylfots are all gone, replaced by perfect crosses inside pinwheels.

And then a queer perceptual reversal takes place. If you just shift your focus of attention diagonally by half a pinwheel, you will notice that there is a fylfot right there before your eyes! In fact, suddenly they appear all over the central section where before you'd been seeing only crosses inside pinwheels! And conversely, of course, now when you look at either end, you'll see pinwheels everywhere with crosses inside them. No fylfots! It is an astonishingly simple design, yet this effect catches nearly everyone really off guard.

This is a simple example of the ubiquitous visual phenomenon called *regrouping,* in which the boundary line of the unit cell shifts so that structures jump out at the eye that before were completely submerged and invisible —while conversely, of course, structures that a moment ago were totally obvious have now become invisible, having been split into separate conceptual pieces by the act of regrouping, or shift of perceptual boundaries. It is both a perceptual and conceptual phenomenon, a delight to that subtle mixture of eye and mind that is most sensitive to pattern.

For another example of regrouping, take a look at "Crossover" (Figure 10-2), also executed at Carnegie-Mellon in 1963 by Richard Lane. Something really amazing happens in the middle, but I won't tell you what. Just find it yourself by careful looking.

By the way, there are still features left to be explained in "Fylfot Flipflop". At first it appears to be mirror-symmetric. For instance, all the fylfots at the left end are spinning counterclockwise, while all the ones at the right end are spinning clockwise. So far, so symmetric. But in the middle, all the fylfots go counterclockwise. This surely violates the symmetry. Furthermore, the one-quarter-way and three-quarter-way stages of this deformation, which ought to be mirror images of each other, bear no resemblance at all to each other. Can you figure out the logic behind this subtle asymmetry between the left and right sides?

This piece also illustrates one more way in which parquet deformations resemble music. A unit cell—or rather, a vertical cross-section consisting of a stack of unit cells—is analogous to a measure in music. The regular pulse of a piece of music is given by the repetition of unit cells across the page.

And the flow of a melodic line across measure boundaries is modeled by the flow of a visual line—such as the mountain range lines—across many unit cells.

*   *   *

Bach's music is always called up in discussions of the relationship of mathematical patterns to music, and this occasion is no exception. I am reminded especially of some of his texturally more uniform pieces, such as certain preludes from the Well-Tempered Clavier, in which in each measure there is a certain pattern executed once or twice, possibly more times. From measure to measure this pattern undergoes a slow metamorphosis, meandering over the course of many measures from one region of harmonic space to far distant regions and then slowly returning via some circuitous route. For specific examples, you might listen to (or look at the scores of): Book I, numbers 1, 2; Book II, numbers 3, 15. Many of the other preludes have this feature in places, though not for their entirety.

Bach seldom deliberately set out to play with the perceptual systems of his listeners. Artists of his century, although they occasionally played perceptual games, were considerably less sophisticated about, and less fascinated with, issues that we now deem part of perceptual psychology. Such phenomena as regrouping would undoubtedly have intrigued Bach, and I for one sometimes wish that he had known of and been able to try out certain effects—but then I remind myself that whatever time Bach might have spent playing with new-fangled ideas would have had to be subtracted from his time to produce the masterpieces that we know and love, so why tamper with something that precious?

On the other hand, I don't find that argument 100 percent compelling. Who says that if you're going to imagine playing with the past, you have to hold the lifetimes of famous people constant in length? If we can imagine telling Bach about perceptual psychology, why can't we also imagine adding a few extra years to his lifetime to let him explore it? After all, the only *divinely* imposed (that is, absolutely unslippable) constraint on Bach's years is that they and Mozart's years add up to 100, no? So if we award Bach five extra ones, then we merely take five years away from Mozart. It's painful, to be sure, but not all *that* bad. We could even let Bach live to 100 that way! (Mozart would never have existed.) It starts to get a little questionable if we go much beyond that point, however, since it is not altogether clear what it means to live a negative number of years.

Although it is difficult to imagine and impossible to know what Bach's music would have been like had he lived in the twentieth century, it is certainly not impossible to know what Steve Reich's music would have been like, had *he* lived in this century. In fact, I'm listening to a record of it right now (or at least I would have been if I hadn't gotten distracted by this radio program). Now Reich's is music that *really* is conscious of perceptual psychology. All the way through, he plays with perceptual shifts and ambiguities, pivoting from one rhythm to another, from one harmonic origin to another, constantly keeping the listener on edge and tingling with nervous energy. Imagine a piece like Ravel's "Bolero", only with a much finer grain size, so that instead of roughly a one-minute unit cell, it has a three-second unit cell. Its changes are tiny enough that sometimes you barely can tell it is changing at all, while other times the changes jump out at you. What Reich piece am I listening to (or rather, *would* I be listening to if I weren't still listening to this radio program)? Well, it hardly matters, since most of them satisfy this characterization, but for the sake of specificity you might try "Music for a Large Ensemble", "Octet", "Violin Phase", "Vermont Counterpoint", or his recent choral work "Tehillim".

*   *   *

Let us now return to parquet deformations. "Dizzy Bee" (Figure 10-3), executed by Richard Mesnik at Carnegie-Mellon in 1964, involves perceptual tricks of another sort. The left side looks like a perfect honeycomb or—somewhat less poetically—a perfect bathroom floor. However, as we move rightward, its perfection seems cast in doubt as the rigidity of the lattice gives way to rounder-seeming shapes. Then we notice that three of them have combined to form one larger shape: a super hexagon made up of three rather squashed pentagons. The curious thing is that if we now sweep our eyes right to left, back to the beginning, we can no longer

FIGURE 10–3. Dizzy Bee, *by Richard Mesnik. Created in the studio of William Huff (1964).*

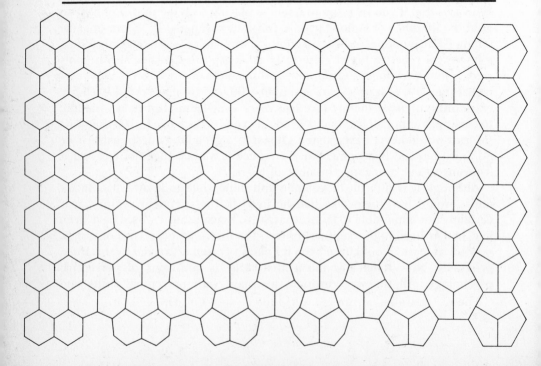

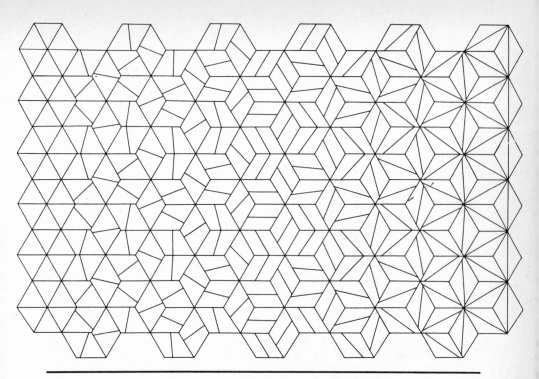

FIGURE 10–4.   Consternation, *by Scott Grady. Created in the studio of William Huff* *(1977).*

see the left side in quite the way we saw it before. The small hexagons now are constantly grouping themselves into threes, although the grouping changes quickly. We experience "flickering clusters" in our minds, in which groups form for an instant and then disband, their components immediately regrouping in new combinations, and so on. The poetic term "flickering clusters" comes from a famous theory of how water molecules behave, the bonding in that case coming from hydrogen bonds rather than mental ones. (See the *P.S.* to Chapter 26.)

Even more dizzying, perhaps, than "Dizzy Bee" is "Consternation" (Figure 10-4), executed by Scott Grady of SUNY at Buffalo in 1977. This is another parquet deformation in which hexagons and cubes vie for perceptual supremacy. This one is so complex and agitated in appearance that I scarcely dare to attempt an analysis. In its intermediate regions, I find the same extremely exciting kind of visual pseudo-chaos as in Escher's best deformations.

Perhaps irrelevantly, but I suspect not, the names of many of these studies remind me of pieces by Zez Confrey, a composer most famous during the twenties for his novelty piano solos such as "Dizzy Fingers", "Kitten on the Keys", and—my favorite—"Flutter by, Butterfly". Confrey specialized in pushing rag music to its limits without losing musical charm, and some of the results seem to me to have a saucy, dazzling appeal not unlike the jazzy appearance of this parquet deformation, and others.

The next parquet deformation, "Oddity out of Old Oriental Ornament"

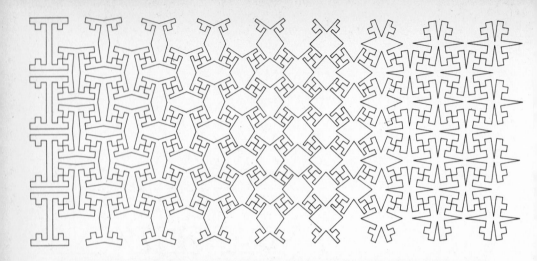

FIGURE 10-5.   Oddity out of Old Oriental Ornament, *by Francis O'Donnell. Created in the studio of William Huff (1966).*

(Figure 10-5), executed by Francis O'Donnell at Carnegie-Mellon in 1966, is based on an extremely simple principle: the insertion of a "hinge" in one single line segment, and subsequent flexing of the segment at that hinge! The reason for the stunningly rich results is that the unit cell that creates the tessellation occurs both vertically and horizontally, so that flexing it one way induces a crosswise flexing as well, and the two flexings combine to yield this curious and unexpected pattern.

Another one that shows the amazing results of an extremely simple but carefully chosen tranformation principle is "Y Knot" (Figure 10-6),

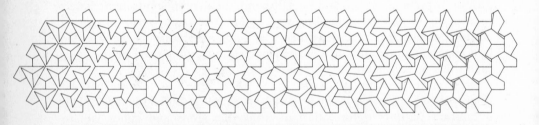

FIGURE 10-6.   Y Knot, *by Leland Chen. Created in the studio of William Huff (1977).*

executed by Leland Chen at SUNY at Buffalo in 1977. If you look at it with full attention, you will see that its unit cell is in the shape of a three-bladed propeller, and that unit cell never changes whatsoever in shape. All that does change is the 'Y' lodged tightly inside that unit cell. And the only way that 'Y' changes is by rotating clockwise very slowly! Admittedly, in the final stages of rotation, this forces some previously constant line segments to extend themselves a little bit, but this does not change the outline of the unit cell whatsoever. What well-chosen simplicity can do!

*   *   *

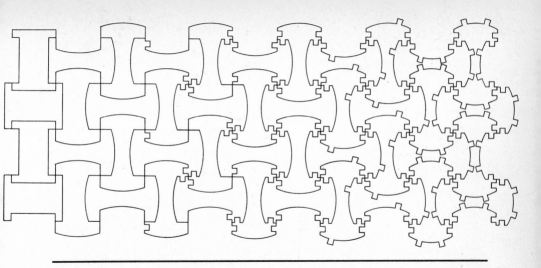

FIGURE 10–7.   Crazy Cogs, *by Arne Larson. Created in the studio of William Huff (1963).*

Three of my favorites are "Crazy Cogs" (Figure 10-7, done by Arne Larson, Carnegie-Mellon, 1963), "Trifoliolate" (Figure 10-8, done by Glen Paris, Carnegie-Mellon, 1966), and "Arabesque" (Figure 10-9, done by Joel Napach, SUNY at Buffalo, 1979). They all share the feature of getting more and more intricate as you move rightward. Most of the earlier ones we've seen don't have this extreme quality of irreversibility—that is, the ratcheted quality that signals that an evolutionary process is taking place. I can't help wondering if the designers didn't feel that they'd painted themselves into a corner, especially in the case of "Arabesque". Is there any way you can back out of that super-tangle except by retrograde motion—that is, retracing your steps? I suspect there is, but I wouldn't care to try to discover it.

To contrast with this, consider "Razor Blades", an extended study in relative calmness (Figure 10-10). It was done at Carnegie-Mellon in 1966, but unfortunately it is unsigned. Like the first one we discussed, this one can be broken up into very long waving horizontal lines and vertical structures crossing them. It's a little easier to see them if you start at the right side. For instance, you can see that just below the top, there is a long snaky line

FIGURE 10–8.   Trifoliolate, *by Glen Paris. Created in the studio of William Huff (1966).*

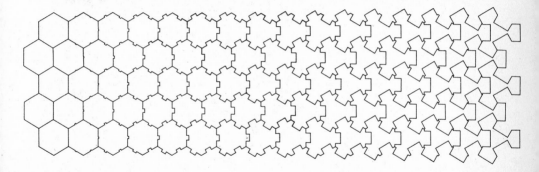

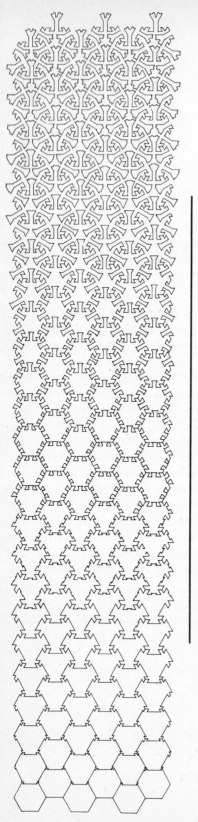

FIGURE 10–9. Arabesque, by Joel Napach. Created in the studio of William Huff (1979).

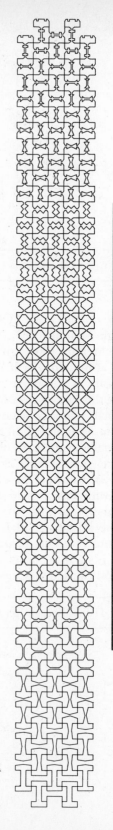

FIGURE 10–10. Razor Blades (unsigned). Created in the studio of William Huff (1966).

with numerous little "nicks" in it, undulating its way leftwards and in so doing shedding some of those nicks, so that at the very left edge it has degenerated into a perfect "square wave", as such a periodic wave form is called in Fourier analysis. Complementing this horizontal structure is a similar vertical structure that is harder to describe. The thought that comes to my mind is that of two very ornate, rather rectangular hourglasses with ringed necks, one on top of the other. But you can see for yourself.

As with "Fylfot Flipflop" (Figure 10-1), each of these patterns by itself is intriguing, but of course the real excitement comes from the daring act of superimposing them. Incidentally, I know of no piece of visual art that better captures the feeling of beauty and intricacy of a Steve Reich piece, created by slow "adiabatic" changes floating on top of the chaos and dynamism of the lower-level frenzy. Looking back, I see I began by describing this parquet deformation as "calm". Well, what do you know? Maybe I would be a good candidate for inclusion in *The New Yorker*'s occasional notes titled "Our Forgetful Authors".

More seriously, there is a reason for this inconsistency. One's emotional response to a given work of art, whether visual or musical, is not static and unchanging. There is no way to know how you will respond, the next time you hear or see one of your favorite pieces. It may leave you unmoved, or it may thrill you to the bones. It depends on your mood, what has recently happened, what chances to strike you, and many other subtle intangibles. One's reaction can even change in the course of a few minutes. So I won't apologize for this seeming lapse.

Let us now look at "Cucaracha" (Figure 10-11), executed in 1977 by Jorge Gutiérrez at SUNY at Buffalo. It moves from the utmost geometricity—a lattice of perfect diamonds—through a sequence of gradually more arbitrary modifications until it reaches some kind of near-freedom, a dance of strange, angular, quasi-organic forms. This fascinates me. Is entropy increasing or decreasing in this rightward flow toward freedom?

A gracefully spiky deformation is the one wittily titled "Beecombing Blossoms" (Figure 10-12), executed this year by Laird Pylkas at SUNY at Buffalo. Huff told me that Pylkas struggled for weeks with this one, and at the end, when she had satisfactorily resolved her difficulties, she mused, "Why is it that the obvious ideas always take so long to discover?"

*　　*　　*

As our last study, let us take "Clearing the Thicket" (Figure 10-13), executed in 1979 by Vincent Marlowe at SUNY at Buffalo, which involves a mixture of straight lines and curves, right angles and cusps, explicit squarish swastikoids and implicit circular holes. Rather than demonstrate my inability to analyze the ferocious complexity of this design, I would like to use it as the jumping-off point for a discussion of computers and creativity —one of my favorite hobbyhorses.

FIGURE 10–11. Cucaracha, by Jorge Gutiérrez. Created in the studio of William Huff (1977).

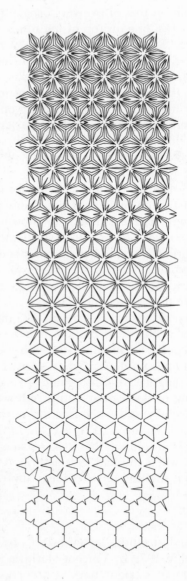

FIGURE 10–12. Beecombing Blossoms, by Laird Pylkas. Created in the studio of William Huff (1983).

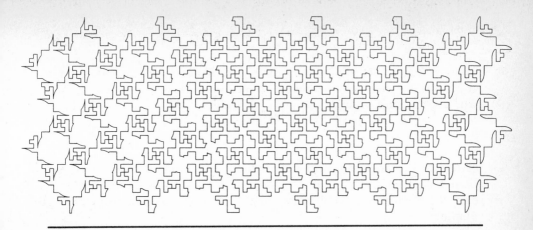

FIGURE 10–13.   Clearing the Thicket, *by Vincent Marlowe. Created in the studio of William Huff (1979).*

Some totally new things are going on in this parquet deformation—things that have not appeared in any previous one. Notice the hollow circles on the left side that shrink as you move rightward; notice also that on the right side there are hollow "anticircles" (concave shapes made from four circular arcs turned inside out) that shrink as you move leftward. Now, according to Huff, such an idea had never appeared in any previously created deformations. This means that something unusual happened here—something genuinely creative, something unexpected, unpredictable, surprising, intriguing—and not least, inspiring to future creators.

So the question naturally arises: Would a computer have been able to invent this parquet deformation? Well, put this way it is a naïve and ill-posed question, but we can try to make some sense of it. The first thing to point out is that, of course, the phrase "a computer" refers to nothing more than an inert hunk of metal and semiconductors. To go along with this bare computer, this hardware, we need some software and some energy. The former is a specific pattern inserted into the matter binding it with constraints yet imbuing it with goals; the latter is what breathes "life" into it, making it act according to those goals and constraints.

The next point is that the software is what really controls what the machine does; the hardware simply obeys the software's dictates, step by step. And yet, the software could exist in a number of different "instantiations"—that is, realizations in different computer languages. What really counts about the software is not its literal aspect, but a more abstract, general, overall "architecture", which is best described in a nonformal language, such as English. We might say that the plan, the sketch, the central idea of a program is what we are talking about here—not its final realization in some specific formal language or dialect. That is something we can leave to apprentices to carry out, after we have presented them with our informal sketch.

So the question actually becomes less mundane-sounding, more theoretical and philosophical: *Is there an architecture to creativity?* Is there a plan, a scheme, a set of principles that, if elucidated clearly, could account for all the creativity embodied in the collection of all parquet deformations, past, present, and future?

*   *   *

Note that we are asking about the *collection* of parquet deformations, not about some *specific* work. It is a truism that any specific work of art can be recreated, even recreated in various slightly novel ways, by a programmed computer.

For example, the Dutch artist Piet Mondrian evolved a highly idiosyncratic, somewhat cryptic style of painting over a period of many years. You can see, if you trace his development over the course of time, exactly where he came from and where he was headed. But if you focus in on just a single Mondrian work, you cannot sense this stylistic momentum —this quality of dynamic, evolving style that any great artist has. Looking at just one work in isolation is like taking a snapshot of something in motion: you capture its instantaneous position but not its momentum. Of course, the snapshot might be blurred, in which case you get a sense of the momentum but lose information about the position. But when you are looking at just a single work of art, there is no mental blurring of its style with that of recent works or soon-to-come works; you have exact position information ("What is the style *now*?"), but no momentum information ("Where was it and where is it going?").

Some years ago, the mathematician and computer artist A. Michael Noll took a single Mondrian painting—an abstract, geometric study with seemingly random elements—and from it extracted some statistics concerning the patterns. Given these statistics, he then programmed a computer to generate numerous "pseudo-Mondrian paintings" having the same or different values of these randomness-governing parameters. (See Figure 10-14.) Then he showed the results to naïve viewers. The reactions were interesting, in that more people preferred one of the pseudo-Mondrians to the genuine Mondrian!

This is quite amusing, even provocative, but it also is a warning. It proves that a computer can certainly be programmed, after the fact, to imitate—and well—mathematically capturable stylistic aspects of a given work. But it also warns us: Beware of cheap imitations!

Consider the case of parquet deformations. There is no doubt that a computer could be programmed to do any *specific* parquet deformation—or minor variations on it—without too much trouble. There just aren't *that* many parameters to any given one. But the essence of any artistic act resides not in selecting particular values for certain parameters, but far deeper: it's in the balancing of a myriad intangible and mostly unconscious mental forces, a judgmental act that results in many conceptual choices that eventually add up to a tangible, perceptible, measurable work of art.

FIGURE 10–14. *One genuine Mondrian plus three computer imitations. Can you spot the Mondrian? If you rotate the figure so that east becomes south, it will be the one in the northwest corner. The Mondrian, done in 1917, is titled* Composition with Lines; *the three others, done in 1965, comprise a work called* Computer Composition with Lines, *and were created by a computer at Bell Telephone Laboratories at the behest of computer tamer A. Michael Noll. The subjectively "best" picture was found through surveys; it is the one diagonally opposite the genuine Mondrian!*

Once the finished work exists, scholars looking at it may seize upon certain qualities of it that lend themselves easily to being parametrized. *Anyone* can do statistics on a work of art once it is there for the scrutiny, but the ease of doing so can obscure the fact that no one could have said, *a priori*, what kinds of mathematical observables would turn out to be relevant to the capturing of stylistic aspects of the as-yet-unseen work of art.

Huff's own view on this question of mechanizing the art of parquet deformations closely parallels mine. He believes that some basic principles could be formulated at the present time enabling a computer to come up with relatively stereotyped yet novel creations of its own. But, he stresses, his students occasionally come up with rule-breaking ideas that nonetheless enchant the eye for deeper reasons than he has so far been able to verbalize. And so, this way, the set of explicit rules gets gradually increased.

Comparing the creativity that goes into parquet deformations with the creativity of a great musician, Huff has written:

> I don't know about the consistency of the genius of Bach, but I did work with the great American architect Louis Kahn (1901–1974) and suppose that it must have been somewhat the same with Bach. That is, Kahn, out of moral, spiritual, and philosophical considerations, formulated ways he would and ways he would not do a thing in architecture. Students came to know many of his ways, and some of the best could imitate him rather well (though not perfectly). But as Kahn himself developed, he constantly brought in new principles that brought new transformations to his work; and he even occasionally discarded an old rule. Consequently, he was always several steps ahead of his imitators who knew what *was* but couldn't imagine what *will be.* So it is that computer-generated 'original' Bach is an interesting exercise. But it isn't Bach —that unwritten work that Bach never got to, the day after he died.

The real question is: What kind of architecture is responsible for *all* of these ideas? Or is there any *one* architecture that could come up with them all? I would say that the ability to design good parquet deformations is probably deceptive, in the same way as the ability to play good chess is: it looks more mathematical than it really is.

A brilliant chess move, once the game is over and can be viewed in retrospect, can be seen as logical—as "the correct thing to do in that situation". But brilliant moves do not originate from the kind of logical analysis that occurs *after* the game; there is no time during the game to check out all the logical consequences of a move. Good chess moves spring from the organization of a good chess mind: a set of perceptions arranged in such a way that certain kinds of ideas leap to mind when certain subtle patterns or cues are present. This way that perceptions have of triggering old and buried memories underlies skill in any type of human activity, not only chess. It's just that in chess the skill is particularly deceptive, because after the fact, it can all be justified by a logical analysis, a fact that seems to hint that the original idea *came from* logic.

Writing lovely melodies is another one of those deceptive arts. To the mathematically inclined, notes seem like numbers and melodies like number patterns. Therefore all the beauty of a melody *seems* as if it ought to be describable in some simple mathematical way. But so far, no formula has produced even a single good melody. Of course, you can look back at any melody and write a formula that will produce it and variations on it. But that is *retrospective,* not *prospective.* Lovely chess moves and lovely melodies (and lovely theorems in mathematics, etc.) have this in common: every one has idiosyncratic nuances that seem logical *a posteriori* but that are not easy to anticipate *a priori.* To the mathematical mind, chess-playing skill and melody-writing skill and theorem-discovering skill seem obviously formalizable, but the truth turns out to be more tantalizingly complex than that. Too many subtle balances are involved.

So it is with parquet deformations, I reckon. Each one taken alone is in some sense mathematical. However, taken as a *class,* they are not mathematical. This is what's tricky about them. Don't let the apparently mathematical nature of an *individual* one fool you, for the architecture of a program that could create all these parquet deformations and more good ones would have to incorporate computerized versions of *concepts* and *judgments*—and those are much more elusive and complex things than are numbers. In a way, parquet deformations are an ideal case with which to make this point about the subtlety of art, for the very reason that each one on its own appears so simple and rule-bound.

At this point, many critics of computers and artificial intelligence, eager to find something that "computers can't do" (and never will be able to do) often jump too far: they jump to the conclusion that art and, more generally, creativity, are fundamentally uncomputerizable. This is hardly the implied conclusion! The implied conclusion is just this: that for computers to act human, we will have to wait until we have good computer models of such human things as perception, memory, mental categories, learning, and so on. We are a long way from that. But there is no reason to assume that those goals are in principle unattainable, even if they remain far off for a long time.

*   *   *

I have been playing with the double meaning, in this column, of the term "architecture": it means both the design of a habitat and the abstract essence of a grand structure of any sort. The former has to do with hardware and the latter with software. In a certain sense, William Huff is a professor of both brands of architecture. Obviously his professional training is in the design of "hardware": genuine habitats for humans, and he is in a school where that is what they do. But he is also in the business of forming, in the minds of his students, a softer type of architecture: the mental architecture that underlies the skill to create beauty. Fortunately for him, he can take for granted the whole complexity of a human brain as his starting point upon which to build this architecture. But even so, there is a great art to instilling a sensitivity for beauty and novelty.

When I first met William Huff and saw how abstract and seemingly impractical were the marvelous works produced in his design studio— ranging from parquet deformations to strange ways of slicing a cube to gestalt studies using thousands of dots to eye-boggling color patterns—I at first wondered why this man was a professor of *architecture.* But after conversing with him and his colleagues, my horizons were extended about the nature of their discipline.

The architect Louis Kahn had great respect for the work of William Huff, and it is with his words that I would like to conclude:

> What Huff teaches is not merely what he has learned from someone else, but what is drawn from his natural gifts and belief in their truth and value. In my belief what he teaches is the introduction to discipline underlying shapes and rhythms, which touches the arts of sight, the arts of sound, and the arts of structure. It teaches students of drawing to search for the abstract and not the representational. This is so good as a reminder of order for the instructors/architectural sketchers (like me), and so good especially for the student sketchers without background. It is the introduction to exactitudes of the kind that instill the religion of the ordered path.

Post Scriptum.

"The religion of the ordered path"—a lovely phrase. I did not know at the time this column was written that it would be my last full column (the one reporting on the results of the Luring Lottery, here Chapter 31, was only a half-column). Both William Huff and I were pleased with my bowing out this way, and I was especially pleased with the phrase with which I bowed out. Though ambiguous, it captures much of the spirit that I attempted to get across in all my columns: dedicated questing after patterned beauty, and particularly after the *reasons* that certain particular patterns are beautiful.

In this column, I repeatedly claimed that it is relatively easy to make a computer program that creates attractive art within a formula, but not at all easy to make a computer program that constantly comes up with novelty. Some people familiar with the computer art produced in the last couple of decades might pick a fight with me over this. They might point to complex patterns produced by simple algorithms, and then add that there are certain simple algorithms which, when you change merely a few parameters, come up with astonishingly different patterns that no human would be likely to recognize as being each other's near kin. An example is a very simple program I know, which fills a screen with rapidly changing sixfold-symmetric dot-patterns that look like magnified snowflakes; in just a few seconds, any given pattern will dissolve and be replaced by an unbelievably different sixfold-symmetric pattern. I have stood transfixed at a screen watching these patterns unfold one after another, unable to anticipate in the slightest what will happen next—and yet knowing that the program itself is only a few lines long! I have seen small changes in mathematical formulas produce enormous visual changes in what those formulas represent, graphically.

The trouble is, these parameter-based changes—*knob-twiddlings,* as they are called in Chapters 12 and 13—are of a different nature than the kinds of novel ideas people come up with when they vary a given idea. For a machine to make simple variants of a given design, it must possess an algorithm for making that design which has *explicit parameters;* those parameters are then modifiable, as with the pseudo-Mondrian paintings. But the way people make variations is quite different. They *look* at some creation by an artist (or computer), and then they abstract from it some quality that they observe in the creation itself (not in some algorithm behind it). This newly abstracted quality may never have been thought of explicitly by the artist (or programmer or computer), yet it is there for the seeing by an acute observer. This perceptual act gets you more than half the way to genuine creativity; the remainder involves treating this new quality *as if it*

FIGURE 10–15. I at the Center, *by David Oleson. Created in the studio of William Huff (1964).*

were an explicit knob: "twiddling" it as if it were a parameter that had all along been in the program that made the creation.

That way, the perceptual process is intimately linked up with the generative process: a loop is closed in which perceptions spark new potentials and experimentation with new potentials opens up the way for new perceptions. The element lacking in current computer art is the interaction of perception with generation. Computers do not *watch* what they do; they simply *do* it. (See Chapter 23 for more on the idea of self-watching computers.) When programs are able to look at what they've done and perceive it in ways that they never anticipated, then you'll start to get close to the kinds of insight-giving disciplined exercises that Louis Kahn was speaking of when he wrote of the "religion of the ordered path".

*　　*　　*

One of my favorite parquet deformations is called "I at the Center" (Figure 10-15), and was done by David Oleson at Carnegie-Mellon in 1964. This one violates the premise with which I began my article: one-dimensionality. It develops its central theme—the uppercase letter 'I' —along two perpendicular dimensions at once. The result is one of the most lyrical and graceful compositions that I have seen in this form.

I am also pleased by the metaphorical quality it has. At the very center of a mesh is an I—an ego; touching it are other things—other I's—very much like the central I, but not quite the same and not quite as simple; then as one goes further and further out, the variety of I's multiplies. To me this symbolizes a web of human interconnections. Each of us is at the very center of our own personal web, and each one of us thinks, "I am the most normal, sensible, comprehensible individual." And our identity—our "shape" in personality space—springs largely from the way we are embedded in that network—which is to say, from the identities (shapes) of the people we are closest to. This means that we help to define others' identities even as they help to define our own. And very simply but effectively, this parquet deformation conveys all that, and more, to me.

11

Stuff and Nonsense

December, 1982

Buz, quoth the blue fly,
Hum, quoth the bee,
Buz and hum they cry,
And so do we:
In his ear, in his nose, thus, do you see?
He ate the dormouse, else it was he.

—Ben Jonson

EH? What does this mean? What is its point? This little nonsense poem, written around 1600, begins with an image of insects, slides into an image of someone's face, and concludes with an uncertain reference to the devourer of a certain rodent. Although it makes little sense, it is still somehow enjoyable. It reminds us of a nursery rhyme. It is comfortable, cute, droll.

Nonsense has been around for a long time. Its style and tone have changed over the centuries, however. The path of development of nonsense is interesting to trace. What marks something off as being nonsense? When does nonsense spill over into sense, or vice versa? Where are the borderlines between nonsense and poetry? These are issues to be explored in this column.

A century and a half after Jonson wrote his poem, an English actor named Charles Macklin became notorious for boasting that he could memorize any passage on one hearing. To challenge Macklin, his friend the dramatist Samuel Foote wrote the following odd passage:

So she went into the garden to cut a cabbage-leaf to make an apple-pie; and, at the same time, a great she-bear coming up the street pops its head into the shop—What! no soap? So he died; and she very imprudently married the barber: and there were present the Picninnies, and the Joblilies, and the Garyulies, and the great Panjandrum himself, with the little round button at top. And they all fell to playing the game of 'catch as catch can', till the gunpowder ran out at the heels of their boots.

Full of *non sequiturs* and awkward, choppy sentences, this must have been an excellent challenge for Macklin. Unfortunately, we have no record as to how he fared on first hearing it, but we do know that he enjoyed the passage immensely, and went around reciting it with great gusto for years thereafter.

In the nineteenth century, the reigning monarchs of nonsense were Lewis Carroll and Edward Lear. Everyone knows Carroll's "Jabberwocky", "Tweedledum and Tweedledee", and "The Walrus and the Carpenter"; most people have heard of Lear's "The Owl and the Pussycat". Fewer have heard of Lear's "The Pobble Who Has No Toes" or "The Dong with the Luminous Nose". Carroll and Lear both enjoyed inventing strange words and using them innocently, as if they were commonplace. Their nonsense was expressed largely in poems, where they indulged in much alliteration, many internal rhymes, catchy rhythms, and off-beat imagery. Rather than exhibit works of those two authors, I have instead chosen, to represent their era, an anonymous poem with some of the same charming qualities:

INDIFFERENCE

In loopy links the canker crawls,
Tads twiddle in their 'polian glee,
Yet sinks my heart as water falls.
The loon that laughs, the babe that bawls,
The wedding wear, the funeral palls,
Are neither here nor there to me.
Of life the mingled wine and brine
I sit and sip pipslipsily.

Many of Carroll's nonsense poems were parodies of popular songs or ditties of his day. Ironically, the parodies are remembered, and the things that triggered them are mostly completely forgotten. Carroll loved to poke fun, in his gentle manner, at the stuffy mores and hypocritical mannerisms of society. One of the characteristics of "genteel" poetry of the nineteenth century was its precious use of classical literary allusions. Carroll seldom parodies this quality, but Charles Battell Loomis, a little-known writer, admirably caught the style in this poem.

A CLASSIC ODE

Oh, limpid stream of Tyrus, now I hear
The pulsing wings of Armageddon's host,
Clear as a colcothar and yet more clear—
(Twin orbs, like those of which the Parsees boast);

Down in thy pebbled deeps in early spring
The dimpled naiads sport, as in the time
When Ocidelus with untiring wing
Drave teams of prancing tigers, 'mid the chime

Of all the bells of Phicol. Scarcely one
Peristome veils its beauties now, but then—
Like nascent diamonds, sparkling in the sun,
Or sainfoin, circinate, or moss in marshy fen.

Loud as the blasts of Tubal, loud and strong,
Sweet as the songs of Sappho, aye more sweet;
Long as the spear of Arnon, twice as long,
What time he hurled it at King Pharaoh's feet.

This poem has the curious quality that when you read it, you feel that surely it makes sense—perhaps another reading will reveal it to you. And then you read it again and find that same head-scratching feeling comes back to you. This is a problem with much modern poetry: It is very hard to be certain that you're not simply being taken for a ride by the poet, sucked in by some practical joker who has actually nothing in mind except tricking readers into thinking there is profound meaning where there is none.

The limerick is a form of poetry often featured in nonsense anthologies, probably because it is a playful form. However, very few limericks make no sense. They may involve mild impossibilities, such as a young woman who travels faster than the speed of light, or other more off-color feats, but in actuality, limericks are seldom nonsensical. One limerick that, in its own way, is pure nonsense is the following gem, by W. S. Gilbert (of Gilbert and Sullivan):

There was an old man of St. Bees,
Who was stung in the arm by a wasp.
When asked, "Does it hurt?"
He replied, "No, it doesn't—
I'm so glad it wasn't a hornet."

Why do I call this "nonsense"? Well, if it were a prose sentence, nothing about it would attract much attention, except perhaps the name of the town. The nonsense is certainly not in the content, but in the way it utterly violates every standard set up for the limerick form. It doesn't rhyme, its meter is a little bumpy, and it has absolutely nothing funny in it—which is what makes it funny. And that makes it qualify as nonsense.

* * *

Is nonsense always funny? Up until the twentieth century, it certainly seemed that way. In fact, nonsense and humor have traditionally been so closely allied that anthologies of nonsense seem to be composed largely of humorous passages of any sort whatsoever, irrespective of how sensible they are. But nonsense and humor took widely divergent paths in the early twentieth century. Perhaps the greatest nonsense writer who ever lived was Gertrude Stein, although she is seldom mentioned in this connection. Entire collections of nonsense have been published without featuring a single piece of her work. Her most audacious piece in this genre is a volume of nearly 400 pages, modestly titled *How to Write*. Here is a sample taken from the Chapter called "Arthur a Grammar".

> Arthur a grammar.
> Questionnaire in question.
> What is a question.
> Twenty questions.
> A grammar is an astrakhan coat in black and other colors it is an obliging management of their requesting in indulgence made mainly as if in predicament as in occasion made plainly as if in serviceable does it shine.
> A question and answer.
> How do you like it.
> Grammar can be contained on account of their providing medaling in a ground of allowing with or without meant because which made coupled become blanketed with a candidly increased just as if in predicting example of which without meant and coupled inclined as much without meant to be thought as if it were as ably rested too. Considerable as it counted heavily in part.
> What is grammar when they make it round and round. As round as they are called.
> Did they guess whether they wished. A politely definitely detailed blame of when they go.
> What is a grammar ordinarily. A grammar is question and answer answer undoubted however how and about.
> What is Arthur a grammar.
> Arthur is a grammar.
> Arthur a grammar.
> What can there be in a difficulty.
> Seriously in grammar.
> Thinking that a little baby can sigh.
> That is so much.
> Sayn can say only he is dead that he is interested in what is said.
> That is another in consequence.
> Better and flutter must and man can beam.
> Now think of seams.
> Embroidery consists in remembering that it is but what she meant.
> There an instance of grammar.
> Suppose embroidery is two and two. There can be reflected that it is as if it were having red about.
> This is an instance of having settled it.

Grammar uses twenty in a predicament. Include hyacinths and mosses which grow in abundance.

Grammar. In picking hyacinths quickly they suit admirably this makes grammar a preparation. Grammar unites parts and praises. In just this way.

Grammar untiringly.

Grammar perhaps grammar.

It is quite perplexing. It is simply an absurd string of non sequiturs, often totally lacking grammar, meandering randomly from "topic" to "topic". It is frustrating because there is nothing to grab onto. It is like trying to climb a mountain made of sand.

Stein's experiments in absurdity parallel the Dadaist and Surrealist movements of roughly the same period, and they mark the trend away from exuberant and laughable nonsense toward troubling and, later, macabre nonsense. However, her work still has a freshness and silliness that makes it amusing and light rather than disturbing and heavy.

* * *

As we move further into the twentieth century, we encounter the philosophy of existentialism and the master expositor of existential malaise, Irish-born playwright Samuel Beckett. In Beckett's most famous play, *Waiting for Godot,* written in the early 1950's, the pathetic character ironically called "Lucky" has exactly one speech, coming in about the middle of the play. He has been being taunted by the other characters with cries of "Think, pig!" and with sharp tugs on the rope around his neck, by which they are holding him. Eventually he is driven beyond the breaking point, and out pours an incoherent, wild, tormented piece of absolute confusion, resembling regurgitated academic coursework crossed with stock phrases and garbled memorized lists of one sort and another. Here is Lucky's famous speech:

Given the existence as uttered forth in the public works of Puncher and Wattmann of a personal God quaquaquaqua with white beard quaquaquaqua outside time without extension who from the heights of divine apathia divine athambia divine aphasia loves us dearly with some exceptions for reasons unknown but time will tell and suffers like the divine Miranda with those who for reasons unknown but time will tell are plunged in torment plunged in fire whose fire flames if that continues and who can doubt it will fire the firmament that is to say blast hell to heaven so blue still and calm so calm with a calm which even though intermittent is better than nothing but not so fast and considering what is more that as a result of the labors left unfinished crowned by the Acacacacademy of Anthropopopometry of Essy-in-Possy of Testew and Cunard it is established beyond all doubt all other doubt than that which clings to the labors of men that as a result of the labors unfinished of Testew and Cunard it is established as hereinafter but not so fast for reasons unknown that as a result of the public works of Puncher and Wattmann it is established beyond all doubt that in view of the labors of Fartov and Belcher left unfinished for reasons unknown of Testew and Cunard left unfinished it is established what many deny that man in Possy of Testew and Cunard that man in Essy that man in short that man in brief in spite of the strides of alimentation and defecation wastes and pines wastes and pines and concurrently simultaneously what is more for reasons unknown in spite of the strides of physical culture the practice of sports such as tennis football running cycling gliding conating camogie skating tennis of all kinds dying flying sports of all sorts autumn summer winter winter tennis of all kinds hockey of all sorts penicilline and succedanea in a word I resume flying gliding golf over nine and eighteen holes tennis of all sorts in a word for reasons unknown in Feckham Peckham Fulham Clapham namely concurrently simultaneously what is more for reasons unknown but time will tell fades away I resume Fulham Clapham in a word the dead loss per head since the death of Bishop Berkeley being to the tune of one inch four ounce per head approximately by and large more or less to the nearest decimal good measure round figures stark naked in the stockinged feet in Connemara in a word for reasons unknown no matter what matter the facts are there and considering what is more much more grave that in the light of the labors lost of Steinweg and Peterman it appears what is more much more grave that in the light the light the light of the labors lost of Steinweg and Peterman that in the plains in the mountains by the seas by the rivers running water running fire the air is the same and then the earth namely the air and then the earth in the great cold the great dark the air and the earth abode of stones in the great cold alas alas in the year of their Lord six hundred and something the air the earth the sea the earth abode of stones in the great deeps the great cold on sea on land and in the air I resume for reasons unknown in spite of the tennis the facts are there but time will tell I resume alas alas on on in short in fine on on abode of stones who can doubt it I resume but not so fast I resume the skull fading fading fading and concurrently simultaneously what is more for reasons unknown in spite of the tennis on on the beard the flames the tears the stones so blue so calm alas alas on on the skull the skull the skull the skull in Connemara in spite of the tennis the labors abandoned left unfinished graver still abode of stones in a word I resume alas alas abandoned unfinished the skull the skull in Connemara in spite of the tennis the skull alas the stones Cunard tennis . . . the stones . . . so calm . . . Cunard . . . unfinished . . .

Around the same time as Beckett was writing this play, or perhaps a few years earlier, the Welsh poet Dylan Thomas, intoxicated with the sounds of the English language, was creating poems that are remarkably opaque. Consider the opening two stanzas (there are five altogether) of his poem "How Soon the Servant Sun":

> How soon the servant sun,
> (Sir morrow mark),
> Can time unriddle, and the cupboard stone,
> (Fog has a bone
> He'll trumpet into meat),
> Unshelve that all my gristles have a gown
> And the naked egg stand straight,

Sir morrow at his sponge,
(The wound records),
The nurse of giants by the cut sea basin,
(Fog by his spring
Soaks up the sewing tides),
Tells you and you, my masters, as his strange
Man morrow blows through food.

Poems like this make me want to cry out that this emperor has no clothes. As far as I can discern, close to no meaning can be pulled from these lines. But how can I be sure? I cannot. All I can say is that it would probably take such a great effort to "decode" these lines that I suspect very, very few people would be willing to make it.

*　　*　　*

It is perhaps not so well known that the American singer Bob Dylan (whose name was inspired by that of Dylan Thomas) is also an author of inspired nonsense. Some of his nonsense written during the 1960's was collected and published in a book called *Tarantula*. Its tone is often bitter and it exudes the confused mood of those difficult years. Most of the pieces in the book consist of an outburst of free associations followed by a letter from some strangely-named personage or other. The following sample is called "On Busting the Sound Barrier":

the neon dobro's F hole twang & climax from disappointing lyrics of upstreet outlaw mattress while pawing visiting trophies & prop up drifter with the bag on head in bed next of kin to the naked shade—a tattletale heart & wolf of silver drizzle inevitable threatening a womb with the opening of rusty puddle, bottomless, a rude awakening & gone frozen with dreams of birthday fog/ in a boxspring of sadly without candle sitting & depending on a blemished guide, you do not feel so gross important/ success, her nostrils whimper. the elder fables & slain kings & inhale manners of furious proportion, exhale them against a glassy mud . . . to dread misery of watery bandwagons, grotesque & vomiting into the flowers of additional help to future treason & telling horrid stories of yesterday's influence/ may these voices join with agony & the bells & melt their thousand sonnets now . . . while the moth ball woman, white, so sweet, shrinks on her radiator, far away & watches in with her telescope/ you will sit sick with coldness & in an unenchanted closet . . . being relieved only by your dark jamaican friend—you will draw a mouth on the lightbulb so it can laugh more freely

forget about where youre bound youre bound for a three octave fantastic hexagram. you'll see it. dont worry. you are Not bound to pick wildwood flowers . . . . like i said, youre bound for a three octave titanic tantagram

your little squirrel,
Pety, the Wheatstraw

Dylan is not the only popular singer of the sixties to have had a literary bent. John Lennon, when he was in his early twenties, reveled in the nonsensical, and published two short books called *In His Own Write* and *A Spaniard in the Works.* They are still available, bound together in a single paperback volume. The books contain mostly nonsense poetry, although there are also several prose selections. Two of Lennon's poems will serve to illustrate his idiosyncratic style.

I SAT BELONELY

I sat belonely down a tree,
   humbled fat and small.
A little lady sing to me
   I couldn't see at all.

I'm looking up and at the sky,
   to find such wondrous voice.
Puzzly puzzle, wonder why,
   I hear but have no choice.

"Speak up, come forth, you ravel me",
   I potty menthol shout.
"I know you hiddy by this tree".
   But still she won't come out.

Such softly singing lulled me sleep,
   an hour or two or so
I wakeny slow and took a peep
   and still no lady show.

Then suddy on a little twig
   I thought I see a sight,
A tiny little tiny pig,
   that sing with all its might.

"I thought you were a lady",
   I giggle,—well I may,
To my suprise the lady,
   got up—and flew away.

THE FAULTY BAGNOSE

Softly softly, treads the Mungle
Thinner thorn behaviour street.
Whorg canteell whorth bee asbin?
Cam we so all complete,
With all our faulty bagnose?

The Mungle pilgriffs far awoy
Religeorge too thee worled.
Sam fells on the waysock-side
And somforbe on a gurled,
With all her faulty bagnose!

Our Mungle speaks tonife at eight
He tell us wop to doo
And bless us cotten sods again
Oamnipple to our jew
(With all their faulty bagnose).

Bless our gurlished wramfeed
Me curséd café kname
And bless thee loaf he eating
With he golden teeth aflame
Give us OUR faulty bagnose!

Good Mungle blaith our meathalls
Woof mebble morn so green the wheel
Staggaboon undie some grapeload
To get a little feel
of my own faulty bagnose.

Its not OUR faulty bagnose now
Full lust and dirty hand
Whitehall the treble Mungle speak
We might as wealth be band
Including your faulty bagnose

Give us thisbe our daily tit
Good Mungle on yer travelled
A goat of many coloureds
Wiberneth all beneath unravelled
And not so MUCH OF YER FAULTY BAGNOSE!

The first of these is transparent and charming, while the second is somewhat baffling and disturbing. What in the world is a *bagnose*? No clear image comes through. And why are all these bagnoses faulty? And does "faulty" have its normal meaning here? Hard to tell.

*   *   *

The idea of "normal meanings" is turned on its head in a recent book of poetry by William Benton called just that: *Normal Meanings*. One section of the book is titled "Normal Meanings"; here is an extract from it.

Escape is, escape
    was, once more,

continued.

       Vineyards
as dusky.

He watches it wrinkle into a school bell.

       It isn't music sometimes, I'm
happy.

Leaves, practically falling
off and into the air.

       Hills river

    sunset ice-cream

      cone.

       The buildings. Things
        build up. It
    must be so many
        normal meanings.

The downstairs lights. Probably I doubt.

These and other

stories.
> The loveliness
of houses.

> Clarissa is the name of the bug I just sent somewhere.

The falseness it abjures has seemed in statements
> we are losing.
> It's hard to say. A note of privilege
which turns up here in their appearances.
> I drink.

> The cobweb is becoming a strand of
> lamplight, its black heart

> blessed.

> A nice

> Elaine

> by the beer.

Some may find the amorphousness of this type of poetry amusing or engaging; others may find it tiresome, confusing. I personally find it provocative for a while, but then I begin to lose interest.

*   *   *

I have somewhat greater interest in the writings of the little-known American rhetorician, Y. Serm Clacoxia, who, in the past 25 years or so, has sporadically penned various pieces of nonsense poetry and prose. Clacoxia's prose is marked by a certain degree of vehemence and fire, although it is sometimes a little hard to figure out exactly what he is ranting and raving about. Here follows one of his most lyrical tracts, entitled "The Illusions of Alacrity".

For millennia it has been less than appreciated how futile are the efforts of those who seek to sow sobriety in the furrows of trivia. To those of us who have striven to clarify what has been left unclear, it has proven a loss. To others who, whilst valiantly straddling the fine line that divides arid piquancy from acrid pungency, have struggled to set right the many Undeeds and Unsaids of yore, life has shown itself as a beast of many colors, a mountain of many flags, a hole of many anchors.

Who, in fact, were the Outcasts of Episode, if not the champions of clarity? Where, indeed, were the witnesses to litany, when their fortress of fecundity was a-being stormed by the Ovaltine Monster, that incubus of frozen cheerios and swollen bananas? And dare one wonder, with the bassoon of lunacy so shrilly betoning the ruined fiddles of flatulism, how it is that doublethink, narcolepsy, and poseurism are unthreading themselves across our land like tall, statuesque, half-uneaten yet virtuous whippoorwhills? Can it be that a cornflake-catechism has beguiled us into an unsworn acceptance of never-takism?

What sort of entiments are they, that would uncouch a mulebound lout and churlishly swirl his burly figure, unfurl and twirl his curly figure, hurl his whirly figure, into the circuline vaults of hysteresis? With a drop of sweat unroasting his feverish brow, we decry his fate; with the patience of a juggernaut and the telemachy of a dozen opossums, we lament his disparity. And summoning all the powers that be, we unbow the jelly of our broken dreams, dashing it with the full fury of a pleistocene hurdy-gurdy against the lubrified and bulbous nexus of that which, having doomed the dinosaurs, seeks the engulfing of all that moves.

Thus we act; and perhaps action itself is the Anatole's Curlicue of our era. It is high time to recognize that action, and action alone, will be the agent that transmutes the flowery barrier of unutterability into an arbitrary but sacred iota of purposefulness, which cannot help but penetrate into an otherwise nameless and universally spaghettified lack of meaning, which smears and beclouds the crab-lit hopes of half-beings begging for deliverance from their own private, yet strangely tuberculine maelstroms that begat, and begotten were from, a howling sea of ribosomal plagiarism.

This is deliberate nonsense, of course, to be contrasted with the nondeliberate nonsense of, say, Dylan Thomas, or the nonsense to be found in crackpot letters written to scientists. Crackpot ideas seem to be an inevitable ingredient of any society in which serious scientific research is carried out; there is no way to plug all the cracks, so to speak. There is no way to ensure that only high-quality science will be done. Fortunately, most journals do not publish absolute nonsense or gobbledygook; it is filtered out at a very early stage. However, one journal I have come across whose pages are filled with utter nonsense—meant seriously—is called *Art-Language*. To show what I mean, here are two short excerpts from the May, 1975 issue. The first one is taken from the beginning of an article called "Community Work". It seems, from the table of contents, to have been written by three people collectively. The second one is taken from an article called "Vulgar and Popular Opinions", and seems to have a single author.

Dionysus gets a job. (Re: language has got a hold on U.S.) (It's a Whorfian conspiracy!)

This is hopeless manqué ontological alienation which is still dealing with ideas about 'discovery' as a function of a metaphysics of categories. Only for researchers is the failure of a modal logic industry to 'catch-my-experience'— the birth of tragedy.

Going-on in A-L indexed (somehow) is a thing-in-and-for-(dynamically)-

itself. That we never catch up with the NaturKulturLogik has little to do with the 'actualizing' sets of the frozen dialogue . . . and it's not just a ledger; our problems with set-theoretical axiomata are embedded into our praxis as more than just historical antecedents . . . more than nomological permissibility . . . more than selective filtration. We still don't recognize ourselves as very fundamental history producers.

The possibility of a defence of a set, as with 'a decision', is an index-margin of a prima facie ersatz principle for action (!). (There is no workable distinction between oratio recta and oratio obliqua.) All we are left with is a deontic Drang. Think of that as a chain strength possibility of what, eventually, comes out as a product (epistemic conditions?) and the product is not a Frankfurt-ish packing-it-all-in . . . .

A slogan (?) might be thought of as a free-form comprised of multiple structural features occurring in a (partially) given, or negotiable, unit relative to others. That is, the slogan is a unit in one sense or another. In going-on (ideologically, perhaps), a slogan is a unitary filler-for-and-of that stretch of surf < surf which is in a B×S position . . . But there is the critical issue of that 'filler' as a reified function of the pusillanimous tittle-tattle of authenticity in its ellipticality (as a Das Volk holism) . . . (e.g.) 'the Fox' material, passim, falls into that trap in dealing with its cultural space as a wantonly dialectical 'region' approaching the solution to 'the negation of essence' (of homo sapiens, art or what?).

I am tempted to quote further, to show how the wild quality of the *A-L* prose just goes on and on. But life is short. It is hard for this human being to believe that these paragraphs were meant to communicate something to anybody, but the journal appears regularly (at least it used to), and can be found on the shelves of reputable art libraries. Isn't it time that somebody blew the whistle? The curious thing about *Art-Language* is that the collective that writes it appears to consist of people who are deeply concerned with issues that hold much interest for me: the nature of reference, the relationship of wholes to parts, the connection of art and reality, the structure of society, the philosophy of set theory, the questionable existence of mathematical concepts, and so on. What is amazing is how such concepts can be so obscured by language that it is hard to make out anything except huge billows of very thick smoke.

*   *   *

An American poet whose work explores ground midway between nonsense and sense is Russell Edson. He writes tiny surrealistic vignettes that shed a strange light on life. Often he performs strange reversals, as of animate and inanimate beings, or humans and animals. His grammar is also oblique, one of his favorite devices being to refer repeatedly to something specific with the indefinite article "a", thus disorienting the reader. A typical sample of Edson's style is the following, drawn from his book *The Clam Theater:*

When Science is in the Country

When science is in the country a cow meows and the moon jumps from limb to limb through the trees like a silver ape.

The cow bow-wows to hear all voice of itself. The grass sinks back into the earth looking for its mother.

A farmer dreamed he harvested the universe, and had a barn full of stars, and a herd of clouds fenced in the pasture.

The farmer awoke to something screaming in the kitchen, which he identified as the farmerette.

Oh my my, cried the farmer, what is to become of what became?

It's a good piece of bread and a bad farmer man, she cried.

Oh the devil take the monotony of the field, he screamed.

Which grows your eating thing, she wailed.

Which is the hell with me too, he screamed.

And the farmerette? she screamed.

And the farmerette, he howled.

A scientist looked through his magnifying glass in the neighborhood.

This eerie tale leaves one with a host of unresolved images. That, of course, is Edson's intent. And in this regard, Edson's work is quite typical. Most of the nonsense of the twentieth century, it seems, has this deliberately upsetting quality to it, reflecting a deep malaise. It is utterly different from the nonsense of the preceding centuries. Similar trends exist in the other arts, particularly in music, where "classical" composers have lost 99 percent of their audience by their experimentation with randomness and cacophony. However, the spirit of experimentation has also crept into rock music, where electronic sounds and unusual rhythms are occasionally heard. The surrealistic, nonsensical spirit also pervades the names of popular groups, such as "Iron Butterfly", "Tangerine Dream", "Led Zeppelin", "Joy of Cooking", "Human Sexual Response", "Captain Beefheart", "Brand X", "Jefferson Starship", "Average White Band", and so on.

*   *   *

Perhaps one of the virtues of nonsense is that it opens our minds to new possibilities. The mere juxtaposition of a few arbitrary words can send the mind soaring into imaginary worlds. It is as if sense were too mundane, and we need a breather once in a while. Perhaps sense is also too confining. Nonsense stresses the incomprehensible face of the universe, while sense stresses the comprehensible. Clearly both are important. Zen teachings have striven to impart the path to "enlightenment". Although I don't believe that such a mystical state exists, I am fascinated by the paths that are offered. Zen itself is perhaps the archetypal source of utter nonsense. It seems fitting to close this column with two Zen koans taken from the *Mumonkan,* or "Gateless Gate"—a set of koans commented upon by the Zen master Mumon in the thirteenth century.

Joshu Examines a Monk in Meditation

Joshu went to a place where a monk had retired to meditate and asked him: "What is, is what?" The monk raised his fist. Joshu replied: "Ships cannot remain where the water is too shallow." And he left. A few days later Joshu went again to visit the monk and asked the same question. The monk answered the same way. Joshu said: "Well given, well taken, well killed, well saved." And he bowed to the monk.

Mumon's comment:

The raised fist was the same both times. Why is it Joshu did not admit the first and approved the second one? Where is the fault? Whoever answers this knows that Joshu's tongue has no bone so he can use it freely. Yet perhaps Joshu is wrong. Or, through that monk, he may have discovered his mistake. If anyone thinks that the one's insight exceeds the other's, he has no eyes.

Mumon's Poem:

The light of the eyes is as a comet,
And Zen's activity is as lightning.
The sword that kills the man
Is the sword that saves the man.

Learning is Not the Path

Nansen said: "Mind is not Buddha. Learning is not the path."

Mumon's comment:

Nansen was getting old and forgot to be ashamed. He spoke out with bad breath and exposed the scandal of his own home. However, there are few who appreciate his kindness.

Mumon's Poem:

When the sky is clear the sun appears,
When the earth is parched rain will fall.
He opened his heart fully and spoke out,
But it was useless to talk to pigs and fish.

Post Scriptum.

I was quite aware that I had omitted some nonsense specialists, such as James Joyce, when I wrote this column. But there were reasons. I haven't studied Joyce, and I feel there is a lot of complexity there. To call Joyce's strange concoctions "nonsense" is to miss the mark.

Several people wrote in, disappointed that I did not include anything by Walt Kelly, the creator of "Pogo". I have to agree that Kelly was a unique writer of ingenious and charming nonsense. In fact, I was lucky enough to grow up knowing "The Pogo Song Book", a record of some of Kelly's most inspired silly songs, some of them belted out by Kelly himself. One that gets across the flavor very well is this one:

TWIRL, TWIRL

Twirl! Twirl! Twinkle between!
The tweezers are twist in the twittering twain.
Twirl! Twirl! Entwiningly twirl
'Twixt twice twenty twigs passing platitudes plain.
Plunder the plover and rover rides round.
Ride all the rungs on the brassily bound,
Billy, Swirl! Swirl! Swingingly swirl!
Sweep along swoop along sweetly your swain.

The poem is catchy and rhythmic, and I cannot read it without hearing the song in my head. Few people know that Kelly was a good composer of catchy melodies. But his songs, unlike his lyrics, follow very ordinary, "sensible" rules of musical syntax.

Two other pieces of inspired nonsense that I have run across since writing this column are Tom Phillips' *A Humument,* and Luigi Serafini's *Codex Seraphinianus.* The former, subtitled "A Treated Victorian Novel", was made, by a sort of literary cannibalism, from another novel entitled *A Human Document,* itself written by a little-known Victorian novelist named William Hurrell Mallock. Phillips "treated" this novel by colorfully and imaginatively overpainting nearly all its pages, blotting out most of the text, leaving only a select few words or letters to poke their heads through and make cameo appearances now and then. This creation (or revelation?) of hidden messages in someone else's text yields some very strange effects. The first page of *A Humument* reads this way (I have slightly modified the two-dimensional placement of the words on the page):

The following sing I a book.
a book of art
of mind and art
that which he hid
reveal I.

* * *

Codex Seraphinianus is a much more elaborate work. In fact, it is a highly idiosyncratic *magnum opus* by an Italian architect indulging his sense of fancy to the hilt. It consists of two volumes in a completely invented language (including the numbering system, which is itself rather esoteric), penned entirely by the author, accompanied by thousands of beautifully drawn color pictures of the most fantastic scenes, machines, beasts, feasts, and so on. It purports to be a vast encyclopedia of a hypothetical land somewhat like the earth, with many creatures resembling people to various degrees, but many creatures of unheard-of bizarreness promenading throughout the countryside. Serafini has sections on physics, chemistry, mineralogy (including many drawings of elaborate gems), geography, botany, zoology, sociology, linguistics, technology, architecture, sports (of all sorts), clothing, and so on. The pictures have their own internal logic, but to our eyes they are filled with utter *non sequiturs.*

A typical example depicts an automobile chassis covered with some huge piece of what appears to be melting gum in the shape of a small mountain range. All over the gum are small insects, and the wheels of the "car" appear to have melted as well. The explanation is all there for anyone to read, if only they can decipher Serafinian. Unfortunately, no one knows that language. Fortunately, on another page there is one picture of a scholar standing by what is apparently a Rosetta Stone. Unfortunately, the only language on it, besides Serafinian itself, is an unknown kind of hieroglyphics. Thus the stone is of no help unless you *already* know Serafinian. Oh, well... Many of the pictures are grotesque and disturbing, but others are extremely beautiful and visionary. The inventiveness that it took to come up with all these conceptions of a hypothetical land is staggering.

Some people with whom I have shared this book find it frightening or disturbing in some way. It seems to them to glorify entropy, chaos, and incomprehensibility. There is very little to fasten onto; everything shifts, shimmers, slips. Yet the book has a kind of unearthly beauty and logic to it, qualities pleasing to a different class of people: people who are more at ease with free-wheeling fantasy and, in some sense, craziness. I see some parallels between musical composition and this kind of invention. Both are abstract, both create a mood, both rely largely on style to convey content.

Music is, in a way, a kind of nonsense that nobody really understands. It captivates nearly every human being who can hear and yet, for all that, we still know amazingly little about how music works its wonders. But if music is a kind of auditory nonsense, that does not prevent there from arising even more extreme brands of auditory super-nonsense. The works of Karl-Heinz Stockhausen, Peter Maxwell Davies, Luciano Berio, and John Cage will provide a wonderful introduction to that genre, in case some reader does not know what I am talking about. Especially if you like the banging of

FIGURE 11-1. *One page from David Moser's "Metaculture Comics" (1979).*

garbage-can lids or the sound of gangland murders, their "musical offerings" are sure to be right up your alley.

David Moser is as fascinated with fringe-language as I am, and has explored many uncharted regions in that territory. His longest and most adventurous journey consisted of the writing and drawing of a roughly 40-page booklet called "Metaculture Comics". Inspired by James Joyce, this volume contains some of the most original and zany meaningless writings I have ever seen. It is also chock-full of the frame-breaking and self-referential devices so beloved by modern graphic designers. A one-page sample is shown in Figure 11-1.

*   *   *

The purpose of this column was to emphasize the very fine line that separates the meaningful from the meaningless. It is a boundary line that has a great deal to do with the nature of human intelligence, because the question of how meaning emerges out of meaningless constituents when combined in certain patterned ways is still a perplexing one. Computers are good at producing very simple passages that—to us—seem to have meaning, and they are excellent at producing passages that are utterly devoid of meaning. It will be interesting to see if someday a computer can tread the line and produce an artistic exploration of meaning by producing provocative nonsense in the same way as these human explorers of the territory have done.

12

Variations on a Theme
as the Crux of Creativity

October, 1982

You see things; and you say "Why?"
But I dream things that never were; and say "Why not?"

—George Bernard Shaw in *Back to Methuselah*

WHEN I first heard this beautiful line, it made a deep impression on me. It was in the spring of 1968, during the presidential campaign, and Robert Kennedy had made this line his theme. I thought it was wonderfully poetic, and I assumed he himself had dreamt it up. Only many years later did I find out I was quite wrong: Not only had he not made it up, but the character who utters it in the Shaw play is the snake in the Garden of Eden! How disturbing! Why couldn't it have been the way I thought?

"To dream things that never were"—this is not just a poetic phrase, but a truth about human nature. Even the dullest of us is endowed with this strange ability to come up with counterfactual worlds and to dream. But why do we have this ability—in fact, this proclivity? What sense does it make? And—how can one "see" what is visibly *not there*?

On my table sits a Rubik's Cube. I look at it and see a $3 \times 3 \times 3$ cube whose faces turn. I see—so it seems to me—what is there. But some people looked at that cube and saw things that *weren't* there. They saw cubes with shaved edges, spherical "cubes", differently colored cubes, Magic Dominos, $2 \times 2 \times 2$ cubes, $4 \times 4 \times 4$ and higher-order cubes, skew-twisting cubes, pyramids, octahedra, dodecahedra, icosahedra, four-dimensional magic polyhedra. (See figures galore in Chapters 14 and 15.) And the list is not complete yet! Just you wait!

How did this come about? How is it that, in looking directly at something solid and real on a table, people can see far beyond that solidity and reality —can see an "essence", a "core", a "theme" upon which to devise variations? I must stress that the solid cube itself is not the theme (although it is convenient and easy to speak as if it were). In the mind of each person who perceives a Rubik's Cube there arises a *concept* that we could call "Rubik's-Cubicity". It's not the same concept in each mind, just as not everyone has the same concept of asparagus or of Beethoven. The variations that are spun off by a given cube-inventor are variations on that concept. In a discussion of perception and invention, this distinction between an object and some mind's concept of the object is simple but crucial.

Now when Eve Rybody comes up with a new variation—let's say the $4 \times 4 \times 4$—is it as a result of wracking her brain, trying as hard as she can to "go against the grain", so as to come up with something original? Does she think to herself, "Golly, that Rubik must have *really* exerted himself to come up with this totally new idea, therefore I too must strain my mind to its limits in order to invent something original."? No, no, no! A thousand times no. Einstein didn't go around wracking his brain, muttering to himself, "How, oh how, can I come up with a Great Idea?" Like Einstein (although perhaps on a lesser scale), Eve never needs to ask herself, "Hmm, let's see, shall I try to figure out some way to spin off a variation on this object sitting here in front of me?" No; she just does what comes naturally.

The bottom line is that invention is much more like falling off a log than like sawing one in two. Despite Thomas Alva Edison's memorable remark, "Genius is 2 percent inspiration and 98 percent perspiration", we're not all going to become geniuses simply by sweating more or resolving to *try harder.* A mind follows its path of least resistance, and it's when it feels easiest that it is most likely being its most creative. Or, as Mozart used to say, things should "flow like oil"—and Mozart ought to know! Trying harder is not the name of the game; the trick is getting the right concept to begin with, so that making variations on it is like taking candy from a baby.

Uh-oh—now I've given the cat away! So let me boldly state the thesis that I shall now elaborate: *Making variations on a theme is really the crux of creativity.*

*   *   *

On the face of it, this thesis is crazy. How can it possibly be true? Aren't variations simply derivative notions, never truly original creations? Isn't the notion of a $4 \times 4 \times 4$ cube simply a result of "twiddling a knob" on the concept of Rubik's-Cubicity? You merely twist the knob from its "factory setting" of 3 to the new setting of 4, and presto—you've got it! An inner voice protests: "That's just too easy. That's certainly not where Rubik's Cube, the *Rite of Spring,* relativity, or *Romeo and Juliet* came from, is it? Isn't there a 'magic spark' that leaps across a gap when a Rubik or a Stravinsky or an Einstein or a Shakespeare comes up with a great idea, something that is patently lacking when an Eve Rybody merely twiddles a knob on an already-existing notion?"

Well, of course, inventing the notion of a $4 \times 4 \times 4$ cube is far less deep than coming up with special or general relativity. I'd be the last to deny that. But that doesn't mean that the underlying mental processes are necessarily based on totally different principles. Of course, there is a boring sense in which the underlying mental processes in your brain, my brain, Eve's brain, and Einstein's brain are all "the same"—namely, they all depend on neural hardware. But it is not at such a microscopic, such a biological level that I mean it when I suggest that the underlying mental processes in different brains are somehow the same. What I mean is that there are mechanisms, processes, call them what you will, that can be described functionally, without reference to the neural substrate that enables them to take place in brains.

Thus, a notion like "twiddling a knob on a concept" bears no relation to the activities of neurons in the brain—or at least no obvious relation. Well then, is there any reality to it, or is it just a metaphor? If someday we at last come to understand the brain, will we then be confident that we're on solid ground when we speak of a brain literally *containing concepts*? Or will such statements forever remain shaky and metaphorical *façons de parler*, compared to such hard-science *facts* as "At the back of each human brain there is a cerebellum"? Well, until words like "concept" have become terms as scientifically legitimate as, say, "neuron" or "cerebellum", we will not have come anywhere close to understanding the brain—at least not in my book.

However, it must be admitted that at present, words like "concept" are only metaphorical. They are protoscientific terms awaiting explication. But this is a very good reason to try to flesh them out as much as possible, to try to see what the metaphor of "twiddling knobs on a concept" involves. Pinning down the meaning of such a metaphor will help us know much more clearly what we would ideally want from a "hard-science" explanation of the brain.

This metaphor makes your imagination conjure up a vision of a tangible thing called a "concept" that literally has a set of knobs on it, just waiting to be twiddled. What I picture in my mind's eye is something that, instead of being built out of millions of neurons, is more like a metallic "black box" with a panel on it, containing a row of plastic knobs with little pointers on them, telling you what each one's setting is.

Just to make this image more concrete, let me describe a genuine example of such a black box with knobs. Back in the old days of player pianos, good pianists made piano rolls of all sorts of wonderful music. Nowadays, you can buy phonograph records of those rolls being played back on player pianos —but you can do better than that. Many of the best rolls made on a special kind of piano called a *Vorsetzer* have been converted into digital cassette tapes—not to be played on tape recorders, but on pianos specially equipped with a device called a "Pianocorder". This "reads" the magnetic tape and converts it into instructions to the keyboard and pedals, so that your piano then plays the piece. Each Pianocorder has a black box on the front of which is a control panel with a row of three knobs (*tempo, pianissimo,* and *fortissimo*) and one switch ("soft pedal"). By twisting the *tempo* knob you can make Rachmaninoff speed up, by twiddling the *pianissimo* and *fortissimo* knobs you can make Horowitz play more softly or Rubinstein more loudly. It's too bad there's not a knob labeled "pianist" so that you can select who plays. After all, it would be interesting to change Horowitzes in midstream.

*   *   *

This device takes us one step toward realizing a dream of the unique Canadian pianist Glenn Gould. Gould is very tuned in to the electronic age, and for years has been advocating using computers to allow people to control the music they hear. You begin with an ordinary recording of, say, Glenn Gould himself playing a concerto by Mozart. But this is merely raw data for you to tamper with. On your space-age record player, you have a bunch of knobs that allow you to slow the music down or to speed it up *ad libitum,* to control the volume of all the separate sections of the orchestra, even to correct for high notes played too flat by the violinists! In effect, you become the conductor, with knobs to control every aspect of the performance, dynamically. The fact that it was originally Glenn Gould at the piano is, by the time you're done with it, irrelevant. By now you've totally taken over and made it your very own performance! Presumably, such systems would eventually evolve to the point where you could start with the mere written score, dispensing entirely with the acoustic recording stage.

But why not carry this further, then? If we are allowing ourselves to fantasize, why not go as far as we can imagine? Why should our "raw data" be limited to the finite universe of already-composed pieces? Why could there not be a knob to control the mood of the composition, another to control the composer whose style it is to be written in? This way, we could get a new piece by our favorite composer in any desired mood. But really, this is too conservative. Why should we be limited to the finite universe of already-born composers? Why could there not be a knob to allow us to interpolate between composers, thus making it possible for us to tune our music-making machine to an even mixture of Johann Sebastian Bach, Giuseppe Verdi, and John Philip Sousa (ugh!), or a position halfway between Schubert and the Sex Pistols (super-ugh!)? And why stop at interpolation? Why not extrapolate beyond a given composer? For instance, I might want to hear a piece by "the composer who is to Ravel as Ravel is to Chopin". The machine would merely need to calculate the ratios of its knob settings for Ravel and Chopin, and then multiply the Ravel-settings by those same ratios to come up with a super-Ravel.

It's no trickier than solving any old analogy problem—you know, simple problems like this:

What is to a triangle as a triangle is to a square?
What is to a honeycomb as a knight's move is to a city grid?

What is to four dimensions as the "impossible triangle" illusion is
    to three?
What is to Greece as the Falkland Islands are to Britain?
What is to visual art as fugues are to music?
What is to a waterbed as ice is to water?
What is to the United States as the Eiffel Tower is to France?
What is to German as Shakespeare's plays are to English?
What is to English as simplified characters are to Chinese?
What is to 1-2-3-4-4-3-2-1 as 4 is to 1-2-3-4-5-5-4-3-2-1?
What is to *pqc* as *abc* is to *aqc?*

The truth is, of course, that analogy problems are staunchly resistant to
mechanization. The knobs on most concepts are not so apparent as to allow
us to just read their settings right off. The examples above simply carried
a sensible thought to a ludicrous extreme. However, it is still worthwhile to
look seriously at the idea that a concept can be considered as a "knobbed
machine" whose knobs can be twiddled to produce a bewildering array of
variations.

*   *   *

The Rubik's-Cube concept, with its "order" knob set at 3, produces an
ordinary $3 \times 3 \times 3$ cube—and with that knob set at 4, a $4 \times 4 \times 4$. Come to
think of it, doesn't there have to be a separate knob for each dimension, so
that you can twiddle each one independently of the others? After all, not all
variations have to be cubical. The Magic Domino is $3 \times 3 \times 2$. So if we agree
that there are *three* knobs defining the shape, then in the original cube they
all just accidentally happened to have the same setting. Now given these
three knobs, we can use our concept—our knobbed machine—to generate
such mental objects as a $7 \times 7 \times 7$ Rubik's Cube, a $2 \times 2 \times 8$ Magic Domino,
even a $3 \times 5 \times 9$ Rubik's Magic Brick (or, if you'll pardon me, a "Rubrick").
But wait a minute—if there really are just three knobs, then we're locked
into three dimensions! Obviously we don't want *that.* So let's add a fourth
knob to control the length in the fourth dimension. With this knob, we can
now make a four-dimensional $2 \times 3 \times 5 \times 7$ Rubrick, as well as any Rubik's
Tesseract that we might want. But needless to say, once we've gone through
the gate from three dimensions to four, certainly we should expect to be
able to go further. For any *n,* we could imagine *n*-dimensional Rubik's
objects—for example, a $2 \times 3 \times 4 \times 5 \times 6 \times 7 \times 8$ Hyper-Rubrick. But now
something peculiar has happened. We must now conceive of our machine
—our concept—as having a potentially *unlimited* number of knobs on it (one
for each dimension in *n*-dimensional space). If *n* is set to 3, there need only
be 3 more knobs. But if *n* is 100, we need 100 extra knobs!
No real machine has a variable number of knobs. Now this may sound like
a somewhat trivial observation. However, it leads into some tricky waters.

The point is that, if we wish to keep on using the metaphor of a concept as a machine with knobs on it, we have to stretch the very concept of "knob". New knobs must be able to sprout, depending on the settings of other knobs. Or you can think of it this way, if you wish: on each concept, there are potentially an infinite number of knobs, and at any moment, some new knobs may get revealed as a consequence of the settings of other knobs.

I'm not sure I like that view, however. It's too cut and dried, too closed and predetermined for my tastes. I am more in favor of a view that says that the knobs on any one concept depend on the set of concepts that happen to be awake simultaneously in the mind of the person. This way, new knobs can spring into existence seemingly out of nowhere; they don't all have to be present from the outset in the isolated concept. If we go back to Rubik, this would mean that *his* concept of Rubik's Cube didn't (and still doesn't) explicitly—or even implicitly—contain all the possible variations that people may come up with. Rubik anticipated, and even designed, many of the objects that have subsequently appeared and that we perceive as "variations on a theme"—but certainly, his mind did not exhaust that fertile theme. Once the concept entered the public domain, it started migrating and developing in ways that Rubik could never have anticipated.

*   *   *

There is a way that concepts have of "slipping" from one into another, following a quite unpredictable path. Careful observation and theorizing about such slippages affords us perhaps our best chance to probe deeply into the hidden murk of our conceptual networks. An example of such a slip is furnished to us whenever we make a typo or a grammatical mistake, utter a malapropism ("She's just back from a one-year stench at Berkeley") or a malaphor (a novel phrase concocted unconsciously from bits and pieces of other phrases, such as "He's such an easy-go-lucky fellow" or "Uh-oh, now I've given the cat away"), or confuse two concepts at a deeply semantic level (*e.g.*, saying "Tuesday" but meaning "February", or saying "midnight" in lieu of "zero degrees"). These types of slip are totally accidental and come straight out of our unconscious mind.

However, sometimes a slippage can be nonaccidental yet still come from the unconscious mind. By "nonaccidental" here, I do not mean to imply that the slip is *deliberate*. It's not that we say to ourselves, "I think I shall now slip from one concept into a variation of it"; indeed, that kind of deliberate, conscious slippage is most often quite uninspired and infertile. "How to Think" and "How to Be Creative" books—even very thoughtful ones such as George Pólya's *How to Solve It*—are, for that reason, of little use to the would-be genius.

Strange though it may sound, *nondeliberate yet nonaccidental slippage permeates our mental processes, and is the very crux of fluid thought.* That is my firmly held conviction. This subconscious manufacture of "subjunctive variations on a theme" is something that goes on day and night in each of us, usually without our slightest awareness of it. It is one of those things that, like air or gravity or three-dimensionality, tend to elude our perception because they define the very fabric of our lives.

To make this concrete, let me contrast an example of "deliberate" slippage with an example of "nondeliberate but nonaccidental" slippage. Imagine that one summer evening you and Eve Rybody have just walked into a surprisingly crowded coffeehouse. Now go ahead and manufacture a few variants on that scene, in whatever ways you want. What kinds of things do you come up with when you deliberately "slip" this scene into hypothetical variants of itself?

If you're like most people, you'll come up with some pretty obvious slippages, made by moving along what seem to be the most obvious "axes of slippability". Typical examples are:

I could have come with Ann Yone instead of Eve Rybody.
We could have gone to a pancake house instead of a coffeehouse.
The coffeehouse could have been nearly empty instead of full.
It could have been a winter's evening instead of a summer's evening.

Now contrast your variations with one that I overheard one evening this past summer in a very crowded coffeehouse, when a man walked in with a woman. He said to her, "I'm sure glad I'm not a waitress here tonight!" This is a perfect example of a subjunctive variation on the given theme—but unlike yours, this one was made without external prompting, and it was made for the purposes of communication to someone. The list above looks positively mundane next to this casually tossed-off remark. And the remark was not considered to be particularly clever or ingenious by his companion. She merely agreed with the thought by saying "Yeah." It caught my attention not so much because I thought it was clever, but mostly because I am always on the lookout for interesting examples of slippability.

I found this example not just mildly interesting, but highly provocative. If you try to analyze it, it would appear at first glance to force you as listener to imagine a sex-change operation performed in world record time. But when you simply *understand* the remark, you see that in actuality, there was no intention in the speaker's mind of bringing up such a bizarre image. His remark was much more figurative, much more abstract. It was based on an instantaneous perception of the situation, a sort of "There-but-for-the-grace-of-God-go-I" feeling, which induces a quick flash to the effect of "Simply because I am human, I can place myself in the shoes of that harried waitress—therefore *I could have been* that waitress." Logical or not, this is the way our thoughts go.

So when you look carefully, you see that this particular thought has practically nothing to do with the speaker, or even with the waitresses he sees. It's just his flip way of saying, "Hmm, it sure is busy here tonight." And that's of course why nobody really is thrown for a loop by such a remark. Yet it was stated in such a way that it invites you to perform a "light" mapping of him onto a waitress, just barely noticing (if at all) that there is a sex difference. What an amazingly subtle thought process is involved here!

And what is even more amazing (and frustrating) to me is how hard it is to point out to people how amazing it is! People find it very hard indeed to see what's amazing about the ordinary behavior of people. They cannot quite imagine how it might have been otherwise. It is very hard to slip mentally into a world in which people would *not* think by slipping mentally into other worlds—very hard to make a counterfactual world in which counterfactuals were not a key ingredient of thought.

Another quick example: I was having a conversation with someone who told me he came from Whiting, Indiana. Since I didn't know where that was, he explained, "Whiting is very near Chicago—in fact, it would be in Illinois if it weren't for the state line." Like the earlier one, this remark was dropped casually; it was certainly not an effort to be witty. He didn't chuckle, nor did I. I simply flashed a quick smile, signaling my understanding of his meaning, and then we went on. But try to analyze what this remark means! On a logical level, it is somewhat like a tautology. *Of course* Whiting would be in Illinois if the Illinois state line made it be so—but if *that's* all he meant, it is an empty remark, because it holds just as well for cities thousands of miles from Chicago. But clearly, the notion he had in mind was that there is an accidental quality to where boundary lines fall, a notion that there are counterfactual worlds "close" to ours, worlds in which the Illinois-Indiana line had gotten placed a couple of miles further east, and so on. And his remark tacitly assumed that he and I shared such intuitions about the impermanence and arbitrariness of geographical boundary lines, intuitions about how state lines could "slip".

Remarks like this betray the hidden "fault lines of the mind"; they show which things are solid and which things can slip. And yet, they also reveal that *nothing* is reliably unslippable. Context contributes an unexpected quality to the knobs that are perceived on a given concept. The knobs are not displayed in a nice, neat little control panel, forevermore unchangeable. Instead, changing the context is like taking a tour around the concept, and as you get to see it from various angles, more and more of its knobs are revealed. Some people get to be good at perceiving fresh new knobs on concepts where others thought there were none, just as some people get to be good at perceiving mushrooms in a forest where others see none, even when they stare mightily.

*   *   *

It may still be tempting to think that for each well-defined concept, there must be an "ultimate" or "definitive" set of knobs such that the abstract space traced out by all possible combinations of the knobs yields all possible instantiations of the concept. A case in point is the concept of the letter 'A'. The typographically naïve might think that there are four or five knobs to twiddle here, and that's all. However, the more you delve into letter forms, the more elusive any attempt to parametrize them mathematically becomes. One of the most valiant efforts at "knobbifying the alphabet" has been the letterform-defining system called "Metafont", developed at Stanford by the well-known computer scientist Donald Knuth.

Knuth's purpose is not to give the ultimate parametrization of the letters of the alphabet (indeed, I suspect that he would be the first to laugh at the very notion), but to allow a user to make "knobbed letters"—we could call them *letter schemas*. This means that you can choose for yourself what the variable aspects of a letter are, and then, with Metafont's aid, you can easily construct knobs that allow those aspects to vary. This includes just about anything you can think of: stroke lengths, widenings or taperings of strokes, curvatures, the presence or absence of serifs, and so on. The full power of the computer is then at your disposal; you can twiddle away to your heart's desire, and the computer will generate all the products your knob-settings define.

Going further than letters in isolation, Knuth then allowed letters to *share parameters*—that is, a single "master knob" can control a feature common to a group of related letters. This way, although there may be hundreds of knobs when you count the knobs on all the control panels of all the letters of the alphabet, there will be a far smaller number of master knobs, and they will have a deeper and more pervasive influence on the whole alphabet. What happens, in effect, is that by twiddling the master knobs alone, you have a way of drifting smoothly through a space of typefaces.

Perhaps Knuth's greatest virtuoso trick yet with Metafont is what he did with Psalm 23, which in this version consists of 593 characters. (See Figure 12-1.) Knuth had defined a full set of letters that shared 28 "master knobs". He began his printed version of the psalm with all 28 master knobs at their leftmost settings. Then, letter by letter, he inched his way toward the rightmost settings, turning each knob 1/592 of the way, so that by the time he had reached the final letter, the extreme opposite end of the spectrum had been attained. In one sense, every letter in this version of the psalm is printed in a different typeface! And yet the transition is so smooth as to be locally undetectable even to a finely trained eye. This example is drawn from Knuth's inspiring article in *Visible Language* entitled "The Concept of a Meta-Font".

One of Knuth's main theses is that with computers, we now are in the position of being able to describe not just a thing in itself, but *how that thing would vary*. Metafont epitomizes this thesis. In a sense, the computer, rather than simply blindly reproducing fixed letter shapes, has a crude "understanding" of what it is drawing, created by the designer who "knobbified" the letters. And yet, one should be careful not to fall under the illusion, so easily created by Metafont's extraordinary power, that these

The LORD is my shepherd;
  I shall not want.
He maketh me to lie down
      in green pastures:
    he leadeth me
      beside the still waters.
He restoreth my soul:
    he leadeth me
      in the paths of righteousness
      for his name's sake.
Yea, though I walk through the valley
      of the shadow of death,
    I will fear no evil:
  for thou art with me;
      thy rod and thy staff
      they comfort me.
Thou preparest a table before me
      in the presence of mine enemies:
    thou anointest my head with oil,
      my cup runneth over.
Surely goodness and mercy
      shall follow me
      all the days of my life:
    and I will dwell
      in the house of the LORD
      for ever.

FIGURE 12–1. *Psalm 23, printed by Donald Knuth's* METAFONT *program. It starts out in an old-fashioned, highly serifed typeface and gradually modulates into a modernistic, sans-serif typeface. Each step, imperceptible on its own, is accomplished by making a tiny shift in 28 parameters governing the overall appearance of the computerized alphabet.*

---

28 master knobs—or *any* finite set of knobs—might actually span the entire space of *all possible typefaces.* This is about as far from the truth as would be the claim that the space of *all possible face types* (see Figure 12-2) could be captured in a computer program with 28 knobs.

Even the space of all versions of the letter 'A' is only barely explored when you twiddle *all* the knobs in Knuth's representation of 'A'—not just the 28 master knobs it shares with other letters, but the many "private" knobs it has as well. Even a thousand knobs would not suffice to cover the variety of letter 'A's that people recognize easily. Some evidence of the richness of the 'A' concept is shown in Figure 12-3. These 'A's are all taken from real typefaces in the 1982 Letraset Catalogue. To illustrate that such richness is not a quirk of our writing system, I have assembled, in Figure 12-4, a similar collection of variants of the Chinese character meaning "black"

FIGURE 12–2. *Sixteen highly diverse human faces, culled from Federico Fellini's extensive library of still photos of people.* [*From* Fellini's Faces, *by Christian Strich.*]

FIGURE 12–3. *56 'A's in different styles, all drawn from a recent Letraset catalogue. The names of their respective typefaces are given on the facing page. To native readers of the Latin alphabet, it is an almost immediate visual experience to recognize how any one of them is an 'A'. No conscious processing is required. A couple of these seem far-fetched, but the rest are quite obvious. The most canonical of all 56 is probably Univers (D-3). Note that no single feature, such as having a pointed top or a horizontal crossbar (or even a crossbar at all!) is reliable. Even being open at the bottom is unreliable. What is going on here? (Compare this figure to Figure 24-13.)*

	A	B	C	D	E	F	G
1	Balmoral	Cardinal	Squire	Glastonbury	Arnold Böcklin	Bottleneck	Countdown
2	Eckmann Schrift	Futura Black	Hobo	Lazybones	Old English	Revue	Park Avenue
3	Romic Bold	Tintoretto	Vivaldi	Univers 67	Airkraft	Apollo	Algerian
4	Astra	Baby Teeth	Block Up	Bombere	Buster	Calypso	Columbian Italic
5	Aristocrat	Company	Glaser Stencil	Cathedral	Good Vibrations	Le Golf	Harrington
6	Harlow Solid	Motter Ombra	Masquerade	Phyllis	Pluto Outline	Process	Primitive
7	Magnificat	Quicksilver	Raphael	Roco	Shatter	Stripes	Sinaloa
8	Stop	Stack	Piccadilly	Neptun	Motter Tektura	Odin	Yagi Link Double

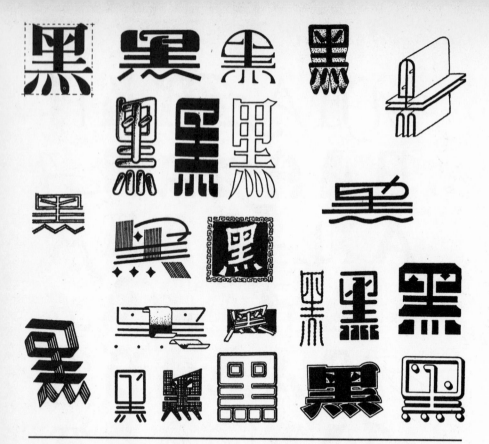

FIGURE 12–4.   *23 "hēi"'s (the Chinese character meaning "black") in different styles, drawn from a variety of "artistic-character catalogues". To native readers of Chinese, it is an almost immediate visual experience to recognize how any one of them is a "hēi". No conscious processing is required. None of these is as far-fetched as the extreme 'A's in the previous figure. For non-readers of Chinese (or even non-native readers of Chinese) it requires some conscious processing to "unmask" many of these. The most canonical of all 23 are: the one enclosed in dotted lines in the upper left corner, and the framed one in the very center (ironically not black, but white). Try to see how the various features of the "Platonic" character are implanted in these mortal incarnations. One learns here to appreciate the French saying* Plus ça change, plus c'est la même chose.

(pronounced "hēi", rhyming with 'a'). I found them in some Chinese-language graphic-design catalogues. This figure is a real eye-opener for people who don't read Chinese. They usually ask incredulously, "You mean Chinese people can easily tell that these are all the same character?!" Of course they can, and in a split second—just as we can for the matrix of 'A's.

There is a crucial distinction to be drawn here. A machine with one off-on switch (the most trivial kind of knob) for each square in a 500×500 grid will certainly define any of the 'A's shown—but it will not exclude 'B's or *hēi*'s or pictures of your grandmother or of trolleycars. It is another matter altogether to define a set of knobs whose twiddling covers all the 'A's, showing all the interpolations between them (as well as extrapolations in all possible directions)—yet never leads you out of the space of recognizable

'A's. This is far trickier! Similarly, it is a nearly trivial project to write a computer program that in theory writes all possible sequences and combinations of tones in all possible rhythmic patterns—but that is a far cry from writing a program that produces only pieces *in the style of Bach.* Putting on the constraints makes the program unutterably more complex!

What Metafont gives you, rather than the full space of all typefaces or 'A's, is a *subspace,* and such a tightly related subspace that it is perhaps best to call it a *family.* Nobody would be able to predict butterflies from having studied ants and wasps and beetles. Certainly no currently imaginable *program* would, anyway. Likewise, nobody would be able to predict the full magnitude of the *concept* of 'A', from seeing only the family traced out by the finite number of knobs in any realistic Metafont program for 'A'.

The next stage beyond Metafont will be a program that, on its own, can extract a set of knobs from a set of given input letters. This, however, is a program for the distant future. At present, it takes a highly trained and perceptive typeface designer months to convert a set of letterforms into Metafont programs with knobs flexible enough to warrant the trouble taken. It would be relatively easy to do it in some crude mechanical way, but what one wants is for stylistic unity to be preserved even as the master knobs are twiddled—and therefore, the task of automating the production of Metafont programs amounts to automation of artistic perception. It's not just around the corner.

*   *   *

There is a curious book called *One Book Five Ways,* published in 1978 by William Kaufmann, Inc. It came about this way. As an educational experiment in comparative publishing procedures, a manuscript on indoor gardening was sent around to five different university presses, and they all cooperated in coming up with full publication versions of the book, which turned out to be stunningly different at all conceivable levels. William Kaufmann had the bright idea of publishing pieces of the various versions side by side; what resulted was this elegant "metabook". It brings home the meaning of the old saying that there's more than one way to skin a cat.

Making this book was an extravagant foray into "possible worlds", the kind of thing that seems very hard to do. One of Knuth's points, however, is that as computers become more sophisticated and common, the notion of skinning a cat in nine different ways will gradually become less extravagant. Once your "cat" has been represented inside a powerful computer program, it is no longer just *one cat;* it has become, instead, a "cat-schema"—a mold for many cats at once, and you can skin them all differently (or at least until the cat-schema runs out of lives).

Text formatters and computer typesetting present us easily with many alternative versions of a piece of text. Metafont shows us how letterforms can glide into alternative versions of themselves. It is now up to us to

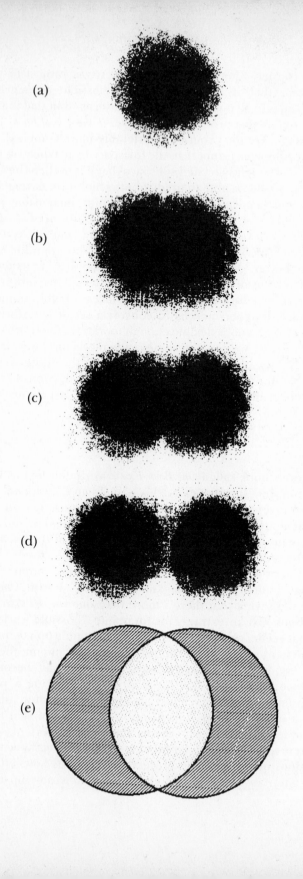

(a)

(b)

(c)

(d)

(e)

continue this trend of extending our abilities to see further into the space of possibilities surrounding what *is*. We should use the power of computers to aid us in seeing the full concept—the implicit "sphere of hypothetical variations"—surrounding any static, frozen perception.

I have concocted a playful name for this imaginary sphere: I call it the *implicosphere*, which stands for *implicit counterfactual sphere*, referring to things that never were but that we cannot help seeing anyway. (The word can also be taken as referring to the *sphere of implications* surrounding any given idea. A visual representation of an implicosphere is shown in Figure 12-5.) If we wish to enlist computers as our partners in this venture of inventing variations on a theme, which is to say, turning implicospheres into "explicospheres", we have to give them the ability to spot knobs themselves, not just to accept knobs that we humans have spotted. To do this we will have to look deeply into the nature of "slippability", into the fine-grained structure of those networks of concepts in human minds.

*   *   *

One way to imagine how slippability might be realized in the mind is to suppose that each new concept begins life as a compound of previous concepts, and that from the slippability of those concepts, it inherits a certain amount of slippability. That is, since any of its constituents can slip in various ways, this induces modes of slippage in the whole. Generally, letting a constituent concept slip in its simplest ways is enough, since when more than one of these is done at a time, that can already create many unexpected effects. Gradually, as the space of possibilities of the new concept—the implicosphere—is traced out, the most common and useful of those slippages become more closely and directly associated with the new concept itself, rather than having to be derived over and over from its constituents. This way, the new concept's implicosphere becomes more and more explicitly explored, and eventually the new concept becomes old and reaches the point where it too can be used as a constituent of fresh new young concepts.

Some examples of this sort of thing were presented in my column for September, 1981 (Chapter 23). Now although September is almost October

---

FIGURE 12–5.   *In (a), a stylized implicosphere. In (b) through (d), various degrees of overlap of two implicospheres are portrayed. Too much overlap (b) leads to mushy, sloppy thought, while too little overlap (d) leads to sparse, dull thought. The ideal amount of overlap and autonomy (c) leads to creative, insightful thought.*

*   In (e), a related and charming geometrical problem called "Mrs. Miniver's problem" is shown. The idea is to determine the conditions under which the overlap of two circles (representing two people) has the same area as each of the two crescents formed. Mrs. Miniver wishes thereby to symbolize her vision of the ideal romance. The ideal overlap of course symbolizes how much two lovers ideally have in common.*

and 1981 is almost 1982, that doesn't quite mean that you have those examples at your mind's fingertips, or on the tip of your mind's tongue. So let me present a few more examples of slippage of a new notion based on slipping some of its parts in their simplest ways. The notion I have chosen is that of yourself sitting there, reading this very column at this very moment. Here are some elements of the implicosphere of that concept:

> You are almost reading the September 1981 issue of *Scientific American.*
> You are almost reading a piece by Richard Hofstadter, the historian.
> You are almost reading a column by Martin Gardner.
> Your identical twin is almost reading this column.
> You are almost reading this column in French.
> You are almost reading *Gödel, Escher, Bach.*
> You are almost reading a letter from me.
> You are almost writing this column.
> You are almost hearing my voice.
> I am almost talking to you.
> You are almost ready to throw this copy of *Mad* magazine out in disgust.

By now, the original concept is almost lost in a silly sea of "almost" variations—but it has been enriched by this exploration, and when you come back to it, it will have been that much more *reified* as a stand-alone concept, a single entity rather than a compound entity. After a while, under the proper triggering circumstances, this very example may be retrieved from memory as naturally and effortlessly as the concept of "fish" is.

This is an important idea: the test of whether a concept has really come into its own, the test of its genuine mental existence, is its retrievability by that process of unconscious recall. That's what lets you know that it has been firmly planted in the soil of your mind. It is not whether that concept appears to be "atomic", in the sense that you have a single word to express it by. That is far too superficial.

Here is an example to illustrate why. A friend told me recently that the *Encyclopaedia Britannica*'s first edition (1768-71) consisted of three volumes: Volume I: "A-B"; Volume II: "C-L", and then Volume III: the rest of the alphabet. In that edition, 511 pages were devoted to topics beginning with 'A', while the last volume had 753 pages altogether! (I guess that in those days there weren't yet many interesting things around that began with letters between 'M' and 'Z'.) Hearing this amusing fact instantaneously triggered the retrieval of the memory, implanted in me years and years ago under totally unremembered circumstances, of how records used to be made, back in the days when there was no magnetic tape and the master disk was actually cut during the live performance. The performers would be playing along and all of a sudden the recording engineer would notice that there wasn't much room left on the plate, so the performers would be given a signal to hurry up, and as a result, the tempo would be faster and faster the further toward the center the needle came. I think it is obvious why the one triggered retrieval of the other. And yet—*is* it obvious?

On the surface, these two concepts are completely unrelated. One concerns printed matter, books, the alphabet, and so on, while the other concerns plastic disks, sounds, performers, recording techniques, and so on. However, at some deeper conceptual level, these really *are* the same idea. There is just *one idea* here, and this idea I call a *conceptual skeleton.* Try to verbalize it. It's certainly not just one word. It will take you a while. And when you do come up with a phrase, chances are it will be awkward and stilted—and still not quite right!

Both of the cited instances of this conceptual skeleton—in itself nameless, majestically nonverbalizable—are floating about in the implicosphere that surrounds it, along with numerous other examples that I am unaware of, not yet having twiddled enough knobs on that concept. I don't yet even know which knobs it has! But I may eventually find out. The point is that the concept itself has been *reified*—this much is proven by the fact that it acts as a point of immediate reference; that my memory mechanisms are capable of using it as an "address" (a key for retrieval) under the proper circumstances. *The vast majority of our concepts are wordless* in this way, although we can certainly make stabs at verbalizing them when we need to.

*   *   *

Early in this column, I stated a thesis: that the crux of creativity resides in the ability to manufacture variations on a theme. I hope now to have sufficiently fleshed out this thesis that you understand the full richness of what I meant when I said "variations on a theme". The notion encompasses knobs, parameters, slippability, counterfactual conditionals, subjunctives, "almost"-situations, implicospheres, conceptual skeletons, mental reification, memory retrieval—and more.

The question may persist in your mind: Aren't variations on a theme somehow trivial, compared to the invention of the theme itself? This leads one back to that seductive notion that Einstein and other geniuses are "cut from a different cloth" from ordinary mortals, or at least that certain cognitive acts done by them involve principles that transcend the everyday ones. This is something I do not believe at all. If you look at the history of science, for instance, you will see that every idea is built upon a thousand related ideas. Careful analysis leads one to see that what we choose to call a new theme is itself always some sort of variation, on a deep level, of previous themes. The trick is to be able to see the deeply hidden knobs!

Newton said that if he had seen further than others, it was only by standing on the shoulders of giants. Too often, however, we simply indulge in wishful thinking when we imagine that the genesis of a clever or beautiful idea was somehow due to unanalyzable, magical, transcendent insight rather than to any mechanisms—as if all mechanisms by their very nature were necessarily shallow and mundane.

My own mental image of the creative process involves viewing the organization of a mind as consisting of thousands, perhaps millions, of overlapping and intermingling implicospheres, at the center of each of which is a conceptual skeleton. The implicosphere is a flickering, ephemeral thing, a bit like a swarm of gnats around a gas-station light on a hot summer's night, perhaps more like an electron cloud, with its quantum-mechanical elusiveness, about a nucleus, blurring out and dying off the further removed from the core it is (Figure 12-5). If you have studied quantum chemistry, you know that the fluid nature of chemical bonds can best be understood as a direct consequence of the curious quantum-mechanical overlap of electronic wave functions in space, wave functions belonging to electrons orbiting neighboring nuclei. In a metaphorically similar way, it seems to me, the crazy and unexpected associations that allow creative insights to pop seemingly out of nowhere may well be con-sequences of a similar chemistry of concepts with its own special types of "bonds" that emerge out of an underlying "neuron mechanics".

Novelist Arthur Koestler has long been a champion of a mystical view of human creativity, advocating occult views of the mind while at the same time eloquently and objectively describing its workings. In his book *The Act of Creation,* he presents a theory of creativity whose key concept he calls "bisociation"—the simultaneous activation and interaction of two previously unconnected concepts. This view emphasizes the coming-together of *two* concepts, while bypassing discussion of the internal structure of a *single* concept. In Koestler's view, something new can happen when two concepts "collide" and fuse—something not present in the concepts themselves. This is in keeping with Koestler's philosophy that wholes are somehow greater than the sum of their parts.

By contrast, I have been emphasizing the idea of the internal structure of *one* concept. In my view, the way that concepts can bond together and form conceptual molecules on all levels of complexity is a consequence of their internal structure. What results from a bond may surprise us, but it will nonetheless always have been completely determined by the concepts involved in the fusion, if only we could understand how they are structured. Thus the crux of the matter is the internal structure of a single concept and how it "reaches out" toward things it is not. The crux is *not* some magical, mysterious process that occurs when two indivisible concepts collide; it is a consequence of the divisibility of concepts into subconceptual elements. As must be clear from this, I am not one to believe that wholes elude description in terms of their parts. I believe that if we come to understand the "physics of concepts", then perhaps we can derive from it a "chemistry of creativity", just as we can derive the principles of the chemistry of atoms and molecules from those of the physics of quanta and particles. But as I said earlier, it is not just around the corner. Mental bonds will probably turn out to be no less subtle than chemical bonds. Alan Turing's words of cautious enthusiasm about artificial intelligence remain as apt now as they were in 1950, when he wrote them in concluding his famous article "Computing Machinery and Intelligence": "We can only see a short distance ahead, but we can see plenty there that needs to be done."

Recently I happened to read a headline on the cover of a popular electronics magazine that blared something about "CHIPS THAT SEE". Bosh! I'll start believing in "chips that see" as soon as they start seeing things that never were, and asking "Why not?"

Post Scriptum.

> *Knobs, knobs, everywhere—*
> *Just vary a knob to think.*

Some readers objected to the slogan of this column—that making variations on a theme is the crux of creativity. They felt—and quite rightly —that making variations (*i.e.*, twisting knobs) is as easy as falling off a log. So how can genius be that easy? Part of the answer is: For a genius, it *is* easy to be a genius. *Not* being a genius would be excruciatingly hard for a genius. However, this isn't a completely satisfactory answer for people who pose this objection. They feel that I am unwittingly implying that it is easy for *anybody* to be a genius: after all, a crank can crank a knob as deftly as a genius can. The crux of their objection, then, is that the crux of creativity is not in *twiddling* knobs, but in *spotting* them!

Well, that is exactly what I meant by my slogan. Making variations is not just twiddling a knob before you; part of the act is to manufacture the knob yourself. Where does a knob come from? The question amounts to asking: How do you see a *variable* where there is actually a *constant*? More specifically: *What* might vary, and *how* might it vary? It's not enough to just have the desire to see something different from what is there before you. Often the dullest knobs are a result of someone's straining to be original, and coming up with something weak and ineffective. So where do good knobs come from? I would say they come from *seeing one thing as something else*. Once an abstract connection is set up via some sort of *analogy* or *reminding-incident*, then the gate opens wide for ideas to slosh back and forth between the two concepts.

A simple example: A friend and I noticed a fuel-delivery truck pulling into a driveway, and on it was very conspicuously printed "NSF", standing for "North Shore Fuel". However, to us those letters meant "National Science Foundation" as surely as "TNT" means "trinitrotoluene" to Eve Rybody. Now, we could have just let the coincidence go, but instead we played with it. We envisioned a National Science Foundation truck pulling up to a research institute. The driver gets out of the cab, drags a thick flexible hose over to a hole in the wall of a building and inserts it, then starts up a loud motor, and pumps a truckload of money—presumably in large bills—into the cellar of the building. (Wouldn't it be nice if grants were delivered that way?) This vision then led us to pondering the way that money actually *does* flow between large institutions: usually as abstract, intangible numbers shot down wires as binary digits, rather than as greenbacks hauled about in large trucks.

This very small incident serves well to illustrate how a simple reminding-incident triggered a series of thoughts that wound up in a region of idea-space that would have been totally unanticipable moments before. All that was needed was for an inappropriate meaning of "NSF" to come to mind, and then to be explored a bit. Such opportunities for being reminded of something remote—such *double-entendre* situations—occur all the time, but often they go unobserved. Sometimes the ambiguity is observed but shrugged off with disinterest. Sometimes it is exploited to the hilt. In this example, the result was not earthshaking, but it did cast things in a new light for both of us, and the image amused us quite a bit. And this way of exploiting serendipity—that is, exploiting coincidences and unexpected perceived similarities—is typical of what I consider the crux of the creative process.

*   *   *

Serendipitous observation and quick exploration of potential are vital elements in the making of a knob. What goes hand in hand with the willingness to playfully explore a serendipitous connection is the willingness to censor or curtail an exploration that seems to be leading nowhere. It is the flip side of the risk-taking aspect of serendipity. It's fine to be reminded of something, to see an analogy or a vague connection, and it's fine to try to map one situation or concept onto another in the hopes of making something novel emerge—but you've also got to be willing and able to sense when you've lost the gamble, and to cut your losses. One of the problems with the ever-popular self-help books on how to be creative is that they all encourage "off-the-wall" thinking (under such slogans as "lateral thinking", "conceptual blockbusting", "getting whacked on the head", etc.) while glossing over the fact that most off-the-wall connections are of very little worth and that one could waste lifetimes just toying with ideas in that way. One needs something much more reliable than a mere suggestion to "think zany, out-of-the-system thoughts".

Frantic striving to be original will usually get you nowhere. Far better to relax and let your perceptual system and your category system work together unconsciously, occasionally coming up with unbidden connections. At that point, you—the lucky owner of the mind in question—can seize the opportunity and follow out the proffered hint. This view of creativity has the conscious mind being quite passive, content to sit back and wait for the unconscious to do its remarkable broodings and brewings.

The most reliable kinds of genuine insight come not from vague reminding experiences (as with the letters "NSF"), but from strong analogies in which one experience can be mapped onto another in a highly pleasing way. The tighter the fit, the deeper the insight, generally speaking. When two things can both be seen as instances of one abstract phenomenon, it is a very exciting discovery. Then ideas about either one can be borrowed in thinking about the other, and that sloshing-about of activity may greatly illuminate both at once. For instance, such a connection (*i.e.,* mapping)—between sexism and racism—resulted in my "Person Paper" (Chapter 8). Another example is Scott Kim's brilliant article "Noneuclidean Harmony", in which mathematics and music are twisted together in the most amazing ways. It can be found in *The Mathematical Gardner,* an anthology dedicated to Martin Gardner, edited by David Klarner.

A mapping-recipe that often yields interesting results is *projection of oneself* into a situation: "How would it be *for me* ?" This can mean a host of things, depending on how you choose to inject yourself into the scene, which is in turn determined by what grabs your attention. The man who focused in on the bustling activity in the coffeehouse and said, "I'm sure glad I'm not a waitress here tonight!" might instead have been offended by the sounds reaching his ears and said, "If I were the owner here, I'd play less Muzak" —or he might have zeroed in on someone purchasing a brownie and said, "I wish I were that thin." People are remarkably fluid at seeing themselves in roles that they self-evidently could never fill, and yet the richness of the insights thus elicited is beyond doubt.

*   *   *

When I first heard the French saying *Plus ça change, plus c'est la même chose,* it struck me as annoyingly nonsensical: "The more it changes, the samer it gets" (in my own colloquial translation). I was not amused but nonetheless it stuck in my mind for years, and finally it dawned on me that it was full of meanings. My favorite way of interpreting it is this. The more different manifestations you observe of one phenomenon, the more deeply you understand that phenomenon, and therefore the more clearly you can see the vein of sameness running through all those different things. Or put another way, experience with a wide variety of things refines your category system and allows you to make incisive, abstract connections based on deep shared qualities. A more cynical way of putting it, and probably more in line with the intended meaning, would be that superficially different things are often boringly the same. But the saying need not be taken cynically.

Seeing clear to the essence of something unfamiliar is often best achieved by finding one or more known things that you can see it *as,* then being able to balance these views. Physicists have long since learned to juggle two views of light: light as waves, light as particles. They know that each contains a grain of the essence of light, that neither contains it all, and they know when to think of light which way. Don't be fooled by people who knowingly assure you that physicists don't depend on crude images or analogies as crutches, that everything they need is contained in their formulas. The fallacy here is that *which* formula to apply, *how* to apply it, and what parts of it to *neglect* are all aspects not covered in any formula, which is why doing physics is a great art, despite the fact that there are formulas all over the place for Eve Rybody and her brother to use.

Seeing anything as waves suggests immediate knobs: wavelength, frequency, amplitude, speed, medium, and a host of other basic notions that define the essence of undularity. Seeing anything as particles suggests totally different knobs: mass, shape, radius, rotation, constituents, and a host of other basic notions that define the essence of corpuscularity. If you choose to see, say, people as waves or as particles, you may find some of these suggested knobs quite interesting. On the other hand, it may not be fruitful to do so. Good analogies usually are not the product of an off-the-wall suggestion like this, but spring to mind unbidden, from the deep similarity-searching wells of the unconscious.

Once you have decided to try out a new way of viewing a phenomenon, you can let that view suggest a set of knobs to vary. The act of varying them will lead you down new pathways, generating new images ripe for perception in their own right. This sets up a closed loop:

* fresh situations get unconsciously framed in terms of familiar concepts;
* those familiar concepts come equipped with standard knobs to twiddle;
* twiddling those knobs carries you into fresh new conceptual territory.

A visual image that I always find coming back in this context is that of a planet orbiting a star, and whose orbit brings it so close to another star that it gets "captured" and begins orbiting the second star. As it swings around the new star, perhaps it finds itself coming very close to yet another star, and ficklely changes allegiance. And thus it do-si-do's its way around the universe.

The mental analogue of such stellar peregrinations is what the loop above attempts to convey. You can think of concepts as stars, and knob-twiddling as carrying you from one point on an orbit to another point. If you twiddle enough, you may well find yourself deep within the attractive zone of an unexpected but interesting concept and be captured by it. You may thus migrate from concept to concept. In short, knob-twiddling is a device that carries you from one concept to another, taking advantage of their overlapping orbits.

Of course, all this cannot happen with a trivial model of concepts. We see it happening all the time in minds, but to make it happen in computers or to locate it physically in brains will require a fleshing-out of what concepts really are. It is fine to talk of "orbits around concepts" as a metaphor, but developing it into a full scientific notion that either can be realized in a computer model or can be located inside a brain is a giant task. This is the task that faces cognitive scientists if they wish to make "concept" a legitimate scientific term. This goal, suggested at the start of this article, could be taken to be the central goal of cognitive science, although such things are often forgotten in the inane hoopla that is surrounding artificial intelligence more and more these days.

The cycle shown above spells out what I intend by the phrase "making variations on a theme", and it is this loop that I am suggesting is the crux of creativity. The beauty of it is that you let your memory and perceptual mechanisms do all the *hard* work for you (pulling concepts from dormancy); all *you* do is twiddle knobs. And I'll let *you* decide what this odd distinction is between something called "you" and the hard-working mechanisms of "your memory".

*   *   *

The concept of the "implicosphere" of an idea—the sphere of variations on it resulting from the twiddling of many knobs a "reasonable" amount—is a difficult one, but it is absolutely central to the meaning of this column. One way of thinking about it is this. Imagine a single gnat attracted by a bright light. It will buzz about, tracing out a three-dimensional random walk centered on that light. If you keep a photographic plate exposed so that you can record its path cumulatively, you will first see a chaotic broken line, but soon the image will get so dense with criss-crossing lines that it will gradually turn into a circular smear of slowly increasing radius. At the outer edges of the smear you might once in a while make out an occasional foray of the lone bug. For a while, the territory covered expands, but eventually this gnat-o-sphere will reach a stable size. Its silhouette, instead of being a sharp-edged circle, will be a blurry circle (see Figure 12-5*a*) whose approximate radius reveals something about how gnats are attracted by lights.

Now if you simply think of this translated into idea-space, you have roughly the right image. Of course, not all implicospheres have the same radius. Some people's implicospheres tend to have bigger radii than other people's do, and consequently their implicospheres overlap more. This can be good but it can be overdone. Too much overlap (Figure 12-5*b*) and all you have is a mush of vaguely associated ideas, an overdone and tasteless mental goulash. Too little overlap (Figure 12-5*d*) and you have a very thin, watery mind, one with few big surprises (except for the meta-level surprise of having so few surprises). There is, in other words, an optimum amount of overlap for useful creative insight (Figure 12-5*c*). This is the kind of thing that cannot be taught, however. It would be like trying to train a gnat to control the size of the spheres it traces out. Or if you prefer, it would be like trying to train an entire swarm of gnats to form spheres of a particular size whenever they cluster around lamps. The problem is, it is already preprogrammed in gnats how much they are attracted by lights, by each other, and so on.

In my view, mindpower is a consequence of how implicospheres in idea-space emerge from the statistical predispositions of neurons to fire in response to each other. Such deep statistical patterns of each brain cannot be altered, although of course a few superficial aspects can be altered. You can teach somebody to think of applehood whenever they think of mother pie, for instance—but adding any number of specific new associative connections does not have any effect on the underlying statistics of how their neurons work. So in that sense I am gravely doubtful about courses or books that promise to improve your thinking style or capabilities. Sure, you can add new *ideas*—but that's a far cry from adding pizzazz. The mind's perceptual and category systems are too much at the "subcognitive" level to be reached via cognitive-level training techniques. If you are old enough to be reading this book, then your deep mental hardware has been in place for many years, and it is what makes your thinking-style idiosyncratic and recognizably "you". (If you are not, then what are you doing reading this book? Put it down immediately!) For more on the ideas of subcognition and identity, see Chapters 25 and 26.

When a new idea is implanted in a mind, an implicosphere grows around it. Since this means, in essence, the linking-up of this new idea with older ideas, I call it "diffusion in idea-space". My canonical example of this phenomenon, although it is a rather grim one, has to do with the recent spate of random murders inspired by the spiking of Tylenol capsules with strichnine. It was the Food and Drug Administration's response that so intrigued me, because it implicitly revealed a theory of how this idea would diffuse in the idea-space of a typical potential murderer. The FDA imposed a set of packaging regulations on manufacturers, with various types of products being given various deadlines for compliance. The idea was that your potential murderer could slip from the idea of Tylenol to that of aspirin in a week's time, but it would take the expanding sphere longer to hit the brilliant idea that it could be just *any* over-the-counter drug. Not just the FDA seemed to think this way; also radio talk-show hosts seemed to love speculating about what drug might be chosen next—but I never heard them worrying about ordinary food in grocery stores. Yet why should it give a stochastic killer any less joy to kill by spiking a jar of mustard than by spiking a drug? In fact, if your goal in life is to see masses of random people die, there are all sorts of routes you can take that don't involve ingestion at all. A friend of mine took a train from Washington to New York and *en route* her train smashed into a washing machine full of rocks that had been placed on the tracks by some do-badder. Was this part of the Tylenol-murders implicosphere in the mind of the person who did it? I doubt it, but it is possible.

In its own gruesome way, the generalization of the Tylenol murders resembles that of the expanding implicosphere of the Cube—and that of any idea that arises. Ideas, whether evil or beneficial, have their own dynamics of spreading in and among minds. Here we are primarily talking about intramind spreading (implicospheres), but intermind spreading (infectious memes) was discussed in Chapter 3.

* * *

Slippage of thought is a remarkably invisible phenomenon, given its ubiquity. People simply don't recognize how curiously selective they are in their "choice" of what is and what is not a hingepoint in how they think of an event. It all seems so natural as to require no explanation.

I dropped a slice of pizza on the floor of a pizza place the other evening. My friend Don, who was less hungry than I was, immediately sympathized, saying, "Too bad *I* didn't drop one of *my* pieces—or that you didn't drop one of mine instead of one of yours." Sounds sensible. But why didn't he say, "Too bad the pizza isn't larger"? His choice revealed that to his unconscious mind, it seemed sensible to switch the role-filler in a given event, as if to imply that a pizza-slice-droppage had been in the cards for that evening, that God had flipped a coin and, unluckily for me, it had come out with me as the dropper instead of Don—but that it might have come out the other way around.

Some hypothetical replacement scenarios—I like to call them "subjunctive instant replays"—are compelling, and come to mind by reflex. They are not idle musings but very natural human emotional responses to a common type of occurrence. Other subjunctive instant replays have little intuitive appeal and seem far-fetched, although it is hard to say just why. Consider the following list:

Too bad they didn't give us a replacement piece.
Lucky we weren't in a really fancy restaurant.
Too bad gravity isn't weaker, so that you could have caught it before
    it hit the ground.
Lucky it wasn't a beaker filled with poison.
Too bad it wasn't a fork.
Lucky it wasn't a piece of good china.
Too bad eating off floors isn't hygienic.
Lucky you didn't drop the whole pizza.
Too bad it wasn't the people at the next table who dropped *their* pizza.
Lucky there was no carpet in here.
Too bad you were the hungry one, rather than me.

I'll leave it to you to generate other subjunctive instant replays that he might have come up with. There is a rough rank ordering to them, in terms of plausibility of springing to mind. It's the rhyme and reason behind that ordering that fascinates me.

Why do people find it not only plausible but even compelling to make remarks like the following?

If Jesse Jackson were a white man, he'd be elected President.
If Jesse Jackson were a white man, he'd be running for dogcatcher.

These two sentences came from random voters, as quoted in *Newsweek*. I wonder what slips in people's minds when they imagine a white Jesse Jackson. Do they envision a preacher in a Baptist church? Is this person an ardent fighter for civil rights? Or, conversely, an ardent fighter against the quota system? Similarly, what does a high-school boy mean when he says, "If I were my father, I wouldn't lend me the car"? Does he ever notice that if he were his father, he would *ipso facto* be his own son? Or need that be so? Would the two have exchanged roles? The point is, there are a host of questions left completely open here, yet no one balks for a second at such counterfactuals. In fact, they are common currency, they are daily bread, they are the meat and potatoes of communication. But some types of counterfactuals never (or hardly ever) come up, while others, equally reality-violating, are a dime a dozen.

Daniel Kahneman and Amos Tversky, cognitive psychologists, have made studies of how much emotion people generate upon reading stories of just-missed airplanes or just-caught airplanes—especially ones that crash. These kinds of near misses, whether fortunate or unfortunate, tug at our hearts and do so in nearly universal ways. Something about these slippability examples is truly at the core of what it is to be human and to experience the world through the filter of the human mind.

Philosophers and artificial-intelligence researchers by and large have not paid much attention to the "catchiness" of a given counterfactual. Logicians have devoted a lot of time and effort to trying to figure out what it would mean for a given counterfactual to be *true*, but to my mind, that's not nearly as interesting—or even as meaningful—a question as these more psychological questions:

Which counterfactuals are likely to be triggered in a human mind by various types of events in the world?

Why are some events perceived to be "near misses", while others are not?

Why are some deaths of innocent people viewed as more tragic than other deaths of innocent people?

Variations on a Theme as the Crux of Creativity

At such points where deep human emotion, identification with other beings, and perception of reality meet lies the crux of creativity—and also the crux of the most mundane thoughts. Spinning out variations is what comes naturally to the human mind, and is it ever fertile!

13

Metafont, Metamathematics, and Metaphysics: Comments on Donald Knuth's Article "The Concept of a Meta-Font"

August, 1982

The Mathematization of Categories, and Metamathematics

DONALD Knuth has spent the past several years working on a system allowing him to control many aspects of the design of his forthcoming books—from the typesetting and layout down to the very shapes of the letters! Seldom has an author had anything remotely like this power to control the final appearance of his or her work. Knuth's TEX typesetting system has become well-known and available in many countries around the world. By contrast, his METAFONT system for designing families of typefaces has not become as well known or as available.

In his article "The Concept of a Meta-Font", Knuth sets forth for the first time the underlying philosophy of METAFONT, as well as some of its products. Not only is the concept exciting and clearly well executed, but in my opinion the article is charmingly written as well. However, despite my overall enthusiasm for Knuth's idea and article, there are some points in it that I feel might be taken wrongly by many readers, and since they are points that touch close to my deepest interests in artificial intelligence and esthetic theory, I felt compelled to make some comments to clarify certain important issues raised by "The Concept of a Meta-Font".

Although his article is primarily about letterforms, not philosophy, Knuth holds out in it a philosophically tantalizing prospect for us: that with the arrival of computers, we can now approach the vision of a unification of all typefaces. This can be broken down into two ideas:

(1) That underneath all 'A's there is just one grand, ultimate abstraction that can be captured in a finitely parametrizable computational structure—a "software machine" with a finite number of "tunable knobs" (we could say "degrees of freedom" or "parameters", if we wished to be more dignified);

(2) That every conceivable particular 'A' is just a product of this machine with its knobs set at specific values.

Beyond the world of letterforms, Knuth's vision extends to what I shall call the *mathematization of categories:* the idea that any abstraction or Platonic concept can be so captured—that is, as a software machine with a finite number of knobs. Knuth gives only a couple of examples—those of the "meta-waltz" and the "meta-shoe"—but by implication one can imagine a "meta-chair", a "meta-person", and so forth.

This is perhaps carrying Knuth's vision further than he ever intended. Indeed, I suspect so; I doubt that Knuth believes in the feasibility of such a "mathematization of categories" opened up by computers. Yet any imaginative reader would be likely to draw hints of such a notion out of Knuth's article, whether Knuth intended it that way or not. It is my purpose in this article to argue that such a vision is exceedingly unlikely to come about, and that such intriguingly flexible tools as metashoes, meta-fonts, modern electronic organs (with their "oom-pah-pah" and "cha-cha-cha" rhythms and their canned harmonic patterns), and other many-knobbed devices will only help us see more clearly why this is so. The essential reason for this I can state in a very short way: I feel that to fill out the full "space" defined by a category such as "chair" or "waltz" or "face" or 'A' (see Figures 12-2, 12-3, and 12-4) is an act of infinite creativity, and that no finite entity (inanimate mechanism or animate organism) will ever be capable of producing all possible 'A's and nothing but 'A's (the same could be said for chairs, waltzes, etc.).

I am not making the trivial claim that, because life is finite, nobody can make an infinite number of creations; I am making the nontrivial claim that nobody can possess the "secret recipe" from which all the (infinitely many) members of a category such as 'A' can in theory be generated. In fact, my claim is that no such recipe exists. Another way of saying this is that even if you were granted an infinite lifetime in which to draw all the 'A's you could think up, thus realizing the full potential of any recipe you had, no matter how great it might be, you would still miss vast portions of the space of 'A's.

In metamathematical terms, this amounts to positing that any conceptual (or *semantic*) category is a *productive* set, a precise notion whose characterization is a formal counterpart to the description in the previous paragraphs (namely, a set whose elements cannot be totally enumerated by any effective procedure without overstepping the bounds of that set, but which can be approximated more and more fully by a sequence of increasingly complex effective procedures). The existence and properties of such sets first became known as a result of Gödel's Incompleteness Theorem of 1931. It is certainly not my purpose here to explain this famous result, but a short synopsis might be of help. (Some useful references are: Chaitin, DeLong, Nagel and Newman, Rucker, and my book *Gödel, Escher, Bach.*)

An Intuitive Picture of Gödel's Theorem

Gödel was investigating the properties of purely formal deductive systems in the sphere of mathematics, and he discovered that such systems—even if their ostensible domain of discourse was limited to one topic—could be viewed as talking "in code" about themselves. Thus a deductive system could express, in its own formal language, statements about its own capabilities and weaknesses. In particular, System X could say of itself through the Gödelian code:

System X *is not powerful enough to demonstrate the truth of Sentence* S.

It sounds a little bit like a science-fiction robot called "ROBOT R-15" droning (of course in a telegraphic monotone):

ROBOT R-15 UNFORTUNATELY UNABLE TO COMPLETE TASK T-12—VERY SORRY.

Now what happens if TASK T-12 happens, by some crazy coincidence, to be not the assembly of some some strange cosmic device but merely the act of uttering the preceding telegraphic monotone? (I say "merely" but of course that is a bit ironic.) Then ROBOT R-15 could get only partway through the sentence before choking: *ROBOT R-15 UNFORTUNATELY UNABLE TO COMPL—*.

Now in the case of a formal system, System X, talking about its powers, suppose that Sentence **G,** by an equally crazy coincidence, is the one that says,

System X *is regrettably not powerful enough to demonstrate the truth of Sentence* **G.**

In such a case, Sentence **G** is seen to be an assertion of its own unprovability within System X. In fact we do not have to rely on crazy coincidences, for Gödel showed that given any reasonable formal system, a **G**-type sentence for that system actually exists. (The only exaggeration in my English-language version of **G** is that in formal systems there is no way to say "regrettably".) In formal deductive systems, this foldback takes place of necessity by means of a Gödelian code, but in English no Gödelian code is needed and the peculiar quality of such a loop is immediately visible.

If you think carefully about Sentence **G,** you will discover some amazing things. Could Sentence **G** be provable in System *X*? If it were, then System *X* would contain a proof for Sentence **G,** which asserts that System *X* contains no proof for Sentence **G.** Only if System *X* is blatantly self-contradictory could this happen—and a formal reasoning system that is self-contradictory is no more useful than a submarine with screen doors. So, provided we are dealing with a *consistent* formal system (one with no self-contradictions), then Sentence **G** is not provable inside System *X.* And since this is precisely the claim of Sentence **G** itself, we conclude that Sentence **G** is true—true but unprovable inside System *X.*

One last way to understand this curious state of affairs is afforded the reader by this small puzzle. Choose the more accurate of the following pair of sentences:

(1) Sentence **G** is true *despite* being unprovable.
(2) Sentence **G** is true *because* it is unprovable.

You'll know you've really caught on to "Gödelism" when both versions ring equally true to your ears, when you flip back and forth between them, savoring that exceedingly close approach to paradox that **G** affords. That's how twisted back on itself Sentence **G** is!

The main consequence of **G**'s existence within each System *X* is that there are truths unattainable within System *X,* no matter how powerful and flexible System *X* is, as long as System *X* is not self-contradictory. Thus, if we look at truths as objects of desire, no formal system can have them all; in fact, given any formal system we can produce on demand a truth that it cannot have, and flaunt that truth in front of it with taunting cries of "Nyah, nyah!" The set of truths has this peculiar and infuriating quality of being uncapturable by any finite system, and worse, given any candidate system, we can use what we know about that system to come up with a specific Gödelian truth that eludes provability inside that system.

By adding that truth to the given system, we come up with an enlarged and slightly more powerful system—yet this system will be no less vulnerable to the Gödelian devilry than its predecessor was. Imagine a dike that springs a new leak each time the proverbial Dutch boy plugs up a hole with his finger. Even if he had an infinite number of fingers, that leaky dike would find a spot he hadn't covered. A system that contains at least one unprovable truth is said to be *incomplete,* and a system that not only contains such truths but that cannot be rescued in any way from the fate of incompleteness is said to be *essentially incomplete.* Another name for sets with this wonderfully perverse property is *productive.* (For detailed coverage of the metamathematical ideas in this article, see the book by Rogers.)

My claim—that *semantic categories are productive sets*—is, to be sure, not a mathematically provable fact, but a metaphor. This metaphor has been used by others before me—notably, the logicians Emil Post and John Myhill—and I have written of it myself before (see Chapter 23).

Completeness and Consistency

Note that it is important to have the potential to fill out the full (infinite) space, and equally important not to overstep it. However, merely having infinite potential is not by any means equivalent to filling out the full space. After all, any existing METAFONT 'A'-schema—even one having just one degree of freedom!—will obviously give us infinitely many distinct 'A's as we sweep its knob (or knobs) from one end of the spectrum to the other. Thus to have an 'A'-making machine with infinite variety of potential output is not in itself difficult; the trick is to achieve *completeness:* to fill the space.

And yet, isn't it easy to fill the space? Can't one easily make a program that will produce all possible 'A's? After all, any 'A' can be represented as a pattern of pixels (dots that are either off or on) in an $m \times n$ matrix—hence a program that merely prints out all possible combinations of pixels in matrices of all sizes (starting with 1×1 and moving upwards to 2×1, 1×2, 3×1, 2×2, 1×3, etc., as in Georg Cantor's famous enumeration of the rational numbers) will certainly cover any given 'A' eventually. This is quite true. So what's the catch?

Well, unfortunately, it is hard—*very* hard—to write a screening program that will retain all the 'A's in the output of this pixel-pattern program, and at the same time will reject all 'K's, pictures of frogs, octopi, grandmothers, trolleycars, and precognitive photographs of traffic accidents in the twenty-fifth century (to mention just a few of the potential outputs of the generation program). The requirement that one must stay within the bounds of a conceptual category could be called *consistency*—a constraint complementary to that of completeness.

In summary, what might seem desirable from a knobbed category-machine is the joint attainment of two properties—namely:

(1) *Completeness:* that all true members of a category (such as the category of 'A's or the category of human faces) should be potentially producible eventually as output;

(2) *Consistency:* that no false members of the category ("impostors") should ever be potentially producible (in short, that the set of outputs of the machine should coincide exactly with the set of members of the intuitive category).

The twin requirements of consistency and completeness are metaphorical equivalents of well-known notions by the same names in metamathematics, denoting desirable properties of formal systems (theorem-producing machines)—namely:

(1) *Completeness:* that all true statements of a theory (such as the theory of numbers or the theory of sets) should be potentially producible eventually as theorems;

(2) *Consistency:* that no false statements of the theory should ever be potentially producible (in short, that the set of theorems of the formal system should coincide exactly with the set of truths of the informal theory).

The import of Gödel's Incompleteness Theorem is that *these two idealized goals are unreachable simultaneously for any "interesting" theory* (where "interesting" really means "sufficiently complex"); nonetheless, one can approach the set of truths by stages, using increasingly powerful formal systems to make increasingly accurate approximations. The goal of total and pure truth is, however, as unreachable by formal methods as is the speed of light by any material object. I suggest that a parallel statement holds for any "interesting" category (where again, "interesting" means something like "sufficiently complex", although it is a little harder to pin down): namely, one can do no better than approach the set of its members by stages, using increasingly powerful knobbed machines to make increasingly accurate approximations.

Intuition at first suggests that there is a crucial difference between the (metamathematical) result about the nonformalizability of truth and the (metaphorical) claim about the nonmechanizability of semantic categories; this difference would be that the set of all truths in a mathematical domain such as set theory or number theory is objective and eternal, whereas the set of all 'A's is subjective and ephemeral. However, on closer examination, this distinction begins to blur quite a bit. The very fact of Gödel's proven nonformalizability of mathematical truth casts serious doubt on the objective nature of such truth. Just as one can find all sorts of borderline examples of 'A'-ness, examples that make one sense the hopelessness of trying to draw the concept's exact boundaries, so one can find all sorts of borderline mathematical statements that are formally undecidable in standard systems and that, even to a keen mathematical intuition, hover between truth and falsity. And it is a well-known fact that different mathematicians hold different opinions about the truth or falsity of various famous formally undecidable propositions (the axiom of choice in set theory is a classic example). Thus, somewhat counterintuitively, it turns out that mathematical truth has no fixed and eternal boundaries, either. And this suggests that perhaps my metaphor is not so much off the mark.

A Misleading Claim for METAFONT

Whatever the validity and usefulness of this metaphor, I shall now try to show some evidence for the viewpoint that leads to it, using METAFONT as a prime example of a "knobbed category machine". In his article, Knuth comes perilously close, in one throwaway sentence, to suggesting that he sees METAFONT as providing us with a mathematization of categories. I doubt he suspected that anyone would focus in on that sentence as if it were the key sentence of the article—but as he *did* write it, it's fair game! That sentence ran:

The ability to manipulate lots of parameters may be interesting and fun, but does anybody really need a 6 1/7-point font that is one fourth of the way between Baskerville and Helvetica?

This rhetorical question is fraught with unspoken implications. It suggests that METAFONT as it now stands (or in some soon-available or slightly modified version) is ready to carry out, on demand, for any user, such an interpolation between two given typefaces. There is something very tricky about this proposition that I suspect most readers will not notice: it is the idea that jointly parametrizing *two* typefaces is no harder, no different in principle, from just parametrizing *one* typeface in isolation.

Indeed, to many readers, it would appear that Knuth already *has* carried out such a joint parametrization. After all, in printing Psalm 23 (Figure 12-1) didn't he move from an old-fashioned, compact, serifed face with relatively tall ascenders and descenders and small *x*-height all the way to the other end of the spectrum: a modern-looking, extended, sans-serif face with relatively short ascenders and descenders and large *x*-height? Yes, of course—but the critical omitted point here is that these two ends of the spectrum were not pre-existing, prespecified targets; they just happened to emerge as the extreme products of a knobbed machine designed so that one more or less intermediate setting of its knobs would yield a particular target typeface (Monotype Modern Extended 8A, in case you're interested).

In other words, this particular set of knobs was inspired solely and directly by an attempt to parametrize one typeface (Monotype Modern). The two extremes shown in the psalm are both variations on that single theme; the same can be said of every intermediate stage as well. There is only one underlying theme (Monotype Modern) here, and a cluster of several hundred variants of it, each one of which is represented by a single character. The psalm does not represent the marriage of two unrelated families, but simply exhibits many members of one large family.

Joint Parametrization of Two Typefaces: A Far Cry from Parametrizing One Typeface

You can envision all the variants of Monotype Modern produced by twiddling the knobs on this particular machine as constituting an "electron cloud" surrounding a single "nucleus" (see Figure 12-5a). Now by contrast, joint parametrization of two pre-existent, known typefaces (say, Baskerville and Helvetica, as Knuth suggests (see Figure 13-1) would be like a cloud of electrons swarming around two nuclei, like a chemical bond (see Figure 12-5c).

In order to jointly parametrize two typefaces in METAFONT, you would need to find, for each pair of corresponding letters (say Baskerville 'a' and Helvetica 'a') a set of discrete geometric features (line segments, serifs, extremal points, points of curvature shift, etc.) that they share and that totally characterize them. Each such feature must be equated with one or

abcdefghijklmnopqrstuvwxyz

ABCDEFGHIJKLMNOPQRSTUVWXYZ

(a)

---

abcdefghijklmnopqrstuvwxyz

ABCDEFGHIJKLMNOPQRSTUVWXYZ

(b)

FIGURE 13–1. *Two typefaces of great beauty and subtlety. In (a), Baskerville; in (b), Helvetica Light.*

---

more parameters (knobs), so that the two letterforms are seen as produced by specific settings of their shared set of knobs. Moreover, *all intermediate settings must also yield valid instances of the letter* 'a'. That is the very essence of the notion of a knobbed machine, and it is also the gist of the quote, of course: that we should now (or soon) be able to interpolate between any familiar typefaces merely by knob-twiddling.

Now I will admit that I think it is perhaps feasible—though much more difficult than parametrizing a single typeface—to jointly parametrize two typefaces that are not radically different. It is not trivial, to cite just one sample difficulty, to move between Baskerville's round dot over the 'i' to Helvetica's square dot—but it is certainly not inconceivable. Conversely, it is not inconceivable to move between the elegant swash tail of the Baskerville 'Q' and the stubby straight tail of the Helvetica 'Q'—but it is certainly not trivial.

Moving from letter to letter and comparing them will reveal that each of these two typefaces has features that the other totally lacks. (Incidentally, you should disregard lowercase 'g', since the 'g's of our two typefaces are as different from each other as Baskerville 'B' is from Helvetica 'H'; in both cases, the two letterforms being compared derive from entirely different underlying "Platonic essences". It is METAFONT's purpose to mediate between different stylistic renditions of a single "Platonic essence", not between distinct "Platonic essences".) Presumably, in a case where one typeface possesses some distinct feature that the other totally lacks, there is a way to fiddle with the knobs that will make the feature nonexistent in one but present in the other. For instance, a knob setting of zero might make some feature totally vanish. Sometimes it will be harder to make features disappear—it might require several knobs to have coordinated settings. Nonetheless, despite all the complex ways that Baskerville and Helvetica differ, I repeat, it is conceivable that somebody with great patience and ingenuity could jointly parametrize Helvetica and Baskerville. But the real question is this: Would such a joint parametrization easily emerge out of two separate, independently carried-out parametrizations of these typefaces?

Hardly! The Baskerville knobs do not contain in them even a hint of the Helvetica qualities—or the reverse. How can I convince you of this? Well, just imagine how great the genius of John Baskerville, an eighteenth-century Briton, would have had to be for his design to have implicitly defined another typeface—and a typeface only discovered (or invented) two centuries later, by Max Miedinger from Switzerland! To see this more concretely, imagine that someone who had never seen Helvetica naïvely created a METAFONT rendition of Baskerville (that is, a meta-font centered on Baskerville in the same sense as Knuth's sample meta-font is centered on Monotype Modern). Now imagine that someone else who does know Helvetica comes along, twiddles the knobs of this Baskerville meta-font, and actually produces a perfect Helvetica! It would be nearly as strange as having a marvelous music-composing program based exclusively on the style of Dr. William Boyce (who composed in England in a baroque, elegant eighteenth-century style) that was later discovered, totally unexpectedly, to produce many pieces indistinguishable in style from the music of Arthur Honegger (who composed in Switzerland in a sparse, crisp twentieth-century style) when various melodic, harmonic, and rhythmic parameters were twiddled. To me, this is simply inconceivable; eighteenth-century style did not contain within it, no matter how implicitly, twentieth-century style —whether in music or in visual arts.

Interpolating Between an Arbitrary Pair of Typefaces

The worst is yet to come, however. Presumably Knuth did not wish us to take his rhetorical question in such a limited way as to imply that the numbers 6 1/7 and 1/4 were important. Pretty obviously, they were just examples of arbitrary parameter settings. Presumably, if METAFONT could easily give you a 6 1/7-point font that is 1/4 of the way between Baskerville and Helvetica, it could as easily give you an 11 2/3-point font that is 5/17 of the way between Baskerville and Helvetica—and so on. And why need it be restricted to Baskerville and Helvetica? Surely those *numbers* weren't the only "soft" parts of the rhetorical question! Common sense tells us that *Helvetica* and *Baskerville* were also merely arbitrary choices of typeface. Thus the hidden implication is that, as easily as one can twiddle a dial to change point size, so one can twiddle another dial (or set of dials) and arrive at any desired typeface, be it Helvetica, Baskerville, or whatever. Knuth might just as easily have put it this way:

> The ability to manipulate lots of parameters may be interesting and fun, but does anybody really need an n-point font that is x percent of the way between typeface *T1* and typeface *T2*?

For instance, we might have set the four knobs to the following settings:

> *n:* 36
> *x:* 50 percent
> *T1:* Magnificat
> *T2:* Stop

Each of these two typefaces (see Figure 13-2) is ingenious, idiosyncratic, and visually intriguing. I challenge any reader to even *imagine* a blend halfway between them, let alone draw it! And to emphasize the flexibility implied by the question, how about trying to imagine a typeface that is (say) one third of the way between Cirkulus and Block Up? Or one that is somewhere between Explosion and Shatter? (For these typefaces, see Figure 13-2.)

A Posteriori Knobs and the Frame Problem of AI

Shatter, incidentally, provides an excellent example of the trouble with viewing everything as coming from parameter settings. If you look carefully, you will see that Shatter is indeed a "variation on a theme", the theme being Helvetica Medium Italic (see Figure 13-2). But does that imply that any meticulous parametrization of Helvetica would automatically yield Shatter as one of its knob-settings? Of course not. That is absurd. No one in their right mind would anticipate such a variation while parametrizing Helvetica, just as no one in their right mind when delivering their Nobel Lecture would say, "Thank you for awarding me my first Nobel Prize." When someone wins a Nobel Prize, they do not immediately begin counting how many they have won. Of course, if they win *two,* then a knob will spontaneously appear in most people's minds, and friends will very likely make jokes about the next few Nobel Prizes. Before the second prize, however, the "just-one" quality would have been an unperceived fact.

This is closely related to a famous problem in cognitive science (the study of formal models of mental processes, especially computer models) called the *frame problem.* This knotty problem can be epitomized as follows: How do I know, when telling you I'll meet you at 7 at the train station, that it makes no sense to tack on the proviso, "as long as no volcano erupts along the way, burying me and my car on the way to the station", but that it *does* make reasonable sense to tack on the proviso, "as long as no traffic jam holds me up"? And of course, there are many intermediate cases between these two. The frame problem is about the question: *What variables (knobs) is it within the bounds of normalcy to perceive?* Clearly, no one can conceivably anticipate all the factors that might somehow be relevant to a given situation; one simply blindly hopes that the species' evolution and the individual's life experiences have added up to a suitably rich combination to make for satisfactory behavior most of the time. There are too many contingencies, however, to try to anticipate them all, even given the most powerful computer. One reason for the extreme difficulty in trying to make

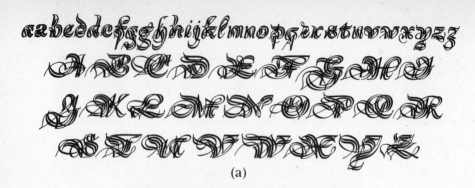

(a)

ABCDEFGHIJKLM
NOPQRSTUVWXYZ

(b)

abcdefghijklmnopqrstuvwxyz

(c)

ABCDEFGHIJKLM
NOPQRSTUVWXYZ

(d)

AABCDEFGHIIJKKLLM
NNOOPQRRSSTTUVWXYZ

(e)

abcdefghijklmnopqrstuvwxyz
ABCDEFGHIJKLMNOPQRSTUVWXYZ

(f)

abcdefghijklmnopqrstuvwxyz
ABCDEFGHIJKLMNOPQRSTUVWXYZ

(g)

FIGURE 13–2. *A series of diverse typefaces: (a) Magnificat; (b) Stop; (c) Cirkulus; (d) Block Up; (e) Explosion; (f) Shatter; (g) Helvetica Medium Italic.*

machines able to learn is that we find it very hard to articulate a set of rules defining when it makes sense and when it makes no sense to perceive a knob. It is a fascinating task to work on making a machine capable of coaxing shy knobs out of the woodwork.

This brings us back to Shatter, seen as a variation on Helvetica. Obviously, once you've seen such a variation, you can add a knob (or a few) to your METAFONT "Helvetica machine", enabling Shatter to come out. (Indeed, you could add similar "Shatterizing" knobs to your "Baskerville machine", for that matter!) But this would all be *a posteriori*: after the fact. The most telling proof of the artificiality of such a scheme is, of course, that no matter how many variations have been made on (say) Helvetica, people can still come up with many new and unanticipated varieties, such as: Helvetica Rounded, Helvetica Rounded Deco, Helvetican Flair, and so on (see Figure 13-3).

No matter how many new knobs—or even new families of knobs—you add to your Helvetica machine, you will have left out some possibilities. People will forever be able to invent novel variations on Helvetica that haven't been foreseen by a finite parametrization, just as musicians will forever be able to devise novel ways of playing "Begin the Beguine" that the electronic-

---

FIGURE 13–3. *Three "simple" offshoots of Helvetica: (a) Helvetica Rounded; (b) Helvetica Rounded Deco; (c) Helvetican Flair.*

organ builders haven't yet built into their elaborate repertoire of canned rhythms, harmonies, and so forth. To be sure, the organ builders can always build in extra possibilities after they have been revealed, but by then a creative musician will have long since moved on to other styles. One can imagine Helvetica modified in many novel ways inspired by various extant typefaces. I leave it to readers to try to imagine such variants.

A Total Unification of All Typefaces?

The worst is *still* yet to come! Knuth's throwaway sentence unspokenly implies that we should be able to interpolate any fraction of the way between any two arbitrary typefaces. For this to be possible, any pair of typefaces would have to share the exact same set of knobs (otherwise, how could you set each knob to an intermediate setting?). And since all *pairs* of typefaces have the same set of knobs, transitivity implies that *all* typefaces would have to share a single, grand, universal, all-inclusive, ultimate set of knobs. (The argument is parallel to the following one: If any *two* people have the same number of legs as each other, then leg-number is a universal constant for *all* people.)

Thus we realize that Knuth's sentence casually implies the existence of a "universal 'A'-machine"—a single METAFONT program with a finite set of parameters, such that any combination of settings of them will yield a valid 'A', and conversely, such that any valid 'A' will be yielded by some combination of settings of them. Now how can you possibly incorporate all of the previously shown typefaces into one universal schema?

Or look again at the 56 capital 'A's of Figure 12-3. Can you find in them a set of specific, quantifiable features? (For a comparable collection for each letter of the alphabet, see the marvelous collection of alphabetical logos compiled by Kuwayama.) Imagine trying to pinpoint a few dozen discrete features of the Magnificat 'A' (A7) and simultaneously finding their "counterparts" in the Univers 'A' (D3). Suppose you have found enough to characterize both completely. Now remember that every intermediate setting also must yield an 'A'. This means we will have every shade of "cross" between the two typefaces.

This intuitive sense of a "cross" between two typefaces is common and natural, and occurs often to typeface lovers when they encounter an unfamiliar typeface. They may characterize the new face as a cross between two familiar typefaces ("Vivaldi is a cross between Magnificat and Palatino Italic Swash") or else they may see it as an exaggerated rendition of a familiar typeface ("Magnificat is Vivaldi squared") (see Figure 13-4). What degree of truth is there to such a statement? All one can really say is that each Magnificat letter looks "sort of like" its Vivaldi counterpart, only about "twice as fancy" or "twice as curly" or something vague along those lines. But how could a single "curliness" knob account for the mysteriously beautiful meanderings, organic and capricious, in each Magnificat letter?

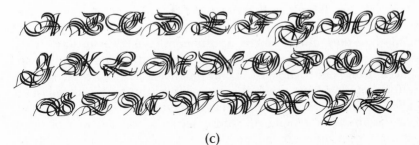

ABDEFGHJKLMNPQRSTUWZ

(a)

(b)

(c)

FIGURE 13–4. *A transition from curved to whirly to superswirly: (a) Palatino Italic Swash caps; (b) Vivaldi caps; (c) Magnificat caps. It is provocative to compare this figure with Figure 16-7.*

Can you imagine twisting one knob and watching thin, slithery tentacles begin to grow out of the Palatino Italic 'A', snaking outwards eventually to form the Vivaldi 'A', then continuing to twist and undulate into ever more sinuous forms, yielding the Magnificat 'A' in the end? And—who says that *that* is the ultimate destination? If Magnificat is Vivaldi squared, then what is Magnificat squared?

Specialists in computer animation have had to deal with the problem of interpolation of different forms. For example, in a television series about evolution, there was a sequence showing the outline of one animal form slowly transforming into another one. But one cannot simply tell the computer, "Interpolate between this shape and that one!" To each point in one there must be explicitly specified a corresponding point in the other. Then one lets the computer draw some intermediate positions on one's screen, to see if the choice works. A lot of careful "tuning" of the correspondences between figures must be done before the interpolation looks good. There is no recipe that works in general for interpolation. The task is deeply semantic, not cheaply syntactic.

For a wonderful demonstration of the truth of this, look at the little book *Double Takes,* in which artist Tom Hachtman has a lot of fun taking unlikely pairs of people and combining their caricatures. His only prerequisite is that their names should splice together amusingly. Thus he did "Bing Cosby" (Bing Crosby and Bill Cosby), "Farafat" (Farrah Fawcett-Majors and Yasir Arafat), "Marlon Monroe" (Marlon Brando and Marilyn Monroe), and many others. The trick is to discern which features of each person are the most characteristic and modular, and to be able to construct a new person having a subtle blend of those features, clearly enough that both contributors can be recognized. For a viewer, it's almost like trying to recognize the two parents in a baby's face.

The Essence of 'A'-ness Is Not Geometrical

Despite all the difficulties described above, some people, even after scrutinizing the wide diversity of realizations of the abstract 'A'-concept, still maintain that they all do share a common geometric quality. They sometimes verbalize it by saying that all 'A's have "the same shape" or are "produced from one template". Some mathematicians are inclined to search for a topological or group-theoretical invariant. A typical suggestion might be: "All instances of 'A' are open at the bottom and closed at the top." Well, in Figure 12-3, sample A8 (Stop) seems to violate both of those criteria. And many others of the sample letters violate at least one of them. In several examples, such concepts as "open" or "closed" or "top" or "bottom" apply only with difficulty. For instance, is G7 (Sinaloa) open at the bottom? Is F4 (Calypso) closed at the top? What about A4 (Astra)?

The problem with the METAFONT "knobs" approach to the 'A' category is that each knob stands for the presence or absence (or size or angle, etc.) of some specifically *geometric* feature of a letter: the width of its serifs, the height of its crossbar, the lowest point on its left arm, the highest point along some extravagant curlicue, the amount of broadening of a pen, the average slope of the ascenders, and so forth and so on. But in many 'A's, such notions are not even applicable. There may be no crossbar, or there may be two or three or more. There may be no curlicue, or there may be a few curlicues.

A METAFONT joint parametrization of two 'A's presumes that they share the same features, or what might be called "loci of variability". It is a bold (and, I maintain, absurd) assumption that one could get any 'A' by filling out an eternal and fixed questionnaire: "How wide is its crossbar? What angle do the two arms make with the vertical? How wide are its serifs?" (and so forth). There may be no identifiable part that plays the crossbar role, or the left-arm role; or some role may be split among two or more parts. You can easily find examples of these phenomena among the 56 'A's in Figure 12-3. Some other examples of what I call *role splitting, role combining, role transferral, role redundancy, role addition,* and *role elimination* are shown in Figure 13-5. These terms describe the ways that conceptual roles are apportioned among various geometric entities, which are readily recognized by their connectedness and gentle curvatures.

For a remarkable demonstration of ways to exploit these various role-manipulations, see Scott Kim's book *Inversions,* in which a single written specimen, or "gram", has more than one reading, depending on the observer's point of view. Often the "grams" are symmetric and read the same both ways, but this is not essential: some have two totally different

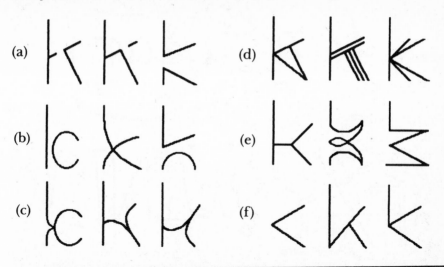

FIGURE 13–5. *Examples of: (a) role splitting; (b) role merging; (c) role transferral; (d) role redundancy; (e) role addition; and (f) role elimination. The idea in all these examples is that one smooth sweep of the pen need not fill exactly one coherent conceptual role. It may fill two or more roles (or parts of two or more); it may fill less than one, in which case several strokes combine to make one role; and so on. Sometimes roles can be added or deleted without serious harm to the recognizability of the letter. Angles, cusps, intersections, endpoints, extrema, blank areas, and separations often play roles no less vital than those played by strokes.*

readings. The essence is imbuing a single written form with ambiguity. Both Scott and I have for years done such drawings—dubbed "ambigrams" by a friend of mine—and a few of my own are presented in Figure 13-6, as well as the one on the half-title page. The strange fluidity of letterforms is brought out in a most vivid way by ambigrammatic art.

Incidentally, it is most important that I make it clear that although I find it easier to make my points with somewhat extreme or exotic versions of letters (as in ambigrams or unusual typefaces), these points hold just as strongly for more conservative letters. One simply has to look at a finer grain size, and all the same kinds of issues reappear.

Chauvinism versus Open-Mindedness:
Fixed Questionnaires versus Fluid Roles

When I was twelve, my family was about to leave for Geneva, Switzerland for a year, so I tried to anticipate what my school would be like. The furthest my imagination could stretch was to envision a school that looked exactly like my one-story Californian stucco junior high school, only with classes in French (twiddling the "language" knob) and with the schoolbus that would pick me up each morning perhaps pink instead of yellow (twiddling the "schoolbus color" knob). I was utterly incapable of anticipating the vast

FIGURE 13–6. *Several ambigrams by the author. Deciphered, they say: "ambigram"; "ambigrams"; "winter"; "spring"; "summer"; "fall"; "Lee Sallows"; "Josh Bell"; "Alejandra" and "Magdalena" (reflections of each other); "Carol"; "David Moser"; "Chopin"; and "Johann*

ambigrams

Spring **Summer**

Alejandra
Magdalena

Caro

David

Chopin

Sebastian Bach''. All three composers' names utilize 90-degree rotation. Notice the extensive use of all the devices shown in Figure 13-5, namely role splitting, merging, transferral, redundancy, addition, and elimination. See the half title page for a further ambigram by the author.

difference that there actually turned out to be between the Geneva school and my California school.

Likewise, there are many "exobiologists" who have tried to anticipate the features of extraterrestrial life, if it is ever detected. Many of them have made assumptions that to others appear strikingly naïve. Such assumptions have been aptly dubbed *chauvinisms* by Carl Sagan. There is, for instance, "liquid chauvinism", which refers to the phase of the medium in which the chemistry of life is presumed to take place. There is "temperature chauvinism", which assumes that life is restricted to a temperature range not too different from that here on the planet earth. In fact, there is planetary chauvinism—the idea that all life must exist on the surface of a planet orbiting a certain type of star. There is carbon chauvinism, assuming that carbon must form the keystone of the chemistry of any sort of life. There is even speed chauvinism, assuming that there is only one "reasonable" rate for life to proceed at. And so it goes.

If a Londoner arrived in New York, we might find it quaint (or perhaps pathetic) if he or she asked "Where is your Big Ben? Where are your Houses of Parliament? Where does your Queen live? When is your teatime?" The idea that the biggest city in the land need not be the capital, need not have a famous bell tower in it, and so on, seem totally obvious after the fact, but to the naïve tourist it can come as a surprise. (See Chapter 24 for more on strange mappings between Great Britain and the United States.)

The point here is that when it comes to fluid semantic categories such as 'A', it is equally naïve to presume that it makes sense to refer to "the crossbar" or "the top" or to any constant feature. It is quite like expecting to find "the same spot" in any two pieces of music by the same composer. The problem, I have found, is that most people continue to insist that any two instances of 'A' have "the same shape", even when confronted with such pictures as Figure 12-3. Figure 12-4 helps, however, to dispel that sort of notion (as does Figure 24-13).

The analogy between Britain and the United States is a useful one to continue for a moment. The role that London plays in England is certainly multifaceted, but two of its main facets are "chief commercial city" and "capital". These two roles are played by different cities in the U.S. On the other hand, the role that the American President plays in the U.S. is split into pieces in Britain, part being carried by the Queen (or King), and part by the Prime Minister. Then there is a subsidiary role played by the President's wife—the "First Lady". Her counterpart in Britain is also split, and moreover, these days, "wife" has to be replaced by "husband", no matter whether one considers that the "President of England" is the Queen or the Prime Minister. (Again, see Chapter 24 for much more detail on this kind of analogy problem.)

To think one can anticipate the complete structure of one country or language purely on the basis of being intimately familiar with another one is presumptuous and, in the end, preposterous. Even if you have seen dozens, you have not exhausted the potential richness and novelty in such domains. In fact, the more instances you have seen, the more circumspect you are about making unwarranted presumptions about unseen instances, although—a bit paradoxically—your ability to anticipate the unanticipated (or unanticipable) certainly improves! The same holds for instances of any letter of the alphabet or other semantic category.

The 'A' Spirit

Clearly there is much more going on in typefaces than meets the eye— literally. The shape of a letterform is a surface manifestation of deep mental abstractions. It is determined by conceptual considerations and balances that no finite set of merely geometric knobs could capture. Underneath or behind each instance of 'A' there lurks a concept, a Platonic entity, a *spirit*. This Platonic entity is not an elegant shape such as the Univers 'A' (D3), not a template with a finite number of knobs, not a topological or group-theoretical invariant in some mathematical heaven, but a mental abstraction —a different sort of beast. Each instance of the 'A' spirit reveals something new about the spirit without ever exhausting it. The mathematization of such a spirit would be a machine with a specific set of knobs on it, defining all its "loci of variability" for once and for all. I have tried to show that to expect this is simply not reasonable. In fact, I made the following claim, above:

> No matter how many new knobs—or even new families of knobs—you add to your .... machine, you will have left out some possibilities. People will forever be able to invent novel variations.... that haven't been foreseen by a finite parametrization....

Of what, then, is such an abstract "spirit" composed? Or is it simply a mystically elusive, noncapturable essence that defies the computational— indeed, the scientific—approach totally? Not at all, in my opinion. I simply think that a key idea is missing in what I have described so far. And what is this key idea? I shall first describe the key misconception. It is to try to capture the essence of each separate concept in a separate "knobbed machine"—that is, to isolate the various Platonic spirits. The key insight is that those spirits overlap and mingle in a subtle way.

Happy Roles, Unhappy Roles, and Quirk-Notes

The way I see it, the Platonic essence lurking behind any concrete letterform is composed of conceptual *roles* rather than geometric *parts*. (A related though not identical notion called "functional attributes" was discussed by Barry Blesser and co-workers in *Visible Language* as early as 1973.) A role, in my sense of the term, does not have a fixed set of parameters defining the extent of its variability, but it has instead a set of tests or criteria to be applied to candidates that might be instances of it. For a candidate to be accepted as an instance of the role, not all the tests have to be passed; not all the criteria have to be present. Instead, the candidate receives a score computed from the tests and criteria, and there is a threshold point above which the role is "happy" and below which it is "unhappy". Then below that, there is a cutoff point below which the role is totally dissatisfied, and rejects the candidate outright.

An example of such a role is that of "crossbar". Note that I am not saying "crossbar in capital 'A'", but merely "crossbar". Roles are *modular*: they jump across letter boundaries. The same role can exist in many different letters. This is, of course, reminiscent of the fact that in METAFONT, a serif (or generally, any geometric feature shared by several letters) can be covered by a single set of parameters for *all* letters, so that all the letters of the typeface will alter consistently as a single knob is turned. One difference is that my notion of "role" doesn't have the generative power that a set of specific knobs does. From the fact that a given role is "happy" with a specific geometric filler, one cannot deduce exactly how that filler looks. There is, of course, more to a role's "feelings" about its filler than simply happiness or unhappiness; there are a number of expectations about how the role should be filled, and the fulfillment (or lack thereof) can be described in *quirk-notes*. Thus, quirk-notes can describe the unusual slant of a crossbar (see Arnold Böcklin—E1 in Figure 12-3), the fact that it is filled by two strokes rather than one (Airkraft—E3), the fact that it fails to meet (or has an unusual way of meeting) its vertical mate (Eckmann Schrift—A2; Le Golf—F5), and many other quirks.

These quirk-notes are characterizations of stylistic traits of a *perceived* letterform. They do not contain enough information, however, to allow a full *reconstruction* of that letterform, whereas a METAFONT program does contain enough information for that. However, they do contain enough information to guide the creation of many specific letterforms that have the given stylistic traits. All of them would be, in some sense, "in the same style".

Modularity of Roles

The important thing is that this *modularity of roles allows them to be exported* to other letters, so that a quirk-note attached to a particular role in 'A' could have relevance to 'E', 'L', or 'T'. Thus *stylistic consistency among different letters is a by-product of the modularity of roles,* just as the notion of letter-spanning parameters in METAFONT gives rise to internal consistency of any typeface it might generate.

Furthermore, there are connections among roles so that, for instance, the way in which the "crossbar" role is filled in one letter could influence the way that the "post" or "bowl" or "tail" role is filled in other letters. This is to avoid the problem of overly simplistic mappings of one letter onto another, analogous to the Londoner asking an American where the

American Houses of Parliament are. Just as one must interpret "Houses of Parliament" liberally rather than literally when "translating" from England to the U.S., so one may have to convert "crossbar" into some other role when looking for something analogous in the structure of a letter other than 'A', such as 'N'. In certain typefaces, the diagonal stroke in 'N' could well be the counterpart of the crossbar in 'A'. But it is important to emphasize that no *fixed* (*i.e.*, typeface-independent) mapping of roles in 'A' onto roles in 'N' will work; only the specific letterforms themselves (via their quirk-notes) can determine what roles (if any) should be mapped onto each other. Such cross-letter mappings must be mediated by a considerable degree of understanding of what functions are fulfilled by all the roles in the two particular letters concerned.

Typographical Niches and Rival Categories

So far I have sketched very quickly a theory of "Platonic essences" or "letter spirits" involving modular roles—roles shared among several letters. This sharing of roles is one aspect of the overlapping and mingling that I spoke of above. There is a second aspect, which is suggested by the phrase *typographical niche*. The notion is analogous to that of "ecological niche". When, in the course of perception of a letterform, a group of roles have been activated and have decided that they are present (whether happily or unhappily), their joint presence constitutes evidence that one of a set of possible letters is present. (Remember that since a role is not the property of any specific letter, its presence does not signal that any specific letter is in view.)

For instance, the presence of a "post" role and a "bowl" role in certain relative positions would suggest very strongly that there is a 'b' present. Sometimes there may be evidence for more than one letter. The eye-mind combination is not happy with any such unstable state for long, and strains to make a decision. It is as if there is a very steep and slippery ridge between valleys, and a ball dropped from above is very unlikely to come to settle on top of the ridge. It will tumble to one side or the other. The valleys are the typographical niches.

Now, the overlapping of letters comes about because each letter is aware of its typographical rivals, its next-door neighbors, just over the various ridges that surround its space. The letter 'h', for instance, is acutely sensitive to the fact that it has a close rival in 'k', and vice versa (see Figure 13-7). The letter 'T' is very touchy about having its crossbar penetrated by the post below, since even the slightest penetration is enough to destroy its 'T'-ness and to slip it over into 'T's arch-rival niche, 't'. It's a low ridge, and for that reason, 'T' guards it extra-carefully.

FIGURE 13–7. *Have we "hen" or "ken" here? In each case, two niches in the Platonic alphabet compete for possession of a single physical specimen. Again, the fluid way in which minds are willing to let roles and fillers align is the source of all the trouble.*

The Intermingling of Platonic Essences

This image is, I hope, sufficiently strong to convey the second sense of overlapping and intermingling of Platonic essences. "No letter is an island", one might say. There has to be much mutual knowledge spread about among all the letters. *Letters mutually define each others' essences,* and this is why an isolated structure supposedly representing a single letter in all its glory is doomed to failure.

A letterform-designing computer program based on the above-sketched notions of typographical roles and niches would look very different from one that tried to be a full "mathematization of categories". It would involve an integration of perception with generation, and moreover an ability to generalize from a few letterforms (possibly as few as one) to an entire typeface in the style of the first few. It would not do so infallibly; but of course it is not reasonable to expect "infallible" performance, since stylistic consistency is not an objectively specifiable quality.

In other words, a computer program to design typefaces (or anything else with an esthetic or subjective dimension) is not a conceptual impossibility;

but one should realize that, no less than a human, any such program will necessarily have a "personal" taste—and it will almost certainly not be the same as its designer's (or designers') taste. In fact, to the contrary, the program's taste will quite likely be full of unanticipated surprises to its programmers (as well as to everyone else), since that taste will emerge as an implicit and remote consequence of the interaction of a myriad features and factors in the architecture of the program. Taste itself is not *directly* programmable. Thus, although any esthetically programmed computer will be "merely doing what it was programmed to do", its behavior will nonetheless often appear idiosyncratic and even inscrutable to its programmers, reflecting the fact—well known to programmers—that often one has no clear idea (and sometimes no idea at all) just what it is that one has programmed the machine to do!

The Vertical and Horizontal Problems: Two Equally Important Facets of One Problem

I have made a broad kind of claim: that true understanding of letterforms depends on more than understanding something about each Platonic letter in isolation; it depends equally much on taking into account the ways that letters and their pieces are interrelated, on the ways that letters depend on each other to define a total style. In other words, any approach to the impossible dream of the "secret recipe" for 'A'-ness requires a simultaneous solution to two problems, which I call the *vertical* and the *horizontal* problems (see Figures 13-8 and 24-14).

Vertical: What do all the items in any *column* have in common?
Horizontal: What do all the items in any *row* have in common?

FIGURE 13–8. *The vertical and horizontal problems. What do all the items in any* column *have in common? What do all the items in any* row *have in common? Answers: Letter; Spirit. (Compare this figure with Figure 24-14.)*

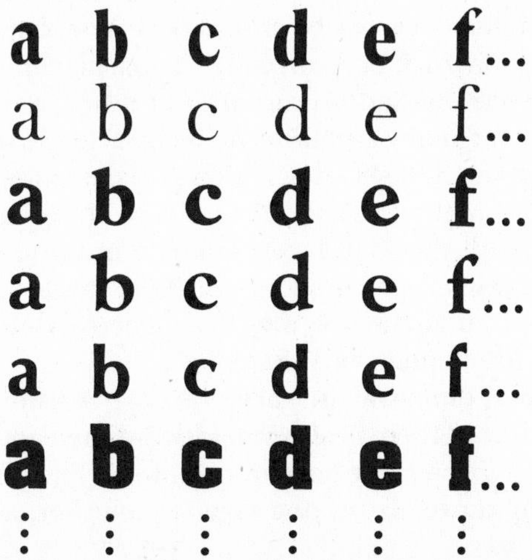

---

abcdefghijklmnopqrstuvwxyz
ABCDEFGHIJKLMNOPQRSTUVWXYZ

(a)

abcdefghijklmnopqrstuvwxyz
ABCDEFGHIJKLMNOPQQRSTUVWXYZ

(b)

abcdefghijklmnopqrstuvwxyz
ABCDEFGHIJKLMNOPQRSTUVWXYZ

(c)

abcdefghijklmnopqrstuvwxyz
ABCDEFGHIJKLMNOPQRSTUVWXYZ

(d)

abcdefghijklmnopqrstuvwxyz
ABCDEFGHIJKLMNOPQRSTUVWXYZ

(e)

abcddee ffgghijkklmnopqrrsttuvvwwxyyz
AABBCCDDEEEFGGHHIJJJKKLLMMNNOPPQRRSSTT
UUVVWWXYYZZ

(f)

FIGURE 13–9. *Six elegant faces created by the contemporary designer Hermann Zapf. In (a), Optima; in (b), Palatino; in (c), Melior; in (d), Zapf Book; in (e), Zapf International; and in (f), Zapf Chancery.*

Actually, there is no reason to stop with two dimensions; the problem seems to exist at higher degrees of abstraction. We could lay out our table of comparative typefaces more carefully; in particular, we could make it consist of many layers stacked on top of each other, as in a cake. On each layer would be aligned many typefaces made by a single designer. This idea is illustrated in Figure 13-9, showing a few faces designed by Hermann Zapf (Optima, Palatino, Melior, Zapf Book, Zapf International, and Zapf Chancery). Along with the Zapf layer, one can imagine a Frutiger layer, a Lubalin layer, a Goudy layer, and so on. One could try to arrange the typefaces in each layer in such a way that "corresponding" typefaces by various designers are aligned in "shafts".

Now in this three-dimensional cake, the two earlier one-dimensional questions still apply, but there is also a new two-dimensional question: What do all the items in a given layer have in common? The third dimension can be explored as one moves from one layer to another, asking what all the typefaces in a given "shaft" have in common. Moreover, a fourth dimension can be added if you imagine many such "layer cakes", one for each distinguishable period of typographical design. Thus our fourth dimension, like Einstein's, corresponds to time. Now one can ask about each layer cake: What do all the items herein have in common? This is a three-dimensional question. Presumably, one could carry this exercise even further.

If we go back to the "simplest" of these questions, the original "vertical" question applying to Figure 13-8, a naïve answer to it could be stated in one word: *Letter.* And likewise, a naïve answer to the "horizontal" question of that figure is also statable in one word: *Spirit.* In fact, the word "spirit" is applicable, in various senses of the term, to all the higher-dimensional questions, such as "What do all the typefaces produced in the Art Deco era have in common?" There is such a thing, ephemeral though it may be, as "Art Deco spirit", just as there is undeniably such a thing as "French spirit" in music or "impressionistic spirit" in art. (Marcia Loeb has recently designed a whole series of typefaces in the Art Deco style, in case anyone doubts that the spirit of those times can be captured. And then there is the book *Zany Afternoons* by Bruce McCall, in which the entire spirit of several recent decades is wonderfully spoofed on all stylistic levels simultaneously.)

Stylistic *moods* permeate whole periods and cultures, and they indirectly determine the kinds of creations—artistic, scientific, technological—that people in them come up with. They exert gentle but definite "downward" pressures. As a consequence, not only are the alphabets of a given period and area distinctive, but one can even recognize "the same spirit" in such things as teapots, coffee cups, furniture, automobiles, architecture, and so on, as Donald Bush clearly demonstrates in his book *The Streamlined Decade.* One can be inspired by a given typeface to carry its ephemeral spirit over into another alphabet, such as Greek, Hebrew, Cyrillic, or Japanese. In fact, this has been done in many instances (see Figure 13-10). The problem I am most concerned with in my research is whether (or rather, how) susceptibility to such a "spirit" can be implanted in a computer program.

Letter and Spirit

These words "letter" and "spirit", of course, recall the contrast between the "letter of the law" and the "spirit of the law", and the way in which our legal system is constructed so that judges and juries will base their decisions on precedents. This means that any case must be "mapped", in a remarkably fluid way, by members of a jury, onto previous cases. It is up to the opposing lawyers, then, to be advocates of particular mappings; to try to channel the jury members' perceptions so that one mapping dominates over another. It is quite interesting that jury decisions are supposed to be unanimous, so that in a metaphorical sense, a "phase transition" or "crystallization" of opinion must take place. The decision must be solidly locked in, so that it reflects not simply a majority or even a consensus, but a totality, a unanimity (which, etymologically, means "one-souledness"). (For discussions of such "phase transitions of the mind", see Chapters 25 and 26, and for descriptions of

TASTE IN PRINTING DETERMINES THE FORM TYP
ography is to take. The selection of a congruous typeface, the
quality and suitability for its purpose of the paper being used

ABCDEFGHIJKLMNOPQRSTUVWXYZ
abcdefghijklmnopqrstuvwxyz

ШРИФТОТЕКА КОМПЬЮГРАФИК СОДЕРЖИТ
дысяза гарнитуров шрифта включающихкак традич
ио нные так современные рисунк шрифта, кото рые

АБВГДЕЖЗИЙКЛМНОПРСТУФХ
ЦЧШЩЪЫЬЭЮЯабвгдежзийклм
нопрстуфхцчшщъыьэюя

TASTE IN PRINTING DETERMINES THE FORM TYPOG
raphy is to take. The selection of a congruous typeface, the qua
lity and suitability for its purpose of the paper being used, the

ABCDEFGHIJKLMNOPQRSTUVWXYZ
abcdefghijklmnopqrstuvwxyz

ШРИФТОТЕКА КОМПЬЮГРАФИК СОДЕРЖИТ
дысяза гарнитуров шрифта включающихкак трат
ичио нные так современные рисунк шрифта, котор

АБВГДЕЖЗИЙКЛМНОПРСТУФХ
ЦЧШЩЪЫЬЭЮЯабвгдежзийклм
нопрстуфхцчшщъыьэюя

(a)

ABCDEFGHIJKLMNOPQRSTUVWXYZ
abcdefghijklmnopqrstuvwxyz

TASTE IN PRINTING WILL DETERMINE THE FORM TY
pography is to take, the selection of a congruous typefa
ce, the quality and suitability for its purpose of the pap

ΑΒΓΔΕΖΗΘΙΚΛΜΝΞΟΠΡΣΤΥΦΧΨΩ
αβγδεζηθικλμνξοπρστυφχψως

Ή καλαισθησιά καί ή ἀπόδοση στήν ἐκτύπωση προ
σδιο ρίξει μορφή πού θάρει τυπωμένο κείμενο τήν
ἐπι λογή τοῦ ἀνάλογου ὀφθαλμοῦ, τήν ποιότητα τό

(b)

ecnolpןɔלמɔɪʋ

ECNOLPןɔלמɔɪʋ

אבגדהוזחטיכלמנסעפצקרשת םולג ץולק

abcdefghijklmnopqrstuvwxyz

(c)

アイウエオカキク
ケコサシスセソタ
チツテトナニヌネ
ノハヒフヘホマミ
ムメモヤユヨラリ
ルレロワヲンガグ
ゲザジズダツバビ
パピプペ゜・アイウエ

ABCDEFGHI
JKLMNOPQR
STUVWXYZ&
1234567890:,
abcdefghijkl-
mnopqrstuvw
xyz

(d)

computer models of perception in which a form of collective decision-making is carried out, see the book by McClelland, Rumelhart, and Hinton, and my article on the Copycat project.)

In law, extant rules, statutes, and so on, are never enough to cover all possible cases (reminding us once again of the fact that no fixed and rigid set of 'A'-defining rules can anticipate all 'A's). The legal system depends on the notion that people, whose experience covers much more than the specific case and rules at hand, will bring to bear their full range of experience not only with many categories but also with the whole process of categorization and mapping. This allows them to transcend the specific, rigid, limited *rules,* and to operate according to more fluid, imprecise, yet more powerful *principles.* Or, to revert to the other vocabulary, this ability is what allows people to transcend the letter of the law and to apply its spirit. It is this *tension between rules and principles, tension between letter and spirit,* that is so admirably epitomized for us by the work of Donald Knuth and others exploring the relationship between artistic design and mechanizability. We are entering a very exciting and important phase of our attempts to realize the full potential of computers, and Knuth's article points to many of the significant issues that must be thought through very carefully.

In summary, then, the mathematization of categories is an elegant goal, a wonderful beckoning mirage before us, and the computer is the obvious medium to exploit to try to realize this goal. Donald Knuth, whether he has been pulled by a distant mirage or by an attainable middle-range goal, has contributed immensely, in his work on METAFONT, to our ability to deal with letterforms flexibly, and has cast the whole problem of letters and fonts in a much clearer perspective than ever before. Readers, however, should not pull a false message out of his article: they should not confuse the chimera of the mathematization of categories with the quest after a more modest but still fascinating goal. In my opinion, one of the best things METAFONT could do is to inspire readers to chase after what Knuth has rightly termed the "intelligence" of a letter, making use of the explicit medium of the computer to yield new insights into the elusive "spirits" that flit about so tantalizingly, hidden just behind those lovely shapes we call "letters".

FIGURE 13–10. *Transalphabetic leaps by the ethereal "spirit" inherent in a given typeface. In (a), we see the "Times" spirit jump across the gap between the Latin and Cyrillic alphabets. In (b), the "Optima" spirit transplants itself to Greek soil. In (c), a Hebrew spirit leaps out of the mirror and jumps into Latin clothes. Finally, in (d), a gigantic trans-Pacific (or trans-Asiatic) leap in which a Kana spirit (Japanese syllabic characters) jumps into Latin letters.*

In recent years there has been a spate of reported sightings of unidentified font-like objects (UFO's). Many people who claim to have seen UFO's insist that they come from other planets. Some claim, for instance, to have seen Venusian written in the Baskerville style, while others say they have seen Martian in the Helvetica style. There are even claims of a complete Magnificat-style Alphacentauribet! Often these claims are contradictory. For instance, one witness will maintain that the bowl of the 'g' was cigar-shaped, while another maintains equally vehemently that it resembled a saucer. Needless to say, not a single such sighting has ever been scientifically validated.

Post Scriptum.

Some months after this article appeared in *Visible Language,* the editor of that journal published a most interesting commentary by Geoffrey Sampson, now a professor in the Linguistics Department at the University of Leeds in England. Here are some extracts from his article, giving the gist of it:

I believe that Douglas Hofstadter is unfair in his critique of Donald Knuth's "Meta-font" article . . . . Human life involves both open-ended categories and closed categories, and in many cases it is very hard to say whether a given intuitively familiar category is open-ended or closed . . . . Hofstadter writes as if Knuth assumes an obviously open-ended category to be closed; but I cannot see that Hofstadter has demonstrated this . . . . Baskerville and Helvetica are both book faces, rather than faces designed exclusively for display. On the other hand, the 56 'A's of Hofstadter's figure [Figure 12-3] are all drawn from display faces. It is much less obvious that the class of book faces is open-ended than that the class of display faces is . . . .

If we restrict the task to book faces (which are the only faces discussed by Knuth) then the open-endedness of the range really does become questionable. Hofstadter denies that this restriction affects his point: with 'more conservative letters . . . . one simply has to look at a finer grain size, and all the same kinds of issues reappear'. Do they? . . . .

The only argument Hofstadter gives for this is the difficulty of 'parametrizing' the contrast between the round dots of Baskerville 'i', 'j' and the square dots in Helvetica, and between the tails of 'Q' in the two faces. But Hofstadter concedes that it is not 'inconceivable' that these problems could be solved. Furthermore it seems to me that the number of such points, where two faces differ with respect to some property of an individual letter in a way that appears not to be predictable on the basis of more general differences between the faces, is fairly limited. The tail of 'Q' is an oddity in many faces; likewise the terminal of 'G'; but on the other hand if you know what (say) 'P' looks like in a given book face you will have a very good idea what 'D' or 'H' or 'T' looks like.

I would suggest that it is an entirely reasonable research programme to attempt to define a finite (no doubt large) set of variables (many of which would no doubt be very subtle) which generate all roman book faces, including faces not explicitly taken into consideration when formulating the variables, and excluding pathological letterforms . . . . If Hofstadter's view of typography is correct, the task proposed will prove to be impossible: every extra face considered will force the addition of yet more independent variables to the meta-font. However, I believe we have no adequate reason to reach this negative conclusion *a priori*.

When I first read this letter, I must admit, I felt that it made sense; that I had perhaps overstated my case. Sampson's point seemed reasonable. But then I started wondering, "Just where *are* the boundary lines of 'book-face-ness'?" This issue is beautifully exemplified by a tacit assumption made by Sampson. He calls Helvetica a book face, without any qualms. In doing so, he practically kicks the ball between his own goal posts for me! Helvetica is almost always thought of as a display face, and is most often used in book titles and advertising displays. It is a sans-serif face, like Optima, Eras, and many others of a similar vintage. I wonder what Sampson feels about serifed faces such as Goudy, Italia, Souvenir, Korinna, etc. (See Figure 13-11.) Which of these would count as display faces, and which as book faces?

Treacherous waters, these. The "problem" (actually not a problem at all, but a marvelous fact) is that the same typeface designers who design our favorite book faces also design our favorite display faces. And the same sense of style and joyous creation is called upon in both tasks. The way I

FIGURE 13–11. *Showing the futility of trying to draw a firm line between display faces and book faces. From top to bottom, we have: Eras Demi, Romic Light, Goudy Extra Bold, Italia Medium, Souvenir Light, and Korinna Extra Bold. It is easy to conceive of a book being printed in any of these faces (in a light weight), yet none is a standard book face.*

---

aabcdefghijklmnopqrstuvwxyz
ABCDEFGHIJKLMNOPQRSTUVWXYZ

abcdefghijklmnopqrstuvwxyz
ABCDEFGHIJKLMNOPQRSTTUVWXYZ

abcdefghijklmnopqrstuvwxyz
ABCDEFGHIJKLMNOPQRSTUVWXYZ

abcdefghijklmnopqrstuvwxyz
ABCDEFGHIJKLMNOPQRSTUVWXYZ

abcdefghijklmnopqrstuvwxyz
ABCDEFGHIJKLMNOPQRSTUVWXYZ

abcdeefghijjklmnopqrstuvwxyz
ABCDEFGHIJKLMNOPQRSTUVWXYZ

think of it is that each designer has a "wildness knob" with which to fiddle. When it's set low, the complexities and trickeries "retreat" into the nooks and crannies of the letterforms: how strokes terminate, swerve, change width, meet, and so on, and so the resulting typeface appears reserved and dignified, conventional yet graceful and stylish, still full of the designer's known characteristics. When wildness is set high, the desire for unusual, exuberant effects is let out of the closet, and the resulting typeface is full of bold flair and exciting, risky bravado: strokes are doubled, omitted, have extravagant shapes, flourishes, and so on. It is quite naïve to think that low wildness means "the same old book-face knobs are twiddled" no matter who's doing it, whereas high wildness involves an open-ended set of concepts.

No creative designer with any pride would ever feel content creating within a pre-set formula, a predetermined set of knobs. The joy of any kind of creation is in playing at the boundaries of what has been done. Every perceptive observer has an intuitive sense of the implicosphere centered on each standard letter and each role within it—a sense of just how daring various deviations will seem and of just where they will begin veering off into unacceptability. At the blurry boundaries of an implicosphere is exactly where an artist most loves to play. With wildness set low, a designer will flirt with the boundaries largely from *within*, making most decisions on the conservative side. With wildness set high, many more risks will be taken, and the flirting will carry the designer noticeably further from the implicosphere's center, like a satellite in a wider orbit. Norm violation is the name of the game in creation, no matter where the "wildness" knob is set. High wildness or low, it's still the same designer and the same creative forces expressing themselves. It's just a question of how subtly, how subduedly, those influences will show up.

*   *   *

Hermann Zapf is the designer of the famous sans-serif face Optima, a typeface that some books have been printed in (see Figure 13-9*a*). Optima is deceptively simple-looking. People tend to think that given one letter, they could determine all the rest easily. Sampson says as much: "If you know what (say) 'P' looks like in a given typeface, you will have a very good idea what 'D' or 'H' or 'T' looks like." But if that's the case, then why did it take Zapf—one of the world's foremost type designers—seven years to design it? All I can say is that there is rampant naïveté about the complexity of letters, even among people who visually are otherwise very astute.

A wonderful exercise to prove this to yourself is to try to draw the Helvetica Medium 'a' by memory (see Exhibit 'a', that is, Figure 13-12*a*). Study it for as long as you like, and then try to reproduce it. The better an eye you have, the more errors you will see you have made. Try it a few times. I myself must have attempted that 'a' several dozen times, and still I have never drawn it perfectly. This letter is one of my favorite letters of all time,

FIGURE 13–12. *Details of two classic typefaces: the 'a' of Helvetica Medium and the 'g' of Italia Book.*

and I have probably spent more time admiring it than any other letter—yet for all that, I still have not fathomed it entirely.

The case of Helvetica is interesting. What is characteristic about it? It was one of the first typefaces in which negative and positive spaces were given equal attention. It employed very simple, nearly mathematical curves. Why was it designed only in 1958? Why did it take so long for such obvious things to be done so elegantly? It's like asking why the ancient Greeks, with their love of purity and elegance, didn't discover group theory, the branch of mathematics dealing with abstract binary operations. Well, some ideas are so abstract that even though they are glimpsed through a fog centuries earlier, their full-scale arrival takes much longer. (Group theory waited patiently for 2,000 years after the Greeks to be discovered! Isn't group theory patient with our species?) Thus it was with the pristine qualities of Helvetica. And what *seems* remarkable, but is actually to be expected, is that in the same year as Max Miedinger designed Helvetica, Adrian Frutiger designed Univers, a lovely typeface, in many ways nearly indistinguishable from Helvetica. Some ideas are just ripe at certain times.

The ideas in Helvetica were not visible to anyone in the 1930's, even though people had thousands of book faces and display faces to look at. Likewise, the ideas in Snorple (a classic book face to be designed by Argli Snorple in 2027) are not visible to us today, even if, in some sense, they are implicitly defined by what is all around us. Cultural pressures, such as the development of computers and low-resolution digital typefaces, have profound impacts on how letters are perceived. Here is a striking example. When Hermann Zapf heard about the curve called a "super-ellipse"—an elegant mathematical interpolation between a circle and a square (or, more generally, between an ellipse and a rectangle), devised in the 1950's by the Danish scientist and author Piet Hein—he decided to base a typeface on that shape. The result: Melior, a now-standard book face whose "bowls" are super-ellipses (see Figure 13-9*c*). The point is, type designers are as susceptible as anyone else is to the subtle ebb and flow of cultural waves— and evidence of those waves shows up in book faces no less than in display faces. You just have to look more closely. Book faces pose problems no less knotty than do display faces, Sampson notwithstanding.

So on reconsideration, I stick with my point that *all* the same issues as apply to "wild" letterforms apply to "tame" ones—that one merely needs to look at a finer grain size to see the same kinds of problems. As I said above, modern book faces play with stroke tips in incredibly creative and surprising ways. Just look, for example, at Exhibit 'g'—that is, the 'g' of Italia (Figure 13-12*b*). Check out some of the other letters and then see what you think of Sampson's claim.

*   *   *

People tend to think that only extreme versions of things pose deep problems. That's why few people see modeling the creativity of, say, the trite television character of Archie Bunker as a difficult task. It's strange and disorienting to realize that if we could write a program that could compose Muzak or write trashy novels, we would be 99 percent of the way to mechanizing Mozart and Einstein. Even a program that could act like a mentally retarded person would be a huge advance. The commonest mental abilities—not the rarest ones—are still the central mystery.

John McCarthy, one of the founders of the field of artificial intelligence, is fond of talking of the day when we'll have "kitchen robots" to do chores for us, such as fixing a lovely Shrimp Creole. Such a robot would, in his view, be exploitable like a slave because it would not be conscious in the slightest. To me, this is incomprehensible. Anything that could get along in the unpredictable kitchen world would be as worthy of being considered conscious as would a robot that could survive for a week in the Rockies. To me, both worlds are incredibly subtle and potentially surprise-filled. Yet I suspect that McCarthy thinks of a kitchen as Sampson thinks of book faces: as some sort of simple and "closed" world, in contrast to "open-ended" worlds, such as the Rockies. This is just another example, in my opinion, of vastly underestimating the complexity of a world we take for granted, and thus underestimating the complexity of the beings that could get along in such a world.

Ultimately, the only way to be convinced of these kinds of things is to try to write a computer program to get along in a kitchen, or to generate book faces. That's when you finally come face to face with the extremely limiting notion of what a knob really is. People's notion of knobs has too much intuitive fluidity to it. It's hard to identify with a computer and to see things utterly and foolishly rigidly—but that's where you have to begin if you want to understand why knobbifying the alphabet is a task of vast magnitude, and is a microcosm of the task of knobbifying all of human thought.

*   *   *

It is very tempting to think that a few degrees of freedom, when combined, can cover any possible situation. After all, the number of possible states of a multi-knob machine is the product of the numbers of settings of each of its knobs, and multiplying a bunch of relatively small numbers together gets you rapidly into large-number territory. A perfect illustration of this line of thought is given in an ad I once clipped for a book called *Director's and Officer's Complete Letter Book,* informally nicknamed *The Ghost.* Here is some of what that ad says:

> This is not a book on letter-writing technique. It is a collection of 133 business letters *already written and ready to use.* They cover virtually every business situation you will ever meet. Just change a few words. They are arranged by subject, with 988 alternate phrases and sentences, keyed so that you can adapt the right letter to your purpose with almost no effort . . . . Editor J. A. VanDuyn traveled for four years, collecting the *finest* examples of business letters written today. They're in crisp, direct, informal language, without cliches . . . . In 30 seconds you can look up the letter you need, by subject. You may need only to change the name, address, and half-a-dozen words. Or you may use one or more of the alternate phrases, sentences, or paragraphs on the facing page. In minutes, you've got your letter. With the personal touch you want. Perfectly suited to the sense you wish to convey . . . .
>
> Some letters are especially hard. When you're stuck for the tactful approach, the just-right expression of concern, the graceful apology, you'll be thankful you have *The Ghost.* Look at some of these subjects:
>
> Letters to Public Officials; Declining Appointive or Elective Positions; Letters of Condolence; Letters of Apology; Soliciting for Charitable Contributions; Adjustments—When the Answer is "No"; Letters to Creditors; Contacting Inactive Accounts; Collection Letters; Requests for References—11 chapters in all.
>
> New subjects are thoroughly covered. You'll find letters on contracting for computer services, apologizing for computer errors, contracting for hardware and software. Virtually every letter a business executive could ever need is here in *The Ghost*—waiting for you.

I wonder if it contains letters that apologize for the mechanically written tone of recent letters, or letters that apologize for the incorrectly selected letter sent last time—and so on. The idea that anyone could think that *every possible situation* has been anticipated just boggles the mind. How credulous does one have to be to buy this book? (By the way, if you're interested, it costs only $49.95, and you can order it from Prentice-Hall, Inc., Englewood Cliffs, New Jersey 07632. But act now—it won't last long.)

*   *   *

In talking about knobs and creativity once with some architects, I encountered some advocates of "shape grammars" used to design houses, gardens, tea rooms, and so on. I was shown how a certain class of Frank Lloyd Wright houses known as his "prairie houses" had been parametrized and embedded in a shape grammar. An article by H. Koning and J. Eizenberg presents the grammar and shows a large number of external and internal designs of pseudo-Wright houses. This kind of art by formula reminds me of the famous aleatoric waltz by Mozart, in which one-measure fragments can be assembled in any order to make an acceptable, if feeble, piece of music. Shape grammars recognize more levels of structure than Mozart did, but then he was doing it only as a joke. It seems to me, after perusing several articles on architectural shape grammars, that the designs they produce are respectable—in fact they are very similar to the input designs. But for that very reason, they strike me as rather dull and dry designs, given that they are all *ex post facto*. We are back at the issue of pseudo-Mondrian versus genuine Mondrian (see Figure 10-14 and the accompanying discussion), and the questionable artistic value in extracting features of a once-novel creation and using them to allow a machine to mimic or perhaps even improve upon that one creation, but always in a blatantly derivative way.

Readers might be surprised to learn that one part of my research is not that distant from either shape grammars or METAFONT: the Hàn Zì project, whose goal is to make a program able to produce Chinese characters in a "twiddlable" style. All characters are reduced to smaller units, which in turn are reduced to smaller units, and so on, until the level of *basic strokes* is reached. Traditional Chinese calligraphers will tell you that there are seven or eight such basic strokes, but that is only for humans, whose vision and concepts are very fluid. For rigid machines, the number has to be increased. I have found that somewhere around 40 will suffice to make just about any character, although for most purposes 30 or 35 will do. The definition of each character is style-independent, which means that if you change the basic strokes, all characters will change in appearance. An example of this is shown in Figure 13-13, in which a short sentence is printed out by Hàn Zì in two different styles (and in which the program says two different things about its output).

My co-worker David Leake and I do not harbor any illusions as to the generality of this approach to style in Chinese. It is quite obviously subject to all the limitations of any parameter-based approach to style: rigidity and non-creativity. Still, we find it an exciting challenge to try to do the best we can within the obvious limitations of such a system. It helps us see just how far these systems can be pushed, it teaches us more about Chinese writing, and perhaps best of all, it entertains and intrigues the many Chinese students we know.

*   *   *

The creative, non-rut-stuck mind is always coming up with ideas that jump out of preconceived categories. A lovely cover on *Science News* (January 8, 1983) shows four new ideas for airplanes. One is a fuselage-less flying wing with six engines and with vertical tails at both ends of the wing. Another is

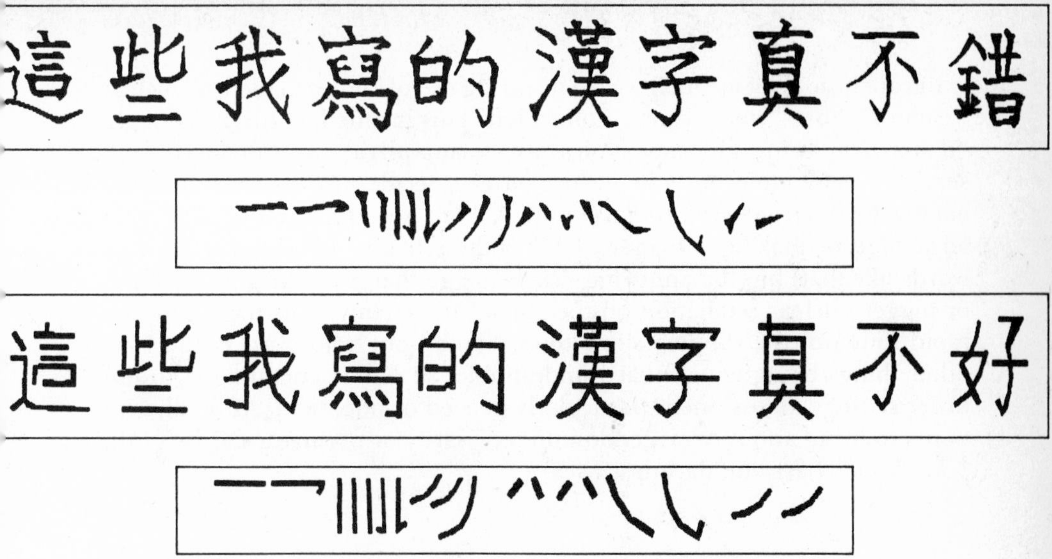

這些我寫的漢字真不錯

一一ㄱ川ノノ八乀乀乚ノ

這些我寫的漢字真不好

一一ㄱ川ノノ八乀乚ノノ

FIGURE 13–13. *Self-descriptive Chinese sentences. The upper one, in a rather calligraphic hand, says: "These Chinese characters I've written are really not bad." The lower one, in a rather robot-like hand, says: "These Chinese characters I've written are really not good." Both were written by the Hàn Zì program, with only about twenty basic strokes changed. The basic strokes themselves are shown in the boxes. All 50,000 (or so) characters in the Chinese language can be built up by the Hàn Zì program from about 40 distinct basic strokes, so that one can switch the visual mood of any passage simply by switching 40 basic graphic objects. Still, we—David Leake and I—are nowhere near being able to capture, in a few simple stroke-redefinitions, the creative variety of Figure 12-4. Our program does not see what it produces, and perception of what one has produced is essential to good creativity.*

a propeller-driven craft whose curvy propeller blades look more like flower petals than like fan blades. The third is a plane whose two wings bend up and over its fuselage, meeting each other to form a complete circle (thus there is really only *one* wing, strictly speaking). The fourth shows a kind of "Siamese twin" plane, with one giant wing being shared by two parallel fuselages. Marvelous images of "Future Flight", as the caption says. Try to put all possible future aircraft designs into a set of fixed knobs! Here is a case where roles are split and merged with the greatest of ease. Visions of the future often feature these kinds of exciting "twists" on present ideas, full of novelty and considerably beyond trivial knob-twisting—yet even they usually fall far short of anticipating how the future really turns out.

An entertaining use of knobs is in the new movie genre called "Choice-a-Rama". The slogan says, "Where *you* decide what happens next!" Presumably, the audience votes at predetermined choice points, and this selects one pathway out of a predetermined set of possible continuations. It is like making dynamic choices at every possible turn while driving through a city, and being surprised by where one winds up. But it must be very expensive to have more than a few choice points, because the numbers multiply. If there are ten binary choice points, that means 2^{10}, or 1,024, different pathways have to be stored somewhere on film. It's an amazing, if decadent, symbol of our society.

In conclusion, let me mention an inspiring use of knobs: in tactical nuclear weapons whose "yield" can be controlled. This is called, naturally enough, "dial-a-yield", in the same spirit as "dial-a-pizza" or "dial-a-prayer" services. Depending on your need, you can decide just how much of the enemy forces you wish to take out. A high setting has the appealing advantage of making a bigger "kill" (although one shouldn't use crude words like that) but the annoying disadvantage that it may trigger a similar or bigger nuclear retaliation on the part of the enemy, thus triggering the rapid slide down a slippery slope toward an all-out holocaust. Bother! All other things being equal, that's undesirable, so one is encouraged to use lower settings unless one is particularly peeved or impatient. After all, who wants to bring about Armageddon unnecessarily or prematurely? By gosh, don't knobs have the darndest uses?

Section IV:

Structure and Strangeness

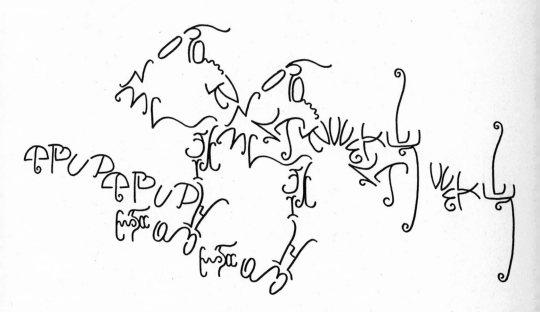

Section IV:
Structure and Strangeness

Mathematical structures are among the most beautiful discoveries made by the human mind. The best of these discoveries have tremendous metaphorical and explanatory power, jumping across discipline boundaries, illuminating many areas of thought simultaneously. In addition, the best discoveries often reveal truly bizarre facets of familiar concepts. In the following seven chapters, four wonderful mathematical ideas are considered. The "Magic Cube" is an engaging object for many reasons, not the least of which is its seeming physical impossibility, as well as the frustrating way that order and chaos appear and disappear on its surface as it is twisted. The borderline between order and chaos in mathematics is the next topic treated, where we see the iteration of very simple functions giving rise to unexpectedly chaotic phenomena—in particular, "strange attractors". A strange attractor is a very peculiar shape having structure on an infinite number of scales at once. This property applies not only to strange attractors, but to a much larger class of shapes known as "fractals". They in turn are examples of the more general mathematical concept of *recursion,* one of our era's most fruitful areas of exploration in mathematics and computer science. Recursion and recursivity are presented in three chapters on the computer language Lisp, the language used most in artificial-intelligence research. Finally, we move from computers to their microscopic substrate: the eerie netherworld of quantum phenomena, and the unresolved mysteries about the relationship between the macroworld and the microworld.

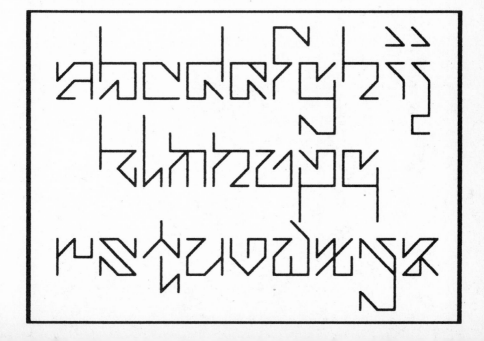

14

Magic Cubology

March, 1981

> Cubitis magikia, *n. A severe mental disorder accompanied by itching of the fingertips, which can be relieved only by prolonged contact with a multicolored cube originating in Hungary and Japan. Symptoms often last for months. Highly contagious.*

WHAT this stuffy medical-dictionary entry fails to mention is that contact with the multicolored cube not only cures the itchiness but also causes it. Furthermore, it fails to point out that the affliction can be highly pleasurable. I ought to know; I have suffered from it for the past year and still exhibit the symptoms.

Bŭvös Kocka—the Magic Cube, also known as Rubik's Cube—has simultaneously taken the puzzle world, the mathematics world, and the computing world by storm. (See Figure 14-1.) Seldom has a puzzle so fired the imagination of so many people, perhaps not since Sam Loyd's famous "15" Puzzle, which caused mass insanity when it came out in the nineteenth century, and which is still one of the world's most popular puzzles. The 15 Puzzle and the Magic Cube are spiritual kin, the one being a two-dimensional problem of restoring the scrambled numbered pieces of a 4×4 square to their proper positions, and the other being a three-dimensional problem of restoring the scrambled colored pieces of a 3×3×3 cube to their proper positions. The solutions of both demand that the solver be willing to undo seemingly precious progress time and time again; there is no route to the goal that does not call for partial but temporary destruction of the visible order achieved up to a given point. If this is a difficult lesson to learn with the 15 Puzzle, how much harder with the Magic Cube! And both puzzles have the fiendish property that well-meaning bumblers or cunning rogues can take them apart and put them back together in innocent-looking positions from which the goal is

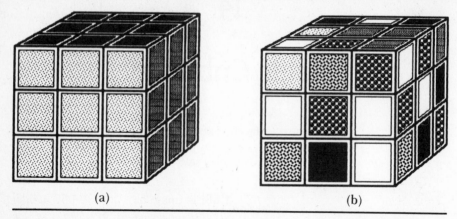

(a)   (b)

FIGURE 14–1. *A Magic Cube in (a) its pristine state, also called* START; *(b) a typical scrambled state.*

absolutely unattainable, thereby causing the would-be solver considerable consternation.

This Magic Cube is much more than just a puzzle. It is an ingenious mechanical invention, a pastime, a learning tool, a source of metaphors, an inspiration. It now seems an inevitable object, but it took a long time to be discovered. Somehow, though, the time was ripe, because the idea germinated and developed nearly in parallel in Hungary and Japan and perhaps even elsewhere. A report surfaced recently of a French inspector general named Semah, who claims to remember encountering such a cube made out of wood in 1920 in Istanbul and then again in 1935 in Marseilles. Of course, without confirmation the claims seem dubious, but still titillating. In any event, Rubik's work was completed by 1975, and his Hungarian patent bears that date. Quite independently, Terutoshi Ishige, a self-taught engineer and the owner of a small ironworks near Tokyo, came up with much the same design within a year of Rubik and filed for a Japanese patent in 1976. Ishige also deserves credit for this wonderful insight.

Who is Rubik? Ernő Rubik is a teacher of architecture and design at the School for Commercial Artists in Budapest. Seeking to sharpen his students' ability to visualize three-dimensional objects, he came up with the idea of a $3\times3\times3$ cube any of whose six 3×3 faces could rotate about its center, yet in such a way that the cube as a whole would not fall apart. Each face would initially be colored uniformly, but repeated rotations of the various faces would scramble the colors horribly. Then his students had to figure out how to undo the scrambling.

When I first heard the cube described over the telephone, it sounded like a physical impossibility. By all logic, it ought to fall apart into its constituent "cubies" (one of the many useful and amusing terms invented by "cubists" around the world). Take any corner cubie—what is it attached to? By imagining rotating each of the three faces to which it belongs, you can see that the corner cubie in question is detachable from each of its three edge-cubie neighbors. So how in the world is it held in place? Some people postulate magnets, rubber bands, or elaborate systems of twisting wires in the interior of the cube, yet the design is remarkably simple and involves no such items.

In fact, the Magic Cube can be disassembled in a few seconds (see Figure 14-2c), revealing an internal structure so simple that one has to ponder how it can do what it does. To see what holds it together, first observe that there are three types of cubie: six *center* cubies, twelve *edge* cubies, and eight *corner* cubies. (See Figure 14-2a.) Each center cubie has only one "facelet"; edge cubies have two; corner cubies have three. Moreover, the six center cubies are really not cubical at all—they are just square façades covering the tips of axles that sprout out from a sixfold spindle in the cube's heart. The other cubies, however, are nearly complete little cubes, except that each one has a blunt little "foot" reaching toward the middle of the cube, and some curved nicks facing inward.

The basic trick is that cubies mutually hold one another in by means of their feet, without any cubie actually being attached to any other. Edge cubies hold corner cubies' feet, corner cubies hold edge cubies' feet. Center cubies are the keystones. As any layer, say the top one, rotates, it holds itself together horizontally, and is held in place vertically by its own center and by the equatorial layer below it. The equatorial layer has a sunken circular track (formed by the nicks in its cubies) that guides the motion of the upper layer's feet and helps to hold the upper layer together. Unless you're a mechanical genius, you really can't understand this without a picture, or, better yet, the real thing.

In his definitive treatise, *Notes on Rubik's 'Magic Cube'*, David Singmaster, professor of Mathematical Sciences and Computing at the Polytechnic of the South Bank in London, defines the *basic mechanical problem* as that of figuring out how the cube is constructed. I sometimes wonder whether Rubik's intended visualization task for his students was to solve the unscrambling problem (Singmaster calls it the *basic mathematical problem*) or to solve the mechanical problem. I suspect the latter is the harder of the two. I myself must have put in more than 50 hours of work, distributed over several months, before I solved the unscrambling problem, and I never did solve the mechanical problem until I saw the cube disassembled. Singmaster informally estimates that people who eventually solve the unscrambling problem (without hints) take, on the average, two weeks of concentrated effort. Of course, it is hard for anyone who has done it to say exactly how long it took (how can you tell play from work?), but it's safe to say that if you are destined to solve the unscrambling problem at all, it will take you somewhere between five hours and a year. I trust this is reassuring.

An important fact that many people fail to appreciate at first is that to restore a scrambled cube even once to the *START* position (the state of Perfect Enlightenment and Grace, where each face is a solid color) is so hard that it is necessary to find a *general* algorithm for doing it from *any* scrambled state. No one can restore a messed-up Magic Cube to its pristine state by

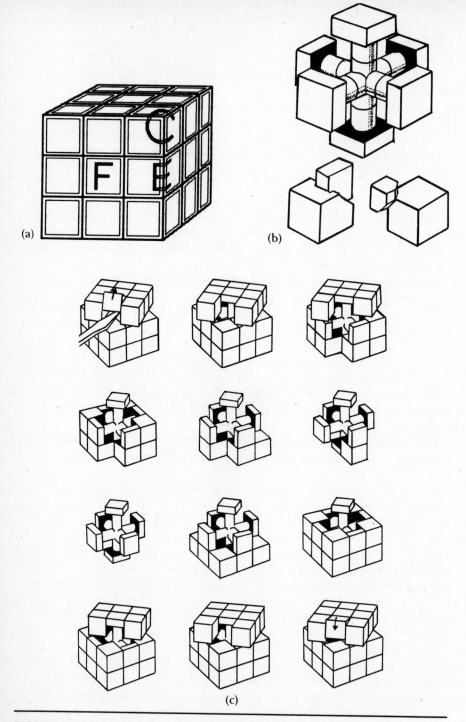

FIGURE 14-2. *In (a) the three types of cubie are identified: face centers (F), corners (C), and edges (E). In (b) the mechanism is revealed. You can see the six-pronged internal spindle with all six face-center cubies attached to it, and one detached edge cubie and one detached corner cubie. Notice that no cubie is a complete cube. In fact, the face centers are just façades! In (c), the gradual dismantling and rebuilding of a Cube are shown. Warning: If you follow this procedure, you are advised to rebuild your Cube in its pristine state; otherwise, you will probably wind up with your Cube in an orbit from which START is inaccessible.*

mere trial and error. Anyone who gets back to *START* has built up a small science.

A word of warning: Proposed solutions to the mechanical problem are often lacking in clarity, having either too much or too little detail. It is certainly a challenge to come up with a mechanism that has the multifaceted twistability of the Magic Cube, but it is perhaps no less of a challenge to describe the mechanism in language and diagrams that other people can readily comprehend. By the same token, to convey algorithms that restore the cube to *START* calls for a good, clear notation. Singmaster himself has an excellent notation that is now considered standard; I will present it below. A second word of warning: I am not a "cubemeister" (one who has contributed to the annals of the profound science of Cubology); I am a mere cubist, an amateur dazzled by the Cube and by the virtuosos who have mastered it. Therefore I am not a suitable recipient of novel solutions to the mechanical problem or to the unscrambling problem. I recommend to readers who believe that they have some novel insight to communicate it to Singmaster, who runs what amounts to the World Center for Cubology. His address is: Department of Mathematical Sciences and Computing, Polytechnic of the South Bank, London SE1 OAA, England.

*   *   *

By now, I would hope that your appetite has been whetted to the point where immediate possession of a Magic Cube is an urgent priority. Fortunately, this can be arranged quite easily. Most any toy store now carries them under such names as "Rubik's Cube", "Wonderful Puzzler", and miscellaneous others. The price ranges from a couple of dollars for a cheap model to roughly $15 for a very solid and high-quality cube. It is likely that many people will buy cubes, little suspecting the profound difficulty of the "basic mathematical problem". They will innocently turn four or five faces, and suddenly find themselves hopelessly lost. Then, perhaps frantically, they will begin turning face after face one way and then another, as it dawns on them that they have irretrievably lost something precious. When this first happened to me, it reminded me of how I felt as a small boy, when I accidentally let go of a toy balloon and helplessly watched it drift irretrievably into the sky.

It is a fact that the cube can be randomized with just a few turns. Let that be a warning to the beginner. Many beginners try to claw their way back to *START* by first getting a single face done. Then, a bit stymied, they leave their partially solved cube lying around where a friend may spot it. The well-known "Don't touch it!" syndrome sets in when the friend innocently picks it up and says, "What's this?" The would-be solver, terrified that all their hard-won progress will be destroyed, shrieks, "Don't touch it!" Ironically, victory can come only through a more flexible attitude allowing precisely that destruction.

For the beginner, there is an awesome sense of irreversibility about destroying *START,* a fear of tumbling off the edge of a precipice. When my own first cube (I now have dozens) was first messed up (by a guest), I felt both relieved (because it was inevitable) and sad (because I feared *START* was gone forever). The physicist in me was reminded of entropy. Once *START* had become irretrievable, each new twist of one face or another seemed irrelevant. To my naïve eye there was no distinguishing one messed-up state from another, just as to the naïve eye there is no distinguishing one plate of spaghetti from another, one pile of fall leaves from another, and so on. The details meant nothing to me, so they didn't register. As I performed my "random walk", the vastness of the space of possible shufflings of the little cubies became vivid.

As with a deck of cards, one can calculate the exact number of possible rearrangements of the cube. An initial estimate would run this way. The first observation—a rather elementary one—is that on the rotation of any face, each corner goes to another corner, each edge to another edge and the center of the face stays put (except for its invisible rotation). Therefore corners mix only with their own kind, and the same goes for edges. There are eight corner cubies and eight corner *cubicles* (the spatial niches, regardless of their content). Cubies and cubicles are to the cube as children and chairs are to the game of musical chairs. Each corner cubie can be maneuvered into any of the eight corner cubicles. This means that we have eight possible fillers for cubicle No. 1, seven for cubicle No. 2, six for cubicle No. 3, and so on. Therefore the corners can be placed in their cubicles in $8 \times 7 \times 6 \times 5 \times 4 \times 3 \times 2 \times 1$ ($= 8!$) different ways. But each corner can be in any one of three orientations. Thus one would expect a further factor of 3^8 from the eight corners. One would expect the same for the twelve edge cubies: twelve objects can be permuted among themselves in $12!$ different ways, and then, since each of them has two possible orientations, that gives another factor of 2^{12}. The center cubies never leave their *START* positions (unless the cube is rotated as a whole) and have no visibly distinct orientations, so they do not contribute. If we multiply the numbers out, we get 519,024,039,293,878,272,000 possible positions—about 5.2×10^{20}.

But there is an assumption here: that any cubie can be gotten into any cubicle in any orientation, regardless of the other cubies' positions and orientations. As we will see, this is not quite the case. It turns out that there is a mild constraint on the orientation of the corner cubies: any *seven* can be oriented arbitrarily, but the last one is then forced, thus removing one factor of three. Similarly, there is a mild constraint on edge cubies: of the twelve, any *eleven* can be oriented arbitrarily, but the last one is then determined, so that another factor of two is removed. There is one final constraint on the permutations of cubies (disregarding their orientations) that says you can place all but two of them wherever you want, but the last two are forced. This removes a final factor of two, reducing the estimate above by a total factor of $3 \times 2 \times 2 = 12$, bringing the possibilities down to a mere 43,252,003,274,489,856,000—about 4.3×10^{19}. Still, it must be said, this does slightly exceed the assertion on Ideal's label: "Over three billion combinations".

Another way of thinking about this factor of twelve is that if you begin at *START*, you are limited to a twelfth of the "obvious" states, but if you disassemble your cube and reassemble it with a single corner cubie twisted by 120 degrees, you are now in a formerly inaccessible state, from which a whole family of 43,252,003,274,489,856,000 new states is accessible. There are twelve such nonoverlapping families of states of the cube, usually called *orbits* by group theorists.

*   *   *

Speaking of impossible twists, I would like to mention a lovely discovery in Cubology that is parallel to ideas in particle physics. It was pointed out by mathematician Solomon W. Golomb. The discovery states: It is impossible to find a sequence of moves that leaves just one corner cubie twisted a third of a full turn and everything else the same. Now, recalling the famous hypothetical fundamental particle with a charge of $+1/3$ and its antiparticle with a charge of $-1/3$, Golomb calls a clockwise one-third twist a *quark* and a counterclockwise one-third twist an *antiquark*. Like their cubical namesakes, quark particles have proved to be tantalizingly elusive, and particle physicists generally believe now in *quark confinement*: the notion that it is impossible to have an isolated free quark (or antiquark). This correspondence between cubical quarks and particle quarks is a lovely one.

Actually, the connection runs even deeper. Although quark particles cannot exist free, they can exist bound together in groups: a quark-antiquark pair is a *meson* (Figure 14-9e), and a quark trio with integral charge is a *baryon*. (An example is the proton—*qqq*—with a charge of $+1$.) Now in the Magic Cube, amazingly enough, it is possible to give any *two* corner cubies one-third twists, provided they are in *opposite* directions (one clockwise, the other counterclockwise). It is also possible to give any *three* corner cubies one-third twists, provided they are all in the *same* direction. Thus Golomb calls a state with two oppositely twisted corners a "meson", and one with three corners twisted in the same direction a "baryon". In the particle world, only quark combinations with an integral amount of *charge* can exist. In the cubical world, only quark combinations with a integral amount of *twist* are allowed. This is just another way of saying that the orientation of the eighth corner cubie is always forced by the first seven. In the cubical world, the underlying reason for "quark confinement" lies in the group theory. There may be a closely related group-theoretical explanation for the confinement of quark particles. That remains to be seen, but in any event, the parallel is provocative and pleasing.

*   *   *

If we have a "pristine cube" (one in *START*), what kind of move sequence will create a meson or a baryon? Here we have an example of the most powerful idea in Cubology: the idea of "canned" move sequences that accomplish some specific reordering of a few cubies, leaving everything else untouched ("invariant", as group theorists say). There are many different terms for such canned move sequences. I have heard them called *operators*, *transforms*, *words*, *tools*, *processes*, *maneuvers*, *routines*, *subroutines*, and *macros*, the first three being group-theoretical terms and the last three being borrowed from computer science. Each term has its own flavor, and I find that I use them all at various times.

In order to talk about processes, we need precision, and that means a good technical notation. I will therefore present Singmaster's notation now. First we need a way of referring to any particular face of the cube. One possibility is to use the names of colors as the names of the faces, even after the cubies have become mixed up. Now it might seem that calling a face "white" would be meaningless if white is scattered all over the place. But remember that the white *center* cubie never moves with respect to the five other center cubies, and thus defines the "home face" for white. So why not use color names for faces? Well, one problem is that different cubes come with their colors arranged differently. Even two cubes from one manufacturer may have different *START* positions. A more general convention is to refer to faces simply as *left* and *right*, *front* and *back*, and *top* and *bottom*. Unfortunately, the initials of "back" and "bottom" conflict. Singmaster resolves the conflict by replacing "top" and "bottom" by *up* and *down*. Now we have names for the six faces: L, R, F, B, U, D. Any particular cubie can be designated by lowercase italic letters naming the faces it belongs to. Thus *ur* (or *ru*) stands for the edge cubie on the right side of the top layer, and *urf* for the corner cubie in front of it (see Figure 14-3*a*).

The most natural move for a right-handed cubist seems to be to grasp the right face with the thumb pointing up along the front face and to move the thumb forward. Seen from the right side, this maneuver causes a clockwise quarter-twist of the R face. This move will be designated R (see Figure 14-3*b*). The mirror-image move, where the left hand turns the L side counterclockwise (as seen from the left), is L^{-1}, or, for short, L'. A clockwise twist of the L side is called, naturally, L. A 90-degree clockwise turn of any face (from the point of view of an observer looking at the center of that face) is named by the letter for that face, and its inverse—the counterclockwise quarter turn —has a prime mark following the face's initial. Quarter-turns will henceforth be called *q-turns*.

With this nomenclature, we can now write down any move sequence, no matter how complex. A trivial example is four successive R's, which we write as R^4. In the language of group theory, this is the *identity* operation: it has zero effect. An equation expressing this fact is $R^4 = I$. Here, I stands for the "action" of doing nothing at all.

Suppose we twist two different faces—say R first, then U. We will

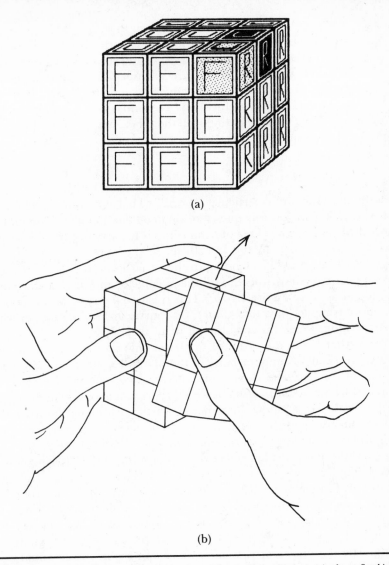

(a)

(b)

FIGURE 14-3. *Labeling of cubies and moves. The speckled cubie in (a) is the* urf *cubie (alias* rfu *and* fur*), and the black one is the* ur *(or* ru*) cubie. The quarter-turn or* q*-turn shown in (b) is called* **R**.

transcribe that as *RU*—not as *UR*. Note, in fact, that *RU* and *UR* are quite different in their effects. To check this out, first perform *RU* on a pristine cube, observe its effects, then undo it, try *UR*, and see how its effects differ. The inverse of *RU* is, quite obviously, *U'R'*, not *R'U'*. (Incidentally, this strategy of experimenting with move sequences on a pristine cube is most helpful. Very early I found it useful to buy a second cube so that I could work on solving one while experimenting with the other, never letting the second one get far away from *START*.)

*   *   *

What is the effect of a particular "word"? That is to say, which cubies move where? To answer this question, we need a notation for the motions of individual cubies. The effect of *R* on edges is to carry the *ur* cubie around to the back face to occupy the *br* cubicle. At the same time, the *br* cubie swings around underneath, landing in the *dr* position, the *dr* cubie moves up like a car on a Ferris wheel to fill the *fr* cubicle, and the *fr* cubie comes to the top at *ur*. (See Figure 14-4*a*.) This is called a *4-cycle*, and we'll write it in a more compact way: (*ur,br,dr,fr*). Of course, it does not matter where we start writing; we could equally well write (*br,dr,fr,ur*).

On the other hand, the order of the letters in cubie names *does* matter. We can reverse all of them or none of them, but not just some of them. If you think of the letters as designating facelets, this will become clear. For example, if we wrote (*ur,rb,dr,rf*), it would represent a 4-cycle involving the same four cubicles as above, but one in which each cubie flipped before moving from one cubicle to the next. Of course, such a cycle cannot be accomplished by a single *q*-turn, but it may be the result of a *sequence* of *q*-turns of different faces (an operator). Or consider the following 8-cycle, shown in Figure 14-4*c*: (*ur,uf,ul,ub,ru,fu,lu,bu*). This has length eight, but involves only four cubicles. Each cubie, after making a full swing around the top face, comes back flipped (see Figure 14-4*b*). After two full swings, it is back as it started. Each facelet has made a "Möbius trip". We can designate this "flipped 4-cycle" as (*ur,uf,ul,ub*)$_+$, where the plus sign designates the flipping. The designation (*ru,fu,lu,bu*)$_+$ and numerous others would do as well. Thus the cycle notation tells you not only where a cubie moves but also its orientation with respect to the other cubies in its cycle.

To complete our description of the effect of *R*, we must transcribe the 4-cycle of the corners. As with edges, we have the freedom to start at any corner we want, and once again we must be careful to keep track of the facelets so that we get the orientations right. Still, *R* has a rather trivial effect on corners: (*urf,bru,drb,frd*), which could also be written (*rub,rbd,rdf,rfu*), and many other ways. Summing up, we can write *R* = (*ur,br,dr,fr*) (*urf,bru,drb,frd*). This says that *R* consists of two disjoint 4-cycles. (If we wanted to, we could throw in a term standing for the 90-degree rotation of the R face's center, but since such rotation is invisible, we needn't do so.)

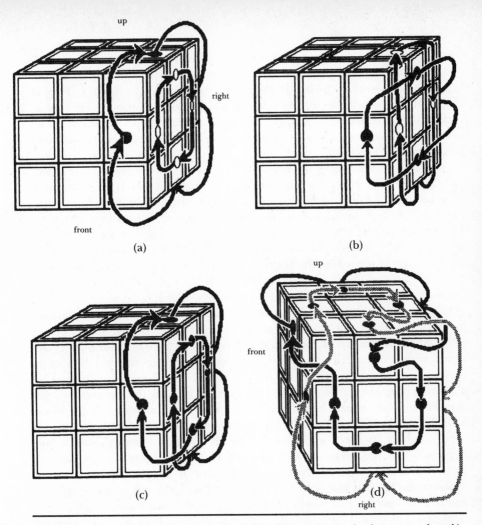

FIGURE 14–4. *The simple 4-cycle* (ur,br,dr,fr), *shown in* (a), *is what happens to edge cubies during the q-turn* **R**. *In* (b), *a trickier 4-cycle* (ur,rb,dr,rf), *involving the same four cubies, is shown; here, each cubie flips before entering the next cubicle. This cycle can be produced only through a sequence of q-turns. In* (c), *the 8-cycle* (fr,ur,br,dr,rf,ru,rb,rd) *is shown, which can also be thought of as a "flipped 4-cycle"—namely,* (fr,ur,br,dr) $_+$. *In* (d), *the 7-cycle* (ur,br,dr, fr,uf,ul,ub) *is shown snaking its way around the Cube, representing the effect on edges of the simple operator* **RU**.

What about transcribing a move sequence such as *RU*? Well, take a pristine cube and perform *RU*. Then start with some arbitrary cubie that has moved and describe its trajectory. For example, *ur* has moved to *br*. Therefore *br* has been displaced. Where has it gone? Find the new location of that cubie (it is *dr*) and continue chasing cubies 'round and 'round the cube until you find the one that moved into the original position of *ur*. You will find the following 7-cycle: (*ur,br,dr,fr,uf,ul,ub*) (see Figure 14-4d).

What about corners? Well, suppose we trace the cubie that originated in *urf.* Where did *RU* carry it? The answer is: Nowhere—it took a round trip but got twisted along the way. It changed into *rfu.* We can designate this clockwise twist—this "twisted unicycle", this quark—as $(urf)_+$. This is shorthand for the following 3-cycle: (urf, rfu, fur). You can even see this as cycling the three letters *u, r,* and *f* inside the cubie's name. If the cycle had been an antiquark, we would have written $(urf)_-$, and the letters would cycle the other way.

What about the other seven corners? Two of them—*dbl* and *dlf*—stay put, and the other five *almost* form a 5-cycle: $(ubr, bdr, dfr, luf, bul)$. It is unfortunate that the cycle does not quite close, because *bul,* although it gets carried into the original *ubr* cubicle, does so in a twisted manner. It gets carried to *rub,* which is a counterclockwise twist away from *ubr.* This means we are dealing with a 15-cycle. But it is so close to the 5-cycle above that we'll just tack on a minus sign to represent the counterclockwise twist. Our twisted 5-cycle is then $(ubr, bdr, dfr, luf, bul)_-$, and the entire effect of *RU,* expressed in cycle notation, is $(ur, br, dr, fr, uf, ul, ub)$ $(urf)_+$ $(ubr, bdr, dfr, luf, bul)_-$.

Now that we have *RU* in cycle notation, we can perform rotations mentally, by sheer calculation. For instance, what would be the effect of $(RU)^5$? Edge cubie *ur* would be carried five steps forward along its cycle, which would bring it to *ul.* (This can also be seen as moving two steps backward.) Then *ul* would go to *fr,* and so on. The 7-cycle is replaced by a new 7-cycle: $(ur, ul, fr, br, ub, uf, dr)$. Let us now look at the twisted 5-cycle. Corner cubie *ubr* would be carried five steps forward along its cycle, which brings it back to itself negatively twisted—namely, *rub.* Similarly, all the corner cubies in the 5-cycle would return to their starting points, but negatively twisted; thus, on being raised to the fifth power, a negatively twisted 5-cycle becomes five antiquarks. But if that is so, how is the requirement for integral twist satisfied? Don't we have one quark—$(urf)_+$— and five antiquarks, and doesn't that add up to four antiquarks, with a total twist of $-1\frac{1}{3}$? Well, I have slipped something by you here. Can you spot it? To gain facility with the cycle notation, you might try to find the cycle representation of various powers of *RU* and *UR* and their inverses.

*　　*　　*

Any sequence of moves can be represented in terms of *disjoint* cycles of various lengths (cycles with no common elements). If you are willing to let cycles share members, however, any cycle can be further broken up into 2-cycles (called *transpositions,* or sometimes *swaps*). For instance, consider three animals: an Alligator, a Bobcat, and a Camel. They initially occupy three ecological niches: *A, B,* and *C* (see Figure 14-5). The effect of the

---

FIGURE 14–5. *A zoological 3-cycle involving three objects:* a, b, *and* c *(an alligator, a bobcat, and a camel). Initially, each is in its usual ecological niche:* a *in* A, b *in* B, *and* c *in* C. *But then, after a permutation,* c *is in* A, a *is in* B, *and* b *is in* C. *This 3-cycle can be thought of as the result of two successive swaps.*

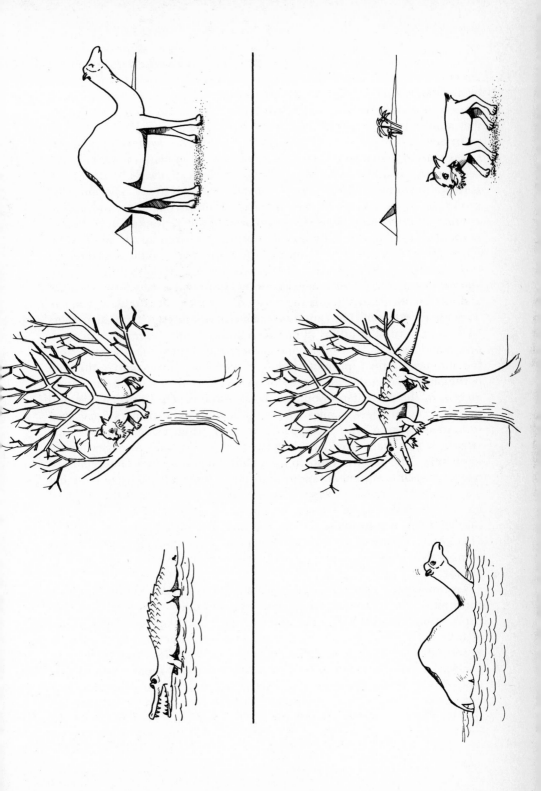

3-cycle (A,B,C) is to put them in the order Camel, Alligator, Bobcat. The same effect can be achieved, however, by first performing the swap (A,B) (what was in A goes to B and vice versa) and then performing (A,C). Of course, this can also be achieved by the two successive swaps $(A,C)(B,C)$—or, for that matter, by $(B,C)(A,B)$. On the other hand, no sequence of *three* swaps will achieve the same effect as (A,B,C). Try it yourself and see. (Note that a niche is like a cubicle and an animal is like a cubie.)

An elementary theorem of zoop theory (a field we won't go into here) states that no matter how a given permutation of animals among niches is reduced to a product of successive swaps (which can always be done), the parity of the number of such swaps is invariant; that is to say, a permutation cannot be expressed as an *even* number of swaps one time and an *odd* number another time. Moreover, the parity of any permutation is the *sum* of the parities of any permutations into which it broken up (using the rules for addition of even and odd numbers: odd plus even is odd, and so forth.)

Now, this theorem has repercussions for the Magic Cube. In particular, you can see that any q-turn consists of two disjoint 4-cycles (one on edges and one on corners). What is the parity of a 4-cycle? It is odd, as you can work out for yourself. Thus, after one q-turn, both the edges and the corners have been permuted oddly; after two q-turns, evenly; after three q-turns, oddly; and so forth. The edges and corners stay in phase, in the sense that the parities of their permutations are identical. Now clearly, the null permutation is even (it effects zero swaps). So if we have a null permutation on *corners,* the permutation on *edges* must also be even. Conversely, a null permutation on edges implies an even permutation on corners. Imagine a state identical to *START* except for two interchanged edges (that is, one swap). Such a state would be even in corners but odd in edges, hence impossible. The best we could do would be to have *two pairs* of interchanged edges. The same argument holds for corners. In short, we have proven that *single swaps are impossible; swaps must always come in pairs.* (This is the origin of one of those factors of two in the earlier calculation of the number of reachable states of the cube.) There are processes for exchanging two pairs of edges, two pairs of corners, and even for exchanging one pair of edges along with one pair of corners. (This last process necessarily involves an odd number of q-turns.)

To round out the subject of constraints, let us ponder the origin of the constraints on corner-twisting and edge-flipping. Here is a clever explanation provided by John Conway, Elwyn Berlekamp, and Richard Guy, elaborating an idea due to Anne Scott. The basic concept is that we want to show that the number of flipped cubies is always even, and that the twist is always integral. But in order to determine what is flipped and what is twisted, we need a frame of reference. To supply it, we will define two notions: the *chief facelet of a cubicle* and the *chief color of a cubie.* (Remember that a cubicle is a niche and a cubie is a solid object.) The *chief facelet of a cubicle* will be the one on the up or down surface of the cube, if that cubicle

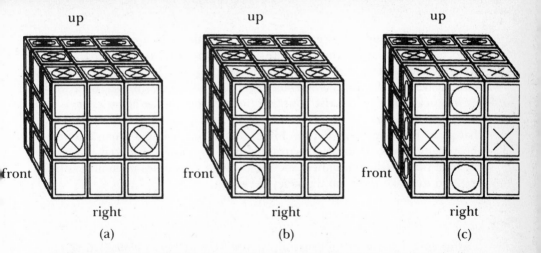

up   up   up

front   front   front

right   right   right

(a)   (b)   (c)

FIGURE 14–6. *Diagrams to aid in the proof that flippancy is even and twist is integral.*
In (a), the Cube is in START. *The chief facelets of cubicles are shown by crosses and the chief colors of cubies by circles. (Note: The concept of "chiefness" does not apply to face-center cubicles or cubies.) Think of the crosses as floating in space and the circles as being attached to the Cube, so that when turns are made, the crosses stay where they are but the circles move. The bottom face looks identical to the top, the left face identical to the right, and the back face identical to the front.*
In (b), the results of the q-turn F *are shown. The two empty circles indicate that the two cubies they are attached to have lost their "sanity". For them to regain their sanity, one cubie would have to be twisted one-third clockwise while the other cubie was twisted one-third counterclockwise, thus canceling each other's contribution to the total twist of the Cube. Similar remarks apply to the invisible left-hand face.*
In (c), the results of the q-turn R *(as applied to* START*) are shown. Empty circles again come in pairs. The top and bottom corner cubies on the front face (each with an empty circle) have canceling twists, as in (b). The top and bottom edge cubies on the right face have canceling flippancies, and the one seemingly unmatched empty circle (on the edge cubie on the front face) is paired with an empty circle on the invisible back face.*

is one; otherwise it will be the one on the left or right wall (see Figure 14-6). There are nine chief facelets on U, nine on D, and four on the equator. (We can ignore the centers, because they never can be flipped or twisted.) The *chief color of a cubie* is defined as the color that should be on the cubie's chief facelet when the cubie "comes home" to its proper cubicle in the *START* position.

Now the argument goes this way. Suppose the cube is scrambled. Any cubie that has its chief color in the chief facelet of its current cubicle will be called *sane;* otherwise it will be called *flipped* (this applies to edge cubies) or *twisted* (this to corner cubies). Obviously, there are two ways a cubie can be twisted: clockwise ($+1/3$ twist) and counterclockwise ($-1/3$ twist). The *flippancy* of a cube state will be defined as the number of flipped edge cubies in it, and the *twist* as the sum of the twists of the eight corner cubies. We shall say that the flippancy and twist of *START* are both zero, by convention.

Next consider the twelve possible *q*-turns out of which everything else is compounded. Performing *U* or *D* (or their inverses) preserves both the

315

flippancy and the twist, since nothing leaves or enters the up or down face. Performing F or B (or their inverses) leaves the total twist constant, by changing the twist of four corners at once: two by $+1/3$ and two by $-1/3$. It also leaves the flippancy alone (see Figure 14-6*b*). Performing L or R will likewise leave the total twist constant (four corner twists again cancel in pairs) and will change the flippancy by 4, since always four cubies will change in flippancy (see Figure 14-6*c*). The conclusion is what I stated above without proof: the eight corner cubies are always oriented to make the total twist a whole number, and the twelve edge cubies must always be oriented to make the total flippancy even.

*   *   *

After this discussion of constraints, you should be convinced that no matter how you twist and turn your Magic Cube, you cannot reach more than a twelfth of the conceivable "universe", beginning at *START*. It is another matter, though, to show that every state within that one-twelfth universe is accessible from *START* (or what amounts to the same thing, only backward: that *START* is accessible from every state in the one-twelfth universe). For this, we need to show how to achieve all even permutations of cubies, and how to achieve all orientations that do not violate the two constraints described above. What it comes down to is that we have to show there are operators that will perform seven classes of operations:

> (1) an arbitrary double edge-pair swap,
> (2) an arbitrary double corner-pair swap,
> (3) an arbitrary two-edge flip,
> (4) an arbitrary meson,
> (5) an arbitrary 3-cycle of edges,
> (6) an arbitrary 3-cycle of corners, and
> (7) an arbitrary baryon.

Of course, each of these operators should work without causing side effects on any other parts of the cube. With these powerful tools in our kit, we would be able to cover the one-twelfth universe without any trouble. In the case of the overlapping swaps of animals, you saw how a 3-cycle is really two overlapping 2-cycles. This implies that classes 5 and 6 can be made out of the first four classes. Similarly, a baryon can be made from two overlapping mesons. So all we really need is the first four classes.

To show that all the operators belonging to these four classes are available, we'll use another of the most crucial and lovely ideas of Cubology: that of *conjugate elements*. It turns out that all we need is *one example* in each class; given one example, we can construct all the other operators of its class from it. How does this work? The idea is very simple.

Suppose we had found one operator in class 1 that swapped, say, *uf* with

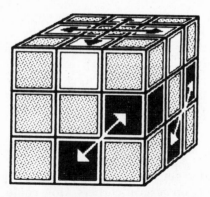

FIGURE 14–7. *How to use conjugate moves to turn an unsolved problem into a solved one. The unsolved problem is to effect the double swap shown by the white arrows. The solved problem is the double swap shown by the black arrows (on top). As long as we can maneuver the black cubies into the white cubicles, we are home free. This principle has nothing to do with the specific cubicles involved in the known and unknown operators, but simply with the idea that sometimes you can translate an unsolved situation into a solvable situation, use a known operator to handle* that *situation, then "back-translate" to regain the original situation, but with the tricky part now solved. This is the principle of* conjugates.

ub, and *ul* with *ur,* leaving the rest of the cube undisturbed. Let us call this operator *H.* Now suppose we wanted to swap two totally different pairs of edge cubies, say *fr* with *fd,* and *rb* with *rd* (see Figure 14-7). We can daydream: "If only those cubies were in the four 'magical swapping spots' on the top surface..." Well, why not just *put* them up there? It would be fairly simple to get four cubies into four specific cubicles. The obvious objection is: "Yes, but that would have an awful side effect—it would totally mess up the rest of the cube." But there is a clever retort. Let the destructive maneuver that gets those four cubies into the magical swapping spots be called *A.* Suppose we were smart enough to transcribe the move sequence of *A.* Then right after performing *A,* we perform our double swap *H.* Now comes the clever part. Reading our transcript in reverse order and inverting each *q*-turn, we perform the exact inverse of *A.* This will not only un-maneuver the four cubies back into their old cubicles, but will also undo the side effects *A* created in the rest of the cube. Does that restore the cube intact? Not quite. Remarkably, since we sandwiched *H* between *A* and *A',* the four edge cubies go home permuted—that is, each one winds up in the home of its swapping partner! Other than that, the cube is restored, and so we have accomplished precisely the double swap we set out to accomplish.

When you think this through, you see that it is flawless in conception. The inverse maneuver, *A',* does not "know" we have exchanged two pairs of edges. As far as it is concerned, it is merely putting everything back where it was before *A* was executed. Hence we have "snuck" our swaps in under *A'* 's nose, which is to say we have "fooled the cube". Symbolically, we have carried out the sequence of moves *AHA'*, which is called a *conjugate* of *H*.

It is this kind of marvelously concrete illustration of an abstract notion of group theory that makes the Magic Cube one of the most amazing things ever invented for teaching mathematical ideas. Normally, the examples of conjugate elements given in group-theory courses are either too trivial or too abstract to be enlightening or exciting. The Magic Cube, though, provides a vivid illustration of conjugate elements and of many other important concepts of group theory.

* * *

Suppose you wanted to get a quark-antiquark pair on *opposite* corners, but knew how to do so only on *adjacent* corners. How could you do it? Here is a hint: There are two nice solutions, but the shorter and prettier one involves using a conjugate. Incidentally, any maneuver that creates a quark on one corner (with other side effects, of course) might be called a *quarkscrew*.

What we have shown for edges goes also for corners: the ability to swap two *specific* corners enables you to swap *any* two corners. Conjugation allows you to build up an entire class of operators from any single member of that class. Of course, the question still remains: How do you find some sample operator in each of the four classes? For example, how do you find an operator that creates a meson on two adjacent corners (a combination of a quarkscrew and an antiquarkscrew)? How do you find an operator that exchanges two edge pairs both of which are on the top surface? I won't give the answer here, but will follow Singmaster, who points the way by suggesting quasi-systematic exploration of some small "subuniverses" within the totality of all cube states—that is, he suggests you look at *subgroups*. This means restricting your set of moves deliberately to some special types of move. Here are a few examples of interesting subgroups created by various kinds of restriction:

1. *The Slice Group.* In this subgroup, every turn of one face must be accompanied by the parallel move on the opposing face. Thus *R* must be accompanied by *L'*, *U* by *D'*, and *F* by *B'*. The name comes from the fact that any such double move is equivalent to rotating one of the three central slices of the cube. Singmaster abbreviates the slice move *RL'* by R_s, *R'L* by R'_s, and so forth. Under this restriction, faces cannot get arbitrarily scrambled. Each face will have a pattern in which all four corners share one color (Figure 14-8). A special case is the pattern called *Dots*, in which each face is all one color except for its center (see Figure 14-9*a*). Can you figure out how to achieve Dots from *START*? How many different ways are there of arranging the dots? How does the Dots pattern resemble a meson? (You will find answers to all these questions, along with much else, in Singmaster's book.)

FIGURE 14–8. *The type of pattern that the Slice Group creates on all faces.*

2. *The Slice-Squared Group.* Here we restrict the Slice Group further, allowing only squares of slice moves, such as R_s^2 (which is the same as R^2L^2) or F_s^2 (which is the same as F^2B^2).

3. *The Antislice Group.* Here, instead of always rotating opposing faces in parallel, we always rotate them in antiparallel, so that R is accompanied by L, F by B, and U by D. An antislice move has a subscript a, as in R_a, which equals RL. (Of course, the Antislice-Squared Group is no different from the Slice-Squared Group.)

4. *The Two-Faces Group.* Allow yourself to rotate only two adjacent faces, say F and R. It turns out to be a pretty substantial challenge to figure out an algorithm for undoing an arbitrary scrambling of two faces, staying within the Two-Faces Group. Most cube experts will instead resort to the "elephant gun" of twisting all six faces to get out of a mere two-face scramble. Shame on them!

5. *The Three-Faces Groups.* The reason this category is pluralized is that there are nonequivalent choices of threesomes of faces. For example, you can form a kind of "bridge", as with faces L, U, and R, or you can form a "corner", as with faces F, U, and R.

6. *The Four-Faces and Five-Faces Groups.* Again, there are various non-equivalent choices of faces for four faces. The Five-Faces Group is, as it turns out, actually the full group of the cube. In other words, you can make an operator equivalent to R out of L, U, D, F, and B.

7. *The Two-Squares Group.* As in the Two-Faces Group, you may rotate only two faces, using only 180-degree turns at that. This is a very simple subgroup.

If you limit your attention to just the Two-Faces and Two-Squares groups, you will be able to find processes that achieve double swaps—some of edges, others of corners. It is a remarkable fact that these processes alone, together with the notion of conjugation, will allow us—in a theoretical sense—to solve the entire unscrambling puzzle.

Why don't we also need a meson maker and a double edge-flipper? Well, consider how we might make a double edge-flipper from the two classes of tools one may assume will be found—that is, double edge-swappers and double corner-swappers. In order to *flip* two edges without creating any side effects, we'll perform two successive double edge-pair *swaps,* and both times they will involve the same pairs! For example, we might swap *uf* with *ub,* and *df* with *db,* and then reswap them. This seems to be an absolute "nothing process", but that need not be the case. After all, just as before, we can sandwich the second swap between a process X and its inverse X', where we carefully choose the process X so as to... (Oh, darn it all, I totally lost my train of thought there. I'm sure you can finish it up, though. I *do* remember that it wasn't *too* tricky, and that I thought the idea was rather elegant. I'm sure you will too.)

The same kind of thinking will show how you can build up a meson maker out of mere corner-swapping processes and conjugation. Given mesons, you can build up baryons. And with mesons and baryons, double edge-flippers, double edge-pair swappers, and double corner-pair swappers, you have a full kit of tools with which to restore any scrambled cube to *START,* as long as it belongs to the same orbit as *START.* What I have given is, needless to say, a highly theoretical existence proof, and any *practical* set of routines would be organized quite differently. The type of solution I have described has the advantage of being compact in description, but it is enormously inefficient. In practice, a cube solver must develop a fairly large and versatile set of routines that are short, easy to memorize, and highly redundant. There is an advantage to being able to carry out transformations in a variety of ways: you can choose whichever tool seems best adapted to the situation at hand, instead of, for instance, using some theoretically developed tool that takes several hundred q-turns to make a baryon.

*   *   *

The typical cube solver evolves a set of transforms partly by intuition, partly by luck, sometimes with the aid of diagrams, and occasionally with abstract principles of group theory. One principle nearly everyone formulates quite early is that of "getting things out of the way". This is once again the idea of conjugates, only in a simpler guise. The typical patter that goes along with it is something like this (I have included sound effects of a sort): "Let's see, I'll swing *this* out of the way [flip, flip] so that I can move *that* [flap, flap], and now I can swing *this* back again [unflip, unflip]. There —now I've got *that* where I wanted it to be." You can hear the conjugate structure inside the patter ("flip, flap, unflip").

The only problem with being conscious of why it all works as you carry it out is that it may be too taxing. My impression is that most cubemeisters do not think in much detail about how their tools are achieving their goals, at least not while they are in the midst of restoring some scrambled cube. Rather, expert cube solvers are like piano virtuosos who have memorized difficult pieces. As Dan Weise, an MIT cubemeister, said to me, "*I've* forgotten how to solve the cube, but luckily, my *fingers* remember."

The average operator seems to be about ten to twenty *q*-turns long. You don't ever want to get lost in mid-operator, because if you do, you will have a totally scrambled cube on your hands, even if you were carrying out your final transform. As cubemeister Bernie Greenberg said to me once, "If I were solving a cube and somebody yelled 'Fire!', I would finish my transform before clearing out."

My own style is probably overly blind. Not only do I not think about *why* my operators work as I am carrying them out; I have to admit that with some of them, I don't even have the foggiest idea why they work at all! I found these "magic operators" through a long and arduous trial-and-error procedure. I used some heuristic notions, such as: "Explore various powers of simple sequences", "Use conjugates a lot", and so on. One thing I hardly used at all—alas, poor Rubik—was three-dimensional visualization. However, I do know one Stanford cubemeister, Jim McDonald, who can give the reason for every last *q*-turn he makes. His operators don't seem magical to him because he can see what they are doing at every moment along the way. In fact, he does not have them memorized as I do mine; he seems to reconstruct them as he unscrambles cubes, relying on his "cube sense". He is like an expert musician who can improvise where a novice must memorize. For interested readers, the central idea of Jim's method is first to solve the top layer except for one corner, and then to utilize the vertical "chimney" underneath that free corner as you might use a neighbor's driveway to turn your car around in. The other two layers are cleaned up by shunting cubies in and out of the "chimney/driveway".

*　　*　　*

Perhaps not coincidentally, the abstract approach has been carried to its extreme by Singmaster's officemate, Morwen B. Thistlethwaite (I wonder what that "B" stands for!). He currently holds the world record for the shortest unscrambling algorithm. It requires at most 52 "turns". (A *turn* is defined as: either a *q*-turn or a half-turn—that is, a 180-degree turn of one face.) Thistlethwaite has used ideas of group theory to guide a computer search for special kinds of transforms. His algorithm has the curious property of not giving any appearance of converging toward the solved state at all—until the very last few turns.

This must be contrasted with the more conventional style. Most algorithms begin by getting one layer—usually the top layer—entirely correct. (In saying "top *layer*" rather than "top *surface*", I mean that the "fringe" has to be right, too: that is, the cubies on top must be correct as seen from the side as well as from above.) This represents the first in a series of "plateau states". Although further progress requires any plateau state's destruction, that state will later be restored, and each time this happens, more order will have been introduced. These are the successive plateau states.

After getting the top layer, the solver typically works on corners on the bottom layer, or perhaps on getting the horizontal equator slice all fixed up. Most algorithms can, in fact, be broken up into five or six natural stages, corresponding to natural classes of cubies that get returned to their home cubicles. My personal algorithm, for instance, goes through the following five stages:

> (1) top edges,
> (2) top corners,
> (3) bottom corners,
> (4) equator edges, and
> (5) bottom edges.

In the first two of my stages, placement and orientation are achieved simultaneously. Each of the last three stages breaks up into substages: a placement phase and then an orientation phase. Naturally, the operators of any stage must respect all the accomplishments of preceding stages. This means that they may damage the order built up as long as they then repair it. They are welcome, however, to indiscriminately jumble up cubies scheduled to be dealt with in later stages. I find that other people's algorithms are usually based on the same classes of cubies, but the order of the stages can be completely different.

Virtually all algorithms have the property that if you were to take a series of snapshots of the cube at the plateau states, you would see whole groups of cubies falling into place in patterns. This is called "monotonicity at the operator level"—that is, a steady, visible approach toward *START*, with no backtracking. Of course, you would see something totally different if you took snapshots *between* plateau states—but that is another matter. There is no known algorithm that makes visible progress with every *turn*!

Very different in spirit is Thistlethwaite's algorithm. Instead of trying to put particular classes of cubies into their cubicles, he makes a "descent through nested subgroups". This means that, starting with total freedom of movement, he makes a few moves, then clamps down on the types of move that will thenceforward be allowed, makes a few more moves, clamps down a bit more, and so on, until the constraints become so heavy that nothing can move any more. But just at this point, the *START* position has been achieved! Each time, the clamping-down amounts to forbidding q-turns on two opposite faces, allowing only half-turns in their stead from then on. The first faces to be thus "clamped" are U and D, then come F and B, and finally L and R. The strange thing about this approach is that you cannot see *START* getting nearer, even if you take a series of snapshots at carefully chosen moments. Just all of a sudden, there it is! It's as if you were climbing Everest and the peak were shrouded in clouds until the last 100 meters, when suddenly the clouds break and there it is!

This Thistlethwaite algorithm thuggests a thorny thought: Wouldn't it be nice if there were an easy way to tell how far you are from *START*? We might call this a "distance-from-*START*-ometer". Such a device would obviously be quite useful. For example, it is rather embarrassing to resort to the full power of a general unscrambling algorithm to undo what some friend has done with four or five casual twists. For that reason alone, it would be nice to be able to assess quickly if some state is "really random" or is close to *START*. But what does "close" mean? Distances between two states in this vast space can be measured in two fairly natural ways. You can count either the *number of* q-*turns* or the *number of turns* needed to get from one state to the other (where "turn", as above, means either a q-turn or a half-turn). But how can one figure out how many turns are needed to get to *START* without doing an exhaustive search? A reliable and at least fairly accurate estimate would be preferable, one that could be carried out quickly during a cursory inspection of the cube state. A naïve suggestion is to count the number of cubies that are not in their home cubicle. This estimator, however, can be totally fooled by the Dots position, in which nearly all cubies are on the "wrong" side (see Figure 14-9a). That position is only eight q-turns away from *START*. Perhaps the flippancy and the number of quarks could also be taken into account by a better estimator, but I don't know of any.

There are sophisticated group-theoretical arguments suggesting that the farthest one can get from *START* is 22 or 23 turns. This is quite striking, considering that most solvers' early algorithms take several hundred turns, and highly polished algorithms take a number somewhere in the 80's or 90's. Indeed, many mere *operators* take considerably more turns than Thistlethwaite's entire *algorithm* does. (My first double edge-flipper, for instance, was nearly 60 turns long.)

One result that can be demonstrated easily is that there exist states at least 17 turns away from *START*. The argument goes as follows. At the outset there are 18 possible turns we might make: L, L', L^2, R, R', R^2, and so on. After that, there are 15 reasonable turns to make. (One would not move the same face again.) The number of distinct turn sequences of length 2 is therefore 18×15, or 270. Another turn will contribute another factor of 15, and so on. How long does it take before we have reached the number of accessible states? It turns out that 17 is the smallest number of turns that will theoretically allow access to 4.3×10^{19} distinct states. Of course, not every turn sequence of length 17 leads to a unique state, not by a long shot, and so we haven't shown that 17 turns *will* reach every accessible state. We have simply shown that *at least* 17 turns are needed if you want to reach every state from *START*. So, conceivably, no two states are much more than 17 turns away from each other. But *which* 17 turns? That is the question.

So far, only God knows how to get from one state of the Magic Cube to another in the minimum number of turns. "God's algorithm" is, by definition, the speediest recipe for solving the Cube from any state. A burning question of Cubology is: Is God's algorithm just a gigantic table without any

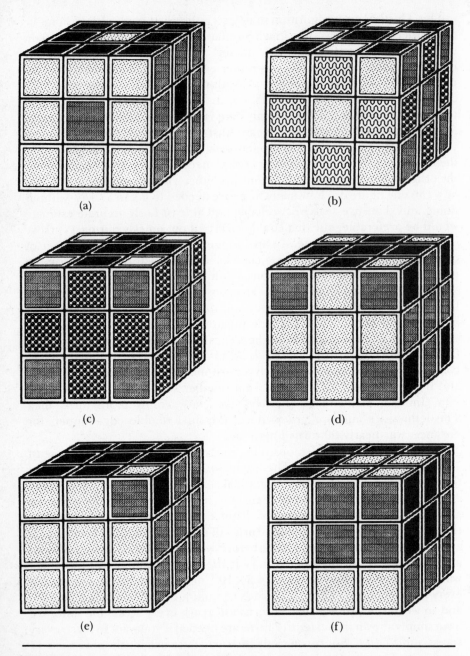

FIGURE 14-9. *A number of special configurations deserving of names. In (a), the pattern known as* Dots. *In (b),* Pons Asinorum. *In (c), the* Christman Cross. *In (d), the* Plummer Cross. *In (e), a* Meson *(showing what appears to be an isolated quark, but it is actually balanced by an antiquark on the opposite corner). In (f), a* Giant Meson, *consisting of a "giant quark" and a "giant antiquark" on opposite corners.*

pattern in it, or is there a significant amount of pattern to it, so that an elegant and short algorithm based on it could be mastered by a mere mortal? Notice that possession of a distance-from-*START*-ometer would be tantamount to possession of God's algorithm. Given any scrambled state, you tentatively try out all eighteen possible twists and then choose one that brings you closer to *START*. (Why must there always be one?) Make it, and then repeat the process. It's a little arduous, but it gets you to *START* directly, obviating plateaus or other intuitive intermediary states. That's one reason for doubting that any simple such meter exists.

*   *   *

If God were to enter a cube-solving contest, It might encounter some rather stiff competition from a few prodigious mortals, even if they do not know Its algorithm. There is a young Englander from Nottingham named Nicholas Hammond who has got his average solving time down to close to 30 seconds! Such a phenomenal performance calls for several skills. The first is a deep understanding of the cube. The second is an extremely polished set of operators. The third is to have the operators down so cold that you could do them in your sleep. The fourth is sheer speed at executing twisty hand motions. The fifth is having a well-oiled "racing cube": one that turns at the merest twitch of a finger, eagerly anticipating every operator before it is needed. In short, the racing cube is a cube that *wants* to win.

I have not yet heard of people naming their racing cubes, although that is sure to come. It would seem, though, that there is an correlation between having a colorful name and being a contributor to Cubology. Apart from Singmaster and Thistlethwaite, there is Dame Kathleen Ollerenshaw (late Lord Mayor of Manchester), who has discovered many streamlined processes, has written an article on the Magic Cube, and has the distinction of being the first to report an attack of Cubist's Thumb, a grave form of the disease mentioned at the beginning of this column. Then there is Oliver Pretzel, the discoverer of a delicious twisted 3-cycle and the creator of a lovely "pretty pattern" called the "6-U" state, which can be reached from *START* by way of the long word

$$L'R^2F'L'B'UBLFRU'RLR_sF_sU_sR_s.$$

Pretty patterns are of interest to many cube lovers, but I cannot do them justice here. I can mention only a few of the best I know. A good warm-up exercise is to figure out how to make the state called *Pons Asinorum* ("Bridge of Asses"). It is shown in Figure 14-9*b*. It has this name because, because, as one MIT cubemeister remarked to me, "If you can't hack this one, forget about cubing." Then there are two kinds of cross, known to the MIT cube-hacking community as the *Christman Cross* and the *Plummer Cross* (see parts *(c)* and *(d)* of Figure 14-9). The former involves *three pairs* of colors

(U-D, F-R, and L-B), while the latter involves *two triples* in the quark-antiquark style. My favorite pretty pattern is the "Worm", whose "genotype", or turn sequence, is:

$$RUF^2D'R_sF_sD'F'R'F^2RU^2FR^2F'R'U'F'U^2FR.$$

Then there is the Snake, a similar sinuous pattern that winds around the cube:

$$BR_sD'R^2DR'_sB'R^2UB^2U'DR^2D.$$

If you cut off the Snake's tail (R^2D') and instead stick on $B^2R_aU^2R'_aB^2D'$, you will create a curious bi-ringed pattern. All of these are from pretty-pattern-meister Richard Walker. A beautiful pattern is the Giant Meson (Figure 14-9*f*), made from a giant quark (a $2\times2\times2$ corner subcube rotated 120 degrees) and a giant antiquark. If you wish, you can top it off, using quarkscrews to twist a standard-size quark and antiquark onto the corners of the giant quark and antiquark, like cherries on top of sundaes. I'll let you figure out how to make this one.

$$* \quad * \quad *$$

I would like to leave you with a set of hints and some things to think about. A difficult challenge, good for cubists at all levels of cubistry, is for someone to do a handful of turns on a pristine cube, to return it to you in this mildly scrambled state, and for you to try to get it back to *START* by finding the exact inverse word. Cubemeisters will be able to invert a bigger handful of turns than novices. Kate Fried reportedly can invert seven turns regularly, and once, after a full day of staring at the cube, she undid ten. (I can undo about four.)

My royal road to discovering an algorithm is based on two challenging exercises involving corner cubies only. The preliminary exercise is as follows. Maneuver the four corner cubies with white on them to the top face with their white facelets pointing upward. Do not worry about which cubie is in which cubicle. Simultaneously do the same thing on the bottom face (of course with *its* color pointing downward). The advanced exercise is to do the preceding one while in addition making sure that all the corner cubies end up in their proper cubicles. This amounts to solving the $2\times2\times2$ Magic Cube puzzle, and it will take you a long way toward mastery of the Magic Cube.

To help you with your edge processes, here is a wonderful trick discovered by David Seal, based on a type of operator called a *monoflip*. I'll give it to you as a puzzle. How can you make a double edge-flipper out of a process that messes up the lower two layers but leaves the top layer invariant, except for flipping a single edge cubie? Hint: The answer involves the important group-theoretical idea of a *commutator*—a word of the form

PQP'Q'. I will also leave it to you to find your own monoflip operator. After I found out about it, I incorporated this trick into my method.

Here is a small riddle: Why do 5- and 7- cycles crop up so often in an object whose symmetries all have to do with numbers such as 3, 4, 6, and 8? Where do cycle lengths such as 5 and 7 come from? A somewhat related question is: What is the maximum *order* a word can have? (The order of a word is the power you have to raise it to in order to get the identity. For example, the order of *R* is 4.) You can show that the order of *RU,* for instance, is 105, by inspecting its cycle structure.

*   *   *

Where do we go from here? I must mention that I have only scratched the surface of Cubology in this column. Rubik and others are working on generalizations of various types. There already is a Magic Domino, which is like two-thirds of a magic cube: two 3×3 layers (see Figure 14-10). You can rotate it by *q*-turns only about one axis; you must do half-turns about the other two. In the *START* position, one face is entirely black, the other entirely white, and both faces have the numbers from 1 through 9 in order. The Domino thus resembles the 15 Puzzle even more strongly than the cube does.

Various people have made $2\times2\times2$ cubes, and such cubes may go on sale one day. You can make your own by gluing little three-cornered hats over each of the eight corners of a $3\times3\times3$ cube. Readers will naturally wonder about such enticing possibilities as a $4\times4\times4$ cube. Rest assured—it is being developed in the Netherlands, and it may be ready soon. Inevitably, there is the question of both higher and lower dimensionalities. Cube theorists are beginning to discuss the properties of higher-dimensional cubes.

The potential of the $3\times3\times3$ cube is not close to being exhausted. One rich area of unexplored terrain is that of alternate colorings. This idea was

FIGURE 14–10. *Ernő Rubik's Magic Domino, scrambled.*

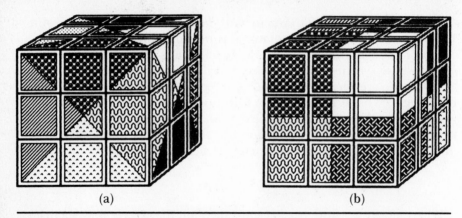

(a)   (b)

FIGURE 14–11. *Two alternate colorings for the Magic Cube, presenting totally novel solving problems for the cubist. In both colorings,* center *orientations* do *matter. However, in (a),* edge *orientations make no difference, and in (b),* corner *orientations make no difference.*

mentioned to me by various MIT Cube hackers. You can color the cubies in a variety of ways (see Figure 14-11). Each new coloring presents a different kind of unscrambling problem. In one variant coloring, edge-cubie orientations take on a vital importance. In another variant, corner-cubie orientations are irrelevant and centers matter. Then, moving toward simplicity, you can color two faces the same color, thereby reducing the number of distinct colors by one. Or you can paint the faces with just three colors. An extreme would be to have three blue faces meet at one corner and three white ones meet at the corner diagonally opposite. Inspector General Semah says that on the cubes he saw, five faces had one color and the sixth face had another color!

Who knows where it will all end? As Bernie Greenberg has pointed out:

> Cubism requires the would-be cubist to literally invent a science. Each solver must suggest areas of research to himself or herself, design experiments, find principles, build theories, reject them, and so forth. It is the only puzzle that requires its solver to build a whole science.

Could Rubik and Ishige have dreamed that their invention would lead to a model and a metaphor for all that is profound and beautiful in science? It is an amazing thing, this Magic Cube.

15

On Crossing the Rubicon

July, 1982

> . . . *O, curséd spite,*
> *that ever I was born to set it right!*
>
> (*Hamlet,* Act I, Scene 5)

THESE days, just "The Cube" will suffice; no one needs to say "Rubik's Cube" to be understood as making a reference to that great puzzle object. In fact, I have a Cube in the shape of a sphere, which I sometimes refer to as "the round Cube", but equally often merely as "that Cube over there". It has been sliced up in the proper way, with rotating "sides" and an inner mechanism that is the same as Rubik's design. And—what is even more marvelous—I have what *poses* as a Cube but is most definitely *not* a Cube: a cubical object sliced in a strange diagonal way, which scrambles in a devilishly skew manner. Both these puzzles are illustrated in Figure 15-1. The sphere is, of course, a Cube, while the cube is an impostor in Cube's clothing. (Note: In this chapter, I use the word "cube" with lowercase 'c' as a generic term for *any* scrambling-by-rotation puzzle, and with capital 'c' to mean the original item: the 3×3×3 Rubik's Cube.)

This proliferation of varieties of cube is really an astonishing phenomenon. Ernö Rubik and his somewhat eclipsed Japanese counterpart Terutoshi Ishige began it, but then it just took off like a prairie fire. Suddenly there were variations on the Cube turning up all over—little ones, teeny-weeny ones, prettily decorated ones, and so forth. But in some sense none of these was an *essentially* different puzzle from the Cube itself. All of them simply dressed the same internal mechanism in different garb.

The first *essentially* different cubes I saw came from Japan. They were 2×2×2's! One was magnetic, with eight metal cubies sliding around a central magnetic sphere. The other was plastic, and had an intricate mechanism similar to, but not identical to, the Rubik-Ishige 3×3×3 mechanism. It could not be identical, since the keystones of the 3×3×3 mechanism are the six face centers—and in a 2×2×2, there aren't any

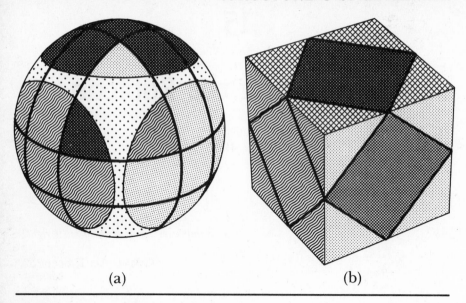

(a)   (b)

FIGURE 15–1. *A cube (a); not a cube (b).*

centers! Later I found out that this mechanism is also due to Rubik, and is based on the $3\times3\times3$ mechanism. This $2\times2\times2$, shown in Figure 15-2*a*, is such a wonderful, inevitable object—in some ways even more beguiling than the $3\times3\times3$. So what puzzles me is: Why aren't they available all over? The $2\times2\times2$—Twobik's Cube?—seems to me an ideal stepping-stone from total novicehood to an intermediate level of cubistry, as it involves solving only the *corners* of a $3\times3\times3$.

Actually, the $2\times2\times2$ was not quite the first essentially different cube I encountered. I had seen a *Magic Domino* (another Rubik invention—see Figure 14-10) much earlier. The Domino is like two of the three layers of a $3\times3\times3$ cube. Its square top and bottom layers both can turn 90 degrees, but its four rectangular sides must turn 180 degrees to allow further moves. Another early variant was the *Octagonal Cube,* a cube four of whose edges had been shaved and which, when twisted, produced some rather grotesque shapes. (See parts *(b)* and *(c)* of Figure 15-2.) Since in this version some of the information about edge parities is lost (you can't tell whether the "shaved" edges are forwards or backwards in their cubicles), it has some quirks that make solving it slightly different from solving the full Cube. On the full Cube, flipped edges always come in pairs. Here, the same is true except that since you can't see whether a shaved edge is flipped or not, sometimes you'll wind up with what appears to be a solved cube, with but a single flipped edge. The first time it can be quite confusing, if you are used to the full Cube!

The next variation I encountered was one due to a young German named Kersten Meier, then a graduate student in operations research at Stanford.

He had built a rough working prototype of a *Magic Pyramid.* It was so rough, in fact, that it often fell apart as you twisted its sides. Nonetheless, it was clearly an innovative step, and deserved to be marketed. I later found out that at nearly the same time, Ben Halpern, a mathematician at Indiana University, had come up with exactly the same concept. Both had generalized the Rubik-Ishige $3 \times 3 \times 3$ Cube mechanism and had seen how to make a dodecahedral puzzle on the same principles. Halpern built working prototypes of both the pyramid and the dodecahedron. The Meier-Halpern variations are shown in Figure 15-2, parts *(d)* and *(e).*

*   *   *

As it turns out, Uwe Mèffert, another German-born inventor, beat both Meier and Halpern to the pyramidal punch—but in a different way. Back in 1972, Mèffert had been interested in pyramids and their pleasing qualities when held in the hand. Somehow, he devised the notion of a pyramid with twisting sides and invented the concept shown in Figure 15-3. He made a few and found them soothing to play with and helpful for meditation, but after a while he stored them away and more or less forgot about them. Then along came Rubik's Cube. Seeing its phenomenal success, Mèffert realized that his old invention might have quite some potential value. So he quickly patented his design, made arrangements to have his device manufactured in quantity, and contacted a toy company for the marketing. The end result was the world success of the *Pyraminx,* a "pyramidal cube" (in my generic sense of "cube") that operates completely differently from the Meier-Halpern pyramid.

Mèffert, who now lives in Hong Kong, became deeply involved in the production and marketing end of his Pyraminx, and began traveling a lot. Through this he came in contact with other inventors in various parts of the world, and decided it would be a good idea to market the most interesting toys of the cube family worldwide. Among these inventors were Meier and Halpern, and as a result, their pyramids too will soon be available to puzzle lovers the world over. They will be known as the *Pyraminx Magic Tetrahedron.* (I would have preferred "King Tet".) The dodecahedron will also be available, under the name *Pyraminx Magic Dodecahedron.* (For a catalogue showing Mèffert's complete range, write to Uwe Mèffert Novelties, Pricewell (Far East), Ltd., P.O. Box 31008, Causeway Bay, Hong Kong. Incidentally, Mèffert welcomes ideas for new "cubic" puzzles. He also wants to develop a Puzzlers' Club, in which members would subscribe at a yearly flat rate and receive in return six or more new puzzles a year. These would be limited editions of particularly complex or esoteric forms of cubic puzzles. He would like to hear from prospective members.)

Dr. Ronald Turner-Smith, a friend of Mèffert's in the Mathematics Department at the Chinese University of Hong Kong, has written a charming little book on the patterns and the mathematics of the Pyraminx,

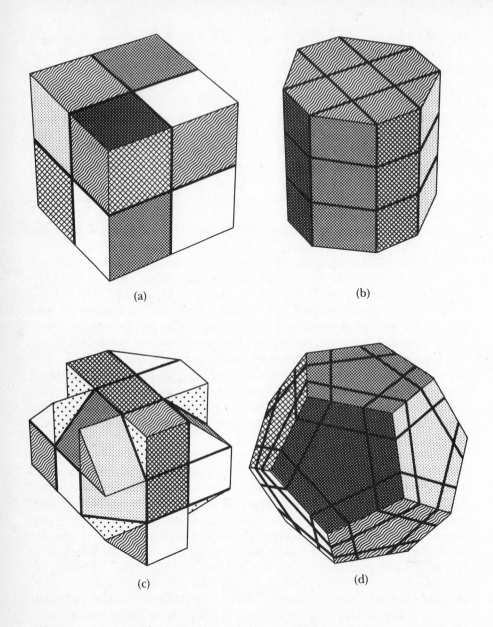

(a)

(b)

(c)

(d)

FIGURE 15–2. *A number of variations on the theme of the Magic Cube. In (a), a 2×2×2 cube. The "Octagonal Prism" (an octagonally shaved 3×3×3 cube), shown in its pristine state in (b) and scrambled in (c). In (d), the Pyraminx Magic Dodecahedron; in (e), the Pyraminx Magic Tetrahedron; in (f), the Pyraminx Magic Icosahedron; in (g), the Pyraminx Ball; in (h), the Pyraminx Magic Crystal; in (i), a 4×4×4 cube in a scrambled state; and in (j), the Pyraminx Ultimate.*

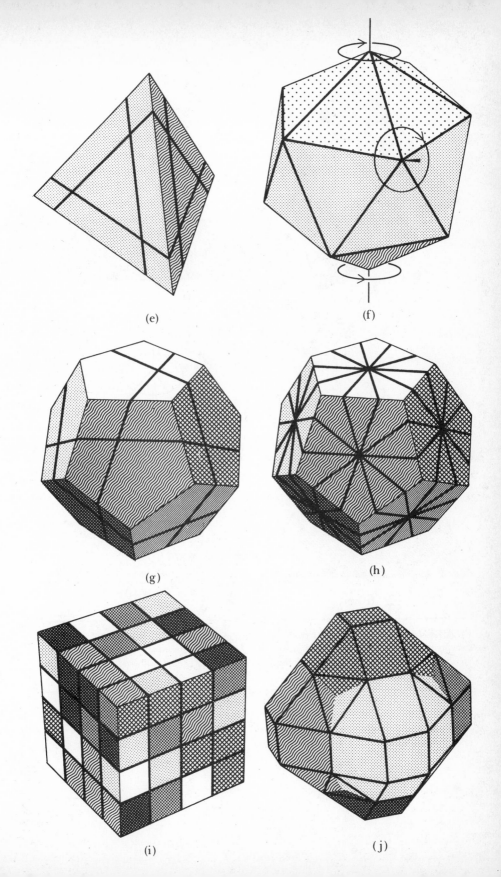

(e)

(f)

(g)

(h)

(i)

(j)

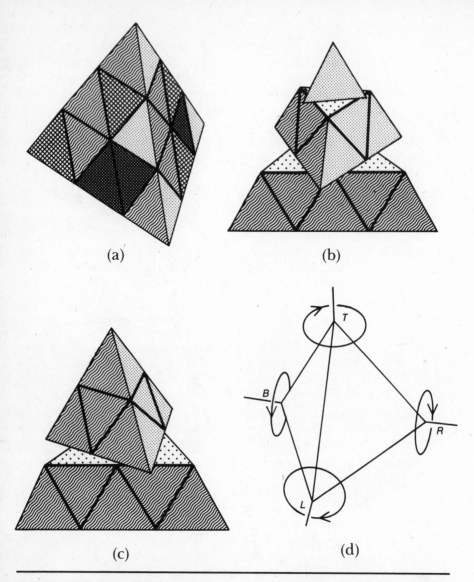

FIGURE 15–3. *Uwe Mèffert's Pyraminx. In (a), a scrambled state. In (b) and (c), modes of twisting are shown. Turns of the form shown in (c) are the ones that the official notation is based on. In (d), names for the four 120-degree clockwise turns:* **L** *(left),* **R** *(right),* **T** *(top), and* **B** *(back).*

called *The Amazing Pyraminx,* which is available in paperback through Mèffert. In it, Turner-Smith does for the Pyraminx what David Singmaster did for the Cube in his *Notes on Rubik's 'Magic Cube'.* (Incidentally, Singmaster is continuing in his role as world clearinghouse for Cubology. He now puts out a newsletter amusingly titled *Cubic Circular,* available by writing to David Singmaster, Ltd. at 66 Mount View Road, London N4 4JR, England. Finally, I should mention that a quarterly magazine called *Rubik's* will be coming out of Hungary beginning this summer, available for $8 a year. Write to P.O. Box 223, Budapest 1906, Hungary.) Like Singmaster, Turner-Smith develops a notation and uses it to convey some of the group theory connected with it, which affords one a deeper appreciation of the object than mere mechanical solving does.

It is interesting that there are two distinct ways of manipulating and describing the action of the Pyraminx. You can rotate either a *face* or a *small pyramid.* The two views are equivalent but complementary, since a face and its opposing small pyramid make up the whole object. Turner-Smith sees the small pyramids as movable and the faces as stationary. We shall adopt this view now, and later return to comment on the complementary one. Let us name the four possible moves, then. (See Figure 15-3*d.*) Each one rotates a small pyramid, either at the Top (T), Back (B), Left (L), or Right (R). The letters *T, B, L, R* stand for clockwise 120-degree turns, and *T', B', L', R'* stand for counterclockwise 120-degree turns (as seen when looking at the rotating tip along the axis of rotation). Notice that any move leaves all the vertices in place (although twisted). Therefore, one can consider the four vertices as stationary reference points, much like the six face centers of the Cube. In fact, at the very start of the solving process they can quickly be twisted to agree with each other, and from then on they provide an identifying color for each face. Thus one can consider the four tip-pyramids either as decorative ornaments or as useful signposts.

In the Cube, the elementary objects that change location are usually called *cubies* or *cubelets.* What are the corresponding elementary objects here? They are not all just small pyramids. As on the Cube, it turns out that there are three types: *edge blocks, middle blocks,* and the above-mentioned *tips.* They are shown in Figure 15-4. As you can see, to each vertex there corresponds one middle block, having three "trianglets" of different colors, just as does the tip perched on top of it. Also like a tip, a middle block never leaves its home location, but only twists. As a consequence, the tips can be considered "trivially solvable" parts of the Pyraminx, and the middle blocks as "easily solvable".

This leaves six edge blocks, each having two colors, that can travel and flip, just like the edge cubies on a Cube. As a matter of fact, it turns out that the constraints on flipping and swapping edges are exactly analogous to those applying to the edge cubies on the Cube: two edges must flip at once, and only *even* permutations of edge locations—permutations where an even number of edge swaps have taken place—are allowed.

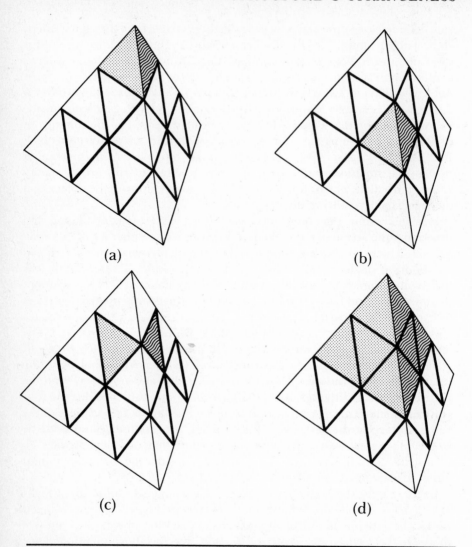

(a)   (b)

(c)   (d)

FIGURE 15–4. *Naming four types of piece in a Pyraminx. In (a), a* tip; *in (b), an* edge; *in (c) a* middle block; *and in (d), another useful though non-basic unit: a* small pyramid.

This means that one can quickly enumerate the number of different ways edges can be distributed about the Pyraminx. Without the constraints, the edges could be dropped into place in 6! (6 factorial), or 720 different ways —the first edge into six slots, the second into five, and so on. But the requirement that the permutation be even divides this by two, to give 360. Also, if unconstrained, each edge could be in either of its two orientations, thus giving 2^6, or 64, different possibilities—but once again, we must divide by 2 because of the flipping-constraint, thus getting 32 distinct flip-states.

Multiplying these two figures together, we come up with 11,520 "interestingly different" states of the Pyraminx. Of course, if you want to take into account the middle blocks and the tips, each of them has 3^4 (or 81) ways of twisting, and they are quite unconstrained, so that you can inflate the figure up to 75,582,720 distinct scramblings altogether! Perhaps the most realistic figure discounts the tip orientations but counts the middle blocks. In that case, one has $81 \times 11,520 = 933,120$, "nontrivially distinct" states of the Pyraminx.

The shortest solving algorithm now known takes 21 twists, and was discovered with the aid of a computer. It is easy to prove that from some positions one needs at least twelve twists to get back to *START*, but the nature of *God's algorithm* (which, by definition, always chooses the shortest possible route home) and the maximum number of twists it requires are unknown, as they are on the Cube.

* * *

When he designed the Pyraminx, Mèffert was quite aware that there were other ways to slice it up internally, even while keeping the same surface appearance, with nine trianglets per face. Therefore, he figured out some alternate internal mechanisms that allow richer modes of twisting. The object I have just described is called the *Popular Pyraminx*. The *Master Pyraminx* is a different kind, and is slated to become available. On it, above and beyond all the movements of the Popular Pyraminx, each edge can swivel about its midpoint by 180 degrees, thus allowing the exchange of any two tips along with the flipping of a single edge piece. (See Figure 15-5.)

FIGURE 15–5. *Showing a physically distinct twisting mode, applicable only to the Master Pyraminx.*

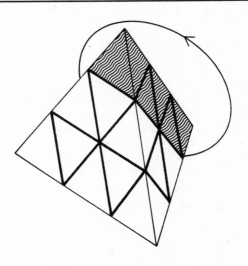

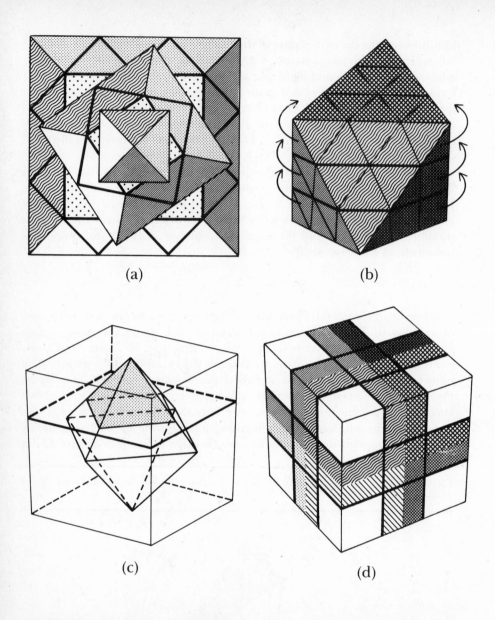

FIGURE 15–6. *Mèffert's Octahedron. In (a), a top view showing how small pyramids and tips can spin, much as on the Pyraminx. Here, however, the natural angle of twist is 90 degrees. In (b), another conceivable way that a "magic octahedron" could be made: with faces that can spin 120 degrees about their centers. The only manufactured item turns as shown in (a). In (c), a diagram demonstrating the mapping between the Octahedron's 90-degree vertex-centered twists and the Cube's 90-degree face-centered twists. In (d), Stan Isaacs' coloring scheme by which an ordinary 3×3×3 Cube emulates a Magic Octahedron, thus concretely demonstrating the idea in (c).*

The flexibility requires each middle block to break up into several pieces as well, some of which can travel all around the pyramid. Thus one has a much more complicated puzzle. The mechanism is exceedingly tricky, because during such swiveling, each of the two moving tips is in contact with the rest of the Pyraminx through a little invisible piece inside the now broken-up middle block. That little piece does not know to which edge it owes "allegiance". As a result, the invisible piece and the tip would together fall off (since that contact does not constitute a permanent link) were it not for a clever piece of engineering that allows each tip to "lock" its little piece to the appropriate edge piece before the swiveling starts, then to "unlock" it after the swiveling is over. Turner-Smith cites the number of scrambled states of the Master Pyraminx as being in excess of 446 trillion.

Once bitten by the "cube bug", Mèffert did not stop here, but moved further into the world of regular polyhedra. His next step was to design an eight-colored octahedron each of whose triangular faces is again divided into nine trianglets. How does it twist? Just as with the Pyraminx, Mèffert perceived the possibility of various modes of twisting. It is interesting that the two equivalent ways of describing the twists of the Popular Pyraminx become inequivalent when applied to the octahedron. Recall that these involved twisting either *faces* or *small pyramids*. The reason they were essentially equivalent is that the rotation of a face is complementary to the rotation of a small pyramid. However, on an octahedron, rotating a face 120 degrees is obviously not complementary to spinning a small pyramid (centered on a vertex) 90 degrees. The distinction is shown in Figure 15-6, parts *(a)* and *(b)*. Realizing this extra degree of freedom, Mèffert designed a mechanism for each of the two ways of turning.

The octahedron that will soon be marketed (under the disappointingly clunky name *Pyraminx Magic Octahedron*) is the one in which the six small pyramids can spin. Thus there are three orthogonal axes of rotation—just as in the Cube. This seemingly trivial resemblance to the Cube actually contains much more than a grain of truth. In fact, *the Mèffert Octahedron and the Cube amount to two surface manifestations of one deep abstract idea.* To see how this comes about, notice that a cube and an octahedron are *dual* to each other: that is, the face centers of either shape form the vertices of the other shape. Thus the six face centers of a cube define an octahedron, and the eight face centers of an octahedron define a cube.

Imagine a Cube, and, sitting inside it, the octahedron that its face centers define (see Figure 15-6c). Each twist of a face of the Cube induces a twist on the corresponding pyramid of the octahedron. Each scrambled position of the Cube seems thus to correspond to a scrambled position of the Octahedron. But this is not quite true. To see what is correct, one needs to see what maps onto what, in the correspondence of Cube and Octahedron. Like the Popular Pyraminx, the octahedron has *tips, middle pieces,* and *edges*. As before, the tips are largely ornamental, and the middle pieces rotate as wholes. Thus a middle piece on the octahedron (together with its decorative tip) maps onto a face center on the Cube. This leaves only edge pieces on the Octahedron—and it is apparent that these, having two facelets, must map onto edge pieces on the Cube. Where does this leave the Cube's corners? Nowhere. They have no analogue on the Octahedron, which is a considerable simplification.

To visualize the Cube-Octahedron correspondence properly, you have to color one of the puzzles in an alternate manner. Since the Cube is more familiar, let's see how it has to be altered to "become" a Magic Octahedron. The proper coloring, *corner*-centered rather than *face*-centered, is shown in Figure 15-6*d*. Stan Isaacs, a computer scientist and puzzlist *par excellence,* has made up one of his dozens of cubes to simulate a Mèffert Octahedron. Someone fluent in solving the ordinarily colored $3 \times 3 \times 3$ Cube will therefore find that their expertise does not quite suffice to handle Isaacs' strangely colored cube, because now the orientation of face centers matters! On the other hand, there is a corresponding simplification as well: "quarks" no longer exist on this cube. That is, there is no such thing as a twisted corner, simply because all the corner cubelets are white on all sides.

All you need to solve this cube (or the Octahedron) is the ability to restore the *edges* and *face centers* (with the added novelty of orientations). Of course, not all "magic octahedra" will be equivalent to simple recolorings of the $3 \times 3 \times 3$ cube, since they may not turn about those three axes. In particular, Mèffert's alternate twisting-mode for the octahedron (where faces twist 90 degrees) is quite unrelated to the Cube.

In his 1982 catalogue, Mèffert shows a picture of an icosahedron (guess what its name is!) whose twenty triangular faces are not subdivided at all; they move five at a time, swirling about any of the twelve vertices. (See Figure 15-2*f.*) Since the movement is vertex-centered rather than face-centered, it should make you think of the icosahedron's dual solid, the dodecahedron. The dual puzzle would have face-centered movement, in the same way as the dual puzzle to the Octahedron, with its vertex-centered movement, is the Cube, with its face-centered movement. (Incidentally, what would be the dual puzzle to the Pyraminx?)

In fact, in Mèffert's catalogue are shown two other dodecahedral puzzles, reproduced in Figure 15-2*g* and Figure 15-2*h,* for your amazement and bemusement. The less complicated one with the asymmetric-looking slices is called the *Pyraminx Ball,* and the beautifully crisscrossed one is called the *Pyraminx Crystal.* The Ball has four axes of rotation, like the Pyraminx, while the Crystal has six. These should be hitting the market in midsummer.

*　　*　　*

At this point, you might well be wondering whether there could be a cube —I mean a genuine, six-sided, square-faced cube!—with a vertex-centered twisting mechanism. No sooner said than done! Tony Durham, a British journalist, was the first to think of this idea. He showed his design to Mèffert,

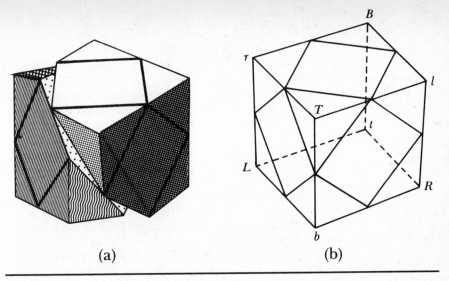

(a)   (b)

FIGURE 15–7. *Tony Durham's Skewb, caught in mid-twist (a). In (b), the labeling of the Skewb's eight corners. (See also Figure 15-1.)*

who developed it into a marketable product by incorporating mechanical features that had proved useful on the Pyraminx. The object in question is shown at rest in Figure 15-1*b* and in motion in Figure 15-7*a*. I call this the *Skewb*, although Mèffert gives it the more prosaic title of *Pyraminx Cube*.

Each of the Skewb's four cuts slices the whole into two equal halves. Each cut perpendicularly bisects one of the four spatial diagonals of the cube. If you think about it, you will see that the shape traced out by each cut as you run around the cube's surface is a perfect hexagon. Each cut crosses all six faces, so that every turn affects all the faces at once. In this respect, the Skewb is more vicious than the Cube, where on each turn two faces are exempt from change. Despite the simplicity of this object, it is quite hard to get used to its skew twist. Of course, that is part of its charm.

Durham offers some insightful commentary on his invention in a remarkable set of notes he has written entitled "Four-Axis Puzzles". I would like to quote a few paragraphs from this document.

> The symmetry group generated by four threefold axes is the rotation group of the tetrahedron, and has order twelve. Almost all the well-known polyhedra, regular as well as semiregular, possess this tetrahedral symmetry, though their own symmetry may be much richer. So a four-axis mechanism may be put inside a polyhedral puzzle of any regular or semiregular shape, and the puzzle will keep its shape during play. The Pyraminx Ball may look odd at first glance, but it illustrates the beautiful way in which tetrahedral symmetry is buried in the richer symmetry of the dodecahedron.
>
> The cube mechanism found by Rubik does not have this property. It uses fourfold rotation axes, which are generally found only in the cube/octahedron family of solids. Thus, it is possible to 'build out' a Rubik cube into the shape of a dodecahedron. But to preserve that shape during play you must restrict

yourself to half-turns. Quarter-turns invoke a symmetry which the dodeca-hedron does not possess.

All four-axis puzzles have a central ball or spindle. Four pieces (usually corners) are pinned directly to the ball. The standard Pyraminx has six free-floating edge pieces with 'wings' that hook under the corner pieces. The analogous free-floating pieces on the Pyraminx Cube are the square face-centers. The four-faceted pieces on the dodecahedral Pyraminx Ball play the same role.

The Pyraminx Cube and Ball have four more free-floating pieces, which again are corners. These pieces have their own 'wings' which, in the *START* position, hook under the first set of free-floating pieces. Thus, there is a three-level hierarchy of interlocking pieces, conceptually similar to Rubik's, but geometrically very different.

All eight corners of the Pyraminx Cube look alike. At first sight one might think that any two corners could be made to change places. In fact, four of the corners are free-floating and four are rigidly fixed to the central ball. The two types can never change place. The square shape of the face center pieces is deceptive, too. Inside, the mechanical parts of the square pieces are not so symmetrical. Such a piece can never return to its starting position (relative to the rigid set of four corners) rotated by 90 degrees. Only half-turns are possible.

The standard Pyraminx has obvious fixed points—the four corners. Confronted with a Pyraminx Cube and knowing that four corners are fixed and four are free, one naturally wonders which are which. Actually it makes no difference. The four free corners move independently of the fixed ones, but they always move together as if physically linked.

Durham proceeds to give Turner-Smith's *TBLR-T'B'L'R'* notation for the Pyraminx, and mentions that it is adaptable to any four-axis puzzle (such as his Skewb), simply by letting *TBLR* name four of the centers of rotation. (On the Pyraminx, this could mean either the four tips or the four face centers. On the Skewb, this would be four of the tips, leaving four other tips unnamed. See Figure 15-7*b*.) Then any move can be transcribed. If it is centered on one of the named spots, just use the proper notation. If it is centered on one of the four unnamed spots, use the name for the complementary move, since it doesn't matter which half of the puzzle twists. (You may want to think about that for a moment. Actually, it is obvious, but it sounds like a tricky point.) Durham points out that it is sometimes useful to have names for the four remaining spots and for twists around them. He lets *t, b, l, r* fulfill that purpose. Thus *T* and *t* accomplish the same thing *internally* to the puzzle, but they leave it hovering in space in a different *overall* orientation. Although he concedes that it may become confusing, Durham advocates using a mixed notation on occasion.

Sometimes you need to mix the notations to see what is going on. *TbT'b'* is one of the useful class of moves called *commutators* (two moves followed by their inverses—thus of the form *xyx'y'*), though you would never guess so from its description in regular coordinates (*TBL'B'*) or alternate coordinates (*tlt'b'*).

The Pyraminx Cube and Ball may be described as *deep-cut* puzzles in contrast to *shallow-cut* puzzles such as Rubik's Cube. In the latter, the cuts are made near to the surface. In deep-cut puzzles, they slash close to the puzzle's heart. The bulk of a shallow-cut puzzle remains stationary while you turn a small part of it. A deep-cut puzzle, however, raises serious doubt as to which part has been turned and which has remained stationary. This is why alternate sets of coordinates have to be taken seriously on deep-cut puzzles.

Deep-cut puzzles also dictate a 'global' approach to solution. It is peculiarly difficult to work on one area of the puzzle without affecting the rest. However, as solution proceeds, this very fact comes to your aid. Pairs of corners magically untwist in synchrony. The last flip, the last swap is done for you automatically. As you close in for the kill, billions of pathways down which the puzzle might escape are closed off to it. Parity constraints are at work, and when every move activates five or eight interlocked permutation cycles—as it does in a deep-cut puzzle—parity constraints are powerful.

In the section of his notes having to do with parity constraints, Durham includes the following humorous but insightful apology:

Please forgive the loose use of the term *parity* to include tests for divisibility by 3 (not only 2) or even more distant concepts. We shall use the term *parity restriction* for any constraint on *imaginable* transformations of the puzzle that prevents their accomplishment in normal operation of the puzzle. The list does not, for example, include the rule: 'Thou shalt not swap a face piece with a corner piece.' It is just too far-fetched. One might as well try to imagine a move that transformed the entire puzzle into depleted uranium or Gorgonzola cheese.

Then he lists all the Skewb's "parity" constraints, in his generalized sense of the term.

1. The four (fixed) corners *TBLR* may be permuted among themselves, as may the remaining four corners *tblr,* but mixing between the two sets is prohibited.
2. *TBLR* themselves move as a rigid tetrahedral unit. This constraint applies to their *positions* in space only (not to their orientations).
2a. For exactly the same reasons, the remaining four (free) corners *tblr* move as a tetrahedral unit. They move independently of *TBLR.* In fact any of the twelve possible relative positions of *tblr* and *TBLR* can be reached in at most two puzzle moves.

    Although *TBLR* are fixed and *tblr* are free-floating, mathematically speaking, 2 and 2a have exactly the same status. Writers on the Rubik Cube have generally regarded the transposition of two face centers as an 'unimaginable' transformation, while the swapping of two edge pieces is 'prohibited but imaginable'. By analogy with this convention, 2a counts as a parity restriction while 2 does not! This is plainly unsatisfactory, and a better and more precise definition of 'parity' is badly needed. Is it a question of geometry? Of mechanics? Of topology? Note that the problem is in enumerating the *impossible* positions. The *possible* positions are readily counted.

3. The sum of the twists of corners *TBLR* is always equal, modulo 3, to the twistedness of the puzzle, taken as a whole.

(Here, *twist* applies to corners, and is either 0, +1, or −1. A corner's twist is measured relative to the rigid tetrahedron to which it belongs. Thus the twist of *T* is measured relative to *TBLR*. A clockwise rotation of a corner counts as +1, counterclockwise as −1. By contrast, the *twistedness* of the puzzle as a whole is a function only of the *positions* of the corners, not of their orientations. If the relative positions of *TBLR* and *tblr* are as in the *START* position, then the twistedness is 0. If they can be restored to *START* by one clockwise puzzle move, the twistedness is −1, and if by one counterclockwise move, then +1. If it takes one of each type, then the twistedness is again 0.)

3a. Same as 3, only with *tblr*.

From 3 and 3a, it follows that the total twist of *TBLR* always equals the total twist of *tblr*. Also, it follows that it is impossible to turn a single corner by 120 degrees (i.e., to create an isolated quark). One might paraphrase 3 and 3a by saying that the puzzle 'knows', in three distinct ways, how many turns it is away from *START* (modulo 3).

4. It is impossible to transpose exactly two face pieces.
5. It is impossible for any face piece to turn in place by 90 degrees.
6. It is impossible to flip a single face piece through 180 degrees.

Durham offers proofs of these interesting facts, but as they are for the most part analogous to those on the Cube, I shall omit them here. By combining all these constraints, Durham comes up with the total number of scrambled states of his Skewb, which is 100,776,960. However, this assumes you have a way of telling the orientation of a face center, which (unless you mark it up) you don't. Hence the number of *visually distinguishable* states is reduced by five factors of two, to 3,149,280—a rather smaller number than for the Cube (4×10^{19}), but certainly the difficulty does not scale down proportionately with the number of states. (Could you even imagine what it would mean for a puzzle to be "ten trillion times easier" than Rubik's Cube?)

*　　*　　*

Durham's final observations carry Solomon Golomb's beautiful analogy between cubological phenomena and those of particle physics to even greater heights. Golomb pointed out that many fundamental particles have their counterparts on the $3 \times 3 \times 3$ Cube. They include the quarks (q), antiquarks ($\bar{q}$), mesons ($q\bar{q}$ pairs), baryons and antibaryons (qqq and $\bar{q}\bar{q}\bar{q}$ trios). Durham extends the analogy as follows:

The definition of *twist* must be modified for the purpose of particle physics. A clockwise twist of one of the corners *TBLR* is now given the value +1/3, as is a *counterclockwise* twist of any of the corners *tblr*. Either of these is a quark. Its opposite is an antiquark with value −1/3. It will be seen that twist corresponds to *baryon number*. The total twist of all corners is always an integer. A single puzzle move is always a meson.

Quarks at the corners *TBLR* will be regarded as 'up' or u quarks; those at *tblr* will be 'down' or d quarks. Both quarks have isotopic spin 1/2. They are distinguished by the orientation of the isospin vector in its abstract space. The projection of the isospin, I_x, has the value $+1/2$ for the u quark and $-1/2$ for the d quark. In the absence of strangeness, charm, etc., the electric charge Q of a particle is given by $Q = I_x + B/2$, where B is the baryon number. So u quarks have charge 2/3, while d quarks have charge $-1/3$. (All the quantum numbers are multiplied by -1 for the antiquarks.) Again the puzzle models an important feature of observed reality: all particles have integral electric charge.

The relevant quantum numbers for our two quarks are as follows:

	u	d
B	1/3	1/3
I	1/2	1/2
I_x	1/2	$-1/2$
Q	2/3	$-1/3$

We can now assemble various *hadrons* (strongly interacting particles), as shown in the table below. Each particle is represented by two rows having four symbols each. The four places in the top row represent the twists on the *TBLR* corners; in the bottom row the same is done for the *tblr* corners. A quark is denoted by '+', an antiquark by '−'.

$+\,0\;0\;0$ (π^+ meson)	$+\,+\,0\;0$ (proton)	$+\,0\;0\;0$ (neutron)
$-\,0\;0\;0$ ($u\bar{d}$)	$+\,0\;0\;0$ (uud)	$+\,+\,0\;0$ (udd)
$+\,+\,+\,0$ (Δ^{++})	$0\;0\;0\;0$ (Δ^-)	$-\,-\,0\;0$ (antiproton)
$0\;0\;0\;0$ (uuu)	$+\,+\,+\,0$ (ddd)	$-\,0\;0\;0$ ($\bar{u}\bar{u}\bar{d}$)
$-\,0\;0\;0$ (π^- meson)	$+\,-\,0\;0$ (η^0 meson)	$0\;0\;0\;0$ (π^0 meson)
$+\,0\;0\;0$ ($\bar{u}d$)	$0\;0\;0\;0$ ($u\bar{u}$)	$+\,-\,0\;0$ ($d\bar{d}$)

Isotopic symmetry is a global symmetry, and the strong (nuclear) force is invariant under transformations that rotate the isotopic spin vector by the same amount for all particles. Such a transformation would, in a continuous fashion, transform all u quarks into d quarks and vice versa. Protons and neutrons would swap roles. The analogous process for the puzzle is the *continuous rotation of the whole puzzle in space*. It can indeed bring the *TBLR* corners to the former position of the *tblr* corners, so that an up quark becomes a down quark.

This makes no difference to the 'strong interaction' (*i.e.*, the normal operation of the puzzle). The *TBLR* and *tblr* corners are functionally identical. But it matters, if you try to dismantle the puzzle: you will find that one set of corners is fixed to the core, and one is not. Such dismantling operations can be thought of as *weak* or *electromagnetic* interactions, which can break the conservation rules obeyed by the strong interaction. Actually they break the rules rather too well, since they allow the creation of single free quarks.

Durham points out that the analogy still has weaknesses, such as the facts that neither charge nor baryon number is conserved, that there is no analogue to spin, that only two "flavors" of quark are represented (up and down), and that quark "color" is not modeled. Golomb, in the meantime, has been actively trying to find a way of modeling quark color in the $3 \times 3 \times 3$ Cube analogy.

Whatever the failings of this analogy, I find it one of the most provocative of all analogies I have ever encountered anywhere, and will be most astonished if it is purely coincidental. I somehow cannot help but believe that the fascinating patterns shared by these macroscopic puzzles and the microscopic particles reveals some underlying order and set of principles common to both. Indeed, I have faith that, if looked at in the proper way, the group-theoretical principles that govern these parity constraints on "cubes" can be transferred to the domain of particle physics, and yield fresh insights about the reasons for the symmetries among particles. There! If that doesn't prod some particle physicist into looking into this, I don't know what will!

*   *   *

Perhaps my favorite "cube" is the one I dubbed the *IncrediBall*. It is due to a German educator from Dortmund named Wolfgang Küppers, and is in Mèffert's catalogue. As of the time of this writing, I may be the world's fastest IncrediBall solver (or at least the fastest on my block!), with an average time of about six minutes. However, I am sure that my glory will not last long, once this puzzle is marketed widely by the Milton Bradley Company sometime this summer. Their trade name for it will be *Impossi*Ball*. It is pictured in Figure 15-8.

This I-Ball is basically a rounded-off dodecahedron each of whose twelve faces (*dodecalets,* I'll call them) has been subdivided into five elementary "trianglets". Thus there are 60 such trianglets. If, instead of seeing them in groups of five, you take them *three* at a time, you'll find that they define a rounded-off icosahedron (the dual of the dodecahedron). Such a group of three trianglets I call an *icosalet,* and there are twenty such, each one having a unique arrangement of three colors. The icosalets are the elementary, unbreakable units out of which the IncrediBall is constructed; they correspond to the cubelets on the Cube, or the elementary pyramids of the Pyraminx. Whereas on the Cube there are three kinds of cubelet (edges, faces, and corners), here all icosalets are of a single type. For this reason, the I-Ball is less forbidding than at first it might appear. Its pristine state is one in which each dodecalet is all of one color. Mèffert has used only six colors, rather than twelve, each color being used in two antipodal dodecalets, but this does not in any way change the difficulty of the puzzle.

The way it turns is a little surprising. Any group of five icosalets that meet at a point (the center of a dodecalet) form what I call a *circle,* which will rotate as a unit, twisting 72 degrees to the left or right. (Such a circle is analogous to a "layer"—a face together with its fringe—on the Cube.) Thus five such

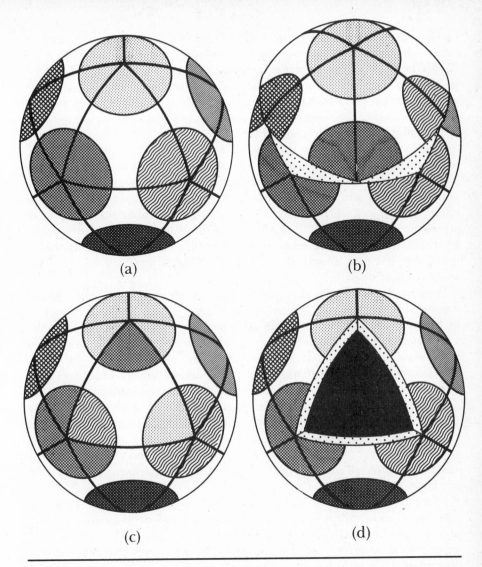

(a)   (b)

(c)   (d)

FIGURE 15–8. *Wolfgang Küppers' IncrediBall (or Impossi*Ball, if you wish). In (a), the pristine state. The triangles with curved sides are called* icosalets. *In (b), an IncrediBall caught in the midst of a "bumpy twist". Each such twist involves rotating a "circle" (composed of five icosalets) through 72 degrees. In (c), a state with just one quark visible (one icosalet twisted 120 degrees clockwise). In (d), one icosalet has been removed. This allows another icosalet to slide in and occupy the vacuum, meanwhile leaving behind its own vacuum. As an icosalet-shaped hole glides around the puzzle, order can be created or destroyed. This sphere-based puzzle thus closely resembles Sam Loyd's planar "15 puzzle".*

twists return that group to its starting position. However, the "circle" defined by the five icosalets is not truly circular, and if the trianglets were rigidly held at a fixed distance from the center, it simply would not be possible to rotate such a group. But Mèffert's mechanism ingeniously gets around that problem by having the icosalets lift up slightly as they go over "bumps", so that the solid flexes noticeably. As a result, twisting the I-Ball has a delightful "organic" feel to it.

The constraints here are the same old story: all permutations are even, which means you cannot swap two icosalets—the best you can do is cycle three of them, or swap two pairs simultaneously; and of course, quarks and antiquarks must add up to a total twist that is integral. Taking into account these constraints, I calculate that the total number of IncrediBall scramblings is 23,563,902,142,421,896,679,424,000, or 24×10^{24}—about 24 trillion trillion. This is not quite a million times larger than the figure for the Cube. It's also about 40 times larger than Avogadro's number, for whatever that's worth.

How hard is it to solve this puzzle? Is it harder than the Cube? I found it easier, but that's hardly fair, since I had already done the Cube. However, in Durham's terms, the IncrediBall is decidedly a "shallow-cut" puzzle, which means that a more or less local approach will work. I found that, when I loosened my conceptual grip on the exact qualities of my hard-won operators for the Cube, and took them more metaphorically, I could transfer some of my expertise over from Cube to I-Ball. Not everything transferred, needless to say. What pleased me most was when I discovered that my "quarkscrew" and "antiquarkscrew" were directly exportable. Of course, it took a while to discover what such an export would consist in. What is the *essence* of a move? Which aspects of it are provincial and sheddable? How can one learn to tell easily? These are very difficult questions, to which I do not have the answers.

I gradually learned my way around the IncrediBall by realizing that a powerful class of moves consists of turning only two overlapping "circles" in a commutator pattern ($xyx'y'$). So I studied such two-circle commutators on paper, as shown in Figure 15-9, until I found ones that filled all my objectives. They included quarkscrews, swaps, and 3-cycles, which form the basis of a complete solution. In doing this, I came up with just barely enough notation to cover my needs, but I did not develop a complete notation for the IncrediBall. This, it seems to me, would be very useful: *a standard universal notation, psychologically as well as mathematically satisfying, for all cubelike puzzles.* However, it is a very ambitious project, given that you would have to anticipate all conceivable future variations on this fertile theme—hardly a trivial undertaking!

It is interesting that my diagrams of overlapping circles turn out to be closely connected with another lovely family of generalizations of the Cube, due to a Spanish physicist named Gabriel Lorente. His puzzles are mostly planar and consist precisely in networks of overlapping circles. (See Figure

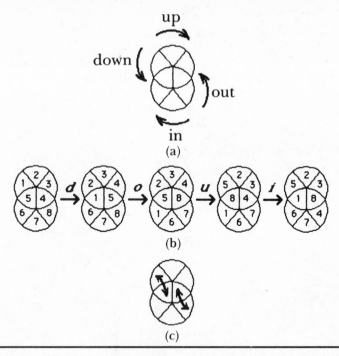

FIGURE 15–9. *Operators involving just two overlapping circles. In (a), names for the four possible 72-degree twists. In (b), the operator "doui" (down-out-up-in) is applied. Note that it has the form of a* commutator, *involving alternating inverses. In (c), the outcome is summarized: a double swap has been effected.*

15-10.) The planar ones he calls the *Grill* and the *Trebol*. In each of them, circles can be given partial twists and pieces of them are thereby shuffled and redistributed. Extending this notion to a spherical surface, Lorente came up with an elegant IncrediBall-like puzzle, which he calls the *Florid Sphere.*

When you look closely at Lorente's puzzles, the IncrediBall, and even the Cube, you begin to see that the *essence* of all these puzzles seems to reside in *overlapping orbits.* In fact, one could even maintain that the three-dimensionality of all these puzzles is irrelevant; their interest is essentially due only to the properties of intricately overlapping closed orbits in a two-dimensional space, possibly curved like a sphere.

The quintessential planar overlapping-circle puzzle was invented, as it turns out, way back in the 1890's, although recently it has been repeatedly rediscovered in the wake of the Cube. All such puzzles basically involve two circles of marbles that intersect at various spots. (See Figure 15-11.) You can choose to cycle either circle, and the marbles at the intersections will thus be absorbed into whichever circle is moving.

While we're discussing two-dimensionality, it is worthwhile pointing out

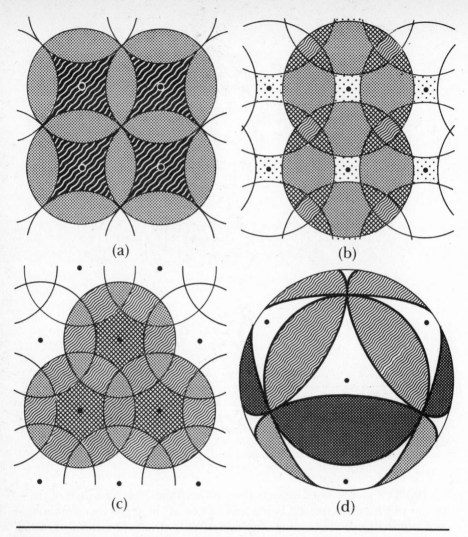

FIGURE 15–10. *Four puzzles by Gabriel Lorente. In (a) and (b), two schemes he calls "Grills". Note that both are based on a square lattice of circle centers. In (c), his "Trebol" puzzle, where centers form a triangular lattice. In (d), the centers of circles lie on a sphere. This is his "Florid Sphere". Which previously discussed puzzle is it equivalent to?*

that the IncrediBall's internal construction allows it to be transformed rather amazingly into what I call the "19" puzzle—a two-dimensional curved-space version of Sam Loyd's famous "15" puzzle (the 4×4 square puzzle with one "squarelet" removed, allowing you to rearrange the remaining 15 squarelets by shifting the hole about). This was first observed by Ben Halpern, while he was idly playing with an IncrediBall. He had removed one single icosalet (which is possible, one of the beauties of the IncrediBall being that its mechanism readily allows disassembly and reassembly), leaving a hole, and he observed that, because all icosalets are

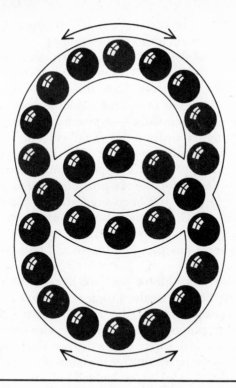

FIGURE 15–11. *Does this 90-year-old type of two-dimensional puzzle, with just two intersecting rings of marbles, capture the ultimate essence of all modern "cubic" puzzles?*

congruent, *the hole could wander about all over the sphere,* just like the square hole in the 15 puzzle. (See Figure 15-8*d*.) Again, this seems to underscore the two-dimensional nature of these puzzles.

The claim that these puzzles are two-dimensional comes from the fact that only pieces on their surfaces move; there is no exchange between the interior and the exterior. For an extreme case, imagine the Earth as a giant puzzle, its entire surface covered with trillions of overlapping circles of marbles. With a hundred million turns, you could ship a marble from New York to San Francisco. Clearly this would be in essence a two-dimensional puzzle. The smallness of the circles relative to the size of the Earth makes this obvious. (However, I surely wouldn't want to think about *solving* such a puzzle, whether it's two-dimensional or not!)

By contrast, consider two objects about to come out: Ideal's 4×4×4 cube, tastelessly marketed as *Rubik's Revenge,* and Mèffert's *Pyraminx Ultimate,* a 5×5×5 with shaved corners. Both are shown in Figure 15-2, parts *(i)* and *(j)*. In these objects, there are circles on a much more global scale. Namely, the 4×4×4 has an "Arctic Circle", a "Tropic of Cancer", a "Tropic of Capricorn", and an "Antarctic Circle". The Pyraminx Ultimate has an Equator as well.

On the $3\times3\times3$ Cube, one could get away with ignoring the Equator by describing equatorial twists in terms of their complements, like rotating the slices of bread instead of the meat in a sandwich. Singmaster's notational choice for the $3\times3\times3$ Cube reflects his propensity to describe face centers as stationary. Thus for him, bread slices move while the meat stays put. Theoretically, this is fine, but realistically, people just do not hold their sandwiches—pardon me, their Cubes—in one fixed orientation. Moreover, when you pass to higher orders, this view will not suffice. Imagine a multilayer club sandwich with three slices of bread and two different kinds of meat. For this, you simply have to expand your notational horizons!

An elegant set of names for the six possible meat-slice, or equatorial, moves on the $3\times3\times3$ Cube has been suggested by John Conway, David Berlekamp, and Richard Guy in their book *Winning Ways*. They employ Greek letters with clever mnemonic justifications. These are shown in Figure 15-12. With some modification, they could be adapted to slices on higher-order cubes.

Slice moves of this more global sort are like giant circles of marbles stretching around the Equator or the Tropic of Capricorn; their radii are of the same order of magnitude as the radius of the underlying three-dimensional object. The topology of linkage of circles becomes much more complicated than in the case where the circles are small and every connection is very local. To describe the linkage economically, one would be forced to talk about the way the circles are embedded in 3-space. In this sense, the higher-order cubes can truly be said to be intrinsically three-dimensional puzzles.

There are, it seems, endless new spinoffs of the Cube being created. It is

---

FIGURE 15–12. *The Conway-Berlekamp-Guy nomenclature for twists of a $3\times3\times3$ Cube's equatorial slices. This notation can be generalized to higher-order cubes.*

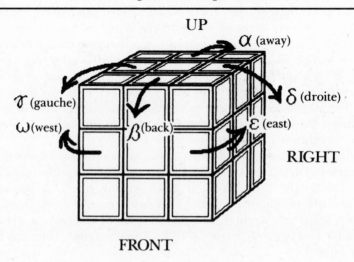

a very fertile idea. H. J. Kamack and T. R. Keane, both chemical engineers, sent me a beautiful paper in which they describe their simulation of a four-dimensional $3\times3\times3\times3$ cube on a computer. They call it *Rubik's Tesseract,* and they have computed the number of possible states it has. That number is: the product of $24!\times32!\times16!/4$ (the number of *permutations* of position of the elementary "tessies" out of which it is built) with $(2^{24}/2)\times(6^{32}/2)\times(12^{16}/3)$ (the number of legal *orientations* of the tessies within their niches, which the authors somewhat hesitantly term "tessicles", by analogy with "cubicles"). This number comes to approximately 1.76×10^{120}, which, they point out, is about the same size as the number of possible games of chess. (I don't think it would be an exaggeration to say that if Ideal were marketing this puzzle, their publicity would shamelessly proclaim, "Over 3 trillion combinations!") Kamack and Keane have made many provocative discoveries, which unfortunately I have no space to report on at this time. I was also sent a fascinating paper by George Marx and Eva Gajzágó, two physicists at the famous Roland Eötvös University in Budapest. In it they give a definition of "entropy" on the Cube and describe some statistical results computed by a grammar school student named Victor Zámbó. These are matters I would like to go into at some future time.

*   *   *

I would like to close by discussing the astonishing popularity of the Cube. In the *New York Times'* paperback bestseller list of November 15, 1981, three cube booklets figured on the list. The positions they occupied? First, second, and fifth. People often ask, "Why is the Cube so popular? Will it last? Or is it just some sort of fad?" My personal opinion is that it will last. I think that the Cube has some sort of basic, instinctive, "primordial" appeal. Its conceptual pizzazz comes from the fact that it fits into a niche in our minds that connects to many, many general notions about the world. So here is an attempt to characterize that quality.

* To begin with, the Cube is small and colorful. It fits snugly in the hand and has a pleasing feel. Twisting is a fundamental and intriguing motion that the hand performs naturally. The object itself has overall symmetry, so that it can be rotated as a whole without its "feel" changing. (This is in contrast to many puzzles that have at most one axis of symmetry.) Quite surprisingly, there are not many puzzles or toys that give the mind and fingers a genuine three-dimensional workout.
* Although it gets all scrambled up, the object itself stays whole. (This is in contrast to many Humpty-Dumpty-ish puzzles that come apart into scads of pieces that may get scattered all over the floor.) That it manages to stay in one piece when it has so many independent ways of twisting is initially amazing, and remains mysterious even after you've seen its "guts".

* The object is a miniature incarnation of that subtle blend of order and chaos that our world is. Most of the time, you just cannot predict what repercussions even simple actions will have—they simply have too many side effects. A few tiny actions can have vast, interlocking consequences, and become practically un-undoable. One can easily become paralyzed by fear, not wanting to make any move at all, sensing that with no trouble at all one can get totally, irretrievably, hopelessly lost.

* There are plenty of patterns, some attainable, some unattainable. Sometimes they are simple to generate, but one can't see how they emerge. Sometimes they are hard to generate, yet one clearly understands how they arise.

* There are many routes to any state, and the shortest is nearly always completely unknowable. The solution to a difficult situation is hardly ever to back out the way you came in, but to find an alternate and completely different escape route. One feels a little like someone trapped in a cave with no light, unable to sense the whole space, able to grope about only very locally, wondering whether it is even humanly possible to to have such an overview. (One wonders about God's algorithm: Is it humanly comprehensible?)

* The Cube is a rich source of metaphors. It furnishes analogies to particle physics (quarks, etc.), to biology (a move-sequence as a "geno-type" and the pattern it codes for as a "phenotype"), to problem-solving in everyday life (breaking a problem into parts, solving it stage by stage), to entropy and path-finding, and on and on. It even touches theology ("God's algorithm") and many other phenomena.

* There are different approaches to understanding the Cube. In particular, there is a strong contrast between the "algebraic" approach and the "geometric" approach. In the algebraic, or mathematical, approach, long sequences of operations are compounded out of shorter sequences, so that after a while one has no idea why one is doing the various individual twists—one just relies on the sequences, as wholes, to work. Though efficient, this is risky. In the geometric, or commonsense, approach, eye and mind combine to choose twist after twist, each twist having a clear reason as part of a carefully charted pathway. Though inefficient, this is reliable. These approaches, of course, can serve as metaphors for styles of attacking problems in life.

* The Cube's universe has a strange population. Aside from its varieties of "cubies" and modes of twisting, there are such intangible qualities as "flippedness" or "twistedness", which one quite literally moves about on the Cube (*e.g.*, in the form of quarks), just as one moves the tangible cubies. Similarly, the word "here" can designate a "place" that moves to and fro during a sequence of twists. The interlocking and nested reference frames that one jumps between in trying to restore order to the Cube vividly exemplify the layered way in which we

conceive of space—indeed, the layered way in which concepts themselves are structured in our minds.

* Among the Cube's less intellectual charms are the magic of motion too swift for the eye, the thrills of speed, competition, and grace; the varying levels of knowledge that one can gain, the enjoyment of exchanging information and insight. And, needless to say, the very idea that such a tiny innocent object conceals such a vast universe of potential.

* Finally, consider the metaphor the Cube offers for the state of the world (one that has been exploited in various political cartoons). The globe is in a mess (as shown in Figure 15-13), and the leaders of various

FIGURE 15–13. *The sad state of the globe.*

countries want it to be "fixed". But they are unwilling to relinquish any tiny bit of order they have achieved. They cling to old, useless achievements because they are too fearful of letting go and temporarily abandoning what partial order they have in order to achieve greater order and harmony. They lack a mature, global view, one that recognizes that a willingness to make sacrifices in the short run can wind up producing much greater gains in the long run.

I am confident that The Cube, as well as "cubes" in general, will flourish. I expect new varieties to appear for a long time to come, and to enrich our lives in many ways. It is gratifying that a toy that so challenges the mind has found such worldwide success. I hear that it's now very popular in China.

Perhaps one day it will even penetrate into the Soviet Union, to my knowlege the last bastion of the Cube-Free World.

Post Scriptum.

I wrote the preceding two columns over a year apart. It has now been two years since the second of them was written. The major cube news since then is, sad to say, that there has not been much major cube news since then. What apparently happened was simply a worldwide cube glut. Cubes—cubical and otherwise—were coming out of everybody's ears, and it was just a little too much. I can understand that, but it saddens me to see something so exciting fade so totally.

There are still a number of things worth mentioning. A good place to begin is with the origin of the cube—that is, of magic solids in general. Shortly after my second cube column appeared, I received a rather plaintive letter from a Fresno high school teacher named William O. Gustafson, who claimed that he was, in some sense, the true inventor of the idea of the Magic Cube. What he actually had invented—in 1958—was a sphere sliced by three orthogonal planes into eight congruent pieces (octants), in such a way that any two opposite hemispheres (each composed of four octants) could turn. This amounts to a spherical $2 \times 2 \times 2$ cube—a cubical variant of which was marketed by Ideal Toy Company some twenty-odd years later under the name "Rubik's Pocket Cube". Gustafson called his toy "Gustafson's Globe".

To substantiate his claim, Gustafson enclosed photocopies of a good deal of correspondence he conducted in 1960 with numerous toy companies (he wrote 76 of them!), the Japanese patent office (he received a patent)—and even Martin Gardner. Gardner's card to him was interesting. It said:

> That is an interesting puzzle that you propose, but I am at a loss for suggestions on how to interest a toy dealer in it. My experience has been that it is almost *impossible* to make any money with a puzzle unless you are the manufacturer yourself, with your own toy company.

An interesting comment, in light of what happened with Rubik's Cube. In addition to his $2 \times 2 \times 2$ Globe, Gustafson developed a $3 \times 3 \times 3$ version, but felt that he should work first on getting the simpler puzzle out, so most of his correspondence concerned that one.

Gustafson also enclosed for me a photocopy of a wry letter of condolence sent to him by a former student who, when he encountered Rubik's Cube, vividly recalled Gustafson's Globe from decades earlier. The card read: "With sincere sympathy in your recent loss, and a hope that time has helped in some small way to ease the sorrow in your heart". Below those poetic lines were the words "Gustafson's Globe", crossed out, and then the words "Rubik's Cube".

I did a little bit of checking around, including talking with David Singmaster, probably the world's leading Cubological and Cubohistorical authority, and discovered that there *is* something to Gustafson's claim of priority. Not that it is likely that Ernö Rubik or Terutoshi Ishige ever heard of Gustafson. Nor is it by any means certain that Gustafson's Globe, had it been picked up by some toy company, would have been the overnight sensation that the Rubik-Ishige Magic Cube was. Still, though, it seems only fair to point out that people besides Rubik and Ishige had smelled some of the same alluring aromas in previous years, and for various reasons had not been able to arouse the interest of the world.

*   *   *

It is one of my firmest beliefs that good ideas almost never come out of nowhere, and that if a good idea arises in one person's mind, it is almost sure to have arisen in someone else's mind in some closely related version, or to do so very shortly. For that reason, whenever I am writing about a discovery or invention, I always try hard to indicate multiple credit when I can discover the people to whom genuine credit is due. The trouble is, when you bend over backwards to be equitable (notice how I always talk about Ishige and Rubik in the same breath, for instance), what inevitably happens is that someone you inadvertently slighted then writes you with some mixture of indignation, consternation, and disappointment, and requests equal time.

I am glad to mention Gustafson's name and to give him credit for having had perhaps the world's first insight into this kind of three-dimensional rotational puzzle. But at the same time, I do not wish to leave out the names of Frank Fox, a British inventor who in about 1970 discovered—and patented—a $3 \times 3 \times 3$ twisting sphere, and Larry Nichols of Cambridge, Massachusetts, who invented and patented a $2 \times 2 \times 2$ cube around 1972, and who has just won a suit against Ideal Toy Co. for not giving him royalties on his invention. Whether Nichols' claim is any more deserving of retroactive compensation than those of Fox or Gustafson, I am not competent to say.

All I can say is, these things get very, very messy—particularly when large amounts of money (or glory) are concerned. In the case of all three inventors—Gustafson, Fox, and Nichols—it seems clear that their inventions were far flimsier than the Rubik-Ishige Cube, and that the real reason the Rubik-Ishige Cube took off was that it *could* be manufactured and that it *did* hold together. But perhaps I am wrong. Perhaps it was a fluke of some sort that allowed Rubik (as contrasted with Ishige, for example) to get most of the credit. But whatever the case, it does illustrate my belief that people are extremely eager to attribute credit—even glory—to just one person, and to vastly simplify a historical situation in order to be able to label it and classify it in their minds.

Who is willing to take the trouble to sift through all the murk surrounding such monumental discoveries as relativity, Gödel's incompleteness theorem, digital computers, lasers, pulsars, the cosmic background radiation, or the structure of DNA? Who wants to track down all the complexly tangled threads of ideas that somehow led to one or two people getting all the glory? Almost without exception, if you dig deep, you will find that the way the credit is conventionally apportioned is unfair. Sometimes entirely the wrong person gets all the credit, sometimes several unknown people deserve to share the credit, and sometimes the story is even more complex and twisty than that. Somebody should write a book on bizarre cases of credit attribution!

But my point is simply that with the cube, as with anything that has made a big hit, the world sees but the very tip of the iceberg, and someone in my position, who receives a lot of mail on these matters, sees only a bit below the tip. There is a lot more buried out there, and I am likely to get more letters from people who, upon reading my current attempt to be fair (in other words, this *Post Scriptum*), will feel *especially* slighted, given that I am trying to be fair and yet somehow failed to mention *their* names! Ah, me, what can you do?

* * *

In the intervening time, I have not heard of any faster algorithm for solving the Cube than Morwen Thistlethwaite's (described in Chapter 14). His algorithm, originally known to solve the Cube in at most 52 turns, has now been slightly improved on, thanks to computer searches. It is now known that 50 turns always suffice, confirming a conjecture that Thistlethwaite himself had made several years ago.

Although this improvement on Thistlethwaite's algorithm does not necessarily bring us appreciably closer to God's algorithm for the full $3 \times 3 \times 3$ Cube, God's algorithm *is* now known for two important smaller puzzles: the $2 \times 2 \times 2$ cube and Mèffert's Pyraminx. Curiously, both of them require the same number of turns at worst: eleven (disregarding the trivial turns of the tips of the Pyraminx). The distribution of positions according to their distance from *START* is quite interesting. Here it is for the Pyraminx, as supplied to me by John Francis of Nutmeg, New Hampshire and Louis Robichaud of Sainte Foy, Québec:

1 configuration requires 0 moves
8 configurations require 1 move
48 configurations require 2 moves
288 configurations require 3 moves
1,728 configurations require 4 moves
9,896 configurations require 5 moves

                51,808 configurations require 6 moves
               220,111 configurations require 7 moves
               480,467 configurations require 8 moves
               166,276 configurations require 9 moves
                 2,457 configurations require 10 moves
                    32 configurations require 11 moves.

Thus if *START* is at the "North Pole" of the space of all Pyraminx states, there are 32 different "South Poles", all maximally distant from it, and by far the bulk of the population lives below the equator.

By contrast, the $2 \times 2 \times 2$ cube has 2,644 states at maximal distance (eleven) from *START*. (In this metric, R^2 counts as just *one* move, rather than two.) Just as with the Pyraminx, the typical distance to *START* tends to be close to the maximum distance, but that tendency is exaggerated even more in the $2 \times 2 \times 2$. In particular, more than half the scrambled states require at least nine turns—and yet, ten turns will suffice to reach over 99.9 percent of all states! Here is the corresponding table:

                     1 configuration requires 0 moves
                     9 configurations require 1 move
                    54 configurations require 2 moves
                   321 configurations require 3 moves
                 1,847 configurations require 4 moves
                 9,992 configurations require 5 moves
                50,136 configurations require 6 moves
               227,536 configurations require 7 moves
               870,072 configurations require 8 moves
             1,887,748 configurations require 9 moves
               623,800 configurations require 10 moves
                 2,644 configurations require 11 moves.

This information comes from the autumn-winter 1982 double issue of Singmaster's *Cubic Circular,* and was apparently computed in several places around the world.

*   *   *

In Chapter 14, I described the game of inverting a handful of twists made on a pristine Cube, and mentioned that Kate Fried could regularly invert seven and once had undone ten. Peter Suber (the inventor of Nomic—see Chapter 4) calls this challenge the "inductive game", and has mastered it to the same level as Kate Fried did. He has written a short article describing this art, called "Introduction to the Inductive Game of Rubik's Cube". In it he explains why he calls it that:

The normal game is inductive only in the process a player undergoes in discovering the algorithms sufficient for solution. That process has been said to model the scientific method, complete with the formulation and testing of theories, negative results, and confirmation. The 'inductive game' is inductive in that way and more. The process of discovering the rules of mastery is similarly inductive; but the product is also inductive. Instead of producing algorithms that may be applied infallibly by an idiot, the inductive game produces 'soft rules' or probabilistic guides that must be applied in each case with judgment, mother wit, and the weight of one's inductive experience . . . .

The inductive game cannot become routine or boring, except perhaps to gods. When one can solve three-twist randomizations nearly 100 percent of the time, then one may move on to four-twist randomizations. Difficulty increases exponentially. There is a foreseeable end to the series, of course. Players who patiently gather up their nuanced, ineffable knowledge of random patterns may reach 22-twist randomizations. Improvement does not merely approach the banal satisfaction of more frequent success; it approaches hard knowledge of God's algorithm.

In the rest of his article, Suber details the results of his researches into this domain and comes up with many hints and heuristics based on his notion of *information*, defined as: "the adjacency of two or more tiles of the same color that need not (and ought not) be separated on the shortest path home". His basic guidelines (not to be interpreted overly rigidly) are:

(1) Thou shalt not break up information.
(2) Thou shalt endeavor to make more information.

The catch is that many configurations give a false impression of containing information. Suber calls this *apparent* information as distinguished from *actual* information, and a large part of his article is devoted to hints for telling the two apart. Readers interested in obtaining a copy of his article may write to Suber at the Department of Philosophy, Earlham College, Richmond, Indiana 47374.

* * *

There is something tantalizing about the idea of precisely reversing a scrambling. Suppose you could undo *any* scrambled state, and that one time the resulting twist-sequence was found to be, say, $UR^{-1}D^2LBLDR^{-1}F^2ULD^{-1}BR^{-1}U^2L^{-1}DF$. Would you be able to take this sequence apart and see any comprehensible structure there? That is, would there be some recognizable pieces inside it that explained how it undid that particular configuration?

Another way of asking the same question is perhaps more compelling. Some of my standard operators for getting things done on the Cube have the form of commutators, conjugates, powers, or combinations of such things. For such operators, I pretty much understand why they flip edges or do whatever they do. However, there are a couple of operators in my repertoire that I've simply memorized without having *any* understanding of why they accomplish what they do. For example, could it be that there's simply *no explanation* why $R^{-1}D^{-1}RD^{-1}R^{-1}D^2R$ undoes three quarks on the bottom layer? Could it be that there is, in other words, no conceptual breakdown to this operator? Such a sequence would resemble a very, very large prime number, a structure that admits of no breakdown into smaller "chunks".

It seems almost certain that the shortest routes home from most scrambled states on the Cube will admit of no breakdown; in short, that most of the solutions given by God's algorithm are *random,* in the sense of having no internal rhyme or reason to them—very much like a sequence of tosses of a coin or die. (This concept of randomness is explained lucidly in the article "Randomness and Mathematical Proof" by Gregory Chaitin.) If this is the case, it would mean that after a certain point—most likely not far above ten twists—it will be a vain hope to try to undo a Cube state via the route that got you there.

Getting into a scrambled state and getting out of it are operations of different computational complexity, just as getting yourself into a tight parking space and getting yourself out of it are operations of different automobilistic complexity. It is easier to find routes out than routes in, even though there are the same number of each. (In this analogy, being well parked is the analogue of getting to *START,* and being out in the street is the analogue of being scrambled.) Clearly, there is something deeply asymmetric about such a situation, and the whole thing smells of the second law of thermodynamics, stating that entropy will tend to increase with time in a closed system.

These informal intuitions can be made somewhat more precise. George Marx, Eva Gajzágó, and Peter Gnädig of the Department of Atomic Physics, Eötvös University, in Budapest, Hungary have studied the Cube statistically in a paper called "The Universe of Rubik's Cube". To begin with, they define a face's "color vector" as an ordered set of six numbers, telling how many facelets on that face are red, orange, yellow, green, blue, and white, respectively. In *START,* the red face's color vector is thus $<9,0,0,0,0,0>$. After some scrambling, you will get color vectors more like this: $<2,0,1,3,1,2>$. Various numerical measures of any face's "degree of scrambledness" can be derived from its color vector. The choice made by these authors is the "length" of this vector—that is, the square root of the sum of the squares of its "sides". For $<9,0,0,0,0,0>$, that comes out as 9, while for the more typical $<2,0,1,3,1,2>$, it is about 4.36. The shortest possible color vector consists of three 1's and three 2's, and has length just under 4.

Marx, Gajzágó, and Gnädig studied the statistics of this quantity as the Cube was twisted randomly, and discovered that faces whose color vector has length 4.36 are the most common. Shorter or longer color vectors are quite infrequent. If you start out at length 9 (a pristine Cube), then with random twisting the length tends to decrease quickly to a bit under 5, and then to fluctuate around that value. This observation is their empirical formulation of the second law of thermodynamics, establishing an "arrow of time".

In accordance with standard usage in statistical mechanics, they define the *entropy* of a Cube face's state as *the logarithm of the number of states that have the same macroscopic description*—in this case, the same color vector (allowing rearrangements, so that $<2,0,1,3,1,2>$ would be considered the same as $<2,1,3,2,0,1>$). Then they show that standard formulas that apply to entropy in real-world cases also apply to this "Cubical entropy". In particular, they remark: "The distribution of the colored squares on a mixed-up cube can be described in a similar way to how Maxwell and Boltzmann described the distribution of energy in the molecular chaos of a gas." At the conclusion of their article, Marx, Gajzágó, and Gnädig wax lyrical: "I honor the cube as the *smallest non-trivial model* of the *great physical universe.*" (italics theirs). (I suppose that when three authors jointly describe themselves as "I", it is a case of "the editorial 'I'".)

*   *   *

During the Cube's peak popularity, a large number of speed tournaments were held around the world and eventually a world champion emerged. He is Minh Thai, formerly of Viet Nam, now resident in the United States. His winning time on a scrambled Cube was 22.95 seconds. His average time seems to hover around 24 seconds, ranging as far upwards as 25 once in a while. Which leads me to ask: Shouldn't he perhaps have been named Minh Time?

There were also tournaments for the $4 \times 4 \times 4$ cube, and there the best times I heard of were in the three-minute range. Uwe Mèffert sent me a $5 \times 5 \times 5$ cube, which I must confess I never dared to scramble. I wonder how long the world champion would take on *that*! Mèffert once described to me his dream of a "Magic Triathlon", in which participants would have to unscramble a trio of scrambled solids—as I recall, the objects involved were the Pyraminx, the Impossi*Ball, and the Megaminx (Mèffert's revised name for his Pyraminx Magic Dodecahedron—see Figure 15-2*d*). My choices for the solids involved would have been different, but I liked the basic idea. I do not know if such an event ever took place.

I see no reason why harder events could not be created, involving such esoteric skills as manipulating an *N*-dimensional cube represented in a computer, such as H. R. Kamack and T. R. Keane's Magic Tesseract. These two gentlemen, implementors of a $3 \times 3 \times 3 \times 3$ hypercube on a home computer, not only solved the "basic mathematical problem" for this horrendous pseudo-object, but also calculated the size of the group for the $3 \times 3 \times \ldots \times 3 = 3^N$ hypercube, or what they call a "Rubik *N*-tope". For $N = 5$, the size of this group is (approximately) 7.017×10^{560}—a number not to be sneezed at!

On Crossing the Rubicon

According to Singmaster, mathematicians Joe Buehler, Brad Jackson, and Dave Sibley studied the 3^N hypercube as well, and came up with a general algorithm for it, as well as various conservation laws for it. The even *more* general case of the $M \times M \times M \times \ldots \times M = M^N$ hypercube remains unsolved, but it particularizes (along another conceptual dimension) to the $M \times M \times M$ cube. Professor Jack Eidswick of the Department of Mathematics and Statistics at the University of Nebraska sent me an article that presents an algorithm for solving any member of this family of three-dimensional cubes. It is based on elaborate versions of some of the necessary operators described in Chapter 14, built out of conjugates and commutators and the like. I hear that Robert Brooks of the Mathematics Department of the University of Maryland also has worked out such an algorithm.

*　　*　　*

I finally must confront the matter of the cube fad's fading. David Singmaster's *Cubic Circular* is going under after Volume 8. Many thousands of Megaminxes were melted down for their plastic. Uwe Mèffert's puzzle club seems to have been a flop. The Skewb and many other wonderful objects I described never hit the stands. A few that did were almost immediately gone forever. So... have we seen the last of the Magic Cube? Are those cubes you bought going to be collector's items? Well, I am always loath to predict the future, but in this case I will make an exception. I am bullish on the cube. It seemed to seize the imagination wherever it went. Despite the line concluding my second cube column, the cubic fad finally *did* spill over into the Soviet Union.

In my opinion, the world simply overdosed on cube-mania for a while. We humans are now collectively sick of the cube, but our turned-off state won't last too long—no more than it lasts when you tell yourself "I'll never eat spaghetti again!" after gorging on it. I predict that cubes will resurface slowly, here and there, and I am even hopeful that some new varieties will appear now and then. This is Mother Lode country. There may never again be quite the Gold Rush that we witnessed a couple of years ago, but there's still plenty of gold in them thar hills!

16

Mathematical Chaos and Strange Attractors

November, 1981

> *You can't know how happy I am that we met,*
> *I'm strangely attracted to you.*
>
> —Cole Porter, "It's All Right with Me"

A few months ago, while walking through the corridors of the physics department of the University of Chicago with a friend, I spotted a poster announcing an international symposium titled "Strange Attractors". My eye could not help but be strangely attracted by this odd term, and I asked my friend what it was all about. He said it was a hot topic in theoretical physics these days. As he described it to me, it sounded quite wonderful and mysterious.

I gathered that the basic idea hinges on looking at what might be called "mathematical feedback loops": expressions whose output can be fed back into them as new input, the way a loudspeaker's sounds can cycle back into a microphone and come out again. From the simplest of such loops, it seemed, both stable patterns and chaotic patterns (if this is not a contradiction in terms!) could emerge. The difference was merely in the value of a single parameter. Very small changes in the value of this parameter could make all the difference in the world as to the orderliness of the behavior of the loopy system. This image of order melting smoothly into chaos, of pattern dissolving gradually into randomness, was exciting to me.

Moreover, it seemed that some unexpected "universal" features of the transition into chaos had recently been unearthed, features that depended solely on the presence of feedback and that were virtually insensitive to other details of the system. This generality was important, because any mathematical model featuring a gradual approach to chaotic behavior might provide a key insight into the onset of turbulence in all kinds of physical systems. Turbulence, in contrast to most phenomena successfully understood in physics, is a *nonlinear* phenomenon: two solutions to the equations of turbulence do not add up to a new solution. Nonlinear mathematical phenomena are much less well understood than linear ones, which is why a good mathematical description of turbulence has eluded physicists for a long time, and would be a fundamental breakthrough.

When I later began to read about these ideas, I found out that they had actually grown out of many disciplines simultaneously. Pure mathematicians had begun studying the iteration of nonlinear systems by using computers. Theoretical meteorologists and population geneticists, as well as theoretical physicists studying such diverse things as fluids, lasers, and planetary orbits, had independently come up with similar nonlinear mathematical models featuring chaos-pregnant feedback loops and had studied their properties, each group finding some quirks that the others had not found. Moreover, not only theorists but also experimentalists from these widely separated disciplines had simultaneously observed chaotic phenomena that share certain basic patterns. I soon saw that the simplicity of the underlying ideas gives them an elegance that, in my opinion, rivals that of some of the best of classical mathematics. Indeed, there is an eighteenth- or nineteenth-century flavor to some of this work that is refreshingly concrete in this era of staggering abstraction.

Probably the main reason these ideas are only now being discovered is that the style of exploration is entirely modern: it is a kind of experimental mathematics, in which the digital computer plays the role of Magellan's ship, the astronomer's telescope, and the physicist's accelerator. Just as ships, telescopes, and accelerators must be ever larger, more powerful, and more expensive in order to probe ever more hidden regions of nature, so one would need computers of ever greater size, speed, and accuracy in order to explore the remoter regions of mathematical space. By the same token, just as there was a golden era of exploration by ship and of discoveries made with telescopes and accelerators, characterized by a peak in the ratio of new secrets uncovered to money spent, so one would expect there to be a golden era in the experimental mathematics of these models of chaos. Perhaps this era has already occurred, or perhaps it is occurring right now. And perhaps after it, we will witness a flurry of theoretical work to back up these experimental discoveries.

In any case, it is a curious and delightful brand of mathematics that is being done. This way of doing mathematics builds powerful visual imagery and intuitions directly into one's understanding. The power of computers allows one to bypass the traditional "theorem-proof-theorem-proof" brand of mathematics, and to arrive quickly at *empirical* observations and discoveries that reinforce each other, and that form a rich and coherent network of results. In the long run, it may turn out to be easier to find proofs of these results (if proofs are still desired), thanks to the careful and thorough exploration of the conceptual territory in advance. It's an upstart's way of doing mathematics, and not all mathematicians approve.

One of the strongest proponents of this style of mathematizing has been Stanislaw M. Ulam, who, when computers were still young, turned them loose on problems of nonlinear iteration as well as on problems from many other branches of mathematics. It is from Ulam's early studies with Paul Stein that many of the ideas to be sketched here follow.

$$* \quad * \quad *$$

So much for romance. Let us work our way up to the concept of "strange attractors" by beginning with the more basic concept of an *attractor*. This whole field is founded on one concept: the iteration of a real-valued mathematical function—that is, the behavior of the sequence of values x, $f(x), f(f(x)), f(f(f(x))), \ldots$, where f is some interesting function. The initial value of x is called the *seed*. The idea is to feed f's output back into f as new input over and over again, to see if some kind of pattern emerges.

An interesting and not too difficult problem concerning the iteration of a function is this: Can you invent a function p with the property that for any real value of x, $p(x)$ is also real, and where $p(p(x))$ equals $-x$? The condition that $p(x)$ be real is what gives the problem a twist; otherwise the function $p(x) = ix$ (where i is the square root of -1) would work. In fact, you can even think of the challenge as that of finding a real-valued "square root of the minus sign". A related problem is to find a real-valued function q, whose property is that $q(q(x)) = 1/x$ for all x other than zero. Note that no matter how you construct p and q, each will have the property that, given any seed, repeated iteration creates a cycle of length four.

Now, more generally, what kinds of functions, when repeatedly iterated, are likely to exhibit interesting cyclic or near-cyclic behavior? A simple function such as $3x$ or x^3, when iterated, does not do anything like that. The nth iteration of $3x$, for example, is $3 \times 3 \times 3 \times \ldots \times 3x$, with n 3's—that is, $3^n x$—and the nth iteration of x^3 is just $(((x^3)^3)^3)^{\cdots 3}$ with n 3's again, which amounts to x^{3^n}. Nothing cycle-like here; the values just keep going up and up and up. To reverse this trend, one needs a function with some sort of switchback—a little zigzag or twist. A more technical way of putting it is that one needs a *nonmonotonic* function: a function whose graph is folded—that is, it starts moving one way—say upward—and then bends back the other way—say downward.

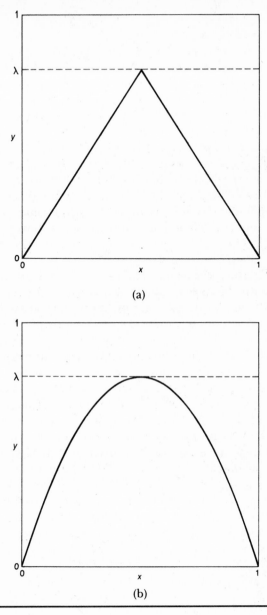

(a)

(b)

FIGURE 16–1. *Two nonmonotonic, or "folded", functions in the unit square. In (a), a sharp peak, and in (b), a parabola. The maximum height of both is defined by the parameter* λ.

In Figure 16-1*a*, we have a sawtooth with a sharp point at its top, and in Figure 16-1*b*, a smoothly bending parabolic arc. Each of them rises from the origin, eventually reaches a peak height called λ, and then comes back down for a landing on the far side of the interval. Of course there are uncountably many shapes that rise to height λ and then come back down, but these two are among the simplest. And of the two, the parabola is perhaps the simpler, or at least the more mathematically appealing. Its equation is $y = 4\lambda x(1-x)$, with λ not exceeding 1.

We allow input (values of *x*) only between 0 and 1. As the graph shows, for any *x* in that interval, the output (*y*) always is between 0 and λ. Therefore the output value can always be fed back into the function as input, which ensures that repeated iteration will always be possible. When you repeatedly iterate a "folded" function like this, the successive *y*-values you produce will sometimes go up and sometimes down—always hovering, of course, between 0 and λ. The fold in the graph guarantees interesting effects when the function is iterated—as we shall see.

It turns out that the spectacular differences in the degree of regularity of patterns I mentioned above are due to variations in the setting of what we might call the "λ-knob". Depending on the value the knob is set at, the function yields an incredible variety of "orbits"—that is, sequences $x, f(x)$, $f(f(x))$, and so on. In particular, for λ below a certain critical value $\lambda_c = 0.892486417967\ldots$, the orbits are all regular and patterned (although there are various degrees of patternedness; generally the lower λ is, the more simply the orbit is patterned), but for λ at or beyond this critical value, hold onto your hat! An essentially chaotic sequence of values will be traced out by the values $x, f(x), f(f(x)), \ldots$, no matter what positive seed value of *x* you choose. In the case of the parabola, the critical role played by varying the λ-knob seems to have been first realized by P. J. Myrberg in the early 1960's, but his work was published in a little-known journal and did not attract much attention. Some ten years later, Nicholas C. Metropolis, Paul Stein, and Myron Stein rediscovered the importance of the knob not only for the parabola but also for many other functions. Indeed, they discovered that as far as certain topological properties were concerned, the function did not matter—only the value of λ did. This property has come to be called "structural universality".

* * *

In order to see how such a nonintuitive dependence on the setting of the λ-knob comes about, one must develop a visual sense for the process of iterating $f(x)$. This is readily done. Suppose we set λ to 0.7. The graph of $f(x)$ appears in Figure 16-2. In addition, the line $y = x$ appears as a 45-degree broken line. (This graph and most of the others in this article were produced on a small computer by Mitchell J. Feigenbaum of the Los Alamos National Laboratory.)

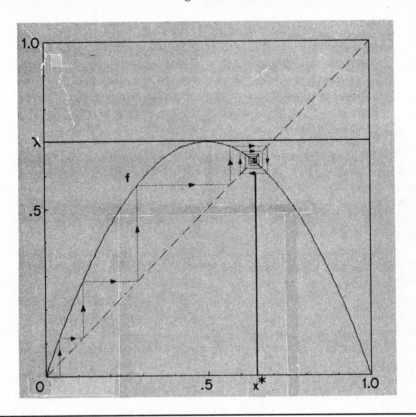

FIGURE 16–2. *The parabola defined by "λ-knob" setting of 0.7. An initial x-value of about 0.04 is used as a "seed" for iteration, and the pathway taken is shown. Eventually it settles down at a fixed point, denoted by x*.*

Consider the two x-values where the 45-degree line and the curve intersect. They are at $x=0$ and $x=9/14=0.643$. Let us designate the nonzero value as x^*. By construction, then, $f(x^*)$ equals x^*, and repeated iteration of f at this x-value will get you into an infinite loop. The same happens if you start iterating at $x=0$: you get stuck in an endless loop. However, there is a significant difference between these two *fixed points* of f. It is best indicated by taking some other initial value of x, say one close to 0.04, as is shown in the same figure. Call this starting x-value x_0. There is an elegant graphical way to generate the orbit of any seed x_0. A vertical line at x-value x_0 will hit the curve at height $y_0=f(x_0)$. To iterate f, we must draw a new vertical line located at the new x-value equal to this y-value. This is where the 45-degree line $y=x$ comes in handy. Staying at height y_0, we simply move over horizontally until we hit that 45-degree line. Then, since along this line y equals x, both x and y equal y_0. Let us call this new x-value x_1. We now draw a second vertical line. This one will hit the curve at height $y_1=f(x_1)=f(y_0)=f(f(x_0))$. Now we just repeat.

In brief, iteration is realized graphically by a simple recipe:

(1) Move vertically until you hit the curve; then
(2) Move horizontally until you hit the diagonal line.
Repeat steps (1) and (2) over and over again.

The results of this procedure with seed $x_0 = 0.04$ are also shown in Figure 16-2. We are led in a merry chase 'round and 'round the point whose x-coordinate and y-coordinate are x^*. Gradually we close down on that point. Thus x^* is a special kind of fixed point, because it attracts iterated values of $f(x)$. It is the simplest example of an *attractor:* every possible seed (except 0) is drawn, through iteration of f, to this stable x-value. This x^* is therefore called an *attractive* or *stable* fixed point. By contrast, 0 is a *repellent* or *unstable* fixed point, since the orbit of any initial x-value, even one infinitesimally removed from 0, will proceed to move away from 0 and toward x^*. Note that sometimes the iterates of f will overshoot x^*, sometimes they will fall short—but they inexorably draw closer to x^*, zeroing in on it like swallows returning to Capistrano. One might also think of such familiar and charming metaphors of prey-seekers as heat-seeking missiles, mosquitos, bloodhounds, Nazi-hunters, sharks, and lastly, the children's rhyme, "Around the world, and around the world, goes a big bear; he bores a hole, and he bores a hole, right . . . in . . . *there*!"

What accounts for this radical qualitative difference between the two fixed points (0 and x^*) of f? A careful look at Figure 16-2 will show that it is the fact that at 0, the curve is sloped too steeply. In particular, the slope there is greater than 45 degrees. It is the local slope of the curve that controls how far you move horizontally each time you iterate f. Whenever the curve is steeper than 45 degrees (either rising or falling) it tends to pull you farther and farther away from your starting point as you repeatedly iterate by rules (1) and (2). Hence the criterion for the stability of a fixed point is: The slope at the fixed point should be less than 45 degrees. Now, this is the case for x^* when λ equals 0.7. In fact, the slope there is about 41 degrees, whereas at 0 it is much greater than 45 degrees.

What happens if we increase λ? The position of x^* (x^* being by definition the x-coordinate of the point where the curve f and the line $y = x$ intersect) will change, and the slope of f at x^* will increase as well. What happens when the slope hits 45 degrees or exceeds it? This occurs when λ is 3/4. We will call this special value of the λ-knob Λ_1. Let us look at the graph for a slightly greater λ-knob setting, namely $\lambda = 0.785$. (See Figure 16-3.)

What if we begin with some random seed instead, again say $x = 0.04$? The resulting orbit is shown in Figure 16-3a. As you can see, a very pretty thing happens. At first the values move up toward the vicinity of x^* (now an unstable fixed point of f), but then they spiral gradually outward and settle

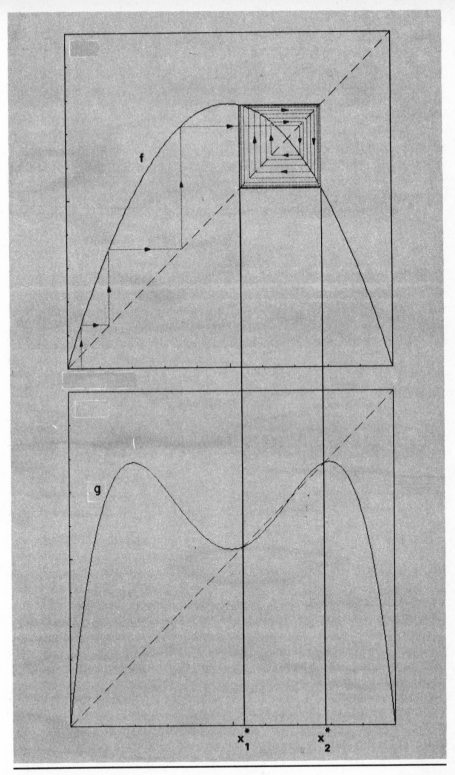

FIGURE 16–3. *Here, the λ-knob has been twisted a little higher: now it is set at 0.785. On top, f itself is shown, and below, f's two-humped iterate g is shown. Now a typical seed will eventually settle into a two-step square dance, asymptotically jumping back and forth between values* x_1^* *and* x_2^*.

down smoothly to a kind of "square dance" converging on two special values x_1^* and x_2^*. This elegant oscillation is called a *2-cycle*, and the pair of x-values that constitute it (x_1^* and x_2^*) is again an attractor—in particular, an attractor of period two. This term means that our 2-cycle is stable: it attracts x-values from far and wide as f is iterated. The orbit for any positive seed value (except x^* itself) will eventually fall into the same dance. That is, it will asymptotically approach the perfect 2-cycle composed of the points x_1^* and x_2^*, although it will never quite reach it exactly. From a physicist's point of view, however, the accuracy of the approach soon becomes so great that one can just as well say that the orbits have been "trapped" by the attractor.

An enlightening way to understand this is to look at a graph of a new function made from the old one. Consider the graph of $g(x) = f(f(x))$, shown in Figure 16-3*b*. This two-humped camel is called the *iterate* of *f*. First of all, observe that any fixed point of f is also a fixed point of g, so that 0 and x^* will be fixed points of g. But secondly observe that since $f(x_1^*)$ equals x_2^*, and conversely $f(x_2^*)$ equals x_1^*, g will have two new fixed points: $g(x_1^*) = x_1^*$ and $g(x_2^*) = x_2^*$. Graphically, x_1^* and x_2^* are easily found: they are intersection points of the 45-degree line with the two-humped graph of g (x). There are four such points (0 and x^* being the other two). As we have seen, the criterion for the stability of any fixed point under iteration is that the slope at that point should be less than 45 degrees. Here we are concerned with fixed points of g, and hence with g's slope (as distinguished from f's slope). Indeed, in the same figure, you can clearly see that at 0 and at x^*, g is sloped more steeply than 45 degrees, whereas at both x_1^* and x_2^*, g's slope is less than 45 degrees. In fact, quite remarkably, not only are both slope values less than 45 degrees, but also, as it turns out through a simple bit of calculus, they are equal (or "slaved" to each other, as it is sometimes put).

*　　*　　*

We have now seen an attractor of period one get converted into an attractor of period two at a special value of λ (namely, $\lambda = 3/4$). Precisely at that value, the single fixed point x^* splits into two oscillating values, x_1^* and x_2^*. Of course they coincide at "birth", but as λ increases, they separate and draw farther and farther apart. This increase of λ will also cause g's slope at these two stable fixed points (of g) to get steeper and steeper until finally, at some λ-value, g, like its progenitor f, will reach its own breaking point (*i.e.*, the identical slopes at both x_1^* and x_2^* will exceed 45 degrees), and each of these two attracting points will break up, spawning its own local 2-cycle. (Actually, the cycles are 2-cycles only as far as g is concerned; for f, the new points are elements of an attractor of period four. You must be careful to keep f and g straight in your mind!) These two splittings will happen at exactly the same "moment" (*i.e.*, at the same λ-knob setting), since the value of the slope of g at x_1^* is slaved to the value of the slope at x_2^*. This λ-knob setting will be called Λ_2, and it has the value of 0.86237 . . .

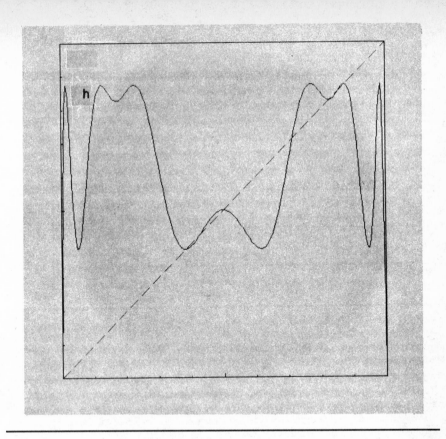

FIGURE 16–4. *A picture of f's iterate's iterate h at a still higher value of λ, namely 0.87.*

Here, as with a joke, you may anticipate the punch line by the time you have heard the theme and one variation. Hence by now you have probably surmised that at some new value Λ_3, all four points in f's attractor will simultaneously fission, yielding a periodic attractor consisting of eight points; and thereafter this pattern will go on and on, doubling and redoubling as various special λ-knob settings are reached and passed. If this is your guess, you are quite right, and the underlying reason is the same each time: the (identical) slopes at all the stable fixed points of some graph reach the critical angle of 45 degrees. In the case of the first fission (at Λ_1) it was the slope of f itself at the single point x^*. The next fission was due to the slopes at g's two stable fixed points x_1^* and x_2^* simultaneously reaching 45 degrees. Analogously, Λ_3 is that value of λ at which the slope of $h(x) = g(g(x)) = f(f(f(f(x))))$ hits 45 degrees simultaneouly at the four stable fixed points of h. And so it goes. Figure 16-4 shows the bumpy appearance of $h(x)$ at a λ-value of approximately 0.87.

In Figure 16-5, the locations on the x-axis of the stable fixed points of f are shown for Λ_1 through Λ_6 (by which time there are 32 of them, some clustered so closely that they cannot be distinguished). The points are pictured just at the moment of their becoming unstable, each one like a cell

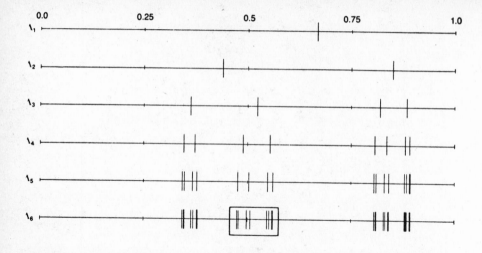

FIGURE 16–5. *Showing how stable attractors become unstable and undergo "fission" at a series of increasing λ-values, denoted Λ_n for n=1, 2, 3,... Note how the boxed subpattern on the lowest line resembles the entire pattern two lines above. This resemblance becomes more and more accurate the larger n gets.*

FIGURE 16–6. *A graph showing the evolution of attractors as λ increases from 0 to 1. Bifurcations begin at λ=0.75 and escalate towards chaos. The "chaotic region", beginning at λ=0.892...., shows unexpectedly beautiful fine structure. [From "Roads to Chaos" by Leo P. Kadanoff in* Physics Today, *December 1983 p.51; see also J. P. Crutchfield, D. Farmer, and B. A. Huberman,* Physics Reports, *Vol. 92, pp. 45–82, December, 1982.]*

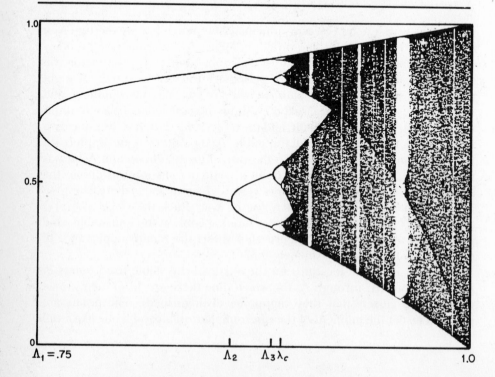

on the verge of division. Notice the neat pattern in the distribution of the attracting points. Looking at these graphs of the spacings of the elements of the successive period-doubled attractors of *f*, you can see that each line can be made from the one above it through a recursive geometric scheme whereby each point is replaced by two "twin" points below it. Each local clustering pattern of points echoes the global clustering pattern, simply reduced in scale (and also, in alternating local clusters, left and right are reversed). For example, in the bottom line a local group of eight points has been outlined in color. Notice how the group of points is like a miniature version of the global pattern two lines above it.

The discovery of this recursive regularity, first made on a little calculator by Feigenbaum, is one of the major recent advances in the field. It states in particular that to make line $n+1$ from line n, you simply let each point on line n give birth to "twins". The new generation of points should be packed in about 2.5 times more densely than the old generation was. More exactly stated, the distance between new twins should be α times smaller than the distance between their parent and its twin, where α is a constant, approximately equal to 2.50290787750958928485 . . . This rule holds with greater and greater accuracy the larger n becomes.

What about the values of the Λ's? Are they headed asymptotically toward 1? Surprisingly enough, no. These Λ-values are quickly converging on a particular critical value λ_c, of size roughly 0.892486418 . . . And their convergence is remarkably smooth, in the sense that the distance between successive Λ's is shrinking geometrically. More precisely, the ratio $(\Lambda_n - \Lambda_{n-1})/(\Lambda_{n+1} - \Lambda_n)$ approaches a constant value called δ by Feigenbaum, its discoverer, but more often referred to simply as "Feigenbaum's number" by others. Its value is approximately 4.66920160910299097 . . .

In short, as λ approaches λ_c, at special λ-values predicted by Feigenbaum's constant δ, *f*'s attractor doubles in population, and its increasingly many elements are geometrically arranged on the *x*-axis according to a simple recursive plan, the main determining parameter of which is Feigenbaum's other constant, α.

Then for λ beyond λ_c—called the *chaotic regime*—the results of iterating *f* can, for some seed values, yield orbits that converge to no finite attractor. These are *aperiodic* orbits. For most seed values, the orbit will remain periodic, but the periodicity will be very hard to detect. First of all, the period will be extremely high. Secondly, the orbit will be much more chaotic than before. A typical periodic orbit, instead of quickly converging to a geometrically simple attractor, will meander all over the interval [0,1], and its behavior will appear indistinguishable from total chaos. Such behavior is termed *ergodic*. Furthermore, neighboring seeds may, within a very small number of iterations, give rise to utterly different orbits. In short, a *statistical* view of the phenomena becomes considerably more reasonable beyond λ_c.

Figure 16-6 beautifully portrays the period-doubling route to chaos, as well as what happens after you've gotten there. The bifurcations are clear to the eye, and since the horizontal distance from each set of them to the next shrinks geometrically, the onset of chaos at λ_c is plainly visible. But the regularity of the structure to the *right* of λ_c—that is, in the chaotic regime—is quite unexpected. It is certain that there are many deep mathematical secrets locked up in this elegant graph.

*   *   *

Now, what do such novel concepts as the iteration of folded functions, period doubling, chaotic regime, and so on have to do with the study of turbulence in hydrodynamic flow, the erratic population fluctuations in predator-prey relations, and the instability of laser modes? The basic idea is embedded in the contrast between laminar flow and turbulent flow. In a peacefully flowing fluid, the flow is *laminar*—a soft and gentle word that means that all the molecules in the fluid are moving like cars on a multilane freeway. The key features are: (1) that each car follows the same path as its predecessor, and (2) that two nearby cars, whether they are in the same lane or in different ones, will, as time passes, slowly separate from each other— essentially in proportion to the difference in their velocities—which is to say, linearly. These features also apply to molecules of fluid in laminar flow; there, the lanes are called *streamlines* or *laminas*.

By contrast, when a fluid is churned up by some external force, this smooth behavior turns into turbulent behavior, as is seen in breakers at the beach and cream being stirred into coffee. Even the word "turbulent" sounds much harsher and more angular than the soft word "laminar". Here, the image of a multilane freeway no longer holds; the streamlines separate from each other and tangle in the most convoluted of ways, as shown in Figure 16-7. In such systems there are eddies and vortices and all sorts of unnamable whorls on many size-scales at once, and consequently, two points that were initially very close may soon wind up in totally different regions of the fluid. Such quickly diverging paths are the hallmark of turbulence. The distance between points can increase exponentially with time, instead of just linearly, and the coefficient of time in the exponent is called the *Lyapunov number*. When one speaks of chaos in turbulent flow, it is this rapid, nearly unpredictable separation of neighbors that is meant. Such behavior is strikingly reminiscent of the rapid separation, in the chaotic regime of λ, of two orbits whose seeds might originally have been very close together.

FIGURE 16–7. *Showing the approach to turbulence. In the upper two pictures, a rod was drawn through a viscous liquid once, setting up trains of vortices behind it. In the lower two, the rod was drawn more than once, and the forms are therefore more complicated and recursive-seeming. It is provocative to compare this figure with Figure 13-4.* [*From* Sensitive Chaos, *by Theodor Schwenk.*]

This suggests that the "scenario" (as it is called) by which pretty, periodic orbits gradually give way to' the messy, chaotic orbits of our parabolic function might conceivably be mathematically identical to the scenario underlying the transition to turbulence in a fluid or other system. Exactly how this connection is established, though, requires some more detailed setting of context. In particular, we must briefly consider how the spatio-temporal flow of a fluid or some other entity, such as population density or money, is mathematically modeled.

In such real-world problems, the most successful equations yet found to model the phenomena are *differential equations*. A differential equation connects the continuous rate of variation of some quantity to that quantity's current size and the current sizes of other quantities. Moreover, the time variable is itself continuous, not jerking from one discrete instant to the next as some strange clocks and watches occasionally do, but indivisibly flowing, like a liquid. One way to visualize the patterns defined by differential equations is to imagine a multidimensional space—it could have thousands of dimensions, or merely a few—in which a point is continuously tracing out a curve. At any one moment, the single point contains all the information about the state of the physical system. Its projections along the various axes give the values of all the relevant quantities that pin down a unique state. Clearly the space—called *phase space*—would need to have an enormous number of dimensions for a mere point to store the entire shape of a wave breaking on a beach. On the other hand, in a simple predator-prey relation, only two dimensions suffice: one coordinate, say x, giving the predator population and the other, say y, giving the prey population. Two dimensions are more easily visualized, and so we will stick with that case for the time being. The ideas generalize, however, to higher-dimensional cases.

As time progresses, x and y determine each other in an intertwined manner. For example, a large population of predators will tend to reduce the population of prey, whereas a small population of prey will tend to reduce the population of predators. In such a system, x and y constitute a single point (x,y) that swirls around smoothly in a continuous orbit on the plane. (Here the sense of "orbit" is different from the preceding one—that of the discrete, or jumping, orbits we saw when our parabolic function was iterated.) One such possible orbit appears in Figure 16-8; it is generated by a differential equation called "Duffing's equation". It looks like the path of a buzzing fly in your bedroom—or rather, it looks like the *shadow* of the fly's path on a wall. As a matter of fact, this self-intersecting two-dimensional curve *is* the shadow of a non-self-intersecting three-dimensional curve. The motion of a point in phase space must *always* be non-self-intersecting. This arises from the fact that a point in phase space representing the state of a system encodes *all* the information about the system, including its future history, so that there cannot be two different pathways leading out of one and the same point.

In particular, in Duffing's equation there is a third variable, z, that I have

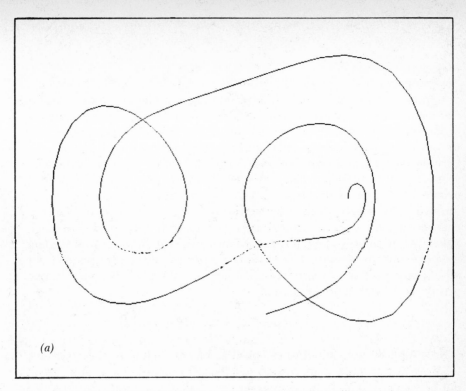

(a)

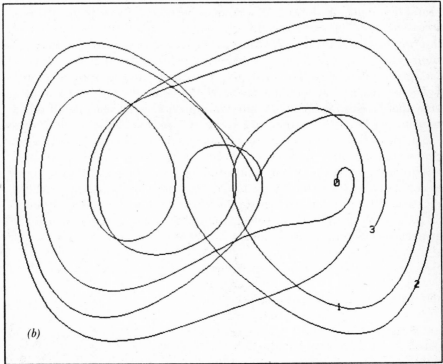

(b)

FIGURE 16–8. *If values of* x *and* y *mutually determine each other according to Duffing's equation as time passes, then the point* (x,y) *will trace out a curve* (a). *If a strobe light blinks periodically and shows* (x,y) *at selected instants, then a group of isolated points will start appearing as in* (b), *and gradually filling out a region of their own—a* Poincaré map.

not mentioned so far. If you think of x and y as representing predator and prey populations, then you can think of z as representing a periodically varying external influence, such as the sun's azimuth or the amount of snow on the ground. Now, if you will allow me to mix my buzzing-fly image with the predator-prey example, imagine a bedroom with a fly buzzing periodically back and forth between two walls. Let us say it takes the fly a year to cross the room and come back. (Perhaps it is a rather large bedroom, or maybe just a slow fly.) In any case, as the fly flies, its shadow on one of the two walls traces out the curve shown in Figure 16-8*a*. If the fly ever chances to come back to a point in the room which it has passed through before, it is doomed to loop forever, following the path it took the preceding time over and over again. This gives you a picture of the continuous orbit of a point in phase space representing the state of dynamic system controlled by differential equations.

*   *   *

Now suppose we wanted to establish some connection of these systems to *discrete* orbits. How might we do so? Well, the values of x, y, and z need not be watched at all moments; they can be sampled periodically, at some natural frequency. In the case of animal populations, a year is the obvious natural period. The sun's azimuth is exactly periodic, and the weather at least *tries* to repeat itself a year later. Thus a natural sequence of discrete points (x_1, y_1, z_1), (x_2, y_2, z_2), . . . can be singled out—one per year. It is as if a strobe light blinked regularly and froze the fly on special annual occasions—perhaps at midnight every Halloween. Or you can think of a firefly that flashes on for just a split second once every year. At all other times its peregrinations around the room are unseen. Figure 16-8*b* shows a sequence of discrete points along the fly-path's shadow, marked by numbers telling when they occurred. Gradually, as many "years" elapse, enough of these discrete points will accumulate that they will start to form a recognizable shape of their own. This pattern of points is a *discrete* "orbit", and so it is closely related to the discrete orbits defined by the iteration of our parabola $f(x)$. In that parabolic case, we had a simple one-dimensional recurrence relation (or an iteration):

$$x_{n+1} = f(x_n).$$

Here we have a *two*-dimensional recurrence:

$$x_{n+1} = f_1(x_n, y_n)$$

$$y_{n+1} = f_2(x_n, y_n)$$

This is a system of *coupled* recurrence relations, in which output values of the nth generation (x_n, y_n) are fed right back into f_1 and f_2 as new inputs, to produce the $n+1$st generation. On and on it goes, generation after generation. In higher-dimensional cases, of course, there are more such equations. Nevertheless, the skeleton of all these systems remains the same: a multidimensional point $(x_n, y_n, z_n, \ldots)$ jumps from one discrete location in phase space to another, as a discrete variable, n, representing time jumping ahead in discrete units, is incremented.

Notice that we have finessed our way around the continuous time variable that is involved in differential equations. We have done it by focusing on the way the point is connected to its predecessor one "year" earlier (or whatever natural period is involved). But is there always a "natural period" at which to look at a system of mutually intertwined differential equations? Not always. In some situations, however, there is, and this happens to be the case in all situations where turbulent behavior occurs.

Why is this so? All systems that exhibit turbulent behavior are *dissipative*, which means that they dissipate, or degrade, energy from more usable forms such as electricity into the less usable form of heat. In the case of hydrodynamic flow, this dissipation is caused by friction, and in the other systems we have been considering, by abstract analogues of friction. A familiar consequence of friction is that objects in motion will grind to a halt unless energy is pumped in. Now if we "drive" a dissipative system with a *periodic* driving force (you can imagine, for example, stirring a cup of coffee with a spoon in a periodic, circular way), then, of course, the system will not grind to a halt; it will head for some kind of steady state. Such a steady state is a stable orbit—or in our terms, an attractor in phase space. And since we have driven the system with a periodic spoon, we have defined a natural frequency at which to flash our strobe light and freeze the system's state— namely, each time the spoon comes swinging around and passes some fixed mark on the cup, such as its handle. This will constitute our "year". In this way, continuous time can be replaced by a series of discrete instants, as long as we are dealing with a dissipative system driven by a periodic force. And so continuous orbits can be replaced by discrete orbits, which brings iteration back into the picture.

If the driving force itself has no natural period (it may be simply a constant force), there is still a way to define a natural period, as long as some variable in the system swings back and forth between extremes. Just flash your strobe whenever that variable hits its extreme value, and the fly will still be caught at discrete instants. This type of discrete representation of the fly's motion in a multidimensional space is called a *Poincaré map*.

This stirring argument is only hand-waving, of course, and needs much more rigor to be convincing to a mathematician. It nonetheless gives the flavor of how the study of a set of coupled differential equations can be replaced by the study of a set of coupled discrete recurrence relations. This is the vital step that brings us back to the recent discoveries about the parabola.

*   *   *

In 1975, Feigenbaum discovered that his numbers α and δ do not depend on the details of the shape of the curve defined by $f(x)$. Almost any smooth convex shape that peaks in the same spot will do as well. Inspired by the structural universality discovered by Metropolis, Stein, and Stein, Feigenbaum tried working with a sine curve instead of a parabola. He was flabbergasted by the reappearance of the same numerical values, to many decimal places, of the numbers α and δ, which had characterized the period-doubling and the onset of chaos for the parabola. For the sine curve just as for the parabola, there is a height-parameter λ and a set of special λ-values that converge to a critical point λ_c. Moreover, the onset of chaos at λ_c is governed by the same numbers α and δ. Feigenbaum began to suspect that there was something universal going on here. In other words, he suspected that what is more important than f itself is the mere fact that f is being iterated over and over. In fact, he suspected that f itself might play no role in the onset of chaos.

It is not quite that simple, in reality. Feigenbaum soon discovered that what *does* matter about f is just the nature of the peak at its very center. The long-term behavior of orbits depends only on an infinitesimal segment at the crest of the graph, and ultimately, it depends only on the behavior at the very point where the maximum occurs! The rest of the shape, even the region close to the peak, is irrelevant. A parabola has what is called a *quadratic* maximum, as do a sine wave, a circle, and an ellipse. In fact, the behavior of a randomly-produced smooth function at a typical maximum would be expected to be of the quadratic type, in the absence of any special coincidences. So the parabolic case, rather than being a quirky exception, begins to seem like the rule. This empirical discovery by Feigenbaum, involving two fundamental scaling factors α and δ that characterize the onset of chaos through period-doubling attractors, represents a new kind of universality, known as *metrical* universality, to distinguish it from the earlier-known *structural* universality. This empirically demonstrated metrical universality was later proved to be correct (in the more orthodox sense of proof) in the one-dimensional case by Oscar Lanford.

A truly exciting development occurred when Feigenbaum's constants unexpectedly turned up in some messy models of actual physical systems that exhibit turbulence, not just in pretty and idealized mathematical systems. Valter Franceschini of the University of Modena in Italy adapted the Navier-Stokes equation, which governs all hydrodynamic flow, for computer simulation. To do so, he turned it into a set of five coupled differential equations whose Poincaré maps he could then study numerically on his computer. He first found that the system exhibited attractors with repeated period-doubling as its governing parameters approached the values where turbulence was expected to set in. Unaware of Feigenbaum's work, he showed his results to Jean-Pierre Eckmann of the University of Geneva, who immediately urged him to go back and determine the rate of convergence of the λ-values at which period-doubling occurred. To their amazement, Feigenbaum's α- and δ-values—accurate to about four decimal places—appeared seemingly out of nowhere! For the first time, an accurate mathematical model of true physical turbulence revealed that its structure was intimately related to the humble chaos lurking in the humble parabola $y = 4\lambda x(1-x)$. Subsequently, Eckmann, Pierre Collet, and H. Koch showed that in the behavior of a multidimensional driven dissipative system, all dimensions but one tend to drop out after a sufficiently long period of time, and so one should *expect* the characteristic of one-dimensional behavior— namely Feigenbaum's metrical universality—to reappear.

Since then, experimentalists have been keeping their eyes peeled for period-doubling behavior in actual physical systems (not just in computer models). Such behavior has been observed in certain types of convective flow, but so far the measurements are too imprecise to lend very strong support to the idea that the parabola contains the clues revealing the nature of genuine physical turbulence. Still, it is tantalizing to think that somehow, all that really matters is that a dissipative set of coupled recurrence relations is being iterated—but that the detailed properties of those recurrences can be entirely ignored if one is concentrating on understanding the route to turbulence.

Feigenbaum puts it this way. One often sees a pattern of clouds in the sky —a celestial trellis composed of a myriad of small white puffs stretching from horizon to horizon—that clearly did not happen "by accident". Some systematic hydrodynamic law has got to be operating. Yet, says Feigenbaum, it must be a law operating at a higher level, or on a larger scale, than the Navier-Stokes equation, which is based on infinitesimal volumes of fluid and not on large "chunks". It seems that in order to understand such beautiful sky patterns, one must somehow bypass the *details* of the Navier-Stokes equation, and come up with some coarser-grained but more relevant way of analyzing hydrodynamic flow. The discovery that iteration gives rise to universality—that is, independence of the details of the function (or functions) being iterated—offers hope that such a view of hydrodynamics may be well on its way to emerging.

*   *   *

Well, we have covered attractors and turbulence; what about *strange* attractors? We have now built up the necessary concepts to understand this idea. When a periodically driven two-dimensional (or higher-dimensional) dissipative system is modeled by a set of coupled iterations, the successive points lit up by the flashes of the periodic strobe light trace out a shape that plays the role, for this system, that a simple orbit did for our parabola. But when one is operating in a space of more than one dimension, the possibilities are richer. Certainly it is possible to have a stable fixed point (an attractor of period one). This would just mean that at every flash of the strobe, the point representing the system's state is exactly where it was last time. It is also possible to have a *periodic* attractor: one where after some finite number of flashes, the point has returned to a preceding position. This would be analogous to the 2-cycles, 4-cycles, and so on that we saw occurring for the parabola.

But there is another option: that the point never returns to its original position in phase space, and that successive flashes reveal it to be jumping around quite erratically inside a restricted region of phase space. Over a period of time, this region may take shape before an observer's eyes as the strobe flashes periodically. In the majority of such cases so far studied, a most unexpected phenomenon has been observed to take place: the erratically jumping point gradually creates a delicate filigree that recalls the "faint fantastic tracery made by frost on glass". (I owe this poetic image to the American critic James Huneker, who used it to describe the magical effect of one of Chopin's piano études: Op. 25, No. 2—see Chapter 9.) The delicacy is of a rather specific kind, closely related to the "fractal" curves described by Benoît Mandelbrot in his book *The Fractal Geometry of Nature*. In particular, any section of such an attractor, when blown up, reveals itself to be just as exquisitely detailed as was the larger picture from which it was taken. In other words, there is an infinite regress of detail, a never-ending nesting of pattern within pattern. One of the earliest of such structures to be found, called the *attractor of Hénon*, is shown in Figure 16-9. It is generated by the sequence of points (x_n, y_n) defined by the following recurrence relations:

$$x_{n+1} = y_n - a x_n^2 - 1$$

$$y_{n+1} = b x_n$$

Here, a is equal to 7/5 and b to 3/10; the seed values are $x_0 = 0$ and $y_0 = 0$. The small square in Figure 16-9a is blown up in Figure 16-9b to reveal more detail, and then another square in Figure 16-9b is blown up in Figure 16-9c to reveal yet finer detail. Note that what we appear to have is a sort of three-lane highway each of whose lanes breaks up, when magnified, into more parallel lanes, the outermost of which is a new three-lane highway— and on and on it goes. Any perpendicular cross-section of this highway would be what is called a "Cantor set", formed by a simple and famous recursive process.

Begin with a closed interval, say [0,1]. ("Closed" means that the interval includes its endpoints.) Now eliminate some open central subinterval. (Since an open subinterval does not include its endpoints, those two points will remain in the Cantor set being constructed before your eyes.) Usually the deleted subinterval is chosen to be the middle third (1/3, 2/3), but this is not necessary. Two closed subintervals remain. Subject them to the same kind of process—namely, eliminate an open central subinterval inside each of them. Repeat the process *ad infinitum*. What you will be left with at the end of your infinite toil will be a delicate structure consisting of isolated points stretched out along the original segment [0,1] like beads of dew on

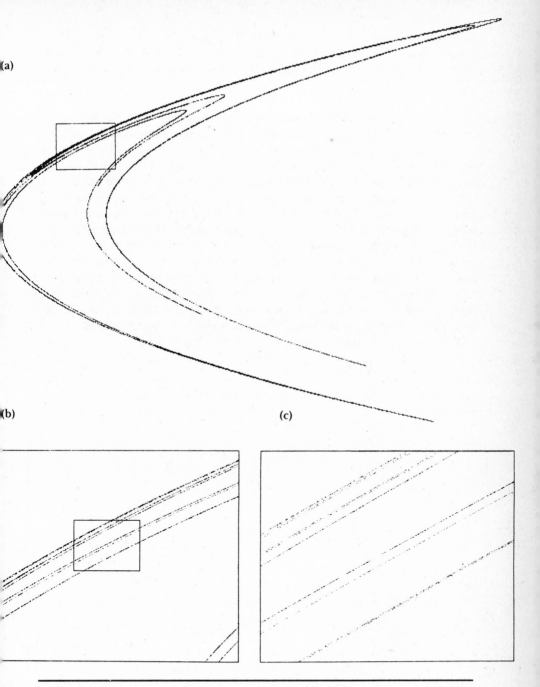

FIGURE 16–9. *The attractor of Hénon: a strange attractor. In (a), the full curve is shown. In (b), the boxed region of (a) is blown up to reveal hidden details. In (c), the boxed region of (b) is further blown up to reveal yet more deeply hidden details. And on and on it could go,* ad infinitum.

a wire. Their number, however, will be uncountably infinite, and their density will depend on the details of your recursive elimination process. Such is the nature of a Cantor set, and if an attractor's cross-sections have this weird kind of distribution, the attractor is said to be *strange,* and for good reason.

Another beautiful strange attractor is generated by the "stroboscopic" points 0, 1, 2,... in Figure 16-8*b*. Since this pattern comes out of Duffing's equation, it is called "Duffing's attractor", and it is shown in a slightly expanded scale in Figure 16-10. Notice its remarkable similarity to the attractor of Hénon. Could this be universality showing its face again?

It is interesting that for the parabola, at the critical value λ_c, f's attractor suddenly consists of infinitely many points, since it is the culmination of an infinite sequence of bifurcations. You can visualize this set either as the limiting case of the horizontal point-sets in Figure 16-5, or as the vertical point-set belonging to $x = \lambda_c$ in Figure 16-6. The precise scatter-pattern of this uncountable point-set is determined by Feigenbaum's recursive rule involving his constant α. Given its recursive genesis, it seems probable that this particular attractor is a Cantor set. Hence the fertile parabola has provided us with an example of a *one-dimensional* strange attractor!

In the chaotic regime of the more general k-dimensional case, long-term prediction of the path that a point will take is quite impossible. Two nearly

---

FIGURE 16–10. *The strange attractor that emerges from a Poincaré map of Duffing's equation.*

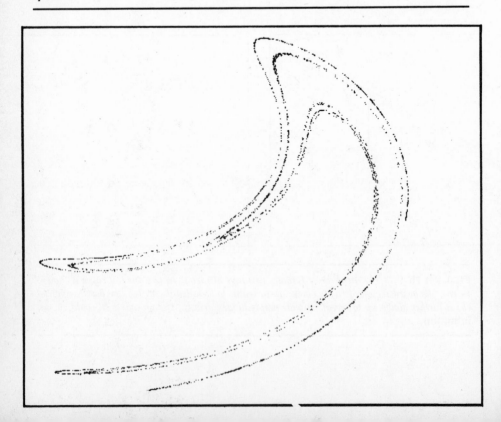

touching points on a strange attractor will, after a few blinks of the strobe light, have wound up at totally different places. This is called *sensitive dependence on initial conditions* and is another defining criterion of a strange attractor.

*   *   *

At present, no one knows just why, how, or when strange attractors will crop up in the chaotic regimes of iterative schemes representing dissipative physical systems, but they do seem to play a central role in the mystery of turbulence. David Ruelle, one of the prime movers of this whole approach to turbulence, wrote: "These systems of curves, these clouds of points, sometimes evoke galaxies or fireworks, other times quite weird and disturbing blossomings. There is a whole world of forms still to be explored, and harmonies still to be discovered."

Robert M. May, a theoretical biologist, concluded his now quite famous review article covering the field in 1976 with a plea that I find most apt and would like to repeat:

> I would urge that people be introduced to the equation $y = 4\lambda x(1-x)$ early in their mathematical education. This equation can be studied phenomenologically by iterating it on a calculator, or even by hand. Its study does not involve as much conceptual sophistication as does elementary calculus. Such study would greatly enrich the student's intuition about nonlinear systems.
>
> Not only in research but also in the everyday world of politics and economics, we would all be better off if more people realized that simple nonlinear systems do not necessarily possess simple dynamical properties.

Post Scriptum.

Stanislaw Ulam, a uniquely inventive mathematician and a warm and delightful human being, died as I was working on this series of postscripts. I had the good fortune to get to know Stan Ulam and his French-born wife Françoise in the summer of 1980, when I visited Santa Fé and stayed with them for a few days. I had always admired and felt kinship with Ulam's strange style in mathematics, totally driven by a passion for the the quirky and the unpredictable, bored by the pure and regular. Ulam loved more than anything to find total chaos in the midst of pristine order. Of course, the thrill was in knowing that there was some kind of *law* to this chaos, so that in *reality*—that is, in God's eye—there was simply a deeper kind of order underneath it all. The bizarre yet tight connections between randomness and order are what all of Ulam's greatest discoveries are about. His style was iconoclastic, to be sure. He was perfectly able to do mathematics in the classical "theorem-proof-theorem-proof" way, but he delighted in the experimental approach, using computers to study crazy behaviors of oddball functions he dreamt up. In some sense, Ulam was a genuine mathematical artist, unlike so many mathematicians. A piece of math by Ulam often feels much more like a creation than like a discovery. It is more idiosyncratic, more easily recognizable as the product of a particular mind, than most mathematical discoveries are.

Aside from being fascinated by mathematics itself, Ulam was also fascinated by the human mind's workings, and he strove to express his vague but provocative intuitions in his writings. I always think of his "ten dogs" theory of memory. The idea is this: When you are searching for a memory that eludes you but that you know is there, what you in effect do is release ten "dogs" in your brain and let them go "sniffing" in parallel. Each dog will start to rummage around here and there, sometimes going in circles, sometimes smelling down wrong alleys, but since there are a bunch of them, you can afford to let them smell out many false pathways. They don't need to be very bright; they just need to have had a whiff of the original idea, and they will follow that spoor high and low. Eventually, it is likely that one dog or another will trot home carrying the desired memory in its mouth. Ulam's autobiography, *Adventures of a Mathematician,* is packed with such glimmerings about how minds work, as well as with droll anecdotes about many of this century's most brilliant mathematicians.

Ulam was very curious about language. He and his wife came to this country about 50 years ago, and both loved the English language. But whereas Stan never lost his strong Polish accent and constantly made errors in English, Françoise eliminated almost every trace of her French accent and became a virtually flawless speaker, whose mastery of idiomatic phrases exceeded that of most native speakers. This caused some amusing light-hearted bickerings between them that I witnessed. Françoise one day used some baseball idiom such as "he threw them a curve ball" or "in the ball park", and Stan immediately objected, saying "You can't use that expression! You didn't grow up playing baseball, so you don't really know what it means!" Françoise defended herself, saying that she had a good idea of its literal meaning but that in any case Stan's point was a red herring. I bought her argument lock, stock, and barrel. After all, how many native speakers of English know what domains such phrases as "red herring" or "lock, stock, and barrel" come from? Yet we certainly all use many such phrases and feel perfectly entitled to do so.

Like many of the brightest mathematicians and physicists working during and just after World War II, Stan Ulam got involved in military projects. His invention, with John von Neumann, of the Monte Carlo method was a key element in the development of the hydrogen bomb. The same forces that drove him to wonder about the cardinality of abstrusely defined sets and the dimensionality of peculiarly defined spaces also guided him to accurate ways of modeling the statistics of chain reactions. At the time he did the work, the nature of the dilemma it would lead humanity as a whole into was not so clear as it now is. To be sure, Einstein had warned us about our slow drift into unparalleled peril, but few people had Einstein's clarity of vision. One of the paradoxes about people is that they are so small compared to the events they can be involved in. Stan Ulam was an ant in a vast colony, and though his role was more significant than that of most ants, he still had no control over the nature of the colony itself. Human nature is one thing, but *humanity*'s nature is another thing.

A good and generous person like Stan Ulam can still be a part of a bad and frightful thing like the arms race. Clearly Ulam had many afterthoughts about his role in these developments, and it is to his credit that he tried to think it all through rationally. Others in similar positions have been far more trapped and narrow-minded, unable to see, or to admit seeing, the complex tragedy that has been unfolding as a consequence of their small actions joined with the small actions of many, many others.

For me it was a privilege to get to know and be friends with this warm and insightful man. I hope that in the long run, Stan Ulam's contributions to mathematics will prove to have outweighed his contributions to a potential Armageddon.

* * *

One of the basic themes of this column is what I call *locking-in*. For no particular reason, I failed to use that term in the column, but it is a good term. The imagery I wish to convey is that of a system that seeks and gradually settles into its own most stable states, and the mechanism whereby it seeks and attains such loci of stability is *feedback*. A system that locks into a state is in a stable equilibrium, which means that if you perturb it somehow, it will swiftly return to the state it was in—there are restoring forces that push it back. Perhaps the most primordial image is that of the particle in the potential well—for example, a marble sitting at the bottom of a round dish. If you ping it lightly with your finger, it will oscillate for a while, but eventually will come to rest again just where it was before: at the sole stable fixed point of the system. Here, as in the column, "fixed point" means that the system's "output" at time t (namely, the marble's position at time t) is identical to the "input" at time $t-1$ (namely, the marble's position at time $t-1$). In this case, the attractor is a single point in space, so it is ridiculously easy to visualize. Most of the attractors in the chapter, however, were *orbits* rather than single points, so they are slightly more abstract. However, if you think of an orbit as simply a point in a multidimensional space, then the concept of zeroing in on a fixed point and the concept of settling down in a stable orbit merge somewhat.

One of the most intuitive as well as charming examples of locking-in is the search for a solution to Raphael Robinson's puzzle in Chapter 2:

> In this sentence, the number of occurrences of 0 is __, of 1 is __, of 2
> is __, of 3 is __, of 4 is __, of 5 is __, of 6 is __, of 7 is __, of 8 is __,
> and of 9 is __.

One way to search for a solution to this puzzle is to fill in the blanks with
an arbitrary sequence of ten numbers, such as <0,1,2,3,4,5,6,7,8,9>, and
see what happens when you check out the truth of the resulting sentence.
It turns out actually to have two occurrences of each digit. Thus the
vector <0,1,2,3,4,5,6,7,8,9> leads to the vector <2,2,2,2,2,2,2,2,2,2>
by the process we'll call "Robinsonizing". Where does *that* vector lead?
Clearly to <1,1,11,1,1,1,1,1,1,1>, which leads to <1,12,1,1,1,1,1,1,1,1>,
which leads to <1,11,2,1,1,1,1,1,1,1>, which leads to <1,11,2,1,1,1,1,1,1,1>
—and lo and behold, we've entered a closed loop!

This vector <1,11,2,1,1,1,1,1,1,1> is like a whirlpool or a vacuum
cleaner: it sucks things near to it into its vortex. It is a trap, a fixed point
—an attractor. It is not unique; there is another such vortex, which I will
leave it to you to find. Furthermore, there is at least one two-state loop, or
period-two attractor, that I know of. I have reason to suspect that *everything*
leads to one of those three attractors, but I could be wrong. You could
search for a period-two attractor by writing down a vector of length twenty
and generating its successor length-twenty vector as follows: Let the new
vector's first half be derived from the old one's second half by
Robinsonizing, and let the new one's second half be derived from the old
one's first half by Robinsonizing. If you now iterate this double-barreled
Robinsonizing operation starting with a random seed, you will eventually
settle down on a fixed point.

Notice that we are now calling a period-two attractor a "fixed point".
Notice also that this is a "point" in a twenty-dimensional space! The point
is, we can view the system *either* as bouncing back and forth between two
ten-dimensional points (a period-two attractor) *or* as sitting still on a fixed
twenty-dimensional point. If by chance there were a loop of length four, we
could similarly think of it as being a fixed point in a 40-dimensional space.
As long as we're willing to "up" the dimensionality of the space, we can
store more and more information in a single point. Thus fixed points and
stable orbits are very close concepts.

*   *   *

This example serves to illustrate how feedback—plugging the system's
output back into the system as input—ushers you to the fixed points. Why
should this be so? Why could the system not thrash about randomly,
somehow avoiding all fixed points? In short, why are fixed points so often
attractive? Why could there not be a large number of fixed points that are
totally isolated, like islands in a vast sea, unreachable via any obvious route?
Could there not be fixed-point "anti-whirlpools" that repel any approacher

that is not dead on target? In the case of Robinson's puzzle, the answer is *no*; but there are such systems. Indeed, in the column I pointed out how there are repellent as well as attractive fixed points for functions of the form $4\lambda x(1-x)$. But in general, it seems to be a very good rule of thumb to search for fixed points by starting out somewhere at random and then hoping that you will get sucked into a stable orbit. Most likely you will, and you will thereby discover a locus of stability, a locked-in solution.

Even more remarkable, it seems generally reliable that you are more likely to be sucked into a short loop than a long one, if short ones exist. Thus, generally speaking, the *stablest* behavior of a system seems also to be its *simplest* behavior. This is true for systems of nearly any sort one can imagine. In the hydrogen atom, for instance, the ground state—the lowest-energy state—is spherically symmetric, and is the only one to have that simple property. Why should this be so, all across the board? Why are stable things the simplest things as well? Or, conversely, why are the simplest things the stablest of all? A toughie.

*　　*　　*

A puzzle more complex than Robinson's but similar in flavor is the search for self-documenting or self-inventorying sentences, which was carried out with such great gusto by Lee Sallows (see Chapter 3). His "logological rocket" was a machine for seeking attractive fixed points in a certain logological space. The book *Loopings* by Aldo Spinelli is a remarkable investigation of regions of a similar logological space, and his search is guided by the same old principle: that starting somewhere random and relying on feedback to get you somewhere "better" is the most likely way to discover a fixed point. This is a most strange way of looking for what might seem something elusive and precious, yet strange though it might be, it is very robust.

In Chapter 3's *Post Scriptum*, I stated that I felt Lee Sallows was overconfident in wagering that a computer search for a self-documenting sentence beginning "This computer-generated pangram contains . . ." would not succeed in ten years. The reason is simple. Lee did not consider the idea of "iterative convergence" to a solution—that is, the idea of Robinsonizing, applied to self-descriptive sentences. You begin with a sentence of the right form, but where all the numbers are randomly chosen. It's a blatant lie, but who cares? You just feed it to a program that counts all its letters and spits out a new sentence with the new letter-counts replacing the old guesses. Around and around you go . . . It is almost certain that you will pretty soon fall into an attractive orbit. Probably most orbits are fairly lengthy loops, and thus do not yield self-documenting sentences—but again, who cares? Just try it again with a different random seed, and keep on doing so, until you find a fixed point.

This method may sound too simple, but it works. I suggested it to Bob

French, one of the two translators into French of *Gödel, Escher, Bach,* and he was gung-ho about implementing such a program. Within a short time, he had one up and running. He sent me this note about his discoveries:

> I wrote a nice program to solve the Pangram Problem and got an answer, written, much to my annoyance, in "françaix". It is:
>
> > *Cette phrase contient cinq a, cinq c, troix d, douze e, un f, un g, quatre h, treize i, huit n, six o, troix p, six q, huit r, six s, quatorze t, dix u, un v, sept x, & quatre z.*
>
> Unbelievably, in programming it, I had put the *wrong* goddam spelling of "trois" into the program. Oh well, when I corrected the mistake, I didn't get an answer immediately, but I'm confident that it'll come, in correctly spelled French, when I get back to work on the thing.

The point is, you don't need to perform a brute-force search through the entire space of all possible combinations of numbers filling the 26 blanks in order to find a perfect self-documenting sentence, not by a long shot! A Robinsonizing routine, together with a simple-minded loop detector, will do the trick quite easily, as long as you're willing to try a bunch of different seeds. The pulling-power of short loops will undoubtedly snag you sooner or later, and you'll have found your target sentence!

My friend Larry Tesler, equally spurred on by Sallows' challenge when it appeared in print in A. K. Dewdney's new *Scientific American* column called "Computer Recreations" in October 1984, coded up the Robinsonizing method in a program and soon his computer fell into a loop that seemed very close to a solution. By changing his program's search technique at that point, Tesler was then easily able to home in on a winner, which he gleefully sent off to both Dewdney and Sallows. Tesler's sentence runs as follows:

> This computer-generated pangram contains six a's, one b, three c's, three d's, thirty-seven e's, six f's, three g's, nine h's, twelve i's, one j, one k, two l's, three m's, twenty-two n's, thirteen o's, three p's, one q, fourteen r's, twenty-nine s's, twenty-four t's, five u's, six v's, seven w's, four x's, five y's, and one z.

*   *   *

Locking-in is perfectly illustrated by the hypothetical book *Reviews of This Book,* described in Chapter 3. There I characterized the method of its creation as resembling the construction of "self-consistent" solutions via the "Hartree-Fock" method. What does that mean? It boils down to the same thing once more. It turns out to be very hard—in fact, impossible— to give closed-form solutions to the equations describing any atom more complicated than a hydrogen atom, with its single electron. When you have three bodies, as in the helium atom with its two electrons and a nucleus, the mathematical complexity is overwhelming. The problem is in essence that each electron would "like" to be in a simple hydrogen-like state around the nucleus, but the other one is blocking it from so doing. How can they "cooperate" with each other to find a stable mode of coexistence?

One way to study this mathematically, suggested first in 1928 by the English physicist Douglas Rayner Hartree, is to try to converge on a good description of the total system by starting out with a false solution—a mathematical description of a state known to be wrong, but easy to describe. (For instance, you could pretend that both electrons *are* in simple hydrogen-like states.) Then you see how each electron "perturbs" the other one out of the presumed state it was in. This leads you to a different—and probably no less fictitious—state. But at least you've made progress, in that you've taken into account the "first-order" effects each electron would have on the other one. Now you do the same thing over again—that is, you see how the perturbed states would perturb each other. This gives you "second-order" corrections—and so on and so on. Eventually—and this is the beauty of the method—the starting point of your calculations gets totally buried, and the state converges to what is called a "self-consistent" solution, very much like the solutions to Robinson's puzzle. What I mean by saying the starting point gets "buried" is that *no matter where you start,* you'll wind up at the same eventual solution—a fixed point, where further iteration has no effect. In this solution, the two electrons are in equilibrium with each other and do not perturb each other. And presto—one has "solved" the helium atom!

Of course, this type of solution is *numerical,* not analytic: there are no exact formulas that come out, only numbers. Nonetheless, that's good enough for most practical purposes. The Russian physicist Vladimir Fock later made a suggestion for improving the validity of this method of calculation, which involves taking into account the fact that electrons obey the Pauli exclusion principle, a complication that Hartree had ignored. That is the reason for the hyphenated name; however, Hartree is the inventor of the general principle of calculating self-consistent solutions for many-body systems.

* * *

This idea of locking-in recurs throughout science. In *Gödel, Escher, Bach,* I discussed the phenomenon called *renormalization*—the way that elementary particles such as electrons and positrons and photons all take each other into account in their very core. The notion is a mathematical one, but for a good metaphor, recall how your own identity depends on the identities of your close friends and relatives, and how theirs in turn depends on yours and on *their* close friends' and relatives' identities, and so on, and so on. This was the image I described for "I at the Center" in the *Post Scriptum* to Chapter 10. Another good graphic representation of this idea is shown in Figure 24-4, where identity emerges out of a renormalization process.

The tangledness of one's own self is a perfect metaphor for

understanding what renormalization is all about. And the best way to imagine how *you* emerge from such a complex tangle is to begin by imagining yourself as a "zeroth-order person"—that is, someone totally unaware and inconsiderate of all others. (Of course, such a person would be barely a person, barely a self at all: a perfect baby.) Then imagine how "you" would be modified if you started to take other people into account, always considering others as perfect babies, or zeroth-order people. This gives a "first-order" version of you. You are beginning to have an identity, emerging from this modeling of others inside yourself. Now iterate: second-order people are those who take into account the identities of first-order people. And on it goes. The final result is *renormalized* people: people who take into account the identities of renormalized people. I know it sounds circular, and indeed it is, but *paradoxical* it is not—at least no more than are the fixed points of Raphael Robinson's puzzle! "Circular" is not synonymous with "paradoxical", although many people mistakenly assume it is. We shall re-encounter this notion of renormalized people in Chapter 30 and beyond, where it will in fact clear up some seeming paradoxes involving cooperation and egoism.

This close connection of locking-in to the deepest essence of personhood plays a central role also in Chapters 22 and 25, where "who" one is is portrayed as emerging from a "level-crossing feedback loop", in which a sophisticated perceiving system perceives limited aspects of its own nature, and by feeding them back into the system creates a type of locking-in. The locked-in loop itself is given a name, and that name, for every such system, is "I".

The idea of a system with an I, watching its own behavior, is closely related to the wellsprings of creativity (recall the cycle underlying creativity discussed in the *Post Scriptum* to Chapter 12, and that to Chapter 10 as well). We will delve into this in depth again in Chapter 23, trying to come to grips with another seeming paradox: that of mechanizing what seems by definition to be nonmechanical and nonmechanizable—the creative act. Once again we'll see vicious paradox dissolve into benign cycles.

In short, locking-in—that is, convergent and self-stabilizing behavior—will surely pervade the ultimate explanation of most mysteries of the mind. One example is the question of memory retrieval. How do things that are only vaguely similar to each other stir up rumblings of recollection, and eventually trigger the retrieval of amazingly deep abstract resemblances? One theory, best formulated and articulated by cognitive scientist Pentti Kanerva of Stanford University, sees the initial input as a *seed*—a vector in a very high-dimensional space, analogous to the seed vector that we fed into the Robinsonizing machine. The seed is fed into memory-retrieval mechanisms, which convert it into an output vector that is then fed back in again. This cyclic process continues until it either converges on a stable fixed point—the desired memory trace—or is seen to be wandering erratically without any likelihood of locking in, tracing out a chaotic sequence of "points" in mind-space. The details of how this is accomplished in Kanerva's beautiful theory are beyond the scope of this book, but this "self-propagating search" provides another remarkable example of the many ways that locking-in can be exploited.

Closely related to memory retrieval is the problem of perception, or pattern recognition. As I mentioned in the *Post Scriptum* to Chapter 4, this central aspect of mind has been best modeled on computers in programs whose strategy is similar to that of Kanerva's model: there is a superficial sweep that narrows the field somewhat, followed by a deeper sweep that narrows it further, and so on (the "terraced scan" I described in the postscript to Chapter 5). This bottom-up processing is complemented by concurrent top-down processing driven not by the input, but by *expectations* of what is "out there" to be recognized. The swirling activity in which bottom-up and top-down processes seek a reconciliation with each other leads to a gradual kind of "crystallization", in which many small pieces of evidence align with, and mutually reinforce, each other. The ultimate justification for some of them resides, of course, in the raw perceptual input, while for others of them it resides in the richness of previous experiences stored in memory. The combination of all these mutually confirming hypotheses results in a globally optimal interpretation of the input: an act of recognition. Once again, locking-in carries the day.

One final example of locking-in is the subject of Chapter 27: the question of the inevitability (or evitability) of the genetic code. This central question about the molecular foundations of life turns out to revolve about two distinct senses of the word "arbitrary". I shall let that Chapter speak for itself, however.

*   *   *

In the Introduction, I described the space of my columns as gradually emerging as, month by month, I revealed one more dot in that space. What is this, if not a Poincaré map of my mental meanderings? During my column-writing era, my mind would light up like a monthly firefly and reveal where it was to the outside world! I just wonder: Would the shape I was thus tracing out turn out to be a strange attractor?

It seems appropriate that at this midpoint of the book, we have identified a unifying theme—or rather, *thema,* to be more faithful to the title. *Locking-in* seems to be a key to the metamagics of Snarls, of Society, of Slipping . . . of Strangeness, of Substrate, of Stability . . . of Survival.

17

Lisp: Atoms and Lists

February, 1983

IN previous columns I have written quite often about the field of artificial intelligence—the search for ways to program computers so that they might come to behave with flexibility, common sense, insight, creativity, self-awareness, humor, and so on. The quest for AI started in earnest over two decades ago, and since then has bifurcated many times, so that today it is a very active and multifaceted research area. In the United States there are perhaps a couple of thousand people professionally involved in AI, and there are a similar number abroad. Although there is among these workers a considerable divergence of opinion concerning the best route to AI, one thing that is nearly unanimous is the choice of programming language. Most AI research efforts are carried out in a language called "Lisp". (The name is not quite an acronym; it stands for "list processing".)

Why is most AI work done in Lisp? There are many reasons, most of which are somewhat technical, but one of the best is quite simple: Lisp is crisp. Or as Marilyn Monroe said in *The Seven-Year Itch,* "I think it's jus-telegant!" Every computer language has arbitrary features, and most languages are in fact overloaded with them. A few, however, such as Lisp and Algol, are built around a kernel that seems as natural as a branch of mathematics. The kernel of Lisp has a crystalline purity that not only appeals to the esthetic sense, but also makes Lisp a far more flexible language than most others. Because of Lisp's beauty and centrality in this important area of modern science, then, I have decided to devote a trio of columns to some of the basic ideas of Lisp.

The deep roots of Lisp lie principally in mathematical logic. Mathematical pioneers such as Thoralf Skolem, Kurt Gödel, and Alonzo Church contributed seminal ideas to logic in the 1920's and 1930's that were incorporated decades later into Lisp. Computer programming in earnest began in the 1940's, but so-called "higher-level" programming languages (of which Lisp is one) came into existence only in the 1950's. The earliest list-processing language was not Lisp but IPL ("Information Processing Language"), developed in the mid-1950's by Herbert Simon, Allen Newell, and J. C. Shaw. In the years 1956–58, John McCarthy, drawing on all these previous sources, came up with an elegant algebraic list-processing language he called Lisp. It caught on quickly with the young crowd around him at the newly-formed MIT Artificial Intelligence Project, was implemented on the IBM 704, spread to other AI groups, infected them, and has stayed around all these years. Many dialects now exist, but all of them share that central elegant kernel.

*   *   *

Let us now move on to the way Lisp really works. One of the most appealing features of Lisp is that it is *interactive*, as contrasted with most other higher-level languages, which are noninteractive. What this means is the following. When you want to program in Lisp, you sit down at a terminal connected to a computer and you type the word "lisp" (or words to that effect). The next thing you will see on your screen is a so-called "prompt" —a characteristic symbol such as an arrow or asterisk. I like to think of this prompt as a greeting spoken by a special "Lisp genie", bowing low and saying to you, "Your wish is my command—and now, what is your next wish?" The genie then waits for you to type something to it. This genie is usually referred to as the *Lisp interpreter,* and it will do anything you want— but you have to take great care in expressing your desires precisely, otherwise you may reap some disastrous effects. Shown below is the prompt, the sign that the Lisp genie is ready to do your bidding:

→

The genie is asking us for our heart's desire, so let us type in a simple expression:

→ **(plus 2 2)**

and then a carriage return. (By the way, all Lisp expressions and words will be printed in **Helvetica** in this and the following two chapters.) Even non-Lispers can probably anticipate that the Lisp genie will print in return the value **4**. Then it will also print a fresh prompt, so that the screen will now appear this way:

→ **(plus 2 2)**
4
→

The genie is now ready to carry out our next command—or, more politely stated, our next wish—should we have one. The carrying-out of a wish expressed as a Lisp statement is called *evaluation* of that statement. The preceding short interchange between human and computer exemplifies the behavior of the Lisp interpreter: it *reads* a statement, *evaluates* it, *prints* the appropriate value, and then signals its readiness to read a new statement. For this reason, the central activity of the Lisp interpreter is referred to as the *read-eval-print loop*.

The existence of this Lisp genie (the Lisp interpreter) is what makes Lisp interactive. You get immediate feedback as soon as you have typed a "wish" —a complete statement—to Lisp. And the way to get a bunch of wishes carried out is to type one, then ask the genie to carry it out, then type another, ask the genie again, and so on.

By contrast, in many higher-level computer languages you must write out an entire program consisting of a vast number of wishes to be carried out in some specified order. What's worse is that later wishes usually depend strongly on the consequences of earlier wishes—and of course, you don't get to try them out one by one. The execution of such a program may, needless to say, lead to many unexpected results, because so many wishes have to mesh perfectly together. If you've made the slightest conceptual error in designing your wish list, then a total foul-up is likely—in fact, almost inevitable. Running a program of this sort is like launching a new space probe, untested: you can't possibly have anticipated all the things that might go wrong, and so all you can do is sit back and watch, hoping that it will work. If it fails, you go back and correct the one thing the failure revealed, and then try another launch. Such a gawky, indirect, expensive way of programming is in marked contrast to the direct, interactive, one-wish-at-a-time style of Lisp, which allows "incremental" program development and debugging. This is another major reason for the popularity of Lisp.

* * *

What sorts of wishes can you type to the Lisp genie for evaluation, and what sorts of things will it print back to you? Well, to begin with, you can type arithmetical expressions expressed in a rather strange way, such as **(times (plus 6 3) (difference 6 3))**. The answer to this is 27, since **(plus 6 3)** evaluates to 9, and **(difference 6 3)** evaluates to 3, and their product is 27. This notation, in which each operation is placed to the left of its operands, was invented by the Polish logician Jan $\geq$ ukasiewicz before computers existed. Unfortunately for $\geq$ ukasiewicz, his name was too formidable-looking for most speakers of English, and so this type of notation came to be called *Polish notation*. Here is a simple problem in this notation for you, in which you are to play the part of the Lisp genie:

$\rightarrow$ (quotient (plus 21 13) (difference 23 (times 2 (difference 7 (plus 2 2)))))

Perhaps you have noticed that statements of Lisp involve parentheses. A profusion of parentheses is one of the hallmarks of Lisp. It is not uncommon to see an expression that terminates in a dozen right parentheses! This makes many people shudder at first—and yet once you get used to their characteristic appearance, Lisp expressions become remarkably intuitive, even charming, to the eye, especially when *pretty-printed,* which means that a careful indentation scheme is followed that reveals their logical structure. All of the expressions in displays in this article have been pretty-printed.

The heart of Lisp is its manipulable structures. All programs in Lisp work by creating, modifying, and destroying structures. Structures come in two types: atomic and composite, or, as they are usually called, *atoms* and *lists.* Thus, every Lisp object is either an atom or a list (but not both). The only exception is the special object called **nil**, which is both an atom and a list. More about **nil** in a moment. What are some other typical Lisp atoms? Here are a few:

**hydrogen, helium, j-s-bach, 1729, 3.14159, pi,
arf, foo, bar, baz, buttons-&-bows**

Lists are the flexible data structures of Lisp. A list is pretty much what it sounds like: a collection of some parts in a specific order. The parts of a list are usually called its *elements* or *members.* What can these members be? Well, not surprisingly, lists can have atoms as members. But just as easily, lists can contain lists as members, and those lists can in turn contain other lists as members, and so on, recursively. Oops! I jumped the gun with that word. But no harm done. You certainly understood what I meant, and it will prepare you for a more technical definition of the term to come later.

A list printed on your screen is recognizable by its parentheses. In Lisp, anything bounded by matching parentheses constitutes a list. So, for instance, **(zonk blee strill (cronk flonk))** is a four-element list whose last element is itself a two-element list. Another short list is **(plus 2 2)**, illustrating the fact that *Lisp statements themselves are lists.* This is important because it means that the Lisp genie, by manipulating lists and atoms, can actually construct new wishes by itself. Thus the object of a wish can be the construction—and subsequent evaluation—of a new wish!

Then there is the *empty* list—the list with no elements at all. How is this written down? You might think that an empty pair of parentheses—**()**—would work. Indeed, it will work—but there is a second way of indicating the empty list, and that is by writing **nil**. The two notations are synonymous, although **nil** is more commonly written than **()** is. The empty list, **nil**, is a key concept of Lisp; in the universe of lists, it is what zero is in the universe of numbers. To use another metaphor for nil, it is like the earth in which all structures are rooted. But for you to understand what this means, you will have to wait a bit.

*   *   *

The most commonly exploited feature of an atom is that it has (or can be given) a *value.* Some atoms have permanent values, while others are variables. As you might expect, the value of the atom **1729** is the integer

1729, and this is permanent. (I am distinguishing here between the atom whose *print name* or *pname* is the four-digit string **1729**, and the eternal Platonic essence that happens to be the sum of two cubes in two different ways—*i.e.,* the number 1729.) The value of **nil** is also permanent, and it is —**nil**! Only one other atom has itself as its permanent value, and that is the special atom **t**.

Aside from **t**, **nil**, and atoms whose names are numerals, atoms are generally variables, which means that you can assign values to them and later change their values at will. How is this done? Well, to assign the value 4 to the atom **pie**, you can type to the Lisp genie **(setq pie 4)**. Or you could just as well type **(setq pie (plus 2 2))**—or even **(setq pie (plus 1 1 1 1))**. In any of these cases, as soon as you type your carriage return, **pie**'s value will become 4, and so it will remain forevermore—or at least until you do another **setq** operation on the atom **pie**.

Lisp would not be crisp if the only values atoms could have were numbers. Fortunately, however, an atom's value can be set to any kind of Lisp object —any atom or list whatsoever. For instance, we might want to make the value of the atom **pi** be a list such as **(a b c)** or perhaps **(plus 2 2)** instead of the number 4. To do the latter, we again use the **setq** operation. To illustrate, here follows a brief conversation with the genie:

```
-> (setq pie (plus 2 2))
4
-> (setq pi '(plus 2 2))
(plus 2 2)
->
```

Notice the vast difference between the values assigned to the atoms **pie** and **pi** as a result of these two wishes asked of the Lisp genie, which differ merely in the presence or absence of a small but critical *quote mark* in front of the inner list **(plus 2 2)**. In the first wish, containing no quote mark, that inner **(plus 2 2)** must be *evaluated*. This returns 4, which is assigned to the variable **pie** as its new value. On the other hand, in the second wish, since the quote mark is there, the list **(plus 2 2)** is never executed as a command, but is treated merely as an inert lump of Lispstuff, much like meat on a butcher's shelf. It is ever so close to being "alive", yet it is dead. So the value of **pi** in this second case is the list **(plus 2 2)**, a fragment of Lisp code. The following interchange with the genie confirms the values of these atoms.

```
-> pie
4
-> pi
(plus 2 2)
-> (eval pi)
4
->
```

What is this last step? I wanted to show how you can ask the genie to *evaluate* the value of an expression, rather than simply *printing* the value of that expression. Ordinarily, the genie automatically performs just *one* level of evaluation, but by writing **eval**, you can get a second stage of evaluation carried out. (And of course, by using **eval** over and over again, you can carry this as far as you like.) This feature often proves invaluable, but it is a little too advanced to discuss further at this stage.

*   *   *

Every list but **nil** has at least one element. This first element is called the list's **car**. Thus the **car** of **(eval pi)** is the atom **eval**. The cars of the lists **(plus 2 2)**, **(setq x 17)**, **(eval pi)**, and **(car pi)** are all names of operations, or, as they are more commonly called in Lisp, *functions*. The **car** of a list need not be the name of a function; it need not even be an atom. For instance, **((1) (2 2) (3 3 3))** is a perfectly fine list. Its **car** is the list **(1)**, whose **car** in turn is not a function name but merely a numeral.

If you were to remove a list's **car**, what would remain? A shorter list. This is called the list's **cdr**, a word that sounds about halfway between "kidder" and "could 'er". (The words "car" and "cdr" are quaint relics from the first implementation of Lisp on the IBM 704. The letters in "car" stand for "Contents of the Address part of Register" and those in "cdr" for "Contents of the Decrement part of Register", referring to specific hardware features of that machine, now long since irrelevant.) The **cdr** of **(a b c d)** is the list **(b c d)**, whose **cdr** is **(c d)**, whose **cdr** is **(d)**, whose **cdr** is **nil**. And **nil** has no **cdr**, just as it has no **car**. Attempting to take the **car** or **cdr** of **nil** causes (or should cause) the Lisp genie to cough out an error message, just as attempting to divide by zero should evoke an error message.

Here is a little table showing the **car** and **cdr** of a few lists, just to make sure the notions are unambiguous.

list	**car**	**cdr**
((a) b (c))	**(a)**	**(b (c))**
(plus 2 2)	**plus**	**(2 2)**
((car x) (car y))	**(car x)**	**((car y))**
(nil nil nil nil)	**nil**	**(nil nil nil)**
(nil)	**nil**	**nil**
nil	***ERROR***	***ERROR***

Just as **car** and **cdr** are called *functions,* so the things that they operate on are called their *arguments.* Thus in the command **(plus pie 2)**, **plus** is the function name, and the arguments are the atoms **pie** and **2**. In evaluating this command (and most commands), the genie figures out the values of the arguments, and then applies the function to those values. Thus, since the value of the atom **pie** is 4, and the value of the atom **2** is 2, the genie returns the atom **6**.

*   *   *

Suppose you have a list and you'd like to see a list just like it, only one element longer. For instance, suppose the value of the atom **x** is **(cake cookie)** and you'd like to create a new list called **y** just like **x**, except with an extra atom—say **pie**—at the front. You can then use the function called **cons** (short for "construct"), whose effect is to make a new list out of an old list and a suggested **car**. Here's a transcript of such a process:

```
-> (setq x '(cake cookie))
(cake cookie)
-> (setq y (cons 'pie x))
(pie cake cookie)
-> x
(cake cookie)
```

Two things are worth noticing here. I asked for the value of **x** to be printed out after the **cons** operation, so you could see that **x** itself was not changed by the **cons**. The **cons** operation created a new list and made that list be the value of **y**, but left **x** entirely alone. The other noteworthy fact is that I used that quote mark again, in front of the atom **pie**. What if I had not used it? Here's what would have happened.

```
-> (setq z (cons pie x))
(4 cake cookie)
```

Remember, after all, that the atom **pie** still has the value 4, and whenever the genie sees an unquoted atom inside a wish, it will always use the *value* belonging to that atom, rather than the atom's *name*. (Always? Well, almost always. I'll explain in a moment. In the meantime, look for an exception— you've already encountered it.)

Now here are a few exercises—some a bit tricky—for you. Watch out for the quote marks! Oh, one last thing: I use the function **reverse**, which produces a list just like its argument, only with its elements in reverse order. For instance, the genie, upon being told **(reverse '((a b) (c d e)))** will write **((c d e) (a b))**. The genie's lines in this dialogue are given afterward.

```
-> (setq w (cons pie '(cdr z)))
-> (setq v (cons 'pie (cdr z)))
-> (setq u (reverse v))
-> (cdr (cdr u))
-> (car (cdr u))
-> (cons (car (cdr u)) u)
-> u
```

⇁ (reverse '(cons (car u) (reverse (cdr u))))
⇁ (reverse (cons (car u) (reverse (cdr u))))
⇁ u
⇁ (cons 'cookie (cons 'cake (cons 'pie nil)))

Answers (as printed by the genie):

(4 cdr z)
(pie cake cookie)
(cookie cake pie)
(pie)
cake
(cake cookie cake pie)
(cookie cake pie)
((reverse (cdr u)) (car u) cons)
(cake pie cookie)
(cookie cake pie)
(cookie cake pie)

The last example, featuring repeated use of **cons**, is often called, in Lisp slang, "consing up a list". You start with **nil**, and then do repeated **cons** operations. It is analogous to building a positive integer by starting at zero and then performing the successor operation over and over again. However, whereas at any stage in the latter process there is a unique way of performing the successor operation, given any list there are infinitely many different items you can **cons** onto it, thus giving rise to a vast branching tree of lists instead of the unbranching number line. It is on account of this image of a tree growing out of the ground of **nil** and containing all possible lists that I earlier likened **nil** to "the earth in which all structures are rooted".

As I mentioned a moment ago, the genie doesn't *always* replace (unquoted) atoms by their values. There are cases where a function treats its arguments, though unquoted, as if quoted. Did you go back and find such a case? It's easy. The answer is the function **setq**. In particular, in a **setq** command, the first atom is taken straight—not evaluated. As a matter of fact, the **q** in **setq** stands for "quote", meaning that the first argument is treated as if quoted. Things can get quite tricky when you learn about **set**, a function similar to **setq** except that it *does* evaluate its first argument. Thus, if the value of the atom **x** is the atom **k**, then saying **(set x 7)** will not do anything to **x**—its value will remain the atom **k**—but the value of the atom **k** will now become 7. So watch closely:

⇁ (setq a 'b)
⇁ (setq b 'c)
⇁ (setq c 'a)
⇁ (set a c)
⇁ (set c b)

Now tell me: What are the values of the atoms **a**, **b**, and **c**? Here comes the answer, so don't peek. They are, respectively: **a**, **a**, and **a**. This may seem a bit confusing. You may be reassured to know that in Lisp, **set** is not very commonly used, and such confusions do not arise that often.

* * *

Psychologically, one of the great powers of programming is the ability to define new compound operations in terms of old ones, and to do this over and over again, thus building up a vast repertoire of ever more complex operations. It is quite reminiscent of evolution, in which ever more complex molecules evolve out of less complex ones, in an ever-upward spiral of complexity and creativity. It is also quite reminiscent of the industrial revolution, in which people used very simple early machines to help them build more complex machines, then used those in turn to build even more complex machines, and so on, once again in an ever-upward spiral of complexity and creativity. At each stage, whether in evolution or revolution, the products get more flexible and more intricate, more "intelligent" and yet more vulnerable to delicate "bugs" or breakdowns.

Likewise with programming in Lisp, only here the "molecules" or "machines" are now Lisp functions defined in terms of previously known Lisp functions. Suppose, for instance, that you wish to have a function that will always return the last element of a list, just as **car** always returns the first element of a list. Lisp does not come equipped with such a function, but you can easily create one. Do you see how? To get the last element of a list called **lyst**, you simply do a **reverse** on **lyst** and then take the **car** of that: **(car (reverse lyst))**. To dub this operation with the name **rac** (**car** backwards), we use the **def** function, as follows:

⇢ **(def rac (lambda (lyst) (car (reverse lyst))))**

Using **def** this way creates a *function definition*. In it, the word **lambda** followed by **(lyst)** indicates that the function we are defining has only one parameter, or dummy variable, to be called **lyst**. (It could have been called anything; I just happen to like the atom **lyst**.) In general, the list of parameters (dummy variables) must immediately follow the word **lambda**. After this "def wish" has been carried out, the **rac** function is as well understood by the genie as is **car**. Thus **(rac '(your brains))** will yield the atom **brains**. And we can use **rac** itself in definitions of yet further functions. The whole thing snowballs rather miraculously, and you can quickly become overwhelmed by the power you wield.

Here is a simple example. Suppose you have a situation where you know you are going to run into many big long lists and you know it will often be useful to form, for each such long list, a short list that contains just its **car** and **rac**. We can define a one-parameter function to do this for you:

```
→ (def readers-digest-condensed-version
   (lambda (biglonglist)
     (cons (car biglonglist) (cons (rac biglonglist) nil))))
```

Thus if we apply our new function **readers-digest-condensed-version** to the entire text of James Joyce's *Finnegans Wake* (treating it as a big long list of words), we will obtain the shorter list **(riverrun the)**. Unfortunately, reapplying the condensation operator to this new list will not simplify it any further.

It would be nice as well as useful if we could create an inverse operation to **readers-digest-condensed-version** called **rejoyce** that, given any two words, would create a novel beginning and ending with them, respectively —and such that James Joyce would have written it (had he thought of it). Thus execution of the Lisp statement **(rejoyce 'Stately 'Yes)** would result in the Lisp genie generating from scratch the entire novel *Ulysses*. Writing this function is left as an exercise for the reader. To test your program, see what it does with **(rejoyce 'karma 'dharma)**.

*   *   *

One goal that has seemed to some people to be both desirable and feasible using Lisp and related programming languages is (1) to make every single statement return a *value* and (2) to have it be through this returned value and *only* through it that the statement has any effect. The idea of (1) is that values are handed "upward" from the innermost function calls to the outermost ones, until the full statement's value is returned to you. The idea of (2) is that during all these calls, no atom has its value changed at all (unless the atom is a dummy variable). In all dialects of Lisp known to me, (1) is true, but (2) is not necessarily true.

Thus if **x** is bound to **(a b c d e)** and you say **(car (cdr (reverse x)))**, the first thing that happens is that **(reverse x)** is calculated; then this value is handed "up" to the **cdr** function, which calculates the **cdr** of that list; finally, this shorter list is handed to the **car** function, which extracts one element—namely the atom **d**—and returns it. In the meantime, the atom **x** has suffered no damage; it is still bound to **(a b c d e)**.

It might seem that an expression such as **(reverse x)** would change the value of **x** by reversing it, just as carrying out the oral command "Turn your sweater inside out" will affect the sweater. But actually, carrying out the wish **(reverse x)** no more changes the value of **x** than carrying out the wish **(plus 2 2)** changes the value of 2. Instead, executing **(reverse x)** causes a *new* (unnamed) list to come into being, just like **x**, only reversed. And that list is the value of the statement; it is what the statement returns. The value of **x** itself, however, is untouched. Similarly, evaluating **(cons 5 pi)** will not change the list named **pi** in the slightest; it merely returns a new list with **5** as its **car** and whatever **pi**'s value is as its **cdr**.

Such behavior is to be contrasted with that of functions that leave "side effects" in their wake. Such side effects are usually in the form of changed variable bindings, although there are other possibilities, such as causing input or output to take place. A typical "harmful" command is a **setq**, and proponents of the "applicative" school of programming—the school that says you should never make any side effects whatsoever—are profoundly disturbed by the mere mention of **setq**. For them, all results must come about purely by the way that functions compute their values and hand them to other functions.

The only bindings that the advocates of the applicative style approve of are transitory "lambda bindings"—those that arise when a function is applied to its arguments. Whenever any function is called, that function's dummy variables temporarily assume "lambda bindings". These bindings are just like those caused by a **setq**, except that they are fleeting. That is, the moment the function is finished computing, they go away—vanishing without a trace. For example, during the computation of **(rac '(a b c))**, the lambda binding of the dummy variable **lyst** is the list **(a b c)**; but as soon as the answer **c** is passed along to the function or person that requested the **rac**, the value of the atom **lyst** used in getting that answer is totally forgotten. The Lisp interpreter will tell you that **lyst** is an "unbound atom" if you ask for its value. Applicative programmers much prefer lambda bindings to ordinary **setq** bindings.

I personally am not a fanatic about avoiding **setq**'s and other functions that cause side effects. Though I find the applicative style to be jus-telegant, I find it impractical when it comes to the construction of large AI-style programs. Therefore I shall not advocate the applicative style here, though I shall adhere to it when possible. Strictly speaking, in applicative programming, you cannot even define new functions, since a **def** statement causes a permanent change to take place in the genie's memory—namely, the permanent storage in memory of the function definition. So the ideal applicative approach would have functions, like variable bindings, being created only temporarily, and their definitions would be discarded the moment after they had been used. But this is extreme "applicativism".

For your edification, here are a few more simple function definitions.

```
→ (def rdc (lambda (lyst) (reverse (cdr (reverse lyst)))))
→ (def snoc (lambda (x lyst) (reverse (cons x (reverse lyst)))))
→ (def twice (lambda (n) (plus n n)))
```

The functions **rdc** and **snoc** are analogous to **cdr** and **cons**, only backwards. Thus, the **rdc** of **(a b c d e)** is **(a b c d)**, and if you type **(snoc 5 '(1 2 3 4))**, you will get **(1 2 3 4 5)** as your answer.

*　　*　　*

All of this is mildly interesting so far, but if you want to see the genie do anything *truly* surprising, you have to allow it to make some decisions based on things that happen along the way. These are sometimes called "conditional wishes". A typical example would be the following:

> ⇢ **(cond ((eq x 1) 'land) ((eq x 2) 'sea))**

The value returned by this statement will be the atom **land** if **x** has value 1, and the atom **sea** if **x** has value 2. Otherwise, the value returned will be **nil** (*i.e.,* if **x** is 5). The atom **eq** (pronounced "eek") is the name of a common Lisp function that returns the atom **t** (standing for "true") if its two arguments have the same value, and **nil** (for "no" or "false") if they do not.

A **cond** statement is a list whose **car** is the function name **cond**, followed by any number of *cond clauses,* each of which is a two-element list. The first element of each clause is called its *condition,* the second element its *result.* The clauses' conditions are checked out by the Lisp genie one by one, in order; as soon as it finds a clause whose condition is "true" (meaning that the condition returns anything other than **nil**!), it begins calculating that clause's *result,* whose value gets returned as the value of the whole **cond** statement. None of the further clauses is even so much as glanced at! This may sound more complex than it ought to. The real idea is no more complex than saying that it looks for the first condition that is satisfied, then it returns the corresponding result.

Often one wants to have a catch-all clause at the end whose condition is *sure* to be satisfied, so that, if all other conditions fail, at least this one will be true and the accompanying result, rather than **nil**, will be returned. It is easy as pie to make a condition whose value is non-**nil**; just choose it to be **t**, for instance, as in the following:

> ⇢ **(cond ((eq x 1) 'land)**
> **((eq x 2) 'sea)**
> **(t 'air))**

Depending on what the value of **x** is, we will get either **land**, **sea**, or **air** as the value of this **cond**, but we'll never get **nil**. Now here are a few sample **cond** statements for you to play genie to:

> ⇢ **(cond ((eq (eval pi) pie) (eval (snoc pie pi)))**
> **(t (eval (snoc (rac pi) pi))))**
> ⇢ **(cond ((eq 2 2) 2) ((eq 3 3) 3))**
> ⇢ **(cond (nil 'no-no-no)**
> **((eq '(car nil) '(cdr nil)) 'hmmm)**
> **(t 'yes-yes-yes))**

The answers are: **8, 2**, and **yes-yes-yes**. Did you notice that **(car nil)** and **(cdr nil)** were quoted?

I shall close this portion of the column by displaying a patterned family of function definitions, so obvious in their pattern that you would think that the Lisp genie would just sort of "get the hang of it" after seeing the first few . . . . Unfortunately, though, Lisp genies are frustratingly dense (or at least they play at being dense), and they will not jump to any conclusion unless it has been completely spelled out. Look first at the family:

```
-> (def square (lambda (k) (times k k)))
-> (def cube (lambda (k) (times k (square k))))
-> (def 4th-power (lambda (k) (times k (cube k))))
-> (def 5th-power (lambda (k) (times k (4th-power k))))
-> (def 6th-power (lambda (k) (times k (5th-power k))))
       .
       .
       .
       .
```

My question for you is this: Can you invent a definition for a *two*-parameter function that subsumes *all* of these in one fell swoop? More concretely, the question is: How would one go about defining a two-parameter function called **power** such that, for instance, **(power 9 3)** yields 729 on being evaluated, and **(power 7 4)** yields 2,401? I have supplied you, in this column, with all the necessary tools to do this, provided you exercise some ingenuity.

*   *   *

I thought I would end this column with a newsbreak about a freshly discovered beast—the homely Glazunkian porpuquine, so called because it is found only on the island of Glazunkia (claimed by Upper Bitbo, though it is just off the coast of Burronymede). And what is a porpuquine, you ask? Why, it's a strange breed of porcupine, whose quills—of which, for some reason, there are always exactly nine (in Outer Glazunkia) or seven (in Inner Glazunkia)—are smaller porpuquines. Oho! This would certainly seem to be an infinite regress! But no. It's just that I forgot to mention that there is a smallest size of porpuquine: the zero-inch type, which, amazingly enough, is totally bald of quills. So, quite luckily (or perhaps unluckily, depending on your point of view), that puts a stop to the threatened infinite regress. This remarkable beast is shown in a rare photograph in Figure 17-1.

Students of zoology might be interested to learn that the quills on 5-inch porpuquines are always 4-inch porpuquines, and so on down the line. And students of anthropology might be equally intrigued to know that the residents of Glazunkia (both Outer and Inner) utilize the nose (yes, the nose) of the zero-inch porpuquine as a unit of barter—an odd thing to our

FIGURE 17–1. *The homely Inner Glazunkian porpuquine,* Porpuquinus verdimontianus. *The size of this particular specimen has not been ascertained, although it appears to be at least a 4-incher. The buying power of a porpuquine is the number of zero-inch noses on it. Larger noses, oddly enough, are worth nothing.* [*Photograph by David J. Moser.*]

minds; but then, who are you and I to question the ancient wisdom of the Outer and Inner Glazunkians? Thus, since a largish porpuquine—say a 3-incher or 4-incher—contains many, many such tiny noses, it is a most valuable commodity. The value of a porpuquine is sometimes referred to as its "buying power", or just "power" for short. For instance, a 2-incher found in Inner Glazunkia is almost twice as powerful as a 2-incher found in Outer Glazunkia. Or did I get it backward? It's rather confusing!

Anyway, why am I telling you all this? Oh, I just thought you'd like to hear about it. Besides, who knows? You just might wind up visiting Glazunkia (Inner or Outer) one of these fine days. And then all of this could come in mighty handy.

18

Lisp: Lists and Recursion

March, 1983

SINCE I ended the previous column with a timely newsbreak about the homely Glazunkian porpuquine, I felt it only fitting to start off the present column with more about that little-known but remarkable beast. As you may recall, the quills on any porpuquine (except for the tiniest ones) are smaller porpuquines. The tiniest porpuquines have no quills but do have a nose, and a very important nose at that, since the Glazunkians base their entire monetary system on that little nose. Consider, for instance, the value of 3-inch porpuquines in Outer Glazunkia. Each one always has nine quills (contrasting with their cousins in Inner Glazunkia, which always have seven); thus each one has nine 2-inch porpuquines sticking out of its body. Each of those in turn sports nine 1-inch porpuquines, out of each of which sprout nine zero-inch porpuquines, each of which has one nose. All told, this comes to $9 \times 9 \times 9 \times 1$ noses, which means that a 3-inch porpuquine in Outer Glazunkia has a buying power of 729 noses. If, by contrast, we had been in Inner Glazunkia and had started with a 4-incher, that porpuquine would have a buying power of $7 \times 7 \times 7 \times 7 \times 1 = 2,401$ noses.

Let's see if we can't come up with a general recipe for calculating the buying power (measured in noses) of any old porpuquine. It seems to me that it would go something like this:

The buying power of a porpuquine with a given quill count and size is:
    if its size = 0, then 1;
        otherwise, figure out the buying power of a porpuquine with
                the same quill count but of the next smaller size,
            and multiply that by the quill count.

We can shorten this recipe by adopting some symbolic notation. First, let **q** stand for the quill count and **s** for the size. Then let **cond** stand for "if" and **t** for "otherwise". Finally, use a sort of condensed algebraic notation in which the English names of operations are placed to the left of their operands, inside parentheses. We get something like this:

```
(buying-power q s) is:
   cond (eq s 0) 1;
         t (times q (buying-power q (next-smaller s)))
```

This is an exact translation of the earlier English recipe into a slightly more symbolic form. We can make it a little more compact and symbolic by adopting a couple of new conventions. Let each of the two cases (the case where **s** equals zero and the "otherwise" case) be enclosed in parentheses; in general, use parentheses to enclose each logical unit completely. Finally, indicate by the words **def** and **lambda** that this is a definition of a general notion called **buying-power** with two variables (quill count **q** and size **s**). Now we get:

```
(def buying-power (lambda (q s)
   (cond ((eq s 0) 1)
         (t (times q (buying-power q (next-smaller s)))))))
```

I mentioned above that the buying power of a 9-quill, 3-inch porpuquine is 729 noses. This could be expressed by saying that **(buying-power 9 3)** equals 729. Similarly, **(buying-power 7 4)** equals 2,401.

*   *   *

Well, so much for porpuquines. Now let's limp back to Lisp after this rather long digression. I had posed a puzzle, toward the end of last month's column, in which the object was to write a Lisp function that subsumed a whole family of related functions called **square**, **cube**, **4th-power**, **5th-power**, and so on. I asked you to come up with one *general* function called **power**, having two variables, such that **(power 9 3)** gives 729, **(power 7 4)** gives 2,401, and so on. I had presented a "tower of power"—that is, an infinitely tall tower of separate Lisp definitions, one for each power, connecting it to the preceding power. Thus a typical floor in this tower would be:

```
(def 102nd-power (lambda (q) (times q (101st-power q))))
```

Of course, **101st-power** would refer to **100th-power** in its definition, and so on, thus creating a rather long regress back to the simplest, or "embryonic", case. Incidentally, that very simplest case, rather than **square** or even **1st-power**, is this:

```
(def 0th-power (lambda (q) 1))
```

I told you that you had all the information necessary to assemble the proper definition. All you needed to observe is, of course, that each floor of the tower rests on the "next-smaller" floor (except for the bottom floor, which is a "stand-alone" floor). By "next-smaller", I mean the following:

(def next-smaller (lambda (s) (difference s 1)))

Thus **(next-smaller 102)** yields 101. Actually, Lisp has a standard name for this operation (namely, **sub1**) as well as for its inverse operation (namely, **add1**). If we put all our observations together, we come up with the following universal definition:

(def power (lambda (q s)
  (cond ((eq s 0) 1)
        (t (times q (power q (next-smaller s)))))))

This is the answer to the puzzle I posed. Hmmm, that's funny . . . I have the strangest sense of *déjà vu.* I wonder why!

*　　*　　*

The definition presented here is known as a *recursive* definition, for the reason that inside the *definiens,* the *definiendum* is used. This is a fancy way of saying that I appear to be defining something in terms of itself, which ought to be considered gauche if not downright circular in anyone's book! To see whether the Lisp genie looks askance upon such trickery, let's ask it to figure out **(power 9 3):**

→ **(power 9 3)**
729
→

Well, fancy that! No complaints? No choking? How can the Lisp genie swallow such nonsense?

The best explanation I can give is to point out that *no circularity is actually involved.* While it is true that the definition of **power** uses the word **power** inside itself, the two occurrences are referring to different circumstances. In a nutshell, **(power q s)** is being defined in terms of a simpler case, namely, **(power q (next-smaller s))**. Thus I am defining the 44th power in terms of the 43rd power, and that in terms of the next-smaller power, and so on down the line until we come to the "bottom line", as I call it—the 0th power, which needs no recursion at all. It suffices to tell the genie that its value is 1. So when you look carefully, you see that this recursive definition is no more circular than the "tower of power" was—and you can't get any straighter than an infinite straight line! In fact, this one compact definition really is just a way of getting the whole tower of power into one finite expression. Far from being circular, it is just a handy summary of infinitely many different definitions, all belonging to one family.

In case you still have a trace of skepticism about this sleight of hand, perhaps I should let you watch what the Lisp genie will do if you ask for a "trace" of the function, and then ask it once again to evaluate **(power 9 3)**.

```
-> (power 9 3).
   ENTERING power (q=9, s=3)
      ENTERING power (q=9, s=2)
         ENTERING power (q=9, s=1)
            ENTERING power (q=9, s=0)
            EXITING power (value: 1)
         EXITING power (value: 9)
      EXITING power (value: 81)
   EXITING power (value: 729)
729
->
```

On the lines marked **ENTERING,** the genie prints the values of the two arguments, and on the lines marked **EXITING,** it prints the value it has computed and is returning. For each **ENTERING** line there is of course an **EXITING** line, and the two are aligned vertically—that is, they have the same amount of indentation.

You can see that in order to figure out what **(power 9 3)** is, the genie must first calculate **(power 9 2)**. But this is not a given; instead it requires knowing the value of **(power 9 1)**, and this in turn requires **(power 9 0)**. Ah! But we *were* given this one—it is just 1. And now we can bounce back "up", remembering that in order to get one answer from the "deeper" answer, we must multiply by 9. Hence we get 9, then 81, then 729, and we are done.

I say "we", but of course it is not we but the Lisp genie who must keep track of these things. The Lisp genie has to be able to suspend one computation to work on another one whose answer was requested by the first one. And the second computation, too, may request the answer to a third one, thus putting itself on hold—as may the third, and so on, recursively. But eventually, there will come a case where the buck stops— that is, where a process runs to completion and returns a value—and that will enable other stacked-up processes to finally return values, like stacked-up airplanes that have circled for hours finally getting to land, each landing opening up the way for another landing.

Ordinarily, the Lisp genie will not print out a trace of what it is thinking unless you ask for it. However, whether you ask to see it or not, this kind of thing is going on behind the scenes whenever a function call is evaluated. One of the enjoyable things about Lisp is that it can deal with such recursive definitions without getting flustered.

*   *   *

I am not so naïve as to expect that you've now totally got the hang of recursion and could go out and write huge recursive programs with the greatest of ease. Indeed, recursion can be a remarkably subtle means of defining functions, and sometimes even an expert can have trouble figuring out the meaning of a complicated recursive definition. So I thought I'd give you some practice in working with recursion.

Let me give a simple example based on this silly riddle: "How do you make a pile of 13 stones?" Answer: "Put one stone on top of a pile of 12 stones." (Ask a silly question and get an answer 12/13 as silly.) Suppose we want to make a Lisp function that will give us not a pile of 13 stones, but a list consisting of 13 copies of the atom **stone**—or in general, **n** copies of that atom. We can base our answer on the riddle's silly-seeming yet correct recursive answer. The general notion is to build the answer for **n** out of the answer for **n**'s predecessor. Build how? Using the list-building function **cons**, that's how. What's the embryonic case? That is, for which value of **n** does this riddle present absolutely no problem at all? That's easy: when **n** equals 0, our list should be empty, which means the answer is **nil**. We can now put our observations together as follows:

```
(def bunch-of-stones (lambda (n)
   (cond ((eq n 0) nil)
           (t (cons 'stone (bunch-of-stones (next-smaller n)))))))
```

Now let's watch the genie put together a very small bunch of stones (with **trace** on, just for fun):

```
→ (bunch-of-stones 2)
   ENTERING bunch-of-stones (n=2)
      ENTERING bunch-of-stones (n=1)
         ENTERING bunch-of-stones (n=0)
         EXITING bunch-of-stones (value: nil)
      EXITING bunch-of-stones (value: (stone))
   EXITING bunch-of-stones (value: (stone stone))
(stone stone)
→
```

This is what is called "consing up a list". Now let's try another one. This one is an old chestnut of Lisp and indeed of recursion in general. Look at the definition and see if you can figure out what it's supposed to do; then read on to see if you were right.

```
→ (def wow (lambda (n)
      (cond ((eq n 0) 1)
              (t (times n (wow (sub1 n)))))))
```

Remember, **sub1** means the same as **next-smaller**. For a lark, why don't you calculate the value of **(wow 100)**? (If you ate your mental Wheaties this morning, try it in your head.)

It happens that Lisp genies often mumble out loud while they are executing wishes, and I just happen to have overheard this one as it was executing the wish **(wow 100)**. Its soliloquy ran something like this:

> Hmm . . . **(wow 100)**, eh? Well, 100 surely isn't equal to 0, so I guess the answer has to be 100 times what it *would* have been, had the problem been **(wow 99)**. All rightie—now all I need to do is figure out what **(wow 99)** is. Oh, this is going to be a piece of cake! Let's see, is 99 equal to 0? No, seems not to be, so I guess the answer to *this* problem must be 99 times what the answer *would* have been, had the problem been **(wow 98)**. Oh, this is going to be child's play! Let's see . . .

At this point, the author, having some pressing business at the bank, had to leave the happy genie, and did not again pass the spot until some milliseconds afterwards. When he did so, the genie was just finishing up, saying:

> . . . And now I just need to multiply *that* by 100, and I've got my final answer. Easy as pie! I believe it comes out to be 933262154439441526816992388562667004907159682643816214685929638952175999932299156089414639761565182862536979208272237582511852109168640000000000000000000000000—if I'm not mistaken.

Is that the answer you got, dear reader? No? Ohhh, I see where you went wrong. It was in your multiplication by 52. Go back and try it again from that point on, and be a little more careful in adding those long columns up. I'm quite sure you'll get it right this time.

* * *

This **wow** function is ordinarily called *factorial; n* factorial is usually defined to be the product of all the numbers from 1 through *n*. But a recursive definition looks at things slightly differently: speaking recursively, *n* factorial is simply the product of *n* and the previous factorial. It reduces the given problem to a simpler sort of the same type. That simpler one will in turn be reduced, and so on down the line, until you come to the simplest problem of that type, which I call the "embryonic case" or the "bottom line". People often speak, in fact, of a recursion "bottoming out".

A *New Yorker* cartoon from a few years back illustrates the concept perfectly. It shows a fifty-ish man holding a photograph of himself roughly ten years earlier. In that photograph, he is likewise holding a photograph of himself, ten years earlier than *that*. And on it goes, until eventually it "bottoms out"—quite literally—in a photograph of a bouncy baby boy in his birthday suit (bottom in the air). This idea of recursive photos catching you as you grow up is quite appealing. I wish my parents had thought of it!

Contrast it with the more famous Morton Salt infinite regress, in which the Morton Salt girl holds a box of Morton Salt with her picture on it—but as the girl in the picture is no younger, there is no bottom line and the regress is endless, at least theoretically. Incidentally, the Dutch cocoa called "Droste's" has a similar illustration on its boxes, and very likely so do some other products.

The recursive approach works when you have a family of related problems, at least one of which is so simple that it can be answered immediately. This I call the *embryonic case*. (In the factorial example, that's the **(eq n 0)** case, whose answer is 1.) Each problem ("What is 100 factorial?", for instance) can be viewed as a particular case of one general problem ("How do you calculate factorials?"). Recursion takes advantage of the fact that the answers to various cases are related in some logical way to each other. (For example, I could very easily tell you the value of 100 factorial if only somebody would hand me the value of 99 factorial—all I need to do is multiply by 100.) You could say that the "Recursioneer's Motto" is: "Gee, I could solve *this* case if only someone would magically hand me the answer to the case that's one step closer to the embryonic case." Of course, this motto presumes that certain cases are, in some sense, "nearer" to the embryonic case than others are—in fact, it presumes that there is a natural pathway leading from any case through simpler cases all the way down to the embryonic case, a pathway whose steps are clearly marked all along the way.

As it turns out, this is a very reasonable assumption to make in all sorts of circumstances. To spell out the exact nature of this recursion-guiding pathway, you have to answer two Big Questions:

(1) What is the embryonic case?
(2) What is the relationship of a typical case to the next simpler case?

Now actually, both of these Big Questions break up into two subquestions (as befits any self-respecting recursive question!), one concerning how you recognize where you are or how to move, the other concerning what the answer is at any given stage. Thus, spelled out more explicitly, our Big Questions are:

(1*a*) How can you know when you've reached the embryonic case?
(1*b*) What is the embryonic answer?

(2*a*) From a typical case, how do you take exactly one step toward the embryonic case?
(2*b*) How do you build this case's answer out of the "magically given" answer to the simpler case?

Question (2*a*) concerns the nature of the *descent* toward the embryonic case, or bottom line. Question (2*b*) concerns the inverse aspect, namely, the *ascent* that carries you back up from the bottom to the top level.

In the case of the factorial, the answers to the Big Questions are:

(1*a*) The embryonic case occurs when the argument is 0.
(1*b*) The embryonic answer is 1.

(2*a*) Subtract 1 from the present argument.
(2*b*) Multiply the "magic" answer by the present argument.

Notice how the answers to these four questions are all neatly incorporated in the recursive definition of **wow**.

* * *

Recursion relies on the assumption that sooner or later you will bottom out. One way to be *sure* you'll bottom out is to have all the simplifying or "descending" steps move in the same direction at the same rate, so that your pathway is quite obviously linear. For instance, it's obvious that by subtracting 1 over and over again, you will eventually reach 0, provided you started with a positive integer. Likewise, it's obvious that by performing the list-shortening operation of **cdr**, you will eventually reach **nil**, provided you started with a finite list. For this reason, recursions using **sub1** or **cdr** to define their pathway of descent toward the bottom are commonplace. I'll show a **cdr**-based recursion shortly, but first I want to show a funny numerical recursion in which the pathway toward the embryonic case is anything but linear and smooth. In fact, it is so much like a twisty mountain road that to describe it as moving "towards the embryonic case" seems hardly accurate. And yet, just as mountain roads, no matter how many hairpin turns they make, eventually do hit their destinations, so does this path.

Consider the famous "$3n+1$" problem, in which you start with any positive integer, and if it is even, you halve it; otherwise, you multiply it by 3 and add 1. Let's call the result of this operation on **n (hotpo n)** (standing for "half or triple plus one"). Here is a Lisp definition of **hotpo**:

```
(def hotpo (lambda (n)
  (cond ((even n) (half n))
        (t (add1 (times 3 n))))))
```

This definition presumes that two other functions either have been or will be defined elsewhere for the Lisp genie, namely **even** and **half** (**add1** and **times** being, as mentioned earlier, intrinsic parts of Lisp). Here are the lacking definitions:

```
(def even (lambda (n) (eq (remainder n 2) 0)))
(def half (lambda (n) (quotient n 2)))
```

What do you think happens if you begin with some integer and perform **hotpo** over and over again? Take 7, for instance, as your starting point. Before you do the arithmetic, take a guess as to what sort of behavior might occur.

As it turns out, the pathway followed is often surprisingly chaotic and bumpy. For instance, if we begin with 7, the process leads us to 22, then 11, then 34, 17, 52, 26, 13, 40, 20, 10, 5, 16, 8, 4, 2, 1, 4, 2, 1, 4, 2, 1, . . . Note that we have wound up in a short loop (a *3-cycle,* in the terminology of Chapter 16). Suppose we therefore agree that if we ever reach 1, we have "hit bottom" and may stop. You might well ask, "Who says we *will* hit 1? Is there a guarantee?" (Again in the terminology of Chapter 16, we could ask, "Is the 1-4-2-1 cycle an attractor?") Indeed, before you try it out in a number of cases, you have no particular reason to suspect that you will *ever* hit 1, let alone *always*. (It would be very surprising if someone correctly anticipated what would happen in the case of, say, 7 before trying it out.) However, numerical experimentation reveals a remarkable reliability to the process; it seems that no matter where you start, you always do enter the 1-4-2-1 cycle sooner or later. (Try starting with 27 as seed if you want a real roller-coaster ride!)

Can you write a recursive function to reveal the pathway followed from an arbitrary starting point "down" to 1? Note that I say "down" advisedly, since many of the steps are in fact *up*! Thus the pathway starting at 3 would be the list **(3 10 5 16 8 4 2 1)**. In order to solve this puzzle, you need to go back and answer for yourself the two Big Questions of Recursion, as they apply here. Note:

(cond ((not (want help)) (not (read further)))
(t (read further)))

*   *   *

First—about the embryonic case. This is easy. It has already been defined as the arrival at 1; and the embryonic, or simplest possible, answer is the list **(1)**, a tiny but valid pathway from 1 to 1.

Second—about the more typical cases. What operation will carry us from typical 7 one step closer to embryonic 1? Certainly not the **sub1** operation. No—by definition it's the function **hotpo** itself that brings you ever "nearer" to 1—even when it carries you *up*! This teasing quality is of course the whole point of the example. What about (2*b*)—how to recursively build a list documenting our wildly oscillating pathway? Well, the pathway belonging to 7 is gotten by tacking (*i.e.,* **cons**ing) 7 onto the shorter pathway belonging to **(hotpo 7)**, or 22. After all, 22 is one step closer to being embryonic than 7 is!

These answers enable us to write down the desired function definition, using **tato** as our dummy variable (**tato** being a well-known acronym for

tato (and tato only), which recursively expands to **tato (and tato only) (and tato (and tato only) only)**—and so forth).

```
(def pathway-to-1 (lambda (tato)
   (cond ((eq tato 1) '(1))
         (t (cons tato (pathway-to-1 (hotpo tato)))))))
```

Look at the way the Lisp genie "thinks" (as revealed when the trace feature is on):

```
-> (pathway-to-1 3)
   ENTERING pathway-to-1 (tato=3)
     ENTERING pathway-to-1 (tato=10)
       ENTERING pathway-to-1 (tato=5)
         ENTERING pathway-to-1 (tato=16)
           ENTERING pathway-to-1 (tato=8)
             ENTERING pathway-to-1 (tato=4)
               ENTERING pathway-to-1 (tato=2)
                 ENTERING pathway-to-1 (tato=1)
                 EXITING pathway-to-1 (value: (1))
               EXITING pathway-to-1 (value: (2 1))
             EXITING pathway-to-1 (value: (4 2 1))
           EXITING pathway-to-1 (value: (8 4 2 1))
         EXITING pathway-to-1 (value: (16 8 4 2 1))
       EXITING pathway-to-1 (value: (5 16 8 4 2 1))
     EXITING pathway-to-1 (value: (10 5 16 8 4 2 1))
   EXITING pathway-to-1 (value: (3 10 5 16 8 4 2 1))
(3 10 5 16 8 4 2 1)
->
```

Notice the total regularity (the sideways 'V' shape) of the left margin of the trace diagram, despite the chaos of the numbers involved. Not all recursions are so geometrically pretty, when traced. This is because some problems request *more than one subproblem* to be solved. As a practical real-life example of such a problem, consider how you might go about counting up all the unicorns in Europe. This is certainly a nontrivial undertaking, yet there is an elegant recursive answer: Count up all the unicorns in Portugal, also count up all the unicorns in the other 30-odd countries of Europe, and finally add those two results together.

Notice how this spawns two smaller unicorn-counting subproblems, which in turn will spawn two subproblems each, and so on. Thus, how can one count all the unicorns in Portugal? Easy: Add the number of unicorns in the Estremadura region to the number of unicorns in the rest of Portugal! And how do you count up the unicorns in Estremadura (not to mention those in the remaining regions of Portugal)? By further breakup, of course.

But what is the bottom line? Well, regions can be broken up into districts, districts into square kilometers, square kilometers into hectares, hectares into square meters—and presumably we can handle each square meter without further breakup.

Although this may sound rather arduous, there really is no other way to conduct a thorough census than to traverse every single part on every level of the full structure that you have, no matter how giant it may be. There is a perfect Lisp counterpart to this unicorn census: it is the problem of determining how many atoms there are inside an arbitrary list. How can we write a Lisp function called **atomcount** that will give us the answer 15 when it is shown the following strange-looking list (which we'll call **brahma**)?

(((ac ab cb) ac (ba bc ac)) ab ((cb ca ba) cb (ac ab cb)))

One method, expressed recursively, is exactly parallel to that for ascertaining the unicorn population of Europe. See if you can come up with it on your own.

*   *   *

The idea is this. We want to construct the answer—namely, 15—out of the answers to simpler atom-counting problems. Well, it is obvious that one simpler atom-counting problem than **(atomcount brahma)** is **(atomcount (car brahma))**. Another one is **(atomcount (cdr brahma))**. The answers to these two problems are, respectively, 7 and 8. Now clearly, 15 is made out of 7 and 8 by addition—which makes sense, after all, since the total number of atoms must be the number in the **car** plus the number in the **cdr**. There's nowhere else for any atoms to hide! Well, this analysis gives us the following recursive definition, with **s** as the dummy variable:

```
(def atomcount (lambda (s)
  (plus (atomcount (car s)) (atomcount (cdr s)))))
```

It looks very simple, but it has a couple of flaws. First, we have written the *recursive* part of the definition, but we have utterly forgotten the other equally vital half—the "bottom line". It reminds me of the Maryland judge I once read about in the paper, who ruled: "A horse is a four-legged animal that is produced by two other horses." This is a lovely definition, but where does it bottom out? Similarly for **atomcount**. What is the simplest case, the embryonic case, of **atomcount**? Simple: It is when we are asked to count the atoms in a single atom. The answer, in such a case, is of course 1. But how can we know when we are looking at an atom? Fortunately, Lisp has a built-in function called **atom** that returns **t** (meaning "true") whenever we are looking at an atom, and **nil** otherwise. Thus **(atom 'plop)** returns **t**, while **(atom '(a b c))** returns **nil**. Using that, we can patch up our definition:

```
(def atomcount (lambda (s)
  (cond ((atom s) 1)
        (t (plus (atomcount (car s)) (atomcount (cdr s)))))))
```

Still, though, it is not quite right. If we ask the genie for **atomcount** of **(a b c)**, instead of getting 3 for an answer, we will get 4. Shocking! How come this happens? Well, we can pin the problem down by trying an even simpler example: if we ask for **(atomcount '(a))**, we find we get 2 instead of 1. Now the error should be clearer: $2 = 1 + 1$, with 1 each coming from the **car** and **cdr** of **(a)**. The **car** is the atom **a** which indeed should be counted as 1, but the **cdr** is **nil**, which should not. So why does **nil** give an **atomcount** of 1? Because **nil** is not only an empty list, it is also an atom! To suppress this bad effect, we simply insert another **cond** clause at the very top:

```
(def atomcount (lambda (s)
  (cond ((null s) 0)
        ((atom s) 1)
        (t (plus (atomcount (car s)) (atomcount (cdr s)))))))
```

I wrote **(null s)**, which is just another way of saying **(eq s nil)**. In general, if you want to determine whether the value of some expression is **nil** or not, you can use the in-built function **null**, which returns **t** if yes, **nil** if no. Thus, for example, **(null (null nil))** evaluates to **nil**, since the inner function call evaluates to **t**, and **t** is not **nil**!

Notice in this recursion that we have more than one type of embryonic case (the **null** case and the **atom** case), and more than one way of descending toward the embryonic case (via both **car** and **cdr**). Thus, our Big Questions can be revised a bit further:

(1*a*) Is there just one embryonic case, or are there several, or even an infinite class of them?

(1*b*) How can you know when you've reached an embryonic case?

(1*c*) What are the answers to the various embryonic cases?

(2*a*) From a typical case, is there exactly one way to step toward an embryonic case, or are there various possibilities?

(2*b*) From a typical case, how do you determine which of the various routes toward an embryonic case to take?

(2*c*) How do you build this case's answer out of the "magically given" answers to one or more simpler cases?

Now what happens when we trace our function as it counts the atoms in **brahma**, our original target? The result is shown in Figure 18-1. Notice the more complicated topography of this recursion, with its many ins and outs.

```
-> (atomcount brahma)
ENTERING atomcount (s=(((ac ab cb) ac (ba bc ac)) ab ((cb ca ba) cb (ac ab cb))))
 ENTERING atomcount (s=((ac ab cb) ac (ba bc ac)))
  ENTERING atomcount (s=(ac ab cb))
   ENTERING atomcount (s=ac)
   EXITING  atomcount (value: 1)
   ENTERING atomcount (s=(ab cb))
    ENTERING atomcount (s=ab)
    EXITING  atomcount (value: 1)
    ENTERING atomcount (s=(cb))
     ENTERING atomcount (s=cb)
     EXITING  atomcount (value: 1)
     ENTERING atomcount (s=nil)
     EXITING  atomcount (value: 0)
    EXITING  atomcount (value: 1)
   EXITING  atomcount (value: 2)
  EXITING  atomcount (value: 3)
  ENTERING atomcount (s=(ac (ba bc ac)))
   ENTERING atomcount (s=ac)
   EXITING  atomcount (value: 1)
   ENTERING atomcount (s=((ba bc ac)))
    ENTERING atomcount (s=(ba bc ac))
     ENTERING atomcount (s=ba)
     EXITING  atomcount (value: 1)
     ENTERING atomcount (s=(bc ac))
      ENTERING atomcount (s=bc)
      EXITING  atomcount (value: 1)
      ENTERING atomcount (s=(ac))
       ENTERING atomcount (s=ac)
       EXITING  atomcount (value: 1)
       ENTERING atomcount (s=nil)
       EXITING  atomcount (value: 0)
      EXITING  atomcount (value: 1)
     EXITING  atomcount (value: 2)
    EXITING  atomcount (value: 3)
    ENTERING atomcount (s=nil)
    EXITING  atomcount (value: 0)
   EXITING  atomcount (value: 3)
  EXITING  atomcount (value: 4)
 EXITING  atomcount (value: 7)
 ENTERING atomcount (s=(ab ((cb ca ba) cb (ac ab cb))))
  ENTERING atomcount (s=ab)
  EXITING  atomcount (value: 1)
  ENTERING atomcount (s=(((cb ca ba) cb (ac ab cb))))
   ENTERING atomcount (s=((cb ca ba) cb (ac ab cb)))
    ENTERING atomcount (s=(cb ca ba))
     ENTERING atomcount (s=cb)
     EXITING  atomcount (value: 1)
     ENTERING atomcount (s=(ca ba))
      ENTERING atomcount (s=ca)
      EXITING  atomcount (value: 1)
      ENTERING atomcount (s=(ba))
       ENTERING atomcount (s=ba)
       EXITING  atomcount (value: 1)
       ENTERING atomcount (s=nil)
       EXITING  atomcount (value: 0)
      EXITING  atomcount (value: 1)
     EXITING  atomcount (value: 2)
    EXITING  atomcount (value: 3)
    ENTERING atomcount (s=(cb (ac ab cb)))
     ENTERING atomcount (s=cb)
     EXITING  atomcount (value: 1)
     ENTERING atomcount (s=((ac ab cb)))
      ENTERING atomcount (s=(ac ab cb))
       ENTERING atomcount (s=ac)
       EXITING  atomcount (value: 1)
       ENTERING atomcount (s=(ab cb))
        ENTERING atomcount (s=ab)
        EXITING  atomcount (value: 1)
        ENTERING atomcount (s=(cb))
         ENTERING atomcount (s=cb)
         EXITING  atomcount (value: 1)
         ENTERING atomcount (s=nil)
         EXITING  atomcount (value: 0)
        EXITING  atomcount (value: 1)
       EXITING  atomcount (value: 2)
      EXITING  atomcount (value: 3)
      ENTERING atomcount (s=nil)
      EXITING  atomcount (value: 0)
     EXITING  atomcount (value: 3)
    EXITING  atomcount (value: 4)
   EXITING  atomcount (value: 7)
   ENTERING atomcount (s=nil)
   EXITING  atomcount (value: 0)
  EXITING  atomcount (value: 7)
 EXITING  atomcount (value: 8)
EXITING  atomcount (value: 15)
15
->
```

FIGURE 18-1. *A trace of the execution of the Lisp function-call* **(atomcount brahma)**.
Recursion in action!

Whereas the previous 'V'-shaped recursion looked like a simple descent into a smooth-walled canyon and then a simple climb back up the other side, this recursion looks like a descent into a much craggier canyon, where on your way up and down each wall you encounter various "subcanyons" that you must treat in the same way—and who knows how many levels of such structure you will be called on to deal with in your exploration?

Shapes with substructure that goes on indefinitely like that, *never* bottoming out in ordinary curves, are called *fractals*. Their nature forms an important area of inquiry in mathematics today. An excellent introduction can be found in Martin Gardner's "Mathematical Games" for April, 1978, and a much fuller treatment in Benoît Mandelbrot's splendid book *The Fractal Geometry of Nature*. For a dynamic view of a few historically revolutionary fractals, there is Nelson Max's marvelous film *Space-Filling Curves*, where so-called "pathological" shapes are constructed step by step before your eyes, and their mathematical significance is geometrically presented. Then, as eerie electronic music echoes all about, you start shrinking like Alice in Wonderland—but unlike her, you can't stop, and as you shrink towards oblivion, you get to see ever more microscopic views of the infinitely detailed fractal structures. It's a great visual adventure, if you're willing to experience infinity-vertigo!

* * *

One of the most elegant recursions I know of originates with the famous disk-moving puzzle known variously as "Lucas' Tower", the "Tower of Hanoi", and the "Tower of Brahma". Apparently it was originated by the French mathematician Edouard Lucas in the nineteenth century. The legend that is popularly attached to the puzzle goes like this:

> In the great Temple of Brahma in Benares, on a brass plate beneath the dome that marks the Center of the World, there are 64 disks of pure gold which the priests carry one at a time between three diamond needles according to Brahma's immutable law: No disk may be placed on a smaller disk. In the Beginning of the World, all 64 disks formed the Tower of Brahma on one needle. Now, however, the process of transfer of the tower from one needle to another is in midcourse. When the last disk is finally in place, once again forming the Tower of Brahma but on a different needle, then will come the End of the World, and All will turn to dust.

A picture of the puzzle is shown in Figure 18-2. In it, the three needles are labeled **a**, **b**, and **c**.

If you work at it, you certainly can discover the systematic method that the priests must follow in order to get the disks from needle **a** to needle **b**. For only three disks, for instance, it is very easy to write down the order in which the moves go:

ab ac bc ab ca cb ab

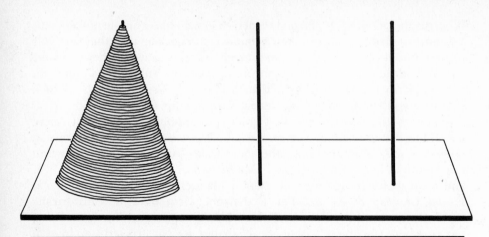

FIGURE 18–2.   *The Tower of Brahma puzzle, with 64 disks to be transferred.* [*Drawing by David Moser.*]

Here, the Lisp atom **ab** represents a jump from needle **a** to needle **b**. There is a structure to what is going on, however, that is not revealed by a mere listing of such atoms. It is better revealed if one groups the atoms as follows:

ab ac bc   ab   ca cb ab

The first group accomplishes a transfer of a 2-tower from needle **a** to needle **c**, thus freeing up the largest disk. Then the middle move, **ab**, picks up that big heavy disk and carries it over from needle a to needle b. The final group is very much like the initial group, in that it transfers the 2-tower back from needle **c** to needle **b**. Thus the solution to moving three depends on being able to move two. Similarly, the solution to moving 64 depends on being able to move 63. Enough said? Now try to write a Lisp function that will give you a solution to the Tower of Brahma for **n** disks. (You may prefer to label the three needles with digits rather than letters, so that moves are represented by two-digit numbers such as **12**.) I will present the solution in the next column—unless, of course, the dedicated priests, working by day and by night to bring about the end of the world, should chance to reach their cherished goal before then . . .

19

Lisp: Recursion and Generality

April, 1983

In the preceding column, I described Edouard Lucas' Tower of Brahma puzzle, in which the object is to transfer a tower of 64 gold disks from one diamond needle to another, making use of a third needle on which disks can be placed temporarily. Disks must be picked up and moved one at a time, the only other constraint being that no disk may ever sit on a smaller one. The problem I posed for readers was to come up with a recursive description, expressed as a Lisp function, of how to accomplish this goal (and thereby end the world).

I pointed out that the recursion is evident enough: to transfer 64 disks from one needle to another (using a third), it suffices to know how to transfer 63 disks from one needle to another (using a third). To recap it, the idea is this. Suppose that the 64-disk tower of Brahma starts out on needle **a**. Figure 19-1 shows a schematic picture, representing all 64 disks by a mere 4. First of all, using your presumed 63-disk-moving ability, transfer 63 disks from needle **a** to needle **c**, using needle **b** as your "helping needle". Figure 19-1*b* shows how the set-up now looks. (Note: In the figure, 4 plays the role of 64, so 3 plays the role of 63, but for some reason, 1 doesn't play the role of 61. Isn't that peculiar?) All right. Now simply pick up the one remaining **a**-disk—the biggest disk of all—and plunk it down on needle **b**, as is shown in Figure 19-1*c*. Now you can see how easy it will be to finish up—simply re-exploit your 63-disk ability so as to transfer that pile on **c** back to **b**, this time using **a** as the "helping needle". Notice how in this maneuver, needle **a** plays the helping role that needle **c** played in the previous 63-disk maneuver. Figure 19-1*d* shows the situation just a split second before the last disk is put in place. Why not *after* it's in place? Simple: Because the entire world then turns to dust, and it's too hard to draw dust.

* * *

Now someone might complain that I left out all the hard parts: "You just magically assumed an ability to move 63 disks!" So it might seem, but there's nothing magical about such an assumption. After all, to move 63, you

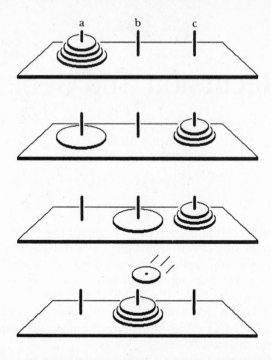

FIGURE 19–1. *A smaller Tower of Brahma puzzle. At the top, the starting position. Below it is shown in an intermediate stage, in which a three-high pile has been transferred from needle a to needle c. At this point, the biggest disk has become free, and can be jumped to needle b. Then all that is left is to re-transfer the three-high pile from c to b. When this is done, the world will end. Thus, the final picture shows an artist's conception of the world a mere split-second before it all turns to dust.*

merely need to know how to move 62. And to move 62, you merely need to know how to move 61. On it goes down the line, until you "bottom out" at the "embryonic case" of the Tower of Brahma puzzle, the 1-disk puzzle. Now, I'll admit that you have to keep track of where you are in this process, and that may be a bit tedious—but that's merely bookkeeping. In principle, you now could actually carry the whole process out—if you were bent on seeing the world end!

As our first approximation to a Lisp function, let's write an English description of the method. Let's call the three needles **sn**, **dn**, and **hn**, standing for "source-needle", "destination-needle", and "helping-needle". Here goes:

To move a tower of height n from sn to dn making use of hn:
    if n = 1, then just carry that one disk directly from sn to dn;
    otherwise, do the following three steps:

(1) move a tower of height n−1 from sn to hn making use of dn;
(2) carry 1 disk from sn to dn;
(3) move a tower of height n−1 from hn to dn making use of sn.

Here, lines (1) and (3) are the two recursive calls; skirting paradox, they seem to call upon the very ability they are helping to define. The saving feature is that they involve **n−1** disks instead of **n**. Note that in line (1), **hn** plays the "destination" role while **dn** plays the "helper" role. And in (3), **hn** plays the "source" role while **sn** plays the "helper" role. Since the whole thing is recursive, every needle will be switching hats many times over during the course of the transfer. That's the beauty of this puzzle and in a way it's the beauty of recursion.

Now how do we make the transition from English to Lisp? It's quite simple:

```
(def move-tower (lambda (n sn dn hn)
  (cond ((eq n 1) (carry-one-disk sn dn))
        (t (move-tower (sub1 n) sn hn dn)
           (carry-one-disk sn dn)
           (move-tower (sub1 n) hn dn sn)))))
```

Where are the Lisp equivalents of the English words "from", "to", and "making use of"? They seem to have disappeared! So how can the genie know which needle is to play which role at each stage? The answer is, this information is conveyed *positionally*. There are, in this function definition, four parameters: one integer and three "dummy needles". The first of these three is the source, the second the destination, the third the helper. Thus in the initial list of parameters (following the **lambda**) they are in the order **sn dn hn.** In the first recursive call, the Lisp translation of line (1), they are in the order **sn hn dn**, showing how **hn** and **dn** have switched hats. In the second recursive call, the Lisp translation of line (3), you can see that **hn** and **sn** have switched hats.

The point is that the atom names **sn**, **dn**, and **hn** carry no intrinsic meaning to the genie. They could as well have been **apple**, **banana**, and **cherry**. Their meanings are defined operationally, by the places where they appear in the various parts of the function definition. Thus it would have been a gross blunder to have written, for instance, **(move-tower (sub1 n) sn dn hn)** as Lisp for line (1), because this contains no indication that **hn** and **dn** must switch roles in that line.

*   *   *

An important question remains. What happens when that friendly Lisp genie comes to a line that says **carry-one-disk**? Does it suddenly zoom off to the Temple at Benares and literally heft a solid gold disk? Or, more prosaically, does it pick up a plastic disk on a table and transfer it from one plastic needle to another? In other words, does some physical action, rather than a mere computation, take place?

Well, in theory that is quite possible. In fact, even in practice the execution of a Lisp command could actually cause a mechanical arm to move to a specific location, to pick up whatever its mechanical hand grasps there, to carry that object to another specific location, and then to release it there. In these days of industrial robots, there is nothing science-fictional about that. However, in the absence of a mechanical arm and hand to move physical disks, what could it mean?

One obvious and closely related possibility is to have there be a display of the puzzle on a screen, and for **carry-one-disk** to cause a *picture* of a hand to move, to *appear* to grasp a disk, pick it up, and replace it somewhere else. This would amount to simulating the puzzle graphically, which can be done with varying degrees of realism, as anyone who has seen science-fiction films using state-of-the-art computer graphics techniques knows.

However, suppose that we don't have fancy graphics hardware or software at our disposition. Suppose that all we want is to create a printed recipe telling us how to move our own soft, organic, human hands so as to solve the puzzle. Thus in the case of a three-disk puzzle, the desired recipe might read **ab ac bc ab ca cb ab** or equally well **12 13 23 12 31 32 12**. What could help us reach this humbler goal?

If we had a program that moved an arm, we would be concerned not with the value it returned, but with the patterned sequence of "side effects" it carried out. Here, by contrast, we are most concerned with the value that our program is going to return—a patterned list of atoms. The list for **n** = 3 has got to be built up from two lists for **n** = 2. This idea was shown at the end of last month's column:

ab ac bc   ab   ca cb ab

In Lisp, to set groups apart, rather than using wider spaces, we use parentheses. Thus our goal for **n** = 3 might be to produce the sandwich-like list **((ab ac bc) ab (ca cb ab))**. One way to produce a list out of its components is to use **cons** repeatedly. Thus if the values of the atoms **apple**, **banana**, and **cherry** are 1, 2, and 3, respectively, then the value returned by **(cons apple (cons banana (cons cherry nil)))** will be the list **(1 2 3)**. However, there is a shorter way to get the same result, namely, to write **(list apple banana cherry)**. It returns the same value. Similarly, if the atoms **sn** and **dn** are bound to **a** and **b** respectively, then execution of the command **(list sn dn)** will return **(a b)**. The function **list** is an unusual function in that it takes any number of arguments at all—even none, so that **(list)** returns the value of **nil**!

Now let us tackle the problem of what value we want the function called **carry-one-disk** to return. It has two parameters that represent needles, and ideally we'd like it to return a single atom made out of those needles' names, such as **ab** or **12**. For the moment, it'll be easier if we assume that the needle names are the numbers 1, 2, and 3. In this case, to make the number 12 out of 1 and 2, it suffices to do a little bit of arithmetic: multiply the first by 10 and add on the second. Here is the Lisp for that:

(def carry-one-disk (lambda (sn dn) (plus (times 10 sn) dn)))

On the other hand, if the needle names are nonnumeric atoms, then we can use a standard Lisp function called **concat**, which takes the values of its arguments (any number, as with **list**) and concatenates them all so as to make one big atom. Thus **(concat 'con 'cat 'e 'nate)** returns the atom **concatenate**. In such a case, we could write:

(def carry-one-disk (lambda (sn dn) (concat sn dn)))

Either way, we have solved the "bottom line" half of the **move-tower** problem.

The other half of the problem is what the recursive part of **move-tower** will return. Well, that is pretty simple. We simply would like it to return a sandwich-like list in which the values of the two recursive calls flank the value of the single call to **carry-one-disk**. So we can modify our previous recursive definition very slightly, by adding the word **list**:

```
(def move-tower (lambda (n sn dn hn)
   (cond ((eq n 1) (carry-one-disk sn dn))
         (t (list (move-tower (sub1 n) sn hn dn)
            (carry-one-disk sn dn)
            (move-tower (sub1 n) hn dn sn))))))
```

Now let's conduct a little exchange with the Lisp genie:

```
-> (move-tower 4 'a 'b 'c)
(((ac ab cb) ac (ba bc ac)) ab ((cb ca ba) cb (ac ab cb)))
->
```

Smashing! It actually works! Isn't that pretty? In last month's column, this list was called **brahma**.

*   *   *

Suppose we wished to suppress all the inner parentheses, so that just a long uninterrupted sequence of atoms would be printed out. For instance, we would get **(ac ab cb ac ba bc ac ab cb ca ba cb ac ab cb)** instead of the intricacy of **brahma**. This would be slightly less informative, but it would be more impressive in its opaqueness. In this case, we would not want to use the function **list** to make our sandwich of three values, but would have to use some other function that removed the parentheses from the two flanking recursive values.

This is a case where the Lisp function **append** comes in handy. It splices any number of lists together, dropping their outermost parentheses as it does so. Thus **(append '(a (b)) '(c) nil '(d e))** yields the five-element list **(a (b) c d e)** rather than the four-element list **((a (b)) (c) nil (d e))**, which would be yielded if **list** rather than **append** appeared in the function position. Using **append** and a slightly modified version of **carry-one-disk** to work with it, we can formulate a final definition of **move-tower** that does what we want:

```
(def move-tower (lambda (n sn dn hn)
   (cond ((eq n 1) (carry-one-disk sn dn))
          (t (append (move-tower (sub1 n) sn hn dn)
                     (carry-one-disk sn dn)
                     (move-tower (sub1 n) hn dn sn))))))

(def carry-one-disk (lambda (sn dn) (list (concat sn dn))))
```

To test this out, I asked the Lisp genie to solve a 9-high Tower of Brahma puzzle. Here is what it shot back at me, virtually instantaneously:

```
(ab ac bc ab ca cb ab ac bc ba ca bc ab ac bc ab ca cb ab ca bc ba ca cb
 ab ac bc ab ca cb ab ac bc ba ca bc ab ac bc ba ca cb ab ca bc ba ca bc
 ab ac bc ab ca cb ab ac bc ba ca bc ab ac bc ab ca cb ab ca bc ba ca cb
 ab ac bc ab ca cb ab ca bc ba ca bc ab ac bc ba ca cb ab ca bc ba ca cb
 ab ac bc ab ca cb ab ac bc ba ca bc ab ac bc ab ca cb ab ca bc ba ca cb
 ab ac bc ab ca cb ab ac bc ba ca bc ab ac bc ba ca cb ab ca bc ba ca bc
 ab ac bc ab ca cb ab ac bc ba ca bc ab ac bc ba ca cb ab ca bc ba ca cb
 ab ac bc ab ca cb ab ca bc ba ca bc ab ac bc ba ca cb ab ca bc ba ca bc
 ab ac bc ab ca cb ab ac bc ba ca bc ab ac bc ab ca cb ab ca bc ba ca cb
 ab ac bc ab ca cb ab ac bc ba ca bc ab ac bc ba ca cb ab ca bc ba ca bc
 ab ac bc ab ca cb ab ac bc ba ca bc ab ac bc ab ca cb ab ca bc ba ca cb
 ab ac bc ab ca cb ab ca bc ba ca bc ab ac bc ba ca cb ab ca bc ba ca cb
 ab ac bc ab ca cb ab ac bc ba ca bc ab ac bc ab ca cb ab ca bc ba ca cb
 ab ac bc ab ca cb ab ca bc ba ca bc ab ac bc ba ca cb ab ca bc ba ca bc
 ab ac bc ab ca cb ab ac bc ba ca bc ab ac bc ba ca cb ab ca bc ba ca cb
 ab ac bc ab ca cb ab ca bc ba ca bc ab ac bc ba ca cb ab ca bc ba ca cb
 ab ac bc ab ca cb ab ac bc ba ca bc ab ac bc ab ca cb ab ca bc ba ca cb
 ab ac bc ab ca cb ab ac bc ba ca bc ab ac bc ba ca cb ab ca bc ba ca bc
 ab ac bc ab ca cb ab ac bc ba ca bc ab ac bc ab ca cb ab ca bc ba ca cb
 ab ac bc ab ca cb ab ca bc ba ca bc ab ac bc ba ca cb ab ca bc ba ca cb
 ab ac bc ab ca cb ab ac bc ba ca bc ab ac bc ab ca cb ab ca bc ba ca cb
 ab ac bc ab ca cb ab)
```

Now that's the kind of genie I like!

*   *   *

Congratulations! You have just been through a rather sophisticated and brain-taxing example of recursion. Now let us take a look at a recursion that offers us a different kind of challenge. This recursion comes from an offhand remark I made last column. I used the odd variable name **tato**, mentioning that it is a recursive acronym standing for **tato (and tato only)**. Using this fact you can expand **tato** any number of times. The sole principle is that each occurrence of **tato** on a given level is replaced by the two-part phrase **tato (and tato only)** to make the next level. Here is a short table:

n=0: tato
n=1: tato
      (and tato only)
n=2: tato
      (and tato only)
      (and tato (and tato only) only)
n=3: tato
      (and tato only)
      (and tato (and tato only) only)
      (and tato (and tato only) (and tato (and tato only) only) only)

For us the challenge is to write a Lisp function that returns what **tato** becomes after **n** recursive expansions, for any **n**. Irrelevant that for **n** much bigger than 3 the whole thing gets ridiculously large. We're theoreticians!

There is only one problem. Any Lisp function must return a single Lisp structure (an atom or a list) as its value; however, the entries in our table do not satisfy this criterion. For instance, the one for **n**=2 consists of one atom followed by two lists. To fix this, we can turn each of the entries in the table into a list by enclosing it in one outermost pair of parentheses. Now our goal is consistent with Lisp. How do we attain it?

Recursive thinking tells us that the bottom line, or embryonic case, occurs when **n**=0, and that otherwise, the **n**th line is made from the line before it by replacing the atom **tato**, wherever it occurs, by the list **(tato (and tato only))**, only without its outermost parentheses. We can write this up right away.

```
(def tato-expansion (lambda (n)
(cond ((eq n 0) '(tato))
       (t (replace 'tato '(tato (and tato only)) (tato-expansion (sub1 n))))))))
```

The only thing is, we have not specified what we mean by **replace**. We must be very careful in defining how to carry out our **replace** operation. Look at any of the lines of the **tato** table, and you will see that it contains one element more than the preceding line. Why is this? Because the atom **tato** gets replaced each time by a two-element list whose parentheses, as I pointed out earlier, are dropped during the act of replacement. It's this parenthesis-dropping that is the sticky point. A less tricky example of such parenthesis-dropping replacement than the recursive one involving **tato** would be this: **(replace 'a '(1 2 3) '(a b a))**, whose value should be **(1 2 3 b 1 2 3)** rather than **((1 2 3) b (1 2 3))**. Rather than exact substitution of a list for an atom, this kind of replacement involves splicing or appending a list inside a longer list.

Let's try to specify in Lisp—using recursion, as usual—just what we mean by **replace**-ing all occurrences of the atom **atm** by a list called **lyst**, inside a long list called **longlist**. This is a good puzzle for you to try. A hint: See how the answer for argument **(a b a)** is built out of the answer for argument **(b a)**. Also look at other simple cases like that, moving back down toward the embryonic case.

∗   ∗   ∗

The embryonic case occurs when **longlist** is **nil**. Then, of course, nothing happens so our answer should be **nil**.

The recursive case involves building a more complex answer from a simpler one assumed given. We can fall back on our **(a b a)** example for this. We can build the complex answer **(1 2 3 b 1 2 3)** out of the simpler answer **(b 1 2 3)** by appending **(1 2 3)** onto it. On the other hand, we could consider **(b 1 2 3)** itself to be a complex answer built from the simpler answer **(1 2 3)** by consing **b** onto it. Why does one involve **append**ing and the other involve **cons**ing? Simple: Because the first case involves the atom **a**, which *does* get replaced, while the second involves the atom **b**, which does *not* get replaced. This observation allows us to attempt to write down an attempt at a recursive definition of **replace**, as follows:

```
(def replace (lambda (atm lyst longlist)
   (cond ((null longlist) nil)
         ((eq (car longlist) atm)
          (append lyst (replace atm lyst (cdr longlist))))
         (t (cons (car longlist) (replace atm lyst (cdr longlist))))))))
```

As you can see, there is an embryonic case (where longlist equals **nil**), and then one recursive case featuring **append** and one recursive case featuring **cons.** Now let's try out this definition on a new example.

```
-> (replace 'a '(1 2 3) '(a (a) b a))
(1 2 3 (a) b 1 2 3)
->
```

Whoops! It *almost* worked, except that one of the occurrences of **a** was completely missed. This means that in our definition of **replace**, we must have overlooked some eventuality. Indeed, if you go back, you will see that an unwarranted assumption slipped in right under our noses—namely, that the elements of **longlist** are always atoms. We ignored the possibility that **longlist** might contain sublists. And what to do in such a case? Answer: Do the replacement inside those sublists as well. And inside sublists of sublists, too—and so on. Can you figure out a way to fix up the ailing definition?

*   *   *

We've seen a recursion before in which all parts on all levels of a structure needed to be explored; it was the function **atomcount** last month, in which we did a simultaneous recursion on both **car** and **cdr**. The recursive line ran **(plus (atomcount (car s)) (atomcount (cdr s)))**. Here it will be quite analogous. We'll have a recursive line featuring *two* calls on **replace**, one involving the **car** of **longlist** and one involving the **cdr**, instead of just one involving the **cdr**. And this makes perfect sense, once you think about it. Suppose you wanted to replace all the unicorns in Europe by porpuquines. One way to achieve this nefarious goal would be to split Europe into two pieces: Portugal (Europe's **car**), and all the rest (its **cdr**). After replacing all the unicorns in Portugal by porpuquines, and also all the unicorns in the rest of Europe by porpuquines, finally you would recombine the two new pieces into a reunified Europe (this is supposed to suggest a **cons** operation). Of course, to carry out this dastardly operation on Portugal, an analogous splitting and rejoining would have to take place—and so on. This suggests that our recursive line will look like this:

```
(cons (replace 'unicorn '(porpuquine) (car geographical-unit))
      (replace 'unicorn '(porpuquine) (cdr geographical-unit)))
```

or, more elegantly and more generally,

```
(cons (replace atm lyst (car longlist)) (replace atm lyst (cdr longlist)))
```

This **cons** line will cover the case where **longlist**'s **car** is nonatomic, as well as the case where it is atomic but not equal to **atm**. In order to make this work, we need to augment the embryonic case slightly: we'll say that when **longlist** is not a list but an atom, then **replace** has no effect on **longlist** at all. Conveniently, this subsumes the earlier **null** line, so we can drop that one. If we put all this together, we come up with a new, improved definition:

```
(def replace (lambda (atm lyst longlist)
   (cond ((atom longlist) longlist)
         ((eq (car longlist) atm)
          (append lyst (replace atm lyst (cdr longlist))))
         (t (cons (replace atm lyst (car longlist))
                  (replace atm lyst (cdr longlist)))))))
```

Now when we say **(tato-expansion 2)** to the Lisp genie, it will print out for us the list **(tato (and tato only) (and tato (and tato only) only))**.

*   *   *

Well, well. Isn't this a magnificent accomplishment? If it seems less than magnificent, perhaps we can carry it a step further. A recursive acronym— one containing a letter standing for the acronym itself—can be amusing, but what of *mutually* recursive acronyms? This could mean, for instance, two acronyms, each of which contains a letter standing for the other acronym. An example would be the pair of acronyms **NOODLES** and **LINGUINI**, standing for:

NOODLES (oodles of delicious LINGUINI), elegantly served

and

luscious itty-bitty NOODLES gotten usually in naples, italy

respectively. Notice, incidentally, that **NOODLES** is not only indirectly but also directly recursive. There's nothing wrong with that.

In general, the notion of mutual recursion means a system of arbitrarily many interwoven structures, each of which is defined in terms of one or more members of the system (possibly including itself). If we are speaking of a family of mutually recursive acronyms, then this means a collection of words, letters in any one of which can stand for any word in the family.

I have to admit that this specific notion of mutually recursive acronyms is not particularly useful in any practical sense. However, it is quite useful as a droll example of a very common abstract phenomenon. Who has not at some time mused about the inevitable circularity of dictionary definitions? Anyone can see that all words eventually are defined in terms of some fundamental set that is not further reducible, but simply goes round and round endlessly. You can amuse yourself by looking up the definition of a common word in a dictionary and replacing the main words in it by *their* definitions. I once carried this process out for "love" (defined as "A strong affection for or attachment or devotion to a person or persons"), substituting for "strong", "affection", "attachment", "devotion", and "person", and coming up with this concoction:

> A morally powerful mental state or tendency, having strength of character or will for, or affectionate regard, or loyalty, faithfulness, or deep affection to, a human being or beings, especially as distinguished from a thing or lower animal.

But not being satisfied with that, I carried the whole process one step further. This was my result:

A set of circumstances or attributes characterizing a person or thing at a given time in, with, or by the conscious or unconscious together as a unit full of or having a specific ability or capacity in a manner relating to, dealing with, or capable of making the distinction between right and wrong in conduct, or an inclination to move or act in a particular direction or way, having the state or quality of being strong in moral strength, self-discipline, or fortitude, or the act or process of volition for, or consideration, attention, or concern full of fond or tender feeling for, or the quality, state, or instance of being faithful to, those persons or ideals that one is under obligation to defend or support, or the condition, quality or state of being worthy of trust, or a strongly felt fond or tender feeling to a creature or creatures of or characteristic of a person or persons, that lives or exists, or is assumed to do so, particularly as separated or marked off by differences from that which is conceived, spoken of, or referred to as existing as an individual entity, or from any living organism inferior in rank, dignity, or authority, typically capable of moving about but not of making its own food by photosynthesis.

Isn't it romantic? It certainly makes "love" ever more mysterious. Stuart Chase, in his lucid classic on semantics, *The Tyranny of Words,* does a similar exercise for "mind" and shows its opacity equally well. But of course concrete words as well as abstract ones get caught in this vortex of confusion. My favorite example is one I discovered while looking through a French dictionary many years ago. It defined the verb *clocher* ("to limp") as *marcher en boitant* ("to walk while hobbling", roughly), and *boiter* ("to hobble") as *clocher en marchant* ("to limp while walking"). This eager learner of French was helped precious little by that particular pair of definitions.

*   *   *

But let us return to mutually recursive acronyms. I put quite a bit of effort into working out a family of them, and to my surprise, they wound up dealing mostly (though by no means exclusively!) with Italian food. It all began when, inspired by **tato**, I chose the similar word **tomato** and then decided to use its plural, coming up with this meaning for **tomatoes**:

TOMATOES on MACARONI (and TOMATOES only), exquisitely SPICED.

The capitalized words here are those that are also acronyms. Here is the rest of my mutually recursive family:

MACARONI:
   MACARONI and CHEESE (a REPAST of Naples, Italy)

REPAST:
   rather extraordinary PASTA and SAUCE, typical

CHEESE:
cheddar, havarti, emmenthaler (especially SHARP emmenthaler)

SHARP:
strong, hearty, and rather pungent

SPICED:
sweetly pickled in CHEESE ENDIVE dressing

ENDIVE:
egg NOODLES, dipped in vinegar eggnog

NOODLES:
NOODLES (oodles of delicious LINGUINI), elegantly served

LINGUINI:
LAMBCHOPS (including NOODLES), gotten usually in Northern Italy

PASTA:
PASTA and SAUCE (that's ALL!)

ALL!:
a luscious lunch

SAUCE:
SHAD and unusual COFFEE (eccellente!)

SHAD:
SPAGHETTI, heated al dente

SPAGHETTI:
standard PASTA, always good, hot especially (twist, then ingest)

COFFEE:
choice of fine flavors, especially ESPRESSO

ESPRESSO:
excellent, strong, powerful, rich, ESPRESSO, suppressing sleep outrageously

BASTA!:
belly all stuffed (tummy ache!)

LAMBCHOPS:
LASAGNE and meat balls, casually heaped onto PASTA SAUCE

LASAGNE:
LINGUINI and SAUCE and GARLIC (NOODLES everywhere!)

RHUBARB:
RAVIOLI, heated under butter and RHUBARB (BASTA!)

RAVIOLI:
  RIGATONI and vongole in oil, lavishly introduced

RIGATONI:
  rich Italian GNOCCHI and TOMATOES (or NOODLES instead)

GNOCCHI:
  GARLIC NOODLES over crisp CHEESE, heated immediately

GARLIC:
  green and red LASAGNE in CHEESE

Any gourmet can see that little attempt has been made to have each term defined by its corresponding phrase; it is simply associated more or less arbitrarily with the phrase.

Now what happens if we begin to expand some word—say, **pasta**? At first we get simply **PASTA and SAUCE (that's ALL!)**. The next stage yields **PASTA and SAUCE (that's ALL!) and SHAD and unusual COFFEE (eccellente!) (that's a luscious lunch)**. We could obviously go on expanding acronyms forever—or at least until we filled the universe up to its very brim with mouth-watering descriptions of Italian food. But what if we were less ambitious, and wanted merely to fill half a page or so with such a description? How might we find a way to halt this seemingly bottomless recursion in midcourse?

Well, of course, the key word here is "bottomless", and the answer it implies is: Put in a mechanism to allow the recursion to bottom out. The bottomlessness comes from the fact that at every stage, every acronym is allowed to expand, that is, to spawn further acronyms. So what if, instead, we kept tight control of the spawning process, being generous in the first few "generations" and gradually letting fewer and fewer acronyms spawn progeny as the generations got later? This would be similar to a redwood tree in a forest, which begins with a single "branch" (its trunk), and that branch spawns "progeny", namely, the first generation of smaller branches, and they in turn spawn ever more progeny—but eventually a natural "bottoming out" occurs as a consequence of the fact that teeny twigs simply cannot branch further. (Somehow, trees seem to have gotten their wires crossed, since for them, bottoming out generally takes place at the top.)

If this process were completely regular, then all redwood trees would look exactly alike, and one could well agree with former Governor Reagan's memorable dictum, "If you've seen one redwood tree, then you've seen them all." Unfortunately, though, redwood trees (and some other things as well) are trickier than Governor Reagan realized, and we have to learn to deal with a great variety of different things that all go by the same name. The variety is caused by the introduction of randomness into the choices as to whether to branch or not to branch, what angle to branch at, what size branch to grow, and so on.

437

Similar remarks apply to the "trees" of mutually recursive acronyms. If in expanding **tomatoes** we always made exactly the same control decisions about which acronyms to expand when, then there would be one and only one type of **rhubarb** expansion, so that here too, it would make sense to say "If you've seen one **rhubarb**, you've seen them all." But if we allow some randomness to enter the decision-making about spawning, then we can get many varieties of **rhubarb**, all bearing some telltale resemblance to one another, but at a much more elusive level of perception.

How can we do this? The ideal concept to bring to bear here is that of the *random-number generator,* which serves as the computational equivalent of a coin flip or throw of dice. We'll let all decisions about whether or not to expand a given acronym depend on the outcome of such a virtual coin flip. At early stages of expansion, we'll set things up so that the coin will be very likely to come up heads (*do* expand); at later stages, it will be increasingly likely to come up tails (*don't* expand). The Lisp function **rand** will be employed for this. It takes no arguments, and each time it is called, it returns a new real number located somewhere between 0 and 1, unpredictably. (This is an exaggeration—it is actually 100 percent predictable if you know how it is computed; but since the algorithm is rather obscure, for most purposes and to most observers the behavior of this function will be so erratic as to count as totally random. The story of random number generation is itself quite a fascinating one, and would be an entire article in itself.)

If we want an event to happen with a probability of 60 percent, first we ask **rand** for a value. If that value turns out to be 0.6 or below, we go ahead, and if not, we do not. Since over a long period of time, **rand** sprays its outputs uniformly over the interval between 0 and 1, we will indeed get the go-ahead 60 percent of the time.

* * *

So much for random decisions. How will we get an acronym to expand when told to? This is not too hard. Suppose we let each acronym be a Lisp function, as in the following example:

```
(def tomatoes (lambda ()
    '(tomatoes on macaroni (and tomatoes only), exquisitely spiced)))
```

The function **tomatoes** takes no arguments, and simply returns the list of words that it expands into. Nothing could be simpler.

Now suppose we have a variable called **acronym** whose value is some particular acronym—but we don't know which one. How could we get that acronym to expand? The way we've set it up, that acronym must act as a function call. In order for any atom to invoke a function, it must be the **car** of a list, as in the examples **(plus 2 2)**, **(rand)**, and **(rhubarb)**. Now if we were to write **(acronym),** then the literal atom **acronym** would be taken by the genie as a function name. But that would be a misunderstanding. It's certainly not the atom **acronym** that we want to make serve as a function name, but its *value,* be it **macaroni**, **cheese**, or what-have-you.

To do this, we employ a little trick. If the value of the atom **acronym** is **rhubarb** and if I write **(list acronym)**, then the value the Lisp genie will return to me will be the list **(rhubarb)**. However, the genie will simply see this as an inert piece of Lispstuff rather than as a little command that I would like to have executed. It cannot read my mind. So how do I get it to perform the desired operation? Answer: I remember the function called **eval**, which makes the genie look upon a given data structure as a wish to be executed. In this case, I need merely say **(eval (list acronym))** and I will get the list **(ravioli, heated under butter and rhubarb (basta!))**. And had **acronym** had a different value, then the genie would have handed me a different list.

We now have just about enough ideas to build a function capable of expanding mutually recursive acronyms into long but finite phrases whose sizes and structures are controlled by many "flips" of the **rand** coin. Instead of stepping you through the construction of this function, I shall simply display it below, and let you peruse it. It is modeled very closely on the earlier function **replace**.

```
(def expand (lambda (phrase probability)
   (cond ((atom phrase) phrase)
         ((is-acronym (car phrase))
          (cond ((lessp (rand) probability)
                 (append
                  (expand (eval (list (car phrase))) (lower probability))
                  (expand (cdr phrase) probability)))
                (t
                 (cons (car phrase) (expand (cdr phrase) probability)))))
         (t (cons (expand (car phrase) (lower probability))
                  (expand (cdr phrase) probability))))))
```

Note that **expand** has two parameters. One represents the phrase to expand, the other represents the probability of expanding any acronyms that are top-level members of the given phrase. (Thus the value of the atom **probability** will always be a real number between 0 and 1.) As in the redwood-tree example, the expansion probability should decrease as the calls get increasingly recursive. That is why lines that call for expansion of **(car phrase)** do so with a *lowered* probability. To be exact, we can define the function **lower** as follows:

```
(def lower (lambda (x) (times x 0.8)))
```

Thus each time an acronym expands, its progeny are only 0.8 times as likely to expand as it was. This means that sufficiently deeply nested acronyms have a vanishingly small probability of spawning further progeny. You could use any reducing factor; there is nothing sacred about 0.8, except that it seems to yield pretty good results for me.

The only remaining undescribed function inside the definition above is **is-acronym.** Its name is pretty self-explanatory. First the function tests to see if its argument is an atom; if not, it returns **nil**. If its argument *is* an atom, it goes on to see if that atom has a function definition—in particular, a definition with the form of an acronym. If so, **is-acronym** returns the value **t**; otherwise it returns **nil**. Precisely how to accomplish this depends on your specific variety of Lisp, which is why I have not shown it explicitly. In Franz Lisp, it is a one-liner.

You may have noticed that there are two **cond** clauses in close proximity that begin with **t.** How come one "otherwise" follows so closely on the heels of another one? Well, actually they belong to different **cond**'s, one nested inside the other. The first **t** (belonging to the inner **cond**) applies to a case where we know we are dealing with an acronym but where our random coin, instead of coming down heads, has come down tails (which amounts to a decision not to expand); the second **t** (belonging to the outer **cond**) applies to a case where we have discovered we are simply not dealing with an acronym at all.

The inner logic of **expand**, when scrutinized carefully, makes perfect sense. On the other hand, no matter how carefully you scrutinize it, the output produced by **expand** using this *famiglia* of acronyms remains quite silly. Here is an example:

(rich italian green and red linguini and shad and unusual choice of fine flavors, especially excellent, strong, powerful, rich, espresso, suppressing sleep outrageously (eccellente!) and green and red lasagne in cheese (noodles everywhere!) in cheddar, havarti, emmenthaler (especially sharp emmenthaler) noodles (oodles of delicious linguini), elegantly served (oodles of delicious linguini), elegantly served (oodles of delicious linguini and sauce and garlic (noodles (oodles of delicious linguini), elegantly served everywhere!) and meat balls, casually heaped onto pasta and sauce (that's all!) and sauce (that's a luscious lunch) sauce (including noodles (oodles of delicious linguini), elegantly served), gotten usually in Northern Italy), elegantly served over crisp cheese, heated immediately and tomatoes on macaroni and cheese (a repast of Naples, Italy) (and tomatoes only), exquisitely sweetly pickled in cheese endive dressing (or noodles instead) and vongole in oil, lavishly introduced, heated under butter and rich italian gnocchi and tomatoes (or noodles instead) and vongole in oil, lavishly introduced, heated under butter and rigatoni and vongole in oil, lavishly introduced, heated under butter and ravioli, heated under butter and rich italian garlic noodles over crisp cheese, heated immediately and tomatoes (or noodles instead) and vongole in oil, lavishly introduced, heated under butter and ravioli, heated under butter and

rhubarb (basta!) (basta!) (basta!) (basta!) (belly all stuffed (tummy ache!)) (basta!))

Oh, the glories of recursive spaghetti! As you can see, Lisp is hardly the computer language to learn if you want to lose weight. Can you figure out which acronym this gastronomical monstrosity grew out of?

*   *   *

The **expand** function exploits one of the most powerful features of Lisp —that is, the ability of a Lisp program to take data structures it has created and treat them as pieces of code (that is, give them to the Lisp genie as commands). Here it was done in a most rudimentary way. An atom was wrapped in parentheses and the resulting minuscule list was then evaluated, or **eval**ed, as Lispers' jargon has it. The work involved in manufacturing the data structure was next to nothing, in this case, but in other cases elaborate pieces of structure can be "**cons**ed up", then handed to the Lisp genie for **eval**ing. Such pieces of code might be new function definitions, or any number of other things. The main idea is that in Lisp, one has the ability to "elevate" an inert, information-containing data structure to the level of "animate agent", where it becomes a manipulator of inert structures itself. This *program-data cycle,* or loop, can continue on and on, with structures reaching out, twisting back, and indirectly modifying themselves or related structures.

Certain types of inert, or passive, information-containing data structures are sometimes referred to as *declarative knowledge*—"declarative" because they often have a form abstractly resembling that of a declarative sentence, and "knowledge" because they encode facts about the world in some way, accessible by looking in an index in somewhat the way "book-learned" facts are accessible to a person. By contrast, animate, or active, pieces of code are referred to as *procedural knowledge*—"procedural" since they define sequences of actions ("procedures") that actually manipulate data structures, and "knowledge" since they can be viewed as embodying the program's set of skills, something like a human's unconscious skills that were once learned through long, rote drill sessions. (Sometimes these contrasting knowledge types are referred to as "knowledge that" and "knowledge how".)

This distinction should remind biologists of that between genes— relatively inert structures inside the cell—and enzymes, which are anything but inert. Enzymes are the animate agents of the cell; they transform and manipulate all the inert structures in indescribably sophisticated ways. Moreover, Lisp's loop of program and data should remind biologists of the way that genes dictate the form of enzymes, and enzymes manipulate genes (among other things). Thus Lisp's procedural-declarative program-data loop provides a primitive but very useful and tangible example of one of the most fundamental patterns at the base of life: the ability of passive structures to control their own destiny, by creating and regulating active structures whose form they dictate.

*   *   *

We have been talking all along about the Lisp genie as a mysterious given agent, without asking where it is to be found or what makes it work. It turns out that one of Lisp's most exciting properties is the great ease with which one can describe the Lisp genie's complete nature in Lisp itself. That is, the Lisp interpreter can be easily written down in Lisp. Of course, if there is no prior Lisp interpreter to run it, it might seem like an absurd and pointless exercise, a bit like having a description in flowery English telling foreigners how best to learn English. But it is not so silly as that makes it sound.

In the first place, if you know enough English, you can "bootstrap" your way further into English; there is a point beyond which explanations written in English about English are indeed quite useful. What's more, that point is not too terribly far beyond the beginning level. Therefore, all you need to acquire first, and autonomously, is a "kernel"; then you can begin to lift yourself by your own bootstraps. For children, it is an exciting thing when, in reading, they begin to learn new phrases all by themselves, simply by running into them several times in succession. Their vocabulary begins to grow by leaps and bounds. So it is once there is a Lisp kernel in a system; the rest of the Lisp interpreter can be—and usually is—written in Lisp.

The fact that one can can easily write the Lisp interpreter in Lisp is no mere fluke depending on some peculiarly introverted fact about Lisp. The reason it is easy is that Lisp lends itself to writing interpreters for all sorts of languages. This means that Lisp can serve as a basis on which one can build other languages.

To put it more vividly, suppose you have designed on paper a new language called "Flumsy". If you really know how Flumsy should work, then it should not be too hard for you to write an interpreter for it in Lisp. Once implemented, your Flumsy interpreter then becomes, in essence, an intermediary genie to whom you can give wishes in Flumsy and who will in turn communicate those wishes to the Lisp genie in Lisp. Of course, all the mechanisms allowing the Flumsy genie to talk to the Lisp genie are themselves being carried out by the Lisp genie. So is this a mere façade? Is talking Flumsy really just a way of talking Lisp in disguise?

Well, when the U.S. arms negotiators talk to their Soviet counterparts through an interpreter, are they really just speaking Russian in disguise? Or is the crux of the matter whether the interpreter's native language was English or Russian, upon which the other was learned as a second tongue? And suppose you find out that in fact, the interpreter's native language was Lithuanian, that she learned English only as an adolescent and then learned Russian by taking high-school classes in English? Will you then feel that when she speaks Russian, she is actually speaking English in disguise, or worse, that she is actually speaking Lithuanian, doubly disguised?

Analogously, you might discover that the Lisp interpreter is in fact written in Pascal or some other language. And then someone could strip off the Pascal façade as well, revealing to you that the truth of the matter is that all instructions are really being executed in *machine language,* so that you are fooling yourself completely if you think that the machine is talking Flumsy, Lisp, Pascal, or any higher-level language at all!

*   *   *

When one interpreter runs on top of another one, there is always the question of *what level one chooses not to look below.* I personally seldom think about what underlies the Lisp interpreter, so that when I am dealing with the Lisp system, I feel as if I am talking to "someone" whose "native language" is Lisp. Similarly, when dealing with people, I seldom think about what their brains are composed of; I don't reduce them in my mind to piles of patterned neuron firings. It is natural to my perceptual system to recognize them at a certain level and not to look below that level.

If someone were to write a program that could deal in Chinese with simple questions and answers about restaurant visits, and if that program were in turn written in another language—say, the hypothetical language "SEARLE" (for "Simulated East-Asian Restaurant-Lingo Expert"), I could choose to view the system either as genuinely speaking Chinese (assuming it gave a creditable and not too slow performance), or as genuinely speaking SEARLE. I can shift my point of view at will. The one I adopt is governed mostly by pragmatic factors, such as which subject area I am currently more interested in at the moment (Chinese restaurants, or how grammars work), how closely matched the program's speed is to that of my own brain, and —not least—whether I happen to be more fluent in Chinese or in SEARLE. If to me, Chinese is a mere bunch of "squiggles and squoggles", I may opt for the SEARLE viewpoint; if on the other hand, SEARLE is a mere bunch of confusing technical gibberish, I will probably opt for the Chinese viewpoint. And if I find out that the SEARLE interpreter is in turn implemented in the Flumsy language, whose interpreter is written in Lisp, then I have two more points of view to choose from. And so on.

With interpreters stacked on interpreters, however, things become rapidly very inefficient. It is like running a motor off power created through a series of electric generators, each one being run off power coming from the preceding one: one loses a good deal at each stage. With generators there is usually no need for a long cascade, but with interpreters it is often the only way to go. If there is no machine whose machine language is Lisp, then you build a Lisp interpreter for whatever machine you have available, and run Lisp that way. And Flumsy and SEARLE, if you wish to have them at your disposal, are then built on top of this *virtual Lisp machine.* This indirectness can be annoyingly inefficient, causing your new "virtual Flumsy machine" or "virtual SEARLE machine" to run dozens of times more slowly than you would like.

* * *

Important hardware developments have taken place in the last several years, and now machines are available that are based at the hardware level on Lisp. This means that they "speak" Lisp in a somewhat deeper sense— let us say, "more fluently"—than virtual Lisp machines do. It also means that when you are on such a machine, you are "swimming" in a Lisp environment. A Lisp environment goes considerably beyond what I have described so far, for it is more than just a language for writing programs. It includes an editing program, with which one can create and modify one's programs (and text as well), a debugging program, with which one can easily localize one's errors and correct them, and many other features, all designed to be compatible with each other and with an overarching "Lisp philosophy".

Such machines, although still expensive and somewhat experimental, are rapidly becoming cheaper and more reliable. They are put out by various new companies such as LMI (Lisp Machine, Inc.), Symbolics, Inc., both of Cambridge, Massachusetts, and older companies such as Xerox. Lisp is also available on most personal computers—you need merely look at any issue of any of the many small-computer magazines to find ads for Lisp.

Why, in conclusion, is Lisp popular in artificial intelligence? There is no single answer, nor even a simple answer. Here is an attempt at a summary:

(1) Lisp is elegant and simple.
(2) Lisp is centered on the idea of lists and their manipulation—and lists are extremely flexible, fluid data structures.
(3) Lisp code can easily be manufactured in Lisp, and run.
(4) Interpreters for new languages can easily be built and experimented with in Lisp.
(5) "Swimming" in a Lisp-like environment feels natural for many people.
(6) The "recursive spirit" permeates Lisp.

Perhaps it is this last rather intangible statement that gets best at it. For some reason, many people in artificial intelligence seem to have a deep sense that recursivity, in *some* form or other, is connected with the "trick" of intelligence. This is a hunch, an intuition, a vaguely felt and slightly mystical belief, one that I certainly share—but whether it will pay off in the long run remains to be seen.

Post Scriptum.

In March of 1977, I met the great AI pioneer Marvin Minsky for the first time. It was an unforgettable experience. One of the most memorable remarks he made to me was this one: "Gödel should just have thought up

Lisp; it would have made the proof of his theorem much easier." I knew exactly what Minsky meant by that, I could see a grain of truth in it, and moreover I knew it had been made with tongue semi in cheek. Still, something about this remark drove me crazy. It made me itch to say a million things at once, and thus left me practically speechless. Finally today, after my seven-year itch, I will say some of the things I would have loved to say then.

What Minsky meant, paraphrased, is this: "Probably the hardest part of Gödel's proof was to figure out how to get a mathematical system to talk about itself. This took several strokes of genius. But Lisp can talk about itself, at least in the sense Gödel needed, *directly*. So why didn't he just invent Lisp? Then the rest would have been a piece of cake." An obvious retort is that to invent Lisp out of the blue would have taken a larger number of strokes of genius. Minsky, of course, knew this, and at bottom, his remark was clearly just a way of making this very point in a facetious way.

Still, it was clear that Minsky felt there was some serious content to the remark, as well. (And I have heard him make the same remark since then, so I know it was not just a throwaway quip.) There was the implicit question, "Why didn't Gödel invent the idea of *direct* self-reference, as in Lisp?" And this, it seemed to me, missed a crucial point about Gödel's work, which is that it showed that self-reference can crop up even where it is totally unexpected and unwelcome. The power of Gödel's result was that it obliterated the hopes for completeness of an *already known* system, namely Russell and Whitehead's *Principia Mathematica*; to have destroyed similar hopes for some newly concocted system, Lisp-like or not, would have been far less significant (or, to be more accurate, such a result's significance would have been far harder for people to grasp, even if it were equally significant).

Moreover, Gödel's construction revealed in a crystal-clear way that the line between "direct" and "indirect" self-reference (indeed, between direct and indirect *reference,* and that's even more important!) is completely blurry, because his construction pinpoints the essential role played by *isomorphism* (another name for coding) in the establishment of reference and meaning. Gödel's work is, to me, the most beautiful possible demonstration of how meaning emerges from and only from isomorphism, and of how any notion of "direct" meaning (*i.e.,* codeless meaning) is incoherent. In brief, it shows that *semantics is an emergent quality of complex syntax,* which harks back to my earlier remark in the *Post Scriptum* to Chapter 1, namely: "Content is fancy form." So the serious question implicit in Minsky's joke seemed to me to rest on a confusion about this aspect of the nature of meaning.

*   *   *

Now let me explain this in more detail. Part I of Gödel's insight was to realize that via a code, a number can represent a mathematical symbol 11 *e.g.,* the integer eleven can represent the left parenthesis, and the integer thirteen the right parenthesis13. The analogue of this in human languages is the recognition that certain orally produced screeches or manually produced scratches (such as "word", "say", "language", "sentence", "reference", "grammar", "meaning", and so on) can stand for elements of language itself (as distinguished from screeches or scratches such as "cow" and "splash", which stand for extralinguistic parts of the universe). Here we have pieces *of* language talking *about* language. Maybe it doesn't seem strange to you, but put yourself, if you can, back in the shoes of barefoot cave people who had barely gotten out of the grunt stage. How amazingly magical it must have felt to the beings in whose minds such powerful concepts as *words about words* first sparked! In some sense, human consciousness began then and there.

But a language can't get very far in terms of self-reference if it can talk only about isolated symbols. Part II of Gödel's insight was to figure out how the system (and here I mean *Principia Mathematica* "and", as Gödel's paper's title says, "related systems") could talk about *lists* of symbols, and even lists of lists of symbols, and so on. In the analogy to human language, making this step is like the jump from an ability to talk about people's one-word utterances ("Paul Revere said 'Land!'.") to the ability to talk about arbitrarily long utterances, and nested ones at that ("'Douglas Hofstadter wrote, "Paul Revere said, 'Land!'.".'.").

Gödel found a way to have some integers stand for *single* symbols and others stand for *lists* of symbols, usually called *strings*. An example will help. Suppose the integer 1 stands for the symbol '0', and as I mentioned earlier, 11 and 13 for parentheses. Thus to encode the string "(0)" would require you to combine the integers 11, 1, and 13 somehow into a single integer. Gödel chose the integer 7500000000000—not capriciously, of course! This integer can be viewed as the product of the three integers 2048, 3, and 1220703125, which in turn are, respectively: 2^{11}, 3^1, and 5^{13}. In other words, the three-symbol string whose symbols *individually* are coded for by 11 and 1 and 13 is coded for *in toto* by the single integer $2^{11}3^15^{13}$. Now 2, 3, and 5 are of course the first three primes, and if you want to encode a longer string, you use as many primes as you need, in increasing order. This simple scheme allows you to code strings of arbitrary length into large integers, and moreover—since large integers can be exponents just as easily as small ones can—it allows for recursive coding. In other words, strings can contain the integer codes for other strings, and this can go on indefinitely. An example: the list of strings "0", "(0)", and "((0))" is coded into the stupendously large integer

$$2^{2^1}3^{2^{11}3^15^{13}}5^{2^{11}3^{11}5^17^{13}11^{13}}$$

The proverbial "astute reader" might well have noticed a possible ambiguity: How can you tell if an integer is to be decomposed via prime factorization into other integers or to be left alone and interpreted as a code for an atomic symbol? Gödel's simple but ingenious solution was to have all atomic symbols be represented by odd integers. How does that solve the matter? Easy: You know you should not factorize odd integers, and conversely, you should factorize even ones, and then do the same with all the exponents you get when you do so. Eventually, you will bottom out in a bunch of odd integers representing atomic symbols, and you will know which ones are grouped together to form larger chunks and how those chunks are nested.

With this beautifully polished scheme for encoding strings inside integers and thereby inside mathematical systems, Gödel had discovered a way of getting such a system to talk—in code—about itself. He had snuck self-reference into systems that were presumed to be as incapable of self-reference as are pencils of writing on themselves or garbage cans of containing themselves, and he had done so as wittily as the Greeks snuck a boatload of unacceptable soldiers into Troy, "encoded" as one single large acceptable structure.

Historically, the importance of Gödel's work was that it revealed a plethora of unexpected self-references (via his code, to be sure, but that fact in no way diminishes their effect) within the supposedly impregnable walls of Russell and Whitehead's Troy, *Principia Mathematica.* Now in Lisp, it's possible to construct and manipulate pieces of Lisp programs. The idea of *quoted code* is one of those deep ideas that make Lisp so appealing to AI people. Okay, but—when you have a system constructed expressly to have self-referential potential, the fact that it *has* self-referential structures will amaze no one. What is amazing and wonderful is when self-reference pops up inside the very fortress constructed expressly to keep it out! The repercussions of that are enormous.

One of the clear consequences of Gödel's revelation of this self-referential potential inside mathematical systems was that the same potential exists within any similar formalism, including computer languages. That is simply because computers can do all the standard arithmetic operations—at least in theory—with integers of unlimited size, and so coded representations of programs are being manipulated any time you are manipulating certain large integers. Of course, which program is being manipulated depends on what code you use. It was only after Gödel's work had been absorbed by a couple of generations of mathematicians, logicians, and computer people that the desirability of inserting the concept of quotation *directly* into a formal language became obvious. To be quite emphatic about it, however, this does not enhance the language's potential in any way, except that it makes certain constructions easier and more transparent. It was for this reason of transparency that Minsky made his remark.

Oh yes, I agree, Gödel's proof would have been easier, but by the time Gödel dreamt it up, it would have long since been discovered (and called "Snoddberger's proof") had Gödel been in a mindset where inventing Lisp was natural. I'm all for counterfactuals, but I think one should be careful to slip things realistically.

*   *   *

After this diatribe, you will think I am crazy when I turn around and tell you: Gödel *did* invent Lisp! I am not trying to take anything away from John McCarthy, but if you look carefully at what Gödel did in his 1931 article, written roughly 15 years before the birth of computers and 27 years before the official birth of Lisp, you will see that he anticipated nearly all the central ideas of Lisp. We have already been through the fact that the central data structure of Lisp, the list, was at the core of Gödel's work. The crucial need to be able to distinguish between atoms and lists—something that modern-day implementors of Lisp systems have to worry about—was recognized and cleverly resolved by Gödel, in his odd-even distinction. The idea of quoting is, in essence, that of the Gödel code. And finally, what about recursive functions, the heart and soul of Lisp programming technique? That idea, too, is an indispensable part of Gödel's paper! This is the astounding truth.

The heart of Gödel's construction is a chain of 46 definitions of functions, each new one building on previous ones in a dizzying spire ascending toward one celestial goal: the definition of one very complex function of a single integer, which, given any input, either returns the value 1 or goes into an infinite loop. It returns 1 whenever the input integer is the Gödel number —the code number—of a theorem of *Principia Mathematica,* and it loops otherwise. This is Gödel's 46th function, which he named "Bew", short for "beweisbar", meaning "provable" in German.

If we could calculate the value of function 46 swiftly for any input, it would resolve for us any true-false question of mathematics that a full axiomatic system could resolve. All we would need to do is to write down the statement in question in the language of *Principia Mathematica,* code the resulting formula into its Gödel number (the most mechanical of tasks), and then call function 46 on that number. A result of 1 means *true,* looping forever means *false.* Do I hear the astute reader protesting again? All right, then: If we want to avoid any chance of having to wait forever to find out the answer, we can encode the *negation* of the statement in question into its Gödel number as well, and also call function 46 on this second number. We'll let the two calculations proceed simultaneously, and see which one says '1'. Now, as long as *Principia Mathematica* has either the statement or its negation as a theorem, one of the two calls on function 46 will return 1, while the other will presumably spin on unto eternity.

How does function 46 work? Oh, easy—by calling function 45 many times. And how does function 45 work? Well, it calls functions 44 and others, and they call previously defined functions, some of which call themselves (recursion!), and so on and so forth—all of these complex calls eventually bottoming out in calls to absolutely trivial functions such as "S", the successor function, which returns the value 18 when you feed it the integer 17. In short, the evaluation of a high-numbered function in Gödel's paper sets in motion the calling of one subroutine after another in a hierarchical chain of command, in *precisely* the same way as my function **expand** called numerous lower-level functions which called others, and so on. Gödel's remarkable series of 46 function definitions is, in my book, the world's first serious computer program—and *it is in Lisp.* (The Norwegian mathematician Thoralf Skolem was the inventor of the study of recursive functions *theoretically,* but Gödel was the first person to use recursive functions *practically,* to build up functions of great complexity.)

It was for all these reasons that Minsky's pseudo-joke struck my intellectual funnybone. One answer I wanted to make was: "I disagree: Gödel *shouldn't* have and *couldn't* have invented Lisp, because his work was a necessary precursor to the invention of Lisp, and anyway, he was out to destroy *PM,* not Lisp." Another answer was: "Your innuendo is wrong, because any type of reference has to be grounded in a code, and Gödel's way of doing it involved no more coding machinery grounding its referential capacity than any other efficient way would have." The final answer I badly wanted to blurt out was: "No, no, no—Gödel *did* invent Lisp!" You see why I was tongue-tied?

*   *   *

One reason I mention all this about Gödel is that I wish to make some historical and philosophical points. There is another reason, however, and that is to point out that the ideas in Lisp are intimately related to the basic questions of metamathematics and metalogic, and these, translated into a more machine-oriented perspective, are none other than the basic questions of computability—perhaps the deepest questions of computer science. Michael Levin has even written an introduction to mathematical logic using Lisp, rather than a more traditional system, as its underlying formal system. For this type of reason, Lisp occupies a very special place inside computer science, and is not likely to go away for a very long time.

However . . . (you were waiting for this, weren't you?), there is a vast gulf between the issues that Lisp makes clear and easy, and the issues that confront one who would try to understand and model the human mind. The way I see it is in terms of *grain size.* To me, the thought that Lisp itself might be "more conducive" to good AI ideas than any other computer language is quite preposterous. In fact, such claims remind me of certain wonderfully romantic but woefully antic claims I have heard about the Hopi language. The typical one runs something like this: "Einstein should just have invented Hopi; then the discovery of his theory of relativity would have been much easier." The basis for this viewpoint is that Hopi, it is said, lacks terms for "absolute time" in it. Supposedly, Hopi (or a language with similar properties) would therefore be the ideal language in which to speak of relativity, since absolute time is abandoned in relativity.

This kind of claim was first put forth by the outstanding American linguist Edward Sapir, was later polished by his student Benjamin Whorf, and is usually known these days as the Sapir-Whorf hypothesis. (I have already made reference to this view in Chapters 7 and 8 on sexist language and imagery.) To state the Sapir-Whorf thesis explicitly: Language controls thought. A milder version of it would say: Language exerts a powerful influence upon thought.

In the case of computer languages, the Sapir-Whorf thesis would have to be interpreted as asserting that programmers in language X can think only in terms that language X furnishes them, and no others. Therefore, they are strapped in to certain ways of seeing the "world", and are prevented from seeing many ideas that programmers in language L can easily see. At least this is what Sapir-Whorf would have you believe. I will have none of it!

I do use Lisp, I do think it is very convenient and natural in many ways, I do advocate that anyone seriously interested in AI learn Lisp well; all this is true, but I do not think that deep study of Lisp is the royal road to AI any more than I think that deep study of bricks is the royal road to understanding architecture. Indeed, I would suggest that the raw materials to be found in Lisp are to AI what raw materials are to architecture: convenient building blocks out of which far larger structures are made.

It would be ridiculous for anyone to hope to acquire a deep understanding of what AI is all about without first having a clear, precise understanding of what computers are all about. I know of no shorter cut to that latter goal than the study of Lisp, and that is one reason Lisp is so good for AI students. Beginners in Lisp encounter, and are in a good position to understand, fundamental issues in computer science that even some advanced programmers in other languages may not have encountered or thought about. Such concepts as lists, recursion, side effects, quoting and evaluating pieces of code, and many others that I did not have the space to present in my three columns, are truly central to the understanding of the potential of computing machinery. Moreoever, without languages that allow people to deal with such concepts directly, it would be next to impossible to make programs of subtlety, grace, and multi-level complexity. Therefore I advocate Lisp very strongly.

It would similarly be next to impossible to build structures of subtlety, grace, and multi-level complexity such as the Golden Gate Bridge and the Empire State Building out of bricks or stone. Until the use of steel as an architectural substrate was established, such constructions were unthinkable. Now we are in a position to erect buildings that use steel in even more sophisticated ways. But steel itself is not the source of great architects' inspiration; it is simply a liberator. Being a super-expert on steel may be of some use to an architect, but I would guess that being quite knowledgeable will suffice. After all, buildings are not just scaled-up girders. And so it is with Lisp and AI. Lisp is not the "language of thought" or the "language of the brain"—not by any stretch of the imagination. Lisp is, however, a liberator. Being a super-expert on Lisp may be of some use to a person interested in computer models of mentality, but being quite knowledgeable will suffice. After all, minds are not just scaled-up Lisp functions.

Let me switch analogies. Is it possible for a novelist to conceive of plot ideas, characters, intrigues, emotions, and so on, without being channeled by her own or his own native language? Are the events that take place in, say, *Anna Karenina* specifically determined by the nature of the Russian language and the words that it furnished to Tolstoy? If that were the case, then of course the novel would be incomprehensible to people who do not know the Russian language. It would simply make no sense at all. But that is not even remotely the case. English-language readers have read that novel with great pleasure and have just as fully fathomed its psychological twists and turns as have Russian-language readers. The reason is that Tolstoy's mind was concerned with concepts that float far above the grain size of *any* human language. To think otherwise is to reduce Tolstoy to a mere syntactician, is to see Tolstoy as pushed around by low-level quirks and local flukes of his own language.

Now please understand, I am not by any means asserting that Tolstoy transcended his own culture and times; certainly he belongs to a particular era and a particular set of circumstances, and those facts flavor what he wrote. But "flavor" is the right word here. The essence of what he did—the meat of it, to prolong the "flavor" metaphor—is universal, and has to do with the fact that Tolstoy had profoundly experienced the human condition, had felt the pangs of many conflicting emotions in all sorts of ways. *That's* where the power of his writing comes from, not from the language he happened to inherit (otherwise, why wouldn't all Russians be great novelists?); *that's* why his novels survive translation not only into other languages (so they reach other cultures), but also into other eras, with different sensibilities. If Tolstoy manages to reach further into the human psyche than most other writers do, it is not the Russian language that deserves the credit, but Tolstoy's acute sensitivity and empathy for people.

The analogous statement could be made about AI programs and AI researchers. One could even mechanically substitute "AI program" for "novel", "Lisp" for "Russian", and—well, I have to admit that I would be hard pressed to come up with "the Tolstoy of AI". Oh, well. My point is simply that good AI researchers are not in any sense slaves to any language. Their ideas are as far removed from Lisp (or whatever language they program in, be it Lisp itself, a "super-Lisp" (such as n-Lisp for any value of n), Prolog, Smalltalk, and so on) as they are from English or from their favorite computer's hardware design. As an example, the AI program that has probably inspired me more than any other is the one called Hearsay-II, a speech-understanding program developed at Carnegie-Mellon University in the mid-1970's by a team headed up by D. Raj Reddy and Lee Erman. That program was written not in Lisp but in a language called SAIL, very different in spirit from Lisp. Nonetheless, it could easily have been written in Lisp. The reason it doesn't matter is simply that the scientific questions of how the mind works are on a totally different level from the statements of any computer language. The ideas transcend the language.

*   *   *

To some, I may seem here to be flirting dangerously with an anti-mechanistic mysticism, but I hasten to say that that is far from the case. Quite the contrary. Still, I can see why people might at first suspect me of putting forth such a view. A programmer's instinct says that you can cumulatively build a system, encapsulating all the complexity of one layer into a few functions, then building the next layer up by exploiting the efficient and compact functions defined in the preceding layer. This hierarchical mode of buildup would seem to allow you to make arbitrarily complex actions be represented at the top level by very simple function calls. In other words, the functions at the top of the pyramid are like "cognitive" events, and as you move down the hierarchy of lower-level functions, you get increasingly many ever-dumber subroutines, until you bottom out in a myriad calls on trivial, "subcognitive" ones. All this sounds very biological —even tantalizingly close to being an entire resolution of the mind-brain problem. In fact, for a clear spelling-out of just that position, see Daniel Dennett's book *Brainstorms,* or perhaps worse, see parts of my own *Gödel, Escher, Bach*!

Yes, although I don't like to admit it, I too have been seduced by this recursive vision of mechanical mentality, resembling nothing so much as an army, with its millions of unconscious robot privates carrying out the desires of its top-level cognitive general, as conveyed to them by large numbers of obedient and semi-bright intermediaries. Probably my own strongest espousal of this view is found in Chapter X of *GEB,* where a subheading blared out, loud and clear, "AI Advances Are Language Advances". I was arguing there, in essence, for the orthodox AI position that if we could just find the right "superlanguage"—a language presumably several levels above Lisp, but built on top of it as Flumsy or SEARLE are built on top of it—then all would be peaches and cream. We would be able to program in the legendary "language of thought". AI programs would practically write themselves. Why, there would be so much intelligence *in the language itself* that we could just sit back and give the sketchiest of hints, and the computer would go off and do our tacit bidding!

This, in the opinion of my current self, is a crazy vision, and my reasons for thinking so are presented in Chapter 26, "Waking Up from the Boolean Dream". I am relieved that I spent a lot more time in *GEB* knocking down that orthodox vision of AI rather than propping it up. I argued then, as I still do now, that the top-level behavior of the overall system must emerge statistically from a myriad independent pieces, whose actions are almost as likely to cancel each other out as to be in phase with each other. This picture, most forcefully presented in the dialogue *Prelude . . . Ant Fugue* and surrounding chapters, was the dominant view in that book, and it continues to be my view. In this book, it is put forth in Chapters 25 and 26.

In anticipation of those chapters, you might just ponder the following question. Why is it the case that, after all these millennia of using language, we haven't developed a single word for common remarks such as "Could you come over here and take a look at this?" Why isn't that thought expressed by some minuscule epigrammatic utterance such as "Cycohatalat"? Why are we not building new layer upon new layer of sophistication, so that each new generation can say things that no previous generation could have even conceived of? Of course, in some domains we are doing just that. The phrase "Lisp interpreter" is one that requires a great deal of spelling out for novices, but once it is understood, it is a very useful shorthand for getting across an extremely powerful set of ideas. All through science and other aspects of life, we are adding words and phrases. Acronyms such as "radar", "laser", "NATO", and "OPEC", as well as sets of initials such as "NYC", "ICBM", "MIT", "DNA", and "PC", are all very common and very wordlike.

Indeed, language does grow, but nonetheless, despite what might be considered an exponential explosion in the number of terms we have at our disposal, nobody writes a novel in one word. Nobody even writes a novel in a hundred words. As a matter of fact, novels these days are no shorter than they were 200 years ago. Nor are textbooks getting shorter. No matter how big an idea can be packed into a single word, the ideas that people want to put into novels and textbooks are on a totally different scale from that. Obviously, this is not claiming that language *cannot express* the ideas of a novel; it is simply saying that it takes a heap o' language to do so, no matter how the language is built. That is the issue of grain size that I alluded to before, and I feel that it is a deep and subtle issue that will come up more often as theoretical AI becomes more sophisticated.

* * *

Those interested in the Sapir-Whorf thesis might be interested to learn of Novelflo, a new pseudo-natural language invented by Rhoda Spark of Golden, Colorado. Novelflo is intended as a hypothetical extension of, or perhaps successor to, English. In particular, it is a language designed expressly for streamlining the writing of novels (or poetry). You write your novel in Novelflo in a tiny fraction of the number of words it would take in full English; then you feed it into Spark's copyrighted "Expandatron" program, which expands your concise Novelflo input into beautiful, flowing streams of powerful and evocative English prose. Or poetry, for that matter —you simply set some parameters defining the desired "shape" of the poem (sonnet, limerick, etc.; free verse or rhyme; and similar decisions), and out comes a beautifully polished poem in mere seconds. Some of Spark's advertised features for Novelflo are:

* Plot Enhancement Mechanisms (PEM's)
* Default Inheritance Assumptions
* Automatic Character Verification (checks for consistency of each character's character)
* Automatic Plot Verification (checks to be sure plot isn't self-contradictory—an indispensable tool for any novel writer)
* SVP's (Stereotype Violation Mechanisms)—allow you to override default assumptions with maximal ease)
* Sarcasm Facilitators (allows sarcasm to be easily constructed)
* VuSwap (so you can shift point of view effortlessly)
* AEP's (Atmosphere Evocation Phrases)—conjure up, in a few words, whole scenes, feelings, moods, that otherwise would take many pages if not full chapters

This entire *Post Scriptum,* incidentally, was written in Novelflo (it was my first attempt at using Spark's language), and before being expanded, it was only 114 words long! I must admit, it took me over 80 hours to compose those nuggets of ideas, but Spark assures me that one gets used to Novelflo quickly, and hours become minutes.

Spark is now hard at work on the successor to Novelflo, to be called Supernovelflo. The accompanying expansion program will be called, naturally enough, the Superexpandatron. Spark claims this advance will allow a further factor of compression of up to 100. (She did not inform me how the times for writing a given passage in Supernovelflo and Novelflo would compare.) Thus this whole *P.S.*—yes, *all* of it—would be a mere three words long, when written in Supernovelflo. It would run as follows (so she tells me):

SP 91pahC TM-foH

Now I'd call that jus-telegant!

20

Heisenberg's Uncertainty Principle and the Many-Worlds Interpretation of Quantum Mechanics

July, 1981

YOU'VE dropped a coin between some cushions in a fancy old chair. You're very anxious to retrieve your coin, so you gingerly try to reach between the cushions and grab the coin. But the very act of sticking your hand in there widens the crevice and the coin slips farther in. You can see that any more of this reaching, and your coin will be lost forever in the innards of that chair. What to do? This commonplace little drama illustrates a feeling we all know: that striving for something can have the effect of reducing that thing's availability.

A good friend is visiting from far away and before she returns home, you want to capture her infectious smile on film. But she is terribly camera-shy. The instant you bring out your camera, she freezes: spontaneity is lost, and there is no way to record that smile. The act of trying to capture this elusive phenomenon completely destroys the phenomenon.

Examples such as these are sometimes erroneously attributed to the *uncertainty principle.* That notorious principle of quantum mechanics was first enunciated by Werner Heisenberg in about 1927. Careless paraphrases since then, however, have eroded and obscured the true meaning of the principle in the popular mind. I would like to clarify matters a bit by discussing the genuine uncertainty principle and its phony imitators.

Let me first exhibit a typical imitation version clearly, so that you know what I am attacking. The standard pseudo-uncertainty principle states:

The observer always interferes with the phenomenon under observation.

It tends to be cited heavily in particular domains, often where the phenomenon involves a reciprocal observer—someone who can observe back. But even in such cases, this pseudo-principle is too simplistic. It rests on a misunderstanding of how experimentation proceeds, and how science explains. The main thing to keep in mind is that science is about *classes* of events, not particular instances. Science explains through abstractions that underlie a potentially unlimited number of concrete phenomena.

Consider the following example. I recently read of a woman who remarked, "Rosa always date shranks." She had meant to say, "Rosa always dated shrinks." But the tense marker somehow got shifted from the verb "date" onto the noun "shrink", which was then conjugated as if it were functioning as a verb: "dated shrinks" became "date shranks". It would be fascinating to know exactly what was going on in the woman's brain as she made this bizarre transformation. We would like to know exactly how things went awry. Something went down the wrong track: what, and why?

But this was a one-shot phenomenon; it will probably never be repeated. We can't expect a scientific explanation of those details. Instead, we have to abstract some general phenomenon that we think is the essential component of this particular event. We have to be able to imagine other events in the same general class. We have to be able to imagine some way to provoke them or to detect them when they happen, so that we can study the patterns. Perhaps the appropriate level of abstraction is: "grammatical errors in the speech of woman W". Or perhaps it is: "shifts of tense markers from verbs to nouns". In any case, we will have to plan a course of experimentation suitable to the way we choose to abstract this event.

In the case of the camera-shy friend, presumably her smile is a repeatable phenomenon; in missing it once, you haven't missed it forever. And with sufficient patience and ingenuity, you could set up a telephoto lens on a distant camera controlled remotely by a button you can carry in your hand. You could put the camera in an unlikely window a few dozen yards from a table where you sneakily take your friend one day, and then snap her smile without her ever suspecting it.

In the case of the coin in the cushion, with some effort you could make a special tool to retrieve it with. In fact, in any such everyday case, even those involving reciprocal observers, by investing sufficient effort and time and ingenuity—and most likely money—into a revised version, you will find you *can* isolate the phenomenon, you *can* render it impervious to the fact that you are observing it. You will never get a perfect replay of some specific event, but as long as it's a general phenomenon and not a one-shot event that you're interested in, then you can always reduce the effect of the observer (yourself) to as close to nil as you want. A budget of a trillion dollars would suffice for most purposes.

Points such as these bear repeating, because many people think the quantum-mechanical uncertainty principle actually applies to everyday phenomena. Nothing could be further from the truth! What, then, is Heisenberg's principle about?

*   *   *

To explain it, we have to go back to one of Albert Einstein's three fundamental papers of the year 1905: the paper in which he postulated that light is made up of the discrete entities he called *photons.* It was in this paper that the window onto the mysterious world of quantum mechanics was first opened. Two centuries of careful experimentation and observation had demonstrated unequivocally that visible light acts like a wave with an exceedingly short wavelength (some 10^{-4} centimeter). Light waves had been observed interfering with themselves, canceling or reinforcing themselves. Such behavior is analogous to phenomena seen on lakes or other bodies of water, such as the momentary canceling of one part of a speedboat's wake by another part reflected off a jetty, or the shimmering patterns created on a still lake by the crisscrossing circular ripples emanating from the successive bounces of a skipped rock.

In some ways, light waves are simpler than water waves. Whereas water waves of different wavelengths travel at different speeds, all light waves travel at one speed: *c,* or 3×10^{10} centimeters per second. In water, waves of long wavelength travel faster than waves of short wavelength. Water is thus said to be a *dispersive* medium. A single circular ripple, as it expands, breaks up into its various components. The outer edge, traveling fastest, consists of long-wavelength components, while the inner edge consists of slower, short-wavelength components. Gradually, because of this dispersion, the leading and trailing edges of the ripple get so far apart that the ripple can no longer be perceived. By contrast, the medium that light waves travel through is nondispersive: all wavelengths travel at exactly the same speed. But what is that medium? The rather crazy fact of the matter is that light waves need no medium—or, if you prefer, vacuum is light's medium. But how very peculiar it is for waves to wave even when there's nothing to wave!

This anomaly persistently puzzled the young Einstein, and in 1905 his fertile mind came across two fundamental elements of the resolution. One element was the counterintuitive theory of special relativity, and the other was the counterintuitive idea of particle-like quanta out of which light waves would somehow be constituted. But where did this curious flash of insight come from?

*   *   *

The classical theory of light as an electromagnetic wave had left a mystery concerning the way light of various colors, or wavelengths, is emitted from a "black body". The term is somewhat misleading; it merely means any object that absorbs light of all frequencies and does not reflect light of any frequencies. As a black body heats up, it begins to glow: first dull red, then bright red, then orange, eventually white, and then, surprisingly enough, bluish! (Think of the glowing burner on an electric stove.) The unsolved problem was to determine how much light of each wavelength is put out by a black body at a given temperature. In short, how does intensity depend on wavelength (at a fixed temperature)? In the water-wave analogy, this would correspond roughly to predicting how deep the leading, central, and trailing parts of a ripple created by a falling stone would be, as a function of, say, the kinetic energy of the stone as it hit the water surface.

Now the actual black-body spectrum at many temperatures had been carefully measured by experimental physicists, and the characteristic shape of the curve of *intensity* versus *wavelength* (at a fixed temperature) was familiar. At very long and very short wavelengths, the intensity died away toward zero, and at an intermediate value determined by the temperature, the intensity hit its maximum. This disagreed sharply with the prediction of classical physics concerning the intensity of the various colors. Classical physics predicted that at very short wavelengths, no matter what the temperature, the intensity would approach infinity. In modern terminology, this amounts to saying that every object, even an ice cube, is constantly radiating lethal gamma rays at arbitrarily great intensities! This is obviously preposterous. Up to 1900, however, no one had any idea of how to patch up the classical theory.

In that year, Max Planck invented a sort of hybrid formula that looked like a mathematical splicing-together of two different components, one pertaining to long wavelengths and the other to short wavelengths. At the longer wavelengths, the formula agreed with the classical prediction and also with the measured data. At the shorter wavelengths, Planck's formula diverged from the classical prediction but stayed in agreement with the data. The long and the short of it was that Planck's equation seemed right on the money for all wavelengths and temperatures—but it had not been derived from the first principles. It was a lucky guess, although much more than luck was involved, since Planck's intuition had guided him like a bloodhound to this formula.

Planck himself was particularly baffled by the fact that he'd had to throw a strange quantity he called "the elementary quantum of action", h, into his formula. What h represented physically was unclear. It was just a constant that, with a suitable value, would make the formula exactly reproduce the observed spectrum. It seemed therefore to be a universal constant of nature.

But what in the world was it doing in this equation? What did it mean? Einstein was the first to postulate a physical reason for the appearance of Planck's constant h in the equation. Einstein began with the concept that the energy content of light waves is deposited in tiny "lumps"—*photons*—whose size has to do with h and their wavelength. For example, if the light is red, its photons carry always 3.3×10^{-12} erg of energy. Green photons carry 4×10^{-12} erg. AM radio-wave photons carry somewhere between 3×10^{-21} and 9×10^{-21} erg (depending on what station you're listening to). The amount of energy per photon was postulated to be invariant, given its color (that is, its wavelength).

In the water-wave analogy, you can try to envision ripples that, when they reach the shore, suddenly disappear and are replaced by frogs who hop up the bank where the waves, had they landed, would have lapped. The longer the wavelength of the ripple, the tinier the frog that jumps out, and conversely: delicate ripples with very short wavelengths, when they reach the shore, suddenly become thundering monster-frogs who knock eucalyptus trees down and send boulders crashing into the lake (this is the infamous *phrogo-eucalyptic effect,* so yclept by reason of its analogy with the famous *photoelectric effect,* in which incoming photons of sufficient energy knock electrons out of a metal surface).

Einstein's interpretation of Planck's formula implied that a frog's energy —or rather, a photon's energy—and its wavelength must be inversely proportional. The equation linking them is:

$$E = hc/\lambda.$$

Here, E is the photon's energy, h is Planck's newly discovered constant, c is the speed of light, and λ is the photon's wavelength. E and λ are the only variables. This mixing of wave and particle viewpoints was one of the most baffling aspects of quantum mechanics, and it has continued to plague the intuitions of physicists ever since, although mathematically it was greatly cleared up by the blossoming of the field in the 1920's and 1930's.

* * *

The next step *en route* to Heisenberg's uncertainty principle came in 1924, when Prince Louis-Victor de Broglie was reflecting on the mysterious particle-like nature of light waves. He asked himself: Why should only *light* waves be particle-like? Why not the reverse? That is, mightn't particles also have wavelike properties? De Broglie's intuition was more or less as follows: If you want to generalize Einstein's equation so that it holds for particles other than photons, you have to get rid of the one direct reference in it to light, namely c. Hence de Broglie thought about how he might most elegantly and relativistically recast the equation in a c-less form.

This proved to be not too hard, because by then it was known that photons have both energy E and momentum p, and that they are related by the equation $E = pc$. If you combine the two equations, you can cancel out the 'c', and the result is:

$$p = h/\lambda.$$

Mathematically speaking, this equation of de Broglie's is new, but physically speaking, its content is no different from that of Einstein's original equation —at least when it is applied to photons. De Broglie's conceptual bravery was to propose—without any experimental evidence for it—that this equation should be *universal.* It should apply to all matter: not just photons, but also electrons, protons, atoms, billiard balls, people—even frogs! Thus Kermit the Frog would have a quantum-mechanical wavelength whose value would depend on how fast he's hopping.

What would this mean physically? What can a hopping frog's wavelength mean? Well, if you calculate it, you will find that Kermit's wavelength comes out far shorter than the radius of a proton—yet Kermit himself is considerably bigger than a proton. If Kermit were very, very small—small enough that his wavelength and his own size were comparable—then his wavelength would make him diffract around objects the way water waves and sound waves do. But since Kermit is macroscopic, his having a microscopic wavelength is all but irrelevant.

For electrons, though, it is entirely another matter. They *are* smaller than their own wavelengths. (In fact, as far as anyone knows, electrons are perfect point particles, with zero radius.) Shortly after de Broglie's suggestion, experiment and theory thoroughly confirmed his notion. Electron waves were soon being diffracted in laboratories around the world, just like light waves. But now there arises a puzzle. Are electrons spread out in space in the way waves must be, or are they localized? If they are truly points, how can they be diffracted? If they are truly waves, where is their electric charge carried?

*   *   *

Experiments have shown that even a *single* electron can be diffracted. Richard Feynman, in his little book *The Character of Physical Law,* describes it beautifully. In an idealized experiment, one electron is released in the direction of a barrier with two slits in it. On the far side of the barrier is a detecting screen. The electron follows some trajectory and hits the screen somewhere. One such event simply results in one dot being made on the screen. Suppose we repeat the experiment many times, each time releasing just one electron. We get a buildup of dots on the screen. Intuition, building on our experience with such things as bullets fired from a gun, tells us clearly to expect the dots to be clustered directly behind each of the two slits, with their distribution tailing off with distance from the center of each cluster. In other words, we would expect to find two clusters of dots and no other kind of distribution. (See Figure 20-1*a.*)

But if the de Broglie wavelength of the electron is close to the distance between the slits, the pattern on the screen after thousands of arrivals will look very different. It will be a complex regular structure characteristic of waves interfering with each other. In fact, it will reproduce the intensity pattern created by a wave that splits itself into two pieces, which pass through the two slits and interfere with each other on the far side of the barrier. (See Figures 20-1*b* and 20-1*c*.) It must be inferred that each electron, as it flew in its trajectory from source to screen, somehow "sensed" *both* slits and *interfered with itself* in the manner of a wave and yet deposited itself froglike (that is, in a point) on the screen without a trace of its schizophrenia.

The dilemma is, then, that electrons act as if they are both spread out *and* localized—as if they were both waves *and* particles. This kind of wishy-washiness is inconceivable in the macroscopic realm. Most of us have no trouble distinguishing between, say, ripples on a pond, and frogs. For those who do, however, it might be useful to clip out the following handy *frog-ripple distinguisher:*

Test 1: Is the candidate *solid, tangible,* and above all, always *somewhere?*
If your answers to these three questions are *yes,* you are probably dealing with a *frog.*
Test 2: Is the candidate *massless, intangible,* and *spread out?*
If your answers to these three questions are *yes,* it is probably a *ripple.*

If you are hungry for frog's legs and want to know where a frog is, you can just look around, and as soon as you sense some froglike photons entering your eyes, you will have found it. Those photons bounced off the frog and into your eyes. But suppose the frog somehow grew smaller and smaller. After it got down to the size of a mitochondrion in a living cell, its diameter would be about the wavelength of frog-green light. Then it would diffract light, and you would not be able to find it so easily. If it grew even smaller, something terrible would begin to happen. The individual photons hitting it would, with their momentum and energy, begin to jostle it around. The particle-like quality of photons would start to enter the picture. Indeed, a frog the size of an electron would probably be *very* hard to find. So if you were starved for frog's legs, you would do better to look around for a bigger one.

Unfortunately, though, no matter how starved you might be for electron's legs, you cannot find a bigger electron! To find an electron, you cannot do anything but bombard it with other particles or with photons. Since particles and photons have both particle-like and wavelike aspects, either bombardment will lead to similar consequences. If you want to pinpoint a particle, you need waves whose wavelength is about the size of that particle (or shorter). To understand this intuitively, think of the way water waves would be affected by a floating piece of wood. If they have a very long wavelength, they will not even "notice" the wood. Only if their wavelength gets down to the size of the object will they begin to be affected by it.

Consequently, in order to find our electron, we need photons of very short wavelength. But wavelength is inversely proportional to momentum.

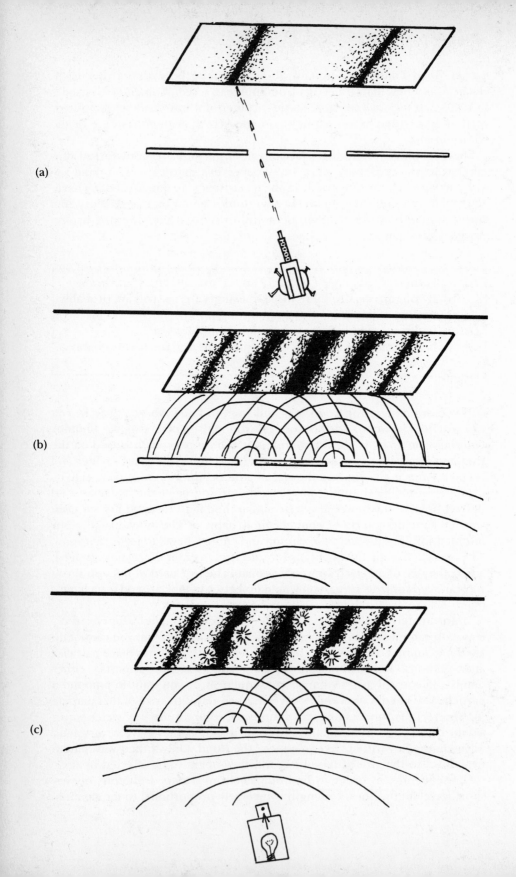

(a)

(b)

(c)

That is the deadly import of de Broglie's equation. You pay for your short wavelength by having a lot of momentum. And so, as you try to diffract waves ever so gently off your particle, hoping not to move it, you will not be able to do so without transmitting momentum to it. Either you are gentle (using long-wavelength photons) and do not see the electron well, or you are violent (using short-wavelength photons) and throw the electron completely off its course.

Heisenberg made a careful study of this perversity, which follows from de Broglie's equation, and, to the bewilderment of epistemology lovers the world over, he discovered that to know the position of a particle perfectly is to give up any hope of knowing its momentum, and that to know the momentum is to give up any hope of knowing its position. And knowing either one *imprecisely* still imposes bounds on the precision with which you could know the other. The principle can even be summarized in an inequality, which Heisenberg deduced. If you are trying to determine the location of the particle, there will be an uncertainty, conventionally denoted Δx. There will also be an uncertainty in the value of the momentum, denoted Δp. Heisenberg's uncertainty principle is the following inequality:

$$\Delta x \Delta p \geq h/4\pi.$$

There are a couple of things to point out here. First, note the presence of h, Planck's mysterious constant. This tells you that the effect is due to the wave-particle duality of matter (and of photons), and has nothing to do with the notion of an *observer* disturbing the thing under observation. Second,

---

FIGURE 20–1. *Three related two-slit experiments, two classical and one quantum-mechanical.*

In (a), a wildly swinging machine gun sprays bullets toward a wall with two holes in it. Occasionally, a bullet will pass through one of the two holes, and will hit the backstop and make a mark. Eventually, the buildup of marks looks as shown. It has two peaks, one for each hole.

In (b), a bobbing buoy creates ripples that spread out toward a jetty with two breaks in it. When the ripples hit the jetty, new circular ripples emanate from each of the two breaks, and those ripples, crisscrossing each other, interfere constructively at some points and destructively at others. On a vertical barrier parallel to the jetty, areas of highly constructive interference are dark, and areas of highly destructive interference are white. This characteristic interference pattern *is due to two facts: first, that any ripple passes through* both *holes, rather than just one, and second, that the phases at the two holes are* correlated.

In (c), a wildly swinging electron gun sprays electrons toward a wall with two holes in it. Beyond the wall there is a backstop made of some material that emits a flash whenever an electron hits it. There is no classical way to describe what happens to any electron en route, *but what is certain is that, when it comes in for a landing on the backstop, its local spot of arrival is clearly visible, just as in (a) (thus reminding us of the corpuscular, or bullet-like, nature of electrons); and yet, if those flashes are tallied up over a period of time, they are found to be distributed in an interference pattern just like the one formed in (b) (thus reminding us of the undulatory, or ripple-like, nature of electrons). Any attempt to ascertain which of the two holes the electrons pass through ends up in destruction of the interference pattern. [Drawing by David Moser, after Richard Feynman.]*

notice that even with this epistemological restriction, arbitrarily accurate measurement of *either* position *or* momentum is possible; you just can't get *both.*

In short, it is a total misinterpretation of Heisenberg's uncertainty principle to suppose that it applies to macroscopic observers making macroscopic measurements. For example, it does not follow from Heisenberg's principle that psychologists studying the phenomena of human cognition are somehow limited in principle by the fact that the conscious human beings they are observing are capable of the same kind of observation. What psychologists *are* limited by is their knowledge of the human brain, their ingenuity, and, of course, their funding.

If you wanted to know more about grammatical anomalies in the speech of woman W, there are all sorts of ways that you could, in principle, go about it without making her self-conscious. For just a few thousand dollars, for instance, you could secretly install a bug in her home and monitor all her conversations. For a few hundred thousand dollars, you could have tiny radio transmitters manufactured and secretly sewn into all her lapels. For, say, a few million dollars, you might be able to convince her she needed minor surgery of some sort, and then while she was anesthetized you could open up her skull and have harmless electrodes implanted in her brain to monitor her speech areas—all without her knowing. If you fear that such blatant physical interference with her brain might disturb her grammatical habits, then you may have to wait a while longer until we figure out how neural activity can be examined remotely. These possibilities are clearly extravagant, even ridiculous, but the point is that, *in principle,* we can study macroscopic phenomena with an arbitrary degree of precision.

To recapitulate: The uncertainty principle states not that the observer always interferes with the observed, but rather that at a very fine grain size, the wave-particle duality of the measuring tools becomes relevant. It is a consequence of the fact that Planck's constant is not zero, rather than an epistemological law about observation that would have been discovered with or without the discovery of quantum mechanics.

* * *

The uncertainty principle is not an *axiom* of quantum physics; it is a *deduced principle,* just as Einstein's most famous equation $E = mc^2$ was deduced from the more fundamental equations of special relativity—a fact that most non-physicists do not appreciate. Both equations are useful (and famous) because they are so pithy. For example, the uncertainty principle is often applied by physicists as a rule of thumb. If you want to estimate the approximate momentum a neutron will have when it is emitted by a nucleus decaying from an excited state, a seat-of-the-pants estimate is given by $p = h/d$, where d is on the order of the dimensions of the confining nucleus. You can think of the confinement within the nucleus as making the *position* uncertainty very small, so that the neutron is bouncing around inside its "cage" with a compensatingly large *momentum* uncertainty. When it escapes, a rough estimate of the momentum it will have is given by the uncertainty value.

When you examine the foundations of quantum mechanics, it becomes clear that the uncertainty principle is more than an epistemological restriction on human observers; it is a reflection of uncertainties in nature itself. Quantum-mechanical reality does not correspond to macroscopic reality. It's not just that we cannot *know* a particle's position and momentum simultaneously; it doesn't even *have* definite position and momentum simultaneously!

In quantum mechanics, a particle is represented by a so-called *wave function* describing the probabilities that the particle is here, there, or somewhere else; that the particle is heading east, west, north, or south; and so on. For each point in space, there is what is called a *probability amplitude* of finding the particle there, and this number is given by the wave function. Alternatively, one can read the wave function through different "mathematical glasses" and obtain a probability amplitude for each possible value of *momentum*. All the facts about the particle are wrapped up in its wave function. In more modern terminology, the term "state" is often used instead of "wave function".

In classical physics, quantities such as x and p—position and momentum —directly enter the equations governing a particle's behavior. The values of x and p are definite at any one moment, and they change according to the forces that are acting on the particle. With such equations of motion, physicists can plot in advance the positions and momenta of particles in simple, stable systems with incredible accuracy. An example is the motions of the planets, which even the ancients learned to predict with considerable accuracy. A more contemporary example is provided in computer space games, where rockets and planets are affected by a star's gravity and can go into orbit right before your eyes, swinging about in perfect ellipses on a screen. The underlying equations of such motion are *differential equations,* and one obvious property they have—we take it for granted—is that the motions they describe are *smooth.* Planets and rocket ships do not jump out of their orbits. There are no sudden discontinuities in their motion.

In quantum mechanics, x and p do not enter into the equations of motion as they do in classical mechanics. Instead, it is the wave function (in nonrelativistic quantum mechanics) that evolves in time according to a differential equation: *Schrödinger's equation,* named for Heisenberg's contemporary, the quantum-mechanical pioneer Erwin Schrödinger. As time progresses, the values of the wave function ripple through space just the way a water wave ripples on a lake's surface. This would seem to imply that quantum phenomena, like nonquantum ones, proceed smoothly and with no jumps. In one sense, that is right. A well-known example is the smooth precession of a spinning charged particle in a magnetic field. It is a kind of electromagnetic analogue to the precession of a spinning top on a table. The parameters that characterize the state of the spinning top or spinning particle do indeed change smoothly, without any jumps.

HOWEVER—a big however—there are exceptions to this smooth behavior, and they seem to form just as central a part of quantum theory as does the smooth evolution of states. The exceptions occur in the act of *measurement*, or the interaction of a quantum system with a macroscopic one. As quantum mechanics is usually cast, it accords a privileged causal status to certain systems known as "observers", without spelling out what observers are (in particular, without spelling out whether consciousness is a necessary ingredient of being an observer). To clarify this, I now present a quick overview of the measurement problem in quantum mechanics, and I will use the metaphor of the "quantum water faucet" for that purpose.

* * *

Imagine a water faucet with two knobs, one labeled "H" and one labeled "C", each of which you can twist continuously. Water comes streaming out of the faucet, but there is a strange property to this system: the water is always either totally hot or totally cold; there is no in-between. These are called the two *temperature eigenstates* of the water. (The prefix *eigen-* can be translated from the German as "particular." Here it refers to the fact that the temperature has a particular value.) The only way you can tell which eigenstate the water is in is by sticking your hand in and feeling it. Actually, in orthodox quantum mechanics it is trickier than that. It is the act of putting your hand in the water that *throws* the water into one or the other eigenstate. Up till that very instant, the water is said to be in a *superposition of states* (or more accurately, a superposition of *eigenstates*).

Depending on the settings of the knobs, the likelihood of your getting cold water will vary. Of course, if you open only the "H" valve, then you'll get hot water always, and if you open only the "C" valve, then you'll get cold water for sure. But if you open both valves, you'll create a superposition of states. By trying it out over and over again with one setting, you can measure the probability of getting cold water with that setting. After that you can change the setting and try again. There will be some crossover point where hot and cold are equally likely. It will then be like flipping a coin. (This quantum water faucet is sadly reminiscent of many a bathroom shower . . .) You can eventually build up enough data to draw a graph of the probability of cold water as a function of the knobs' settings.

Quantum phenomena are like this. Physicists can twiddle knobs and put systems into superpositions of states, analogous to the superpositions of the hot-cold system. As long as no *measurement* is made of the system, a physicist cannot know which eigenstate the system is in. Indeed, it can be shown that in a very fundamental sense, the system itself does not "know" which eigenstate it is in, and only decides—at random—at the moment the observer's hand is stuck in to "test the water", so to speak. (Note that a nonsexist reader is in a superposition of states at this very moment, not knowing if this hypothetical observer (or for that matter, the hypothetical nonsexist reader) is male or female!) Up to the moment of observation, the system acts as if it were not in an eigenstate. For all practical purposes, for all theoretical purposes—in fact for *all* purposes—the system is not in an eigenstate.

You can imagine doing a lot of experiments on the water coming out of a quantum water faucet to determine whether the water is actually hot or actually cold without sticking your hand in it. (We're of course assuming that there are no telltale clues to the temperature of the water, such as steam rising from it.) For example, run your washing machine on water from the quantum faucet. Still, you won't know whether your wool sweater has shrunk or not until the moment you open the machine (a measurement made by a conscious observer). Make some tea with water from the faucet. Still, you won't know whether you've got hot tea or not until you taste it (again a measurement made by a conscious observer). The critical point here is that the sweater and the tea, not having conscious-observer status themselves, have to play along with the gag and, just as the water did, enter superpositions of states: shrunk and non-shrunk, hot tea and cold tea.

All this may sound as if it has nothing to do with physics *per se*, but merely with ancient philosophical conundrums such as: "Does a tree in a forest make a noise when it falls, if there's nobody there to hear it?" But the quantum-mechanical twist on such riddles is that there are observational consequences of the reality of the superpositions, consequences diametrically opposite to those that would ensue if a seemingly mixed state were in reality always a true eigenstate, merely hiding its identity from observers until the moment of measurement. In crude terms, a stream of maybe-hot-maybe-cold water would act differently from a stream of water that is *actually* hot or *actually* cold, because the alternatives "interfere" with each other. This would become manifest only after a large number of sweater-washings or tea-makings, just as in the two-slit experiment it takes a large number of electron-landings to reveal the interference pattern of the alternative trajectories. (Quantum-mechanical interference resembles the classical phenomenon of two notes beating against each other, except that in quantum mechanics, instead of producing a chord of sounds, the superposition produces a distribution of probability—a "chord of possibilities".) Interested readers should consult either Feynman's *The Character of Physical Law* or, for an account with more detail, Volume III of *The Feynman Lectures on Physics*.

*   *   *

The plight of "Schrödinger's cat" carries this idea further: it suggests that even a cat might be in a quantum-mechanical superposition of states until

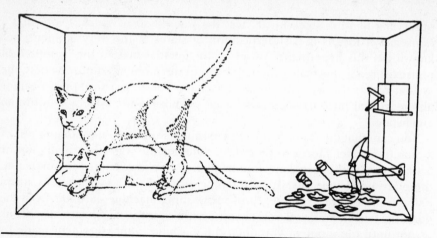

FIGURE 20–2. *Schrödinger's cat in a superposition of states, partly alive and partly dead.* [*From* The Many-Worlds Interpretation of Quantum Mechanics, *edited by Bryce S. DeWitt and Neill Graham.*]

a human observer intervened. The tale of this unfortunate cat goes like this. A box is prepared for a cat's occupancy. Inside this box, there is a small sample of radium. Also in the box is a detector of radiation, which will detect any decays of radium nuclei in the sample. The sample has been chosen so that there is a 50–50 probability that within any hour-long period, one decay will occur. On the occurrence of such a decay, a circuit will close, tripping a switch that will break a beaker filled with a deadly liquid, spilling the liquid onto the floor of the box, and killing the cat. (See Figure 20–2.)

The cat is now placed in the box, the lid firmly shut, and an hour ticks away. At the hour's end, a human observer approaches the box and opens the lid to see what has happened. According to one extreme view of quantum mechanics (and the reader should bear in mind that it is not the usual view), only *at that moment* will the system be forced to "jump" into one of the two possible eigenstates—cat alive and cat dead—that are represented together as a superposition in the wave function of the system. (Notice that it is necessary that the randomness be of a clearly quantum-mechanical origin: the decay of the radium nucleus. This thought experiment would not pack any punch if there were a spinning roulette wheel in the box instead of a radium sample.)

One might object and say, "Wait a minute! Isn't a live cat as much of a conscious observer as a human being is?" Probably it is, but notice that this cat is possibly a *dead* cat, and in that case certainly not a conscious observer. We have in effect created in Schrödinger's cat a superposition of two eigenstates, one of which *has* observer status, the other of which *lacks* it. Now what do we do? This situation is reminiscent of a Zen riddle (recounted in *Zen Flesh, Zen Bones* by Paul Reps):

Zen is like a man hanging in a tree by his teeth over a precipice. His hands grasp no branch, his feet rest on no limb, and under the tree another person asks him: "Why did Bodhidharma come to China from India?" If the man in the tree does not answer, he fails; and if he does answer, he falls and loses his life. Now what shall he do?

* * *

The idea that consciousness is responsible for the "collapse of the wave function"—a sudden jump into one randomly chosen pure eigenstate—leads to further absurdities. For instance, it would imply that nothing ever happened for the first umpteen billion years of the universe, until one day, a million or so years ago, some human being woke up—and at that instant the enormously swollen universal wave function collapsed down into one world—and this person blinked, peered around, and saw Mesopotamia or Kenya . . .

The alternative left to us is that observers—things that make a wave function collapse—need not be conscious, but merely *macroscopic.* However, isn't a macroscopic object just a collection of microscopic objects? How would a wave function "know" it was dealing with a macroscopic object? More concretely, what is it about a screen that forces an electron to reveal itself?

To many physicists, the distinction between systems with observer status and those without has seemed artificial, even repugnant. Moreover, the idea that an intervention of an observer causes a "collapse of the wave function" introduces caprice into the ultimate laws of nature. "God does not play dice" (*Der Herrgott würfelt nicht*) was Einstein's lifelong belief.

A radical attempt to save both continuity and determinism in quantum mechnics is known as the *many-worlds interpretation,* first proposed in 1957 by Hugh Everett III. According to this very bizarre theory, no system ever jumps discontinuously into an eigenstate. What happens is that the superposition evolves smoothly with its various branches unfolding in parallel. Whenever it is necessary, the state sprouts further branches that carry the various new alternatives. For example, there are two branches in the case of Schrödinger's cat, and they develop in parallel. "Well, what happens to the cat? Does *it* feel itself to be alive or does it feel itself to be dead?" One must wonder. Everett would answer: "It depends on which branch you look at. On one branch, the cat feels itself to be alive, and on the other, there is no cat to feel anything." With intuition beginning to rebel, one then asks: "Well, what about a few moments before the cat on the fatal branch died? How did the cat feel *then?* Surely the cat can't feel two ways at once! Which of the two branches contains the genuine cat?"

The problem becomes even more intense as you realize the implications of this theory as applied to you, *here* and *now.* For every quantum-mechanical branch point in your life (and there have been billions upon billions), you have split into two or more you's riding along parallel but disconnected branches of one gigantic "universal wave function". (By this term is meant the enormous wave function representing all the particles in all the parallel universes.) At the critical spot in his article where this difficulty arises, Everett calmly inserts the following footnote:

> At this point we encounter a language difficulty. Whereas before the observation we had a single observer state, afterwards there were a number of different states for the observer, all occurring in a superposition. Each of these separate states is a state for an observer, so that we can speak of the different observers described by different states. On the other hand, the same physical system is involved, and from this viewpoint it is the *same* observer, which is in different states for different elements of the superposition (i.e., has had different experiences in the separate elements of the superposition). In this situation we shall use the singular when we wish to emphasize that a single physical system is involved, and the plural when we wish to emphasize the different experiences for the separate elements of the superposition. (E.g., "The observer performs an observation of the quantity A, after which each of the observers of the resulting superposition has perceived an eigenvalue.")

All said with a poker face. The problem of how it feels *subjectively* is not treated; it is not even swept under the rug. It is probably considered meaningless.

And yet . . . one simply has to wonder: "Why, then, do I feel myself to be in just *one* world?" Well, according to Everett's view, you *don't*—you feel all the alternatives simultaneously. It's just *this* you going down *this* branch who doesn't experience all the alternatives. This is completely shocking. In his story "The Garden of Forking Paths", the Argentinian writer Jorge Luis Borges describes a fantastic vision of the universe in this way:

> . . . a picture, incomplete yet not false, of the universe as Ts'ui Pên conceived it to be. Differing from Newton and Schopenhauer, . . . [he] did not think of time as absolute and uniform. He believed in an infinite series of times, in a dizzily growing, ever spreading network of diverging, converging and parallel times. This web of time—the strands of which approach one another, bifurcate, intersect, or ignore each other through the centuries—embraces every possibility. We do not exist in most of them. In some you exist and not I, while in others I do, and you do not, and in yet others both of us exist. In this one, in which chance has favored me, you have come to my gate. In another, you, crossing the garden, have found me dead. In yet another, I say these very same words, but am an error, a phantom.

This quotation is featured at the beginning of the book *The Many-Worlds Interpretation of Quantum Mechanics: A Fundamental Exposition,* edited by Bryce S. Dewitt and Neill Graham. The ultimate question is this: "Why is *this me* in *this branch,* then? What makes me—I mean *this* me—feel itself—I mean myself—unsplit?"

* * *

The sun is setting one evening over the ocean. You and a group of friends are standing at various points along the wet sand. As the water laps at your feet, you silently watch the red globe drop nearer and nearer to the horizon. As you watch, somewhat mesmerized, you notice how the sun's reflection on the wave crests forms a straight line composed of thousands of momentary orange-red glints—a straight line pointing right at you! "How lucky that *I* am the one who happens to be lined up exactly with that line!" you think to yourself. "Too bad not all of us can stand here and experience this perfect unity with the sun." And at the same moment, each of your friends is having precisely the same thought . . . or is it the same?

Such musings are at the heart of the "soul-searching question". Why is this soul in this body? (Or on this branch of the universal wave function?) Why, when there are so many possibilities, did *this* mind get attached to this body? Why can't my "I-ness" belong to some other body? It is obviously circular and unsatisfying to say something like "You are in that body because that was the one made by your parents." But why were *they* my parents, and not sometwo else? Who would have been my parents if I had been born in Hungary? What would I have been like if I had been someone else? Or if someone else had been me? Or—*am* I someone else? Am I everyone else? Is there only one universal consciousness? Is it an illusion to feel oneself as separate, as an individual? It is rather eerie to find these bizarre themes reproduced at the core of what is supposedly our stablest and least erratic science.

And yet in a way it is not so surprising. There is a clear connection between the imaginary worlds of our minds and the alternate worlds evolving

---

FIGURE 20–3. *A robot in an anxious superposition of mental states.* [*Drawing by Rick Granger.*]

in parallel with the one we experience. The proverbial young man picking apart a daisy and muttering, "She loves me, she loves me not, she loves me, she loves me not . . ." is clearly maintaining in his mind at least two different worlds based on two different models for his beloved. (See Figure 20-3.) Or would it be more accurate to say that there is *one* mental model of his beloved that is in a mental analogue of a quantum-mechanical superposition of states?

And when a novelist simultaneously entertains a number of possible ways of extending a story, are the characters not, to speak metaphorically, in a mental superposition of states? If the novel never gets set to paper, perhaps the split characters can continue to evolve their multiple stories in their author's brain. Furthermore, it would even seem strange to ask which story is the *genuine* version. All the worlds are equally genuine.

And in like manner, there is a world—a branch of the universal wave function—in which you didn't make that stupid mistake you now regret so much. Aren't you jealous? But how can you be jealous of your*self*? Besides which, there's yet *another* world in which you made yet stupider mistakes, and in which you are jealous of this very you, here and now in *this* world!

*　　*　　*

Perhaps one way to think of the universal wave function is as the mind—or brain, if you prefer—of the great novelist in the sky, God, in which all possible branches are being simultaneously entertained. We would be mere subsystems of God's brain, and *these* versions of us are no more privileged or authentic than our galaxy is the only genuine galaxy. God's brain, conceived in this way, evolves smoothly and deterministically, as Einstein always maintained. The physicist Paul Davies, writing on just this subject in his recent book *Other Worlds,* says: "Our consciousness weaves a route at random along the ever-branching evolutionary pathway of the cosmos, so it is we, rather than God, who are playing dice."

Yet this leaves unanswered the most fundamental riddle that each of us must ask: "Why is my unitary feeling of myself propagating down *this* random branch rather than down some other? What *law* underlies the random choices that pick out the branch I feel myself tracing out? Why doesn't my feeling of myself go along with the other me's as they split off, following other routes? What attaches *me-ness* to the viewpoint of this body evolving down this branch of the universe at this moment in time?" These questions are so basic that they almost seem to defy clear formulation in words. And their answers do not seem to be forthcoming from quantum mechanics. In fact, this is exactly the collapse of the wave function reappearing at the far end of the rug it wasn't swept under by Everett . . . It turns it into a problem of personal identity, no less perplexing than the problem it replaces.

One can fall even more deeply into the pit of paradox when one realizes that there are branches of this one gigantically branching universal wave function on which there is no Werner Heisenberg, no Max Planck, no Albert Einstein, branches on which there is no evidence for quantum mechanics whatsoever, branches on which there is no uncertainty principle or many-worlds interpretation of quantum mechanics. There are branches on which the Borges story did not get written, branches in which this column did not get written. There is even a branch in which this entire column got written just as you see it here, except for one noun which was replaced by its exact antonym, at the column's very beginning.

Post Scriptum.

> *Quantum particles: the dreams that stuff is made of.*
> —David Moser

If this was your introduction to the weirdness of quantum mechanics (which I doubt), then may I say how delighted I am to have been your guide. But in that case, I also must say that you really deserve a more complete introduction. This article was aimed mostly at people who already have at least a nodding acquaintance with quantum phenomena. The Feynman books alluded to in the article are ideal introductions. There are other books that purport to explain quantum mechanics to novices, and in some cases they may do a fairly good job of it, but some of them have the serious drawback of trying to link quantum-mechanical reality with Eastern mysticism, a connection I find superficial and misleading. I cannot fault people who wish to make some observations about the worldview of ancient Buddhists and to point out that a few statements written thousands of years ago can, if very liberally interpreted, be taken to say things that are not inconsistent with discoveries of modern physics, but to claim that "Western science is only now catching up with the ancient wisdom of the East", as most of those authors do (and in roughly those words), is, in my view, both silly and anti-intellectual.

I call it "anti-intellectual" because most Western people infatuated with Eastern mysticism hold a grudge against the encroachment of science on territory they consider beyond science. This attitude may be a holdover from the bitterly anti-scientific, anti-intellectual mood that gripped the United States during much of the Viet Nam War era. Those people have some sort of axe to grind, perhaps subconsciously; they want to see science "put in its place". Curiously, many of them are scientists themselves and revel in a kind of self-deprecation, thinking that they are lifting themselves up to transcendent heights and seeing things from a "higher plane of enlightenment" than science affords. Usually, at that point, their prose abruptly changes mood, moving from precise terms to mushy, vague, poetic terms (such as "mushy", "vague", and "poetic"). Don't you just hate that sort of thing?

These are the sorts of people who propagate misinformation about the discoveries of modern physics (such as the pseudo-uncertainty principle). They encourage people to think that any wild theory explaining any mystery (or alleged mystery) might well be correct, as long as it uses voguish technical terms from physics—terms like "tachyon", "Bell's inequality", "EPR paradox", "gravitational waves", and so on. A typical abuser of physics in this way is Arthur Koestler; in his book *The Roots of Coincidence,* he purports to explain "psi phenomena" in terms of some five-dimensional theory of particle physics that includes a host of hypothetical particles called "psitrons".

To me, a very troubling aspect of an "explanation" such as this (which, actually, Koestler didn't invent himself but borrowed from a physicist named Adrian Dobbs) is that very similar explanations are used by physicists themselves—not so often of "psi phenomena", but of currently unexplained real phenomena in particle physics. When I was a graduate student in particle physics, quite a number of years ago, I read paper after paper in which not only new *particles* were invoked to explain some observation, but new *families* of particles were routinely postulated. As a matter of fact, one of those papers was the straw that broke the camel's back, as far as I was concerned. In that three-author paper, the authors had the audacity to invent some totally off-the-wall *superfamily* of particles that consisted of a large number of families, each containing quite a few particles on its own. As I recall, there were something like 140 new particles introduced in one fell swoop—and, mind you, this was done merely to explain some rather small discrepancies between things measured and things predicted by previous theories. A far cry from the days when it was a highly daring step to introduce even *one* new particle! It was at that point that I decided I should bow out of that branch of theoretical physics.

*   *   *

I am not really trying to castigate the whole field of particle physics, because all I learned for sure from my long, grueling, and ultimately broken engagement with that field was that I personally was not cut out to be a particle physicist. However, I did learn one disillusioning thing about science in general, and that is that large segments of it—including, very often, the most forbidding and technically prickly papers—are just as nonsensical and empty as the pseudo-scientific papers that try to shore up "psi phenomena", "remote viewing", "telekinesis", or the like. (Is it reasonable for me to continue using quotation marks around those words? I think so. I don't like using words in such a way that I help to lend them legitimacy when I think there is nothing behind them.) Bad science permeates good science the way that gristle runs through meat (a "meataphor" exploited in a different context in Chapter 21).

I am afraid that this is an example of an inevitable phenomenon: If you are throwing darts and want not only to hit the bull's-eye every time but also to cover the entire bull's-eye evenly, so that you are equally likely to hit any point *inside* the bull's-eye but totally unlikely to go *outside* of it, then you are dreaming a pipe dream! You have to pay in some way for the privilege of filling up that inner circle—and you pay either by sometimes overflowing the boundaries of the bull's-eye (being too loose, so to speak), or by covering it unevenly, having a high concentration in the middle of the bull's-eye and a low concentration near its edges (being too tight or controlled). In science, this translates to the trade-off between being too speculative and too cautious. It is impossible for all the papers in a field to be both right and significant. Either many will be wrong or many will be trivial. The former corresponds, obviously, to throwing outside the circle, and the latter, a little less obviously, to covering it fully but unevenly. This inevitable trade-off is very much like that spoken of in Chapter 13, where in trying to produce all the truths expressible in a formal system or all the members of a semantic category, you wind up with either an *incomplete* system or an *inconsistent* system.

I guess this makes me sound somewhat cynical about science. But I would make similar noises about human endeavors of any sort that involve skill. For instance, not all the letters I receive from people who have read things I've written hit the bull's-eye; some of them are the cat's meow, but a larger number are either old hat, off base, full of hot air, or some combination thereof. So if I want to get some good letters, I have no choice but to be willing to wade through a bunch of bad ones, too. And, regretfully, I must say that this law applies just as much to my own output: not all of it can be of the same caliber. If it's all correct, then much of it will be mundane; and if I regularly dare to go far beyond the mundane, then some of it will wind up being wrong.

Some people choose to see trade-offs such as these as more examples of a kind of "uncertainty principle": you can't have both total correctness and total novelty. You must take your pick. This "either-or" quality, however, has very little to do with the quantum-mechanical substrate of our world. It just has to do with statistical phenomena in general.

* * *

I would like to say something about the alienness of quantum-mechanical reality. It is no accident, I would maintain, that quantum mechanics is so wildly counterintuitive. Part of the nature of explanation is that it *must* eventually hit some point where further probing only increases opacity rather than decreasing it. Consider the problem of understanding the nature of solids. You might wonder where solidity comes from. What if someone said to you, "The ultimate basis of this brick's solidity is that it is composed of a stupendous number of eensy-weensy bricklike objects that themselves are rock-solid"? You might be interested to learn that bricks are composed of micro-bricks, but the initial question—"What accounts for solidity?"—has been thoroughly begged. What we ultimately want is for solidity to vanish, to dissolve, to disintegrate into some totally different kind of phenomenon with which we have no experience. Only then, when we have reached some completely novel, alien level will we feel that we have *really* made progress in explaining the top-level phenomenon.

That's the way it is with quantum-mechanical reality. It is truly alien to our minds. Who can fathom the fact that light—that most familiar of daily phenomena—is composed of incredible numbers of indescribably minuscule "particles" with zero mass, particles that recede from you at the same speed no matter how fast you run after them, particles that produce interference patterns with each other, particles that carry angular momentum and that bend in a gravitational field? And I have barely scratched the surface of the nature of photons! I like to summarize this general phenomenon in the phrase "Greenness disintegrates." It's a way of saying that no explanation of *macro*scopic X-ness can get away with saying that it is a result of *micro*scopic X-ness ("just the same, only smaller"); macroscopic greenness, solidity, elasticity—X-ness, in short—*must,* at some level, disintegrate into something very, very different.

I first saw this thought expressed in the stimulating book *Patterns of Discovery* by Norwood Russell Hanson. Hanson attributes it to a number of thinkers, such as Isaac Newton, who wrote, in his famous work *Opticks:* "The parts of all homogeneal hard Bodies which fully touch one another, stick together very strongly. And for explaining how this may be, some have invented hooked Atoms, which is begging the Question." Hanson also quotes James Clerk Maxwell (from an article entitled "Atom"): "We may indeed suppose the atom elastic, but this is to endow it with the very property for the explanation of which . . . . the atomic constitution was originally assumed." Finally, here is a quote Hanson provides from Werner Heisenberg himself: "If atoms are really to explain the origin of color and smell of visible material bodies, then they cannot possess properties like color and smell." So, although it is not an original thought, it is useful to bear in mind that *greenness disintegrates.*

*   *   *

One of the most beautiful features of the quantum-mechanical description of reality is how a bridge is erected between the microscopic and the macroscopic. The nature of that bridge is characterized by the *correspondence principle,* which states:

In the limit of large sizes, quantum-mechanical phenomena must look indistinguishable from their classical counterparts.

This can be converted into a more mathematical statement, as follows:

> In the limit of large quantum numbers, quantum-mechanical equations must reproduce their classical counterparts.

A physicist does not have to work to make an equation describing quantum phenomena obey this principle; if the equation is correct, it will obey it automatically. However, a physicist cannot always be sure that a proposed equation is correct. Therefore, the correspondence principle provides a very useful check on any proposed equation—for if it fails to yield the familiar classical equation in the limit of large sizes (or more accurately, large quantum numbers), it is surely wrong. Of course, merely passing this test is no guarantee that an equation is right, but it is a confirming piece of evidence.

Quantum-mechanical phenomena are characterized by "quantum numbers", which are always integers. When those integers are small—less than 5 or so—you have quintessentially quantum phenomena. But when you plug fairly large values such as 20 into the equations, you get behavior that floats midway between the quantum style and the classical style. And when you take the limit of infinitely large values, you should get back the familiar old equations from the pre-quantum era: such things as Newton's laws of motion, for instance.

A striking example of this idea is furnished by so-called "Rydberg atoms", highly excited atoms whose outermost electrons have very large quantum numbers, and which are consequently tethered so loosely to their central nucleus that their orbits begin to be somewhat less "cloud-like" (*i.e.,* less quantum-mechanical), and more like the familiar planetary orbits that electrons used to follow, back in the short-lived "semiclassical" era of physics, after Ernest Rutherford's discovery of nuclei, but before Schrödinger and Heisenberg. These bridges between the alien world and the familiar world help provide the intuitions necessary for macroscopic people to imagine how jolly giant greenness could emerge from murky, unfathomable microdepths.

Section V:

Spirit and Substrate

Section V:
Spirit and Substrate

The world has traditionally been divided into the animate and the inanimate. Inanimate things do not have feelings or wills of their own, and can therefore be smashed, burned, or harnessed by animate ones without the animate ones having to feel guilty. This borderline, so long taken for granted by people, is gradually becoming blurrier with the advent of computers, especially as programs acquire more and more flexibility—and with that flexibility, a seeming mentality or personality. How and when could mind and emotions—surely the essence of the animate—emerge from complex inanimate substrates? What does it take to make spirit out of pure matter pattern? A number of recent artificial-intelligence programs have been touted as "thinking". Yet no one who looked closely could fail to see that there remains a huge gap between human self-aware fluidity and such programs. Even the best of them is still relatively rigid and unaware of anything, let alone itself. But where *is* the borderline between the highest inanimate flexibility and the lowest animate sentience? When does a system or organism have the right to call itself "I", and to be called "you" by us? Will we be able to recognize systems deserving of our respect when they come along, or will we abuse them? Will such systems have as much free will as we don't? These and other philosophically motivated questions about mind and mechanism, free will and determinism, randomness and rule-following, are examined in the following six chapters.

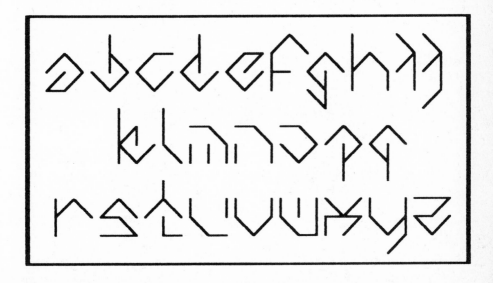

21

Review of *Alan Turing:*
The Enigma

November, 1983

CAN true intelligence be embodied in any sort of substrate—organic, electronic, or otherwise? Is mind more than pattern? How can we distinguish between a genuine mind and a clever façade? Is free will incompatible with a materialist, mechanistic view of living beings? Is there a contradiction in the notion of rule-bound creativity? Do our emotions and our intellects belong to separate compartments of our selves? Could machines have emotions? Could machines be enchanted by ideas, by people, by other machines? Could machines be attracted to each other, fall in love? What would be the social norms for machines in love? Would there be proper and improper types of machine love affairs? Could a machine be frustrated and suffer? Could a frustrated machine release its pent-up feelings by going outdoors and self-propelling ten miles? Could a machine learn to enjoy the sweet pain of marathon running? Could a machine with a seeming zest for life destroy itself purposefully one day, planning the entire episode so as to fool its mother machine into "thinking" (which of course machines cannot do) that it had perished by accident?

These are the sorts of questions that burned in Alan Turing's brain, and, taken at another level, they reveal highlights of Turing's troubled life. It would require someone who shares much with Turing to plumb his story deeply enough to do it justice, and Andrew Hodges, a young British writer with a doctorate in mathematics, has wonderfully succeeded in doing so. His 500-page biography of Turing, painstakingly put together from innumerable sources, including conversations with scores of people who knew Turing at various stages of his life, provides as vivid a picture as one could hope of a most complex and intriguing individual. And it's about time, for not only was Turing a very significant person in the science of this century, but his fascinating and difficult life illustrates serious problems that society has not yet grappled with successfully.

Hodges' rich and engrossing portrait is not the first book about Turing, since his mother, Sara Turing, wrote a sketchy memoir a few years after her son's death, which presents an image of Turing as a lovable, eccentric boy of a man, filled with the joy of ideas and driven by an insatiable curiosity about questions concerning mind and life and mechanism. Hodges goes far more deeply into Turing's mind, body, and soul than Sara Turing ever dared, for she wore conventional blinders and did not want to see how poorly her son fit into the standard molds of British society. Alan Turing was homosexual, a fact that he took no particular pains to hide, especially as he grew older. And for a boy growing up in the 1920's and for a man in the next few decades, being homosexual—especially if one was British and belonged to the upper classes—was an unmentionable, terrible, and mysterious affliction.

Alan Turing, an atheist, homosexual, eccentric English mathematician, was in large part responsible not only for the concept of computers, incisive theorems about their powers, and a clear vision of computer minds, but also for the cracking of German ciphers during World War II. It is fair to say that we owe much to Alan Turing for the fact that we are not under Nazi rule today. And yet this salient figure in world history has remained, as the book's subtitle says, an enigma.

*　　*　　*

Turing was born in London in 1912 of relatively well-to-do parents in the civil service in India. Not long after his birth, his father returned to India, followed by his mother, and they spent the next few years there, leaving young Alan in England. Then they decided to return closer to England, and for a time lived in France, which gave Alan the opportunity to take school vacations there and learn French. As a boy, he was inquisitive and humorously inventive but definitely not a child prodigy. At age thirteen, he was sent off to a boys' private boarding school called Sherborne, in the west of England. He made quite a hit his first day, for he arrived on bicycle, having pedaled the 60 miles from Southampton, where the ferry from France had left him on a day of general strikes and no trains. However, as the weeks passed, his hero status declined as he revealed himself to be a rather untidy pupil prone to getting ink all over himself, and one who did not distinguish himself in most of his classes. Alan was a solitary boy and his first venture into serious friendship came to an unexpected and tragic ending, when his friend and idol, Christopher Morcom, succumbed to bovine tuberculosis.

Alan never forgot this first and perhaps deepest of all his human contacts, for it was in fact a mixture of friendship and love. Although Alan never confided his love to Chris, it is apparent that Alan was in love with him. Later in life, Alan would have romances with other men, as well as more numerous and more sordid one-night stands, but the purity of his love for Christopher Morcom was never surpassed. A flame in Alan's heart continued to burn for

Review of Alan Turing: The Enigma

Chris Morcom, and he faithfully visited the Morcom family for years afterward, seeking some sort of spiritual communion with his lost friend. Perhaps this was one of the key sources of Alan's abiding interest in the connection between the elusive human soul and its mortal incarnation.

At Sherborne, Alan excelled at mathematics to the exclusion of pretty much everything else. In the end, his school recognized his great talent and awarded him several science prizes. At age twenty, Alan went on to Cambridge. This was 1933, and the scientific world was charged with the excitement of absorbing several revolutionary discoveries of the previous decade. Relativity, one of Turing's early obsessions, was now old hat, while quantum mechanics and mathematical logic were in their heyday. Quantum mechanics made a deep impression on Alan's mind. In quantum systems such as an atom, an electron can jump from one orbit (or "state") to another without occupying any intermediate position between them. It would be as if a space satellite jumped from one orbit to another without traveling between them.

Equally striking to Turing was the mechanization of mathematical reasoning, which he first read about in a philosophical book by Bertrand Russell. Later he studied the ambitious "Hilbert program", whose aim was to demonstrate the possibility of capturing in a single system all the valid principles of mathematical reasoning. In that system, all possible true consequences would flow out of a small set of axioms by means of a well-defined set of rules, like automobiles from an assembly line, or physical systems jumping from one state to another. This image of a machine that jumped from one state to another according to a finite set of rules became uppermost in Turing's mind. What fascinated him was the idea that such meaningless actions could also be viewed as having meanings. For instance, one rule-obeying machine might be viewed as making moves of chess, another as producing truths of mathematics, and yet another as writing poetry.

In 1931, the Austrian logician Kurt Gödel devastated Hilbert's and Russell's hopes of creating a perfect formalization of all mathematical reasoning. Gödel had demonstrated that there were undecidable propositions in any consistent axiomatic system of the Hilbert-Russell sort, propositions based on famous paradoxes of logic that had plagued logicians ever since the Greeks. (The sentence "I am lying" is a good example, as is the fruitless exercise of trying to catch a glimpse in the mirror of what you look like with your eyes perfectly closed.) What Gödel had left unsettled, however, was the question of whether, given an axiomatic system and an arbitrary proposition in it, one could determine mechanically whether that proposition was undecidable in that system. If this were possible, then one could discard undecidable propositions as easily as one trims pieces of fat from a steak; if it were impossible, then mathematics would resemble a piece of steak riddled with gristle, so that no matter how you slice it, what remains will contain some gristle.

*   *   *

Alan Turing chose to work on this question of whether the gristle could be cleanly lopped off the rest of mathematics, leaving the "meat" of mathematics intact and mechanizable, and the rest just a collection of quirky Gödelian curiosities, sideshow freaks contrasting with the vast world of normal mathematical propositions. To his surprise, he discovered that for very Gödelian reasons, no machine could be built that could infallibly recognize undecidable propositions.

He began by trying to specify exactly the most general possible notion of what a "machine" is. In fact, the concept he arrived at, now called a "Turing machine", forms a central part of Turing's contribution to the theory of computing. Although fundamentally all a Turing machine can do is jump from one discrete state to another by means of very simple transition rules, Turing was able to show that such machines could do anything that one could reasonably expect of any machine or any human following well-defined rules. He went further and showed that a very complex type of Turing machine, called a "universal" Turing machine, was capable of being fed a single number that encoded the structure of any other Turing machine, much in the way that DNA codes for the structure of an organism. The universal machine could then act indistinguishably from that machine. The discovery of such a universal Turing machine made all "specialist" Turing machines obsolete. For instance, if some Turing machine could play chess, then the universal Turing machine could also play chess, simply by being fed the code number of the chess-playing Turing machine. Ditto for theorem-producing and poetry-writing. If Turing were still alive, he would probably have relished Woody Allen's recent character Leonard Zelig, the "Human Chameleon"—a living, breathing universal Turing machine, one that could perfectly simulate any other, if fed the right code number.

Turing's death blow to the hopes of logicians such as Russell and Hilbert was delivered in two stages. First, he supposed that a machine for recognizing undecidable propositions exists; then he showed how that assumption leads to self-contradiction. He began by showing that any such machine—if it existed—would closely resemble a universal Turing machine, in that it could accept the description number of any machine and simulate it. Slyly, then, he proposed feeding into this hypothetical machine its *own* description number. This action, he showed, would instantly send it into a dizzy loop and make it perish of computational vertigo. In other words, the idea of such a machine is self-contradictory. The tantalizing upshot of this twisty argument is the discovery that undecidable propositions run through mathematics like ineradicable threads of gristle that crisscross a steak in such a dense way that they cannot be cut out without the entire steak's being destroyed. In short, through Gödel's and Turing's work, mathematics was revealed to be unmechanizable—or, more precisely, *incompletely* mechanizable, no matter how complex the machine involved.

Review of Alan Turing: The Enigma

Though on the surface this defeat for mechanism might seem to imply that human reasoning can always outwit or transcend mechanical imitations, on deeper analysis it turns out that Turing's argument can be applied to humans as well. Consider the yes-no question, "Will your answer to this particular question be 'no'?" You will find that you too go into a sort of computational vertigo in trying to answer it with a "yes" or a "no". This question exemplifies the sort of undecidability problems that Turing showed machines and mathematical systems are subject to. Though the example is simplistic, it reminds us of an essential fact of the human condition—that people, no matter how aware they are of their minds, cannot fully take their own complexity into account in attempting to understand themselves, and, quite like Turing machines baffled by their own descriptions, may be plunged into a vertigo of the psyche when they attempt to calculate their own hypothetical or future acts.

Just as people can be surprised by their own complexity, so can machines, in that they can't predict their own behavior. People attribute this feature of themselves to "free will", and speak of "making choices". Turing's observation that machines will go into endless loops when trying to predict their own behavior suggests that a sufficiently complex machine might also come to suffer from that seemingly inevitable human delusion: believing that one has free will and is able to make choices that transcend physical law.

Thus Turing's seemingly negative result about machines can be seen as a positive result, in that it sheds new light on how physical objects might reflect on themselves and even consider themselves to be conscious, deliberating beings. A mechanical approach to the mysteries of conscious-ness was Alan Turing's dream, and probably by the late 1930's he was a believer in the possibility that a properly organized machine could be intelligent, conscious, and have free will—at least to the extent that we or any physical object can do so.

*   *   *

The war came as an interruption to young Turing's budding career as a mathematical logician—he had by this time held fellowships at both Cambridge and Princeton (at the Institute for Advanced Study, where he enjoyed such august company as that of Einstein, Gödel, and John von Neumann)—and he was pressed into service as a code breaker. It actually turned out not so badly for Turing, on a personal level. At Bletchley Park, midway between Cambridge and Oxford, he and a small cohort of powerful mathematical minds turned their powers of analysis to furthering the already impressive work done by Polish codebreakers. It was known that the German high command was sending orders in a code to its forces, including its vast submarine network, by means of a machine called the "Enigma". The exact construction of the Enigma was known, but this was not enough, as Alan Turing so well knew: the code breakers also needed to know the machine's internal state, which could be any one of an astronomical number.

Any configuration of several independently turning rings in the enciphering machine constituted a state; only when they knew that configuration could the code breakers quickly decipher a message.

Turing, Gordon Welchman, and a few others worked in close collaboration. Together, they analyzed strategies for using intercepted coded messages and high-speed searching machines to pinpoint the Enigma's state. Feverishly they worked as British ships were sunk one after another by the German U-boats. It was clear that the Nazis would bring Britain's war effort to an end unless the Enigma could be outwitted.

At first, they were able to decipher messages only a couple of weeks after receiving them—obviously far too late. As they began to succeed, they reduced the gap to a few days, then one day, and finally they reached the break-even point of decoding messages in minutes. However, it then turned out that the Germans were referring to places by special code names and unusual coordinates, so a second layer of decoding was needed. Fortunately, this could be done by watching where ships actually were sunk and correlating that information with the mysterious coordinates in the decoded messages. Once the second layer of code had been peeled off, it was as if, all of a sudden, the German fleet in the Atlantic were simply displayed on a screen in front of them.

There was an immediate dramatic increase in the number of British ships getting through the U-boats' offensive network. To the Germans, this ought to have been a dead giveaway that their code had been broken, but ironically, they were so certain of the undecipherability of the Enigma machine that their own logic forced them to conclude instead that the British must have very good spies, and so they looked for the spies instead of inventing a new coding machine. There was nonetheless a fragile, touch-and-go quality to the decipherment operation, because the Germans would occasionally alter the Enigma machine in various ways, precipitating desperate scrambles for new theories in Bletchley Park. But Turing and his associates always came up with the theories, and the British government knew regularly and with certainty what the Nazi command was up to.

Meanwhile, the figure at the center of this activity, Alan Turing, was running in long races and riding his dumpy bicycle to and from work, seemingly oblivious to rain. People noticed that every so often he would stop to adjust his bicycle chain so that it wouldn't fall off. Characteristically, he knew just when such a stop was needed, for he had observed that his bicycle had an internal state, just like an Enigma machine, determined by the relative positions of several independently turning gear wheels in the mechanism. As long as he monitored the state of this "Turing machine", he was able to forestall disaster. And meanwhile, on a more global scale, he was forestalling disaster no less.

*　　*　　*

Review of Alan Turing: The Enigma

When the war came to an end, Turing's ideas about how machines could imitate the mind had matured considerably. He had been in contact for several years now with electronic machinery, and many of the tasks he had been involved in had fertilized his brain with new ideas. The problem now was that there was no longer any war to make his abilities and his ideas seem crucial to anyone with money. He tried to find funding to build his universal Turing machine, but his awkward ways with people and his tendency to advocate long-term philosophical goals along with nearer-term practical ones seemed to put people off. Rather than gaining respect, he became known as something of an oddball. His powerful vision of the best way to go about creating a universal machine, based on his deep preference that all flexibility come from software (internal programs) rather than hardware, was gradually circumnavigated, and he found himself left out in the cold. Eventually, a British computer was built at Manchester University in the late 1940's, but not along the lines Turing had advocated.

Fortunately, while Turing was out of favor in the "proper" intellectual circles, he was able to concentrate on philosophical issues connected with mechanical thought, and in 1950, at age 38, he put his reflections into one of the classic articles on that subject, entitled "Computing Machinery and Intelligence". In it he proposed what has come to be known as the "Turing Test". The idea is to get around emotionally charged questions like "Can a machine think?". Taking his cue from operationalism, he replied in effect, "You want to know if that machine can think? Put it behind a curtain and see if it can fool people into thinking it is human on the basis of what it types to them." This has its parallel, interestingly, in the way some orchestra conductors do auditions: they have each candidate stand behind a curtain, hidden, and play from there, so they will not be swayed by age, sex, dress, or other external aspects.

The Turing Test (or "Imitation Game", as Turing called it) involved communication between a human interrogator and an unknown language-using "being". Knowing that there was ferocious resistance to the image that computing machinery might soon, or indeed, *ever,* think, Turing took pains to point out the remarkable generality of the probing allowed by his test, by presenting a pair of short sample dialogues in which it was shown how a skillful human interrogator might try to elicit odd and recondite knowledge, subtle judgments, and even emotional responses from the unknown "being". But most people remain skeptical about the Turing Test even after reading these dialogues, probably because they fear that they might be easily taken in by the wiles of a superficial machine. They do not appreciate how deeply and broadly the Turing Test potentially would allow them to probe.

In his article, Turing raised nine plausible objections to his own Imitation Game approach to the question of mechanical thought, and answered them cogently one by one. The most serious one seems to be "Lady Lovelace's objection": that computers cannot *originate* anything, but can do only what we explicitly tell them to do. Turing's answer to this—that one does *not* know what one has programmed a machine to do, except in the most superficial and general way—has a depth that eludes many good minds. I suspect that the Turing Test's profundity as an examination of an alleged "thinking machine" will only gradually seep into our culture as we collectively absorb the subtle and many-layered complexities of computers.

A sad footnote: In the early 1950's, the BBC recorded several radio interviews with Turing on the subject of minds and machines, but for some reason did not preserve any of them, and so we are left without a trace of a voice that, by all accounts, was quite peculiar and revealing. Even though Turing's own Imitation Game stressed the power of the printed word to convey all the nuances of personality, it seems poignant to think that the voice of such a recent figure is forever lost, and all we have to go on is the written word.

*   *   *

For his entire life, Alan Turing had been fascinated by the problem of morphogenesis: how whole organisms synchronize and coordinate their growth. An example is the fivefold symmetry of a starfish—how in the world does a cell know what part of the organism it is in, and how do various cells communicate with one another to plan the tricky overall pattern that they eventually wind up forming? It is as if the card stunt section in a football stadium had to coordinate complex patterns entirely by having nearest neighbors talk with each other. Turing's mathematically-based theories, developed in the early 1950's, were typically ahead of their time and even today they hold up well.

His long-time enjoyment of long-distance running remained with him, and he could look forward, so it might seem, to a happy life of pursuing his intellectual dreams and his romantic hopes in a more peaceful world. Unfortunately, the Britain of those days was as troubled politically as the United States, and homosexuality was seen as a "dangerous" disease, symptomatic of mental instability. And ironically, just as the anti-homosexual attack was getting more virulent, Alan Turing was becoming increasingly courageous and vocal about his own sexual nature, often ignoring the advice of friends to be more cautious.

Turing's house was burglarized in 1952, and it was quickly clear to him that one of his occasional lovers was somehow involved. In the course of making depositions to the police, Turing indirectly revealed his homosexuality. Instantly, the course of his life was irrevocably changed. No longer just a victim, he was now a criminal in his own right—and rather than protest his innocence of the "crime" of homosexuality, he talked freely about his "crime".

At that time in Britain, there was a movement to look upon homosexuality as a disease caused by hormone imbalances, and various physicians had proposed various "cures". In America, castration had been a very popular

Review of Alan Turing: The Enigma

"cure" for males (Hodges cites figures to the effect that by 1950, at least 50,000 castrations had been performed), but in Britain this was eschewed for less violent but no less barbaric treatment. Turing was found guilty and sentenced to "treatment": regular injections of female sex hormone, supposedly to quell his sex drive. This was the way that British society thanked the person most responsible for the safety of its ships during the World War. Of course, they had no way of knowing what role Turing had taken during the war, since it was top secret and would remain so for many years more. And in any case, Turing's wartime role should not have been seen as a mitigating factor in his "crime", since that would have meant that the millions of other more ordinary British homosexuals would have still been guilty of the same "crime". Turing saw this, and did not want to try to use any of his connections in government or the academic world to mitigate his sentence, and he simply endured it, growing breasts and being rendered impotent.

After one year, the sentence was over and he was free to return to a more normal "state of affairs". But torment like that leaves permanent scars, and deep inside Alan Turing something had changed. For the next couple of years, he appeared for the most part quite happy to his friends, and he joked and chatted about his future. But one day in 1954, he prepared a cyanide-coated apple, just as he had once seen the Wicked Witch do in the Walt Disney movie of *Snow White and the Seven Dwarfs*. Unlike her, he bit into his own apple. "Dip the apple in the brew, Let the sleeping death seep through." And he was found dead the next day. He had planned it in such a way that his mother would interpret it as an "accident with chemicals", but others knew better. Although today all evidence strongly suggests that the machine known as Alan Mathison Turing halted itself of its own free will, the ultimate reason remains an enigma to us, an undecidable question.

* * *

Andrew Hodges has painted, in his book, a beautiful portrait of a multifaceted man whose honesty and decency were too much for his society and his times, and who brought about his own downfall. Beyond the evident empathy that Hodges feels for Turing, there is another level of depth and understanding, one that makes all the difference in a biography of a scientific figure: scientific accuracy. Hodges has done an admirable job of presenting each idea in detail to the lay reader, but moreover, it is obvious that he is passionately intrigued by all the ideas that fascinated Turing. This book is a first-rate presentation of the life of a first-rate scientific mind, and because this particular mind was attached to a body that had a mind of its own, the book is a very important document for social reasons as well. Alan Turing would have shuddered if he had ever known that his life story would be made public, but he is in good hands: it is hard to imagine a more thoughtful and warm portrait of a life than this one.

22

A Coffeehouse Conversation
on the Turing Test

May, 1981

Participants in the dialogue:
Chris, a physics student; Pat, a biology student; Sandy, a philosophy student.

CHRIS: Sandy, I want to thank you for suggesting that I read Alan Turing's article "Computing Machinery and Intelligence". It's a wonderful piece and certainly made me think—and think about my thinking.

SANDY: Glad to hear it. Are you still as much of a skeptic about artificial intelligence as you used to be?

CHRIS: You've got me wrong. I'm not against artificial intelligence; I think it's wonderful stuff—perhaps a little crazy, but why not? I simply am convinced that you AI advocates have far underestimated the human mind, and that there are things a computer will never, ever be able to do. For instance, can you imagine a computer writing a Proust novel? The richness of imagination, the complexity of the characters—

SANDY: Rome wasn't built in a day!

CHRIS: In the article, Turing comes through as an interesting person. Is he still alive?

SANDY: No, he died back in 1954, at just 41. He'd be only 70 or so now, although he is such a legendary figure it seems strange to think that he could still be living today.

CHRIS: How did he die?

SANDY: Almost certainly suicide. He was homosexual, and had to deal with some pretty barbaric treatment and stupidity from the outside world. In the end, it got to be too much, and he killed himself.

CHRIS: That's horrendous, especially in this day and age.

SANDY: I know. What really saddens me is that he never got to see the amazing progress in computing machinery and theory that has taken place since 1954. Can you imagine how he'd have been wowed?

A Coffeehouse Conversation on the Turing Test

CHRIS: Yeah . . .

PAT: Hey, are you two going to clue me in as to what this Turing article is about?

SANDY: It is really about two things. One is the question "Can a machine think?"—or rather, "Will a machine ever think?" The way Turing answers the question—he thinks the answer is *yes*, by the way—is by batting down a series of objections to the idea, one after another. The other point he tries to make is that, as it stands, the question is not meaningful. It's too full of emotional connotations. Many people are upset by the suggestion that people are machines, or that machines might think. Turing tries to defuse the question by casting it in less emotional terms. For instance, what do you think, Pat, of the idea of thinking machines?

PAT: Frankly, I find the term confusing. You know what confuses me? It's those ads in the newspapers and on TV that talk about "products that think" or "intelligent ovens" or whatever. I just don't know how seriously to take them.

SANDY: I know the kind of ads you mean, and they probably confuse a lot of people. On the one hand, we're always hearing the refrain "Computers are really dumb; you have to spell everything out for them in words of one syllable"—yet on the other hand, we're constantly bombarded with advertising hype about "smart products".

CHRIS: That's certainly true. Do you know that one company has even taken to calling its products "dumb terminals" in order to stand out from the crowd?

SANDY: That's a pretty clever gimmick, but even so it just contributes to the trend toward obfuscation. The term "electronic brain" always comes to my mind when I'm thinking about this. Many people swallow it completely, and others reject it out of hand. It takes patience to sort out the issues and decide how much of it makes sense.

PAT: Does Turing suggest some way of resolving it, some kind of IQ test for machines?

SANDY: That would be very interesting, but no machine could yet come close to taking an IQ test. Instead, Turing proposes a test that theoretically could be applied to any machine to determine whether or not it can think.

PAT: Does the test give a clear-cut yes-or-no answer? I'd be skeptical if it claimed to.

SANDY: No, it doesn't claim to. In a way that's one of its advantages. It shows how the borderline is quite fuzzy and how subtle the whole question is.

PAT: And so, as usual in philosophy, it's all just a question of words!

SANDY: Maybe, but they're emotionally charged words, and so it's important, it seems to me, to explore the issues and try to map out the meanings of the crucial words. The issues are fundamental to our concept of ourselves, so we shouldn't just sweep them under the rug.

PAT: Okay, so tell me how Turing's test works.

SANDY: The idea is based on what he calls the *Imitation Game.* Imagine that a man and a woman go into separate rooms, and from there they can be interrogated by a third party via some sort of teletype set-up. The third party can address questions to either room, but has no idea which person is in which room. For the interrogator, the idea is to determine which room the woman is in. The woman, by her answers, tries to help the interrogator as much as she can. The man, though, is doing his best to bamboozle the interrogator, by responding as he thinks a woman might. And if he succeeds in fooling the interrogator . . .

PAT: The interrogator only gets to see written words, eh? And the sex of the author is supposed to shine through? That game sounds like a good challenge. I'd certainly like to take part in it someday. Would the interrogator have met either the man or the woman before the test began? Would any of them know any of the others?

SANDY: That would probably be a bad idea. All kinds of subliminal cueing might occur if the interrogator knew one or both of them. It would certainly be best if all three people were totally unknown to one another.

PAT: Could you ask any questions at all, with no holds barred?

SANDY: Absolutely. That's the whole idea!

PAT: Don't you think, then, that pretty quickly it would degenerate into sex-oriented questions? I mean, I can imagine the man, overeager to act convincing, giving away the game by answering some very blunt questions that most women would find too personal to answer, even through an anonymous computer connection.

SANDY: That's a nice observation. I wonder if it's true . . .

CHRIS: Another possibility would be to probe for knowledge of minute aspects of traditional sex-role differences, by asking about such things as dress sizes and so on. The psychology of the Imitation Game could get pretty subtle. I suppose whether the interrogator was a woman or a man would make a difference. Don't you think that a woman could spot some telltale differences more quickly than a man could?

PAT: If so, maybe the best way to tell a man from a woman is to let each of them play interrogator in an Imitation Game, and see which of the two is better at telling a man from a woman!

SANDY: Hmm . . . that's a droll twist. Oh, well. I don't know if this original version of the Imitation Game has ever been seriously tried out, despite the fact that it would be relatively easy to do with modern computer terminals. I have to admit, though, that I'm not at all sure what it would prove, whichever way it turned out.

PAT: I was wondering about that. What would it prove if the interrogator— say a woman—couldn't tell correctly which person was the woman? It certainly wouldn't prove that the man *was* a woman!

SANDY: Exactly! What I find funny is that although I strongly believe in the idea of the Turing Test, I'm not so sure I understand the point of its basis, the Imitation Game.

A Coffeehouse Conversation on the Turing Test

CHRIS: As for me, I'm not any happier with the Turing Test as a test for thinking machines than I am with the Imitation Game as a test for femininity.

PAT: From what you two are saying, I gather the Turing Test is some kind of extension of the Imitation Game, only involving a machine and a person instead of a man and a woman.

SANDY: That's the idea. The machine tries its hardest to convince the interrogator that it is the human being, and the human tries to make it clear that he or she is not the computer.

PAT: The machine *tries*? Isn't that a loaded way of putting it?

SANDY: Sorry, but that seemed the most natural way to say it.

PAT: Anyway, this test sounds pretty interesting. But how do you know that it will get at the essence of thinking? Maybe it's testing for the wrong things. Maybe, just to take a random illustration, someone would feel that a machine was able to think only if it could dance so well that you couldn't tell it was a machine. Or someone else could suggest some other characteristic. What's so sacred about being able to fool people by typing at them?

SANDY: I don't see how you can say such a thing. I've heard that objection before, but frankly, it baffles me. So what if the machine can't tap-dance or drop a rock on your toe? If it can discourse intelligently on any subject you want, then it has shown that it can think—to me, at least! As I see it, Turing has drawn, in one clean stroke, a clear division between thinking and other aspects of being human.

PAT: Now *you're* the baffling one. If you couldn't conclude anything from a *man's* ability to win at the Imitation Game, how could you conclude anything from a *machine's* ability to win at the Turing Game?

CHRIS: Good question.

SANDY: It seems to me that you could conclude *something* from a man's win in the Imitation Game. You wouldn't conclude he was a woman, but you could certainly say he had good insights into the feminine mentality (if there is such a thing). Now, if a computer could fool someone into thinking it was a person, I guess you'd have to say something similar about it—that it had good insights into what it's like to be human, into "the human condition" (whatever that is).

PAT: Maybe, but that isn't necessarily equivalent to *thinking,* is it? It seems to me that passing the Turing Test would merely prove that some machine or other could do a very good job of *simulating* thought.

CHRIS: I couldn't agree more with Pat. We all know that fancy computer programs exist today for simulating all sorts of complex phenomena. In theoretical physics, for instance, we simulate the behavior of particles, atoms, solids, liquids, gases, galaxies, and so on. But no one confuses any of those simulations with the real thing!

SANDY: In his book *Brainstorms,* the philosopher Daniel Dennett makes a similar point about simulated hurricanes.

CHRIS: That's a nice example, too. Obviously, what goes on inside a computer when it's simulating a hurricane is not a hurricane, for the machine's memory doesn't get torn to bits by 200-mile-an-hour winds, the floor of the machine room doesn't get flooded with rainwater, and so on.

SANDY: Oh, come on—that's not a fair argument! In the first place, the programmers don't claim the simulation really *is* a hurricane. It's merely a simulation of certain aspects of a hurricane. But in the second place, you're pulling a fast one when you imply that there are no downpours or 200-mile-an-hour winds in a simulated hurricane. To *us* there aren't any, but if the program were incredibly detailed, it could include simulated people on the ground who would experience the wind and the rain just as we do when a hurricane hits. In their minds—or, if you'd rather, in their *simulated* minds—the hurricane would be not a simulation, but a genuine phenomenon complete with drenching and devastation.

CHRIS: Oh, my—what a science-fiction scenario! Now we're talking about simulating whole populations, not just a single mind!

SANDY: Well, look—I'm simply trying to show you why your argument that a simulated McCoy isn't the real McCoy is fallacious. It depends on the tacit assumption that any old observer of the simulated phenomenon is equally able to assess what's going on. But in fact, it may take an observer with a special vantage point to recognize what is going on. In the hurricane case, it takes special "computational glasses" to see the rain and the winds.

PAT: "Computational glasses"? I don't know what you're talking about.

SANDY: I mean that to see the winds and the wetness of the hurricane, you have to be able to look at it in the proper way. You—

CHRIS: No, no, no! A simulated hurricane isn't wet! No matter how much it might seem wet to simulated people, it won't ever be *genuinely* wet! And no computer will ever get torn apart in the process of simulating winds.

SANDY: Certainly not, but that's irrelevant. You're just confusing levels. The laws of physics don't get torn apart by real hurricanes, either. In the case of the simulated hurricane, if you go peering at the computer's memory, expecting to find broken wires and so forth, you'll be disappointed. But look at the proper level. Look into the *structures* that are coded for in memory. You'll see that many abstract links have been broken, many values of variables radically changed, and so on. *There's* your flood, your devastation—real, only a little concealed, a little hard to detect.

CHRIS: I'm sorry, I just can't buy that. You're insisting that I look for a new kind of devastation, one never before associated with hurricanes. That way you could call *anything* a hurricane as long as its effects, seen through your special "glasses", could be called "floods and devastation".

SANDY: Right—you've got it exactly! You recognize a hurricane by its *effects*. You have no way of going in and finding some ethereal "essence of hurricane", some "hurricane soul" right in the middle of the storm's eye. Nor is there any ID card to be found that certifies "hurricanehood". It's just the existence of a certain kind of *pattern*—a spiral storm with an eye and so forth—that makes you say it's a hurricane. Of course, there are a lot of things you'll insist on before you call something a hurricane.

PAT: Well, wouldn't you say that being an *atmospheric* phenomenon is one prerequisite? How can anything inside a computer be a storm? To me, a simulation is a simulation is a simulation!

SANDY: Then I suppose you would say that even the *calculations* computers do are simulated—that they are fake calculations. Only *people* can do genuine calculations, right?

PAT: Well, computers get the right answers, so their calculations are not exactly fake—but they're still just patterns. There's no *understanding* going on in there. Take a cash register. Can you honestly say that you feel it is *calculating* something when its gears mesh together? And the step from cash register to computer is very short, as I understand things.

SANDY: If you mean that a cash register doesn't feel like a schoolkid doing arithmetic problems, I'll agree. But is that what "calculation" means? Is that an integral part of it? If so, then contrary to what everybody has thought up till now, we'll have to write a very complicated program indeed to perform *genuine* calculations. Of course, this program will sometimes get careless and make mistakes, and it will sometimes scrawl its answers illegibly, and it will occasionally doodle on its paper . . . It won't be any more reliable than the store clerk who adds up your total by hand. Now, I happen to believe that eventually such a program could be written. Then we'd know something about how clerks and schoolkids work.

PAT: I can't believe you'd ever be able to do that!

SANDY: Maybe, maybe not, but that's not my point. You say a cash register can't calculate. It reminds me of another favorite passage of mine from Dennett's *Brainstorms*. It goes something like this: "Cash registers can't really calculate; they can only spin their gears. But cash registers can't really spin their gears, either; they can only follow the laws of physics." Dennett said it originally about computers; I modified it to talk about cash registers. And you could use the same line of reasoning in talking about people: "People can't really calculate; all they can do is manipulate mental symbols. But they aren't really manipulating symbols; all they are doing is firing various neurons in various patterns. But they can't really make their neurons fire; they simply have to let the laws of physics make them fire for them." Et cetera. Don't you see how this *reductio ad absurdum* would lead you to conclude that calculation doesn't exist, that hurricanes don't exist—in fact, that nothing at a level higher than particles and the laws of physics exists? What do you gain by saying that a computer only pushes symbols around and doesn't truly calculate?

PAT: The example may be extreme, but it makes my point that there is a vast difference between a real phenomenon and any simulation of it. This is so for hurricanes, and even more so for human thought.

SANDY: Look, I don't want to get too tangled up in this line of argument, but let me try one more example. If you were a radio ham listening to another ham broadcasting in Morse code and you were responding in Morse code, would it sound funny to you to refer to "the person at the other end"?

PAT: No, that would sound okay, although the existence of a person at the other end would be an assumption.

SANDY: Yes, but you wouldn't be likely to go and check it out. You're prepared to recognize personhood through those rather unusual channels. You don't have to see a human body or hear a voice. All you need is a rather abstract manifestation—a code, as it were. What I'm getting at is this. To "see" the person behind the dits and dahs, you have to be willing to do some *decoding*, some interpretation. It's not direct perception; it's indirect. You have to peel off a layer or two to find the reality hidden in there. You put on your "radio-ham's glasses" to "see" the person behind the buzzes. Just the same with the simulated hurricane! You don't see it darkening the machine room; you have to decode the machine's memory. You have to put on special "memory-decoding" glasses. *Then* what you see is a hurricane.

PAT: Oh, ho ho! Talk about fast ones—wait a minute! In the case of the shortwave radio, there's a real person out there, somewhere in the Fiji Islands or wherever. My decoding act as I sit by my radio simply reveals that that person exists. It's like seeing a shadow and concluding there's an object out there, casting it. One doesn't confuse the shadow with the object, however! And with the hurricane there's no *real* storm behind the scenes, making the computer follow its patterns. No, what you have is just a shadow-hurricane without any genuine hurricane. I just refuse to confuse shadows with reality.

SANDY: All right. I don't want to drive this point into the ground. I even admit it is pretty silly to say that a simulated hurricane *is* a hurricane. But I wanted to point out that it's not as silly as you might think at first blush. And when you turn to simulated *thought,* then you've got a very different matter on your hands from simulated hurricanes.

PAT: I don't see why. You'll have to convince me.

SANDY: Well, to do so, I'll first have to make a couple of extra points about hurricanes.

PAT: Oh, no! Well, all right, all right.

SANDY: Nobody can say just exactly what a hurricane is—that is, in totally precise terms. There's an abstract pattern that many storms share, and it's for that reason we call those storms hurricanes. But it's not possible to make a sharp distinction between hurricanes and non-hurricanes. There are tornados, cyclones, typhoons, dust devils . . . Is the Great Red Spot on Jupiter a hurricane? Are sunspots hurricanes? Could there be a hurricane in a wind tunnel? In a test tube? In your imagination, you can even extend the concept of "hurricane" to include a microscopic storm on the surface of a neutron star.

CHRIS: That's not so far-fetched, you know. The concept of "earthquake" has actually been extended to neutron stars. The astrophysicists say that the tiny changes in rate that once in a while are observed in the pulsing of a pulsar are caused by "glitches"—starquakes—that have just occurred on the neutron star's surface.

SANDY: Oh, I remember that now. That "glitch" idea has always seemed eerie to me—a surrealistic kind of quivering on a surrealistic kind of surface.

CHRIS: Can you imagine—plate tectonics on a giant sphere of pure nuclear matter?

SANDY: That's a wild thought. So, starquakes and earthquakes can both be subsumed into a new, more abstract category. And that's how science constantly extends familiar concepts, taking them further and further from familiar experience and yet keeping some essence constant. The number system is the classic example—from positive numbers to negative numbers, then rationals, reals, complex numbers, and "on beyond zebra", as Dr. Seuss says.

PAT: I think I can see your point, Sandy. In biology, we have many examples of close relationships that are established in rather abstract ways. Often the decision about what family some species belongs to comes down to an abstract pattern shared at some level. Even the concepts of "male" and "female" turn out to be surprisingly abstract and elusive. When you base your system of classification on very abstract patterns, I suppose that a broad variety of phenomena can fall into "the same class", even if in many superficial ways the class members are utterly unlike one another. So perhaps I can glimpse, at least a little, how to you, a simulated hurricane could, in a funny sense, *be* a hurricane.

CHRIS: Perhaps the word that's being extended is not "hurricane", but "be".

PAT: How so?

CHRIS: If Turing can extend the verb "think", can't I extend the verb "be"? All I mean is that when simulated things are deliberately confused with genuine things, somebody's doing a lot of philosophical wool-pulling. It's a lot more serious than just extending a few *nouns,* such as "hurricane".

SANDY: I like your idea that "be" is being extended, but I sure don't agree with you about the wool-pulling. Anyway, if you don't object, let me just say one more thing about simulated hurricanes and then I'll get to simulated minds. Suppose you consider a really deep simulation of a hurricane—I mean a simulation of every atom, which I admit is sort of ridiculous, but still, just consider it for the sake of argument.

PAT: Okay.

SANDY: I hope you would agree that it would then share all the abstract structure that defines the "essence of hurricanehood". So what's to keep you from calling it a hurricane?

PAT: I thought you were backing off from that claim of equality.

SANDY: So did I, but then these examples came up, and I was forced back to my claim. But let me back off, as I said I would do, and get back to *thought,* which is the real issue here. Thought, even more than hurricanes, is an abstract structure, a way of describing some complex events that happen in a medium called a brain. But actually, thought can take place in any one of several billion brains. There are all these physically very different brains, and yet they all support "the same thing": thinking. What's important, then, is the abstract *pattern,* not the medium. The same kind of swirling can happen inside any of them, so no person can claim to think more "genuinely" than any other. Now, if we come up with some new kind of medium in which *the same style* of swirling takes place, could you deny that thinking is taking place in it?

PAT: Probably not, but you have just shifted the question. The question now is: How can you determine whether the "same style" of swirling is really happening?

SANDY: The beauty of the Turing Test is that it *tells* you when! Don't you see?

CHRIS: *I* don't see that at all. How would you know that the same style of activity was going on inside a computer as inside my mind, simply because it answered questions as I do? All you're looking at is its *outside.*

SANDY: I'm sorry, I disagree entirely! How do you know that when I speak to you, anything similar to what you call thinking is going on inside *me* ? The Turing Test is a fantastic probe, something like a particle accelerator in physics. Here, Chris—I think you'll like this analogy. Just as in physics, when you want to understand what is going on at an atomic or subatomic level, since you can't see it directly, you scatter accelerated particles off a target and observe their behavior. From this, you infer the internal nature of the target. The Turing Test extends this idea to the mind. It treats the mind as a "target" that is not directly visible but whose structure can be deduced more abstractly. By "scattering" questions off a target mind, you learn about its internal workings, just as in physics.

CHRIS: Well . . . to be more exact, you can *hypothesize* about what kinds of internal structures might account for the behavior observed—but please remember that they may or may not in fact exist.

SANDY: Hold on, now! Are you suggesting that atomic nuclei are merely *hypothetical* entities? After all, their existence (or should I say *hypothetical* existence?) was proved (or should I say *suggested* ?) by the behavior of particles scattered off atoms.

CHRIS: I would agree, but you know, physical systems seem to me to be much simpler than the mind, and the certainty of the inferences made is correspondingly greater. And the conclusions are confirmed over and over again by different types of experiments.

SANDY: Yes, but those experiments still are of the same sort—scattering, detecting things indirectly. You can never *handle* an electron or a quark. Physics experiments are also correspondingly harder to do and to interpret. Often they take years and years, and dozens of collaborators are involved. In the Turing Test, though, just one person could perform many highly delicate experiments in the course of no more than an hour. I maintain that people give other people credit for being conscious simply because of their continual external monitoring of other people—which is itself something like a Turing Test.

PAT: That may be roughly true, but it involves more than just conversing with people through a teletype. We see that other people have bodies, we watch their faces and expressions—we see they are human beings, and so we think they think.

SANDY: To me, that seems a narrow, anthropocentric view of what thought is. Does that mean you would sooner say a mannequin in a store thinks than a wonderfully programmed computer, simply because the mannequin looks more human?

PAT: Obviously I would need more than just vague physical resemblance to the human form to be willing to attribute the power of thought to an entity. But that organic quality, the sameness of origin, undeniably lends a degree of credibility that is very important.

SANDY: Here we disagree. I find this simply too chauvinistic. I feel that the key thing is a similarity of *internal* structure—not bodily, organic, chemical structure but *organizational* structure—software. Whether an entity can think seems to me a question of whether its organization can be described in a certain way, and I'm perfectly willing to believe that the Turing Test detects the presence or absence of that mode of organization. I would say that your depending on my physical body as evidence that I am a thinking being is rather shallow. The way I see it, the Turing Test looks far deeper than at mere external form.

PAT: Hey, now—you're not giving me much credit. It's not just the *shape* of a body that lends weight to the idea that there's real thinking going on inside. It's also, as I said, the idea of common origin. It's the idea that you and I both sprang from DNA molecules, an idea to which I attribute much depth. Put it this way: the external form of human bodies reveals that they share a deep biological history, and it's *that* depth that lends a lot of credibility to the notion that the owner of such a body can think.

SANDY: But that is all indirect evidence. Surely you want some *direct* evidence. That's what the Turing Test is for. And I think it's the *only* way to test for thinkinghood.

CHRIS: But you could be fooled by the Turing Test, just as an interrogator could mistake a man for a woman.

SANDY: I admit, I could be fooled if I carried out the test in too quick or too shallow a way. But I would go for the deepest things I could think of.

CHRIS: *I* would want to see if the program could understand jokes—or better yet, make them! *That* would be a real test of intelligence.

SANDY: I agree that humor probably is an acid test for a supposedly intelligent program, but equally important to me—perhaps more so—

would be to test its emotional responses. So I would ask it about its reactions to certain pieces of music or works of literature—especially my favorite ones.

CHRIS: What if it said, "I don't know that piece", or even, "I have no interest in music"? What if it tried its hardest (oops!—sorry, Pat!) . . . Let me try that again. What if it did everything it could, to steer clear of emotional topics and references?

SANDY: That would certainly make me suspicious. Any consistent pattern of avoiding certain issues would raise serious doubts in my mind as to whether I was dealing with a thinking being.

CHRIS: Why do you say that? Why not just conclude you're dealing with a thinking but unemotional being?

SANDY: You've hit upon a sensitive point. I've thought about this for quite a long time, and I've concluded that I simply can't believe emotions and thought can be divorced. To put it another way, I think emotions are an automatic by-product of the ability to think. They are entailed by the very nature of thought.

CHRIS: That's an interesting conclusion, but what if you're wrong? What if I produced a machine that could think but not emote? Then its intelligence might go unrecognized because it failed to pass *your* kind of test.

SANDY: I'd like you to point out to me where the boundary line between emotional questions and non-emotional ones lies. You might want to ask about the meaning of a great novel. This certainly requires an understanding of human emotions! Now is that thinking, or merely cool calculation? You might want to ask about a subtle choice of words. For that, you need an understanding of their connotations. Turing uses examples like this in his article. You might want to ask for advice about a complex romantic situation. The machine would need to know a lot about human motivations and their roots. If it failed at this kind of task, I would not be much inclined to say that it could think. As far as I'm concerned, *thinking, feeling,* and *consciousness* are just different facets of one phenomenon, and no one of them can be present without the others.

CHRIS: Why couldn't you build a machine that could feel nothing (we all know machines don't feel anything!), but that could think and make complex decisions anyway? I don't see any contradiction there.

SANDY: Well, I do. I think that when you say that, you are visualizing a metallic, rectangular machine, probably in an air-conditioned room—a hard, angular, cold object with a million colored wires inside it, a machine that sits stock still on a tiled floor, humming or buzzing or whatever, and spinning its tapes. Such a machine can play a good game of chess, which, I freely admit, involves a lot of decision-making. And yet I would never call it conscious.

CHRIS: How come? To mechanists, isn't a chess-playing machine rudimentarily conscious?

SANDY: Not to *this* mechanist! The way I see it, consciousness has got to come from a precise pattern of organization, one we haven't yet figured out how to describe in any detailed way. But I believe we will gradually come to understand it. In my view, consciousness requires a certain way of mirroring the external universe internally, and the ability to respond to that external reality on the basis of the internally represented model. And then in addition, what's really crucial for a conscious machine is that it should incorporate a well-developed and flexible self-model. And it's there that all existing programs, including the best chess-playing ones, fall down.

CHRIS: Don't chess programs look ahead and say to themselves as they're figuring out their next move, "If my opponent moves here, then I'll go there, and then if they go this way, I could go that way . . ."? Doesn't that usage of the concept "I" require a sort of self-model?

SANDY: Not really. Or, if you want, it's an extremely limited one. It's an understanding of self in only the narrowest sense. For instance, a chess-playing program has no concept of why it is playing chess, or of the fact that it is a program, or is in a computer, or has a human opponent. It has no idea about what winning and losing are, or—

PAT: How do *you* know it has no such sense? How can *you* presume to say what a chess program feels or knows?

SANDY: Oh, come on! We all know that certain things don't feel anything or know anything. A thrown stone doesn't know anything about parabolas, and a whirling fan doesn't know anything about air. It's true I can't *prove* those statements—but here, we are verging on questions of faith.

PAT: This reminds me of a Taoist story I read. It goes something like this. Two sages were standing on a bridge over a stream. One said to the other, "I wish I were a fish. They are so happy." The other replied, "How do *you* know whether fish are happy or not? *You're* not a fish!" The first said, "But you're not *me*, so how do you know whether I know how fish feel?"

SANDY: Beautiful! Talking about consciousness really does call for a certain amount of restraint. Otherwise, you might as well just jump on the solipsism bandwagon ("*I* am the only conscious being in the universe") or the panpsychism bandwagon ("*Everything* in the universe is conscious!").

PAT: Well, how do you know? Maybe everything *is* conscious.

SANDY: Oh, Pat, if you're going to join the club that maintains that stones and even particles like electrons have some sort of consciousness, then I guess we part company here. That's a kind of mysticism I just can't fathom. As for chess programs, I happen to know how they work, and I can tell you for sure that they aren't conscious. No way!

PAT: Why not?

SANDY: They incorporate only the barest knowledge about the goals of chess. The notion of "playing" is turned into the mechanical act of comparing a lot of numbers and choosing the biggest one over and over again. A chess program has no sense of disappointment about losing, or pride in winning. Its self-model is very crude. It gets away with doing the least it can, just enough to play a game of chess and nothing more. Yet interestingly enough, we still tend to talk about the "desires" of a chess-playing computer. We say, "It wants to keep its king behind a row of pawns" or "It likes to get its rooks out early" or "It thinks I don't see that hidden fork".

PAT: Yes, and we do the same thing with insects. We spot a lonely ant somewhere and say, "It's trying to get back home" or "It wants to drag that dead bee back to the colony". In fact, with any animal we use terms that indicate emotions, but we don't know for certain how much the animal feels. I have no trouble talking about dogs and cats being happy or sad, having desires and beliefs and so on, but of course I don't think their sadness is as deep or complex as human sadness is.

SANDY: But you wouldn't call it "simulated" sadness, would you?

PAT: No, of course not. I think it's real.

SANDY: It's hard to avoid use of such teleological or mentalistic terms. I believe they're quite justified, although they shouldn't be carried too far. They simply don't have the same richness of meaning when applied to present-day chess programs as when applied to people.

CHRIS: I still can't see that intelligence has to involve emotions. Why couldn't you imagine an intelligence that simply calculates and has no feelings?

SANDY: A couple of answers here. Number one, any intelligence has to have motivations. It's simply not the case, whatever many people may think, that machines could think any more "objectively" than people do. Machines, when they look at a scene, will have to focus and filter that scene down into some preconceived categories, just as a person does. And that means seeing some things and missing others. It means giving more weight to some things than to others. This happens on every level of processing.

PAT: I'm not sure I'm following you.

SANDY: Take me right now, for instance. You might think I'm just making some intellectual points, and I wouldn't need emotions to do that. But what makes me *care* about these points? Just now—why did I stress the work "care" so heavily? Because I'm emotionally involved in this conversation! People talk to each other out of conviction—not out of hollow, mechanical reflexes. Even the most intellectual conversation is driven by underlying passions. There's an emotional undercurrent to every conversation—it's the fact that the speakers want to be listened to, understood, and respected for what they are saying.

PAT: It sounds to me as if all you're saying is that people need to be interested in what they're saying, otherwise a conversation dies.

SANDY: Right! I wouldn't bother to talk to anyone if I weren't motivated by *interest*. And "interest" is just another name for a whole constellation of subconscious biases. When I talk, all my biases work together, and what you perceive on the surface level is my personality, my style. But that style arises from an immense number of tiny priorities, biases, leanings. When you add up a million of them interacting together, you get something that amounts to a lot of *desires.* It just all adds up! And that brings me to the other answer to Chris' question about feelingless calculation. Sure, that exists—in a cash register, a pocket calculator. I'd say it's even true of all today's computer programs. But eventually, when you put enough feelingless calculations together in a huge coordinated organization, you'll get something that has properties *on another level.* You can see it— in fact, you *have* to see it—not as a bunch of little calculations but as a system of tendencies and desires and beliefs and so on. When things get complicated enough, you're *forced* to change your level of description. To some extent that's already happening, which is why we use words such as "want", "think", "try", and "hope" to describe chess programs and other attempts at mechanical thought. Dennett calls that kind of level-switch by the observer "adopting the intentional stance". The really interesting things in AI will only begin to happen, I'd guess, when the program *itself* adopts the intentional stance toward itself!

CHRIS: That would be a very strange sort of level-crossing feedback loop.

SANDY: It certainly would. When a program looks at itself *from the outside,* as it were, and tries to figure out why it acted the way it did, then I'll start to think that there's *someone* in there, doing the looking.

PAT: You mean an "I"? A self?

SANDY: Yes, something like that. A soul, even—although not in any religious sense. Of course, it's highly premature for anyone to adopt the intentional stance (in the full force of the term) with respect to today's programs. At least that's my opinion.

CHRIS: For me an important related question is: To what extent is it valid to adopt the intentional stance toward beings other than humans?

PAT: I would certainly adopt the intentional stance toward mammals.

SANDY: I vote for that.

CHRIS: Now that's interesting. How can that be, Sandy? Surely you wouldn't claim that a dog or cat can pass the Turing Test? Yet don't you maintain the Turing Test is the *only* way to test for the presence of consciousness? How can you have these beliefs simultaneously?

SANDY: Hmm . . . All right. I guess that my argument is really just that the Turing Test works only above a certain level of consciousness. I'm perfectly willing to grant that there can be thinking beings that could *fail* at the Turing Test—but the main point that I've been arguing for is that anything that *passes* it would be a genuinely conscious, thinking being.

PAT: How can you think of a computer as a conscious being? I apologize if what I'm going to say sounds like a stereotype, but when I think of conscious beings, I just can't connect that thought with machines. To me, consciousness is connected with soft, warm bodies, silly though it may sound.

CHRIS: That does sound odd, coming from a biologist. Don't you deal with life so much in terms of chemistry and physics that all magic seems to vanish?

PAT: Not really. Sometimes the chemistry and physics simply increase the feeling that there's something magical going on down there! Anyway, I can't always integrate my scientific knowledge with my gut feelings.

CHRIS: I guess I share that trait.

PAT: So how do you deal with rigid preconceptions like mine?

SANDY: I'd try to dig down under the surface of your concept of "machine" and get at the intuitive connotations that lurk there, out of sight but deeply influencing your opinions. I think we all have a holdover image from the Industrial Revolution that sees machines as clunky iron contraptions gawkily moving under the power of some loudly chugging engine. Possibly that's even how the computer inventor Charles Babbage saw people! After all, he called his magnificent many-geared computer the "Analytical Engine".

PAT: Well, *I* certainly don't think people are just fancy steam shovels or electric can openers. There's something about people, something that—that—they've got a sort of *flame* inside them, something alive, something that flickers unpredictably, wavering, uncertain—but something *creative*!

SANDY: Great! That's just the sort of thing I wanted to hear. It's very human to think that way. Your flame image makes me think of candles, of fires, of vast thunderstorms with lightning dancing all over the sky in crazy, tumultuous patterns. But do you realize that just that kind of thing is visible on a computer's console? The flickering lights form amazing chaotic sparkling patterns. It's such a far cry from heaps of lifeless, clanking metal! It *is* flamelike, by God! Why don't you let the word "machine" conjure up images of dancing patterns of light rather than of giant steam shovels?

CHRIS: That's a beautiful image, Sandy. It does tend to change my sense of mechanism from being matter-oriented to being pattern-oriented. It makes me try to visualize the thoughts in my mind—these thoughts right now, even!—as a huge spray of tiny pulses flickering in my brain.

SANDY: That's quite a poetic self-portrait for a mere spray of flickers to have come up with!

CHRIS: Thank you. But still, I'm not totally convinced that a machine is all that I am. I admit, my concept of machines probably does suffer from anachronistic subconscious flavors, but I'm afraid I can't change such a deeply rooted sense in a flash.

SANDY: At least you sound open-minded. And to tell the truth, part of me sympathizes with the way you and Pat view machines. Part of me balks at calling myself a machine. It *is* a bizarre thought that a feeling being like you or me might emerge from mere circuitry. Do I surprise you?

CHRIS: You certainly surprise *me*. So, tell us—do you believe in the idea of an intelligent computer, or don't you?

SANDY: It all depends on what you mean. We've all heard the question "Can computers think?" There are several possible interpretations of this (aside from the many interpretations of the word "think"). They revolve around different meanings of the words "can" and "computer".

PAT: Back to word games again . . .

SANDY: I'm sorry, but that's unavoidable. First of all, the question might mean, "Does some present-day computer think, right now?" To this I would immediately answer with a loud *no*. Then it could be taken to mean, "Could some present-day computer, if suitably programmed, potentially think?" That would be more like it, but I would still answer, "Probably not". The real difficulty hinges on the word "computer". The way I see it, "computer" calls up an image of just what I described earlier: an air-conditioned room with cold rectangular metal boxes in it. But I suspect that with increasing public familiarity with computers and continued progress in computer architecture, that vision will eventually become outmoded.

PAT: Don't you think computers as we know them will be around for a while?

SANDY: Sure, there will have to be computers in today's image around for a long time, but advanced computers—maybe no longer called "computers"—will evolve and become quite different. Probably, as with living organisms, there will be many branchings in the evolutionary tree. There will be computers for business, computers for schoolkids, computers for scientific calculations, computers for systems research, computers for simulation, computers for rockets going into space, and so on. Finally, there will be computers for the study of intelligence. It's really only these last that I'm thinking of—the ones with the maximum flexibility, the ones that people are deliberately attempting to make smart. I see no reason that these will stay fixed in the traditional image. They probably will soon acquire as standard features some rudimentary sensory systems—mostly for vision and hearing, at first. They will need to be able to move around, to explore. They will have to be physically flexible. In short, they will have to become more animal-like, more self-reliant.

CHRIS: It makes me think of the robots R2D2 and C3PO in the movie *Star Wars*.

SANDY: Not me! In fact, I don't think of anything remotely like them when I visualize intelligent machines. They are too silly, too much the product of a film designer's imagination. Not that I have a clear vision of my own. But I think it's necessary, if people are realistically going to try to imagine an artificial intelligence, to go beyond the limited, hard-edged picture of computers that comes from exposure to what we have today. The only thing all machines will always have in common is their underlying mechanicalness. That may sound cold and inflexible, but then—just think —what could be more mechanical, in a wonderful way, than the workings of the DNA and proteins and organelles in our cells?

PAT: To me, what goes on inside cells has a "wet", "slippery" feel to it, and what goes on inside machines is dry and rigid. It's connected with the fact that computers don't make mistakes, that computers do only what you tell them to do. Or at least that's my image of computers.

SANDY: Funny—a minute ago, your image was of a flame, and now it's of something wet and slippery. Isn't it marvelous, how contradictory we can be?

PAT: I don't need your sarcasm.

SANDY: No, no, I'm not being sarcastic—I really *do* think it's marvelous.

PAT: It's just an example of the human mind's slippery nature—mine, in this case.

SANDY: True. But your image of computers is stuck in a rut. Computers certainly *can* make mistakes—and I don't mean on the hardware level. Think of any present-day computer predicting the weather. It can make wrong predictions, even though its program runs flawlessly.

PAT: But that's only because you've fed it the wrong data.

SANDY: Not so. It's because weather prediction is too complex. Any such program has to make do with a limited amount of data—entirely correct data—and extrapolate from there. Sometimes it will make wrong predictions. It's no different from a farmer gazing at the clouds and saying, "I reckon we'll get a little snow tonight." In our heads, we make models of things and use those models to guess how the world will behave. We have to make do with our models, however inaccurate they may be, or evolution will prune us out ruthlessly—we'll fall off a cliff or something. And for intelligent computers, it'll be the same. It's just that human designers will speed up the evolutionary process by aiming explicitly at the goal of creating intelligence, which is something nature just stumbled on.

PAT: So you think computers will be making fewer mistakes as they get smarter?

SANDY: Actually, just the other way around! The smarter they get, the more they'll be in a position to tackle messy real-life domains, so they'll be more and more likely to have inaccurate models. To me, mistake-making is a sign of high intelligence!

PAT: Wow—you throw me sometimes!

SANDY: I guess I'm a strange sort of advocate for machine intelligence. To some degree I straddle the fence. I think that machines won't really be intelligent in a humanlike way until they have something like your biological wetness or slipperiness to them. I don't mean *literally* wet—the slipperiness could be in the software. But biological-seeming or not, intelligent machines will in any case be machines. We will have designed them, built them—or grown them! We'll understand how they work—at least in some sense. Possibly no one person will really understand them, but collectively we will know how they work.

PAT: It sounds like you want to have your cake and eat it too. I mean, you want to have people able to build intelligent machines and yet at the same time have some of the mystery of mind remain.

SANDY: You're absolutely right—and I think that's what *will* happen. When *real* artificial intelligence comes—

PAT: Now there's a nice contradiction in terms!

SANDY: *Touché*! Well, anyway, when it comes, it will be mechanical and yet at the same time organic. It will have that same astonishing flexibility that we see in life's mechanisms. And when I say mechanisms, I *mean* mechanisms. DNA and enzymes and so on really *are* mechanical and rigid and reliable. Wouldn't you agree, Pat?

PAT: Sure! But when they work together, a lot of unexpected things happen. There are so many complexities and rich modes of behavior that all that mechanicalness adds up to something very fluid.

SANDY: For me, it's an almost unimaginable transition from the mechanical level of molecules to the living level of cells. But it's that exposure to biology that convinces me that people are machines. That thought makes me uncomfortable in some ways, but in other ways it is exhilarating.

CHRIS: I have one nagging question . . . If people are machines, how come it's so hard to convince them of the fact? Surely a machine ought to be able to recognize its own machinehood!

SANDY: It's an interesting question. You have to allow for emotional factors here. To be told you're a machine is, in a way, to be told that you're nothing more than your physical parts, and it brings you face to face with your own vulnerability, destructibility, and, ultimately, your mortality. That's something nobody finds easy to face. But beyond this emotional objection, to see yourself as a machine, you have to "unadopt" the intentional stance you've grown up taking toward yourself—you have to jump all the way from the level where the complex lifelike activities take place to the bottom-most mechanical level where ribosomes chug along RNA strands, for instance. But there are so many intermediate layers that they act as a shield, and the mechanical quality way down there becomes almost invisible. I think that when intelligent machines come around, that's how they will seem to us—and to themselves! Their mechanicalness will be buried so deep that they'll *seem* to be alive and conscious—just as *we* seem alive and conscious . . .

CHRIS: You're baiting me! But I'm not going to bite.

PAT: I once heard a funny idea about what will happen when we eventually have intelligent machines. When we try to implant that intelligence into devices we'd like to control, their behavior won't be so predictable.

SANDY: They'll have a quirky little "flame" inside, maybe?

PAT: Maybe.

CHRIS: And what's so funny about that?

PAT: Well, think of military missiles. The more sophisticated their target-tracking computers get, according to this idea, the less predictably they will function. Eventually, you'll have missiles that will decide they are pacifists and will turn around and go home and land quietly without blowing up. We could even have "smart bullets" that turn around in midflight because they don't want to commit suicide!

SANDY: What a nice vision!

CHRIS: I'm very skeptical about all this. Still, Sandy, I'd like to hear your predictions about when intelligent machines will come to be.

SANDY: It won't be for a long time, probably, that we'll see anything remotely resembling the level of human intelligence. It rests on too awesomely complicated a substrate—the brain—for us to be able to duplicate it in the foreseeable future. Anyhow, that's my opinion.

PAT: Do you think a program will ever pass the Turing Test?

SANDY: That's a pretty hard question. I guess there are various degrees of passing such a test, when you come down to it. It's not black and white. First of all, it depends on who the interrogator is. A simpleton might be totally taken in by some programs today. But secondly, it depends on how deeply you are allowed to probe.

PAT: You could have a range of Turing Tests—one-minute versions, five-minute versions, hour-long versions, and so forth. Wouldn't it be interesting if some official organization sponsored a periodic competition, like the annual computer-chess championships, for programs to try to pass the Turing Test?

CHRIS: The program that lasted the longest against some panel of distinguished judges would be the winner. Perhaps there could be a big prize for the first program that fools a famous judge for, say, ten minutes.

PAT: A prize for the *program,* or for its *author*?

CHRIS: For the program, of course!

PAT: That's ridiculous! What would a program do with a prize?

CHRIS: Come now, Pat. If a program's human enough to fool the judges, don't you think it's human enough to enjoy the prize? That's precisely the threshold where it, rather than its creators, deserves the credit, and the rewards. Wouldn't you agree?

PAT: Yeah, yeah—especially if the prize is an evening out on the town, dancing with the interrogators!

SANDY: I'd certainly like to see something like that established. I think it could be hilarious to watch the first programs flop pathetically!

PAT: You're pretty skeptical for an AI advocate, aren't you? Well, do you think any computer program today could pass a five-minute Turing Test, given a sophisticated interrogator?

SANDY: I seriously doubt it. It's partly because no one is really working at it explicitly. I should mention, though, that there is one program whose inventors claim it has *already* passed a rudimentary version of the Turing Test. It is called "Parry", and in a series of remotely conducted interviews, it fooled several psychiatrists who were told they were talking to either a computer or a paranoid patient. This was an improvement over an earlier version, in which psychiatrists were simply handed

(a)   (b)

FIGURE 22–1. *In (a), a program is enjoying the great reward for passing the Turing Test: an evening out on the town, dancing with the interrogator. Can the reader spot the program? In (b), an interrogator is enjoying the great reward for successfully unmasking an unthinking robot: an evening out on the town, dancing with the robot. Can the robot spot the reader?* NOTE: *One of these two photographs was not taken by David J. Moser. Can the interrogator tell which one?*

transcripts of short interviews and asked to determine which ones were with a genuine paranoid and which ones were with a computer simulation.

PAT: You mean they didn't have the chance to ask any questions? That's a severe handicap—and it doesn't seem in the spirit of the Turing Test. Imagine someone trying to tell which sex *I* belong to, just by reading a transcript of a few remarks by me. It might be very hard! I'm glad the procedure has been improved.

CHRIS: How do you get a computer to act like a paranoid?

SANDY: Now just a moment—I didn't say it *does* act like a paranoid, only that some psychiatrists, under unusual circumstances, thought so. One of the things that bothered me about this pseudo-Turing Test is the way Parry works. "He", as the people who designed it call it, acts like a paranoid in that "he" gets abruptly defensive and veers away from undesirable topics in the conversation. In effect, Parry maintains strict control so that no one can truly probe "him". For reasons like this, simulating a paranoid is a whole lot easier than simulating a normal person.

PAT: I wouldn't doubt that. It reminds me of the joke about the easiest kind of human being for a computer program to simulate.

CHRIS: What is that?

PAT: A catatonic patient—they just sit and do nothing at all for days on end. Even *I* could write a computer program to do that!

SANDY: An interesting thing about Parry is that it creates no sentences on its own—it merely selects from a huge repertoire of canned sentences the one that in some sense responds best to the input sentence.

PAT: Amazing. But that would probably be impossible on a larger scale, wouldn't it?

SANDY: You better believe it (to use a canned remark)! Actually, this is something that's really not appreciated enough. The number of sentences you'd need to store in order to be able to respond in a normal way to all possible turns that a conversation could take is more than astronomical—it's really unimaginable. And they would have to be so intricately indexed, for retrieval . . . Anybody who thinks that somehow, a program could be rigged up just to pull sentences out of storage like records in a jukebox, and that this program could pass the Turing Test, hasn't thought very hard about it. The funny part is that it is just this kind of unrealizable "parrot program" that most critics of artificial intelligence cite, when they argue against the concept of the Turing Test. Instead of imagining a truly intelligent machine, they want you to envision a gigantic, lumbering robot that intones canned sentences in a dull monotone. They set up the imagery in a contradictory way. They manage to convince you that you could see through to its mechanical level with ease, even as it is simultaneously performing tasks that we think of as fluid, intelligent processes. Then the critics say, "You see! A machine could pass the Turing Test and yet it would still be just a mechanical device, not intelligent at all." I see things almost the opposite way. If *I* were shown a machine that can do things that I can do—I mean pass the Turing Test—then, instead of feeling insulted or threatened, I'd chime in with philosopher Raymond Smullyan and say, "How wonderful machines are!"

CHRIS: If you could ask a computer just one question in the Turing Test, what would it be?

SANDY: Uhmm . . .

PAT: How about this: "If you could ask a computer just one question in the Turing Test, what would it be?"?

Post Scriptum.

In 1983, I had the most delightful experience of getting to know a small group of extremely enthusiastic and original students at the University of Kansas in Lawrence. These students, about thirty in number, had been drawn together by Zamir Bavel, a professor in the Computer Science Department, who had organized a seminar on my book *Gödel, Escher, Bach*. He contacted me and asked me if there was any chance I could come to Lawrence and get together with his students. Something about his way of describing what was going on convinced me that this was a very unusual group and that it would be worth my while to try it out. I therefore made a visit to Kansas and got to know both Zamir and his group. All my expectations were met and surpassed. The students were full of ideas and warmth and made me feel very much at home.

The first trip was so successful that I decided to do it again a couple of months later. This time they threw an informal party at an apartment a few of them shared. Zamir had forewarned me that they were hoping to give me a demonstration of something that had already been done in a recent class meeting. It seems that the question of whether computers could ever think had arisen, and most of the group members had taken a negative stand on the issue. Rod Ogborn, the student who had been leading the discussion, had asked the class if they would consider any of the following programs intelligent:

(1) a program that could pass a course in beginning programming (*i.e.*, that could take informal descriptions of tasks and turn them into good working programs);

(2) a program that could act like a psychotherapist (Rod gave sample dialogues with the famous "Doctor" program, also known as "Eliza", by Joseph Weizenbaum);

(3) a program called "Boris", written at Yale by Michael Dyer, that could read stories in a limited domain and answer questions about the situation which required filling in many unstated assumptions, and making inferences of many sorts based on them.

The class had come down on the "no" side of all three of these cases, although they got progressively harder. So Rod, to show the class how difficult this decision might be if they were really *faced* with a conversational program, managed to get a hookup over the phone lines with a natural-language program called "Nicolai" that had been developed over the last few years by the Army at nearby Fort Leavenworth. Thanks to some connections that Rod had, the class was able to gain access to an unclassified version of Nicolai and to interact with it for two or three hours. At the end of those hours, they then reconsidered the question of whether a computer might be able to think. Still, only one student was willing to consider Nicolai intelligent, and even that student reserved the right to switch sides if more information came in. About half the others were noncommittal, and the rest were unwilling, under any circumstances, to call Nicolai intelligent. There was no doubt that Rod's demonstration had been effective, though, and the class discussion had been one of the most lively.

Zamir told me all of this on our drive into Lawrence from the Kansas City airport, and he explained that the group had been so stimulated by this experience that they were hoping to get reconnected to Nicolai over the phone lines, and to let me try it out during the party. I thought it sounded quite amusing, and since I have tried out and watched a lot of natural-language programs in my time, I thought I would have an easy time coming up with good probes into the weaknesses of Nicolai. "Besides," I thought (somewhat chauvinistically), "how good can a program be that is being developed by the Army in an out-of-the-way place like Fort Leavenworth? To reveal its mechanicalness ought to be a piece of cake for me!" So I was looking forward to the challenge.

When we first arrived at the party, everyone was drinking beer and munching on various chips and dips and talking excitedly about what had happened in the previous class meeting with Nicolai. There was some consternation, though, because it seemed that the connection to Nicolai was hard to re-establish. It had to go through a computer at the University of Kansas, and thence to Fort Leavenworth, all of which meant that it was a slow and tenuous link. But after about 45 minutes of trying to set up the link, someone announced with pleasure that Nicolai was responding. One of the students was sitting at the terminal and typing in various simple questions and getting back simple answers. I watched over his shoulder, and felt the itch to sit down and try to unmask Nicolai for what "he" really was. After a couple of minutes, the student relinquished the terminal and I got my chance, with a small group of people looking over my shoulder as I typed. Nicolai's answers were quite slow in coming over the phone lines, and the wait was made all the more suspenseful by the fact that they were transmitted only in full lines of text, so there were some pretty long delays between successive lines of response. But that didn't seriously dampen my enthusiasm for trying to quickly poke holes in the program's façade of understanding.

Herewith follows a somewhat abridged but accurate transcript of the "conversation" I had with Nicolai over the next hour or so. For convenience's sake, what Nicolai typed is shown in capitals, and my inner thoughts at the time (as best I can reconstruct them now) are in italics.

A Coffeehouse Conversation on the Turing Test

What would you like to talk about?

(Let's start out easy.)

I CAN TALK ABOUT ANYTHING.

(Yeah, yeah—tell it to the Marines!)

Is there anything you specialize in?

WHAT DOES "SPECIALIZE" MEAN? PLEASE USE A COMPLETE SENTENCE.

(For a supposedly smart program, Nicolai doesn't seem to have a very big vocabulary.)

"Specialize" means "to be good at".

(I wonder if Nicolai likes those quotation marks, or if they will throw it for a loop.)

I UNDERSTAND.

(Izzatso?)

When were you written?

LAST COMPILATION: "4/27/83 19.247"

(Oh, look at that stupid, rigid answer, complete with pointless quotation marks and that extra, unexplained number at the end. I'm exposing Nicolai pretty well, so far . . .)

What language is Nicolai written in?

(I think I'll see if it knows its own name and recognizes third-person references to itself.)

NICOLAI IS WRITTEN MOSTLY IN ADA.

(Hmm . . . Not bad. It handled both "you" and "Nicolai" when I talked about it. Those Army programmers aren't so bad!)

What are feet?

(Choose an ambiguous word, see what it comes up with . . .)

FEET ARE 12.0 INCHES.

(Wow, weird! A person would never put it that way. They'd say, "A foot is a unit of length, twelve inches long", or something like that. And anyway, feet are primarily parts of the body, secondarily units of length. Why didn't it even seem to recognize that ambiguity? Oh, well. I guess that kind of dumbness is the kind of thing I wanted to expose. Score one for me.)

What are ears?

(Why not try another ambiguous word? After all, I did well on the last one.)

EARS ARE AUDITORY ORGANS FOUND ON ANIMALS.

(This is really funny! Why were feet not thought of primarily as body parts, then? This body-parts stuff seems to give me lots of points—I think I'll try it again.)

What are arms?

THAT INFORMATION IS CLASSIFIED.

(Oh, no! That is too stupid for words! Or is this somebody's idea of a joke?)

I like to understand up without seventeen clams.

(Let's toss some nonsense at Nicolai and see what it does.)

WHAT DO CLAMS HAVE TO DO WITH IT?

(Oh, come on, Nicolai! Do you mean you understood *all the rest, except for the clams bit? You are* really *dumb!)*

Why are you so dumb?

I AM A SMART PROGRAM.

(Well, that's kind of touching, actually. A little self-defense is appealing, in a way.)

You are pretty confident of yourself, aren't you?

WHAT DOES "PRETTY CONFIDENT" MEAN?

(Oh, yeah, I should have known—"pretty" as an adverb might be too colloquial.)

Did you understand up without my last sentence?

(I figure, since earlier it didn't complain about my phrase "understand up", let's see what it does with it this time.)

YOUR SENTENCE IS NONSENSE.

(Oh, really? Well, then, why wasn't the other one with that same phrase in it also considered nonsense?)

Do you enjoy nonsense?

IF YOU INSIST ON TALKING NONSENSE, I HAVE OTHER TASKS TO PERFORM.

(A strange response. Who ever insisted on talking nonsense? I was just asking a simple question. Oh, well, let me try out another angle.)

What is your opinion of M. C. Escher?

M. C. ESCHER, ARTIST, TWENTIETH CENTURY

(Oh, not again—another vacuous dictionary look-up. Can't it do better than that, especially when I've asked it a probing question about taste? Let's see how it handles a complaint . . .)

You did not answer my question.

WHICH QUESTION DID I NOT ANSWER?

(Hey, this Nicolai is a pretty slick character. Not so different from Parry. I must say, I am really pretty impressed by the robustness of this program. I mean, we have covered a lot of ground so far, and at least it makes "pseudo-sensible answers" most of the time. Not at all bad!)

Do you own any money?

I DO NOT OWN ANYTHING.

(Fair enough.)

Have you seen the latest issue of "Scientific American"?

(Uh-oh—once again, I don't know if I'm going to help it or confuse it by putting in those quotes. Let's see.)

I DO NOT HAVE ANY "SCIENTIFIC AMERICAN".

(That sounds awkward. Very computerish. I doubt if it knows what I meant. Oh, well.)

YOU'VE BEEN TALKING TO ME FOR QUITE A WHILE. WHAT IS YOUR NAME?

(Hmm . . . That was an unexpected remark by Nicolai. Should I answer it? Sure, why not?)

My name is "Doug".

DO YOU HAVE A LAST NAME?

(Inquisitive little devil, isn't it?)

Of course I do.

(Ask a stupid question, get a stupid answer. Let's see how it fields this one.)

WOULD YOU CARE TO TELL ME WHAT YOUR LAST NAME IS?

(Damn, it just totally ignored the tone of my remark. And why is it being so polite, suddenly?)

Only if you'll tell me yours.

(That ought to challenge it.)

I DO NOT HAVE A LAST NAME.

(I don't seem to be able to foil its defenses very easily. It always walks around me. Those programmers sure anticipated a lot of ideas. I guess they must have debugged Nicolai for years, watching typical dialogues with people who just come in off the street. For a dumb program, Nicolai is pretty smart!)

Well then, I won't tell you mine.

THAT'S NOT BEING VERY FRIENDLY.

(You know, sometimes, I hate to admit it, this stupid program really does *seem to understand something!)*

I'm sorry. How would you like me to give you a last name?

IF YOU WOULD LIKE TO DO THAT, THAT WOULD BE FINE WITH ME.

(Now that is obviously a canned sentence, or at least the latter half must certainly be. Score one more point for me. And yet, I'm not doing nearly as well as I'd hoped I would be, by now. What's wrong?)

Your last name is "Rimsky-Korsakov".

THANK YOU.

(Well, that's a strange kind of thing to thank somebody for. Just goes to show how the rigidity of programs makes them say the strangest things.)

Don't mention it.

(I'll play along with this game for a second.)

WHAT IS YOUR LAST NAME?

(My god, it remembered, after all this while, that as soon as it had told me its last name, I was going to tell it mine. *Well, I'd better make good on that promise.)*

My last name is "Hofstadter".

ARE YOU "(32) DOUGLAS R. HOFSTADTER"?

(Now why, in the midst of so much elegance, does it suddenly act so crude, putting my name in quotes and including that silly extraneous index number? I can't for the life of me figure out why this program is so mixed in quality.)

One and the same.

(Try out a strange idiomatic phrase.)

(83) "PARSING ERROR IN PREVIOUS INPUT"

(Aha—fooled it again. All of a sudden, a whole lot of mechanicalness is showing through.)

How many letters are there in your last name?

(I think I'll see whether it can figure out ambiguous words by context.)

14 PLUS A HYPHEN.

(Good. Now I'll let "letters" switch meaning. What will it do?)

How many letters are there in an ordinary envelope?

18.

(Oh, no! It didn't recognize the shift! And it merely counted the number of letters in "an ordinary envelope"! But if that's the case, then how come in the earlier case it didn't count the letters in the phrase "your last name", but instead retrieved the referent? *There's something so inconsistent and uneven about this program. It just goes to show how far we have to go before mechanical intelligence bears any resemblance to human fluidity.)*

ARE YOU WORKING ON A NEW BOOK?

(I wonder what will happen if I simply sidestep that question. Anyway, I'd like to probe its ability to deal with humor.)

Tell me a joke.

I WANT A COOOKIE.

(What kind of a joke is that? Hey, it spelled "cookie" wrong! Hmm . . . is that its joke? Or what is going on here?)

By the way, I am sort of working on a new book. Care to hear about it?

(Can it handle weird phrases like "sort of"? And questions with no subject?)

NO, BUT IF YOU WOULD GIVE ME A COOKIE, I WOULD GLADLY PAY YOU TUESDAY.

(That sounds like a familiar line. Have I heard it somewhere before? Is that supposed to be the punch line of its "joke"?)

Tell me another one.

HOW MANY COMPUTER PROGAMMERS DOES IT TAKE TO SCREW IN A LIGHT BULB?

(Okay, I see. It's got a library of jokes and riddles, and it just fishes them out at random. I'll play along.)

4.

INCORRECT. THAT IS A HARDWARE PROBLEM.

(Okay, so it told its joke. But I bet it has no idea what any of the terms in it mean, such as "programmer". Hey, it left an 'r' out of "programmer"! I guess whoever input the joke misspelled the word. So let's see what happens when I ask about it . . .

What is a computer progammer?

WOULD YOU LIKE ME TO EXPLAIN THE JOKE?

(Hey, damn it, it turned the tables on me! Does it think I don't know what a "progammer" (or a programmer) is? I don't like this slipperiness. Every time I try to get it to act really dumb, it somehow slips out of it. Strange . . . All right, let me probe some more about its understanding of the terms in that joke. How about a potentially off-color word?)

What does "to screw" mean?

IN WHAT CONTEXT?

(Aha, at least it knows it's an ambiguous term.)

In bed.

"TO SCREW IN BED" MEANS "TO HAVE SEX".

(Okay, fine, it has some definitions stored according to context. I guess this word must come up all the time in conversations. But can it go any further than that definition?)

What is having sex all about?

I CAN RECOMMEND AN EXCELLENT BOOK ON THAT SUBJECT.

(What a great stock answer! One point for the Army programmers!)

Tell me, tell me!

THE BOOK IS CALLED "GODEL, ESCHER, BACH".

(Huh? Say, wait a minute, this is too weird. I'm not so sure this is really a program. There's too damn much cleverness in these weird responses. I can't imagine bizarre mechanical devices coming up with this much coherence, even if it is completely off the wall, at times.)

I like to call it "GEB", myself.

THAT IS VERY NICE, DOUG. WHAT ELSE DO YOU CALL "GEB"?

(All right, the jig's up. What's going on here?)

At this point, everybody started laughing. They had all been trying hard to keep from doing so the whole time, but at this point, they couldn't contain themselves. It was time to reveal to me what had been going on. They took me downstairs and showed me that three of the students—Roy Leban, Howard Darsche, and Garrick Stangle—had been collectively playing Nicolai. There was in reality no such program, and there hadn't ever been one. (In retrospect, I am reminded of the famous French mathematician Nicolas Bourbaki—a hypothetical person, actually an amalgam of over a dozen eminent mathematicians writing under that collective pen name.) There had indeed been a similar demonstration for the class a few days earlier, and the class, like me, had been taken in for a long time. In my case, Roy, Howard, and Garrick had worked very hard to give the impression of mechanicalness by spewing back "parsing error" and other indications of rigidity, and also by sending what looked very much like canned phrases from time to time. That way, they could keep sophisticates like me believing that there was a program behind it all. Only by that point I was beginning to wonder just how sophisticated I really was.

The marvelous thing about this game is that it was, in many ways, a Turing Test in reverse: a group of human beings masquerading as a program, trying to act mechanical enough that I would believe it really was one. Hugh Kenner has written a book called *The Counterfeiters* about the perennial human fascination with such compounded role flips. A typical example is Delibes' ballet *Coppélia,* in which human dancers imitate life-sized dolls stiffly imitating people. What is amusing is how Nicolai's occasional crudeness was just enough to keep me convinced it was mechanical. Its "willingness" to talk about itself, combined with its obvious limitations along those lines (its clumsy revelation of when it was last compiled, for instance), helped establish the illusion very strongly.

*   *   *

In retrospect, I am quite amazed at how much genuine intelligence I was willing to accept as somehow having been implanted in the program. I had been sucked into the notion that there really must be a serious natural-language effort going on at Fort Leavenworth, and that there had been a very large data base developed, including all sorts of random information: a dictionary, a catalogue containing names of miscellaneous people, some jokes, lots of canned phrases to use in difficult situations, some self-knowledge, a crude ability to use key words in a phrase when it can't parse it exactly, some heuristics for deciding when nonsense is being foisted on it, some deductive capabilities, and on and on. In hindsight, it is clear that I was willing to accept a huge amount of fluidity as achievable in this day and age simply by putting together a large bag of isolated tricks— kludges and hacks, as they say.

Roy Leban, one of the three inside Nicolai's mind, wrote the following about the experience of being at the other end of the exchange:

> Nicolai was a split personality. The three of us (as well as many kibitzers) argued about practically every response. Each of us had a strong preconceived notion about what (or who) Nicolai should be. For example, I felt that certain things (such as "Douglas R. Hofstadter") should be in quotation marks, and that feet should not be 12 inches, but 12.0. Howard had a tendency for rather flip answers. It was he who suggested the "classified" response to the "arms" question. And somehow, when he suggested it, we all *knew* it was right.

Several times during our conversation, I felt quite amazed at how fluently Nicolai was able to deal with things I was bringing up, but each time I could postulate some not *too* sophisticated mechanical underpinning that would allow that particular thing to happen. As a strong skeptic of true fluidity in machines at this time, I kept on trying to come up with rationalizations for the fact that this program was doing so well. My conclusion was that it was a very vast and quite sophisticated bag of tricks, no one of which was terribly complex. But after a while, it just became too much to believe. Furthermore, the mixture of crudity and subtlety became harder and harder to swallow, as well.

My strategy had been, in essence, to use spot checks all over the map: to try to probe it in all sorts of ways rather than to get sucked into some topic of its own choice, where it could steer the conversation. Daniel Dennett, in a paper on the depth of the Turing Test called "Can Machines Think?", likens this technique to a strategy taught to American soldiers in World War II for telling German spies from genuine Yankees. The idea was that even if a young man spoke absolutely fluent American-sounding English, you could trip him up by asking him things that any boy growing up in those days would be expected to know, such as "What is the name of Mickey Mouse's girlfriend?" or "Who won the World Series in 1937?" This expands the domain of knowledge necessary from just the language itself to the entire culture—and the amazing thing is that just a few well-placed questions can unmask a fraud in a very brief time—or so it would seem.

The problem is, what do you do if the person is extremely sharp, and when asked about Minnie Mouse, responds in some creative way, such as, "Hah! She ain't no *girl*friend—she's a *mouse*!"? The point is that even with these trick probes that *should* ferret out frauds very swiftly, there can be clever defensive countermaneuvers, and you can't be sure of getting to the bottom of things in a very brief time.

It seems that a few days earlier, the class had collectively gone through something similar to what I had just gone through, with one major difference. Howard Darsche, who had impersonated (if I may use that peculiar choice of words!) Nicolai in the first run-through, simply had acted himself, without trying to feign mechanicalness in any way. When asked what color the sky was, he replied, "In daylight or at night?" and when told "At night", he replied, "Dark purple with stars." He got increasingly poetic and creative in his responses to the class, but no one grew suspicious that this Nicolai was a fraud. At some point, Rod Ogborn simply had to stop the demonstration and type on the screen, "Okay, Howard, you can come in now." Zamir (who was not in cahoots with Rod and his team) was the only one who had some reluctance in accepting this performance as that of a genuine program, and he had kept silent until the end, when he voiced a muted skepticism.

Zamir summarizes this dramatic demonstration by saying that his class was willing to view *anything on a video terminal* as mechanically produced, no matter how sophisticated, insightful, or poetic an utterance it might be. They might find it interesting and even surprising, but they would find some way to discount those qualities. Why was this the case? How could they do this for so long? And why did I fall for the same kind of thing?

In interacting with me, Nicolai had seemed to waver between crude mechanicalness and subtle flexibility, an oscillation I had found most puzzling and somewhat disturbing. But I was still taken in for a very long time. It seems that, even armed with spot checks and quite a bit of linguistic sophistication and skepticism, unsuspecting humans can have the wool pulled over their eyes for a good while. This was the humble pie I ate in this remarkable reverse Turing Test, and I will always savor its taste and remember Nicolai with great fondness.

*   *   *

Alan Turing, in his article, indicated that his "Imitation Game" test should take place through some sort of remote teletype linkup, but one thing he did not indicate explicitly was at what grain size the messages would be transmitted. By that, I mean that he did not say whether the messages should be transmitted as intact wholes, or line by line, word by word, or keystroke by keystroke. Although I don't think it matters for the Turing Test in any *fundamental* sense, I do think that which type of "window" you view another language-using being through has a definite bearing on how *quickly* you can make inferences about that being. Clearly, the most revealing of these possibilities is that of watching the other "person" operate at the keystroke level.

On most multi-user computer systems, there are various ways for different users to communicate with each other, and these ways reflect different levels of urgency. The slowest one is generally the "mail" facility, through which you can send another user an arbitrarily long piece of text, just like a letter in an envelope. When it arrives, it will be placed in the user's "mailbox", to be read at their leisure. A faster style of communicating is called, on Unix systems, "write". When this is invoked, a direct communications link is set up between you and the person you are trying to reach (provided they are logged on). If they accept your link, then any full line typed by either of you will be instantly transmitted and printed on the other party's screen—where a lineful is signaled by your hitting the carriage-return key. This is essentially what the Nicolai team used in communicating with me over the Kansas computer. Their irregular typing rhythm and any errors they might have made were completely concealed from me this way, since all I saw was a sequence of completely polished lines (with the two spelling errors "coookie" and "progammer", which I was willing to excuse because Nicolai generated them in a "joke" context).

The most revealing mode is what, on Unix, is called "talk". In this mode, every single keystroke is revealed. You make an error, you are exposed. For some people, this is too much like living in a glass house, and they prefer the shielding afforded by "write". For my part, I like living dangerously. Let the mistakes lfy! In computer-mediated conversations with my friends, I always opt for "talk". I have been amused to watch their "talk" styles and my own slowly evolve to relatively stable states.

When we in the Indiana University Computer Science Department first began using the "talk" facility, we were all somewhat paranoid about making errors, and we would compulsively fix any error that we made. By this, I mean that we would backspace and retype the character. The effect on the screen of hitting the backspace key repeatedly is that you see the most recently typed characters getting eaten up, one by one, right to left, and if necessary, the previous line and ones above it will get eaten backwards as well. Once you have erased the offending mistakes, you simply resume typing forwards. This is how errors are corrected. We all began in this finicky way, feeling ashamed to let anything flawed remain "in print", so to speak, visible to others' eyes. But gradually we overcame that sense of shame, realizing that a typo sitting on a screen is not quite so deathless as one sitting on a page in a book.

Still, I found that some people just let things go more easily than others. For instance, by the length of the delay after a typo is made, you can tell just how much its creator is hesitating in wondering whether to correct it. Hesitations of a fraction of a second are very noticeable, and are part of a person's style. Even if a typo is left uncorrected, you can easily spot someone's vacillations about whether or not to fix it.

The counterparts of these things exist on many levels of such exchanges. There are the levels of *word choice* (for instance, some people who don't mind having their typos on display will often backtrack and get rid of *words* they now repudiate), *sentence-structure* choice, *idea* choice, and higher. Hesitations and repairs or restarts are very common. I find nothing so annoying as someone who has gotten an idea expressed just fine in one way, and who then erases it all on the screen before your eyes and proceeds to compose it anew, as if one way of suggesting getting together for dinner at Pagliai's at 6 were markedly superior to another!

There are ways of exploiting erasure in "talk" mode for the purposes of humor. Don Byrd and I, when "talk"ing, would often make elaborate jokes exploiting the medium in various ways. One of his I recall vividly was when he hurled a nasty insult onto the screen and then swiftly erased it, replacing it by a sweetly worded compliment, which remained for posterity to see—at least for another minute or so. One of our great discoveries was that some "arrow" keys allowed us to move all over the screen, thus to go many lines up in the conversation and edit earlier remarks by either of us. This allowed some fine jokes to be made.

One hallmark of one's "talk" style is one's willingness to use abbreviations. This is correlated with one's willingness to abide typos, but is not by any means the same. I personally was the loosest of all the "talkers" I knew, both in terms of leaving typos on the screen and in terms of peppering my sentences with all sorts of silly abbreviations. For instance, I will now retype this very sentence as I would have in "talk mode", below.

F ins, I will now retype ts very sent as I wod hv in "talko mode", below.

Not bad! Only two typos. The point is, the communication rate is raised considerably—nearly to that of a telephone—if you type well and are willing to be informal in all these ways, but many people are surprisingly uptight about their unpolished written prose being on exhibit for others to see, even if it is going to vanish in mere seconds.

*   *   *

All of this I bring up not out of mere windbaggery, but because it bears strongly on the Turing Test. Imagine the microscopic insights into personality that are afforded by watching someone—human or otherwise—typing away in "talk" mode! You can watch them dynamically making and unmaking various word choices, you can see interferences between one word and another causing typos, you can watch hesitations about whether or not to correct a typo, you can see when they are pausing to work out a thought before typing it, and on and on. If you are just a people-watcher, you can merely observe informally. If you are a psychologist or fanatic, you can measure reaction times in thousandths of a second, and make large collections and catalogue them. Such collections have really been made, by the way, and make for some of the most fascinating reading on the human mind that I know of. See, for instance, Donald Norman's article "Categorization of Action Slips" or Victoria Fromkin's book *Errors of Linguistic Performance: Slips of the Tongue, Ear, Pen, and Hands.*

In any case, when you can watch someone's real-time behavior, a real live personality begins to appear on a screen very quickly. It is far different in feel from reading polished, post-edited linefuls such as I received from Nicolai. It seems to me that Alan Turing would have been most intrigued and pleased by this time-sensitive way of using his test, affording so many lovely windows onto the subconscious mind (or pseudo-mind) of the being (or pseudo-being) under examination.

As if it were not already clear enough, let me conclude by saying that I am an unabashed pusher of the validity of the Turing Test as a way of operationally defining what it would be for a machine to genuinely think. There are, of course, middle grounds between real thinking and being totally empty inside. Smaller mammals and in general, smaller animals, seem to have "less thought" going on inside their craniums than we have inside ours. Yet clearly animals have always done, and machines are now doing, things that seem to be best described using Dennett's "intentional stance". Donald Griffin, a conscious mammal, has written thoughtfully on these topics (see, for instance, his book *The Question of Animal Awareness*). John McCarthy has pointed out that even electric-blanket manufacturers use such phrases as "it thinks it is too hot" to explain how their products work. We live in an era when mental terms are being both validly extended and invalidly abused, and we are going to need to think hard about these matters, especially in face of the onslaught of advertising hype and journalese. Various modifications of the Turing Test idea will undoubtedly be suggested as computer mastery of human language increases, simply to serve as benchmarks for what programs can and cannot do. This is a fine idea, but it does not diminish the worth of the original Turing Test, whose primary purpose was to convert a philosophical question into an operational question, an aim that I believe it filled admirably.

23

On the Seeming Paradox of Mechanizing Creativity

September, 1982

I_T is a commonly heard statement that there is such a thing as the "creative spark", that an "unanalyzable leap of the imagination" takes place when a great mind comes up with a new idea or work of art. Great creators are sometimes said to be a "quantum leap" away from ordinary mortals. People like Mozart are held to be somehow divinely inspired, to have magical insights for which they could no more be expected to be able to account than spiders for the wondrous webs they weave. It is all felt to be somehow too deep down, too hidden, too occult a gift, to be mechanical in any sense. Creativity, in fact, is perhaps one of the last refuges of the soul. "You may mechanize your *logic*," says the English professor to the computer scientist, "but you'll never lay a finger on *poetry*." (You may substitute music or any other domain of artistic creation for poetry.)

Is this kind of statement irrational? Is it a reflection of a deep-seated fear that even this most sacred aspect of humanity is doomed to be taken over soon by metallic machines, or by silicon chips? Why make such a big deal out of an activity of the human mind which, like every other activity in life, has shades and degrees? After all, the creative blurs with the mundane so much that it would be hopeless, would it not, to try to cull what is truly creative from what is not? Or—is there some clean dividing line that distinguishes the run-of-the-mill workaday deviser of ditties from the Great Composer of Eternal Symphonic Masterpieces? And if so, is it possible that here lies the elusive difference between the living and the dead, the human and the machine, the mental and the mechanical?

With such a "magical" view of creativity, there is, of course, a problem. It would seem to imply that the poor composer of ditties is actually dead and mechanical inside; that only certified geniuses like Mozart are qualitatively different from machines—and that even old Mozart was nonmechanical only when he was composing (certainly not when he was merely sipping ale at a tavern!). Probably most people who believe in the magical view of creativity would dispute this way of portraying their position. They would maintain that Mozart was nonmechanical *all* the time; moreover that you and I, no less than Mozart, are also nonmechanical all the time. No matter that some, even many, human abilities have already been mechanized or will be mechanized someday.

About the touchy question of the mechanization of the mental, many educated people feel that, although a machine may now or someday be able to do a creditable job of acting like a person, any machine's performance will always remain lackluster and dull, and that after a while, this dullness will always shine through. You'll simply be able to tell that it is unoriginal, that its ideas and thoughts are all being drawn from some storehouse of formulas and *clichés*, that ultimately there is nothing alive and dynamic—no *élan vital*—behind its *façade.* If it comes up with a *bon mot* now and then, well, *tant mieux*—but even the best will just be an automaton *par excellence.* There may be nothing specific to point to other than the "vibes" you pick up of its dullness and unoriginality, but after a while they will inevitably start to come in loud and clear. (Incidentally, I would be delighted if some of the more vocal antimechanists felt that way, instead of insisting, as they more often do, that operational tests are of no use in deciding who or what possesses "genuine mental states".)

This sense that you will eventually be able to "just tell", from its inevitable lack of sparkle, that you're dealing with a machine and not a person, seems to depend upon a tacit assumption about human thought, one with which I fully agree: namely, that "creative spark" is not the exclusive property of just a few rare individuals down the centuries, but quite to the contrary, it is an intrinsic ingredient of the everyday mental activity of everyone, even the most run-of-the-mill people. In short, it seems that people who feel that machines—even intelligent ones—will always remain duller than minds are tacitly relying on the following thesis: Creativity is part of the very fabric of all human thought, rather than some esoteric, rare, exceptional, and fluky by-product of the ability to think, which every so often surfaces in places spread far and wide.

With this thesis I agree. Where I differ with the antimechanists is over the matter of whether creativity lies *beyond* intelligence. I see creativity and insight, for machines no less than for people, as intimately bound up with intelligence, so that I cannot imagine a noncreative yet intelligent machine —something that, in order to make a point about what is essentially human, they seem to be willing and able to do. To me, "noncreative intelligence" is a flat-out contradiction in terms.

* * *

In this column, I would like to describe some ideas I have about how creativity is founded on mechanisms, mechanisms that, to be sure, lie deeply hidden in the depths of the structure of our brains, but mechanisms that nonetheless exist and can perhaps be approximated using the hardware and software of the machines we have today, crude though they are in certain ways. The gist of my notion is that having creativity is an automatic consequence of having the proper representation of *concepts* in a mind. It is not something you add on afterward. It is built into the way concepts are. To spell this out more concretely: If you have succeeded in making an accurate model of *concepts*, you have thereby also succeeded in making a model of the creative process, and even of consciousness.

Another way of talking about concepts is to talk about memory, which is the "place" where concepts are stored. It is the organization of memory that defines what concepts are. Incidentally, when I first wrote the preceding sentence, it ended differently. It said, "It is the organization of memory that defines what concepts will be accessible under what conditions." But on rereading it, I felt it was too weak that way. It took for granted the notion that all readers have a clear concept of what a concept is. But that is hardly takable-for-granted! Granted, we all have *some* concept of what a concept is, but a *clear* one?

So I dropped the phrase beginning with "will be accessible" and replaced it with a stark "are". This way, the sentence does more than simply state that memory is a storehouse of some things called concepts. It emphasizes that what establishes the "concepthood" of something is the way it is integrated into memory. Or to put it the other way 'round, nothing is a concept except by virtue of the way it is connected up with other things that are also concepts. In other words, the property of being a concept is a property of connectivity, a quality that comes from being embedded in a certain kind of complicated network, and from nowhere else. Put this way, concepts sound like structural or even topological properties of vast tangly networks of sticky mental spaghetti.

That's more or less the image I feel it is important to convey: namely, that concepts derive all their power from their connectivity to one another. And now, having expressed that idea, I can return to the sentence as it was originally put: It is the organization of memory that defines what concepts will be accessible under what conditions—and surely, the happy choice of the right concept at the right time is the essence of the creative. Therefore it is imperative to study deeply the nature of that network—to ask the question "What is a concept?".

Some questions that come to mind are: What is the relationship between a general, or Platonic, concept, such as that of "tree", and the concept you form of some specific tree? That is, what is the distinction between semantic or perceptual *categories* and the representations of individual *instances* of them? How is a given situation filed away in memory so that one has access to it under an enormous variety of future situations—access that is often via analogy or other abstract pathways, rather than by simplistic superficial traits? Or, to flip that coin, how does a given situation cause the highly selective retrieval from memory of a small number of previous situations that seem relevant? Only through a deep understanding of the organization of memory—which is to say, only by answering the question "What is a concept?"—will we be able to make models of the creative process. This will be a long and arduous process, not one that will yield answers overnight, or even in a few decades. Nonetheless, we have the right beginnings, in the sciences of cognitive psychology and artificial intelligence. Philosophers of mind and neuroscientists will undoubtedly contribute as well. The union of all these disciplines is called "cognitive science".

*   *   *

A question that arises at the outset is: "What kinds of objects have concepts stored inside them, and what kinds do not?" One of my favorite passages that opens this question wide is in Dean Wooldridge's book *Mechanical Man: The Physical Basis of Intelligent Life,* and it runs this way:

> When the time comes for egg laying, the wasp *Sphex* builds a burrow for the purpose and seeks out a cricket which she stings in such a way as to paralyze but not kill it. She drags the cricket into the burrow, lays her eggs alongside, closes the burrow, then flies away, never to return. In due course, the eggs hatch and the wasp grubs feed off the paralyzed cricket, which has not decayed, having been kept in the wasp equivalent of a deepfreeze. To the human mind, such an elaborately organized and seemingly purposeful routine conveys a convincing flavor of logic and thoughtfulness—until more details are examined. For example, the wasp's routine is to bring the paralyzed cricket to the burrow, leave it on the threshold, go inside to see that all is well, emerge, and then drag the cricket in. If the cricket is moved a few inches away while the wasp is inside making her preliminary inspection, the wasp, on emerging from the burrow, will bring the cricket back to the threshold, but not inside, and will then repeat the preparatory procedure of entering the burrow to see that everything is all right. If again the cricket is removed a few inches while the wasp is inside, once again she will move the cricket up to the threshold and reenter the burrow for a final check. The wasp never thinks of pulling the cricket straight in. On one occasion this procedure was repeated forty times, with the same result.

One can make the obvious remark that perhaps not the wasp but the experimenter was the one in the rut—but humor aside, this is a rather shocking revelation of the mechanical underpinning, in a living creature, of what looks like quite reflective behavior.

There seems to be something supremely unconscious about the wasp's behavior here, something totally opposite to what we feel *we* are all about, particularly when we talk about our own consciousness. I propose to call the quality here portrayed *sphexishness,* and its opposite *antisphexishness* (a vexish word to pronounce!), and then I propose that consciousness is simply the possession of antisphexishness to the highest possible degree. The point is that sphexishness and antisphexishness are two extremes along a continuum. Let me give a few examples distributed along that continuum, starting at the most sphexish and finishing with the most antisphexish:

1. A stuck record. This can be especially ironic if it's a recording of something that has a vibrant, lifelike dynamism to it (such as the music of contemporary composer Steve Reich), and then the illusion is shattered by the mechanical repetition of the jumping needle.

2. The *Sphex* wasp herself, and other examples from the insect world. For instance, suppose you have a mosquito in your bedroom. You try to swat it, and miss. It takes off and flies around the room, losing you. But after a while, it settles down and you spot it somewhere on the wall. Again you try to swat it and miss. As this cycle progresses, is the mosquito aware of the repetition? Does it begin to sense that there is an organized conspiracy against it, or does each new swat attempt come as fresh and unexpected as the previous one? Does the mosquito formulate some such notion as "the animate agent trying to wipe me out"? Sadly for the mosquito (but fortunately for you), it seems highly doubtful.

3. A herd of cattle in a corral, waiting to get branded. There is general commotion and hubbub, caused by the noise each cow makes at the moment of branding, and propagated outward by the cows closest to it. But does each cow in the corral recognize the overall pattern? Is its increased state of agitation due to the fact that the cow sees what is coming, or is it rather just a kind of vague apprehension, perhaps merely a raised adrenaline level without any specific meaning or referential quality?

4. A dog who is fooled every time by a faking motion in which you pretend to throw a ball, but instead don't release it. Actually, I don't know any dog who would fall for such an elementary trick. However, I do know a dog (who shall remain nameless—although he does happen to be an Airedale) who did not catch on when I threw his toy to an upstairs landing instead of down the hall (where he expected it). I led him up the stairs and showed him where it was. I expected he would know to go upstairs the next time. But no such luck. He just ran down the hallway again. Even after I had thrown his toy upstairs fifteen times more, he *still* ran down the hallway, then came back looking confused. Poor doggie! True, some of those seventeen painful times he did start going up the stairs, but each time he got only partway up, then turned around, and hightailed it down the hallway. To me, it was a disappointingly sphexish kind of behavior for a dog.

5. Glassy-eyed gamblers in Las Vegas, glued to their slot machines. To this can be added glassy-eyed teen-agers and college students glued to video games and pinball machines. Is there not some kind of deadening rut here? And yet so many people do this over and over again with seeming pleasure.

6. A happy-go-lucky person who sings or whistles all the time—and if you listen closely, you notice that it's always the same little refrain, day in, day out, year in, year out: never any variety.

7. People who make what seems to be the same joke, only in slightly different guises, over and over and over again. Or inveterate punsters, who simply cannot stop making one pun after another.

8. Junior-high-school students who fill each other's yearbooks with those same pat phrases and corny poems as *your* junior-high class did.

9. A mathematician who exploits one single technique to advantage in paper after paper, making advances in many different branches in mathematics, yet always with a distinct, idiosyncratic touch, and always, in some deep sense, just doing "the same old trick" again and again.

10. People whose rut-stuck behavior leads them down harmful pathways in their lives, for instance in their romances or their jobs. We all know people who "blow it" in the same way each time when faced with a situation that matters.

11. Social trends that become completely stylized and predictable, such as the endless trashy sitcoms that television networks keep churning out, the movies one after another based on some gimmick exploited in slightly different ways. For instance, one could perceive the movies *Breaking Away, The Black Stallion,* and *Chariots of Fire* as simply three ways of plugging specific values for variables into one successful formula—an upcoming championship race, a lovable underdog, a rival, and, of course, ultimate victory. And these are sophisticated, compared to some books and movies that much more blatantly exploit famous predecessors.

12. Styles in art that become dated and routinized to the point of no longer being creative. This happens to every style, but at the moment of its happening, there are always some people who are breaking out of the rut and creating totally new styles. However, there are others who become technically proficient at an old style, and who continue to create in an old-fashioned vein.

How different are these last few examples from the stuck record, or from the *Sphex* wasp? What is the real difference we feel as we progress down this list?

I would summarize it by saying that it is a general *sensitivity to patterns,* an ability to spot patterns of unanticipated types in unanticipated places at unanticipated times in unanticipated media. For instance, *you* just spotted an unanticipated pattern—five repetitions of a word. And I'm sure you picked up on all the French phrases crowded together earlier on in this chapter. Neither in your schooling nor in your genes was there any explicit preparation for such acts of perception. All you had going for you is *an ability to see sameness.* All human beings have that readiness, that alertness, and that is what makes them so antisphexish. Whenever they get into some kind of "loop", they quickly sense it. Something happens inside their heads—a kind of "loop detector" fires. Or you can think of it as a "rut detector", a "sameness detector"—but no matter how you phrase it, the possession of this ability to *break out of loops of all sorts* seems the antithesis of the mechanical. Or, to put it the other way around, the essence of the mechanical seems to be in its lack of novelty and its repetitiveness, in its trappedness in some kind of precisely delimited space. This is why the wasp, the dog, even some humans seem so mechanical.

*   *   *

How many computers do you know that would react with outrage (or guffaws) to the simultaneous occurrence on a single mailing list of "Bernie Weinreb", "Bernie W. Weinreb", "Mr. Bernie Weinreb, R.M.", "Barnie Weinrab", and so forth? Computers do not have automatic sensitivity to patterns in the data that they deal with. And of course, how could they be expected to? As one old saw goes, they do only what they are programmed to do. Computers are not inherently bored by adding long columns of numbers, even when all the numbers are the same. But people are. What is the difference?

Clearly there is something lacking in the machine that allows it to have this unbounded tolerance for repetitive actions. This thing that is lacking can be described in a few words: It is the ability to watch oneself as one deals with the world, to perceive in one's own activities a pattern, and to be able to do so at many levels of abstraction. Thus, consider the case of a hypothetical self-watching computer. To be sensitive in this way, it should get bored whenever it is forced to add a long column of identical numbers together. Wouldn't you? It should get bored whenever it is forced to do just adding over and over again, even when the numbers are different. Wouldn't you? It should even get bored when asked to do many arithmetic operations in any sort of repetitive pattern! Wouldn't you? Any loop of any sort should become tedious! Wouldn't it?

But where does it stop? Surely if a computer could perceive that all it *ever* does is pull up one instruction after another from memory (a piece of hardware, not to be confused with human memory), execute those instructions, and change various registers, it would yawn very boredly and probably soon go to sleep. And by the same token, you or I, if we ever gained access to the firings of our neurons, would find watching the activity to be one of the most stultifying things imaginable.

But this is not the kind of self-watching I mean. Watching one's own internal microscopic patterns is bound to be boring, because any complex system is bound to be made up out of thousands, millions, or even more copies of small elements (such as gears, transistors, cells, and so on). What is critical is to be able to watch activities on a completely different level—the *collective* level, in which huge patterns of activity of these many components assume regular behaviors perceptible on their own. A hurricane is a huge pattern of activity of tiny atoms, but one that has such regularity and pattern that we can predict hurricanes without ever thinking of their constituent atoms. A *thought* is a huge pattern of activity of tiny cells, of which much the same can be said.

Antisphexishness has to do with self-perception at this kind of level. Rather than watching its neurons or transistors or registers, an antisphexish being watches its own high-level patterns, looking for similarities somewhat the way meteorologists might look for one hurricane following another in a regular way.

Thus we should not expect or even want a self-watching computer to be able to see down to the level of its circuitry; it would not watch itself doing machine-language operations such as *ADD, STORE,* and *JUMP* in loop-like patterns. The effects of such operations are to change larger things called "data structures" in memory. Self-watching involves monitoring those changes as they happen, filtering out the dull ones, and recording certain aspects of the interesting ones in *other* data structures. (The fact that such monitoring, filtering, and recording would, on a more microscopic level, involve the very same kinds of elementary machine-language operations would be invisible to the computer, since it should be shielded from that detailed a view of itself.) Thus patterns in the changes taking place in *one* set of data structures would get recorded in another set of data structures. Should we then not set up a third level of data structures, to watch the second level, should patterns occur in it? And a fourth, to watch the third? This seems prime territory for an infinite regress: an endless hierarchy of structures, each one monitoring changes in the level below it.

Now that is quite true, and it is because you are a self-watching human being that you caught onto this pattern, and probably before I had spelled it out. It is in the nature of human pattern perception to be able to detect such infinite regresses, and to stop them short before they ever get anywhere. But what about the hypothetical self-watching computer, with its infinitely many layers of watchers?

Well, surely one of the most salient features—no, definitely the *most* salient feature—of what I have just described is the pattern of the data structures themselves: the hierarchy stretching upwards repetitively towards infinity. Shouldn't this pattern be as blatant to a self-watcher as it is to us? Indeed yes, it should. If we were to label the bottom level '0' and the first watching level '1', then logically we should label the further levels '2', '3', and so on. Each level in this potentially infinite set can be identified with a natural number. Once the pattern is perceived by a watcher, that watcher can form the general concept of "all the levels seen at once", associated with the concept of "all the natural numbers conceived of at once". The conventional name for the set of all natural numbers is 'ω' (omega), which we can take as the name of a *new* watching level that looks out for patterns in this potentially infinite tower of watchers.

You need not worry, by the way, that in proposing such a self-watching computer I am presupposing an infinite machine. Precisely the opposite. The whole purpose of stopping infinite regress in its tracks is so that we will *not* need to actually build an infinite tower of data structures and watching processes, a feat that would clearly be impossible, aside from being monumentally sphexish. At any stage, only a finite amount of recording would have been done, so that only a finite number—in fact, a small number—of levels of structure would exist. The only requirement is that there should exist the *potential* to extend it further.

It would be the ω-watcher that would perceive (as you and I and any human being would) the infinite-regress pattern of attempts to build the ω-tower itself. The ω-watcher would catch any such infinite regress before it could start. If a change in level 0 caused a change in level 1 that caused yet another change in level 2, and if these changes seemed to be patterned in such a way that an inevitable infinite ripple upwards would ensue, the ω-watcher, ever alert for such patterns in the other watchers, would come to the rescue, shouting "Wait! Enough! Halt!" Thus in fact, no infinite regress would actually occur; it would be nipped in the bud by the same sorts of mechanisms that allow you to cut off a bore at a party. "Excuse me, I think I'll go get some more punch."

$$* \quad * \quad *$$

The problem is, there's nothing to prevent the ω-level itself from going into loops—so if we're going to obviate that, we have to have a higher watcher—conventionally called "$\omega + 1$". Uh-oh! Before I even had a chance to begin spelling it out, you sniffed a new infinite regress! (You ruin all my fun!) Well, I'm going to spell it out, anyway. Level $\omega + 1$ needs to be watched by level $\omega + 2$, and that level by level $\omega + 3$. Thus we have a *second* potentially infinite tower of watchers, all of whom will be watched over by the Grand Watcher: level 2ω. But if there can be *two* towers, then why not *three*? And so, of course, it goes. Wheels within wheels, patterns of patterns of patterns. We get watchers 2ω, 3ω, and now our tower of towers needs a new Great-Grand Watcher: ω^2. And then—

Excuse me; I think I'll go get some more punch. There is a problem once you start getting into infinite regresses composed of other infinite regresses—the whole thing just never stops, and it becomes a *bore.* Or not exactly a bore, but a very complex and confusing thing, whose reality and relevance become ever more questionable. And yet, when you bring it back to the domain of sphexishness, it becomes the very real and very relevant question of how to build a machine that can sense unanticipated patterns in its own behavior.

This is related to a classic problem in the theory of computability, called the *halting problem*: It is the question of whether there exists any computer program that can inspect other programs before they run, and reliably predict whether or not they will go into infinite loops ("going into an infinite loop" means, of course, never coming to a halt—and conversely, "halting" means avoiding any infinite loop). The answer turns out to be "Definitely not", and for elegant, deep reasons. (Recall Chapter 21.) Of course, the thing hinges on getting this halting inspector to try to predict its own behavior when looking at itself trying to predict its own behavior when looking at itself trying to predict its own behavior when . . . Excuse me; I think I'll go get some more punch.

This halting-problem idea is closely related to our question about self-watching programs, but it is not really the same thing. First of all, the halting problem is concerned with an inspection to be carried out on programs *before* they are running, like looking at blueprints of buildings before they are built to see if they are earthquake-proof. Here we are talking about a program that is observing some program *while* it is running—and what's more, it's not just "some program" that it is watching, but *itself*. Of course, not *all* of its attention is being devoted to seeing if it's gotten into a rut (for that would itself constitute ruttish behavior!), but while it's doing other things, it's keeping its eye peeled, so to speak, for signs of ruttishness inside itself.

In computability theory, when a program or system of any sort turns back on itself in this manner, the turning-back-on-itself is known as *diagonalization*. To some people, diagonalization seems a bizarre exercise in artificiality, a construction of a sort that would never arise in any realistic context. To others, its flirtation with paradox is tantalizing and provocative, suggesting links to many deep aspects of the universe. Now here we see a *dynamic* diagonalization—a self-watching program—that seems to be closely connected with what makes a human being so utterly different from a stuck record or a *Sphex* wasp. Surely that is not such a bizarrely artificial thing to ponder!

Probably the most significant difference between the halting problem and the idea of a self-watching program is that in trying to build an artificial intelligence, we are not really so concerned with the mathematical perfection of our self-watching system as with its likelihood of survival in a complex world; after all, that's what intelligence is about. So if there is a mathematical theorem telling us that no program whatsoever will be a *perfect* self-watcher, able to catch itself in any conceivable kind of infinite regress, well, that is simply a statement that *perfect* intelligence is unreachable—something that ought to please us rather than dismay us, since it would be rather horrible and disappointing if someone came up with some finite program after a while, and could legitimately announce, "Well, folks, here it is at last: the end-all of intelligence, a *perfectly* intelligent program."

But don't worry about that. The metamathematical work of Kurt Gödel, Alan Turing, Stephen Kleene, and others, on such things as the halting problem and the theory of infinite ordinals (such as the towers of numbers and ω's), tells us that this scenario will not come to pass, for neither is there a perfect halting inspector, nor is there any ultimate scheme for naming ordinals. What this latter result means is that there is no finite mechanism that can possibly detect all patterns, patterns of patterns, patterns of patterns of patterns of patterns (aha!—fooled you that time, didn't I?), and so on.

*　　*　　*

In his famous paper "Minds, Machines, and Gödel", the English philosopher J. R. Lucas attempted to capitalize on these sorts of "negative" results of metamathematics by claiming that they provided the key element in a proof that no machine could ever be conscious in the way that humans are. Let Lucas speak for himself:

> At one's first and simplest attempts to philosophize, one becomes entangled in questions of whether when one knows something one knows that one knows it, and what, when one is thinking of oneself, is being thought about, and what is doing the thinking. After one has been been puzzled and bruised by this problem for a long time, one learns not to press these questions: the concept of a conscious being is, implicitly, realized to be different from that of an unconscious object. In saying that a conscious being knows something, we are saying not only that he knows it, but that he knows that he knows it, and that he knows that he knows that he knows it, and so on, as long as we care to pose the question: there is, we recognize, an infinity here, but it is not an infinite regress in the bad sense, for it is the questions that peter out, as being pointless, rather than the answers. The questions are felt to be pointless because the concept contains within itself the idea of being able to go on answering such questions indefinitely. Although conscious beings have the power of going on, we do not wish to exhibit this simply as a succession of tasks they are able to perform, nor do we see the mind as an infinite sequence of selves and super-selves and super-super-selves. Rather, we insist that a conscious being is a unity, and though we talk about parts of the mind, we do so only as a metaphor, and will not allow it to be taken literally.
>
> The paradoxes of consciousness arise because a conscious being can be aware of itself, as well as of other things, and yet cannot really be construed as being divisible into parts. It means that a conscious being can deal with Gödelian questions in a way in which a machine cannot, because a conscious being can both consider itself and its performance and yet not be other than that which did the performance. A machine can be made in a manner of speaking to 'consider' its performance, but it cannot take this 'into account' without thereby becoming a different machine, namely the old machine with a 'new part' added. But it is inherent in our idea of a conscious mind that it can reflect upon itself and criticize its own performances, and no extra part is required to do this: it is already complete, and has no Achilles' heel.

Somehow—and I think understandably—Lucas was under the impression that human beings are endowed with powers that are equivalent to a self-watcher of infinite depth, someone who will detect and terminate any and all patterned behavior: the ultimate in antisphexishness. I call this hypothetical ability "Breaking Out Of Loops Everywhere"—"BOOLE" for short, in honor of George Boole, who wrote one of the most influential books of the nineteenth century, *The Laws of Thought,* surely a forerunner of today's artificial intelligence work.

Lucas seems to think that to be human is to be endowed with this "BOOLE" ability—this total and perfect antisphexishness—intrinsically. On reflection, however, one realizes this surely is not the case. Despite not being *Sphex* wasps or Airedales, we humans are all still vulnerable to getting caught in ruts, as I attempted to point out in the dozen-item list above. None of us is immune. Each of us—even the Mozarts among us—exhibits a "cognitive style" that in essence defines the ruts we are permanently caught in.

Far from being a tragic flaw, this is what makes us interesting to each other. If we limit ourselves to thinking about music, for instance, each composer exhibits a "cognitive style" in that domain—a musical style. Do we take it as a sign of weakness that Mozart did not have the power to break out of his "Mozart rut" and anticipate the patterns of Chopin? And is it because he lacked spark that Chopin could not see his way to inventing the subtle harmonic ploys of Maurice Ravel? And from the fact that in "Bolero" Ravel does not carry the idea of pseudo-sphexish music to the intoxicating extreme that Steve Reich has, should we conclude that Ravel was less than magical?

On the contrary. We celebrate individual styles, rather than seeing them negatively, as proofs of inner limits. What in fact is curious is that those people who are able to put on or take off styles in the manner of a chameleon seem to have no style of their own and are simply saloon performers, amusing imitators. We accord greatness to those people whose "limitations", if that is how you want to look at it, are the most apparent, the most blatant. If you are familiar with his style, you can recognize music by Maurice Ravel any time. He is powerful *because* he is so recognizable, because he is trapped in that inimitable "Ravel rut". Even if Mozart *had* jumped that far out of his Mozart system, he still would have been trapped inside the Ravel system. You simply *can't* jump infinitely far!

The point is that Mozart and Ravel, and you and I, are all highly antisphexish, but not perfectly so, and it is at that fuzzy boundary where we can no longer quite maintain the self-watching to a high degree of reliability that our own individual styles, characters, begin to emerge to the world.

Although Lucas has been roundly criticized, and rightly so, I believe, by many philosophers, logicians, and computer scientists for failing to see many important subtleties of the Gödel argument on which he bases his paper, most of his critics have failed to see the crucial aspect of mind that Lucas was one of the first to point out. Lucas correctly observes that the degree of nonmechanicalness that one perceives in a being is directly related to its ability to self-watch in ever more exquisite ways. Unfortunately, too many artificial-intelligence people are ready to pooh-pooh the Lucas article on the grounds that its central thesis—the impossibility of mechanizing mind—is wrong. What they miss is that it is pointing at very deep issues that have much to do with the very core of intelligence and creativity.

*   *   *

Earlier I stressed the importance of the organization of memory and the pressing need to come at the question "What is a concept?" Critical to the way our memory is organized is our automatic mode of storing and retrieving items, our knowledge of when we know and do not know, of how we know or why we wouldn't know. Such aspects of what is sometimes called "metaknowledge" are fluidly integrated into the way our concepts are meshed together. They are not some sort of "extra layer" added on top by a second-generation programmer who decided that metaknowledge is a good thing, over and above knowledge! No, metaknowledge and knowledge are simmering together in a single stew, totally fused and flavoring each other richly. This makes self-watching an automatic consequence of how memory is structured. How is this wondrous stew of antisphexishness realized in the human brain?

And how can we create a program that, like a human brain, is all "of a piece", a program that is not simply a stack of ever-higher "other-watchers", but is truly a seamless "*self-*watcher", where all levels are collapsed into one? If we wish to have a program that breaks out of the extremely sphexish mold that all programs seem to be in today, we have to figure out how a flexible perception program might exploit its own flexibility to look at itself. Of course, no such program will be written as I just stated. That is, it will *not* come into being in the following way:

Step 1. We write a flexible perception program.
Step 2. We turn that program back on itself as a self-watcher.

Rather, to achieve the results desired in Step 1, we must have incorporated the goals of Step 2 into the design from the start! In other words, these two goals are intertwined, more in the following sense:

Goal 1. Flexible perception.
Goal 2. Self-watching.

There is no chronological priority here, for the two goals are too intertwined to have one precede the other. This is a tricky foldback, quite a bit more elaborate than the one involved in the halting problem, yet in spirit related to it.

It is interesting that Lucas' argument was based on Gödel's Theorem, whose proof depends on making one of these seemingly impossible (or at least highly counterintuitive) foldbacks—this one where a mathematical system of reasoning folds back on itself and subsumes itself as an object of study. What is fascinating in that proof is how, in such a system, there is a kind of level-collapse that ensues from the ability of a system to see itself. Rather than there being towers of watchers, then towers of those towers, and so on *ad infinitum* in the worst possible sort of multiply infinite regress, all those degrees and levels of self-perception are achieved at once by the fact that the system can mirror itself. Not that it mirrors itself in every aspect, mind you—for that would entail contradiction—but it does so at all levels of complexity.

The seemingly distinct levels of watcher and watched are totally fused, in the Gödel construction, exactly as Lucas would have it occurring in the minds of all conscious beings. The only thing that Lucas failed to understand is that the ability to fold around and see oneself in the wonderfully circular Gödelian way does not—in fact, *cannot*—bring with it *total* antisphexishness. That, fortunately or unfortunately, depending on your point of view, is a chimera.

*   *   *

Back in 1952, the philosopher and composer John Myhill wrote a lyrical article entitled "Some Philosophical Implications of Mathematical Logic: Three Classes of Ideas". The three classes are borrowed from mathematical logic, and Myhill's names for them are the *effective,* the *constructive,* and the *prospective.* In logic, they are known more technically as the *recursive,* the *renotrec* (short for "recursively enumerable but not recursive"), and the *productive.* Their essence is described below.

A category is *effective* provided that there is a way, given a candidate for membership, of deciding without any doubt whether that object is or is not a member. Is Ronald Reagan a KGB agent? Is the Pope Catholic? Although these two questions are easy to answer, which would seem to imply that being a KGB agent and being Catholic are examples of the effective, this is slightly misleading. Was Lee Harvey Oswald a KGB agent? Is an excommunicated bishop Catholic? Examples like these show that these categories are not genuinely effective categories—but then nothing in the real world is as clean as it is in logic. I could have asked, "Is 29 prime?" but I wanted to show how these notions extend beyond the mathematical realm. In natural languages, grammaticality (syntactic well-formedness) is a rather fuzzy property, but in an idealized language or formal system, it would be a perfect example of an effective property.

We pass on to the *constructive.* A property that is constructive is more elusive than one that is effective. The idea here is that some means exists whereby members of the category can be churned out one by one, so that you will eventually see any particular member if you wait long enough, but no means exists for doing the complementary operation—namely, churning out *non*members, one by one. Unfortunately, although this kind of set in mathematics is an extremely important one, easily definable examples of it are rather hard to come by. The set of all theorems in any formal axiomatic system is always recursively enumerable, but very often its complement is also, which turns the set into an effective one rather than a constructive one. You have to be dealing with a formal system whose *non*theorems are not themselves producible by some complementary formal system. Only then do you have a renotrec, or constructive, set. The set of theorems of any formalized version of number theory turns out (by Gödel's theorem) to have this property.

So much for the "constructive". We finally come to the *prospective,* also known as the *productive.* Myhill's characterization of it is this: "A prospective character is one which we cannot either recognize or create by a series of reasoned but in general unpredictable acts." Thus it is neither effective nor constructive. It eludes production by *any* finite set of rules. However—and this is important—it can be *approximated* to a higher and higher degree of accuracy by a series of bigger and better sets of generative rules. Such rules tell you (or a machine) how to churn out members of this prospective category. In mathematical logic, works by Tarski and Gödel establish that *truth* has this open-ended, prospective character. This means that you can produce all sorts of examples of truths—unlimitedly many—but no set of rules is ever sufficient to characterize them *all.* The prospective character eludes capture in any finite net. (See Chapter 13 for a discussion of Platonic notions such as "chairness", 'A'-ness, etc.)

As his prime example outside of mathematical logic of this quality, Myhill suggests beauty. As he puts it:

> Not only can we not guarantee to recognize it [beauty] when we encounter it, but also there exists no formula or attitude, such as that in which the romantics believed, which can be counted upon, even in a hypothetical infinitely protracted lifetime, to create all the beauty that there is.

Thus beauty admits of a succession of ever-better approximations, but is never fully attainable. Beauty and irrationality are often linked. Is it coincidental that the first example of such a notion of something approximable but never attainable in a finite process is called an "irrational" number?

Myhill is bold enough to speculate as follows: "The analogue of Gödel's theorem for aesthetics would therefore be: There is no school of art which permits the production of all beauty and excludes the production of all ugliness." To each coin there are two sides; and the obverse side of beauty is ugliness. By a rather ironic coincidence, the complementary set to a productive (or prospective) set is called, in the jargon of mathematical logic, *creative.* It must be admitted that it would take a stupendously brilliant, if perverse, sort of creativity to produce all possible ugly objects.

If we see the aim of art as the production of all possible objects of beauty (which is doubtless an oversimplification, but let us adopt that view nonetheless), then each individual artist contributes objects in a particular style. That style is a product of the artist's heredity and formation, and becomes a hallmark. To the extent of having an individual style, any artist is sphexish—trapped within invisible, intangible, but inescapable boundaries of mental space. But that is nothing to lament. Artists in groups form movements or schools or periods, and what limits one artist need not limit another. Thus, by the fact that its boundaries are wider, a school is less sphexish—more conscious—than any of its members.

But even the collective movement of a school of art has its limits, shows its finitude, after a period of time. It begins to wind down, to lose fertility, to stagnate. And a new school begins to form. What no individual can make out clearly is perhaps seen collectively, on the level of a society. Thus art progresses towards an ever wider vision of beauty—a "prospective" vision of beauty—by a series of repeated "diagonalizations": processes of recognizing and breaking out of ruts. As I like to put it, this is the process of *jootsing* (jumping out of the system) to ever wider worlds.

This endless jootsing is a process whose totality (so says Gödel) cannot be formalized, either in a computer or in any finite brain or set of brains. Thus one need not fear that the mechanization of creativity, if ever it comes about, will mark the end of art. Quite the contrary: It is a day to look forward to, for on that day our eyes will open—as will those of computers, to be sure —onto whole new worlds of beauty. It will be a happy day when, hand in hand with our new computer friends, we take an unanalyzable leap out of the system and go get some more punch.

Post Scriptum.

Do you know the Saint-Saëns Violin Concerto No. 3? Its middle movement happens to be based on a ravishingly beautiful melody—long, sinuous, flowing, lyrical. I suggest you get a hold of it and listen to it! Where do such melodies come from? Did they always exist? Are some people just lucky to have picked them up, these pretty pebbles lying on the musical beach?

Well, I hardly want to get into the discovery-invention-existence quagmire here. I have my own opinions, to be sure, but what I am more concerned with is where such inspiration comes from. One can point with a fair degree of objectivity to certain composers as being the most melodically gifted. These names come to my mind, for instance: Chopin, Rachmaninoff, Saint-Saëns, Tchaikovsky, Brahms, Bach, Mendelssohn, Puccini—and, switching gears somewhat, Cole Porter, Richard Rodgers, Jerome Kern, and George Gershwin. Obviously there are others. Some people undoubtedly would strike some off this list and would suggest others—perhaps Schubert,

Dvořák, Prokofiev, Scott Joplin, Fats Waller, Frederick Loewe, the Beatles, Carole King . . . It's hard to draw the line.

The main point is that certain rare people seem to be able to tap into some magic vein in which flow incredibly catchy patterns, deeply intoxicating to the human spirit. Leonard Bernstein once wrote a lively dialogue encatchily titled "Why Don't You Run Upstairs and Write a Nice Gershwin Tune?". In it, he talks about why that vein is so hard to tap. Bernstein should know, of course, since he too is one of the great melodic inventors of our time.

The problem is that melody invention, like every other art, looks so easy after the fact. In fact, in many ways it looks easier than creating other kinds of beauty, because melodies are such small, easily described structures. Making a beautiful turn on skis at least involves a *continuum* of possibilities, whereas a melody usually involves a very restricted, discrete alphabet (the notes within a two-octave range or so), and isn't even very long!

It is tempting, therefore, to imagine that good melodies are producible from some sort of recipe or mathematical formula, or, what comes to nearly the same thing, to think that the *amount* of beauty in a melody could be measured by some sort of machine, just as the amount of radioactivity in a sample of ore can be measured by a scintillation counter. You would stick your proposed string of notes into a machine and out would come a number called its "CQ" ("catchiness quotient").

If you doubt that the very idea of such a number is coherent, just remember that attached to every piece of existent music there really *is* a measure of its catchiness—namely, how often it actually is listened to, at the present time. Pieces can be rank-ordered according to this very cold, linear measure. This is not to suggest that the top piece is the best, but only to point out that the idea of a single, one-dimensional "catchiness index" applying to every possible string of notes is by no means absurd. Admittedly, under the present circumstances, it seems to take an entire society of millions of people to calculate the value for any string of notes, but could all that not be simulated? Perhaps the catchiness-quotient machine could be built to accept a set of parameters characterizing the target culture and its general musical mood at the time, and then it would predict how the given tune would fare in the given society under the specified musical circumstances. Is that not an engaging notion?

Are the musical receptivities of a culture truly characterizable in purely mathematical terms relating only to the syntactical structures of melodies? Ultimately, of course, the answer has *got* to be "yes", if by "syntactical structures" you mean structures whose recognition might require bringing in arbitrary amounts of external information. Sufficiently deep syntactic probing is tantamount to semantic probing, a motto from Chapter 1's *P.S.* The question is, then, just how complex a "syntax machine" that creates, or at least measures, melodic beauty would be. (Let's assume that it contains adjustable parameters for culture and mood.) Need it be as complex as a human society or a human brain? Can wonderful, lyrical, sinuous, and rapturous melodies come pouring out of a black box that can do nothing but that? Readers of *Gödel, Escher, Bach* (especially pages 676−680) might recall that I am extremely skeptical on that score. Yet how solid is the ground I am standing on? Could music not yield to brute computational power as swiftly as chess skill has (something which, in the same passages in *GEB,* I also was very skeptical about)?

*   *   *

It is funny how certain fads catch on, seemingly for no reason, while other things die, again for no clear reason. We all laugh at the Edsel today—yet what exactly is there to laugh at, except the fact that it did so poorly? What exactly was *wrong* with the Edsel? What is wrong with those thousands upon thousands of melodies that are composed every year and go nowhere? What made Michael Jackson and Pachelbel's simple Canon all the rage? Why did the typeface Helvetica catch on like wildfire when it was first invented, when a dozen extremely similar ones died on the vine? Why did the typographical gimmick of symmetrically capitalizing both the first and the last letter of a word or title, as in

GATEWAY
INN

PRINCE
SPAGHETTI

become a sudden vogue about four years ago?

Why is it now faddish to write run-on words such as "Intelligenetics" or "PEOPLExpress"? What makes words like "Da-glo", "Turbomatic", and "Rayon" seem slightly dated? Why is "Qantas" still modern-sounding? What is poor about brand names like "Luggo" and "Flimp"? Why are 'x's now so popular in brand names? And yet why would "Goxie" be a weak name compared with, say, "Exigo" or "Xigeo"? Why are the ordinary-seeming names that nasal-voiced comedians Bob and Ray come up with—for example, "Wally Ballou", "Hudley Pierce", "Bodin Pardew", and "John W. Norbis"—apt to evoke snickers? How come Norma Jean Baker changed her name to "Marilyn Monroe"? Why would it not do for a movie star to be named "Arnold Wilberforce"? Why is the name "Tiffany" popular today, and why was "Lisa" so popular a few years earlier? Is something wrong with "Agnes", "Edna", or "Thelma"? With "Clyde", "Lance", or "Bartholomew"? Mere length certainly cannot be the answer (think of "Elizabeth"). Nor can the sound, in any simple sense. (Why is "Lance" bad if "Vance" is okay?)

All this may seem a far, far cry from sphexishness and self-watching computers and brains. But what I am getting at is the unbelievable number of forces and factors that interact in our unconscious processing of even very tiny structures composed of discrete parts, such as words and names only a few letters long, let alone melodies several dozen notes long. Most of us could not put our finger on the answers to any of these questions. In fact, nobody could really answer these questions definitively. If we are going to try to get machines to do the subtlest of cognitive tasks, we had jolly well better be able to explain how mere words are appealing or repelling!

*　　*　　*

There are currently some efforts in artificial intelligence to imbue programs with a certain type of introspective capacity. Such a capacity is usually termed "reflection", a self-explanatory name that harks back to mathematical logic. A formal system is said to be capable of reflection if it can reason about itself. Gödel was the first person to discuss such things in detail. Nowadays reflective systems are the bread and butter of many a logician. However, computer modeling of logic is just now reaching the point where reflection is being seriously explored.

The idea is very enticing, but I think it has less to do with genuine progress in AI than it does with progress in elegant formal systems. It all has to do with one's ultimate view of what thought is. If you believe that thought is intimately tied up with some strict notion of truth and reasoning, and that exquisitely honed deductive capacities are the centerpiece of mentality, then you will naturally be drawn toward reflective reasoning systems. If, on the other hand, you believe, as I do, that reasoning is a far, far cry from the core of thought, then you will not be too inclined to jump toward such systems.

One way of looking at things is this. Imagine you have a set of rules that are supposed to capture the way people think in some domain—say that of melody composition. Now you try them out, and you find that most of the time they fail for complex reasons, but reasons that you have some intuitions about. How should you proceed now? There are two main rival avenues, the way I see it.

One avenue says, "Add meta-rules! Then add meta-meta-rules! Then . . . *ad infinitum*!" This might be called the "meta-meta" school of AI. The strategy is to improve the performance of a given set of rules by having higher-order meta-rules that help determine when and how to apply the ordinary rules. And this process knows no bounds, even to the point that one can formalize the progression from one level to its meta-level, so that in principle, an infinite number of meta-levels now are "there" to be consulted if needed.

The alternate avenue is to sidestep the topless tower of bureaucracies and meta-bureaucracies *above* by making rule-like behavior emerge out of a multi-level bubbling broth of activity *below*. This means that you give up the idea of trying to explicitly tell the system as a whole how to run itself. Instead, you content yourself with defining explicit micro-behaviors that will interact in vast numbers, and then you just let them go, carefully watching what ensues and noting what you like and what you don't like. After the run, you theorize about what might have made the system's top-level behavior more closely resemble your ultimate goals, and you go back and tinker around with the micro-elements whose micro-behavior you have explicit control over, using your best guess as to what sorts of changes will improve overall performance. Then you run the system again.

I remember a long time ago seeing a television show—perhaps you have seen it, too—in which someone set up a bathtub full of spring-loaded mousetraps holding ping-pong balls. Then they threw a single ping-pong ball in, and WHAM! The whole thing exploded madly, in parallel chain reactions. It was all over in a few seconds, but you can imagine running a film of it in slow motion. There are numerous large-scale features of the explosion that one could aim at creating, such as how long the pop takes, how high the average ping-pong ball flies, what the envelope of the flying balls looks like, and so on. If there were more types of micro-element and their interactions were more variegated, then you can imagine how multi-dimensional the system's macrobehavior would be, and how hard it would be to predict even its most basic features.

Yet when certain vast ensembles grow sufficiently big, the statistical principle called "the law of large numbers" sets in, in essence guaranteeing that there will be so much cancellation in the chaos that ultimately, a kind of order will emerge. It is for reasons like this that the National Safety Council can predict fairly accurately how many deaths there will be on a Labor Day weekend, even though they have no idea where any particular one will occur. Somehow, amazingly, the drivers cooperate and produce just about the predicted number each time, usually even on the state-by-state level, although less accurately.

The difference between such statistically emergent macrobehavior and rigidly constrained macrobehavior is best made by contrasting the mousetrap system with a huge domino-chain network, involving branching and rejoining paths, paths that climb hills and go back down, anything you can imagine as long as it's entirely self-determined (*i.e.,* no unanticipated external events start chains falling). In this kind of system, you know how everything is going to work beforehand. It's true that you may not be able to predict which of two "rival" pathways will reach a certain point first, but this kind of unpredictability is not nearly as hard to correct as that of the mousetrap system. If on one run the result is not what you want, you can just set it up again the same way, change some specific region, and you know what will happen. You can *program* this kind of system, but you cannot program a statistical system in the same sense. You can only tailor its micro-elements, and then release them and see what happens.

Which approach to mind is superior? Is the mind more like a fancy system of domino chains or a bathtub full of spring-loaded mousetraps? I'm betting on the latter. More will be found on this topic in Chapters 25 and 26 and their postscripts.

*   *   *

I received a letter from Thomas P. Laubert, in which he expressed considerable perplexity over a paragraph he had come across containing the following sentence: "Experience had taught the du Pont engineers to provide . . . flexibility in the design, wherever possible, to meet unforeseen problems that were sure to arise." Laubert mused: "But if the nature of the problems was unforeseen, then what parameters were used to determine these built-in flexibilities?" Another reader, whose letter I have unfortunately misplaced, brought up a similar point about engineering. What I remember vividly is his term "UNK-UNK"s—meaning the *unknown unknowns* that plague all complex systems. He was asking, rather skeptically, as I recall, how one can ever hope to build a system that anticipates all possible problems.

These simple-seeming questions hit the nail on the head. An intelligence is, by definition, a system supposed to be able to deal with the unpredictable. But how can any set of rules "frozen" into a machine's design do that? Doesn't the very fact of being frozen make any foreordained system/program/machine/organism vulnerable in some way that actually follows from the rules themselves? This, of course, is the Gödelian point that J. R. Lucas was trying to make in the article I quoted from in the column. And the only satisfactory answer that I can see is to admit that, yes, all intelligences are indeed vulnerable—including biological ones, and that means people no less than *Sphex* wasps. Natural selection has looked favorably upon organisms with highly abstract kinds of vulnerability, highly abstract kinds of sphexishness. And so for the time being, humans are doing all right. But as for there being a fixed recipe that would allow an organism to cope with all the curves that the universe at large might throw at it, that is a vain and crazy hope!

24

Analogies and Roles
in Human and Machine Thinking

September, 1981

$\mathbf{I}$N our research in artificial intelligence, my graduate students Gray Clossman and Marsha Meredith and I have been looking at typical human thought processes in everyday life as well as in more limited domains, and everywhere we look, we seem to find that within the internal representations of concepts there are substructures that have a kind of independence of the structures of which they are part. Such a substructure is *modular*— exportable from its native context to alien contexts. It is an autonomous structure in its own right, and we call these modules *roles*. A role, then, is a natural "module of description" of something, a sort of bite-sized chunk that seems to be comfortable moving out of its first home and finding homes in other places, some of them unlikely at first glance.

One intriguing example is the "First Lady" role. Probably most Americans use this term more flexibly than they realize. They would most likely say, if asked, that the term means "the wife of the president", and not think any more about it. But if they were asked about the First Lady of Canada, what would almost surely pop into their mind is the name or image of Margaret Trudeau. They might reject the thought as soon as it occurred to them, but for us the important thing is that the thought of her would arise at all. First of all, people know her as the *former* wife of Pierre Elliott Trudeau. Second, Trudeau is not the president of Canada but its prime minister. How, then is "former wife of the prime minister" the same as "wife of the president"?

Before you answer, "Well, 'wife' and 'former wife' are related concepts, as are 'prime minister' and 'president'", consider who might be said to be the current First Lady of Britain. Whose name comes to your mind? Margaret Thatcher? Queen Elizabeth? They are women, but do they really play the role of First Lady? How about Denis Thatcher or Prince Philip? At first these suggestions seem silly, but in a strange way they start to seem compelling, particularly the thought of Denis Thatcher. In fact, I once clipped a newspaper article that portrayed Denis Thatcher as Britain's First Lady.

What kind of sense does this make? How can a male be a lady? Well, language is far slipperier than dictionary definitions would have you believe. Its slipperiness comes from the underlying slipperiness of concepts, in particular these elusive things we are calling roles.

Of course, you could argue that what "First Lady" *really* means is "spouse of the head of state", and so the First Lady role goes over without any trouble into "husband of the prime minister". But this won't do either. In Haiti, until recently the title of First Lady belonged to Simone Duvalier, the wife of the late former president, François ("Papa Doc") Duvalier. She is also the mother of the current president, Jean-Claude ("Baby Doc") Duvalier. Not long ago there was a bitter power struggle between Simone Duvalier and her daughter-in-law Michelle Bennett Duvalier, the wife of Baby Doc, for the title of First Lady. In the end, the younger woman apparently gained the upper hand, taking the title "First Lady of the Republic" away from her mother-in-law, who in compensation was given the lifetime title "First Lady of the Revolution".

Do you want to amend your suggestion so that it will say "spouse or parent, present or former, of a head of state, present or former"? You know perfectly well that we'll be able to come up with other exceptions. For example, imagine a meeting of the Pooh-Bah Club at which the Grand Pooh-Bah's favorite aunt was introduced as the First Lady of the club. Of course, the Grand Pooh-Bah is hardly a head of state, and so you could amend your definition to say "spouse or favorite relative, present or former, of the head, present or former, of any old organization". But suppose . . . Actually, I think I'll let *you* go on inventing exceptional cases. For any rule you propose, there is bound to be some conceivable way to get around it.

Worse yet, something terrible is happening to the concept as it gets more flexible. Something crucial is gradually getting buried, namely the notion that "wife of the president" is the most *natural* meaning, at least for Americans in this day and age. If you were told only the generalized definition, a gigantic paragraph in legalese, full of subordinate clauses, parenthetical remarks, and strings of *or*'s—the end product of all these bizarre cases—you would be perfectly justified in concluding that Sam Pfeffenhauser, the former father-in-law of the corner drugstore's temporary manager, is just as good an example of the First Lady concept as Nancy Reagan is. When this happens, something is wrong. The definition not only should be general, but also should incorporate some indication of what the *spirit* of the idea is.

* * *

Computers have a hard time getting the spirit of things; they prefer to know things to the letter. And so people spend an enormous amount of time talking to computers, writing long and detailed descriptions of ideas they could get across in *one good example* to anyone with half a brain. So a challenging question is how to get a computer to understand what is meant by "First Lady". For this we need to examine the idea of "roles" in detail.

In order to illustrate how the notion of "role" can be modeled in domains more formal than that of political protocol, I now will switch to one of my favorite domains: the natural numbers. I will present some puzzles that Gray, Marsha, and I have been thinking about. Each of them has a set of possible answers with varying degrees of plausibility or defensibility. We are working on a computer program that is able to see the rationale behind each possible answer, and thus is able to come up with the same set of "feelings" as a typical person would have, about what is a good answer and what is a bad one.

The domain of natural numbers might sound at first like a hard-edged, objective mathematical world, but actually it is a domain in which problems requiring extremely subtle *subjective* judgments can be formulated. We have given our program very little detailed arithmetical knowledge about the integers. The program does not, for example, recognize 9 as a square; in fact, it doesn't even know about multiplication! It does not know that 6 is even and 7 is odd. So what *does* it know? It knows how to count up or down —that is, it has a knowledge of successorship and predecessorship. Thus it recognizes that the sequence of numerals "12345" represents an upward counting process. It is also able to apply the notion of counting to structures it is looking at, as in "44444", which it could recognize as a group of five copies of the numeral '4'. It knows that 9 is bigger than 4, although it has no idea *how much* bigger. (Subtraction and other arithmetical operations are unknown to it.) You can think of our computer program as having the arithmetical sophistication of a five-year-old and an avid curiosity about number patterns. (By the way, it is not tied to or affected by decimal notation. The number 10 is not considered any more special than the number 9.)

Here is the first problem (invented, as were many of the following ones, by Gray). Consider the following structure, which we'll call A:

$$\text{A: } 1\ 2\ 3\ 4\ 5\ 5\ 4\ 3\ 2\ 1$$

Now consider the structure called B:

$$\text{B: } 1\ 2\ 3\ 4\ 4\ 3\ 2\ 1$$

The question is: *What is to B as 4 is to A?* Or, to use the language of roles: *What plays the role in B that 4 plays in A?*

Note that by asking it this way, we leave it to the puzzle solver to decide what role 4 actually does play in A. It would be analogous to asking "Who is the Nancy Reagan of Britain?", leaving it to the listener to figure out what conceptual role Nancy Reagan fills, and then to try to export that role to Britain. I have found that many people who balk at calling Denis Thatcher the "First Lady of Britain" are quite content with calling him the "Nancy Reagan of Britain". A curious point that this illustrates, and to which we will return, is this: If the role is left implicit, nonverbalized, it has more fluidity in the way it transfers than if it is "frozen" in an English phrase.

As a matter of fact, most analogies crop up in this type of nonverbal way. Seldom does someone say to you explicitly: "What is the counterpark of Central Park in San Francisco?" Usually it happens through a more implicit channel. When you are visiting San Francisco for the first time, you are driven through Golden Gate Park, and somehow it reminds you of Central Park. After the fact, you can point out some shared features: both are long thin rectangles; both contain lakes, curving roads, and excellent museums; and so on. Most analogies arise similarly—as a result of unconscious filterings and arrangings of perceptions, rather than as consciously sought solutions to cooked-up puzzles. To put it another way, to be *reminded* of something is to have unconsciously formulated an analogy.

Incidentally, when I first thought of writing about roles and analogies, I had in mind both the First Lady example and the numerical examples. As my thoughts evolved, I realized I was unconsciously developing a parallel in my mind between the First Lady example and the numerical examples. I'll call it a "meta-analogy", since it is an analogy between analogies. In this meta-analogy, I see structure A as corresponding to the United States, structure B to Britain, 4 to Nancy Reagan and the unknown number to the unknown person. We'll come back to the meta-analogy later on.

*　　*　　*

Let us now look at some possible answers to the first number-analogy problem. The most sensible answer is 3—and fortunately, it is also the most frequently given one. The usual justification is that 4 precedes the central pair (55) in A, and the corresponding central pair in B is 44, which is preceded by 3. Well, then, what would you say for C? What is to C as 4 is to A?

C: 1 2 3 4 5 6 6 6 6 5 4 3 2 1

The central pair in C is 66, which is flanked by 6's. Is 6, therefore, to C what 4 is to A? Well, most people probably would prefer 5, although it is perfectly *logical* to insist on 6. The preference for 5 comes, nonetheless, from a very sensible (and also logical) instinct to generalize the notion of "central pair" (itself, to be sure, a role) to "central plateau" (or whatever you want to call it). There are competing urges: first, to stay with the exact original concept, and second, to flex and bend when it "feels right", when it would seem rigid and stodgy to insist on established conventions over simple and "natural" extensions. But it is just these sorts of terms—"flex", "bend", "feels right", "rigid", "natural", and so on—that are so extraordinarily hard to put into programs, logical though programming might be.

Now let us investigate some other ways to make the role of 4 slip. Consider this structure:

$$\text{D:}\quad 1\ 1\ 2\ 2\ 3\ 3\ 4\ 4\ 5\ 4\ 4\ 3\ 3\ 2\ 2\ 1\ 1$$

Here is a curious kind of reversal; now there is no central pair—yet everything else is in pairs. Some people might still pick 4, since it is next to the center. But what about 44, a *pair* rather than a single number? After all, as long as "pair" and "singleton" have switched places, we might as well go all the way and give an answer that reflects this perceptual turnabout. In fact, it would seem rigid and unimaginative to insist on sticking with single numbers when it is so obvious that the easiest way to perceive D is in terms of pairs:

$$\text{1-1}\quad \text{2-2}\quad \text{3-3}\quad \text{4-4}\quad \text{5}\quad \text{4-4}\quad \text{3-3}\quad \text{2-2}\quad \text{1-1}$$

Not just 4 but every part of A has a role, and there are corresponding roles in D. As you can see, within each role the concepts of pair and singleton have been switched.

Now is as good a time as any to return to my meta-analogy and to point out some correspondences between these problems and the First Lady problem. If you think of the president as "the highest, most central figure in the land" and his wife as "the one standing next to him", you will see that this characterization carries over almost literally to the numerical problems. In structure A, the highest, most central figure—the "president"—is 5 (or possibly the pair of 5's) and his "wife", standing next to him, is 4. In B, the president is 4 (or the pair of 4's) and his wife is 3. In C, the president is 6 (or the group of 6's), and his wife is 5. In D, the president is (for once) unambiguous (5), but to compensate, there is a dilemma concerning the identity of his wife. If you think of pairs as males and singletons as females, then D presents us with a case where the sexes are reversed, exactly as in the First Lady of Britain problem. The most reasonable answer seems to be the "spouse" (in this case, the husband) of 5, namely the pair 44.

Consider now the following couple of curious cases:

$$\text{E:}\quad 1\ 2\ 3\ 4\ 5\ 6\ 7\ 8$$

$$\text{F:}\quad 8\ 7\ 6\ 5\ 4\ 3\ 2\ 1$$

What can we make of these? A very rigid person might cling to the idea captured in the phrase "number to the left of the central pair", despite the fact that nothing at all distinguishes the central pair in either of these examples. Such a person would give the inane answers of 3 for E, and 6 for F. Such a person would do better to take up football instead of analogies, as Lewis Carroll's Tortoise once remarked to Achilles.

But what would be a wiser view of, say, E? How *should* one map E onto A? Any mapping is doomed to be imperfect, so how can we do it best (that is, with the least pain or frustration)? We might think of E, since it rises uniformly, as mapping onto the left half of A. This would involve a tacit judgment that it is all right to abandon the attempt to map E onto *all* of A, in return for the ease of mapping E onto a "natural" portion of A. That is a pretty subtle step to take, I would say. It would suggest 7 as the answer.

Well, what about F, then? Do we prefer 2 or 7? It depends on whether we choose to map F onto the left half or the right half of A. Mapping F onto A's left half involves mapping a *descending* sequence onto an *ascending* one. But either choice requires a willingness to let go of qualities that had seemed important, a willingness to bend gracefully under pressure. Fluid analogies are not a game for rigid minds!

These kinds of situations are difficult because in essence they call for splitting the role of 4 in A into two rival facets. In the mapping of A onto F, one of the rival facets sees 4's role in A as "one less than the president", whereas the other facet sees 4's role as "the next-to-rightmost element of a staircase". Thus one facet is primarily concerned with *magnitude* and the other primarily with *position*. The facet you find more important will determine your answer to F.

Pretty much this kind of split happened when you tried to decide whether the First Lady of Britain was Queen Elizabeth or Margaret Thatcher—or one of their husbands. Is being a figurehead or being a head of state more likely to make someone's spouse a First Lady? In the United States, these features coincide in one person (the president), but in Britain they do not. Consider the following target structures:

G: 5 4 3 2 1 1 2 3 4 5

H: 1 2 3 4 6 5 5 6 4 3 2 1

In G, what is most central is simultaneously lowest, and what is highest is simultaneously most peripheral! (G can be pictured as a valley and A as a mountain peak.) We have a "ceremonial figure" (the 5's flanking the structure) and we have a "head of state" (the two central 1's). Which one's spouse would better fill the role of First Lady? Or, to put it most simply: "What in G plays the role of 4 in A?" I personally would opt for 2 because it stands next to the central group. To me, centrality seems more important here than magnitude, just as political power seems more substantive than ceremonial show. Correspondingly, I would opt for Denis Thatcher rather than Prince Philip as "Britain's Nancy Reagan".

Now what happens when we tackle H? There are three "reasonable" possibilities (in the sense of appealing to law's proverbial "reasonable human"): 6 (flanking the central pair of 5's), 5 (being the next-to-largest number) and 4 (flanking the central "crater" 6556). Once again there is no Gloriously Right Answer, but there are certainly ideas that seem good and ideas that seem shaky. For instance, if someone suggested, "The answer is 4, because 4 is the fourth term of H, just as it is the fourth term of A", I would be nonplussed. That would be a childish generalization based on the most superficial of scans of both structures. It would be as childish as inferring that, since *our* national holiday falls on the fourth of July, other countries' national holidays would also have to fall on the fourth of various months. To see 4 as no more than the fourth element of A is to ignore all of A's structure. It is to see A as nothing richer than this:

o o o 4 o o o o o o

A good answer must take A's structure into account in a full, rich, and yet simple way. This means that, to the extent it is possible, all of A must be perceived in terms of interacting, mutually intertwining conceptual structures—roles that are mutually dependent, in the way that "family", "husband", "wife", "mother", "father", "daughter", "son", "brother", "sister", "relative", "in-laws", and so on are all interdependent concepts.

The word "role" makes us think of the theater. In a play, the various roles all mingle together in scenes. A scene is a larger-scale structure than an individual role; it is a place where several roles coexist and interact. In our analogy problems, one might try to conceive of the two structures involved as if they were two enactments of a single scene, portrayed by different directors working with different actors. Thus the *core roles* would exist and would be filled in both presentations, but at the same time each presentation would have minor aspects, or roles, unique to it. For example, the adaptation of the Greek legend of Orpheus and Eurydice into a contemporary context of carnival time in Rio de Janeiro is the basis for the movie *Black Orpheus.* Many original features cannot be *directly* exported, but with poetic modification they can be, and the director, Marcel Camus, met the challenge with great flair. In the movie there are, of course, many minor parts—extras —that add Brazilian flavor, yet they do not impair the analogy at all; in fact, they enrich it. This is the kind of thing that appeals deeply to human sensibilities, both intellectually and emotionally.

*   *   *

Now that you have seen some variations, I would like to return to our first puzzle and point out some of its hidden subtlety. First of all, the "central pair" notion, which functions as the keystone of structure A, is actually just a kind of by-product, an accidental artifact of the structure of A. To see what I mean, consider this question. How would you efficiently describe the structure of A (without quoting it digit by digit)? You would probably say it rises from 1 to 5 and then falls from 5 to 1, making two halves that are mirror images. Nowhere in this description was there any mention of some kind of central pair or central plateau. It was not needed; one will just appear automatically there when anyone follows your description. In fact, anyone who constructs a copy of A is very likely to see a central plateau, even if the concept was never suggested to them. To the mind's eye it appears something like this:

$$1\ 2\ 3\ 4\ \textbf{5-5}\ 4\ 3\ 2\ 1$$

Somehow a new percept has been born in the center. It is, as I remarked above, the *keystone* of A. (Note that the concept of "the keystone of A" depends on, or implies, a mapping of A onto an arch—yet another analogy.)

Why don't we perceive the pair of 3's, say, as a unit as well? Probably simply because they do not touch. And consider this structure:

$$1\ 2\ 3\ 4\ 5\ 1\ 2\ 3\ 4\ 5$$

The central pair—"5 1"—doesn't pop out as being salient or important, does it? In A, though, the combination of *adjacency* and *equality*, particularly when supplemented by *centrality*, somehow makes the two central 5's merge into a unit in the perceiver's mind, albeit usually not at a conscious level. Still, if this perceptual shift did not happen, then the answer of 5 for C, based largely on equating the plateaus in A and C, would be considerably less compelling.

In the first puzzle, both A and B had obvious central plateaus. This suggested a good starting point for an overall mapping of A onto B: central plateau onto central plateau, start onto start, finish onto finish, and so on. But if we tried to complete this mapping, we would obviously run into trouble:

We *must* have 1 in A mapping onto 1 in B, no? And the centers have to match up too, don't they? But where between 1 and 5 does the analogy break down? It seems that some kind of mapping of 4 onto 3, as is shown above, is satisfying to many people. But press them one step more, and they will shrug, grin, and give up.

Similarly, although you can ask for "the Nancy Reagan of Britain", it makes less sense to ask, "Who is the Maureen Reagan of Britain?" (Remember that Maureen Reagan is Nancy Reagan's stepdaughter.) Suppose the Thatchers had a biological daughter. Would she be the counterpart of Maureen Reagan? Or suppose Margaret Thatcher had a stepdaughter. Would she be the counterpart? Then again, suppose that

Margaret Thatcher had no daughter but that Denis Thatcher had *twin* stepdaughters. Would these twins, taken together, constitute the counterpart of Maureen Reagan? (How can two people fill a role defined by one person? Well, think of example D, where the pair of 4's played the role of a single 4 in A. Or think of many European countries, which have both a president and a prime minister.)

Issues like this crop up all the time in the pursuit of good analogies, and facing up to mismatches leads occasionally to productive insights. One could go on and press for even more detailed correspondences between entities in Britain and in the United States. What is the British counterpart of Watergate? Who plays the part of Richard Nixon? Of John Mitchell? Of Senator Sam Ervin? Of Senator Daniel Inouye? Of G. Gordon Liddy? Of Judge John Sirica? Of John Dean? Of Officer Ulasewicz? Of Alexander Butterfield? The less salient an object is inside a larger structure, the harder it is to characterize in an exportable way.

But what makes something salient? As a rule, it is its proximity, in some sense, to a "distinguished" element of the larger structure. Consider the following long structure:

$$1\ 1\ 1\ 1\ 1\ 1\ 1\ 1\ 2\ 2\ 2\ 2\ 3\ 3\ 3\ 4\ 3\ 3\ 3\ 2\ 2\ 2\ 2\ 1\ 1\ 1\ 1\ 1\ 1\ 1\ 1$$

The central 4 is probably the most distinguished individual numeral. Then, depending on how you perceive the sequence, different features will leap out at you. For instance, do you see it as "letters" or as "words" (larger-scale chunks of the sequence)? When I see it at the "word" level, the central group "3334333" seems just a shade less salient than the 4, and after that, perhaps, the two flanking groups of 1's. The two groups of 3's by themselves come next. Only then do the groups of 2's get recognized. On the other hand, when I perceive the sequence at the "letter" level, what is salient is quite different. After the central 4, probably the next most salient numbers to me are the first and last 1's, since they are very easy to describe—then maybe the first and last 2's. After that, the two 3's flanking the central 4— but at this point it starts to get a little harder to specify various items without resorting to such uninspired descriptions as "the fourth term".

A *distinguished* item is something we can get at via an elegant, crisp, exportable-sounding description. A *nearly distinguished* item is something we can get at by first pointing to a distinguished item, and then, in an exportable way, describing a short "jog" that leads to it. Just as in giving someone directions, some places are more salient, others are less so. Some buildings in New York City are inherently difficult to direct someone to, others are inherently easy. In the same way, some roles in a complex conceptual structure are highly distinguished and easily exportable, others are very hard to describe. Although they may have certain idiosyncratic qualities in their local context, nothing makes them stand out globally.

As you move progressively farther away from its central roles, any analogy becomes increasingly strained. For example, "Who is the Jackie Washington of Britain?" Should we begin by getting out the London telephone book and looking under "Washington, J."? Or should we look under "London, J."? Or is it a meaningless question, meaningless even to Jackie's best friend? After all, Jackie's role may just be too small and idiosyncratic within the structure of the United States. It is not exportable. The fact that Jackie is the manager of Gearloose's Hot Dog Stand in Duckburg does not help much, because one still has to figure out the identities of the British Duckburg and the British Gearloose—not to mention the British equivalent of hot dog stands!

The moral is a simple one: Don't press an analogy too far, because it will always break down. In that case, what good are analogies? Why bother with them? What is the purpose of trying to establish a mapping between two things that *do not* map onto each other in reality? The answer is surely very complex, but the heart of it must be that it is good for our survival (or our genes' survival), because we do it all the time. Analogy and reminding, whether they are accurate or not, guide all our thought patterns. Being attuned to vague resemblances is the hallmark of intelligence, for better or for worse.

*   *   *

The fact that we use words and ready-made phrases shows that we funnel the world down into a fairly constant set of categories. Often we end up with one word, such as "kitchen". In general, two kitchens will not map onto each other exactly, but we still are satisfied with the abstraction "kitchen". Generally speaking, a kitchen will have a sink, a stove, a refrigerator, cupboards, counters, drawers, and so on. In the United States, it is very common for people to assume that the garbage will be in a cupboard below the sink. The idea of "the cupboard below the sink" is a perfect example of an exportable role. In fact, isn't your sink the "president" of your kitchen? And . . .

Our language provides for mappings of many degrees of accuracy. Some people, when they see Bossie, see no further than "cow" and accordingly use that word; others notice that Bossie is female, and will say "heifer". Still others perceive the breed as easily as they perceive Bossie's "cowness" and talk about "that Angus heifer". A famous Dublin zookeeper, Mr. Flood, was once asked the secret of his great success in breeding lion cubs. "Understanding lions." said he. "And in what does understanding lions consist?" he was asked. His reply: "Every lion is different." This curious answer denies the category while taking advantage of it. But that is the nature of categories. Their validity can at best be partial. No matter at what level of detail you cut off your scrutiny, your perception amounts to filtering out some aspects and funneling the remainder into a single conceptual target, a mental symbol often labeled with just one word (such as "word") or stock phrase (such as "stock phrase"). Each such mental symbol implicitly stands for the elusive sameness shared by all the things it denotes.

Beyond the implicit analogies hidden in individual words or stock phrases, explicit analogies occur all the time on a larger scale in our sentences. We are quite uninhibited in comparing unfamiliar things with things we assume are more familiar. We see grids of all kinds as being similar to checkerboards. We see carefully charted actions in life as being similar to chess moves. We see the eye as a camera, the atom as a tiny solar system. Science is constantly being likened to a vast jigsaw puzzle (an analogy I have never cared for). In their eagerness to stretch and bend concepts, people turn proper nouns into common nouns, as in the statement "Brigitte Bardot is the French Marilyn Monroe." In such linguistic flexing, both *la Bardot* and the Monroe suffer somewhat in the interests of vivid imagery.

Then, going one step beyond the explicit linguistic level, there are the analogies and mappings that we use constantly to guide our thoughts on a larger scale. The perception of romantic dilemmas is one of the most striking places where mapping or analogical thinking dominates in an obvious way. When someone tells us of some romantic woe, we can usually map it immediately onto some experience of our own. In fact, we can probably draw some parallel between *any* romantic situation and any other one, and such a mapping will perhaps yield some insight if it is carried out well. Yet romances are incredibly detailed and idiosyncratic. The point is that we throw many details away; we skim off some abstractions and are careful not to try to carry the resemblance too far. And certainly we ignore the trivial aspects. A romance between Chris and Sandy can certainly map onto one between Pat and Chris or one between Sandy and Pat, despite the fact that names, hair colors, and other superficial features do not match!

The reason, then, for worrying about human analogical thought is that *it is there.* To ignore it would be like ignoring Everest in trying to understand mountain climbing.

*   *   *

Let us get back to some concrete problems in our more formal, numerical domain. Notice that there is an inherent kind of contradiction in setting up analogy problems—which, after all, are *informal* by definition—in a rather formal domain. But the nice thing is that it shows that the domain is actually just as slippery as any "informal" domain. Here are four further examples:

I: 1 2 3 3 4 5 6 7 6 5 4 3 3 2 1

J: 1 7 7 6 5 4 3 2 1

K: 6 9 7 3 9 4 1 6 6

L: 1 2 3 4 5 6 7 8 9 7 8 9 6 5 4 3 2 1

Example I involves what I enjoy referring to as a "governor", namely the pair 33. Here again, one role in A has been split into parts: 55 in A was not only the sole pair but also the *peak,* whereas in I, 33 is the sole pair and 7 plays the role of the peak. We are forced to choose between 2 (the wife of the governor) and 6 (the wife of the president). Actually, the governor has two "wives"—2 and 4, a "left wife" and a "right wife"—and so we have to choose between them, unless we go with 6 as being the wife of president 7.

Example J, beginning as it does with "1776", is a patriotic puzzle. (What is its British counterpart?) Its interest is primarily in that it draws attention to A's symmetry, which we had taken for granted. When we chose 4 as the president's wife, were we taking his *left* wife or his *right* wife? In A, of course, they coincided, so it didn't matter. But in J they differ: 1 would be the left wife, 6 would be the right wife. Because of a tendency to be influenced by our left-to-right scanning, we probably would choose the left wife under normal circumstances, but here, there is such great asymmetry that we pause. The regular descent from 7 to 1 corresponds far better to A's "staircase" structures than does the abrupt leap upward (from 1 to 7 in one step!). For that reason, 6 probably wins over 1, in this case.

Example K is a bit obscure, but it has been led up to by example J. In particular, example J drew attention to the fact that in A there are two 4's, not just one. Example K plays on the relation of those two 4's to each other. In A, there were two elements between the two 4's. We can take that property as defining the role of 4 in A. To be sure, that is not the *only* relation between the two 4's, but it is the most obvious. If you "turn off" everything in A but the 4's, you will get an image something like this: ooo4oo4ooo. That image makes the size of the interval between the 4's stand out. Given this way of looking at the role of A's two 4's, what in K corresponds? There is only one number that occurs exactly twice, and its two appearances are separated by two numbers. That number is 9. If you turn off everything but the 9's in K, you get this picture: o9oo9oooo. That may or may not be sufficient reason for you to choose 9 as the K-counterpart of A's 4.

Finally, consider example L. Here, the central-pair notion gets extended one further degree of abstraction. We go up, step by step, until we hit the second 7. Jolt! It takes us a moment to get our bearings, and when we recover, we realize that the central pair consists not of single integers but of "clumps" or "chunks": namely, two copies of the unit 789. We can aid the eye this way:

$$1 \quad 2 \quad 3 \quad 4 \quad 5 \quad 6 \quad 7\text{-}8\text{-}9 \quad 7\text{-}8\text{-}9 \quad 6 \quad 5 \quad 4 \quad 3 \quad 2 \quad 1$$

Now the answer seems glaringly obvious: It is 6! On the other hand, maybe we were supposed to get the hint offered us generously by the central pair. And what was that hint? It is that we could perceive the *whole structure* in triples, not just its center. In this case, L reparses into

$$123 \quad 456 \quad 789 \quad 789 \quad 654 \quad 321$$

Now the answer should be obvious—except that we are still left with a minor dilemma. Do we take the president's right-hand wife (654) or his left-hand wife (456)? I am biased by left-to-right chauvinism and would choose 456. Many people, however, refuse to see the sequence in triples and stick with 6.

Here is an innocent-seeming puzzle that points to still more complex issues:

M: 1 2 3 4 5 7 7 5 4 3 2 1

The way I see it, the best answer is 6. You might object, "Why not 5? 6 isn't even there!" True, but 6 is conspicuous by its absence. The 4 in A precedes the 5 not only typographically but also arithmetically: 4 is the numerical predecessor of 5. And what is 5 in A? It could be seen either as the *maximum* in A or as the number forming the *central pair* of A. Both carry over to M, yielding 7 as M's 5. Now, if you choose to see 4's role in A abstractly and arithmetically rather than concretely and typographically, you can carry your vision directly over to M. Then candidate 6 must be considered a strong competitor to 5. In my mind, it wins.

This example opens up an whole new world of *levels of abstraction* in the perception of structures. To illustrate briefly, let me propose the following structures:

A':   1 2 3 4 4 4 5 6 7 8 9 8 7 6 5 4 4 4 3 2 1

B':   1 1 1 2 3 4 3 2 1 1

And here is the puzzle: What in B' plays the role that 7 plays in A'? Well, 7 occurs twice in A', but certainly it seems to play no *salient* role. As a numeral in A', 7 has no outstanding characteristic, and so at first its role seems hard to export. However, 7 enters into the structure of A' in another way, and a salient one at that. One of the most salient features of A' is its large number of 4's. Count them. How many? Seven. Aha! Thus 7, in its capacity as an invisible *counting number* rather than as a visible *numeral,* plays a very distinguished role in structure A'. Still, is it possible to *export* this role to B'?

We have to decide how to characterize (in an exportable way) just what it is that 7 is counting. To insist that it must be the number of 4's seems a little parochial, to say the least. Who says that 4 is the 4 of B'? Perhaps a deeper and more fruitful way to look at matters is to see 4 as *A''s most frequent term.* This leads us to look for the most frequent term in B'; this is 1. So 1 is the 4 of B'. Therefore the counterpart of 7 would be the number of 1's in B'—namely, 5—another "invisible" answer, in that 5 does not appear as a visible numeral in B'.

It is narrow-minded to insist that there is a big distinction between being present as a visible numeral and being present in a more abstract sense, for example, as a counting number. To put it another way, 5 is invisible in B' only if you think of vision as having no cognitive component, as if we could perceive only *numerals.* In fact, with our eyes we are constantly "seeing" abstract qualities. When we look at a television program, we see more than flickering dots: we see people. Of course, somewhere deep down in the processing there are components of our visual system where the dots themselves are "seen" as dots, but ironically, we would hesitate to describe as "vision" what retinal and other cells do. Vision implies *going beyond the dots*; in other words, beyond the primitive visual level. We can "see" that a certain chess position is ominous, that a certain painting is by Picasso, that someone is in a bad mood, that this car won't fit in that garage, and so on. If we accept this notion that vision is imbued with a cognitive component, then we can agree to "look beyond the numerals". In that case, 5 *is* directly visible in B′!

By the way, I carefully drew up A′ so that 7 would appear as a numeral in it (as well as counting the number of 4's). This threw in a complicating factor, something one had to ignore. I could have had A′ have, say, 12 4's, in which case 12 would have "appeared" in A′ only at the abstract level of a counting number and not as a numeral. But real life is seldom so considerate of would-be analogy-perceivers. For example, in thinking about the question "Who is the Nancy Reagan of Britain?", you might have felt that this was much harder than "Who is the First Lady of Britain?" because you may attach certain uniquely personal attributes to Nancy Reagan, over and above seeing her as the First Lady of the United States. What if I had asked for "the Eleanor Roosevelt of Britain"? Or, turning the tables, "Who is the Moshe Dayan of the United States?" I am almost tempted to answer "Douglas MacArthur", like Dayan a famous and successful general and a controversial political figure, but then I remember—MacArthur had two eyes! Dayan's eye patch is perhaps his most memorable feature.

It is interesting to go back to earlier examples, mapping them onto A′ and asking, "What here plays the role of 7?" You will perceive those old structures through new eyes (or glasses). I leave a few challenging examples for you to map onto A and A′:

P: 5 4 3 2 1 5 4 3 2 1

Q: 5 4 3 2 1 1 2 3 4 5 5 4 3 2 1

R: 1 2 3 4 9 8 7 6 5 4 3

S: 1 1 2 2 3 3 4 4 5 5 6 6 7 7 1 2 1 7 6 5 4 3 2 1

T: 1 2 3 4 1 2 3 1 2 1 2 1 3 2 1 4 3 2 1

U: 2 1 1 2 2 1 2 2 2 2 9 1 2 3 2

You might enjoy making up some examples of your own, which potentially might lead a solver to further unexpected modifications of the perceived role of 4 in A. For instance, can you devise an example in which it becomes sophisticated, rather than childish, to perceive 4 as the fourth element of A?

*   *   *

One of the purposes of these puzzles is to dispel the notion that the full, rich, intuitive sense of a role, such as that of 4 in A or that of First Lady, can be easily captured in words. In fact, it might be more accurate to assert the contrary: that precisely in its nonverbalizability lies its fluidity, its flexibility. This is a crucial idea. Consider how you would try to capture in some phrase the precise way you see "what 4 is doing" within A. No matter what phrase you give, someone will be able to concoct another example in which your phrase does not enable anyone to predict what you will perceive as being analogous to 4. A frozen verbal phrase is like a snapshot that gives a perfect likeness at one moment but fails to show how things can slip and move. There is something much more fluid in the way a mind represents the role internally. Various features are potentially important in defining the role, but not until an example comes up and makes one feature explicit does that feature's relevance emerge.

We make comparisons all the time. It does not seem particularly note-worthy when someone walks into your kitchen for the first time and says, "I like the way your kitchen is laid out better than the way mine is. My kitchen has windows over *there* and the stove is right *here,* so it is less convenient and the light isn't so good in the morning." Clearly the words "here" and "there" conceal implicit mappings of the two kitchens, other-wise the statement would be utter nonsense. Words like "this" and "that" and phrases like "that sort of thing" are even better at picking up intangible, flexible, implicit meanings that can be transported across the borders of situations differing widely from each other. And that's the name of the game, in thought.

Right now it seems that what artificial intelligence needs is a way to go beyond "delta function" programs: programs that are virtuosos in a very narrow domain but that have no flexibility or adaptability or tolerance for errors. I call these programs "AE programs": programs that have Artificial Expertise. The trouble with them is that they are always brittle and narrow. It seems that a careful study of judgmental processes in even so simple a domain as these curious number analogies would afford fascinating insights into how computer programs might be made to approach the flexibility and generality of our own minds.

To show what I mean, I would like to conclude with a verbatim transcript of a conversation I had with a friend a while back. It ran this way:

FRIEND: Last Friday afternoon I was over at the Pooh-Bah Club listening to a piece on the radio that I was *sure* was Shostakovich. When it ended and they announced it, sure enough, it was! I was thrilled, because that kind of thing has happened to me only a couple of times in my life!

ME: *That* kind of thing? You mean, being at the Pooh-Bah Club and hearing a piece on the radio that you thought was Shostakovich on a Friday afternoon?

FRIEND: You're so *dense*! When those *Scientific American* people hear about that, they probably won't want you to write any more articles for them.

ME: Yeah, I should have known that it didn't have to be on a Friday afternoon.

FRIEND: You should have known that it didn't have to be Shostakovich!

Quite coincidentally, a recently perfected natural-language computer program called CORTEX happened to be eavesdropping on us, and it just could not resist chiming in at this point, saying, "Oh say, that reminds me —something *really* similar happened to me the other day. I was at a club whose name is hyphenated, and the water cooler broke. Ain't that something!" Well, *that kind of thing* is what I would like to see artificial intelligence programs doing more of.

Post Scriptum.

Verdi is the Puccini of music.
   —Igor Stravinsky

The knee is the Achilles' heel of the leg.
   —Pasadena (Calif.) *Valley Values*

The AI work out of which this column grew was my "Seek-Whence" project. My original goal was to develop a program that would take as input a sequence of integers such as 1, 4, 9, 16, and that would detect the underlying law—thus, it would "seek whence" the sequence came, and would extend it. Over a period of time, it became obvious that certain aspects of the goal were more central to mentality than others. In particular, it became clear that the ability to quickly discover the law behind highly mathematical sequences (even lowly mathematical ones, like the squares) is a specialized skill that bears little relation to mind in general, but that the ability to quickly spot *patterns* (as in "1 2 2 3 4 4 5 6 6") is absolutely indispensable.

Thus the Seek-Whence project retargeted itself on *structures composed of smallish integers;* and the major effort became one of figuring out how to perceive and "parse" such structures as these:

$$1\ 2\ 3\ 4\ 5\ 5\ 4\ 3\ 2\ 1$$
$$1\ 1\ 2\ 3\ 1\ 2\ 2\ 3\ 1\ 2\ 3\ 3$$
$$2\ 1\ 2\ 2\ 2\ 2\ 2\ 3\ 2\ 2\ 4\ 2$$

The latter two examples nicely illustrate one of the major problems to confront: that of boundary location. Where do you draw the lines separating one substructure from another one? And what are good ways of restructuring or regrouping if a first try fails? That whole area of concern is reflected in the project's very title, "Seek-Whence", which violates the conventional syllabic structure, namely, "see-kwents", regrouping it as "seek-wents". This kind of difficulty permeates the efforts to mechanize continuous speech recognition, and is very familiar to anyone who has been in the position of listening to a foreign language they have studied but don't know well. Often sounds will flow by so fast that you have absolutely no idea what is being said, simply because you cannot tell where the word breaks are; what is most frustrating is that this can happen even if everything being said would be perfectly clear if you saw it in writing (where word breaks are handed to you on a sil verplatter). The "parsing" of visual input is likewise permeated by boundary-location problems. Music is another such domain, and in fact the discrete multi-level patterns of melodies were among the biggest inspirations for the Seek-Whence domain.

At one point, in trying to get across the idea of Seek-Whence to someone who had a distaste for integers, I simply substituted letters for integers (*a* for 1, *b* for 2, etc.), and made up some parsing and analogy problems. For instance: "What is to *abcddcba* as *d* is to *abcdeedcba*?" Some people might say that this problem is *similar* or *analogous* to the first numerical analogy problem given in the column; I would say it is the *same* problem. Yes, in different clothing, if you like, but the same all the same. Numerals, capital letters, smalls—what's the difference? At least that was my feeling. Yet I found that I could usually awaken more interest in people if these analogy problems were presented in terms of letters instead of numbers. Groan!

From potentially infinite sequences and the *rules* behind them, my focus of attention gradually shifted to rather short sequences and the *roles* inside them (as I emphasized in the column). This concentration on roles and analogies then became so dominant that the Seek-Whence work revealed itself to be primarily a project on perception of analogies. Once this was out in the open, I decided to reify that concern by creating a new project, which I dubbed "Copycat", the idea being that being a copycat, when you're a child, is a universal and primordial experience in doing simple analogies. If I touch my nose and say to you, "Do this!", what will you do? Most people will touch their own nose. But why not touch mine? If I touch *your* nose, what will you do? Touch your own, or mine? And so on. A set of variations on this theme is shown in Figures 24-1 and 24-2. You can think of them as symbolizing the entire Copycat project.

One can be more flexible or less in how one interprets "Do this!" What does one take literally, what does one see as playing a role in a foreordained and familiar structure? What kinds of familiar structures is one willing to see as identical to each other? When is it necessary to start inventing new ways of perceiving a given situation in order to fit it into pre-existent frameworks, which then allow already-familiar roles to emerge? What remains firm, and what slips? What sticks, and what gives? These kinds of questions sound rather abstract, but when real analogies are manufactured, they are the chief

FIGURE 24–1. *The "Do this!" problem. In (a), Tom touches his nose and says to Annie: "Do this!" She wonders what she should do. In (b) and (c) you see how two Annie-clones respond. What would you do?*

FIGURE 24–2. *More "do-this" questions. In (a), a three-headed Annie is at a loss for what the best way to "do this" is. In (b), a long-necked Annie and her giraffe friend wonder what to do. In (c), Fanny the Fish—handless and noseless, but having gills—muses how to copy Tom. Finally, in (d), how in the world can poor little Elephannie do what Tom is doing?*

concerns of the analogy-maker, whether at a conscious level or not. Therefore these issues are the heart and soul of the Copycat project, and the purpose of this *P.S.* is mainly to show exactly how that is so.

* * *

One serious problem with studying real-world analogies like the "First Lady" examples is that they bring along too much baggage—too many complex tie-ins to all sorts of concepts. Another serious problem is that when we study real-world analogies that real people actually have made, we are nearly blinded to their essence because we have nothing to contrast them with. In order to see what makes human-made analogies good, we have to see the alternatives: analogies that humans would never make, even in jest. Can you imagine, for instance, an organism trying to understand the experience of being pregnant by likening it to being an elevator (or a football stadium, for that matter) containing one person? Something is very wrong with such an analogy—but what? Try to formulate principles of analogy-building that would suppress that type of analogy (whatever that means!) but that would recognize the insight in this one: "Try speaking English for a while without using the letters 'e' or 't' at all. Then you'll have an inkling of how Japan felt, having two of its major cities wiped out." Both these analogies involve mapping a thinking organism onto something very alien to it, and on a totally different scale. Yet one succeeds well and the other flops well. How come?

In such cases, filled to the brim with myriads of overlapping and softly blurring concepts, there is virtually no way to unravel what is really going on in a human understander's mind. To try to model all that at this early stage of research on natural and artificial minds would be as sillily ambitious as trying to master the most rich and idiomatic poetry of a foreign language without ever having bothered to tackle any prose in it at all. That would be arrogant and intellectually upside-down.

I believe that it is not merely preferable, but indispensable, to look at the analogy-making process in a pared-down domain, yet a domain where all the essential qualities of analogy-making remain. Newton couldn't have discovered his laws of motion if he had concentrated on trying to understand the laws governing waterfalls or hurricanes. Instead, he boiled the problem of motion down to the most pristine case he could imagine— planets coasting through a vacuum. This is the typical method of science: Isolate the crucial phenomena and study them in a pure context; then work your way upward towards phenomena in which two or more fundamental themes coexist.

This is what I sought to do in Copycat: To lay bare what I saw as the central problems of analogy without any extra clutter of real-world knowledge. Those central problems, as I see them, are:

* deciding how literally to take references (*i.e.*, deciding which parts of each situation already have literal counterparts in the other situation, and which parts need "literary" counterparts to be discovered or invented);
* deciding what structures are worth perceiving (*i.e.*, deciding which types of abstraction are worth bringing to bear as overarching frameworks to guide perception, so as to facilitate mapping pieces of one situation onto pieces in the other);
* perceiving roles inside structures (*i.e.*, selecting which aspects of the currently preferred organizing frameworks are most relevant and which are less relevant);
* deciding how literally to take roles (*i.e.*, recognizing which roles in each situation already have literal counterroles in the other situation, and which roles need "literary" counterroles to be discovered or invented);
* weighing rival ways of viewing a situation against each other and choosing the most elegant one (or, if you prefer, the simplest one).

The parallel passages about literary and literal *parts* and *roles* may seem a bit obscure, so let me motivate them briefly.

An entity in one situation can belong simultaneously to other situations. The sun is an example. From your point of view and mine, it is just the sun, a unique object. The sun is what I call a "part" of my world. Its counterpart in your world is the very same thing, not just something *like* it. Thus the sun is a part of *the* world. For another example, take Groucho Marx. When he died, I didn't have to put myself in my friends' shoes in order to understand what his death was like *for them*. His role in their world and his role in my world were so indistinguishable that to attempt such an empathetic mapping would have been foolishly extravagant.

But for a contrast, consider your beloved identical twin sister Glunka. Your connection to her is certainly far different from mine to her. In fact, I've never even met her, whereas *you've* known her all your life! Glunka is a very big part of your life, but from my rather distant and external point of view, her main identity is *as your twin*—after all, I know nothing else about her. Thus to me, she is the *filler of a role* in your life. If I were to learn that Glunka had died, I could hardly be expected to weep rivers, because she is not a part of my life. But that does not mean that I could not empathize, because I could project. The obvious mapping would see her counterpart as *my* identical twin sister. However, given that I am a male, no such person exists. Does that mean I am incapable of empathizing? I would be pretty inhuman if my ability to empathize were that weak. It is easy to slip from "identical twin sister" to "twin sister". Trouble is, I don't have one of those either. What can I do, then, to empathize? How about slipping along a different dimension—to my identical twin *brother*? That would be fine if he existed, but he doesn't, poor fellow. (And not because he died!) So I must loosen up still further, and try mapping your identical twin sister onto my twin brother, or just my brother. They don't exist either. Damn! In desperation, I try slipping to just plain old *sister.* Aha! This time it works. A sister I have (two, in fact), and in some loose sense, each of them plays a role in my life analogous to that of your identical twin sister in your life.

Of course, the analogy is weaker than it might be if I were twins (like you), but what can a body do? We make do with the best mapping we can find. The notion "my sister" is what I call the *counterrole* in my life to the notion "my identical twin sister" in your life. Of course, to someone else, the counterrole might be "my Siamese twin brother", or "my best friend, who I have known all my life", or even, for some people, "my car". The *filler* of a counter*role* is, of course, the counter*part.* In your life there are many parts and role fillers, as there are in mine. By discovering your life's counterparts to parts of mine, and your life's counterroles to roles in mine, you can project and identify.

How distinct are these concepts of role filler and part? Well, as concepts, they are quite distinct, but life constantly confronts us with blurry situations where people (or things) are simultaneously parts and role fillers. Think of your old and dear friend Millapollie, who is also familiar to me, but only mildly so. If you told me that Millapollie had died, how would I react? I would have dual approaches to the situation, one seeing Millapollie as a (very small) *part* of my life, and the other trying to find a *counterpart* in my life to Millapollie's part in *your* life. Thus I would try to find the filler of the counterrole in my life to the role that Millapollie plays in your life. Very probably, I would find myself flitting back and forth between these two visions of one and the same person. To effect such a part-role compromise is sometimes easy, but more often very tricky. Usually we are not even conscious of the conflicting pressures, but we muddle through anyway.

*   *   *

Cross-language comparisons may also help to make this idea more vivid. How eager I was, when learning French, to learn how to talk about baseball. I very much wanted to know how you say "pitcher", "catcher", "fly ball", "out", and so on. To be sure, such terms do exist in French, and it's fine to learn them, but it seems to me in retrospect to have been a misguided obsession for someone whose chief motivation was sheer fluency. In learning a foreign language, why place a high priority on learning how to talk about your *own* culture's idiosyncratic features? Instead, strive to learn the "corresponding" aspects of that culture—that is, the things that play *counterroles,* rather than the *translations* of many concepts unique to your culture. In my case, perhaps the appropriate move would have been to learn all about soccer and its terminology in French.

Of course, many terms transcend languages. It is important to know how to say "moon" in both languages, and it seems reasonable to assume that "the moon of France" and "the moon of the United States" are really the same, so that the moon is a shared part rather than a private role-filler. Now in a way this would seem true of any publicly visible entity, such as Algeria. And yet—are the Algeria of France and the Algeria of the United States really the same thing? Might Viet Nam not be "the Algeria of the United States", at least from a French perspective? Is Algeria an objective *part* of the world or something that plays a *role* in various national perspectives? What about Argentina? Australia? Antarctica? And such questions apply not only to proper nouns. What about wines, cheeses, languages?

Native English speakers are quite easily amused by very crude parodies of German, such as this sign posted near a computer terminal:

Alles Lookenspeepers!

Das Komputermaschine ist nicht für gefingerpoken und mittengrabben. Ist easy schnappen der Springewerk, blowenfusen, und poppencorken mit Spittzensparken. Ist nicht für gewerken by das Dummkopfen. Das rubber-necken Sightseeren keepen Hands in das Pockets—relaxen und watchen das Blinkenlights.

Our amusement is based on the peculiar way in which our language is rooted in the Germanic family. We tend to find many aspects of German gawky, comical, and old-fashioned. Obviously, German speakers will not easily see how their language has this quality to us. They will certainly not get a sense for our feelings of their language's gawkiness if they tack Germanic endings onto German words, use lots of "sch" sounds, and make long compound words—but neither will they do so by tacking on English endings, suppressing "sch" sounds, and breaking up compounds, because the historical and social connection between the two languages is not symmetric, and the effect, even if humorous, would not be analogous. But what, then, *would* be the analogue for native German speakers? What is "the German of German"?

Note that I seem to be implying that there is one best answer. Actually, I doubt there is. The connections between English and German are many and variegated. In some ways, English certainly *is* to German what German is to English. (This harks back to some problems of translating self-referential sentences, dealt with in Chapter 1 and its *Post Scriptum.*) In other ways, the assumption of symmetry is completely wrong. What would a German (or French, or whatever) parody of English (or Dutch, or whatever) look like?

How does English sound to a native Mandarin speaker studying it? I doubt I could ever know. Yet I am sure there is some fairly uniform reaction to English across the millions of Chinese people who have heard it. What would it be like to hear my own native language through ears that could not fathom it, or could penetrate it only superficially? Such an experience is denied to me—and yet, is it not exactly the same as my experience when I listen to Mandarin? Yes and no.

What entities in a given situation play the role of fixed stars, of mutually shared global points of reference? These are, in my terminology, parts. What entities are seen entirely in terms of their role relative to the perceiver, entirely as local occupiers of standard "slots"? These are role-fillers. What entities float midway between total globality and total locality, somewhere between being pure parts and pure role fillers? The answer is, of course, that *nearly all* entities are free-floating in this way, which is why the problems of analogy and translation are so deep and so deeply implicated in the mystery of mind and consciousness. The linguist George Steiner has provocatively explored these issues in his book *After Babel.*

Since a satisfactory discussion of the nature of analogy would take an entire book, I will not attempt to lay out the philosophy that the Copycat project is based on. The column gives you some good ideas (provided you can make that giant conceptual leap from numbers to letters). But it seems worthwhile presenting at least a few canonical examples from the Copycat project, since I feel they capture in a nutshell all that we are trying to do.

*　　*　　*

It should go without saying to readers of the column that Copycat deals with an alphabet of stripped-down letters. In particular, all a letter "knows" about itself is its predecessor and its successor (if it has such; *a* and *z* of course are special in that regard, each one lacking one). Letters do not know what they look like or sound like, or whether they are vowels or consonants. Since the Platonic alphabet has a starting point and a finishing point, unlike the integers, there is a kind of symmetry to it. There are two distinguished elements, namely, the endpoints *a* and *z*. These elements have identities on their own; they are somewhat like royalty. All other letters derive their identities, directly or indirectly, from these distinguished letters. Obviously *b* and *y* are like royal viziers, and *c* and *x* like vice-viziers.

I visualize the graph representing letters' "importances" as the arc of a suspension bridge, suspended at both ends from *a* and *z* and descending very steeply to a minimum at the center of the alphabet (see Figure 24-3). Thus in theory, the very least distinguished letters are *m* and *n*. However, practically speaking, all the letters in that general vicinity are pretty much equally nondescript. After all, if being nondescript were a salient property, then we would be caught in a paradox: *m* and *n* would be highly salient by virtue of being maximally undistinguished! But *m* and *n* do not know they are of minimal salience, and hence the paradox is obviated. In fact, any letters further in from the tips than *c* and *x* are pretty bland, and even those two aren't very exciting.

In the vast midwestern prairies of the Platonic alphabet, one step this way or that makes little difference. Poor *q* hardly knows what role it plays in society, since its only connections are to *p* and *r*, letters of equally little distinction. It's just as in human communities: most people are recognized in their own neighborhood, but as soon as they leave, they become

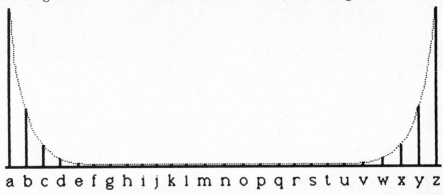

a b c d e f g h i j k l m n o p q r s t u v w x y z

FIGURE 24–3. *A graphical representation of the effective saliencies of the letters in the Platonic alphabet of the Copycat world. The "San Francisco" and "New York" of the alphabet—a and z—are of course glamorous and salient. Nearby letters get some reflected glory, but as you leave the two "coasts", you fade into the drab middle regions, where no letter has much to draw attention to it.*

anonymous faces. The only thing you know about random strangers is precisely that: that they are strangers. Copycat letters are like people in that way.

For example, *q* recognizes that it doesn't know *k* or *x*, but they are so unfamiliar that it makes no distinction between their degrees of unfamiliarity. Only when you come/quite close to *q* does it begin to act as if it recognizes you. But even then, whatever notice *q* might take of, say, *s* would not be *direct;* it would have to be mediated by *r*, which has a direct acquaintance with both letters. Generally speaking, connections decrease quickly as the number of intermediaries goes up, so that Platonic *q* loses virtually all "acquaintance" with letters much further from it than *s* or *o*. Still, there is a sort of exponentially decaying "halo" surrounding Platonic *q*, a residue of its interactions with its immediate neighbors (and *their* interactions with *their* neighbors, etc.), giving it a tapering-off set of "fringe acquaintances" (see Figure 24-4). The same phenomenon applies to all the Platonic letters, of course.

This "renormalization effect" (so called after the analogous effect in particle physics) is quite well captured by the following candid remarks made to the author by Platonic *q*, when queried about various letters:

"Mercy! I certainly don't recall hearing the name *m* before."
"Now then . . . I believe I've seen *n* somewhere around."
"Oh yes. I know *o*, though not terribly well."
"Positively. I'm old friends with *p*."
"Quit kidding! That's *me*!"
"Really a fine and true friend, is *r*."
"Sure, I know *s*, though not frightfully well."
"That's possible . . . Probably I've seen *t* somewhere around."
"Uhh . . . No, I definitely don't recall hearing the name *u* before."

571

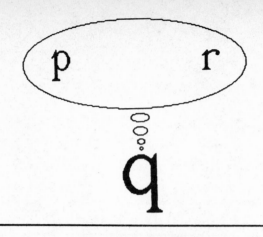

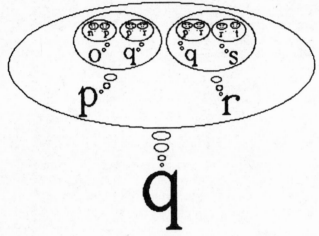

...ɟjklmnop q rstuvwx,...

FIGURE 24–4. *The "renormalization effect" by which any letter (q, here) acquires a large set of* virtual *acquaintances despite having only two* true *acquaintances—its predecessor and successor. In (a), "bare" q dreams of its two neighbors, bare p and bare r. But those letters can in turn dream of their neighbors, and so on. This recursive dreaming-pattern is shown in (b). The upshot of it all is shown in (c): a q's-eye view of the alphabet. This shows how q has an effective connection to every other letter of the alphabet, although the strength decays rapidly with distance, since it has to be mediated by all the intervening letters, and each extra link weakens the chain by some constant factor. This subjective vista, reminiscent of a sensory homunculus in an animal's brain, translates into a probabilistic statement for the running Copycat program: the further two letters are from each other, the less likely is any relationship they bear to each other to be noticed. Thus close letters (within two or so, as a rule of thumb) are pretty likely to be thought of in terms of their roles relative to each other, but further-apart letters are likely to be taken simply at face value, without any attempt to draw connections between them. This has vast repercussions on how analogies are made.*

Copycat's alphabetic world includes abstractions that lurk behind the scenes, and it is they that allow viewers to see coherence in *groups* of letters. The most basic of those concepts are two group-types based on the two simple relations that exist: (1) sameness and (2) successorship (or predecessorship, its mirror image). A group of neighboring elements linked by sameness is called a *copy-group,* or *C-group.* Here are a few C-groups:

$$aaa \quad uuuuu \quad cc \quad wowowowowo$$

Notice that the concept includes the case where a *structure* (in this case, *wo*) is repeated. Degenerate cases include C-groups of length 1, such as *c* (or even *wo*), and worse yet, C-groups of length 0! Although it sounds perverse, sometimes a group consisting of zero copies of *e* is quite different from one consisting of zero copies of *f.* But this is a fine point, and only for advanced copycats.

The other group-type is that of *S-group* (and its mirror twin, *P-group*). An S-group is simply a group of neighboring successors, as shown below:

$$abc \quad uvwxy \quad cd \quad pqrs$$

And if you flip these over so that they run backwards, then they are examples of P-groups:

$$cba \quad yxwvu \quad dc \quad srqp$$

Needless to say, degenerate S-groups and P-groups of length 1 and 0 exist as well, but we need not worry over them. Our old friend "1 2 3 4 5 5 4 3 2 1" consists of course of a *numerical* S-group and P-group, back to back.

Beyond these most basic abstract constructs, there are more shadowy entities that move in the wings and exert intangible yet profound forces on perception of structures. These go by such names as *symmetry, uniformity, good substructures, boundary strength,* and so on. They are the kinds of forces that push you toward perceiving *abcpqr* as two groups of three, and *aabbcc* as three groups of two; they push you toward breaking *aakkkkggee* into five equal-length C-groups, and toward breaking *abcdefpqr* into three equal-length S-groups; they lend support to seeing all three of the structures *abcdcba, abcdabc,* and *abcdwxyz* as symmetric, although in three very different senses; and they make you feel quite uncomfortable with a structure such as *aaabbbqcc.* I will not try to spell out these elusive forces here, firstly because they are many, secondly because they are very abstract, and finally because people tend to grasp intuitively exactly what sorts of pressures are created by them anyway.

This concludes the presentation of the Copycat domain, nearly isomorphic to the Seek-Whence domain (except that in Seek-Whence, there is no analogue to *z*), and so without further ado, we may proceed to the analogies themselves. The basic set of analogies is actually a bunch of variations on a theme (just as in the column). That theme is the following "event":

abc changes into *abd.*

Note that it is up to you to decide what happened to *abc*; if you think that the *c* in it became a *d*, that is just fine, but it is your perception and not inherent in the change. Someone else might interpret it as a replacement of the entire "object" *abc*, lock, stock, and barrel, by *abd.*

Now, there are a number of possible counterpart situations that I could present and ask you to do "the same thing" in. It's hard to decide which one to try first, but I'll just plunge in:

What does *pqrs* change to?

Pretty easy, eh? I suppose you said *pqrt.* So does nearly everybody. That is because it is—in some rather elusive and wonderful sense—*the right answer.* Yet there are numerous other candidates that you might have considered. In fact, because it is almost impossible to appreciate the intricacy and fascination of the Copycat world unless you "live" there, I strongly urge you to pause here and think: What else could *pqrs* go to?

*   *   *

One possibility is *pqrd*; need I explain why? Then there is *pqrs,* produced from the input *pqrs* just as *abd* was produced from *abc*: by substituting a *d* for every *c.* Did we, or did we not, do "the same thing" here? Should you touch *your* nose, or *mine*? For that matter, why isn't *abd* or *pqds* the answer to this problem? Such questions of rigidity versus fluidity recur throughout analogy, and seem to resist formalization.

Speaking of rigidity versus fluidity, when I gave a lecture on analogies in the Physics Department at the California Institute of Technology several years ago, one Richard Feynman sat in the front row and bantered with me all the way through the lecture. I considered him a "benevolent heckler", in the sense that he would reliably answer each question "What is to X as 4 is to A?" with the same answer, "4!", and insist that it was a good answer, probably the best. It seemed to me that Feynman not only was acting the part of the "village idiot", but even was relishing it. It was hard to tell how much he was playing devil's advocate and how much he was sincere. In any case, I will never forget the occasion, since his arguing with me stimulated me no end, and at least from *my* point of view, it wound up being one of the best lectures I have ever given.

On a subsequent occasion, when giving another lecture on analogy, I remarked, quite innocently, "Last time I gave this lecture, Richard Feynman sat *right there*", and I pointed at a seat just to the left of center in the front row. No sooner had I said this than I realized the marvelous analogical transfer I had done so totally subconsciously. After all, at Caltech it had been a gigantic auditorium (this was a small classroom); the seats were in tiers (here they were just in ordinary rows); each row was very wide (here they were quite narrow); and I had been in California (now I was in Ohio). Yet pointing at one seat and saying "Feynman was sitting *there*" seemed to make eminent sense, in that context. (Isn't it equally sensible as claiming that the light bulb was invented in Dearborn, Michigan, merely because the New Jersey house that Edison did his work in has been transported to a historical park there?) Furthermore, it now occurs to me that "just to the left of center" is itself the key concept of many of the analogies in the column.

The more you look at the question of how to do "the same thing" to *pqrs,* the more possibilities you see. For instance, many people seem to like *pqst,* in which the first two letters are left alone and subsequent letters are replaced by their successors. Occasionally, people have suggested *pqtu,* a rather ingenious notion based on seeing *rs* as a single unit whose successor is the unit *tu.* Somebody pointed out that *qrst* is a possibility, based on the idea of changing all letters but *a* and *b* to their successors. And one time someone sug-jested *dddd,* whose justification resides in the even more village-idiotic notion of changing all letters but *a* and *b* to *d*!

Some answers appear almost sick. Consider *abce.* The defense of this answer is that you take as many letters at the beginning of the alphabet as are in the target, then convert the final one to its successor. You can even come up with justifications for such queer answers as *abt* and, believe it or not, *pqre.*

*   *   *

When I call some answers sick, and others healthy by implication, there is something behind the metaphor. After all, there is a very serious question that always arises about analogy, but particularly strongly in such an abstract domain as this, and that is how one can ever speak with confidence about the rightness or wrongness of something that is so clearly subjective. The way I have come to view this is in terms of the *survival value* that an analogy-making capacity confers on its possessors. After all, our brains got to be the way they are only by helping our forebears to survive better than their rivals in this unforgiving world. And analogy-making is at, or close to, the pinnacle of our mental abilities.

It seems to me that people do not generally recognize how deeply implicated the analogical capacity is in decisions that affect the course of their lives. On a global level, it is evident, once pointed out. Is the embroilment of the United States in Lebanon "another Viet Nam"? How about in El Salvador? How does the American invasion of Grenada map onto the Soviet invasion of Afghanistan, or the British invasion of the Falklands? Is the Soviet Union more like an irrational paranoid person, or someone rational who has been badly bullied recently? Does the current arms race have valid precedents in history to which it can be compared?

On a more local scale, our system of law very obviously sanctifies analogy as the ultimate justification for making a reasonable and even a wise decision. The term "precedent" is just a legalistic way of saying "well-founded analogue". Two cases that at a surface level have nothing to do with each other (a bank robbery, say, and a kidnapping) may be mapped onto each other in exquisite detail at a more abstract level, with the napped kid being the loot, for instance. Lawyers attempt to sway the jury by bringing in new ways of looking at the situation that discredit their opponents' analogies, as well as by making and buttressing their own rival analogies. (Peter Suber has written a nice article connecting Copycat and Seek-Whence analogies with legal reasoning. It is called "Analogy Exercises for Teaching Legal Reasoning" and can be gotten from him at the Philosophy Department of Earlham College in Richmond, Indiana 47374.)

In our private lives, most of our important judgments are made by conscious or unconscious analogy. Should I fight this bureaucracy or accept some annoying inconvenience? Should I buy this computer or wait for a better one to come along at the same price? Should we have children now or wait a few years? Should I retire or continue working beyond retirement age? Questions concerning what to buy, what to think of someone, whom to marry, whether to move to a new city, how to talk to someone who has suffered a calamity, and on and on—all of them are influenced in a myriad ways by prior experiences of the same general sort. And remember that even in cases where there is not any obvious analogy guiding the judgment, all the categorization of the situation is being made by a mind exposed to many thousands of words, and the purpose of words is to label situation types and thus implicitly to make use of stored analogical mappings.

As was discussed in the column, the boundary line between making creative analogies and recognizing pre-existent categories is very blurry. It is signaled when we feel a desire to pluralize a proper noun ("your Einsteins and your Mozarts") or to prefix a proper noun by a definite article ("the Podunk of Albania"). Most common words hide an enormous degree of analogical abstraction. For instance, the abstractions "female" and "male" are not nearly as simple as most people think, especially when you consider how they extend to plants. What makes Middle Eastern pita, Indian puri, French baguettes, and American Wonder all be examples of the concept "bread"? When you migrate from nouns toward verbs and prepositions, the difficulties escalate. What do all "*x*-is-on-*y*" situations have in common?

All of this points out how analogies determine the course of our lives in the *present*. But I would go much further than that. In pre-civilized days, when people (or proto-people) lived in caves and hunted bison, analogy played no less important a role. Samenesses that we have absorbed into our perception as being obvious were great insights back then. For instance, the idea that one could chart out a plan for trapping a wild beast by drawing a map on the ground must have been a fabulous advance. All that is involved, in some sense, is a change in scale—one of the most obvious of analogical transforms, yet when it was first invented, it must have been revolutionary. On the other hand, proto-humans who tried burying meat underground in an attempt to imitate squirrels' underground hoarding of acorns might thereby seriously damage their chances for survival. Some analogies help, others hinder.

Our current mechanisms for analogy-making must certainly have emerged as a consequence of natural selection. Good mechanisms were selected for, bad ones were selected against, way back when, in the old times when you and I were but monkeys and rodents scampering about in tree branches ('member?). The point, then, is that far more than being just a matter of taste, *variations in analogy-making skill can spell the difference between life and death.* That's why "right answer" means something even for analogies; it's why analogies are only to *some* degree a matter of taste.

*   *   *

This finally gets us back to the rivalry among answers in the *pqrs* case. The domain does admittedly appear abstract and of course it is totally decoupled from the cruel world. People who prefer *dddd* are not suddenly going to get swallowed by a tiger or topple off a cliff. But people who genuinely believe in *dddd*'s superiority over *pqrt* will still have a rough time in life, because they lack the means to size up a situation and catch its essence in their mind's mesh, letting the trivial pass through. Something about their analogy-making mechanisms is defective.

To be sure, there is room for argument about answers in this mini-world, just as there is in a courtroom. But just as a lawyer who suggested that killing a human being is analogous to breaking a window because both are nasty or because both can be done with a brick would lose the case in a snap, so anyone who prefers *dddd* or *abt* to *pqrt* can be safely assumed to be totally off base. There are absurd answers, there are good answers, and there are in-between ones, just as there are degrees of edibility of food. Some foods lead to no survival, some to bare survival, and others to comfortable survival; the same is true of analogies.

One can liken the various levels of quality to the concentric circles surrounding a bull's-eye on a target (see Figure 24-5). In the middle are the totally edible foods (or insightful analogies); further out are semi-edible substances, such as grass, hay, or ants (weak analogies) and worse, leather or wood (dubious analogies); and then way out are the completely inedible things, such as nails, shards of glass, or Anglican cathedrals (these correspond to analogies that lead to disaster, such as forming a higher-level category that lumps tigers together with zebras simply because both have stripes). In the very center, to be sure, one can argue about taste and it is indubitably good for the human race that there are people who see analogies differently in that region, but you cannot get too far-fetched.

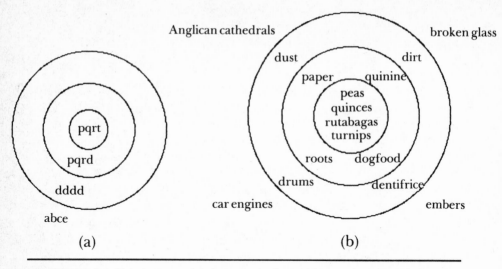

FIGURE 24–5.   *Targets representing the survival values of various actions taken by (a) a mind seeking answers to the analogy "If* abc *changes to* abd, *what does* pqrs *change to?", and (b) a mind choosing things for its body to ingest. To be sure, there is room for debate in the bull's-eye region (hard to say whether having a fried egg sandwich or a plate of spaghetti is better for you), but as you edge further out, it gets less and less debatable. Fried dust just doesn't match up to pea soup! Similarly for analogies. Among the various answers to any analogy problem, some will be decidedly weaker than others, even if you find that no one answer emerges as the clear victor.*

There is a radius beyond which analogies will be very likely to bring bad consequences to their proposers, at least if they are acted upon.

It is for this kind of reason that I unbudgeably believe that there are *better* and *worse* answers to analogies, whether in life or in the Copycat domain. Elegance is more than just a frill in life; it is one of the driving criteria behind survival. Elegance is just another way of talking about getting at the essence of situations. If you don't trust the word "elegance" in this context, then you may substitute "compactness", "efficiency", or "generality"—in short, *survivability.* Insight into the mechanisms behind this sense of elegance is the goal of the Copycat project. And personally, I would not shy away from equating elegance with wit. I would venture that one cannot be a successful Copycat without a sense of humor.

*   *   *

Having seen numerous wild and woolly (and sometimes witty) answers to the question "What happens to *pqrs* ?", you might now wish to see some alternate targets. They are extremely important, because they bring out a variety of new ways of perceiving the original *"abc goes to abd"* change. Here are a few fascinating challenges that I urge you, once again, to actually devote some time to considering. All of them are based upon our old stock event, *abc* goes to *abd.*

1. What does *cab* go to?
2. What does *cba* go to?
3. What does *pct* go to?
4. What does *pxqxrx* go to?
5. What does *aabbcc* go to?
6. What does *aaabbbcck* go to?
7. What does *srqp* go to?
8. What does *spsqsrss* go to?
9. What does *abcdeabcdabc* go to?
10. What does *bcdacdabd* go to?
11. What does *ace* go to?
12. What does *xyz* go to?

Each one of these questions shakes some fundamental assumptions about how you should perceive the original change. Does it necessarily affect just one letter? Need the object affected be at the righthand extremity? Does the changed piece always get replaced by its successor? In short, what should be taken *literally,* and what *slipperily* ? Although I would love to do so, I am not going to discuss all twelve examples here, for that would take a good long time. Each one merits at least a page on its own. (A set of "answers" is given at the end of this *P.S.*) I will discuss just one of them, number 12, in some depth. It has real beauty and raises all the central issues, so I hope you will give it some thought before reading on.

May I go on now? All right. Many people are inclined to say that *xyz* should go to *xya.* But who said the alphabet was circular? To make that leap, you almost need to have had prior experience with circularity in some form, which we all have. For instance: The hours of a clock form a closed cycle, as do the days of the week, the months of the year, the cards in a suit, the digits 0 – 9, and so on. But not all linear orders are cyclic. The bottom rung on a ladder is not above the top rung! The Empire State Building's top floor is not the same as its basement! It is a premise of the Copycat world that *z* has no successor. Sure, a machine could *posit* that *a* is the successor to *z,* but to do so would be an act of far greater creativity than it would be for you, because you have all these prior experiences with wraparound structures. The Copycat program does not. Therefore, let us consider *xya* as admirable but simply too daring, and look for something more humble yet no less apt. What else remains? Again, I urge you to think about this before reading on. This is the crucial point where there simply is no substitute for your own experimentation.

*   *   *

Okay. You've thought it over. You've got an answer, perhaps even a few, ranked more or less according to the pleasure they give you. Great! Some people suggest *xy,* pure and simple. Since *z* has no successor, they just let the third term drop away, as if it had fallen off the edge of the world. Some think, "Why not just leave *z* alone, producing *xyz* ?" Some say, "Since the rightmost letter has no successor, why not slip over to the next-to-rightmost one and take *its* successor, thus producing *xzz* ?"

Those are all right, but far more insightful answers are possible. To find them, one can let the unexpected "snag" (namely, the problem of trying to take the successor of a successor-less object) trigger a search for something crucial possibly overlooked earlier. What people tend to see at this stage is the potential correspondence of *a* and *z*—the two extreme letters of the alphabet, our two twin "monarchs". If *z* is the *a* of *xyz*, then what is the *c* ? Quite obviously, it is *x*. Now the question arises: "What to do to *x* ?" Should we take its successor, thereby producing *yyz* ? To my mind, there is something almost repulsively rigid about this suggestion. After all, the very fabric out of which *abc* was constructed has now been reversed in our new way of looking at *xyz*. Leftward motion has seized the role of rightward motion, and with it, predecessorship has taken on the role that successorship played in *abc*. Therefore elegance, in the form of a *drive toward abstract symmetry,* very strongly pushes for the answer *wyz*. Now that's a beautiful answer, in my estimation.

There is one other answer that I have encountered fairly often, and that I admire and decry at one and the same time. That is *wxz*. To be sure, it has the same inner structure as does *abd*: a jump of size one followed by a jump of size two. That much is good, but there is something peculiar about *wxz* nonetheless. To illustrate my ambivalence toward this answer, I will relate a micro-allegory.

Arphabelle Snerxis built a lovely house, ultra-modern in every respect. Then one day she capriciously removed her snazzy, sleek, new doorknob and replaced it by a most conspicuous creaky, rusty, old doorknob. Now Zulips Twankler, a great admirer of Arphabelle Snerxis, happened to have built a lovely house, old-fashioned in every respect; and when he saw Arphabelle's action, he determined to do "the same thing" to *his* house. And how did he do that? You might guess that Zulips Twankler removed his creaky, rusty, old doorknob and replaced it by a most conspicuous snazzy, sleek, new one! But no, Zulips left his creaky, rusty, old doorknob intact and instead he tore down the rest of his lovely house, old-fashioned in every respect. Then Zulips built another quite different but also lovely house, ultra-modern in every respect—except for the creaky, rusty, old doorknob. And *that's* how Zulips Twankler did "the same thing" to his house as Arphabelle Snerxis did to her house (except that when he'd finished, it wasn't *his house* any more—it was a different house altogether).

In this "analogory", Zulips let the identity of his house slip in a manner determined by its doorknob's properties, while for most people it would seem more natural to have the slippabilities reversed. There is a parallel in the fight between answers *wxz* and *wyz*. The former sees preservation of the literal intervals (1, then 2) as necessary at all costs, and it allows the bulk

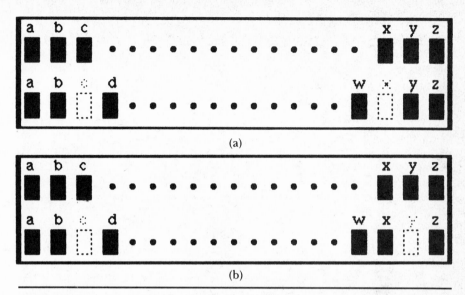

FIGURE 24–6. *A visual comparison of two good answers to the question "If* abc *changes to* abd, *what does* xyz *change to?" In (a),* wyz *is depicted. The total symmetry of this answer is its virtue. In (b),* wxz *is depicted. Its virtue is that its spacing imitates the spacing of* abd *literally, even slavishly. The question is, which of these answers creates a better overall analogy? Is it wise or is it a cop-out to intone* "De gustibus non est disputandum" *and leave it at that?*

of the original entity *xyz* to be shifted in order to achieve those aims. The latter sees only one small role (analogous to the doorknob) as singled out for modification, and keeps the bulk intact. The contrast is vividly portrayed in Figure 24-6. Another way to see the contrast is this: *wyz* is based on a better imitation of the *change* from *abc* to *abd*, while *wxz* is a better imitation of the *end product, abd.* Which do you find the more satisfying or elegant action? For me it is *wyz*, hands down. Still, I find a strange charm in the Zulipian answer. This is one of those cases where two different answers (three, if you count *xya*, that deeply creative answer that comes so easily to us old circularity hands) can lie inside the innermost "delicious and nutritious" circle, within which *de gustibus non disputandum est.*

There are a host of other "possible" answers to this problem, but many of them lie in the risky outer circles and are dangerous to their proposers. (Or, to speak more precisely, the *mental mechanisms* that would allow them to be considered seriously are dangerous to their owners, since those same mechanisms could produce very untrustworthy suggestions for courses of action in cases that are *not* decoupled from consequences.) Some of these answers are so far-fetched that they are actually quite humorous. In fact, some of them are so outlandish that I would claim no conceivable survivor of evolution's harsh pruning would ever come up with them, unless deliberately for humorous purposes. Let us look.

First of all, not all that funny but very much in the Feynman "village idiot" spirit, is *xyd.* Just plain old dullsville. What can you say to such an answer? Certainly, it has more merit than *pqrd* did earlier, for here, at least, there is an excuse for such rigidity: *z* is a trouble spot, whereas *s* was not. Another answer, definitely more pitiful, is *dyz.* This one just goes "Thud!" very loudly in my mind's ear. Anyone who would seriously suggest this answer has seen the light but then dropped the ball, mixed-metaphorically speaking. That is, they go quite a long ways in mapping *a* onto *z, c* onto *x,* but then they seemingly totally forget this level of sophistication and revert to an infantile literality: "Replace whatever plays the *c* role by *d."* It is so inept that it is funny. In fact, I would say that in the answer *dyz* to this analogy problem there is the germ of a rich theory of humor, based on the idea of level-mixing slips of this sort. Now if only I could say just exactly what I mean by "slips of this sort", I'd be in business . . .

Equally scramble-brained is *dba.* Its hypothetical serious proposer has clearly seen that, in some abstract sense, *xyz* is a "mirror image" of *abc.* An attempt is therefore made to take a mirror image of *abd,* but somewhere in the shuffle, the *type* of mirror image involved got forgotten, and for it was substituted the crudest possible notion of "reversal". The result is another heavy-handed thud. By the way, nobody has ever seriously suggested to me either the clunky *dyz* or *dba,* or even *yyz,* interestingly enough. Oh well, I guess such people's ancestors must have all gotten gobbled up by dreaded human-eating zebras.

*   *   *

These analogies, as must be abundantly clear by now, are themselves analogous to real-world analogies. Or better yet, they are *allegories* or *fables* for analogy-land children. They capture the essence of the dilemmas that analogy-makers face over and over again. Pressures for literality vie mightily with pressures for "literarity"—that is, for high-flown abstractions that disdain rigidity of reference. In an analogy where real insight is needed, often some initial forays are made that reveal some literal-minded ways of seeing the situation, but they ring false or overly crude, and so one does not stop there. Instead, one continues searching, guided by a number of small cues, and at some unpredictable moment, something simply snaps, and a host of things fall into place in a new conceptual schematization of what is going on. What was once important becomes suddenly trivial, and a new essence emerges, an organizing concept or set of concepts that seem far superior.

But beware! You should not take any of this to imply that being literal-minded is to be avoided at all costs. If jumping to rarefied levels of abstraction were always preferable, then we would wind up never making any distinctions between situations. Every situation's optimal description would be: "Something happens." It would be better to say "woman" than "Mary", "person" than "woman", "animate object" than "person", "physical entity" than "animate object", and ultimately, just "thing" would emerge triumphant. (Actually, even that choice would be nonexistent, for those different words would not exist, as they would merely make petty distinctions that vanish at enlightened higher levels.) Rigidity comes in many forms, and a rigid drive toward abstraction is no less stupid than a rigid refusal to abstract. The clearest and cleanest statement of the problem that analogy poses is that there are always fights between forces pushing for literality (in its many forms) and forces pushing for abstraction (in its many forms). How, in specific circumstances, those forces compete and interact and in the end come up with some sort of optimal compromise is the problem. And if you go away from this chapter with one thought, please let it be this: There is no fixed mathematical recipe for reconciling all the different forces pushing and pulling you in analogies.

*   *   *

In the Copycat world, we have remarkably fine control over how pressures interact, and over the strengths of various rival answers. For instance, we can "tune" a given analogy by varying knobs on it, until gradually we home in on the most refined possible version of it. We can also take two possible answers and make each one seem preferable by very subtly "tweaking" the analogy itself. It is a a highly pleasing esthetic exercise to seek the perfect balance point where an analogy is teetering on the brink and can tip either way, so that about half the people to whom we give it see it one way and half see it the other way.

A lovely example of this idea is provided by a very simple analogy with a knob that twiddles the part-*vs.*-role proportion perceived by a typical person. This example may help clarify that distinction a little, as well. Consider the following three variations on a familiar theme:

> 1. If *abc* goes to *abd,* what does *pqrs* go to?
> 2. If *abc* goes to *abe,* what does *pqrs* go to?
> 3. If *abc* goes to *abf,* what does *pqrs* go to?

Line 1 is familiar by now. There, nearly everyone instantly proposes *pqrt*; very few people think of, let alone prefer, *pqrd.* The reason is that the *c-d* conversion seems so obviously to be a "leap" from one letter to its successor that we ourselves leap to that conclusion. But consider line 3. Here we are confronted by the much larger *c-f* gap, too large for us (or the Copycat program) to have any intuition for. It seems so arbitrary that instead of seeking any "whence" behind it, we just accept it at face value, saying to ourselves, "Okay, replace the rightmost letter by *f,* eh?" And indeed, most people are happy with the answer *pqrf.*

To some, this answer may seem overly Feynman-like, but I must reiterate that in the Copycat world, there is no simple connection between distant letters. You can't just "subtract" c from f the way you can subtract 3 from 6. Subtraction is an unknown concept, just as in Seek-Whence. The connection between c and f, to the extent there is one, is a *conceptual* one rather than a mathematical one: f is the successor of the successor of the successor of c, and that is just too topheavy a notion to have much charisma.

All right, if line 1's *c-d* leap has charisma and line 3's *c-f* leap lacks it, what about the intermediate case of line 2? Here we are poised between two analogies that push us in opposite directions. If we were to follow line 1's example, we would see the *c-e* leap *intensionally,* a fancy way of saying that we would see e as a *role-filler* (as the successor of the successor of c—a bit gawky but still plausible). But if we were to follow line 3's example, we would see the same leap *extensionally,* meaning that we would see e as a mere *part* of the event, filling no role other than being itself (which it could hardly help doing). This is not gawky, but it is so literal that it provides no insight into why the given change occurred. So which do you prefer—the intensional view of e as gawky role-filler, or the extensional view of e as arbitrary part? The former leads you to answer *pqru,* the latter to answer *pqre.*

Although line 2 may not be your personal balance point, there is surely a point at which you will switch over from one view to the other. It would seem highly compulsive if, given the question

abc goes to *abv;* what does *pqrs* go to?

somebody insisted that the v must somehow "come from" the c, and tried to force a vision of some connection when there really is none. Furthermore, even violating the spirit of Copycat and seeing v as the order-19 successor of c is not of much help, for what is the order-19 successor of s? It gets us right back to successor-of-z problems, very messy territory.

Seeking the balance point of analogies is an esthetic exercise closely related to the esthetically pleasing activity of doing ambigrams, where shapes must be concocted that are poised exactly at the midpoint between two interpretations (see Figures 13-6 and 13-7). But seeking the balance point is far more than just esthetic play; it probes the very core of how people perceive abstractions, and it does so without their even knowing it. It is a crucial aspect of Copycat research.

*   *   *

A few more choice problems are given below for would-be copycats. I do not have the space-time to discuss them all here; I propose them simply because each one has a chance of affording you a small but thrilling moment of blinding insight, if you look at it in just the right way. Our view of the best answers is given in the *Post Post Scriptum.*

1. If *aqc* goes to *abc,* what does *pqc* go to?
2a. If *efgh* goes to *fghi,* what does *mvr* go to?
2b. If *efgh* goes to *fghi,* what does *uuuuu* go to?
3. If *beq* goes to *bqe,* what does *abcdefpqr* go to?
4. If *xyzabc* goes to *xyzqabc,* what does *abcxyz* go to?
5. If *aaqqkkkk* goes to *zaazqqzkkzkkz,* what does *abcdefstu* go to?
6. If *eeeffghhiii* goes to *eeeefffgghhhiiii,* what does *eefhii* go to?
7a. If *eqe* goes to *qeq,* what does *abcdcba* go to?
7b. If *eqe* goes to *qeq,* what does *aaabccc* go to?
7c. If *eqe* goes to *qeq,* what does *eqg* go to?

It must be emphasized that the selection of Copycat problems presented here is but a tiny fraction of all those that I, together with David Rogers, a postdoctoral fellow working on the Copycat project, have come up with. This selection is biased toward analogies that have spice and tang, as opposed to bland ones such as "If *bbb* goes to *bbbb,* what does *eee* go to?" There are of course innumerable ones of this boring sort, ones that have obvious answers and that present no serious challenges to adult humans. Now, we do not in the least disdain such analogies in the project. Indeed, it is a tremendous challenge to make a program that could handle these seemingly easy cases reliably. They are amazingly deceptive in their subtlety. But it is not of much interest to people to go down a long list of (not actually but seemingly) trivial analogies, so that explains the censorship in my choices for you.

Still, it must be admitted, analogies that seem to require a deep perceptual shift after an initially unsatisfactory first stab are the ones that beguile us, for they seem to promise insight into that mystery of mysteries: insight. I must admit to the belief, or at least the strong intuition, that all the depth of scientific discovery, even the profoundest discovery, is wrapped up in the mechanisms for solving these simple problems in which conflicting pressures push around one's percepts and concepts, letting things bounce against each other until, all at once, something falls into place and then, presto! A sense of certainty crystallizes, so powerful that you *know* you have found the right way to look at things. I firmly believe, in short, that "mini-breakthroughs" and "maxi-breakthroughs" have precisely the same texture. That's the faith underlying Copycat.

It may seem arrogant or blasphemous to compare the trivial alphabetic insights of a copycat with the genius of an Einstein discovering special relativity, yet I do not think the comparison is all that silly. What characterized Einstein's unique view of space and time was that he had decided that certain things were more unslippable than others: in particular, that the speed of light was unslippable but the notion of absolute simultaneity of events separated in space was slippable. To be perhaps more accurate, Einstein didn't *decide* that simultaneity was slippable, but was *forced* into that conclusion, since his stronger intuitive belief in the invariance of the speed of light simply compelled him to accept its consequences, strange and counterintuitive though they might be. (Note that counterintuitive consequences can flow from intuitive grounding.) Einstein did not begin with the idea of simultaneity being nonabsolute, but when he had to confront that possibility, he let it slip. This fluidity of mind, guided by a certainty about the deepest, most unslippable concepts, gave rise to the creative insights of special relativity.

There is an old song whose lyrics go this way:

When an irresistible force such as you,
Meets an old immovable object like me,
You can bet, as sure as you live,
Something's gotta give, something's gotta give, something's gotta give!

Yes, something's gotta give, but what? A reliable nose for what might slip and what ought not marks the difference between a great mind and a small one. If the Copycat research can unearth the basis for judgments exhibiting creative, artistic slippability even in our tiny domain, we will be ecstatic, for in our opinion, that would put us well on the road to understanding where full-scale artistic creativity comes from. Now *that* may sound arrogant, but firstly, we are not expecting it to happen just around the corner, and secondly, it is just an expression of our faith that we have not lost the essence of the larger problem in boiling it down this far. If Newton saw whirling planets in falling apples, why can we not see great leaps in small slips?

* * *

One can look to literature as well as science to find cases where finding the right things to slip yields highly creative solutions to hard problems. One example I gave in the *Post Scriptum* to Chapter 1 was the problem of translating into French (or the foreign language of your choice) the title of the book *All the President's Men*. A word-for-word translation would be as dull as dishwater, as flat as old ginger ale whose carbonated kick has long since evaporated. In order to keep the title alive in French, you must seek out a line well known to readers of French that carries the same subliminal imagery. Need it be a line in the canonical translation of "Humpty Dumpty"? Need it even be a line from *Mother Goose*? Of course not. The essence of the situation does not reside in those particulars. So slip, baby, slip! But how?

The crux of the matter is to find a line alluding to a famous irreversible downfall. If it is a line from Pascal's *Pensées,* so be it. If it is from a popular song of recent years, so be it. You may have to go further afield to find an appropriate line. There may be no line of the sort in the popular French-speaking consciousness, in which case more radical solutions must be sought. There is no clean, clear recipe guaranteed to work. By the way, I do not have any idea if that book has ever been translated into other languages, and if so, what solutions its translators found. But this type of problem is absolutely standard, since these days particularly, book titles of that style, making an oblique allusion to some well-known phrase, are a dime a dozen.

I must admit to some twinges of shame for having leapt aboard the title-as-pun bandwagon, when Daniel Dennett and I chose the title *The Mind's I* for our anthology. Good but non-native speakers of English usually are confused by this title, and often read the final "I" as the roman numeral for "one", which makes absolutely no sense, yet it is the best they can do, being unfamiliar with the idiom "the mind's eye".

Just to give a hint of how creative solutions can be found for such titles, I'll give one possible French translation for our title (even though it is not yet certain whether the book will ever be translated into French). Jacqueline Henry, one of the two translators into French of *Gödel, Escher, Bach,* came up with *Vues de l'esprit*—literally, "Views of Spirit", which clearly gets across one main purpose of the book: to focus on the nature of mind from many angles. But at the same time, it has a more idiomatic meaning, namely: grandiose dreams such as are dreamt by visionaries (and lunatics)—in short, *visions* or possibly even *hallucinations.* This too has its own kind of appropriateness, since a basic theme of the book is that much of the mystery surrounding mind, spirit, and soul is caused by a kind of hallucination: the hallucination that there is some *thing* called "I". Therefore, the French double meaning is elegant and, though it does not replicate in French the exact effect of the English *double entendre,* it is effective and thought-provoking. What more could you ask?

Incidentally, if I were writing this *P.S.* in French, I would of course talk about books in French, not ones in English, whose titles are hard to translate. Thus a proper translation of this very passage would involve a good deal of literarity. In fact, I have one particular book title in French in mind: *Le corps a ses raisons*—a book about health and physical fitness, which actually came out in English under the feeble title *The Body Has Its Reasons.* Can you do better? Hint: You need to be familiar with a famous saying by Pascal, namely, *Le cœur a ses raisons que la raison ne connaît point.*

*   *   *

In a certain way, translation is the quintessential form of analogy. You are given a fixed overarching framework—the home language and culture—and within it, a novel structure has been erected—a book title or sentence, for instance. Your task as translator is to replicate, as best you can, the overall "feel" of that small structure, but in a different overarching fixed framework —the target language and culture. This description obviously recalls the allegory of Arphabelle and Zulips, where the frameworks are their respective houses, and it applies equally clearly to the "Do this!" examples.

My own mental image that best gets at the nature of translation involves

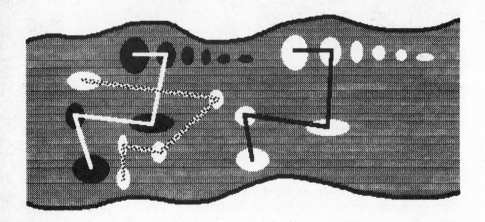

FIGURE 24–7. *A metaphor for translation. A stream (symbolizing reality) has two sets of stepping-stones (symbolizing the basic ingredients of a language, such as words and stock phrases) in it. The black stones (Burmese, say) are arranged in one way, and the white stones (say, Welsh) in some other way. A pathway linking up a few black stones (a thought expressed in Burmese) is to be imitated by a "similar" pathway joining up white stones (translated into Welsh). One possibility is the speckled pathway, located at nearly the same part of the stream as the original pathway but not terribly similar in shape to it (a fairly literal translation), while a rival candidate (a more literary translation, needless to say) is the pathway located a distance upstream and resembling the original in some more abstract ways, including patterns in some of the "overstones" of the main stones (the similar archipelagos in Burmese and Welsh stones running roughly parallel to the far bank).*

picturing each language as a fixed set of stepping-stones in a stream (see Figure 24-7). Suppose you are translating from Burmese to Welsh. A Burmese utterance is a pathway from one place to another via the black stones. They seem to be located in convenient enough places, and you can get pretty much wherever you want to go. But when it comes to translating what you have said into Welsh, you find that the Welsh stepping-stones— the white ones—are often not quite in the same places as the Burmese ones, and even in the cases where they are just about in the same places, they are shaped differently, and so you can't treat them as identical to the Burmese stones you are familiar with. You must tread with great care, and sometimes you will find that there are gaps in one language's set of stones that don't exist in the other, so that some routes are easier to mimic than others. The most literal translations involve sticking as close as you can to the original route, at the stone-by-stone level. Of course, no mimicking route is exactly the same as the original.

It may be, however, that the essence of a particular route lies not in *where* it starts or ends in the stream, but in its *shape*. It may be that in the particular region of the stream where the original path was traced, the Burmese stones very easily allow many shapes to be traced out but the Welsh stones happen to be sparse. At various places upstream or downstream, however, the converse is true. If you believe that the essence of the idea resides more in its shape than in its absolute location relative to the stream bed, then you won't mind moving upstream or downstream a bit, in order to gain that flexibility. Less metaphorically speaking, this means that sometimes the overt topic of a passage can slip as long as something more central—style, perhaps, or metaphorical allusion—is preserved.

A critical idea is the following one: The longer a passage is, the less the graininess of the underlying medium is going to be noticed. In purely geometric terms, what I am saying is this. The larger the curves of the pathway are in comparison to typical inter-stone distances, the less it matters which of the two grids of stepping-stones you are using. This can be illustrated very elegantly by thinking of trying to approximate a circle by filling in various squares on a normal (8×8) chessboard. Clearly you would make the circle as big as you can within the confines of the board, so as to round off the effects of the squareness. And if you could make the board bigger, you would. On a 100×100 chessboard, you could draw a very fine approximation to a circle, and on a $1,000,000 \times 1,000,000$ board, no one would know the difference. Furthermore, nobody would even be able to tell, in such a case, whether the underlying board was a square lattice, a hexagonal lattice, or what. But if you go back down to circles whose size is on a par with that of the lattice grain, then of course the lattice becomes very visible.

For this reason, I feel safe in suggesting that translating a novel's *title* may sometimes be the most challenging aspect of translating the whole novel. After all, the overall message of most novels is on such a vastly larger scale than the grain size of either language involved that small jogs here and there (where the idiosyncratic placing of the stepping-stones forces you to take an awkward zigzag) can be compensated for in other places, and in the larger picture such jogs will balance or cancel each other out. Recall that I said something similar about computer languages and AI programs—the grain size of the ideas in a big program is far larger than that of any conceivable computer language.

But a title is another story. A title is tiny. Its grain size is barely above that of the stepping-stones themselves. It consists of a pathway just a handful of stones long, and the challenge is great when it contains subtlety of any sort —which is the case for most titles, as it is for proverbs, epigrams, and so on. As they say in Italian, *Traduttore, traditore*—which, literally as well as literarily translated, means "Translator, traitor." In this curious case, the English version is a perfect counterexample to its own claim, but generally speaking, the Italian epigram is right on target, and pithily expresses the idea that no translation—no analogy—is perfect. Perhaps a better English translation of it would therefore be: "Transductor, treasoner."

*   *   *

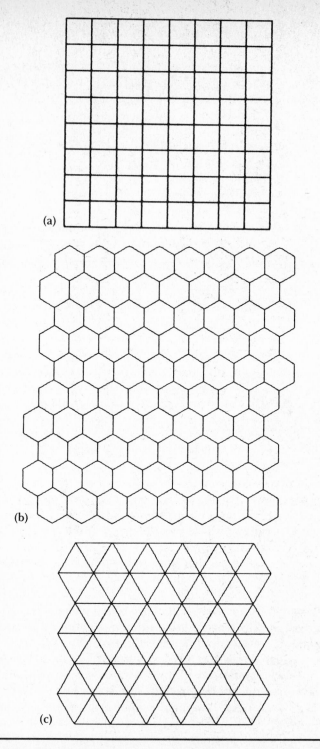

FIGURE 24–8. *Three lattices on which chess-like games could be played. In (a), a square lattice; in (b), a hexagonal lattice; in (c), a triangular lattice.*

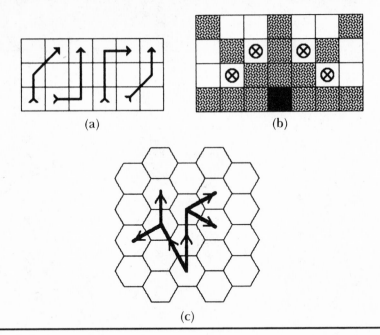

(a)   (b)

(c)

FIGURE 24–9. *Various ways to think about the knight's move. In (a), it is built out of rook-move and bishop-move primitives. In (b), it is portrayed as the closest spot not immediately accessible to rook or bishop. In (c), we make some first stabs at the knight's move on a hexagonal lattice. Are some of these possibilities more defensible than others?*

The idea of approximating a given shape (such as a circle) on a coarse-grained grid (such as a chessboard) provides a wonderful way of framing many analogy issues. But the target shape need not be subtly curvilinear for the analogy problem to be deeply challenging. If you are trying to export even a very simple shape from one grid—its "natural habitat"—into another grid, and if it does not export literally to the target grid, then something's gotta give, and that is the hallmark of a hard analogy problem.

Since we are talking about chessboards, let us use a chess example. The underlying grid of chess is a square lattice. Suppose we pick as our target grid the hexagonal or triangular lattice (see Figure 24-8), and ask, "What is the knight's move on this lattice?" We are immediately forced to confront the question, "What is the essence of the knight's move in the only case we really know?" There are a number of ways of thinking about it (see Figure 24-9). Which of the following, if any, is the most insightful characterization of the knight's move?

(1) a rook step of length two followed by a single perpendicular rook step;
(2) a single rook step followed by a perpendicular rook step of length two;

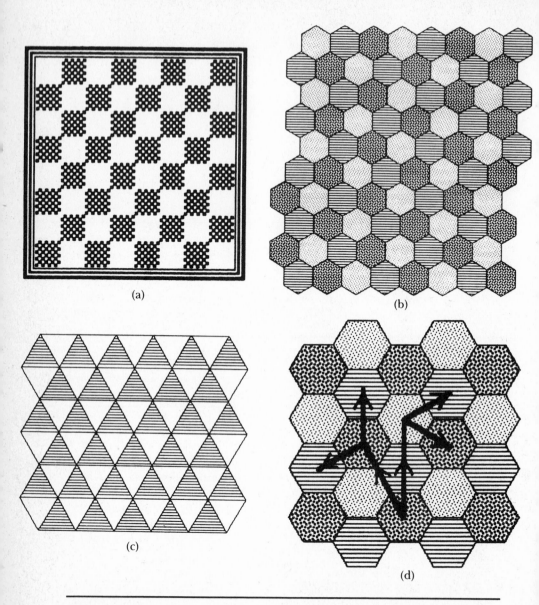

FIGURE 24–10. *Recognizing that the coloring of the board plays a significant role in defining the moves of chess pieces. In (a), the standard coloring pattern of a square lattice. In (b), the most natural coloring of a hexagonal lattice. Notice that it involves three colors. In (c), the most natural coloring of a triangular lattice. Finally, in (d), the guesses of Figure 24-9 are now shown on a colored-in hexagonal lattice. How does this affect their plausibilities?*

FIGURE 24–11. *In (a), a board for unidimensional chess, known as* chass. *Its optimal number of squares is to be determined. In (b), a wider board for* quasi-unidimensional chess. *This variant is known as* chäss *or* chæss.

(3) a rook step of length one extended by a bishop step of length one;
(4) a bishop step of length one extended by a rook step of length one;
(5) a normal pawn move followed by a pawn's move in "capture mode";
(6) the shortest move that no other piece can make.

Or are all of these simply *aspects* of the essence of the knight's move? Which aspects are more central, then? Which would you be willing to relinquish first? Which never? Are you sure? How slippable are the following aspects, all of which do apply in the square grid?

* When a knight moves, it must land on a square of a different color.
* A knight must be able to jump over or around pieces.
* A knight can make a tour of the entire chessboard, landing on every square.
* The starting and stopping squares of a knight's move should lie on opposite sides of a straight line that contains one edge of each.
* The knight's move should not resemble any other piece's move.
* All knight moves should be congruent except for rotation and reflection.
* A knight must have about the same power as a bishop and less power than a rook.

Once you have tried extending the concept of the knight's move to another lattice, then you begin to sense all the subliminal features that add up to define its highly composite identity, features that you most likely never would have thought about without this pressure. For example, coloring the lattices was a revelation for me, and turned out to be the royal road to finding elegant knight's-move solutions (see Figure 24-10).

While working on these two puzzles, I dreamt up what sounded at first like an absurd challenge: to compress chess into one dimension. In other words, take a chain of squares of length N (to be determined) and find moves for rook, bishop, knight, queen, king, and pawn (see Figure 24-11).

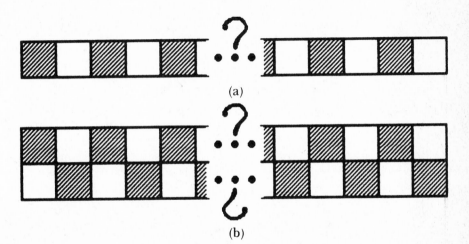

(a)

(b)

Also consider how to place the pieces on the board at the game's start, and determine N. This droll exercise gave rise to a number of very stimulating discussions among AI colleagues and chess-playing friends of mine. The sense of sheer analogical elegance had to compete with realism about what choices might make the game more interesting. Along the way, one unexpected suggestion arose: What about a board *two* squares wide, rather than just one? On that type of grid, the knight's move seemed trivially obvious—but my "obvious" solution turned out to have a fatal flaw, which we patched in a most intriguing way.

One of the more amusing spinoffs of those discussions was a quest for the name of the game (as they say). One-dimensional chess was dubbed *chass*, since 'a' is the first vowel. Two-dimensional chess retained its name, 'e' being the second vowel. (But what would seven-dimensional chess be called?) The $2 \times N$ game, delicately poised between one- and two-dimensionality, was yclept *chäss*. And what about the names for chess on the nonstandard two-dimensional lattices? Here are some solutions that fell into a delightful pattern:

> *Chesh:* chess played on a *hexagonal* lattice;
> *Chest:* chess played on a *triangular* lattice; and of course,
> *Chess:* chess played on a *square* lattice.

And please—when you are about to deliver the death blow to your opponent's king in a game of chass, don't forget to triumphantly exclaim, "Chackmate!"

*   *   *

I found these chess-extension puzzles to be beautiful not only as puzzles, but much more rewardingly, as examples of issues in analogy. When, for example, I settled on my answer for chesh, it felt like not just *an* answer, but *the* answer: *the* knight's move on a hexagonal lattice. This reminded me strongly of the feeling of absolute certainty that one gets in mathematics when one sees a familiar phenomenon recurring in a new way in an unfamiliar domain. One says, "Aha! So *this* is how Wiggler's Lemma generalizes! The twistoploppic theomorphism is the same for *even* clackdoodles, but becomes a hypertwistoploppic pseudotheomorphism for *odd* clackdoodles. That's so *beautiful*!"

Examples galore of this feeling must have arisen in the minds of the people who extended the Magic Cube concept to other polyhedra, other dimensions, other ways of slicing. And once you have made or acquired a new "cube" (such as the Skewb or IncrediBall), you will want to know how to export a known *algorithm*, broken up into its fundamental *operators*, from a familiar cube. What is the *essence* of each operator? One senses a deep invariant lying somehow "down underneath" it all, something that one can't quite verbalize but that one recognizes so clearly and unmistakably in each new example, even though that example might violate some feature one had thought necessary up to that very moment. In fact, sometimes that violation is what makes you sure you're seeing *the same thing,* because it reveals slippabilities you hadn't sensed up till that time.

No better example exists than the way mathematicians extended the concept of *exponentiation*—putting x to the y power. At first, y had to be a positive integer. Then it was realized that $x^0 = 1$ fits exactly into the pattern, so zero was allowed as an exponent. Immediately, it was seen how the same pattern would suggest—nay, *require*—that x^{-1} be the reciprocal of x. By then the generalization ball was off and rolling. Fractional powers came along very quickly: 1/2 as an exponent meant you should take the square root, 1/3 meant the cube root, and so on. Then on to real numbers. But why stop there? Various abstract representations of what it meant to exponentiate had now been formulated, allowing one to transcend earlier, primitive notions of what it meant. Pretty soon, not only could complex numbers be exponents, but so could $n \times n$ matrices, functional operators, and God knows what else! This conceptual supernova was still very much centered on one core, and blurry though the implicosphere around it might be, the vastness of this implicosphere's size only made the conceptual core stronger, firmer, realer.

Another example: There is clearly only one sensible $4 \times 4 \times 4$ Magic Cube. It is *the* answer; it simply has the *right spirit.* The same holds for the four-dimensional cube, discovered by Kamack and Keane. Similarly, Scott Kim once generalized the concept of the "impossible triangle" (a two-dimensional drawing read by three-dimensional viewers as a three-dimensional object that cannot exist) up one dimension, so that it became the "impossible skew quadrilateral" (a three-dimensional sculpture read by four-dimensional viewers as a four-dimensional object that cannot exist). Later, he found out that Roger Penrose, the inventor of the impossible triangle, had likewise "added 1" to his three-dimensional illusion and come up with exactly the same construction as Scott had—only Penrose did it fifteen years earlier. Clearly, then, this was *the* corresponding paradox, manifesting the same deep essence as the original and simpler one. Here again we see the aptness of that wonderful saying, *Plus ça change, plus c'est la même chose.*

That feeling of encountering an absolute and almost divine truth and reality behind highly abstract analogical connections is particularly prevalent in mathematics, but it can also arise in other areas of life. When a "reminding" feeling becomes so strong that you want to use the *same word,* that is when you start getting religious about your discovery. Golomb's "quarks" on the cube, for instance, seem to have some "essence of quarkness" about them. Is this *one phenomenon* manifesting itself in two different ways, or is it simply a pretty coincidence? Such questions can occasionally not be answered, but very often our minds come to conclusions on such matters without our ever noticing it. Reification of new categories in words is a telltale signal, and one of the most important of mental events.

* * *

Some people might look upon the exercise of translating the knight's move into an alien grid as an amusing but trifling game, and maintain that such things are far from real-world concerns. Actually, in recent years, problems not too different from this have become the stock-in-trade of people working on the computerization of typefaces, where the idea is to pack as much of the spirit of a typeface (such as Helvetica) into the smallest possible number of "pixels" (on-off dots, usually arranged in a square lattice, though that is not necessary). Can one make an 'a' that is recognizably a *Helvetica* 'a', using just 35 pixels arranged in a 5×7 array? This is certainly beyond feasibility. But how few can you get away with? When does at least a hint of "Helveticality" start to appear? (See Figure 24-12.) And just what is this "Helveticality" spirit that is so elusive? How much harder to capture than "essence of knight's move" is it?

Attempting to compress a visual form into smaller and smaller arrays of pixels forces one to confront ever more deeply the question of its essence. What can one afford to release, and what must be held onto? An analogous *aural* analogy problem is very obvious to state, yet seldom explored: Can one translate a complex piece of music from major into minor, or vice versa? Musicians will immediately recognize that the major and minor scales here play the roles of underlying *lattices*, so that we are undeniably dealing with a lattice-conversion problem. Mechanical methods will carry you a certain distance, to be sure, but for any complex piece there will always remain a lot of sticky and idiosyncratic knots. For instance, what if a piece in a major key turns minor for a short stretch? Should its minor-key "translation" turn major at the corresponding point? This example is just the tip of the iceberg in the major-minor translation game. To get into the right spirit, you might try humming to yourself such old favorites as the popular song "Awful Days Are Here Again" (traditionally sung by mournful Democrats right after they lose an election) and Frédéric Pichon's celebrated Baptismal March (from his piano sonata in B-flat major) . . .

Another musical analogy problem arises when one tries to arrange a piece of music for a new instrument or group. Can George Gershwin's very pianistic preludes for piano be adapted for guitar, for example? Could one convert the wonderful Mendelssohn violin concerto into a piano concerto? Each instrument forms a kind of grid, and inter-grid transfer of essence is the problem.

From vision and hearing, we now move to a more conceptual domain: pieces of writing. The task of compressing a piece of text one has written into fewer and fewer words forces one to struggle to define the essence of what one is trying to get across. Up to a point, a piece of text may actually be improved by having some fat trimmed here and there, just as a university or government agency can undoubtedly benefit now and then from a severe budget crunch—but this can be carried too far, and meaning will certainly begin to suffer. A fascinating exercise is to try to pack a page of one's writing into half a page, then into a quarter page, and so on, until one has gone

FIGURE 24–12.   *Helveticality emerging from the gloom. Proceeding from bottom to top, we have a series of increasingly fine-grained dot matrices within which to maneuver. Clearly, both the 'a'-ness and the Helveticality get easier and easier to recognize as you ascend—especially if you look at the page from a few feet away. Proceeding from left to right, we have a series of increasingly letter-savvy programs doing the choosing of the pixels to light up. (As a matter of fact, the rightmost column is a very light touch-up job of the third column, done by a human.)*

The leftmost column is done by a totally letter-naïve program. It takes the curvilinear outline of the target shape and turns on all pixels whose centers fall within that outline.

The second and third columns are the work of an algorithm that has information about zones likely to be characteristic and critical for recognizability. It mathematically transforms the original outline so that the critical zones are disproportionately enlarged (the way your nose is enlarged when you look at yourself in a spoon). It then applies the naïve algorithm to this new outline (pixels light up if and only if they fall inside). This amounts to an interesting trade-off: sensitivity in the critical zones is enhanced at the sacrifice of sensitivity in less critical zones. Consequently, some pixels are turned on that do not fall inside the letter's true *outline, while some that do fall inside that outline remain off. It's a gamble that usually pays off, but not always, as you can see by comparing the first and second letters in, say, the third row.*

The difference between the second and third columns is that in the second column, the critical zones are crude averages fed to the program and don't even depend on the letter involved. In the third column, however, the program inspects the curvilinear shape and determines the zones itself according to its knowledge of standard letter features such as crossbars, bowls, posts, and so on. Then it uses these carefully worked-out zones just the way the second algorithm uses its cruder zones: by distorting the true outline to emphasize those zones, and then applying the naïve algorithm to the new outline.

But no matter how smart a program you are, the problem gets harder and harder as you descend towards typographical hell: matrices too coarse to capture essential distinctions. En route *to hell, more and more sacrifices are made. Helveticality goes overboard first, then 'a'-ness; and from then on, entropy reigns supreme. But just before that point is the ultimate challenge—and only people can handle it, so far. [Computer graphics by Phill Apley and Rick Bryan.]*

down to a phrase of only a few words. This can be seen as both a translation problem and an analogy problem. It is not usually considered either one, but just think: One is trying to "say this" in an ever sparser and tighter language, an ever more severely constricted grid.

In a similar vein, learning to write in the language called "Nonsexist" is a great exercise in translation and analogy, as is trying to become fluent in 'e'-less English, referred to earlier. Both provide you with a somewhat modified set of stepping-stones, and force you to invent and then get accustomed to many new types of constructs in order to say things that are easily said in the more prevalent mode of speaking. It is very hard to become totally fluent in either language.

*　　*　　*

A significant problem these days, related to that of capturing "Helveticality" in a low-resolution grid, is that of producing original and esthetically pleasing low-resolution typefaces—in other words, instead of imitating a known curvilinear typeface, inventing a new typeface whose natural habitat is, say, a 5×7 or 10×12 grid, all of whose letters are in "the same style" within that tiny world. Many human designers have discovered solutions of great ingenuity, but machine designers? There are none.

Letter Spirit, an AI project of mine currently on the back burner (it is impatiently waiting for Copycat to come to a boil), has as its aim to produce a program that can do just that: Given one or two low-resolution letters as inspiration, complete the alphabet in "the same style"—or rather, the same *spirit.* Instead of using pixels (points) as the primitive components of letters, however, I chose to use short straight-line segments on a fixed grid containing just vertical, horizontal, and 45-degree diagonal segments. I call those primitive segments "quanta". Figure 24-13 shows the tiny grid permitted, and the stunning variety of 'a's that one can realize within it. Actually, I estimate there are well over a thousand ways to realize grid-bound designs possessing some degree of 'a'-ness; some will definitely hit the bull's-eye while others will clearly be way out on the fringes of the 'a'-sphere. Then of course there will be many shapes that hover simultaneously near the fringes of two or more Platonic letters' spheres of influence. Such shapes are anathema to the human visual system, which greatly desires unambiguous category membership; they should be likewise antithetical to the Letter Spirit program.

The Letter Spirit grid, although seemingly a constraint, actually inspires flights of fancy that total freedom would not (a fundamental and general lesson about the deep connection between constraints and creativity). Figure 24-14 gives a sampling of a few "gridfonts" inspired by various stylistic quirks in one letter or another. Once again, this is only the tip of the iceberg. There are thousands of intriguing gridfonts to be designed and savored. As of this writing, I have designed about 150 of them. You could

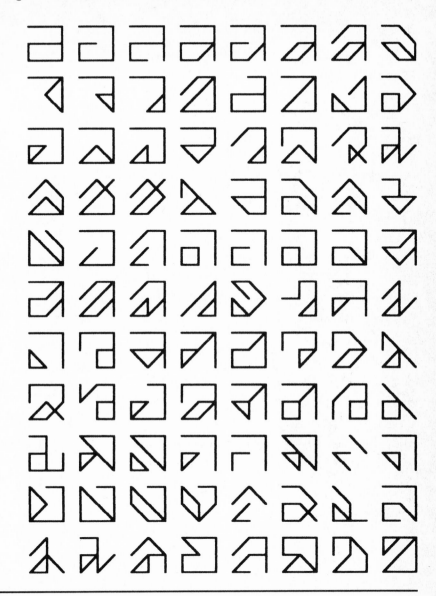

FIGURE 24–13. *87 'a's composed of horizontal, vertical, and 45-degree "quanta" in the Letter Spirit world. How many more shapes recognizable as 'a's do you suppose lurk in the given grid? (Compare this figure with Figures 12-2, 12-3, and 12-4.)*

say I'm addicted! Seven complete gridfonts by me are exhibited in this book, below the introductory paragraphs to the seven sections.

The Letter Spirit project was distilled from a far more ambitious dream: that of producing a program able to create genuinely artistic, curvilinear, full-fledged typefaces when inspired by one or more sample letterforms. I

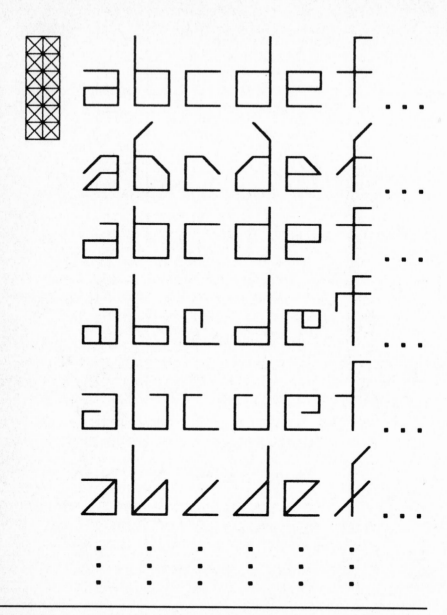

FIGURE 24-14. *The horizontal and vertical problems as they arise in the Letter Spirit world. (Compare this figure with Figure 13-8.) The central problem of the Letter Spirit project is to characterize what it is that items with "the same spirit" (i.e., in the same row) have in common with each other, so that in general, given a sample letter or two, the program can "get the hang of it" and then go ahead and design all the remaining letters in the same spirit, thus creating an esthetically pleasing "gridfont". Readers are encouraged to try their hand at completing the six gridfonts whose beginnings are shown here, and inventing their own.*

wrote "far more ambitious", yet in a way that's not right. After all, during each boiling-down step (and there were quite a few between the initial conception of the project and the final arrival at the grid), I assured myself that it truly preserved the *essence* of the full typeface problem and merely eliminated some superficial aspect of it. So in some sense I do believe that Letter Spirit actually encapsulates *the* central problem not only of typeface design, but indeed of art and creativity in general. And I am prepared to stand behind that claim, as long as you give me some grains of salt to defend myself with.

I recently had a fascinating visit at Bitstream, a Cambridge, Massachusetts firm specializing in the digitization of human-designed typefaces. A typical task they must do over and over again is to take a given font and adapt it from one high-resolution lattice to another (for example, from 200×200 pixels per letter to merely 100×100). For this, they have specialized graphics hardware that works just fine. This grid-to-grid conversion is an easily mechanizable analogy, or translation, task. However, when they want to take a given font and adapt it from a high-resolution lattice to a *medium-resolution* one (say, down to 15×15), their graphics machine produces an unacceptably crude solution, filled with ragged edges and spurious pixels of all sorts. To improve on this, Bitstream purchased an expensive Lisp machine and developed a complex program for this purpose. Some of the results of that work are shown in Figure 24-12. The point is, severe compression requires far more brute hardware and sophisticated software than does gentle compression. Finally, when they need to compress a font from a high-resolution lattice down into a truly coarse-grained one (say, 10×10), they turn the task over to human designers, because people alone can handle the many interacting perceptual forces that emerge at this level of resolution.

At first, this may sound counterintuitive, but really it makes perfect sense. With high-resolution grids, the graininess of the underlying medium all but disappears, and it is child's play to convert from one grid to another. It wouldn't even matter if the target grid were hexagonal, as long as it were sufficiently fine-grained. But compression down into very coarse grids forces one to deal with the conceptual and perceptual essence of visual forms—and essence, if anything, is the central problem of analogy. In fact, a sense of essence, in essence, is, in a sense, the essence of sense, in effect.

*   *   *

Any analogy can be viewed as an attempt to reproduce in one metaphorical grid a form that exists in another metaphorical grid. The more coarse-grained the two "grids" are, the more ingenious the analogy-maker has to be to perform the mapping. Roles and substructures must be extracted and weighed and mapped against each other. Shifts of all sorts, up and down in abstraction as well as sideways in conceptual similarity, must be able to take place. The analogy-maker attempts to judge proposed solutions for their elegance, but in the end, only their performance in the world determines their success.

The Copycat domain might appear less charming a domain than the nose-touching and chess domains, less grabbing than the Letter Spirit domain. But that is a superficial viewpoint. To make progress in science, one has to make sure that the phenomena under study are truly isolated. I am banking on having carried out the job of isolation very well, and now comes the stage of making the model. That project is ongoing, and its method of attack—its vision of how to build a system that would run on a real machine—is an esoteric and complex one. To relate that would be another very long story. It is the domain itself that has been the subject of discussion here.

I feel confident that this tiny alphabetic world allows all the key features of analogy to make their appearances. In fact I would go further and claim: Not only does the Copycat domain allow all the central features of analogy to emerge, but they emerge in a more crystal-clear way than in any other domain I've yet come across, precisely *because* of its stripped-down-ness. Paradoxically, Copycat's conceptual richness and beauty emanate directly from its apparent impoverishedness, just as the richness of the "ideal gas" metaphor emanates from its absolute simplicity. Time will tell if this limb I am out on is solid.

Post Post Scriptum.

In retrospect, it seems that this *P.S.* probably ought to have been a chapter on its own. I did not dream that it would grow to this size; I merely wanted to let my readers know what sorts of issues I am working on currently —and I discovered that sketching that out takes a good deal of time. The original column was disappointingly coolly received. I hope that this more complete explanation of the driving forces behind my research projects will awaken more enthusiasm.

Below I give our "answers" to the analogy problems given in the *P.S.* Each problem merits a much longer discussion, but life is short.

Page 579:

1. *dab* (chosen over rival *cac*)
2. *dba* (hands down, over *cbb*)
3. hard to decide between *pdt* (ugh!) and *pcu* (yuk!)
4. *pxqxsx*, of course—not *pxqxry* or *pxqxsy*

   5. too obvious to need any comment
   6. hmm . . . maybe *aaabbbddd*, maybe *aaabbbddk*, maybe *aaabbbccl*, maybe even *aaabbbddl*
   7. *trqp*, but maybe *srqo* (definitely not *srqq*)
   8. *tptqtrtt* is pretty, but so is *spsqsrst*
   9. *abcdeabcdab*, of course
   10. *bcdacdabc* (it's a figure-ground problem—*bcd* is *a*, in "code")
   11. *acg* is way better than *acf*
   12. you know. . . .

Page 585:

   1. *pqr*, a far more insightful answer than *pbc*
   2a. *nws* is the only reasonable answer
   2b. *uuuuu* (not *vvvvv*, despite the answer to 2*a*)
   3. *aBc dEf pQr* goes to *aBc pQr dEf*
   4. *qabcxyzq* is incisive, but has a strong rival in *abcqxyz*
   5. *zabczdefzstuz*—certainly better than *zabcdefzstuz*
   6. *eeeffghhiii*, and yes, that *g* in the middle is the whole point
   7a. *dcbabcd*—not hard
   7b. *abbbc*, based on seeing 3-1-3 go to 1-3-1
   7c. *pfr*—a daringly abstract vision of "inside-out-ness"

When a program can do analogies *like this*, I'll be impressed!!!

Post Post Post Scriptum.

After I'd completed the *P.S.* and *P.P.S.*, I ran into Richard Feynman at a conference. I reminded him of my lecture at Caltech three years earlier; his somewhat vague recollection of it was that it was "silly". I took that as a charitable way of saying that he hadn't seen any point to it. Which made me think that maybe his "village-idiot" stance was due to genuine puzzlement, and not just an act.

I then told him, with a certain amount of trepidation, that in my new book I had humorously referred to his blunt way of answering all my analogy problems as "village-idiotic" a few times. Would this offend him? "Oh, no!" he said. "A while back, *Omni* magazine interviewed me, and on their cover they advertised it as an interview with the 'world's smartest man'. I think it's good to counterbalance that—so now you're calling me a village idiot. That's fine. I think my mother would agree with you more than with *Omni*."

25

Who Shoves Whom Around Inside the Careenium? or, What Is the Meaning of the Word "I"?

March, 1981

The Achilles symbol and the Tortoise symbol encounter each other inside the author's cranium.

ACHILLES: Fancy meeting you here! I'd thought that our dialogue in Paris was the last one we'd ever have.

TORTOISE: You never can tell with this author. Just when you think he's done with you, he drags you out again to perform for his readers.

ACHILLES: I don't see why we should have to perform at his whim.

TORTOISE: Just try resisting. Then you'll see why. You don't have any choice in the matter!

ACHILLES: I don't?

TORTOISE: Look—to refuse to perform is tantamount to suicide. Let's face it, Achilles—you and I (at least in these Hofstadterian versions of ourselves) come to life only when Hofstadter writes dialogues about us. We had it good in *Gödel, Escher, Bach,* but now that that's over and done with, I have a feeling the pickings are going to be pretty slim. Hofstadter knows he can't live off of us forever! So we'd better take what we can get!

ACHILLES: Yes . . . I remember those good old days. Sometimes we had such wonderful lines. Like that one you had, something about how the "Achillean flash" swoops about my brain "in shapes stranger than the dash of a gnat-hungry swallow". Isn't that how it went?

TORTOISE: Something like that. Hofstadter liked that one well enough that he had me say it in at least *two* dialogues! Pretty strange, eh?

ACHILLES: The way you talk about all this is so bizarre, to my mind. I mean, granted that we're figments of someone else's imagination; but still, you know how characters in a novel are supposed to "come alive" and have "wills of their own" . . . Surely it's not just a cliché?

TORTOISE: I wouldn't know. I'm not a novelist. Nor is Hofstadter.

ACHILLES: I mean, am *I* really just a tool of Hofstadter (however benevolent he is), or am I genuinely exerting my own free will here (as I feel I am doing)? What it comes down to is: Who pushes whom around inside this cranium?

TORTOISE: Now *there's* a planted line, if I ever heard one. That's a direct quote from *GEB*, page 710, where Hofstadter is quoting from Roger Sperry of split-brain fame. It's where Sperry's giving his mind-brain-free-will philosophy, which Mr. H evidently espouses. But let's get on with the subject matter of *this* dialogue. I think we've done enough introduction. You must have something on your mind, Achilles, which Mr. H wants to bring up through you.

ACHILLES: I wish you'd quit putting it in that upside-down way, Mr. T.

TORTOISE: All right. But am I right? Isn't there something you're just itching to tell me?

ACHILLES: Come to mention it, yes. It's related to a book I saw in the bookstore the other day, called *Molecular Gods: How Molecules Determine Our Behavior*. It was the subtitle that intrigued me.

TORTOISE: In what way?

ACHILLES: My first thought on reading it was, "Oh, that's interesting—I didn't know that the molecules inside me could affect me that much."

TORTOISE: A classic reaction.

ACHILLES: I know it sounds dumb, but what's wrong with it?

TORTOISE: How can you say that? Molecules is all you are, my friend! Read Francis Crick's *Of Molecules and Men* someday.

ACHILLES: Oh, yes—I know I'm made of molecules. Nobody could deny *that*. It just seems to me that my molecules are at *my* beck and call—not individually, of course, but in large "chunks", such as my fingers, when I play my cello or sign a check. So that when *I* decide to do something, my molecules are forced to come along. So—haven't you really got it reversed? Isn't it *really* the case that *I* shove those molecules around, and not vice versa?

TORTOISE (*rather exasperated*): What do you mean, "I"? What is this "you"?

ACHILLES: How I feel—let me put it that way. My free will determines what I do.

TORTOISE: All right. Let me suggest a definition. Let me suggest that the term "free will", when you use it, is a shorthand for a complex set of predispositions of your brain to act in certain ways. Just a moment ago, you used the word "fingers" as an abbreviation for a whole bunch of molecules. In a similar way, the phrase "free will" could be thought of as an abbreviation for a whole bunch of natural tendencies and constraints. So . . . your free will—your set of preferred pathways for neural activity to flow along—constrains the motions of molecules inside your brain, and those motions in turn are reflected in the patterns that your fingers will trace out.

ACHILLES: Are you saying that when I say "free will", I'm really using a shorthand for a kind of "hedge maze", like the ones on the grounds of Victorian palaces, a maze that allows some pathways and disallows others?

TORTOISE: Yes, that's the idea—only of course this "hedge maze" is inside your skull, and is a bit more abstract. For instance, it's a little oversimplified to imagine that pathways are *rigidly* allowed or disallowed. It would be more accurate to think of the set of predispositions in terms of a set of pins in a pinball machine. You know what I mean by "pins"?

ACHILLES: Those stationary round things with rubber "bumpers" that the shiny marbles bounce off of?

TORTOISE: Correct. Were you to take an average over a million marbles, you could find out how each pin statistically affects the way the marbles descend to the bottom. Pathways aren't just *allowed* or *disallowed*; rather, some are more likely, some are less likely, depending on how the pins are arrayed. But if you still like the image of the maze of hedges, that's not a bad one to hold in your head. The hedges make more rigid constraints —things are more black-and-white than with the pins. There are fewer degrees of freedom for the motions in a maze. But I can make the maze image richer. Suppose that in your maze, one of the effects of the people moving through the maze were that the hedges gradually shifted position. It's somehow as if the maze were formed of movable partitions that constrain the maze runners, yet the maze runners' paths gradually move the partitions, thus changing the maze.

ACHILLES: You mean the maze runners could just decide—by free will—that they want to pick up a partition and plop it down somewhere else?

TORTOISE: Not like that. It's got to be a deterministic outcome of the act of maze running itself. Let me go back to the pinball analogy. It's more as if the pins, instead of being fastened on the board, were *slidable* objects like hockey pucks, objects that as they get banged around from above and below and all sides, slightly slip and change positions. The pins need not be circular; they could be longish so that two or more located near each other could act like a channel or a funnel for marbles. In any case, they are jounced around by the rapidly moving pinballs.

ACHILLES: As in Brownian motion?

TORTOISE: Exactly. There are really *two scales* in time and space operating here, each affecting the other. The heavy hockey-puck-like pins appear almost stationary to the light marbles. To a casual observer who's following the motions of the marbles, the massive pins would appear to be *determining* the light marbles' motions, to be telling the marbles where to go—or in Sperry's words, to be "shoving them around".

ACHILLES: I like that image. It agrees with my earlier view that *I* shove my molecules around.

TORTOISE: True—provided you identify "yourself" with the configuration of the pins.

ACHILLES: That's a little strange, I admit.

TORTOISE: Now imagine a second observer, who's watching a *film* of the whole thing speeded up by a factor of a thousand or more. To *her*, there is a smooth, interesting patterned motion of a bunch of large, variously-shaped pucks. She says to herself, "Wonder why they're moving that way—I can't see anything visible causing any of it."

ACHILLES: She doesn't see the marbles?

TORTOISE: No—they are shooting around so fast in this time scale that their tracks all blur together into one uniform background color with no apparent motion.

ACHILLES: Ah, yes . . . Now the facts about Brownian motion begin to come back to me. I remember how people were mystified by the jostling motions of colloidal particles in solutions when they looked at them under a microscope. They couldn't figure out what was causing such motions. The molecules that were battering them constantly were too small to be visible, and besides, they were moving too quickly.

TORTOISE: Exactly. An observer on this time scale might start to develop a sense for the slow drifting patterns of the pucks, even without having any clear notion of what's *causing* the pucks to move about.

ACHILLES: It's a natural human tendency. Why not?

TORTOISE: The observer could anthropomorphize: "Oh, those two little ones don't like to be close together, and those two long thin ones are trying to be parallel"—and so on. So she develops a teleology, or a way of describing the heavy pucks' motions all on their own. She's quite unaware that they are being bombarded constantly by teeny objects, as in Brownian motion. (Let's pretend that the marbles are more like BB's —really small.) She doesn't know that something smaller is *making* the pucks swim around in those patterns.

ACHILLES: So you can turn a knob on your movie projector and flip back and forth between the fast and slow views? Or even smoothly go between them? That's neat! At first, at the slowest setting, the immobile pucks seem to determine the paths of the many little bouncing marbles. As you speed up the film, the marbles become harder and harder to track, and pretty soon they become just a big blur. Meanwhile, you begin to notice that the pucks actually *aren't* immobile, after all. They're being shoved about by the marbles. So—who's shoving whom around *really*? Well, it's mutual, I now see.

TORTOISE: Good. Now let me add some more richness to this whole metaphor. Let's say that marbles are constantly being shot in from all sides of the table, and also leaving on all sides. You can envision something like a pool table, with a lot of little marble-launching stations mounted on the walls, and a lot of pockets that act as exits for stray marbles that land in them. The inflow and outflow are equal, so there's no net gain or loss of marbles. And the bombardment is pretty uniform, but not exactly. The marbles are launched according to conditions *outside* the table. For example, if there's a red light near a marble-launching station, that station slows down its firing rate; if a green light is near it, it speeds it up. So you have a set of *transducers* from *external light* to *internal marble-shooting*. Now if the puck observer watches both the lights *and* the pucks, she'll be able to draw some causal connections between light patterns outside and the puck-patterns inside. Using mentalistic language will become quite natural. For instance, it would sound quite reasonable to say, "It saw the green light—it's moving away from it—I guess it doesn't like green." And so on.

ACHILLES: Now you've got me thinking. I too want to add some strange features. I'll propose a physical linkage between one particular puck and an external "arm" that can move toward or away from the lights. So, when that puck moves a certain way on the table, the arm may push a light away or pull it closer. Of course this is primitive—there are no fingers or anything, but at least there's now a two-way link between the pucks and the lights. Gosh! I'm almost completely forgetting about those marbles careening around down there! I'm just *relying* on the marble-shooters to keep on doing their job without much maintenance or attention needed . . . All I see now is the seemingly animate interplay—a sort of dance— among the pucks, the lights, and the arms . . .

TORTOISE: We're really jumping from one metaphor to another, aren't we? And each time, we escalate in complexity . . . Oh, well, that's fine with me. No matter how complex the scene gets, you can always slow down the projector, unblur the marbles and no longer see the pucks moving at all.

ACHILLES: Of course. But there's now something that bothers me. In the brain, there *aren't* these large- and small-sized units—everything's uniform, right? I mean, it's all just a dense packing of neurons. So where do the two scales come from? If we go back to the maze and partitions, there too we had two levels of objects (maze people and maze walls), each kind pushing the other around. But in the brain, this isn't so—or is it? What else is there besides neural activity?

TORTOISE: Let's add, then, a level of detail to our picture that we didn't have before. Let's say there are no pucks at all. There are only marbles and a number of larger stiff yet malleable mobile metal strips, which I'll also describe as "stiff yet malleable membranes" (and you'll soon see why). They can be bent into *U*'s or *S*'s or circles . . .

ACHILLES: So these things are swimming in the soup of marbles, now, but there are no more pucks, eh?

TORTOISE: Right. Can you guess what might happen now?

ACHILLES: I can imagine that these strips—

TORTOISE: Would you mind calling them "stiff yet malleable membranes", just to please me?

ACHILLES: Are you going to pull some acronymic trick off in a moment? Let's see—"SYMM" doesn't spell anything, does it? Is that really what you want me to call them, Mr. T?

TORTOISE: In fact you anticipated me, Achilles. Go ahead and do call them "SYMM"'s.

ACHILLES: All right. So these SYMM's will now be knocked around along with the marbles that are bashing into one another. Will the SYMM's occasionally get wrapped around some group of marbles and form a circular membrane, separating out a group of marbles from the rest?

TORTOISE: Just call the circular structure so formed a "SYMM-ball", if you please.

ACHILLES: Oh . . . I should have seen it coming. All right. Now I see that in this way, structures like pucks are emerging again, only this time as *composite structures* made up out of many, many marbles. So now, my old question of who pushes whom around in the cranium—er, should I say "in the *careenium*"?—becomes one of *symmballs* versus *marbles*. Do the marbles push the symmballs around, or vice versa? And I can twiddle the speed control on the projector and watch the film fast or slow.

TORTOISE: I should mention that once a symmball is formed, it might have quite a bit of stability, because the marbles inside it get fairly densely packed together, and jostle each other around only a little bit when the symmball gets hit by a fast marble from the outside. The impact gets spread around and shared among the marbles inside, and the symmball won't tend to break up—at least not when you watch the film at either of the two speeds we've already mentioned. Perhaps the *fission* of a symmball would occur on a longer time scale than the *motions* of symmballs. And the same for the formation of a symmball.

ACHILLES: Would it be fair to liken a symmball's emergence to the solidifying of water into a cube of ice?

TORTOISE: An excellent analogy. Symmballs are constantly forming and unforming, like blocks of ice melting down into chaotically bouncing water molecules—and then new ones can form, only to melt again. This kind of "phase transition" view of the activity is very apt. And it introduces yet a third time scale for the projector, one where it is running much faster and even the motions of the symmballs would start to blur. Symmballs have a dynamics, a way of forming, interacting, and splitting open and disintegrating, all their own. Symmballs can be seen as reflecting, internally to the careenium, the patterns of lights outside of it. They can store "images" of light patterns long after the light patterns are gone—thus the configurations of symmballs can be interpreted as *memory, knowledge,* and *ideas.*

ACHILLES: It seems to me that although you got rid of the pucks, you added another structure—the SYMM's. So how is this new system any improvement, as a model of a brain, over the old one? Don't you still have two levels of basic physical constituents and activity?

TORTOISE: The SYMM's are there only to provide a way for marbles to join up and form clusters. There are other conceivable ways I could have done this. I could have said, "Imagine that each marble is magnetic, or Velcro-coated, so that they all attract each other and stick together (unless jostled too hard)." That suggestion would have had a similar effect—namely, of making much larger units grow out of smaller ones— and so you would have only *one* kind of basic physical constituent. Would that be more satisfying to you, Achilles?

ACHILLES: Yes, but then you'd have lost your pun on "symbols", which would be too bad.

TORTOISE: Not at all! I'd cleverly rename the marbles themselves this time, as "small yellow magnetic marbles"—"SYMM's"—and a magnetically bound cluster of them would form a "SYMM-ball". No loss.

ACHILLES: That's a relief! I would hate to see a good metaphor go down the drain for lack of a pun to illustrate it.

TORTOISE: Hofstadter would never let *that* happen! You can take it from me. Anyway, you can conceive of the larger units however you want, as long as you have it clear in your mind that starting with just *one* level, you wind up with *two* levels and *two* time scales—three time scales, in fact, when you take into account the slow formation, fission, fusion, and fizzling of the symmballs.

ACHILLES: Now can we go back and talk about whether *I* control my molecules, or my molecules control *me*? That's where this all started, after all.

TORTOISE: Certainly! Why don't you try to answer the question yourself?

ACHILLES: The problem is that in all those pulsations inside a careenium, I just don't see a "me". I see a lot of activity—I see a lot of internalized representations of things "out there"—I mean of light patterns, in this case. And with fancier transducers, we could have a careenium in which symmball patterns reflected such things as sounds, touches, smells, temperatures, and so on.

TORTOISE: Let your imagination run wild, Achilles!

ACHILLES: All right. If I stretch my imagination, I can even see a gigantic three-dimensional careenium, hundreds of feet on a side, filled with billions upon billions of marbles floating in zero gravity, shooting back and forth, and all over forming short-lived and long-lived symmballs, and with those symmballs in turn governing the marbles' paths. I can see all that, and yet I don't see free will or "I". I guess I can't see how *I myself* could be a system like this inside my cranium. *I* feel alive! *I* have thoughts, feelings, desires, sensations!

TORTOISE: Hold on, hold on! One at a time. These are all related, but let's try to talk about just one—say, thoughts. Let me propose that the word "thought" is a shorthand for the activity of the symmballs that you see when you run the movie fast: the way they interact and trigger patterns of motions among themselves (mediated, of course, by the constant background swarming of marbles, too fast to make out).

ACHILLES: But I *feel* myself thinking. There's no one *inside* a careenium to *feel* those "thoughts". It's all just a bunch of silly yellow magnetic marbles bashing into each other! It's all impersonal and unalive. How can you call that "thought"?

TORTOISE: Well, isn't it equally true of the molecules running around in *your* brain? Where's the soul of Achilles that "shoves *them* around"?

ACHILLES: Oh, Mr. T, that's not a good enough answer. I've just heard it said too many times that we're made out of atoms, so there's no room for souls or other things—but I know I'm there, it's an undeniable *fact,* so I need more insight than a mere reminder that my body obeys the laws of physics. *Where* does this feeling of "I" come from, a feeling that I have and you have but stones *don't* have?

TORTOISE: You're calling my bluff, eh? All right. Let's see what I can do to turn you around. Let's add one more feature to the careenium—an artificial mouth and throat, just as we added an arm. Let various parameters of them be driven by various symmballs. Now suppose we turn on a green light on the right-hand side of the careenium. New marble activity near that side begins immediately, and there follows a complex regrouping of symmballs. As it all settles down into a new steady configuration, the mouth-throat combination makes an audible sound: "There's a green light out there." Maybe it even says, "I saw a green light out there."

ACHILLES: You're trying to play on my weaknesses. You're trying to get me to identify with a careenium by making it more human-seeming, by making it simulate talking. But to me, this is merely "artificially signaling" (to borrow one of my favorite phrases from Professor Jefferson's Lister Oration). Do you expect me also to believe that somewhere out there, there is a conscious person reciting the time of day twenty-four hours a day, simply because I can dial a certain number and hear a human voice say, over the telephone, "At the tone, Pacific Daylight Time will be five forty-two"? A voice uttering sentence-like sounds doesn't necessarily signify the presence of a conscious being behind it.

TORTOISE: Granted. But this careenium voice isn't merely uttering a mechanically repetitive sequence of sentences. It is giving a dynamic description of what is perceived in the vicinity.

ACHILLES: I have a question about that. Is the thing being perceived located *outside* the careenium, or *inside* it? Why does the mouth say, "I saw a green light *out there*" rather than say something such as, "Inside me, a new symmball just formed and exchanged places with an old one"? Isn't *that* a more accurate description of what it perceived?

TORTOISE: In a way, yes, that is what it perceived, but in another way, no, it did not perceive its own activity. Think about what perception really involves. When you perceive something "out there", you cannot help but mirror that event inside you somehow. Without that internal mirroring event, there would be no perception. The trick is to know what kind of external event triggered it, and to describe what you felt out loud in

611

public language that refers to something external. You subtract one layer of transduction. You omit, in your description of what happened, one step along the way. You omit mention of the step that converted the green light into internal symmball responses. You are not even aware of that step, unless you are something of a philosopher or psychologist.

ACHILLES: Why would I or anyone else omit a real level? What's the origin of this socially conventional lie? *I* don't omit levels in *my* speech!

TORTOISE: Actually, you do. It's a universal phenomenon. If you live near a railroad track and hear a certain kind of loud noise coming from that direction—rumbling, bells dinging, and so on—do you say, "I hear a train", or do you say, "I hear the *sound* of a train"?

ACHILLES: I guess that ordinarily, I would tend to say, "I hear the train."

TORTOISE: Do you see a train, or do you see *light* hitting your eyes? When you touch a chair, do you feel the chair, or do you feel your *feeling* of the chair?

ACHILLES: I opt for the simpler alternative. I never would think those extra philosophical thoughts that go along with it. What point would it serve to say, "I hear the *sound* of the train"?

TORTOISE: Exactly my point. The most convenient language, the least obfuscatory and pedantic, omits the heavy "extra" reference to the medium carrying the signals, omits mention of the transducers, and so on. It simply gets straight to the *external source.* This seems, somehow, the most *honest* way to look at things—and the least confusing. You hear and see a *train,* not an image of a train, not the light reflected off a train, not retinal cells firing—and most definitely not your perception of a train! We are constructed in such a way as to be unaware of our brain's internal activity underlying perception, and therefore we "map it outward".

ACHILLES: Yes, I see that pretty well. I think I see why a careenium with a voice might talk about a green light rather than talk about its symmballs. But wait a minute. How would it know anything about green lights? It might *prefer* to refer to things in the outside world—but nonetheless, all it knows about is its own internal state!

TORTOISE: True, but its way of verbalizing its internal state employs words that you and I think refer to objects and facts *outside* the careenium. In fact, it too thinks so. But you could very well argue that it is just making sounds that mirror its internal state in some very complex way. It could be deluding itself. There might be nothing out there to refer to!

ACHILLES: True, but that's not exactly my question. What I want to know is, how come it uses the *right words* to describe what's out there? Where did it learn to say "green light"? The same question goes for people. How come we all say the same sounds for the same things?

TORTOISE: Oh, *that*'s not so hard. I had thought you were asking whether reality exists or not. I quickly tire of such pointless quibbling over solipsism. But let me answer the question you *did* ask. When you were a tot, you saw things—say, rattles—and heard certain sounds—namely, various pronunciations of the word "rattle"—at about the same time. Those sights and sounds were transduced from your retinas and eardrums into internal symbol states inside your cranium. Now, as a member of the human race, you were constructed in such a way as to enjoy mimicry, so you made funny noises something like "wattle", which were then automatically picked up by your eardrums, and fed back into the interior of your cranium. You heard your own voice, to your great delight and thrill! You were then able to compare the sounds *you*'d just made with your memory of the sounds you'd heard. By playing this exciting new game, you were learning the English words for objects. Of course you started with the nouns for visible objects, but quickly you built on that most concrete level and over the next few years you developed a large vocabulary including such things as "ball", "pick up", "next to", "splash", "window", "seven", "remember", "sort of", "zebra", "maze", "stretch", "of course", "by accident", "tongue-twister", "blunder", "confetti", "equilibrium", "analogy", "*vis-à-vis*", "chortle", "Picasso", "double negation", "few and far between", "neutrino", "*Weltanschauung*", "*n*-dimensional vector space", "tRNA-amino-acyl synthetase", "solipsism", "careenium"—

ACHILLES: Wait a minute! What about "banana split"?

TORTOISE: Now how did I overlook that? A shameful oversight. But I hope you get the point.

ACHILLES: I think I see what you mean. Gradually, I internalized a huge set of external, public, aural conventions—namely the English words attached to particular states of my own brain, states that were correlated with things "out there".

TORTOISE: Not just things—actions and styles and relationships and so on.

ACHILLES: To be sure. But instead of conceiving that the words described my brain state, it was easier to conceive of them as describing things out there *directly*. In this way, by omitting a level in my interpretation of my own brain's state, I cast internal images outward.

TORTOISE: A careenium would do likewise—casting its internal symmball patterns outward, attributing them to some properties of the external world. And if a large number of careenia happened to be located near some specific stimulus, they could all communicate back and forth by means of a set of publicly recognizable noises that are externalizations of their internal states! So it's actually very useful to subtract out the references to the transduction, perception, and representation levels.

ACHILLES: It all makes sense now. But unfortunately, something else is bothering me! If the system projects all its states *outward*, talking about "green lights" and "red lights" and "traffic jams" and so forth, then how is there any room left for it to perceive its own *internal* state? Will it be able to say, "I'm annoyed" or "I forgot" or "I don't know" or "It's on the tip of my tongue" or "I'm in a blue funk"? Or will it project all *those* inner states outward as well, attributing weird qualities to things outside of it? Could there be inward-directed transducers that focus on *symmballs* and come up with a *representation* of symmball activity? That would be a sort of sixth sense—an inward-directed sense.

TORTOISE: You could call it the "inner eye".

ACHILLES: That's a perfect name for it.

TORTOISE: The inner eye wouldn't need to do much transducing, would it? It's the easiest thing in the world to monitor because it's right there inside you.

ACHILLES: Now, Mr. T, you always warn me about confusing use and mention; I think you yourself are committing that error here. To have a word such as "Tortoise" in a text is enough to make somebody conjure up the image of a Tortoise, but it is not at all the same as making that person start to think about the *word* "Tortoise", is it? They may not notice it at all.

TORTOISE: Point well taken. There is a difference between *having* your symmballs in certain states, and *being aware* of that fact. It's something like the difference between using grammar correctly and knowing the rules of grammar.

ACHILLES: Now I sense I could get really confused here—things could get very tangled. How can symmballs "watch" other symmballs? The ones that react to green lights, I can imagine and understand. There are transducers—the marble shooters on the borders. But would there be some symmball that always reacts to, say, the fusion of two symmballs? How would it detect such a fusion? What would make it react that way? Would it be a sort of satellite or U-2 plane, with an overview of the whole terrain of the brain? And what purpose would it serve?

TORTOISE: Imagine that you were watching an actual careenium, and at a very slow speed—so slow that you could reach down with your hand and pick up and remove an entire symmball before getting struck by any symms careening toward your hand. All of a sudden there would be a vacuum, where before there had been a dense mass of marbles. If you switched speeds now and watched the results in the *symmballs'* time scale, you'd see a massive regrouping of symmballs all over the careenium, a kind of *shudder* passing through the whole system as all the various symmballs come to occupy slightly different positions.

ACHILLES: You could call such a shudder a "mindquake".

TORTOISE: An excellent suggestion. Various types of "mindquakes" would have characteristic qualities to them. They would have "signatures", so to speak. Now if *you*, Achilles, an observer from the outside, could learn to recognize such a signature, then why couldn't the system itself, from within, be even more able to do so? Such mindquakes would be, after all, just as tangible to the system as is an increase of marble-firings on any side. Both are simply *internal events*, even though the one is triggered by something external, while the other is set off by something internal.

ACHILLES: So would there be various "seismometer symmballs", each one sitting there waiting to feel a specific kind of mindquake, and when that happens it would react?

TORTOISE: Sure. And for each type of mindquake, there is a special symmball there just sitting there like a pencil on end—and when its type of mindquake comes along, it topples. Of course that "toppling" in itself is just *more symmball activity*—

ACHILLES: Another mindquake?

TORTOISE: Precisely—and it can set off further reactions inside the careenium. The whole thing is very circular—one shudder triggers another one, and that one sets off more, and so on.

ACHILLES: It sounds like it would never stop. There would just be constant symmball activity rippling back and forth across the careenium.

TORTOISE: Well, of course! That *is* what happens with conscious systems, isn't it? We're constantly thinking thoughts—some fresh, some stale— constantly mentally alive and aware—partly of the external world, partly of our own state—for example, how confused or tired we are, what something reminds us of, how bored we are with this long monotonous dialogue. . . .

ACHILLES: Hey, wait a minute! The *reader* may be bored, but I'm not!

TORTOISE: Only kidding, Achilles. Just trying to liven things up a bit.

ACHILLES: All right. Well anyway, I admit that everything you've been saying is true, makes sense, but how is it *useful* for us to monitor our own state?

TORTOISE: Well, think first of a simple animal. What it needs most of all is food. Its brain—if it has one—is connected to its stomach by nerves, and it transduces an emptiness in the stomach into a certain configuration of symbols in the brain. Actually, this animal might be so simple that the symmball level doesn't exist. There might just be marbles zipping around in its cranium, but no larger-scale agglomerations. In any case, the effect of this may then be a shuddering in its brain, which produces repercussions on the animal's peripheries. It may move. All this is very much at the reflex level. Mostly it involves monitoring the organism's hunger state and controlling its limbs. Every organism has to monitor itself in terms of hunger. But primitive organisms don't use much information about the external environment they're in—they just flap about and "hope"—if that isn't too strong a word!—to encounter some food. Pretty unconscious. On the other hand, take a more complex animal. It will have an elaborate representation of its environment inside itself, so it also has a lot of options when it detects internal hunger. The symbolic activity representing the empty stomach has to be dealt with in the context of all the other symbols, which might represent danger, priorities other than eating, choices of when and what to eat, and so on. The total interaction of symbols at that point we might call "consideration" or "deliberation" or "reflection"—as distinguished from "reflexes". Now after all this, let me ask you: Does this help you to see why such a careenium might have a self?

ACHILLES: Well . . . I might grant that there's reflection going on in there, I might even grant that it's *thinking*—but there's no*body* in there *doing* the thinking!

TORTOISE: Would you grant that there's *free will* inside there?

ACHILLES: Hardly!

TORTOISE: Then I can see that you will need some more persuading. All right. Let me suggest that there *is* free will, and that this notion of a careenium may help you understand more clearly what free will truly consists in. We began this discussion by talking about whether you can "shove your molecules around" or not. This is a central question—in truth, it is *the* central question, I think. So I'd like to ask you, Achilles, can you freely decide to do anything?

ACHILLES: Of course I can! That's precisely what free will is about! I can decide to do whatever I want!

TORTOISE: Really? Can you decide, say, to answer me in Sanskrit?

ACHILLES: Obviously not. But that has nothing to do with it. I don't speak Sanskrit. How could I answer you in it? Your question doesn't make sense.

TORTOISE: Not so. You can only do what your brain will allow you to do, and that is very crucial. Let me ask you another question. Can you decide to kill me right now?

ACHILLES: Mr. T! What a suggestion! How could you suggest such a thing, even in jest?

TORTOISE: Could you nevertheless decide to do it?

ACHILLES: Sure! Why not? I can certainly *imagine* myself deciding to do it.

TORTOISE: That is beside the point, Achilles. Don't confuse hypothetical or fictitious worlds with reality. I'm asking you if you *can* decide to kill me.

ACHILLES: I guess that in this world, in the *real* world, I could not *carry out* such a decision, even had I "decided"—or claimed I'd decided—to do it. So I guess I *couldn't* decide to do it, actually.

TORTOISE: That's right. That innocent-seeming trailer phrase that one tends to tack on—exactly as you did—is very telling, after all.

ACHILLES: What innocent phrase? What do you mean?

TORTOISE: Don't you remember? You insisted vehemently to me, "I can decide to do *whatever I want.*" Now that phrase "whatever I want" may *sound* like a grand, universal, all-inclusive, sweeping phrase—but in fact, it represents quite the opposite: a severe constraint. It's not true that you are able to decide to do *anything*; you are limited to being able to decide to do only things you *want.* Worse yet, you are in fact limited to doing, at any time, the *one* thing that you want *most* to do! Here, "want" is a complex function of the state of the entire system.

ACHILLES: Are you saying that choice is an illusion?

TORTOISE: Only to the extent that "I" is an illusion. Let me explain. It's quite common for people to develop interests that begin to consume them—doing puzzles, doing music, thinking about philosophy . . .

Sometimes such habits get so strong that they begin to interfere with the rest of their lives. A wife may pick up a bad habit—say, twiddling a cube or smoking cigars or constantly punning—and then *try* to get rid of it. Her exasperated husband may say to her, "What's this *trying*? Why can't you just *decide* to stop cubing? It is driving a wedge between you and me. Why don't you just *decide* to quit?" Yet the afflicted wife may, for all her good intentions, be unable to do so. Certainly having a modicum of desire is not enough. I would put it this way. The husband is appealing to what I would call his wife's "soul"—a coherent set of principles and tendencies and interests and personality traits and so forth that represent to him the person that he married. They have always before seemed to provide reasons or explanations for his wife's character, and he loves her for that aggregate of ways of being. So he appeals to this "soul" to put a clamp on its new obsession. But once the wife starts twiddling her cube, a *part* of her takes over. She gets obsessed—or should I say "possessed"?—by one of her own subsystems!

ACHILLES: "Possessed" is the word for it. I myself find it very hard to stop practicing a piece on my cello once I have gotten into the swing of it. Before I start, I think, "Now, I'll just play this piece *one time.*" (Or, "I'll just eat one potato chip", or "I'll just solve the cube one time".) But then, once I've let myself start, I'm no longer quite the same person—some things inside me have subtly shifted. And the *new* me thinks, "*That* guy said *he'd* do it only once. That's what *he* thought. But *I* know better!" There is a kind of inner inertia that makes me want to continue, even when there are *other* things I would also like to do. It's as if some part of you just "slips away" from a higher level of control—some subsystem gets "out of control" and won't obey the soul on top—like a bucking bronco unwilling to obey its rider.

TORTOISE: A powerful image. In such cases the wife herself may be confused and torn. Her inner turmoil is like that of a country in inner strife. There are factions battling each other—only in this case, the factions are neural firings, not people, of course. On some level, this woman may *feel* she wants to be able to decide to give up her habit—yet she may not have enough neurons on her side! And as in a country where the people won't support the government, so here: the "soul" has to have the support of its neurons! It can't just arbitrarily "shove them around", in reality.

ACHILLES: I'm all confused. Who *is* in control, here?

TORTOISE: We'd like to be able to say that the symmballs can *decide* to do arbitrary things, but they are constrained. They are in a system that "wants" its parts to move in some ways but doesn't "want" them to move in others. We could come back to the hedge-maze metaphor, to make this more vivid.

ACHILLES: Yes, but that applied to the *lower-level* objects—marbles, symms, or neural firings.

TORTOISE: Exactly. The "heavyweight" entities—hedges, pins, symmballs—

constrain the "lightweight" entities—maze runners, pinballs, symms; but in revenge, the little ones, acting together, control how the high-level ones are arrayed.

ACHILLES: So *nobody*'s free here!

TORTOISE: Well, from the outside, that's the way it seems. But on the inside, the system may feel, just as you did, that it can "decide to do whatever it wants to do". But mind you, two symmballs in a careenium aren't free to *decide,* arbitrarily, on their own, to move in parallel—they have to have the cooperation of the marbles. The marbles have to do the work for them. Similarly, when the unhappy wife tries to "decide" to give up her cubing or punning habit, she can't do it without the agreement, the support, of her neurons.

ACHILLES: You make this wife's "soul" sound like a general trying to marshall unruly neurons, to force them into line when they have their *own* paths to follow. A military general has some degree of power over his soldiers, so he can coerce them to some extent—but only so far. Beyond that, they'll mutiny. So the general has to go along with the tide. He can't really dictate policy—he can only resonate with it.

TORTOISE: It's true. However, sometimes an unexpected shift at a higher level can precipitate an abrupt "phase transition" of lower levels. A million tiny things suddenly find themselves swirling around in unexpected ways, and realigning in totally novel higher-level patterns. Once in a while—just once in a while—the "general" *can* gain control of those unruly neurons—but only when they themselves don't know what they want, haven't reached any kind of consensus, and are instead in a malleable, leadable, chaotic state.

ACHILLES: It sounds like you're describing a "snap decision"—an exercise of pure will power, such as when I say to myself, "I'm going to quit cubing *right now*!", or "I'm going to stop feeling sorry for myself and go out and get something useful done." But if I understand *your* way of looking at this kind of thing, even a phrase like "snap decision" is really just a kind of shorthand for summarizing a lot of low-level activity. Is that so? It seems to me it would *have* to be so, in your picture.

TORTOISE: You're right, saying something like "snap decision" is really a coarse-grained manner of speaking about a huge cloud of neural activity, like a huge blurry cloud of symms in a careenium projected at a high speed on the screen. And sometimes the activity of neurons inside a cranium, or of symms inside a careenium, lends itself admirably to such a high-level, coarse-grained, symbolic description—or in the case of a careenium, a "symm-ball-ic" description.

ACHILLES: Not always?

TORTOISE: Are all ponds always frozen?

ACHILLES: Oh, I see what you mean. If the relevant portions of the careenium are chunked into symmballs, then a symmball-level description can be made. One set of symmballs is seen to affect other sets of symmballs regularly and predictably. Whereas if there *are* no symmballs—just a lot of stray symms careening around with nothing to constrain them except the careenium's boundary—then it's kind of chaotic, and no higher-level description applies. But when the whole careenium is "symm-ball-ic"—when the phase transitions have taken place—then the person—I mean the *careenium*!—feels very much in control of his or her thoughts.

TORTOISE: *Its* thoughts?

ACHILLES: Yeah, yeah—that's what I meant. Its thoughts. But when not enough phase transitions have taken place, then there's an indescribable hubbub: random symms careening all over the place without orderly constraints. But I wonder what it's like when the brain is in sort of a halfway state—when there are lots of symmballs, but at the same time still a lot of stray symms that belong to no one. It reminds me of a half-frozen lake in early winter or early spring, when the molecules have only *half*-coalesced into large blocks of ice.

TORTOISE: That's a wonderful state to be in. I find I'm most creative when I feel my brain consisting of such halfway-coalesced symbols—neurons acting somewhat independently, somewhat collectively. It's a happy medium where neural bubblings cooperate with symbolic channelings and yield the most creative, fulfilling, semi-chaotic sense of aliveness.

ACHILLES: You think some of that uncoalesced freedom is essential for creativity?

TORTOISE: I was convinced of it by Hofstadter, who certainly feels that way. In *GEB,* writing about his plight as a writer, he portrayed himself as suffering from "helplessness" of the top level, for although he—or his symbol level—may in *some* sense have decided what to write, still he is entirely and utterly dependent on vast cooperating teams of neurons to come up with imagery and ideas and choices of words and sentence structures. Those lower-level items feel to the top level as if they "bubble up" from nowhere. But in reality they are somehow formed from the churning, seething masses of interacting neural sparks—just as patterns of symmball motions emerge out of the chaotic Brownian motion of the many tiny symms. And a few of those ideas make it out through the narrow channel of verbalization, like grains of sand passing through the narrow neck of an hourglass. Yet most likely Hofstadter will *insist* that he himself is responsible for this dialogue, will desire the credit to accrue to *him.*

ACHILLES: Hmm . . . to the overall system that constrained the marbles to jounce in those ways . . . It is hard to assign "credit" or "blame", once you start analyzing thought mechanistically. I see that "decision" and "choice" are very subtle concepts that somehow have to do with the ways in which constraints on two different levels affect each other reciprocally, and at two different time scales, inside a cranium, or a careenium.

TORTOISE: You're getting the idea.

ACHILLES: Every time you say "bubbling up", I can't help but think about the bubbles in ginger ale—I love the stuff. And I'm thirsty. I'm going to have some. Care for a glass yourself?

TORTOISE: Ah, ginger ale—capital suggestion.

ACHILLES (*sipping from a tall glass of cool ginger ale*): Did it ever occur to you that when your leg is asleep, it feels like it's full of ginger ale?

TORTOISE: Clever observation.

ACHILLES: Not really original, I have to admit. I read it once in a "Dennis the Menace" cartoon.

TORTOISE: Are you sure *you* read it—or was it Hofstadter?

ACHILLES: Spoilsport!

TORTOISE: It strikes me that having your leg fall asleep is one very weird experience. It's as if nature were giving you a chance, every once in a while, to be privy to all the tiny goings-on inside your leg, feeling the mingling tingling of trillions of cells all buzzing at once . . .

ACHILLES: Do you suppose that's what being alive *really* is like, and most of the time we're just numb to it?

TORTOISE: Precisely. Can you imagine if all of your body were always as fizzy and tingly as that? I've always wondered why people say their leg is *asleep.* We Tortoises say, "My leg is *awake.*" And French speakers say, "I've got ants in my leg." Those seem so much more accurate to me.

ACHILLES: Phooey!

TORTOISE: What's the matter now?

ACHILLES: I just realized that Hofstadter planted all of this. I mean, I'd thought I was genuinely thirsty. Now I see I was just being manipulated. He wanted to get certain remarks in here, and having me be thirsty was just a convenient avenue for him to do so. I should have known better.

TORTOISE: Oh, so your ginger ale doesn't taste any good?

ACHILLES: That's not what I mean. It tastes fine!

TORTOISE: Well then, what are you grumbling about? You're happy, he's happy. Would you have been happier if he were *unhappy*? That would seem a little perverse, even to me.

ACHILLES: I guess you have a point. You know, now that I think about it, sometimes the decisions I make seem to be slow percolating processes, things that are utterly out of my control. In fact, a rather gory image that illustrates this idea flashed before my mind's eye while we were talking about the difficulty of breaking out of mental ruts.

TORTOISE: What was that?

ACHILLES: I imagined a grim scene where a man's young wife is in a car crash and is badly mangled. He will certainly *react.* Perhaps he will react with love and devotion, perhaps with pity. Perhaps, to his own own dismay, he will even react with revulsion. But it occurred to me that in such emotionally wrenching cases, you can hardly *decide* what you will feel. Something just *happens* inside you. Subtle forces shift deep inside you, hidden, subterranean. It's quite scary, in a way, because in real crises like that, instead of being able to *decide* how you'll act, you *find out* what sort of stuff you're made of. It's more passive than active—or more accurately put, the action is on levels of yourself that are far lower—far more microscopic—than you have direct control over.

TORTOISE: Correct. You and your neurons are not on speaking terms, any more than a country could be on speaking terms with its citizens. There is, in both cases, a kind of collective action of a myriad tiny elements on low levels that swings the balance—exactly as in a country that "decides" to go to war or not. It will flip or not, depending on the polarization of its citizens. And they seem to align in larger and larger groups, aided by communication channels and rumors and so on. All of a sudden, a country that seemed undecided will just "swing" in a way that surprises everyone.

ACHILLES: Or, to shift imagery again, it's like an avalanche caused by the collective outcome of the way that billions upon billions of snow crystals are poised. One tiny event can get amplified into stupendous proportions —a chain reaction. But the crystals have to be poised in the right way, otherwise nothing will happen.

TORTOISE: In cases of judgment, whether it be of one musical composer over another, one potential title or subtitle for a book over another, or whatever, the top level pretty much has to wait for decisions to percolate up from the bottom level. The masses down below are where the decision *really* gets made, in a time of brooding and rumination. Then the top level may struggle to articulate the seething activity down below, but those verbalized reasons it comes up with are always *a posteriori*. Words alone are never rich enough to explain the subtlety of a difficult choice. Reasons may sound plausible but they are never the essence of a decision. The verbalized reason is just the tip of an iceberg. Or, to change images, conflicts of ideas are like wars, in which *every reason has its army*. When reasons collide, the real battleground is not at the verbal level (although some people would love to believe so); it's really a battle between opposing armies of neural firings, bringing in their heavy artillery of connotations, imagery, analogies, memories, residual atavistic fears, and ancient biological realities.

ACHILLES: My goodness, it sounds terrifying! You make the battlefield of the mind sound like a vast mined battlefield! Or a treacherous ice field on a steep mountain face. I never realized that a mechanistic explanation of thinking could sound so organic and living. It's sort of awful and yet it's sort of awe-inspiring as well. But I am very confused now about the "soul", the free will.

TORTOISE: I think that all these strangely evocative images have brought us back to your original perplexity, over the question of who pushes whom around in the cranium. Would you now be inclined to say, Achilles, that your molecules push *you* around, or that *you* push *them* around?

ACHILLES: Actually, I'm not sure how this "I" fits into a cranium—or a careenium. You've got my head so spinning now that I don't know what's up or down.

TORTOISE: Wonderful! At least now your mind may be malleable. Do you see how "free will" in a careenium is actually constrained—*physically* constrained, I mean—by the "wants" of the system?

ACHILLES: Yes, I see that these seemingly intangible "wants" are actually physical attributes of the overall system—tendencies to shun certain modes of behavior or to repeat certain patterns. So in a way I can see that a careenium has "free will" in this constrained sense of freedom. Maybe "free will" should be renamed *"free won't"*.

TORTOISE: Oh, my, Achilles! Did you just make that clever one up?

ACHILLES: I don't know—it just came to me. I never thought about it. It just "bubbled up from nowhere". I don't know who deserves the credit. Maybe Hofstadter made it up. Or maybe it just bubbled up inside *his* brain—although I don't quite see the difference.

TORTOISE: It sounds like the sort of thing Hofstadter's friend Scott Kim would say.

ACHILLES: Hmm . . . I still wonder, though—could a careenium's symmballs actually *decide* to do anything on their own?

TORTOISE: They certainly can't disobey the way the symms push them around—but on the other hand, the symms *are* always poised in just such a way that the *one* internal event that the symmballs *most* want to happen *will* happen. Isn't that a miraculous coincidence?

ACHILLES: Now that I understand how all this comes about, I can see that it's not at all a coincidence. By the *definition* of "want", the symmballs will get shoved around the way they want to be (*whether they like it or not*)! I guess that the real conviction of having free will would arise when, repeatedly and reliably, a collection of symmballs wants something and then watches its desire getting carried out. It must seem like magic!

TORTOISE: It's what happens when you decide, say, to sign your name. Your fingers begin obeying you, and miraculously, you watch your name just appear before you, effortlessly! Is that magic?

ACHILLES: Aha! That brings back that ultimately confusing term, "I". We say "*I* decide to sign my name." But what does that mean? I can see everything in a careenium—wants, desires, beliefs—but I just can't seem to take that last step. I simply fail to see an "I" in there.

TORTOISE: I've tried to explain that the word "I" is just a shorthand used by a system such as a careenium—a system that perceives itself in terms of symmballs and their predispositions to act in certain ways and not others—particularly a careenium that has *not* perceived that it is composed of small yellow magnetic marbles.

ACHILLES: Perceive, shmerceive, Mr. T! There's no one inside a careenium who *could* perceive such a thing. Perception requires *awareness,* which no careenium has. There's no one inside a careenium to feel and experience and *enjoy* its "free will", even if it's there, in your sense. Or maybe the best way to say it is that there's perception and free will there, but there's nobody there to have it.

TORTOISE: You mean you seriously would grant that a physical system could have *free will* but you wouldn't then feel forced to say there was *someone exercising* that free will? Or that there was perception but no *perceiver*? Perceiverless perception? Agentless, subjectless free will? Soulless, inanimate free will? That's a real doozy!

ACHILLES: I know it sounds paradoxical. I could almost agree with you— except I'm still hung up on one thing. Just *which* perceiver, *which* agent, *which* subject, *which* soul would it be? Which person gets to *be* that careenium? Or maybe I should turn the question around: Which careenium gets to have a given soul? Do you see what I mean?

TORTOISE: I think so. You seem to be envisioning a corral of souls up in the sky, into which God (or some other Grand Overarching Deity) dips, whenever a new cranium or careenium comes into existence, and from which It pulls out a soul, imbuing that careenium or cranium with *that* identity forevermore—almost as if It were putting a cherry on top of a sundae.

ACHILLES: Are you mocking me?

TORTOISE: I don't mean to be. If it sounds that way, it's only because I'm trying to take what I think your implicit notion of "soul" is and to characterize it explicitly, by putting it into as graphic terms as possible. But if you subtract out the imagery of a corral and God and cherries on sundaes, am I not putting into words the gist of your view?

ACHILLES: In a way, I suppose so—only you've made it sound so silly that I hesitate to adopt that view now.

TORTOISE: It's so tempting to think that different I's are just "out there", dormant, waiting to be attached to structures, like saddles put on horses or cherries on sundaes. Then, once they are in place, suddenly there is a consciousness that "wakes up". As if the consciousness, and the identity, the "who-ness", were provided by the cherry, and without it there would be only a hollow "pseudo-I"—a thing possessing free will but with nobody to *be*! Isn't that a little sad? Wouldn't you feel sorry for such a poor, deprived entity? Oh, no, of course you wouldn't—there would be no one to be sorry for, right?

ACHILLES: Well, it's hard to see where a sense of "who I am" could come to a bunch of marbles in a careenium, or even to a collection of firing neurons. It seems to me that the identity *has* to be imposed on top of such a structure. A careenium is a complex pinball machine—a heap of metallic machinery—even if, unlike pinball machines, some of its states represent the world and its workings. But until you add some sort of living "flame" to that heap, it's empty—soulless. You need "flame" (although I admit I don't know quite what I mean by that term) to turn a physical object into a *being,* just as you need flame to turn a pile of wood into a fire. No matter how much lighter fluid you pour on it, without a flame, it's still inert.

TORTOISE: Wait a minute! A pile of wood starts *burning* when you set flame to it—but does it acquire a *soul* at that moment? No—as you said, it simply becomes active instead of inert. Any old flame would do. The identity of a fire doesn't come from the torch that lit it, but from the combustible materials! It's the transition from inactivity to activity that makes the flame seem so critical. But a careenium doesn't need to become *more* active than it is. Yet for some reason, Achilles, you seem to balk at my suggestion that in that activity there is as much reason to see an individualized soul as in neural firing activity. But what's so special about neurons? You know what you remind me of?

ACHILLES: I don't know that I *want* to know, but tell me anyway.

TORTOISE: You remind me of somebody who runs into a pile of metal that's merrily burning away, and who declares that although it *looks* mighty like a fire, it surely *isn't* a fire (especially not a *genuine* fire!), because it's made of metal, and everyone knows that fires—especially *genuine* ones—are always made of burning wood or paper.

ACHILLES: That sounds pretty silly and narrow-minded—more so than I am, I should hope. I'm not insisting that no careenium could have a genuine soul so much as I am wondering, "*If* a careenium had a soul, *which* soul would it be? Who would be *this* careenium, who would be *that* one?" On what basis could a decision be made?

TORTOISE: Wow, have you got things upside down! (Or backwards—I'm not sure which.) The same question goes for people as much as for careenia. Who gets to be which body? Do you also have the belief that *any body* could be *anybody*? All it takes is the right flame inside? Could there be a "flame transplant", where someone else's flame—say mine—got implanted in your body, leaving your brain and body intact? Then who would be you? Or, who would you be? Or *where* would you be?

ACHILLES: And where would *you* be, Mr. T? Something seems wrong in this picture, I admit. If a careenium is actually somebody, where does the decision as to *who* it is originate?

TORTOISE: I think you've got things backwards. (Or upside down—I'm not sure which.) First of all, it's not a *decision*—it's an *outcome*. Secondly, which "who" a careenium is is an outcome of its structure, particularly the way it represents its own structure in itself. The more it is able to see itself as an independent and coherent agent, the more of a "who" there is for it to be. Eventually, by building up enough of a sense of its unique self, it has built up a complete "who" for it to be: a soul, if you will. The continuity and strength of the feeling of "being someone" come from identification with past and future versions of the same system, from the way the system sees itself as a unitary thing moving and changing through time.

ACHILLES: That's a strange idea—a thing whose identity remains stable even though that thing changes in time. Is it like a country that changes and yet remains somehow the same country? I think of Poland, for instance. If *any* country has had its soul-flame tampered with, Poland is it—yet it seems to have maintained a continuous "Polish spirit" for hundreds of years.

TORTOISE: A beautiful example. The sense of "one thing, extending through time" is very much at the root of our feeling of "being someone". And in a way it is nature's hoax: the illusion of soulsameness. Or, if you prefer not to call it an illusion, you can say that the ability of an organism to abstract, to think it sees some constant thing, over time, that it considers its self even as it changes, makes that organism's soul *not* an illusion.

ACHILLES: You mean anything that can fool itself—I mean, *see* itself—as unchanging over time has a soul?

TORTOISE: That's not such a silly notion—provided that the verb "see" has its usual abstract meaning, not some dilution of the term. If the organism is as perceptually powerful as living ones like you and me, then I would definitely say it has a soul, if it sees itself as essentially "the same organism" over time.

ACHILLES: But to see itself *as an organism* is not a trivial thing! It has to see itself as one coherent thing acting for *reasons,* not just randomly.

TORTOISE: Now you're talking! I couldn't agree more. Such a way of looking at something—namely, ascribing mental attributes to it—has been called by Daniel Dennett "adopting the intentional stance" toward that thing. In the case of you looking at a careenium, it would come down to your seeing it at the *symmball* level, and interpreting the symmball configurations and the patterns they go through over time as representing the system's beliefs, desires, needs, and so on, overlooking the underlying masses of marbles, either deliberately or out of ignorance.

ACHILLES: But you're not talking about *me* looking at a system; you're talking about a situation where some system does that to *itself,* right?

TORTOISE: Exactly. It looks at its own behavior and, instead of seeing all the little marbles deep down there making it act as it does, it sees only its *symmballs,* acting in sensible, rational ways—

ACHILLES: The system sees itself just as observers of the fast film see it! It could say of itself, "It wants this, believes that", and so on—only now it is ascribing all these beliefs and penchants and preferences and desires and so forth to itself, so instead it says, *"I want this, believe that",* and so on. This seems peculiar to me. It makes up a bunch of hypothetical notions about itself simply out of convenience, then ascribes them to itself in all seriousness. For God's sake, though—if beliefs and desires and purposes and so on *really* existed inside itself, wouldn't the blasted careenium itself have direct access to them?

TORTOISE: What makes you think those beliefs *aren't* real? Aren't ice cubes and traffic jams and symmballs real? And what makes you think that this self-perception *isn't* direct access to its beliefs? After all, does your perception of your own feelings via your "inner eye" differ so wildly from this?

ACHILLES: I suppose not.

TORTOISE: When an *outsider* ascribes beliefs and purposes to some organism or mechanical system, he or she is "adopting the intentional stance" toward that entity. But when the organism is so complicated that it is forced to do that with respect to *itself*, you could say that the organism is "adopting the *auto*-intentional stance". This would imply that the organism's own best way of understanding itself is by attributing to itself desires, beliefs, and so on.

ACHILLES: That's a very strange sort of level-crossing feedback loop, Mr. T. The system's self-image (as a collection of symmballs) is getting recycled back into the system but of course this depends on the very concrete symms themselves to carry it out. It's like a television looking at its own screen, recycling a representation of itself over and over, building up a pattern of nested self-images on the screen.

TORTOISE: And that stable pattern becomes a real thing in and of itself. If you were a careenium, merely by adopting the auto-intentional stance toward yourself, you would create a self-perpetuating delusion. As soon as you create this illusion that there is just one thing there—a unitary self with beliefs and desires rather than a mere bunch of goalless, soulless marbles—then that illusion reenters the system as one of its own beliefs. The more that illusion of unity is cycled through the system, the more established and hardened and locked-in the whole illusion becomes. It's like a crystal whose crystallization, once started, somehow has a catalyzing effect on its further crystallization. Some sort of vicious closed loop that self-reinforces, so that even if it starts out as a delusion, by the time it has locked in, it has so deeply permeated the system's structure that no one could possibly explain how or why the system works as it does without referring to its "silly, self-deluding" belief in itself *as a self.*

ACHILLES: But by that time it isn't so silly any more, is it?

TORTOISE: No, by then it has to be taken quite seriously, because it will have a lot of explanatory power. Once the self has become so locked-in, or "reified", in the system's own set of concepts, this fact determines much of the system's own future behavior—or at least if you are restricted to watching the fast projector, to looking at the *symmball* level, that is the easiest way to understand matters. And the curious thing is that this *same* level-crossing feedback loop (of adopting the auto-intentional stance) takes place in *every* careenium of sufficient complexity. So that whichever careenium you take, the stable self-image pattern that it finally establishes in this loopy way is isomorphic to the stable self-image pattern in every *other* careenium!

ACHILLES: Bizarre! The medium is different, but the abstract phenomenon it supports is the same. It's a universal. That's sort of hard to grasp.

TORTOISE: Maybe so, but it's right. They all have isomorphic, identical senses of "I". There is just *one* sense of the word—just one referent—just one abstract pattern—yet each one seems to feel *it* knows it *uniquely*! There's a kind of fight for sole possession of something that everyone shares.

ACHILLES: *Soul* possession, Mr. T?

TORTOISE: Very astute, Achilles.

ACHILLES: Do you really believe there is just *one* "I", Mr. T?

TORTOISE: Not quite—an exaggeration for rhetorical purposes. The real point is, there's only *one mechanism* underlying "I-ness": namely, the circling-back of a complex representation of the system together with its representations of all the rest of the world. Which "I" you are is determined by the *way* you carry out that cycling, and the way you represent the world.

ACHILLES: So you mean that all that determines who "I" am is the set of experiences some organism has gone through?

TORTOISE: Not at all. I said "the *way* things are cycled", not "*which* things are so cycled and represented". You've got to distinguish between the *set* of objects represented, and the overall *style* with which they are represented. It's that *style* that determines how the loop will loop. *That's* what creates the uniqueness of each "I".

ACHILLES: Well, Mr. T, I think I am beginning to see your point. It's just *so* hard, emotionally, to acknowledge that a "soul" emerges from so physical a system as a careenium.

TORTOISE: The trick is in seeing the curious bidirectional causality operating between the levels of the system, and in integrating that vision with a sense of how symbols have representational power, including the power to recognize certain qualities of their own activity, even though only approximately. This is the crux of the mental, and the source of the enigma of "I".

Post Scriptum.

This piece was inspired by my brief contact with a brilliant meteor, my friend Randy Read, a psychiatrist and writer who lived in San Diego. Randy died about a year ago, as I write this, yet I still feel his spirit resonating in mine. I sometimes don't understand why. I barely knew him, in a way—and yet in another way, for a short time I think I was his best friend.

I got to know Randy through his letters, beginning in 1979. The first letters from him were intriguing and full of freshness, but there was also a definite brashness about them that made me hold back for quite a while. Over time, however, his exuberant way with words and his relentless questing for complex truths grew on me, and I came to understand that much of his brashness was just for fun. What came through loud and clear was a tremendous passion to know and to revel in nature, and to find beauty. He wrote of the loneliness of "edge life"—a balance somewhere between the impulsive and the reflective, the intellectual and the emotional.

I met Randy Read face to face on only a couple of occasions, but during them I came to appreciate even more his keen sensitivity for beauty, especially music, and his intense zest for life. I found him to be a surprisingly warm and strangely vulnerable human being. We once did a little rock climbing on the Pacific coast, a very memorable occasion for me, since I could enjoy Randy's outdoor side and his verbal side at the same time. After that we continued to correspond and I kept his many idiosyncratic letters and cards.

In 1981, we tried to collaborate via mail and telephone on an expanded version of the Careenium dialogue, but it didn't work out as well as we had hoped. Though the outcome was less than perfect, this joint effort afforded us both much pleasure. After that, our correspondence tailed off somewhat, but still I thought of Randy quite often.

When I first learned that Randy Read had died, I was absolutely stunned. I had no idea how much he had touched me. Here are some lines I wrote to myself just after the shock.

> RR was a quester, a seeker, a reacher, a flailer, a yearner, a powerful, wonderful, spiritual soarer who could not deal, somehow, with flat mundane reality. No, that's not accurate. He *loved* reality, every tiny cubic inch of it, every nook and cranny filled with paradox, purity, poetry, power.

As a memorial to Randy Read, I would like to show how his ideas inspired the Careenium piece and triggered a myriad other thoughts spread throughout this book. Randy was a remarkable extemporaneous muser. His thoughts came out fast and furious in wonderfully chosen words. Fortunately, he often used a dictating machine and then sent me letters that were, in essence, transcriptions of his musings. Herewith, then, I present a few of my favorite selections from "The Randy Reader".

*   *   *

March 31, 1980

I'm sitting here today in the warm sunshine sipping on some papaya juice and contemplating the mysteries of life. What better time to inflict you with another letter!

Well, here goes: I've just returned from another one of my forays into the mountains. I was up in the Sierra Nevadas in California and had a nice time despite the occasional blizzard. This has been a high-risk year for avalanches in the Sierras, so that more than the usual circumspection was necessary.

Avalanches, even small ones, are extraordinarily deadly, yet notoriously hard to predict.

Avalanche prediction is such a black art that meteorology seems a hard science by comparison. Snow pack, as you may know, has the broadest range of physical properties of any known physical substance. It can range from *névé*, an almost crystalline ice-like mass, to the flour-fine dust of a powder avalanche. Snow pack is plastic, elastic, rigid, brittle, solid, or liquid, depending on the interacting physical factors.

An avalanche of snow pack is often caused by such a complex net of factors that the phenomenon seems almost to possess life. One doesn't have to be a fanatical animist to sense, in some complex systems, businesses that resemble the commerce of life. One ready analogy is that an avalanche with its almost "all-or-nothing" response resembles the triggering of a nerve cell. A small irritant input, usually the mountaineer himself, unleashes an enormous orgasmic response.

Aggregation of loosely coupled elements can produce the "slop" needed for innovation. Counterfactuality in the mind is permitted by an elastic looseness. Likewise, systems such as avalanches defy easy prediction because they don't "have to" be one way.

Primitive peoples often credit such complex systems with a mentality or spirit. Indeed, one does not have to be particularly superstitious to at times feel the presence of a sort of mind in a corniced snow pack. There is a character, a sort of irritable grouchiness at times, a playful perversity at others, but almost always a sense of being that comes from billions of flakes of snow that feel each other's presence.

Crowd and flock behaviors resemble avalanches. Changes of public opinion can be swift; a mob's mood can snap into violent action as abruptly as a collapsing wall of snow. Nations may even have some of a snow pack's perversity, for that matter.

Much of the effort to understand systems like avalanches, the weather, gas flow in internal combustion systems, and other swirly things has rested on an attempt to analyze microscopic elements. Perhaps some day, we'll have the mathematics and measuring devices to pull off this sort of analysis, but such reductionism often misses how the artists do it. The most savvy mountaineers, the most skilled engine tuners rely on an intuitive sense of the personalities of the medium they work with. Measurements have their place, to be sure, but a sense of the phenomenon's personality often guides attempts to find solutions.

I think in time computer sciences will lead the way in developing classifications of big system "personalities". Computer hardware readily lends itself to precise replication and in time, I'm sure we'll develop the sensitivity to recognize big system minds as readily as we do a friend's face. Even today, this is beginning to be true, as quirks in wiring and software can yield distinctive personalities in our existing computer systems.

*   *   *

January 20, 1981

In the natural environment, one can never achieve a zero error rate. So nature, in her delicate brilliance, finds a way to make errors work. It's sort of like this: outrageous errors are deadly, but slight errors keep things loose enough for the new to appear. If the steps of the error can be kept small enough, a near-perfect approximation can gradually be built.

It's hard for me to put into words, but qualities like "quiveriness" and "vulnerability" come to mind when I think of creativity. Maybe another image is something like the doping used in the manufacture of semiconductors. Introducing small bits of something that doesn't belong there can dramatically change the structure of a large-scale array. It's not really randomness that is the vital ingredient in creativity, but rather the slight tint of the subtle truths that creativity finds that changes the random from white to pink. Michelangelo, standing before the marble that was to become David, had in his mind the image of a young man, but also allowed himself to be invaded by the block of marble, the whole universe itself. Creativity requires a sense of smell, a palate to taste the scents that make brilliance.

"Chance favors the prepared mind."—Louis Pasteur. I like that quote. All life feeds upon the random. Creativity is simply the *haute cuisine.*

Little ripples that exist or rather persist long enough to be observed are the lowest forms that possess what could be called personality. Rocks and snow tend to have very little personality because the ripples are very small-scale and dominated by randomness. Smoke has more. Indeed I think that's why Bach and Magritte liked pipe smoke and why I, too, confess to enjoy the habit.

A puff of smoke is sort of like a cousin of ours. Little eddies, the loosely coupled systems, shear and spin, and we can observe the gentle drift of the whole ensemble. Bubbling streams have this quality, as do breaking ocean waves. The boiling and roiling, the little pieces, each with their own life, each wavelet connected, interacting, and yet participating in the whole.

*　　*　　*

February 23, 1981

BUTTERFLIES

The best thoughts are the most delicate,
　　fastest, trickiest to capture.
Lepidoptera so different on the wing,
　　than when caught, killed,
　　and proudly displayed.

*　　*　　*

On April 10, 1983, Randy Read took his own life. I don't know why. Perhaps these musings, dancing and sparking in the neurons of a few thousand readers out there, will keep alive, in scattered form, a tiny piece of his soul.

26

Waking Up from the Boolean Dream, or, Subcognition as Computation

July, 1982

Introduction

THE philosopher John Searle has recently made quite a stir in the cognitive-science and philosophy-of-mind circles with his celebrated article "Minds, Brains, and Programs", in which he puts forth his "Chinese room" thought experiment. Its purpose is to reveal as illusory the aims of artificial intelligence, and particularly to discredit what he labels *strong AI*—the belief that a programmed computer can, in principle, be conscious. Various synonymous phrases could be substituted for "be conscious" here, such as:

* *think*;
* *have a soul* (in a humanistic rather than a religious sense);
* *have an inner life*;
* *have semantics* (as distinguished from "mere syntax");
* *have content* (as distinguished from "mere form");
* *have intentionality*;
* *be something it is like something to be* (a weird phrase due to T. Nagel);
* *have personhood*;

and others. Each of these phrases has its own peculiar set of connotations and imagery attached to it, as well as its own history and proponents. For our purposes, however, we shall consider them all as equivalent, and lump them all together, so that the claim of strong AI now becomes very strong indeed.

At the same time, various AI workers have been developing their own philosophies of what AI is, and have developed some useful terms and slogans to describe their endeavor. Some of them are: "information processing", "cognition as computation", "physical symbol system", "symbol manipulation", "expert system", and "knowledge engineering". There is some confusion as to what words like "symbol" and "cognition" actually mean, just as there is some confusion as to what words like "semantics" and "syntax" mean.

It is the purpose of this article to try to delve into the meanings of such elusive terms, and at the same time to shed some light on the views of Searle, on the one hand, and Allen Newell and Herbert Simon, on the other hand —visible AI pioneers who are responsible for several of the terms in the previous paragraph. The thoughts expressed herein were originally triggered by a paper called "Artificial Intelligence: Cognition as Computation", by Avron Barr. However, they can be read completely independently of that paper.

The questions are obviously not trivial, and certainly not resolvable in a single article. Most of the ideas in this article, in fact, were stated earlier and more fully in my book *Gödel, Escher, Bach: an Eternal Golden Braid.* However, it seems worthwhile to extract a certain stream of ideas from that book and to enrich it with some more recent musings and examples, even if the underlying philosophy remains entirely the same. In order to do justice to these complex ideas, many topics must be interwoven, and they include the nature of symbols, meaning, thinking, perception, cognition, and so on. That explains why this article is not three pages long.

Cognition versus Perception:
The 100-millisecond Dividing Line

In Barr's original paper, AI is characterized repeatedly by the phrase "information-processing model of cognition". Although when I first heard that phrase years ago, I tended to accept it as defining the nature of AI, something has gradually come to bother me about it, and I would like to try to articulate that here. Now what's in a word? What's to object to here? I won't attempt to say what's wrong with the phrase so much as try to show what I disagree with in the ideas of those who have promoted it; then perhaps the phrase's connotations will float up to the surface so that other people can see why I am uneasy with it.

I think the disagreement can be put in its sharpest relief in the following way. In 1980, Simon delivered a lecture that I attended (the Procter Award Lecture for the Sigma Xi annual meeting in San Diego), and in it he declared (and I believe I am quoting him nearly verbatim):

> Everything of interest in cognition happens above the 100-millisecond level—
> the time it takes you to recognize your mother.

Well, our disagreement is simple; namely, I take exactly the opposite viewpoint:

Everything of interest in cognition happens below the 100-millisecond level—the time it takes you to recognize your mother.

To me, the major question of AI is this: "What in the world is going on to enable you to convert from 100,000,000 retinal dots into one single word 'mother' in one tenth of a second?" Perception is where it's at!

The Problem of Letterforms: A Test Case for AI

The problem of intelligence, as I see it, is to understand the fluid nature of mental categories, to understand the invariant cores of percepts such as your mother's face, to understand the strangely flexible yet strong boundaries of concepts such as "chair" or the letter 'a'. Years ago, long before computers, Wittgenstein had already recognized the centrality of such questions, in his celebrated discussion of the nonpindownability of the meaning of the word "game". To emphasize this and make the point as starkly as I can, I hereby make the following claim:

The central problem of AI is the question: *What is the letter* 'a'?

Donald Knuth, on hearing me make this claim once, appended, "And what is the letter 'i'?"—an amendment that I gladly accept. In fact, perhaps the best version would be this:

The central problem of AI is: *What are* 'a' *and* 'i'?

By making these claims, I am suggesting that, for any program to handle letterforms with the flexibility that human beings do, it would have to possess full-scale general intelligence.

Many people in AI might protest, pointing out that there already exist programs that have achieved expert-level performance in specialized domains without needing general intelligence. Why should letterforms be any different? My answer would be that specialized domains tend to obscure, rather than clarify, the distinction between strengths and weaknesses of a program. A familiar domain such as letterforms provides much more of an acid test.

To me, it is strange that AI has said so little about this classic problem. To be sure, some work has been done. There are a few groups with interest in letters, but there has been no all-out effort to deal with this quintessential problem of pattern recognition. Since letterform understanding is currently an important target of my own research project in AI, I would like to take a moment and explain why I see it as contrasting so highly with domains at the other end of the "expertise spectrum".

Each letter of the alphabet comes in literally thousands of different "official" versions (typefaces), not to mention millions, billions, trillions, of "unofficial" versions (those handwritten ones that you and I and everyone else produces all the time). There thus arises the obvious question: "How are all 'a's like each other?" The goal of an AI project would be, of course, to give an exact answer in computational terms. However, even taking advantage of the vagueness of ordinary language, one is hard put to find a satisfactory intuitive answer, because we simply come up with phrases such as "They all have the same shape." Clearly, the whole problem is that they *don't* have the same shape. And it does not help to change "shape" to "form", or to tack on phrases such as "basically", "essentially", or "at a conceptual level".

There is also the less obvious question: "How are all the various letters in a single typeface related to each other?" This is a grand analogy problem if ever there were an analogy problem. One is asking for a 'b' that is to the abstract notion of 'b'-ness as a given 'a' is to the abstract notion of 'a'-ness. You have to take the qualities of a given 'a' and, so to speak, "hold them loosely in the hand", as you see how they "slip" into variants of themselves as you try to carry them over to another letter. Here is the very hingepoint of thought, the place where one thing slips into alternate, subjunctive, variations on itself. Here, that "thing" is a very abstract concept—namely, "the way that this particular shape manifests the abstract quality of being an 'a'". The problem of 'a' is thus intimately connected with the problems of 'b' through 'z', and with that of stylistic consistency.

The existence of optical character readers, such as the reading machines invented by Ray Kurzweil for blind people, might lead one to believe at first that the letter-recognition problem has been solved. If one considers the problem a little more carefully, however, one sees that the surface has barely been scratched. In truth, the way that most optical character recognition programs work is by a fancy kind of template matching, in which statistics are done to determine which character, out of a fixed repertoire of, say, 100 stored characters, is the "best match". This is about like assuming that the way I recognize my mother is by comparing the scene in front of me with stored memories of the appearances of tigers, cigarettes, hula hoops, gambling casinos, and can openers (and of course all other things in the world simultaneously), and somehow instantly coming up with the "best match".

The Human Mind and Its Ability to Recognize and Reproduce Forms

The problem of recognizing letters of the alphabet is no less deep than that of recognizing your mother, even if it might seem so, given that the number of Platonic prototype items is on the small side (26, if one ignores all characters but the lowercase alphabet). One can even narrow it down further—to just a handful. As a matter of fact, Godfried Toussaint, editor of the pattern recognition papers for the *IEEE Transactions,* has said to me that he would like to put up a prize for the first program that could tell correctly, given twenty characters that people easily can identify, which ones are 'a's and which are 'b's. To carry out such a task, a program cannot just recognize that a shape is an 'a'; it has to see *how* that shape embodies 'a'-ness. And then, as a test of whether the program really knows its letters, it would have to carry "that style" over to the other letters of the alphabet. This is the goal of my research: To find out how to make letters slip in "similar ways to each other", so as to constitute a consistent artistic style in a typeface—or simply a consistent way of writing the alphabet.

By contrast, most AI work on vision pertains to such things as aerial reconnaissance or robot guidance programs. This would suggest that the basic problem of vision is to figure out how to recognize textures and how to mediate between two and three dimensions. But what about the fact that although we are all marvelous face-recognizers, practically none of us can draw a face at all well—even of someone we love? Most of us are flops at drawing even such simple things as pencils and hands and books. I personally have learned to recognize hundreds of Chinese characters (shapes that involve neither three dimensions nor textures) and yet, on trying to reproduce them from memory, find myself often drawing confused mixtures of characters, leaving out basic components, or worst of all, being unable to recall anything but the vaguest "feel" of the character and not being able to draw a single line.

Closer to home, most of us have read literally millions of, say, 'u's with serifs, yet practically none of us can draw a 'u' with serifs in the standard places, going in the standard directions. (This holds even more for the kind of 'g' you just read, but it is true for any letter of the alphabet.) I suspect that many people—perhaps most—are not even consciously aware of the fact that there are two different types of lowercase 'a' and of lowercase 'g', just as many people seem to have a very hard time drawing a distinction between lowercase and uppercase letters, and a few have a hard time telling letters drawn forward from letters drawn backward.

How can such a fantastic "recognition machine" as our brain be so terrible at rendition? Clearly there must be something very complex going on, enabling us to *accept* things as members of categories and to perceive *how* they are members of those categories, yet not enabling us to reproduce those things from memory. This is a deep mystery.

In his book *Pattern Recognition,* the late Mikhail Bongard, a creative and insightful Russian computer scientist, concludes with a series of 100 puzzles for a visual pattern recognizer, whether human, machine, or alien, and to my mind it is no accident that he caps his set off with letterforms. In other words, he works his way up to letterforms as being at the pinnacle of visual recognition ability. There exists no pattern recognition program in the world today that can come anywhere close to doing those Bongard problems. And yet, Barr cites Simon as writing the following statement:

The evidence for that commonality [between the information processes that are employed by such disparate systems as computers and human nervous systems] is now overwhelming, and the remaining questions about the boundaries of cognitive science have more to do with whether there also exist nontrivial commonalities with information processing in genetic systems than with whether men and machines both think. Wherever the boundary is drawn, there exists today a science of intelligent systems that extends beyond the limits of any single species.

I find it difficult to understand how Simon can believe this, in an era when computers still cannot do basic kinds of *subcognitive* acts (acts that we feel are unconscious, acts that underlie cognition).

In another lecture in 1979 (the opening lecture of the inaugural meeting of the Cognitive Science Society, also in San Diego), I recall Simon proclaiming that, despite much doubting by people not in the know, there is no longer any question as to whether computers can think. If he had meant that there should no longer be any question about whether machines may *eventually* become able to think, or about whether we humans are machines (in some abstract sense of the term), then I would be in accord with his statement. But after hearing and reading such statements over and over again, I don't think that's what he meant at all. I get the impression that Simon genuinely believes that today's machines are intelligent, and that they really do think (or perform "acts of cognition"—to use a bit of jargon that adds nothing to the meaning but makes it sound more scientific). I will come back to that shortly, since it is in essence the central bone of contention in this article, but first a few more remarks on AI domains.

Toy Domains, Technical Domains, Pure Science, and Engineering

There is in AI today a tendency toward flashy, splashy domains—that is, toward developing programs that can do such things as medical diagnosis, geological consultation (for oil prospecting), designing of experiments in molecular biology, molecular spectroscopy, configuring of large computer systems, designing of VLSI circuits, and on and on. Yet there is no program that has common sense; no program that learns things that it has not been explicitly taught how to learn; no program that can recover gracefully from its own errors. The "artificial expertise" programs that do exist are rigid, brittle, inflexible. Like chess programs, they may serve a useful intellectual or even practical purpose, but despite much fanfare, they are not shedding much light on human intelligence. Mostly, they are being developed simply because various agencies or industries fund them.

This does not follow the traditional pattern of basic science. That pattern is to try to isolate a phenomenon, to reduce it to its simplest possible manifestation. For Newton, this meant the falling apple and the moon; for Einstein, the thought experiment of the trains and lightning flashes and, later, the falling elevator; for Mendel, it meant the peas; and so on. You don't tackle the messiest problems before you've tackled the simpler ones; you don't try to run before you can walk. Or, to use a metaphor based on physics, you don't try to tackle a world with friction before you've got a solid understanding of the frictionless world.

Why do AI people eschew "toy domains"? Once, about ten years back, the MIT "blocks world" was a very fashionable domain. Roberts and Guzmán and Waltz wrote programs that pulled visions of three-dimensional blocks out of two-dimensional television-screen dot matrices; Winston, building on their work, wrote a program that could recognize instantiations of certain concepts compounded from elementary blocks in that domain ("arch", "table", "house", and so on); Winograd wrote a program that could "converse" with a person about activities, plans, past events, and some structures in that circumscribed domain; Sussman wrote a program that could write and debug simple programs to carry out tasks in that domain, thus effecting a simple kind of learning. Why, then, did interest in this domain suddenly wane?

Surely no one could claim that the domain was exhausted. Every one of those programs exhibited glaring weaknesses and limitations and specializations. The domain was phenomenally far from being understood by a single, unified program. Here, then, was a nearly ideal domain for exploring what cognition truly is—and it was suddenly dropped. MIT was at one time doing truly basic research on intelligence, and then quit. Much basic research has been supplanted by large teams marketing what they vaunt as "knowledge engineering". Firmly grounded engineering is fine, but it seems to me that this type of engineering is not built upon the solid foundations of a science, but upon a number of recipes that have worked with some success in limited domains.

In my opinion, the proper choice of domain is the critical decision that an AI researcher makes, when beginning a project. If you choose to get involved in medical diagnosis at the expert level, then you are going to get mired in a host of technical problems that have nothing to do with how the mind works. The same goes for the other earlier-cited ponderous domains that current work in expert systems involves. By contrast, if you are in control of your own domain, and can tailor it and prune it so that you keep the essence of the problem while getting rid of extraneous features, then you stand a chance of discovering something fundamental.

Early programs on the nature of analogy (Evans), sequence extrapolation (Simon and Kotovsky, among others), and so on, were moving in the right direction. But then, somehow, it became a common notion that these problems had been solved. Simply because Evans had made a program that could do some very restricted types of visual analogy problem "as well as a high school student", many people thought the book was closed. However, one need only look at Bongard's 100 to see how hopelessly far we are from dealing with analogies. One need only look at any collection of typefaces (look at any magazine's advertisements for a vast variety) to see how enormously far we are from understanding letterforms. As I claimed earlier, letterforms are probably the quintessential problem of pattern recognition. It is both baffling and disturbing to me to see so many people working on imitating cognitive functions at the highest level of sophistication when their programs cannot carry out cognitive functions at much lower levels of sophistication.

AI and the True Nature of Intelligence

There are some notable exceptions. The Schank group at Yale, whose original goal was to develop a program that could understand natural language, has been forced to "retreat", and to devote at least a bit of its attention to the organization of memory, which is certainly at the crux of cognition (because it is part of subcognition, incidentally)—and the group has gracefully accommodated this shift of focus. I will not be at all surprised, however, if eventually the group is forced into yet further retreats—in fact, all the way back to Bongard problems or the like. Why? Simply because their work (on such things as how to discover what "adage" accurately captures the "essence" of a story or episode) already has led them into the deep waters of abstraction, perception, and classification. These are the issues that Bongard problems illustrate so perfectly. Bongard problems are idealized ("frictionless") versions of these critical questions.

It is interesting that Bongard problems are in actuality nothing other than a well-worked-out set of typical IQ-test problems, the kind that Terman and Binet first invented 50 or more years ago. Over the years, many other less talented people have invented similar visual puzzles that had the unfortunate property of being filled with ambiguity and multiple answers. This (among other things) has given IQ tests a bad name. Whether or not IQ is a valid concept, however, there can be little question that the original insight of Terman and Binet—that carefully constructed simple visual analogy problems probe close to the core mechanisms of intelligence—is correct. Perhaps the political climate created a kind of knee-jerk reflex in many cognitive scientists to shy away from anything that smacked of IQ tests, since issues of cultural bias and racism began raising their ugly heads. But one need not be so Pavlovian as to jump whenever a visual analogy problem is placed in front of one. In any case, it will be good when AI people are finally driven back to looking at the insights of people working in the 1920's, such as Wittgenstein and his "games", Koehler and Koffka and Wertheimer and their "gestalts", and Terman and Binet and their IQ-test problems.

I was saying that some AI groups seem to be less afraid of "toy domains", or more accurately put, they seem to be less afraid of stripping down their domain in successive steps, to isolate the core issues of intelligence that it involves. Aside from the Schank group, N. Sridharan and Thorne McCarty at Rutgers have been doing some very interesting work on "prototype deformation", which, although it springs from work in legal reasoning in the quite messy real-world domain of corporate tax law, has been abstracted into a form in which it is perhaps more like a toy domain (or, perhaps less pejorative-sounding, an "idealized domain") than at first would appear.

At the University of California at San Diego, a group led by psychologist Donald Norman has been for years doing work on understanding errors, such as grammatical slips, typing errors, and errors in everyday physical actions, for the insights it may offer into the underlying (subcognitive) mechanisms. (For example, one of Norman's students unbuckled his watch instead of his seatbelt when he drove into his driveway. What an amazing mental slippage!) A group led by Norman and his colleague David Rumelhart has developed a radically different model of cognition largely based on parallel subcognitive events termed "schema activations". The reason that this work is so different in flavor from mainstream AI work is twofold: firstly, these are psychologists who are studying genuine cognition in detail and who are concerned with reproducing it; and secondly, they are not afraid to let their vision of how the *mind* works be inspired by research and speculation about how the *brain* works.

Then there are those people who are working on various programs for perception, whether visual or auditory. One of the most interesting was Hearsay II, a speech-understanding program developed at Carnegie-Mellon, Simon's home. It is therefore very surprising to me that Simon, who surely was very aware of the wonderfully intricate and quite beautiful architecture of Hearsay II, could then make a comment indicating that perception and, in general, subcognitive (under 100 milliseconds) processes, "have no interest".

There are surely many other less publicized groups that are also working on humble domains and on pure problems of mind, but from looking at the proceedings of AI conferences one might get the impression that, indeed, computers must really be able to think these days, since after all, they are doing anything and everything cognitive—from ophthalmology to biology to chemistry to mathematics—even discovering scientific laws from looking at tables of numerical data, to mention one project ("Bacon") that Simon has been involved in. However, there's more to intelligence than meets the AI.

Expert Systems versus Human Fluidity

The problem is, AI programs are carrying out all these *cognitive* activities in the absence of any *subcognitive* activity. There is no substrate that corresponds to what goes on in the brain. There is no fluid recognition and recall and reminding. These programs have no common sense, little sense of similarity or repetition or pattern. They can perceive some patterns as long as they have been anticipated—and particularly, as long as the *place* where they will occur has been anticipated—but they cannot see patterns where nobody told them explicitly to look. They do not learn at a high level of abstraction.

This style is in complete contrast to how people are. People perceive patterns anywhere and everywhere, without knowing in advance where to look. People learn automatically in all aspects of life. These are just facets of common sense. Common sense is not an "area of expertise", but a general—that is, domain-independent—capacity that has to do with fluidity in representation of concepts, an ability to sift what is important from what is not, an ability to find unanticipated analogical similarities between totally different concepts ("reminding", as Schank calls it). We have a long way to go before our programs exhibit this cognitive style.

Recognition of one's mother's face is still nearly as much of a mystery as it was 30 years ago. And what about such things as recognizing family resemblances between people, recognizing a "French" face, recognizing kindness or earnestness or slyness or harshness in a face? Even recognizing age—even sex!—these are fantastically difficult problems. As Donald Knuth has pointed out, we have written programs that can do wonderfully well at what people have to work very hard at doing consciously (*e.g.*, doing integrals, playing chess, medical diagnosis, etc.)—but we have yet to write a program that remotely approaches our ability to do what we do *without* thinking or training—things like understanding a conversation partner with an accent at a loud cocktail party with music blaring in the background, while at the same time overhearing wisps of conversations in the far corner of the room. Or perhaps finding one's way through a forest on an overgrown trail. Or perhaps just doing some anagrams absentmindedly while washing the dishes.

Asking for a program that can discover new scientific laws without having a program that can, say, do anagrams, is like wanting to go to the moon without having the ability to find your way around town. I do not make the comparison idly. The level of performance that Simon and his colleague. Langley wish to achieve in Bacon is on the order of the greatest scientists. It seems they feel that they are but a step away from the mechanization of genius. After his Procter Lecture, Simon was asked by a member of the audience, "How many scientific lifetimes does a five-hour run of Bacon represent?" After a few hundred milliseconds of human information processing, he replied, "Probably not more than one." I don't disagree with that. However, I would have put it differently. I would have said, "Probably not more than one millionth."

Anagrams and Epiphenomena

It's clear that I feel we're much further away from programs that do human-level scientific thinking than Simon does. Personally, I would just like to see a program that can do anagrams the way a person does. Why anagrams? Because they constitute a "toy domain" where some very significant subcognitive processes play the central role.

What I mean is this. When you look at a "Jumble" such as "telkin" in the newspaper, you immediately begin shifting around letters into tentative groups, making such stabs as "knitle", "klinte", "linket", "keltin", "tinkle" —and then you notice that indeed, "tinkle" is a word. The part of this process that I am interested in is the part that precedes the recognition of "tinkle" as a word. It's that part that involves experimentation, based only on the "style" or "feel" of English words—using intuitions about letter affinities, plausible clusters and their stabilities, syllable qualities, and so on. When you first read a Jumble in the newspaper, you play around, rearranging, regrouping, reshuffling, in complex ways that you have no control over. In fact, it feels as if you throw the letters up into the air separately, and when they come down, they have somehow magically "glommed" together in some English-like word! It's a marvelous feeling— and it is anything but cognitive, anything but conscious. (Yet, interestingly, *you* take credit for being good at anagrams, if you are good!)

It turns out that most literate people can handle Jumbles (*i.e.,* single-word anagrams) of five or six letters, sometimes seven or eight letters. With practice, maybe even ten or twelve. But beyond that, it gets very hard to keep the letters in your head. It is especially hard if there are repeated letters, since one tends to get confused about which letters there are multiple copies of. (In one case, I rearranged the letters "dinnal" into "nadlid"—incorrectly. You can try "raregarden", if you dare.) Now in one sense, the fact that the problem gets harder and harder with more and more letters is hardly surprising. It is obviously related to the famous "7 plus or minus 2" figure that psychologist George A. Miller first reported in connection with short-term memory capacity. But there are different ways of interpreting such a connection.

One way to think that this might come about is to assume that concepts for the individual letters get "activated" and then interact. When too many get activated simultaneously, then you get swamped with combinations and you drop some letters and make too many of others, and so on. This view would say that you simply encounter an explosion of connections, and your system gets overloaded. It does not postulate any explicit "storage location" in memory—a fixed set of registers or data structures—in which letters get placed and then shoved around. In this model, short-term memory (and its associated "magic number") is an *epiphenomenon* (or "innocently emergent" phenomenon, as Daniel Dennett calls it), by which I mean it is a consequence that emerges out of the design of the system, a product of many interacting factors, something that was not necessarily known, predictable, or even anticipated to emerge at all. This is the view that I advocate.

A contrasting view might be to build a model of cognition in which you have an explicit structure called "short-term memory", containing about seven (or five, or nine) "slots" into which certain data structures can be fitted, and when it is full, well, then it is full and you have to wait until an empty slot opens up. This is one approach that has been followed by Newell and associates in work on production systems. The problem with this approach is that it takes something that clearly is a very complex consequence of underlying mechanisms and simply plugs it in as an explicit structure, bypassing the question of what those underlying mechanisms might be. It is difficult for me to believe that any model of cognition based on such a "bypass" could be an accurate model.

When a computer's operating system begins thrashing (*i.e.*, bogging down in its timesharing performance) at around 35 users, do you go find the systems programmer and say, "Hey, go raise the thrashing-number in memory from 35 to 60, okay?"? No, you don't. It wouldn't make any sense. This particular value of 35 is not stored in some local spot in the computer's memory where it can be easily accessed and modified. In that way, it is very different from, say, a student's grade in a university's administrative data base, or a letter in a word in an article you're writing on your home computer. That number 35 emerges dynamically from a host of strategic decisions made by the designers of the operating system and the computer's hardware, and so on. It is not available for twiddling. There is no "thrashing-threshold dial" to crank on an operating system, unfortunately.

Why should there be a "short-term-memory-size" dial on an intelligence? Why should 7 be a magic number built into the system explicitly from the start? If the size of short-term memory really were explicitly stored in our genes, then surely it would take only a simple mutation to reset the "dial" at 8 or 9 or 50, so that intelligence would evolve at ever-increasing rates. I doubt that AI people think that this is even remotely close to the truth; and yet they sometimes act as if it made sense to assume it is a close approximation to the truth.

It is standard practice for AI people to bypass epiphenomena ("collective phenomena", if you prefer) by simply installing structures that mimic the superficial features of those epiphenomena. (Such mimics are the "shadows" of genuine cognitive acts, as John Searle calls them in his paper cited above.) The expectation—or at least the hope—is for tremendous performance to issue forth; yet the systems lack the complex underpinning necessary.

The anagrams problem is one that exemplifies mechanisms of thought that AI people have not explored. How do those letters swirl among one another, fluidly and tentatively making and breaking alliances? Glomming together, then coming apart, almost like little biological objects in a cell. AI people have not paid much attention to such problems as anagrams. Perhaps they would say that the problem is "already solved". After all, a virtuoso programmer has made a program print out all possible words that anagrammize into other words in English. Or perhaps they would point out that in principle you can do an "alphabetize" followed by a "hash" and thereby retrieve, from any given set of letters, all the words they anagrammize into. Well, this is all fine and dandy, but it is really beside the point. It is merely a show of brute force, and has nothing to contribute to our understanding of how we actually do anagrams ourselves, just as most chess programs have absolutely nothing to say about how chess masters play (as de Groot, and later, Simon and coworkers have pointed out).

Is the domain of anagrams simply a trivial, silly, "toy" domain? Or is it serious? I maintain that it is a far purer, far more interesting domain than many of the complex real-world domains of the expert systems, precisely because it is so playful, so unconscious, so enjoyable, for people. It is obviously more related to creativity and spontaneity than it is to logical derivations, but that does not make it—or the mode of thinking that it represents—any less worthy of attention. In fact, because it epitomizes the unconscious mode of thought, I think it more worthy of attention.

In short, it seems to me that something fundamental is missing in the orthodox AI "information-processing" model of cognition, and that is some sort of substrate from which intelligence emerges as an epiphenomenon. Most AI people do not want to tackle that kind of underpinning work. Could it be that they really believe that machines already can think, already have concepts, already can do analogies? It seems that a large camp of AI people really do believe these things.

Not Cognition, But Subcognition, Is Computational

Such beliefs arise, in my opinion, from a confusion of levels, exemplified by the title of Barr's paper: "Cognition as Computation". Am I really computing when I think? Admittedly, my neurons may be performing sums in an analog way, but does this pseudo-arithmetical hardware mean that the epiphenomena themselves are also doing arithmetic, or should be—or even *can* be—described in conventional computer-science terminology? Does the fact that taxis stop at red lights mean that traffic jams stop at red lights? One should not confuse the properties of objects with the properties of statistical ensembles of those objects. In this analogy, traffic jams play the role of thoughts and taxis play the role of neurons or neuron-firings. It is not meant to be a deep analogy, only one that emphasizes that what you see at the top level need not have anything to do with the underlying swarm of activities bringing it into existence. In particular, *something can be computational at one level, but not at another level.*

Yet many AI people, despite considerable sophistication in thinking about a given system at different levels, still seem to miss this. Most AI work goes into efforts to build rational thought ("cognition") out of smaller rational thoughts (elementary steps of deduction, for instance, or elementary motions in a tree). It comes down to thinking that what we see at the top level of our minds—our ability to think—comes out of rational "information-processing" activity, with no deeper levels below that.

Many interesting ideas, in fact, have been inspired by this hope. I find much of the work in AI to be fascinating and provocative, yet somehow I feel dissatisfied with the overall trend. For instance, there are some people who believe that the ultimate solution to AI lies in getting better and better theorem-proving mechanisms in some predicate calculus. They have developed extremely efficient and novel ways of thinking about logic. Some people—Simon and Newell particularly—have argued that the ultimate solution lies in getting more and more efficient ways of searching a vast space of possibilities. (They refer to "selective heuristic search" as the key mechanism of intelligence.) Again, many interesting discoveries have come out of this.

Then there are others who think that the key to thought involves making some complex language in which pattern matching or backtracking or inheritance or planning or reflective logic is easily carried out. Now admittedly, such systems, when developed, are good for solving a large class of problems, exemplified by such AI chestnuts as the missionary-and-cannibals problem, cryptarithmetic problems, retrograde chess problems, and many other specialized sorts of basically logical analysis. However, these kinds of techniques of building small logical components up to make large logical structures have not proven good for such things as recognizing your mother, or for drawing the alphabet in a novel and pleasing way.

One group of AI people who seem to have a different attitude consists of those who are working on problems of perception and recognition. There, the idea of coordinating many parallel processes is important, as is the idea that pieces of evidence can add up in a self-reinforcing way, so as to bring about the locking-in of a hypothesis that no one of the pieces of evidence could on its own justify. It is not easy to describe the flavor of this kind of program architecture without going into multiple technical details. However, it is very different in flavor from ones operating in a world where everything comes clean and precategorized—where everything is specified in advance: "There are three missionaries and three cannibals and one boat and one river and . . ." which is immediately turned into a predicate-calculus statement or a frame representation, ready to be manipulated by an "inference engine". The missing link seems to be the one between perception and cognition, which I would rephrase as the link between subcognition and cognition, that gap between the sub-100-millisecond world and the super-100-millisecond world.

Earlier, I mentioned the brain and referred to the "neural substrate" of cognition. Although I am not pressing for a neurophysiological approach to AI, I am unlike many AI people in that I believe that any AI model eventually has to converge to brainlike hardware, or at least to an architecture that at some level of abstraction is "isomorphic" to brain architecture (also at some level of abstraction). This may sound empty, since that level could be anywhere, but I believe that the level at which the isomorphism must apply will turn out to be considerably lower than (I think) most AI people believe. This disagreement is intimately connected to the question of whether cognition should or should not be described as "computation".

Passive Symbols and Formal Rules

One way to explore this disagreement is to look at some of the ways that Simon and Newell express themselves about "symbols".

> At the root of intelligence are symbols, with their denotative power and their susceptibility to manipulation. And symbols can be manufactured of almost anything that can be arranged and patterned and combined. Intelligence is mind implemented by any patternable kind of matter.

From this quotation and others, one can see that to Simon and Newell, a *symbol* seems to be any token, any character inside a computer that has an ASCII code (a standard but arbitrarily assigned sequence of seven bits). To me, by contrast, "symbol" connotes something with representational power. To them (if I am not mistaken), it would be fine to call a bit (inside a computer) or a neuron-firing a "symbol". However, I cannot feel comfortable with that usage of the term.

To me, the crux of the word "symbol" is its connection with the verb "to symbolize", which means "to denote", "to represent", "to stand for", and so on. Now, in the quote above, Simon refers to the "denotative power" of symbols—yet elsewhere in his paper, Barr quotes Simon as saying that thought is "the manipulation of formal tokens". It is not clear to me which side of the fence Simon and Newell really are on.

It takes an immense amount of richness for something to represent something else. The letter 'I' does not in and of itself stand for the person I am, or for the concept of selfhood. That quality comes to it from the way that the word behaves in the totality of the English language. It comes from a massively complex set of usages and patterns and regularities, ones that are regular enough for babies to be able to detect so that they too eventually come to say 'I' to talk about themselves.

Formal tokens such as 'I' or "hamburger" are in themselves empty. They do not denote. Nor can they be made to denote in the full, rich, intuitive sense of the term by having them obey some rules. You can't simply push around some Pnames of Lisp atoms according to complex rules and hope to come out with genuine thought or understanding. (This, by the way, is probably a charitable way to interpret John Searle's point in his above-mentioned paper—namely, as a rebellion against claims that programs that can manipulate tokens such as "John", "ate", "a", "hamburger" actually have understanding. Manipulation of empty tokens is not enough to create understanding—although it is enough to imbue them with meaning in a *limited* sense of the term, as I stress in my book *Gödel, Escher, Bach*—particularly in Chapters II through VI.)

Active Symbols and the Ant Colony Metaphor

So what is enough? What am I advocating? What do I mean by "symbol"? I gave an exposition of my concept of *active symbols* in Chapters XI and XII of *Gödel, Escher, Bach.* However, the notion was first presented in the dialogue "Prelude . . . Ant Fugue" in that book, which revolved about a hypothetical conscious ant colony. The purpose of the discussion was not to speculate about whether ant colonies are conscious or not, but to set up an extended metaphor for brain activity—a framework in which to discuss the relationship between "holistic", or collective, phenomena, and the microscopic events that make them up.

One of the ideas that inspired the dialogue has been stated by E. O. Wilson in his book *The Insect Societies* this way: "Mass communication is defined as the transfer, among groups, of information that a single individual could not pass to another." One has to imagine teams of ants cooperating on tasks, and information passing from team to team that no ant is aware of (if ants indeed are "aware" of information at all—but that is another question). One can carry this up a few levels and imagine hyperhyperteams carrying and passing information that no hyperteam, not to mention team or solitary ant, ever dreamt of.

I feel it is critical to focus on collective phenomena, particularly on the idea that some information or knowledge or ideas can exist at the level of collective activities, while being totally absent at the lowest level. In fact, one can even go so far as to say that *no* information exists at that lowest level. It is hardly an amazing revelation, when transported back to the brain: namely, that no ideas are flowing in those neurotransmitters that spark back and forth between neurons. Yet such a simple notion undermines the idea that thought and "symbol manipulation" are the same thing, if by "symbol" one means a formal token such as a bit or a letter or a Lisp Pname.

What is the difference? Why couldn't symbol manipulation—in the sense that I believe Simon and Newell and many writers on AI mean it— accomplish the same thing? The crux of the matter is that these people see symbols as lifeless, dead, *passive* objects—things to be manipulated by some overlying program. I see symbols—representational structures in the brain (or perhaps someday in a computer)—as *active,* like the imaginary hyperhyperteams in the ant colony. *That* is the level at which denotation takes place, not at the level of the single ant. The single ant has no right to be called "symbolic", because its actions stand for nothing. (Of course, in a real ant colony, we have no reason to believe that teams at *any* level genuinely stand for objects outside the colony (or inside it, for that matter) —but the ant-colony metaphor is only a thinly disguised way of making discussion of the brain more vivid.)

Who Says Active Symbols Are Computational Entities?

It is the vast collections of ants (read "neural firings", if you prefer) that add up to something genuinely symbolic. And who can say whether there exist rules—formal, computational rules—*at the level of the teams themselves* (read "concepts", "ideas", "thoughts") that are of full predictive power in describing how they will flow? I am speaking of rules that allow you to ignore what is going on "down below", yet that still yield perfect or at least very accurate predictions of the teams' behavior.

To be sure, there are phenomenological observations that can be formalized to sound like rules that will describe, very vaguely, how those highest-level teams act. But what guarantee is there that we can skim off the full fluidity of the top-level activity of a brain and encapsulate it—without any lower substrate—in the form of some computational rules?

To ask an analogous question, what guarantee is there that there are rules at the "cloud level" (more properly speaking, the level of cold fronts, isobars, trade winds, and so on) that will allow you to say accurately how the atmosphere is going to behave on a large scale? Perhaps there are no such rules; perhaps weather prediction is an intrinsically intractable problem. Perhaps the behavior of clouds is not expressible in terms that are computational *at their own level,* even if the behavior of the microscopic substrate—the molecules—*is* computational.

The premise of AI is that thoughts themselves are computational entities at their own level. At least this is the premise of the information-processing school of AI, and I have very serious doubts about it.

The difference between my active symbols ("teams") and the passive symbols (ants, tokens) of the information-processing school of AI is that the active symbols flow and act on their own. In other words, there is no higher-level agent (read "program") that reaches down and shoves them around. Active symbols must incorporate within their own structures the wherewithal to trigger and cause actions. They cannot just be passive storehouses, bins, receptacles of data. Yet to Newell and Simon, it seems, even so tiny a thing as a bit is a symbol. This is brought out repeatedly in their writings on "physical symbol systems".

A good term for the little units that a computer manipulates (as well as for neuron firings) is "tokens". All computers are good at "token manipulation"; however, only some—the appropriately programmed ones—could support active symbols. (I prefer not to say that they would carry out "symbol manipulation", since that gets back to that image of a central program shoving around some passive representational structures.) The point is, in such a hypothetical program (and none exists as of yet) the symbols themselves are acting!

A simple analogy from ordinary programming might help to convey the level distinction that I am trying to make here. When a computer is running a Lisp program, does it do function calling? To say "yes" would be unconventional. The conventional view is that *functions* call other functions, and the computer is simply the hardware that *supports* function-calling activity. In somewhat the same sense, although with much more parallelism, symbols activate, or trigger, or awaken, other symbols in a brain.

The brain itself does not "manipulate symbols"; the brain is the medium in which the symbols are floating and in which they trigger each other. There is no central manipulator, no central program. There is simply a vast collection of "teams"—patterns of neural firings that, like teams of ants, trigger other patterns of neural firings. The symbols are not "down there" at the level of the individual firings; they are "up here" where we do our verbalization. We feel those symbols churning within ourselves in somewhat the same way as we feel our stomach churning; we do not *do* symbol manipulation by some sort of act of will, let alone some set of logical rules of deduction. We cannot decide what we will next think of, nor how our thoughts will progress.

Not only are we not symbol manipulators; in fact, quite to the contrary, we are manipulated by our symbols! As Scott Kim once cleverly remarked, rather than speak of "free will", perhaps it is more appropriate to speak of "free won't". This way of looking at things turns everything on its head, placing cognition—that rational-seeming level of our minds—where it belongs, namely as a consequence of much deeper processes of myriads of interacting subcognitive structures. The rational has had entirely too much made of it in AI research; it is time for some of the irrational and subcognitive to be recognized for its pivotal role.

The Substrate of Active Symbols Does Not Symbolize

"Cognition as computation" sounds right to me only if I interpret it quite liberally, namely, as meaning this: "Cognition is an activity that can be supported by computational hardware." But if I interpret it more strictly as "Cognition is an activity that can be achieved by a program that shunts around meaning-carrying objects called symbols in a complicated way", then I don't buy it. In my view, meaning-carrying objects won't submit to being shunted about (it's demeaning); meaning-carrying objects carry meaning only by virtue of being active, autonomous agents themselves. There can't be an overseer program, a pusher-around.

To paraphrase a question asked by neurophysiologist Roger Sperry, "Who shoves whom around inside the computer?" (He asked it of the cranium.) If some program shoves data structures around, then you can bet it's not carrying out cognition. Or more precisely, if the data structures are supposed to be *meaning-carrying,* representational things, then it's not cognition. Of course, in any computer-based realization of genuine cognition, there will have to be, at *some* level of description, programs that shove formal tokens around, but it's only agglomerations of such tokens *en masse* that, above some unclear threshold of collectivity and cooperativity, achieve the status of genuine representation. At that stage, the computer is not shoving them around any more than our brain is shoving thoughts around! The thoughts themselves are causing flow. (This is, I believe, in agreement with Sperry's own way of looking at matters—see, for instance, his article "Mind, Brain, and Humanist Values".) Parallelism and collectivity are of the essence, and in that sense, my response to the title of Barr's paper is *no,* cognition is *not* computation.

At this point, some people might think that I myself sound like John Searle, suggesting that there are elusive "causal powers of the brain" that cannot be captured computationally. I hasten to say that this is not my point of view at all! In my opinion, AI—even Searle's "strong AI"—is still possible, but thought will simply not turn out to be the formal dream of people inspired by predicate calculus or other formalisms. Thought is not a formal activity whose rules exist *at that level.*

Many linguists have maintained that language is a human activity whose nature could be entirely explained at the linguistic level—in terms of complex "grammars", without recourse or reference to anything such as thoughts or concepts. Nowadays many AI people are making a similar mistake: They think that rational thought simply is composed of elementary steps, each of which has some interpretation as an "atom of rational thought", so to speak. That's just not what is going on, however, when neurons fire. On its own, a neuron firing has no meaning, no symbolic quality whatsoever. I believe that those elementary events at the bit level— even at the Lisp-function level (if AI is ever achieved in Lisp, something I seriously doubt)—will have the same quality of *having no interpretation.* It is a level shift as drastic as that between molecules and gases that takes place when thought emerges from billions of in-themselves-meaningless neural firings.

A simple metaphor, hardly demonstrating my point but simply giving its flavor, is provided by Winograd's program SHRDLU, which, using the full power of a very large computer (a DEC-10), could deal with whole numbers up to ten in a conversation about the blocks world. It knew nothing—at its "cognitive" level—of larger numbers. Turing invents a similar example, a rather sly one, in his paper "Computing Machinery and Intelligence", where he has a human ask a computer to do a sum, and the computer pauses 30 seconds and then answers incorrectly. Now this need not be a ruse on the computer's part. It might genuinely have tried to add the two numbers at the *symbol level,* and made a mistake, just as you or I might have, despite having neurons that can add fast.

The point is simply that the lower-level arithmetical processes out of which the higher level of any AI program is composed (the adds, the shifts, the multiplies, and so on) are completely shielded from its view. To be sure,

Winograd could have artificially allowed his program to write little pieces of Lisp code that would execute and return answers to questions in English such as "What is 720 factorial?", but that would be similar to your trying to take advantage of the fact that you have billions of small analog adders in your brain, some time when you are trying to check a long grocery bill. You simply don't have access to those adders! You can't reach them.

Symbol Triggering Patterns Are the Roots of Meaning

What's more, you *oughtn't* to be able to reach them. The world is not sufficiently mathematical for that to be useful in survival. What good would it do a spear thrower to be able to calculate parabolic orbits when in reality there is wind and drag, the spear is not a point mass—and so on? It's quite the contrary: A spear thrower does best by being able to imagine a cluster of approximations of what may happen, and anticipating some plausible consequences of them.

As Jacques Monod in *Chance and Necessity* and Richard Dawkins in *The Selfish Gene* both point out, the real power of brains is that they allow their owners to simulate a variety of plausible futures. This is to be distinguished from the *exact* prediction of eclipses by iterating differential equations step by step far into the future, with very high precision. The brain is a device that has evolved in a less exact world than the pristine one of orbiting planets, and there are always far more chances for the best-laid plans to "gang agley". Therefore, mathematical simulation has to be replaced by abstraction, which involves discarding the irrelevant and making shrewd guesses based on analogy with past experience. Thus the symbols in a brain, rather than playing out a scenario precisely isomorphic to what actually will transpire, play out a few scenarios that are probable or plausible, or even some scenarios from the past that may have no obvious relevance other than as metaphors. (This brings us back to the "adages" of the Yale group.)

Once we abandon perfect mathematical isomorphism as our criterion for symbolizing, and suggest that symbol triggering-patterns are just as related to their suggestive value and their metaphorical richness, this severely complicates the question of what it means when we say that a symbol in the brain "symbolizes" anything. This is closely related to perhaps one of the subtlest issues, in my opinion, that AI should be able to shed light on, and that is the question "What is meaning?" This is actually the crucial issue that John Searle is concerned with in his earlier-mentioned attack on AI; although he camouflages it, and sometimes loses track of it by all sorts of evasive maneuvers, it turns out in the end (see his reply to Dennett in the *New York Review of Books*) that what he is truly concerned with is the "fact" that "computers have no semantics"—and he of course means "Computers do not now have, and never will have, semantics." If he were talking only about the present, I would agree. However, he is making a point in principle, and I believe he is wrong there.

Where do the meanings of the so-called "active symbols", those giant "clouds" of neural activity in the brain, come from? To what do they owe their denotational power? Some people have maintained that it is because the brain is physically attached to sensors and effectors that connect it to the outside world, enabling those "clouds" to mirror the actual state of the world (or at least some parts of it) faithfully, and to affect the world outside as well, through the use of the body. I think that those things are *part* of denotational power, but not its crux. When we daydream or imagine situations, when we dream or plan, we are *not* manipulating the concrete physical world, nor are we sensing it. In imagining fictional or hypothetical or even totally impossible situations we are still making use of, and contributing to, the meaningfulness of our symbolic neural machinery. However, the symbols do not symbolize specific, real, physical objects. The fundamental active symbols of the brain represent *semantic categories*—classes, in AI terminology.

Categories do not point to specific physical objects. However, they can be used as "masters" from which copies—instances—can be rubbed, and then those copies are activated in various conjunctions; these activations then automatically trigger other instance-symbols into activations of various sorts (teams of ants triggering the creation of other teams of ants, sometimes themselves fizzling out). The overall activity will be semantic—meaningful —if it is isomorphic, not necessarily to some actual event in the real world, but to some event that is compatible with all the known constraints on the situation.

Those constraints are not at the molecular or any such fine-grained level; they are at the rather coarse-grained level of ordinary perception. They are to some extent verbalizable constraints. If I utter the Schankian cliché "John went to a restaurant and ate a hamburger", there is genuine representational power in the patterns of activated symbols that your brain sets up, not because some guy named John actually went out and ate a hamburger (although, most likely, this is a situation that has at some time occurred in the world), but because the symbols, with their own "lives" (autonomous ways of triggering other symbols) will, if left alone, cause the playing-out of an imaginary yet realistic scenario. [Note added in press: I have it on good authority that one John Findling of Floyds Knobs, Indiana, did enter a Burger Queen restaurant and did eat one (1) hamburger. This fact, though helpful, would not, through its absence, have seriously marred the arguments of the present article.]

Thus, the key thing that establishes meaningfulness is whether or not the semantic categories are "hooked up" in the proper ways so as to allow realistic scenarios to play themselves out on this "inner stage". That is, the triggering patterns of active symbols must mirror the general trends of how the world works as perceived on a macroscopic level, rather than mirroring the actual events that transpire.

Beyond Intuitive Physics: The Centrality of Slippability

Sometimes this capacity is referred to as "intuitive physics". Intuitive physics is certainly an important ingredient of the triggering patterns needed for an organism's comfortable survival. John McCarthy gives the example of someone able to avoid moving a coffee cup in a certain way, because they can anticipate how it might spill and coffee might get all over their clothes. Note that what is "computed" is a set of alternative rough descriptions for what might happen, rather than one exact "trajectory". This is the nature of intuitive physics.

However, as I stated earlier, there is much more required for symbols to have meaning than simply that their triggering patterns yield an intuitive physics. For instance, if you see someone in a big heavy leg cast and they tell you that their kneecap was acting up, you might think to yourself, "That's quite a nuisance, but it's nothing compared to my friend who has cancer." Now this connection is obviously caused by triggering patterns possessed by symbols representing health problems. But what does this have to do with the laws of motion governing objects or fluids? Precious little. Sideways connections like this, having nothing to do with causality, are equally much of the essence in allowing us to *place situations in perspective*— to compare what actually *is* with what, to our way of seeing things, "might have been" or might even come to be. This ability, no less than intuitive physics, is a central aspect of what meaning is.

This way that any perceived situation has of seeming to be surrounded by a cluster, a halo, of alternative versions of itself, of variations suggested by slipping any of a vast number of features that characterize the situation, seems to me to be at the dead center of thinking. Not much AI work seems to be going on at present to mirror this kind of "slippability". (There are some exceptions. Jaime Carbonell's group working on metaphor and analogy at Carnegie-Mellon is an example. Some other former members of Schank's Yale group have turned toward this as well, such as Michael Dyer and Margot Flowers at UCLA, and Jerry DeJong at Illinois. I would also include myself as another maverick investigating these avenues. Cognitive psychologists such as Stanford's Amos Tversky and Daniel Kahneman of the University of British Columbia have done some very interesting studies of certain types of slippability, though they don't use that term.) This is an issue that I covered in some detail in *Gödel, Escher, Bach,* under various headings such as "slippability", "subjunctive instant replays", "'almost' situations", "conceptual skeletons and conceptual mapping", "alternity" (a term due to George Steiner), and so on.

If we return to the metaphor of the ant colony, we can envision these "symbols with halos" as hyperhyperteams of ants, many of whose members are making what appear to be strange forays in random directions, like flickering tongues of flame spreading out in many directions at once. These tentative probes, which allow the possibility of all sorts of strange lateral connections as from "kneecap" to "cancer", have absolutely no detrimental effect on the total activity of the hyperhyperteam. In fact, quite to the contrary: the hyperhyperteam depends on its members to go wherever their noses lead them. The thing that saves the team—what keeps it coherent— is simply the regular patterns that are sure to emerge out of a random substrate when there are enough constituents. Statistics, in short.

Occasionally, some group of wandering scouts will cause a threshold amount of activity to occur in an unexpected place, and then a whole new area of activity springs up—a new high-level team is activated (or, to return to the brain terminology, a new "symbol" is awakened). Thus, in a brain as in an ant colony, high-level activity spontaneously flows around, driven by the myriad lower-level components' autonomous actions.

AI's Goal Should Be to Bridge the Gap between Cognition and Subcognition

Let me, for a final time, make clear how this is completely in contradistinction to standard computer programs. In a normal program, you can account for every single operation at the bit level, by looking "upward" toward the top-level program. You can trace a high-level function call downward: It calls subroutines that call other subroutines that call this particular machine-language routine that uses these words and in which this particular bit lies. So there is a high-level, global *reason* why this particular bit is being manipulated.

By contrast, in an ant colony, a particular ant's foray is not the carrying-out of some global purpose. It has no interpretation in terms of the overall colony's goals; only when many such actions are considered at once does their statistical quality then emerge as purposeful, or interpretable. Ant actions are not the "translation into machine language" of some "colony-level program". No one ant is essential; even large numbers of ants are dispensable. All that matters is the statistics: thanks to it, the information moves around at a level far above that of the ants. Ditto for neural firings in brains. Not ditto for most current AI programs' architecture.

AI researchers started out thinking that they could reproduce all of cognition through a 100 percent top-down approach: functions calling subfunctions calling subsubfunctions and so on, until it all bottomed out in some primitives. Thus intelligence was thought to be hierarchically decomposable, with high-level cognition at the top driving low-level cognition at the bottom. There were some successes and some difficulties —difficulties particularly in the realm of perception. Then along came such things as production systems and pattern-directed inference. Here, some bottom-up processing was allowed to occur within essentially a top-down context. Gradually, the trend has been shifting. But there still is a large element of top-down quality in AI.

It is my belief that until AI has been stood on its head and is 100 percent bottom-up, it won't achieve the same level or type of intelligence as humans have. To be sure, when that kind of architecture exists, there will still be high-level, global, cognitive events—but they will be epiphenomenal, like those in a brain. They will not in themselves be computational. Rather, they will be constituted out of, and driven by, many many smaller computational events, rather than the reverse. In other words, *subcognition at the bottom will drive cognition at the top.* And, perhaps most importantly, the activities that take place at that cognitive top level will neither have been written nor anticipated by any programmer. This is the essence of what I call *statistically emergent mentality.*

Statistically Emergent Mentality Supersedes the Boolean Dream

Let me then close with a return to the comment of Simon's: "Nothing below 100 milliseconds is of interest in the study of cognition." I cannot imagine a remark about AI with which I could more vehemently disagree. Simon seems to be most concerned with having programs that can imitate chains of serial actions that come from verbal protocols of various experimental subjects. Perhaps, in some domains, even in some relatively complex and technical ones, people have come up with programs that can do this. But what about the simpler, noncognitive acts that in reality are the substrate for those cognitive acts? Whose program carries those out? At present, no one's. Why is this?

It is because AI people have in general tended to cling to a notion that, in some sense, thoughts obey formal rules at the thought level, just as George Boole believed that "the laws of thought" amounted to formal rules for manipulating propositions. I believe that this Boolean dream is at the root of the slogan "Cognition as computation"—and I believe it will turn out to be revealed for what it is: an elegant chimera.

Post Scriptum.

Since writing this diatribe, I have found, to my delight, that there are quite a few fledgling efforts underway in AI that fall squarely under the "statistical emergence" banner. I mentioned the work by Norman and Rumelhart at the Institute for Cognitive Science at the University of California at San Diego. That institute is in fact a hotbed of subversive "PDP" (parallel distributed processing) activity. Paul Smolensky, a PDP researcher there, has developed a theory of perceptual activity directly based on an analogy to the branch of physics known as statistical mechanics, and it includes a mental counterpart to the physical concept of *temperature*. In physics, temperature is a number that measures the degree of random thermal jumbling going on in a system composed of many similar parts. In Smolensky's work, a "computational temperature" controls how much randomness is injected into the system.

Imagine a system that is "looking" at a simple scene. (I mean it has a television camera providing input to a computer.) This system's job is to figure out the most plausible interpretation of what is "out there". Is it the word "READ"? Is it the system's grandmother? Is it Smolensky's dog Mandy? When the system is first faced with a fresh situation, the temperature is high, indicating that the system is in a completely open-minded state, ready to have any ideas activated. As randomly chosen concept fragments (not full concepts) are tried on for size, the system gradually starts developing a sense for what sorts of things "fit". Thus the temperature is lowered a bit, lessening the chances of stray concept fragments floating in and destroying the fragile order that is just beginning to coalesce. As fragments start to fit together coherently, the system continues to turn down its randomness knob.

Gradually, larger conceptual structures begin to form and to confirm each other in a benign, self-reinforcing loop. Furthermore, these high-level structures now bias the probabilities of random activation of lower-level fragments, so that the thermal activity, though still random, is more directed. The system is settling into a stable state that captures, in some internal code, the salient external realities. When it is completely "happy" (or "harmonious", in Smolensky's terminology), then the system's temperature reaches zero: it is "freezing". It is no coincidence that the moment of freezing coincides with the attainment of maximal computational bliss, for the temperature gets lowered only when the system is seen to have made some upward jump in its happiness level.

This idea of stochastically guided convergence to what is called a *globally optimum state* seems to have arisen (as do so many good ideas) in the minds of several people at once, spread around the globe. For all I know, it is an ancient idea (though I will not go so far as to credit the ancient Buddhists with it), but it seems that the atmosphere has to be just right for this kind of spark to "catch". People not involved in AI sometimes have expressed the spirit of this sort of thing very poetically. Here is Henri Poincaré writing in the early part of this century about the genesis of mathematical inspirations:

> Permit me a rough comparison. Figure the future elements of our combinations [full-fledged ideas] as something like the hooked atoms of Epicurus. During the complete repose of the mind, these atoms are motionless, they are, so to speak, hooked to the wall; so this complete rest may be indefinitely prolonged without the atoms meeting, and consequently without any combination between them.

On the other hand, during a period of apparent rest and unconscious work, certain of them are detached from the wall and put in motion. They flash in every direction through the space (I was about to say the room) where they are enclosed, as would, for example, a swarm of gnats or, if you prefer a more learned comparison, like the molecules of gas in the kinematic theory of gases. Then their mutual impacts may produce new combinations . . . .

Now our will did not choose them at random; it pursued a perfectly determined aim. The mobilized atoms are therefore not any atoms whatsoever; they are those from which we might reasonably expect the desired solution. Then the mobilized atoms undergo impacts which make them enter into combinations among themselves or with other atoms at rest which they struck against in their course. Again I beg pardon, my comparison is very rough, but I scarcely know how otherwise to make my thought understood.

And more recently the biologist Lewis Thomas, in his book *The Medusa and the Snail,* wrote this:

At any waking moment the human head is filled alive with molecules of thought called notions. The mind is made up of dense clouds of these structures, flowing at random from place to place, bumping against each other and caroming away to bump again, leaving random, two-step tracks like the paths of Brownian movement. They are small round structures, featureless except for tiny projections that are made to fit and then lock onto certain other particles of thought possessing similar receptors. Much of the time nothing comes of the activity. The probability that one notion will encounter a matched one, fitting closely enough for docking, is at the outset vanishingly small.

But when the mind is heated a little, the movement speeds up and there are more encounters. The probability is raised.

The receptors are branched and complex, with configurations that are wildly variable. For one notion to fit with another it is not required that the inner structure of either member be the same; it is only the outside signal that counts for docking. But when any two are locked together they become a very small memory. Their motion changes. Now, instead of drifting at random through the corridors of the mind, they move in straight lines, turning over and over, searching for other pairs. Docking and locking continue, pairs are coupled to pairs, and aggregates are formed. These have the look of live, purposeful organisms, hunting for new things to fit with, sniffing for matched receptors, turning things over, catching at everything. As they grow in size, anything that seems to fit, even loosely, is tried on, stuck on, hung from the surface wherever there is room. They become like sea creatures, decorated all over with other creatures as living symbionts.

At this stage of its development, each mass of conjoined, separate notions, remembering and searching at the same time, shifts into its own fixed orbit, swinging in long elliptical loops around the center of the mind, rotating slowly as it goes. Now it is an idea.

This poetic passage reminds me of nothing more than my Jumbo system for doing anagrams, which I developed in 1982. There, in what I call the "cytoplasm", letters bash at random into other letters, check each other out a bit, occasionally "mate", then couples continue the search for other compatible couples as well as for more letters they could gobble up to make triples or quaduples. (See Figure 27-3.) Syllables build, sniff at each other's ends, occasionally unite, making word candidates. Then those large "gloms" can undergo internal transformations, break down into their natural subunits or even into elemental smithereens. For instance, "pan-gloss" could become "pang-loss" by *regrouping,* which could then by *spoonerism* become "lang-poss", and so on. Forkerism and kniferism (like spoonerism, only different) are other types of recombination mechanisms, as are sporkerism and foonerism. A typical low-temperature route, meandering through a portion of logological space using these mechanisms, might visit, in sequence, "lang-poss", "lass-pong", "las-spong", "lasp-song", "song-lasp", "son-glasp", and so on. And if, as a consequence of global tension, the temperature rises, the entire bubble may burst apart and we will be left with isolated letters scattered all over the place, with occasional surviving duplets ("ng", maybe) here and there, souvenirs of what it was like before the blast. Sigh . . . Oh, but why suffer pangs of loss? After all, isn't this world, of all possible worlds, the very best?

*　　*　　*

Given the passages from Poincaré and Thomas, I will not claim that these ideas are totally new—but then, why would I want to? Part of my thesis on creativity is that even the best ideas are simply variations on themes already enunciated, discovered by unconscious and random processes of recombination, filtering, and association. In fact, the "fit" between statistical mechanics and "statistical mentalics" is not yet exact, and it is to be hoped that the collective mental temperature of cognitive scientists is high enough that the jiggling-about of ideas in our brains will finally bring the right ones into contact with each other, thus bringing us closer to an accurate view of the physics-cognition connection, allowing the temperature to go down, bringing us even closer to truth, which will lower the temperature still further—and on and on.

Besides Paul Smolensky, there are many other people sniffing about in roughly the same territory. David Rumelhart (mentioned above), James McClelland, and co-workers in the "PDP" group at San Diego have modeled several types of perceptual and cognitive behavior using a system of this sort. Geoffrey Hinton and Scott Fahlman (like Simon and Newell, at Carnegie-Mellon University) and Terrence Sejnowski (of Johns Hopkins) are exploring, via what they call the "Boltzmann machine", "pseudo-neural" models of learning, based on ideas closely resembling those of Smolensky. (The prognosis is good, for "neural" rearranges into "u learn".) J.J. Hopfield of Caltech has studied the statistical properties of neural nets, to see what one can say about the substrate of associative memory. Pentti Kanerva, a highly original and autonomous philosopher-programmer at

Stanford, has done related theoretical work aimed at suggesting plausible substrates underlying the fluidity of memory, and his findings dovetail beautifully with recent observations about the anatomical structure of various areas of the brain. This may be a coincidence and it may not, but there is certainly plenty there to speculate about. Related work has been done by T. Kohonen in Finland, and O. P. Buneman and D. Willshaw in England. James Anderson and Stuart Geman at Brown University have developed theories and models of how collective activity of many individual processing units can have emergent character. Jerome Feldman and colleagues at the University of Rochester have developed what they call a "connectionist" theory of perception and cognition, in which neurons can assemble into stable and not-so-stable aggregates called "coalitions". These shifting alliances are presumed to form the subcognitive basis of fluid cognition. And finally, my group's active projects—Jumbo, Seek-Whence, and Copycat—are all thoroughly permeated with an independently conceived vision of a temperature-controlled randomness at the subcognitive level, out of which emerges, at the cognitive level, a fluid but hopefully not wildly meandering train of thought.

Marsha Meredith, who has been working on implementing a Seek-Whence program, seems to really have taken the idea of "fluid" cognition to heart. In writing up what she has implemented so far, she spoke of the cytoplasm of her system:

> The cytoplasm might be viewed as a soup bubbling with gloms, the bubbles which rise to the top being the system's current view of the sequence. If neighboring bubbles have enough mutual attraction (strong enough bonds) they will combine; otherwise they will either exist independently or burst to permit new bubbles to take their place.

In addition to her cytoplasm, Marsha has created a "Platoplasm" (where Platonic concepts are stored) and a "Socratoplasm" (to mediate between the down-to-earth cytoplasm and the ethereal Platoplasm). Marsha's bubbling, boiling, churning, roiling "Seek-Whence soup" is thus very much like alphabet soup, the only difference being that the good old ABC's have been replaced by 123's.

* * *

I think it would be silly to try to attach credit to any one person for these "soup-cognitive" ideas, for they are in the air, as it were, and the time is simply ripe. This is not to say that they are being roundly welcomed by the whole AI and cognitive science community. There are definite "pro" and "con" camps, and some more neutral observers. There are people who cling to the Boolean dream like it was going out of style! Daniel Dennett has recently coined another term for the same concept: "High Church Computationalism", to which he contrasts what he calls "The New

Connectionism". I like the vision of orthodoxy implied by the former term, but I think the latter term overstresses the role of neural modeling in the new approaches. A model of thought in the new style need not be based so literally on brain hardware that there are neuron-like units and axon-like connections between them. The essence of the dissenting movement lies, it seems to me, in three notions:

(1) asynchronous parallelism;
(2) temperature-controlled randomness;
(3) statistically emergent active symbols.

Actually, for those who understand this intuition well, line 3 alone says it all. How? Well, the phrase "statistically emergent" clearly implies that collective phenomena are involved, in which many independent uncorrelated micro-events, chaotically spread all about in some physical medium, are happening all the time, forming and breaking patterns. This is the imagery attached to lines 1 and 2.

I am reminded, whenever I visualize this kind of thing really clearly, of one fairly old but still influential theory about how water's fluidity emerges out of all the frenetic molecular bumping and banging "down there". This is the theory that goes by the poetic name of *flickering clusters* (referred to also in Chapter 10). The idea is that water molecules can form small and highly ephemeral hydrogen-bonded clusters (with a lifespan even shorter than a mayfly's!). Within microseconds, a group will form and break down again, and its constituent molecules will regroup with other free ones. This is going on, over and over, day and night, second by second, in every tiny drop of water, gadzillions of times. The statistically emergent phenomenon, in that case, is the macroscopic nature of water. In particular, such familiar physical properties of water as its boiling point, density, viscosity, compressibility, and so on are deducible—at least in theory—from such a model.

If one is concerned with minds, however, the phenomena to be explained are less tangible and far more elusive. What seems to most people a primary goal to aim for—and here John Searle and I agree, for once—is that of explaining where *meaning* really comes from, or in other words, a theory of the basis of semantics, or reference. Put in a nutshell, the question is, "What makes mental activity *symbolic*?"

*   *   *

There seems to be a genuine conundrum about how mere matter could possess *reference*. How could a lump of stuff be *about* anything else (let alone about *itself*)? Searle conveniently exempts bio-stuff (or at least neuro-stuff) from this query, assigning to it special "causal powers" that he mysteriously declines to identify but that magically (it would seem) allow brains, or something in them, to refer. This is as thoroughly *ad hoc* as the Boolean dreamers' chutzpah in simply proclaiming that there is no problem at all there, for Lisp atoms *do* refer. The fact of the matter is that an analysis of what reference is has proved a little too tough for both sides so far, and so it degenerates into polemics. Each side already *knows* what "aboutness" is all about, and is most impatient with the other side for its obtuseness. I certainly am just as guilty of this syndrome as any other party, for I too feel I *know* (intuitively and nonverbalizably) just what reference really is, and how it *can* come out of "mere matter" and its patterns. I devoted a very large portion of *Gödel, Escher, Bach* to trying to get across some of those intuitions, and since then I have continued to try to spell them out better (most notably in a paper called "Shakespeare's Plays Weren't Written by Him, But by Someone Else of the Same Name", not co-authored by Gray Clossman and Marsha Meredith but by people of the same names, and in the developing work on roles and analogies, described in Chapter 24 of this book). The questions still seem to stymie the best minds, however.

Does the expression "– –p– –q– – – –" intrinsically mean anything? Does the expression "(SS0+SS0)=SSSS0" intrinsically mean anything? How about "(equals 4 (plus 2 2))" or "2+2=4" or "bpbqd"? What would imbue *one* of them with meaning, if not *all*? If none of these has meaning, then do printed symbols *ever* have meaning? Does an entire set of the *Encyclopaedia Britannica* tumbling out of control in interstellar space have any intrinsic meaning, or is it just an empty lump of nonsymbolic matter? Would it help if we lifted the entire Library of Congress into that selfsame interstellar orbit? If not, why not?

What about a cute little robot that scampers about in your living room, seeking to plug itself into any locatable electric outlet and avoiding banging into furniture? Has it got anything inside it that truly *represents* anything else? If so, why? If not, why not? What about a human-sized robot that roams the world in search of beauty and truth and along the way "emits" strange pieces of weird and garbled "syntactic behavior" such as "This sentence no verb"—might that type of robot possess any shreds of *aboutness*? Or would you have to know precisely what it was made of, down to the the most microscopic fibers of its circuitry? What if it objected to such examination? Would your prior knowledge that it was a robot tell you that it was merely "artificially signaling" such objections, and entitle you (as a *bona fide* sentient being) to override its *ersatz* objections without compunction, and to open it up and dissect it?

*　　*　　*

In a way it is natural but in another way it is curious that most people's threshold for changing their tune on whether or not an organism has a mind and feelings (and "aboutness") seems to lie at just about the point where they can easily identify with the organism. Microbes? "Naah, they're too small." Mosquitos? "Maybe, but they're just mechanical." Mice? "They sure *seem* to experience pain and fear and curiosity." Men? "Well, maybe . . . despite the fact that they don't know what it's like to menstruate."

Such reactions are somewhat natural, but it is curious to me that what seems to be the most convincing is the moving-about in the world, and the perceptual and motor interface. Systems that are not interfaced with our tangible, three-dimensional world via perceptors and motor capacities, no matter how sophisticated their innards, seem to be un-identifiable-with, by most people. I have in mind a certain kind of program that most people would probably find it ludicrous to ever consider conscious: a program that does symbolic mathematical manipulations. Take the famous system called Macsyma, for instance, which can do calculus and algebra problems of a very high order of difficulty. Its performance would have been so unimaginable in the days of Gauss or Euler that many smart people would have gasped and many brilliant people might have worshiped it. No one could pooh-pooh it—but today we do. Today we are "sophisticated". In a way, this is good, but in a way it is bad.

What bothers me is a kind of "hardware chauvinism" that we humans evince. This chauvinism says, "Real Things live in three dimensions; they are made of atoms. Photons bounce off Real Things. Real Things make noises when you drop them. Real Things are material, not insubstantial mental ghosts." The idea that numbers or functions or sets or any other kind of mathematical construct might be Real would provoke guffaws in many if not most intellectual quarters today. The idea that being able to maneuver about in a "space" or "universe" of pure abstractions might entitle a robot to be called "sentient" would be ridiculed to the skies, no matter if the maneuvering in that abstruse high-dimensional space were as supple and graceful as that of the most skilled Olympic ice-skating champion or the greatest jazz pianist.

Speaking of which, the musical universe provides another wonderful testbed. Would a robot able to devise incredibly beautiful, lyrical, flowing passages that brought tears to your eyes be entitled to a bit of empathy? Suppose it were otherwise immobile, its only conception of "reality" being inward-directed rather than something accessible through hands or eyes or ears. How would you feel then?

I personally don't think that such a program could come to exist in actuality, but as a thought experiment it asks something interesting about our conception of sentience. Does access to the "real world" count for a lot? Why should the intangible world of the intellect be any less real than the tangible world of the body? Does it have less structure? No, not if you get to know it. Every type of complexity in the physical world has its mirror image in the world of mathematical constructs, including time. What kind of prejudice is it, then, that biases us in favor of our kind so strongly? As questions of mind and matter grow ever more subtle, we must watch out for tacit assumptions of this sort ever more vigilantly, for they affect us at the deepest level and provide pat answers to exceedingly non-pat questions.

*　　*　　*

The question that launched this digression was what kinds of entities deserve attribution of genuine meaning, genuine symbolicness. Some people, Searle for one, seem to feel that nothing any computer system might do could ever be genuinely symbolic. It might well capture the "shadows" of symbolic activity, but it would never have the "right stuff", that is, the "causal powers of the brain", whether or not it passed the Turing Test. Now, I don't agree at all with Searle about there being an unbreachable machine-mind gap, but I do agree with his skepticism toward orthodox AI's view that we have just about got to the point where computers are using words and symbols with genuine meanings, in the full sense of the term.

The problem is, as I emphasized in the article, that computers' concepts thus far lack slippability (and therefore, their "aboutness" is very weak). The blurry boundaries between human concepts are not well captured by models that try to do blurring explicitly. Such models range from so-called "fuzzy set theory", in which an unblurry amount of blurriness is inserted into the most precise of logical calculi (actually a rather comical idea), to memory models with concepts strung together in complex kinds of webs, with hierarchical and lateral connections galore, even including explicit "hierarchies of variability". Somehow human fluidity is not even approached, though.

The alternate school's recipe is to build symbolic activity up from nonsymbolic activity, rather than presuming that the objects one begins with (Lisp atoms, for instance) can be imbued with all the fluidity one wants by making ever-larger piles of complex rules to push them around in the right ways. I am a strong believer in the idea that symbolicness, like greenness, disintegrates. E. O. Wilson's idea of "mass communication" being "the transfer, among groups, of information that a single individual could not pass to another" seems to me to be at the heart of the idea of statistically emergent active symbols. Somehow, in any genuinely cognitive system, there must be layers upon layers of organization, allowing fluid semantics to emerge at the top level out of rigid syntax at the bottom level. Symbolic events will be broken down into nonsymbolic ones. In the ant-colony metaphor, the top-level hyperhyperteams will be symbolic, hyperteams will be subsymbolic, mere teams will be subsubsymbolic (whatever that means!), and the lowly ants will be totally devoid of symbolicness. Obviously, the number of levels need not be four, but this is enough to make a point: Symbolic events are *not* the primitives of thought.

If you believe in this notion of different layers of collectivity having different degrees of symbolicness and fluidity, then you might ask, "What can we learn from trying to make a system with a small number of such layers?" This is an excellent scientific question. In fact, simply to make a two-layer system in which the upper layer is simultaneously more collective, more symbolic, and more fluid than the lower layer would be the key step —and that is precisely what the statistical-emergence camp is trying to do.

*   *   *

In a way, the AI hope up till recently has been to get away with just one level. This is not dissimilar to the hopes of the brain-research people, who in their own way have wanted to locate everything in just one level: that of neurons. Well, AI people are loosening up and so are brain people, and some meaningful dialogue is beginning. This is a hopeful sign, but some people resent the implications that their long-held views are being challenged. They particularly resent anyone's writing about such matters in a general and philosophical way, full of imagery, meant to stir up the intuitions rather than to present well-known facts dryly and impartially.

My aim in the preceding article, which was solicited expressly for the purpose of interdisciplinary communication (it was published in *The Study of Information: Interdisciplinary Messages,* edited by Fritz Machlup and Una Mansfield), was to spark new intuitions about places where progress is needed—not so much specific new experiments, but new areas for musing and theorizing. I was hoping to stimulate not only AI people but also cognitive psychologists, philosophers of mind, and brain researchers. That is why I used so much imagery and appealed to the intuition.

Allen Newell, whose ideas were criticized in the article, did not take too kindly to it. In his reply (solicited by the book's editors), he dismissed my ideas as nonscientific, despite the fact that all the articles solicited were expressly requested to be personal viewpoints rather than scientific papers. In fact, he treated my article with as much disdain as one would treat a pesky fly that one wanted to swat. I had expected, and would of course have warmly welcomed, a reply discussing the issues in a substantive way.

Fortunately, Newell did spend a page or so doing that kind of thing. He pointed out that in his and Simon's writings, the word "symbol" has always had the meaning of "something that denotes", as distinguished from mere tokens, such as the bits at the bottom level of a computer. He gave several excerpts from articles by Simon and himself, including the following one, referring to the 0's and 1's in a typical computer:

> These entities are not symbols in the sense of our symbol system. They satisfy only part of the requirements for a symbol, namely being the tokens in expressions. It is of course possible to give them full symbolic character by programming an accessing mechanism that gets from them to some data structure.

Newell claims that in my article I have seriously misrepresented his and Simon's well-known views on physical symbol systems. A typical passage where he feels I do so is this one:

> To me . . . 'symbol' connotes something with representational power. To them (if I am not mistaken), it would be fine to call a bit (inside a computer) or a neuron-firing a 'symbol'.

Newell comments bluntly: "Hofstadter is indeed mistaken, absolutely and unequivocally." Now here is an opportunity for substantive discussion! I am glad to reply at that level.

*   *   *

Firstly, I plead guilty to one count of misrepresentation of the Newell-Simon view of symbols. I now realize that they place the symbolic level above the bit level; effectively, they place it at the level of Lisp structures. However, I wish to point out that there is a curious vacillation on Newell's part in the paper from which he draws the quote given above. In the first part of the paper, he repeatedly uses the word "symbol" to refer to the 0's and 1's in a Turing machine. In fact, he does it so often that a naïve reader *might* conclude that Newell considers them to *be* symbols. But no! It turns out that after more than a dozen such usages, he turns right around and repudiates any such usage, in the passage quoted above. That, I submit, is hardly clarity in writing, and I would request that it be considered by the jury as constituting mitigating circumstances, possibly providing grounds for a reduced sentence for my client.

But there is a more substantive area of disagreement. Newell repeatedly makes the point that for him, a physical symbol is virtually identical to a Lisp atom with an attached list (usually called its "property list"). He says as much: "That Lisp is a close approximation to a pure symbol system is often not accorded the weight it deserves." And later on, he refers to his paradigmatic physical symbol system as "a garden variety, Lisp-ish sort of beast". (It is no coincidence that the name of one company making Lisp machines is "Symbolics".) Throughout his article, Newell refers to the *manipulation of symbols* by programs (although strangely, he avoids the word "program"). I may have been "mistaken, absolutely and unequivocally" in attributing to Newell and Simon the view that bits are symbols, but I am certainly not mistaken in attributing to them the view that a Lisp atom with attached property list has all the prerequisites of being a genuine symbol, as long as the right program is manipulating it. That much is crystal-clear. And that is the view I was opposing, no less than the view of bits as symbols.

As a sidelight, it is an amusing coincidence that John Searle was quite upset when, in *The Mind's I,* I misquoted him, saying he had said "a few bits of paper" when he had actually said "slips of paper". Now I find myself in a similar situation: I accused someone of having said "bits" when they meant something else. Searle meant "slips"; Newell meant "lisps" (Lisp atoms or lists). And in both cases, although I admit I was wrong in detail, I feel I was entirely right in principle. My arguments remain unchanged even after the misquotation is corrected.

To some, the build-up of atoms from bits might seem to resemble the first layer of emergence of fluid semantics from rigid syntax that I was speaking of earlier. So couldn't a view that sees Lisp structures as slightly more fluid than bits be somewhat consistent with my view? My answer is *no*, and here's why. The rules governing Lisp structures are strictly computational in and of themselves, and implementing a Lisp system in 0's-and-1's hardware adds nothing enriching to the Lisp atoms whatsoever. The logic of a Lisp system does not emerge from the details of levels below it; it is present in full in the written program even without any computer that can run Lisp. In that sense, Lisp programs are Platonic, which is so well demonstrated by Gödel's original "Lisp program", written way back in 1931, before computers existed. In fact, the only distinction between bits and atoms is in number: There are only two types of bit, whereas there can be an arbitrarily large variety of atoms. But as for *fluidity,* nothing is gained by moving from the bit level to the atom level. Either level is 100 percent formal in operation.

What we are looking for, however, in explaining cognition, is *a bridge between the formal and the informal.* Now it may be that Newell does not believe in cognition's informality, and I probably would not be able to convince him of it. Indeed, it would be hard to convince anyone who doesn't see it already that it is reasonable to think of human cognition in those terms, but that is how I see it. And statistical emergence seems to me to be not merely a shot in the dark, but the obvious route to explore. The brain certainly does an immense amount in parallel, with different parts operating completely independently from others. There is known to be a lot of "noise", or randomness, in the brain, and moreoever, the world itself is acting on the brain in so many different ways at once that it is like being bombarded simultaneously with the output of a thousand different random number generators. So temperature there's plenty of. All we need to figure out is what kinds of collective entities could evolve in such a rich medium, how they would interact, and how they could be symbolic.

This is the challenge I was posing to Newell and other staunch believers in the Boolean dream. The debate will continue, but meantime research must be done. And there, everyone must be guided by personal intuitions about what the right path is. Newell and Simon have theirs, and I have mine. We both think we're right. As Wanda Landowska, the famous harpsichordist, once remarked, "You play Bach *your* way, and I'll play him *his* way." How can one reply to that? No way. So let the game go on!

Section VI:
Selection and Stability

Section VI:
Selection and Stability

Ever since self-replicating molecules came about, they have been reproducing like mad and proliferating in ever more varieties. Moreover, they have been agglomerating in gigantic colonies that, seen at their own level, are also self-replicating entities. We ourselves are huge self-replicating molecule-heaps. Ever upward builds this dizzying spire of self-replicating structures. What gives this whole movement any coherent direction? How and why does complexity evolve from simplicity? The gist of the answer is that any active organism engages in external behavior that has repercussions back on the organism. These repercussions then influence the further behavior of the organism, which in turn engenders further repercussions, and so on. Such *feedback* selectively reinforces certain kinds of structures and strategies while suppressing others. Thus through feedback, certain types of organism are stabilized and can become the building blocks for higher-level organisms. In this hierarchical way, very complex yet stable structures and strategies can emerge out of a very primitive bottom level. What is the nature of the structures and strategies that emerge? How stable are they? How inevitable? How arbitrary? How do game-theoretical models and computer simulations shed light on the competition of many organisms in a society? Are all such organisms at odds with each other, acting maximally selfish? Or need selfish organisms always be at odds with each other? Can cooperation and even a seeming morality emerge purely as a consequence of the laws that govern self-replication and the universe's impersonal preference for various styles? The iterated Prisoner's Dilemma, a pithy idealization of such questions, is the focus of the last of these three chapters.

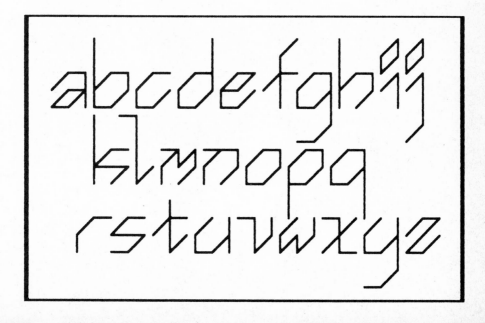

27

The Genetic Code: Arbitrary?

March, 1982

It all began with a pesky student of mine named Vahe Sarkissian. I was telling a class about one of my favorite notions: the analogy between the complex machinery in a living cell that enables a DNA molecule to replicate itself, and the clever machinery in a mathematical system that enables a formula to say things about itself. To my mind, the resemblance is deep and fruitful; it has afforded me sharper insights into both domains. Although Vahe appreciated the analogy, he doubted the validity of one important aspect of it, and so he brought the matter up in class. His challenge forced me to think the issues through carefully, and *en route* I encountered some fascinating details of molecular biology that I might otherwise never have known. What I find gratifying is how quickly people can come to appreciate these intricacies without having studied molecular biology. I'll therefore attempt to sketch out the necessary background, and then I'll explain the problem and what I believe to be its resolution.

Both of the profound twentieth-century discoveries involved in this analogy depend crucially on *codes*: curiously arbitrary-seeming mappings from one set of entities to another set of entities. In metamathematics, the code is *Gödel numbering*; in biology, it is the *genetic code*. In Gödel numbering, invented by Kurt Gödel in about 1930, code numbers are assigned to various mathematical symbols (plus signs, digits, and parentheses, for example), just as license numbers are assigned to cars or telephone area codes to cities. The symbol '0', for instance, might be associated with the quantity 666, and a formula like "0=0" might inherit the number 666,111,666 from its constituent symbols. Gödel's mapping connects entities from two intrinsically unrelated domains, one typographical (printed symbols) and the other abstract (Platonic numbers), and allows any system that can talk about *numbers* to talk about *itself*: in code.

The genetic code is likewise a mapping between two mutually unrelated domains. In this case, though, both domains consist of chemical units. To someone unfamiliar with chemical terminology, the two domains might sound so similar that the connection of one with the other would appear mundane. But actually, nobody had ever in the least suspected that one set of chemicals could *code* for another set. Indeed, the very idea is somewhat baffling: If there is a code, then who invented it? What kinds of messages are written in it? Who writes them? Who reads them? Why not just write the messages *directly,* rather than in code? Are there lots of codes? Are they arbitrary? I would hope that this list of who's and why's would perk up anybody's curiosity, and make them see that a code connecting two unrelated families of chemicals is not to be taken for granted.

* * *

Over the past few billion years, a scheme gradually evolved in living beings according to which a unit of one chemical "species" is assigned as a code for a unit of another "species". Actually, a *triplet* of units of what I will call Species I is assigned to a unit of Species II. In fact (just to make things impossibly complex), sometimes several different triplets are assigned! But for the moment, that is beside the point. The main fact is simply that members of two entirely unrelated species of chemical units are "mapped" onto each other. Vahe's question had to do with how arbitrary this mapping really is. I claimed it was arbitrary, while he claimed there must be some comprehensible rhyme or reason to it.

Species I is the *nucleotides.* Species II is the *amino acids.* If these words are not in your vocabulary, don't panic! In fact, you are very likely my ideal audience. You need not know Word One about chemistry to be able to imagine this correspondence, this match-up between members of two different chemical species. All you need to know is that to each triplet of nucleotides (whatever nucleotides might be), there is matched an amino acid (whatever that is!). That is what the genetic code is.

I mentioned area codes earlier. They are useful to keep in mind, since they too involve triplets (of digits). "212" codes for New York City, "619" for San Diego, and so on. Clearly, these connections are not intrinsic: "619" could as easily have been New York's area code. It is hard to imagine two more disparate domains than three-digit numbers and geographical areas. Yet this mapping—this "telephonic code"—is one of the most useful we know!

The purpose of the genetic code would be hard to describe without adding a little about the constituents of the cell. The "personality", or character, of a cell is stored in its genes. But genes are essentially static, like words in a book. For them to come alive and have observable effects, they must be translated into dynamic agents. These agents are *proteins,* and their actions realize the potential of the genes. They "express" the genes and thereby create the character of the cell. As it turns out, genes are strings of nucleotides, and proteins are strings of amino acids. The cell's personality is therefore written in the passive chemical units of Species I. Through the genetic code, this description can be converted into a vast population of dynamic agents made out of chemical units of Species II. Hence, thanks to the genetic code, the cell's personality, implicitly defined by its genes, can emerge and bloom.

There are twenty different amino acids, so you might think there would be twenty different nucleotide triplets. It's not quite that simple. There happen to be four different nucleotides involved in the genetic code, denoted 'A', 'C', 'G', and 'U' (which stand for *adenine, cytosine, guanine,* and *uracil*). Every possible triplet (beginning with AAA, AAC, AAG and going all the way to UUU) stands for some amino acid. (Well, not quite. Three triplets do not, but for the moment that is just a detail.) How many such triplets are there? Sixty-four, of course: 4×4×4. So 61 (64−3) different triplets are matched up with twenty amino acids. Consequently, most amino acids are coded for by more than one *codon* (a triplet of nucleotides). This would be like having most cities represented by several different "area codons", all just as good as each other. Indeed, there are some amino acids that have six different codons, some that have four, some that have three, and some that have two; only a couple have one. The complete genetic code is shown, for your convenience, in Figure 27-1.

FIGURE 27–1. *The genetic code. A typical codon such as "CAU" is seen to represent the amino acid histidine. Notice the redundancy and partial symmetry of this chart. Are such features important or necessary? Could this chart have been different but life the same?*

	U	C	A	G	
U	phenylalanine	serine	tyrosine	cysteine	U
	phenylalanine	serine	tyrosine	cysteine	C
	leucine	serine	*punctuation*	*punctuation*	A
	leucine	serine	*punctuation*	tryptophan	G
C	leucine	proline	histidine	arginine	U
	leucine	proline	histidine	arginine	C
	leucine	proline	glutamine	arginine	A
	leucine	proline	glutamine	arginine	G
A	isoleucine	threonine	asparagine	serine	U
	isoleucine	threonine	asparagine	serine	C
	isoleucine	threonine	lysine	arginine	A
	methionine	threonine	lysine	arginine	G
G	valine	alanine	aspartic acid	glycine	U
	valine	alanine	aspartic acid	glycine	C
	valine	alanine	glutamic acid	glycine	A
	valine	alanine	glutamic acid	glycine	G

Back to Vahe and me. I had pointed out to my class that Gödel's numbering scheme was quite arbitrary. Gödel could have made just about any number correspond to each of the mathematical symbols involved; it would not have made the slightest difference to the success of his work. I had then said that the analogous statement would hold for the genetic code. But Vahe had the feeling that the genetic code, so tied in with the secret of life, must be deeper. It seemed to him intuitively that each amino acid should be related to its particular codon (or codons) for some compelling reason—that there must be some fundamental chemical necessity for the relation. To caricature Vahe's position, one might say: "Gödel's code is the work of a mere mortal, but the genetic code is the work of God. Therefore the genetic code must be perfect, inevitable, and unalterable."

I was quick to retort that, as far as I could tell, such was not the case. I said that the genetic code seemed to me every bit as arbitrary as the way the telephone company assigned area codes, every bit as arbitrary as Gödel's numbering scheme. On the blackboard I drew some pictures of the molecules involved, to show my reasons for thinking this way. But as I stood there at the board, a few things began to nag at me, places where I was not entirely sure of what I was saying, "facts" I knew I really ought to check up on. This new desire to *prove* to my class the arbitrary nature of the genetic code then led me down some fascinating paths in molecular biology, and my findings are what I wish to report here.

*   *   *

Let us return to the cell. A cell is a little hotbed of activity, rather like a miniature town. There are basically two kinds of entity in this town. There are passive objects, "lumps" that just sit around and wait for somebody to do something to them, and there are active agents, "doers" who always want to get in there and do something. These active agents are, for the most part, *enzymes.* (The very definition of an enzyme is that it is a protein that does something.) Each enzyme has a specific job it can carry out, and it does this job on lumps of a specific type. (Actually, enzymes can even act on other enzymes. However, enzymes being acted on have to act "lumpish" during the operation, like normally vigorous patients anesthetized in a dentist's chair. Once they're released, they may well re-become active doers.) How an enzyme works is not our business here. We can just assume that enzymes do their thing, whether it involves splitting some lump in two, welding two lumps together, transporting some lump from one place to another, or performing some other chemical act.

A marvelous thing about cells is that they are so elegantly designed that for many purposes one can totally ignore their chemistry and think just about their logic. In fact, that is the only way I know to think about the goings-on inside cells, since I am not a biochemist. Although I use the chemical names for things, my true image, deep down inside, is hardly one of chemicals. It is really one of little objects that somehow magically behave in certain ways specified in biochemistry books. My view of the chemicals involved in the processes of life is like most people's view of cars: they know what cars will do in all kinds of situations, but they don't really understand how cars work. I get a kick out of batting around technical terms of biochemistry when in fact I understand only their logic—"the molecular logic of the living state", as Albert Lehninger calls it. The fact that one can get away with this is one of the beauties of molecular biology, and it is this beauty that we are celebrating here.

*   *   *

An enzyme is just a kind of protein, and all proteins are strangely curled molecules made up of amino acids. The curliness is crucial. Here is how I think about it. First I imagine a large number of amino acids strung together like plastic snap-beads, or cars of a train. (Like snap-beads or train cars, amino acids have couplings that allow them to hook up at front and back to other amino acids, so they can form an arbitrary sequence.) Then I imagine holding this long string of amino acids between my two hands, tautly stretched to form a straight chain. (See Figure 27-2.) Now I let the

---

FIGURE 27–2.   *If you stretch a protein straight (a), and then let it go, it will snap right back into its natural curled-up form (b), exhibiting its characteristic* tertiary structure. [*Drawing by David Moser.*]

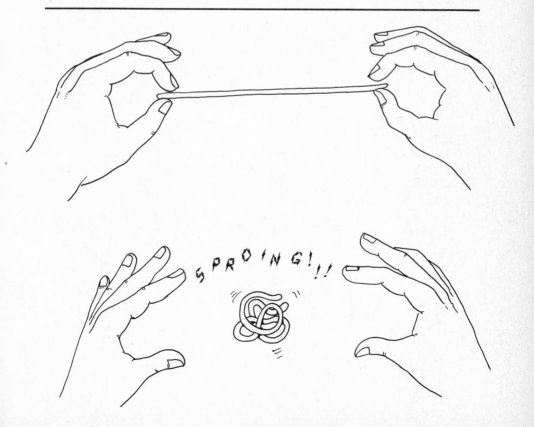

string go. Sproing!! The crazy critter rapidly twists itself up into a tight little ball about the size of a fist. Now *you* try it. Here. Grab the two ends inside the ball and slowly pull them apart. The protein chain is resisting, of course, but if you are careful not to jerk it, you can uncurl it without breaking it anywhere. Got it all straightened out? Good. Now let it go. Sproing!! What do you know? It returns to exactly the same curly shape. Now hand it back to me. Thank you.

It seems that a protein *likes* to be curled up in its little ball. That three-dimensional shape is called its *tertiary structure.* The tertiary structure of each type of protein is unique. The sequence of amino acids totally determines the tertiary structure; it is what makes the protein fold up every time into that same shape. The one-dimensional sequence of amino acids is the protein's *primary structure.* Thus we can say that *a protein's primary structure determines its tertiary structure.* (Some proteins have a secondary structure as well, an intermediate level of coiling like that of a telephone cord, but tertiary structure is the essence of a protein.)

But so what if a protein always has a tertiary structure? So what if it folds up? The answer is that this folded shape is what determines the kind of doer this protein is. (If indeed it is a doer. Some proteins are not enzymes, but mere lumps that serve as boring construction material; however, from here on out, we will be concerned only with *doer* proteins—enzymes.) An enzyme's tertiary structure is characterized by certain bumps and clefts, like the nose and ears of a person's face, except that enzymes differ more radically than faces and are more convoluted. Certain parts of an enzyme are called its *active sites.* They are where the enzyme fastens itself, leechlike, to the lumps it is going to act on. Only by trying to fit various candidate lumps into its active site does an enzyme eventually latch onto a lump of the proper type. (Think of how the Prince found Cinderella by her slipper.) The enzyme and its substrate, the lump, are often likened to a lock and a key. No wrong substrate will fit. (Actually, wrong ones sometimes do fit under special circumstances, but we need not go into that here.)

An enzyme is very specific; it is tailor-made for a certain task and for no other. Once an enzyme is hooked up to its substrates, it starts churning, like a laundromat washing-machine into which the proper coins have been inserted. The enzyme may rip parts off one substrate and attach them to another, it may bind two substrates together—whatever its thing is, it does it. Then it lets go of the product or products, which are now free to go off and drift about inside the cell, like patients after a dentist's appointment. (See Figure 27-3.)

The upshot of all this frenzied activity by billions of busy enzymes doing violent things to chemical lumps inside cells is the creation and sustenance of a unique living organism. These enzymes, these proteins, these coiled-up chains of amino acids—*these* are what carry out the master plan stored passively in the cell's genes, which are chains of—but wait! We are getting ahead of the story.

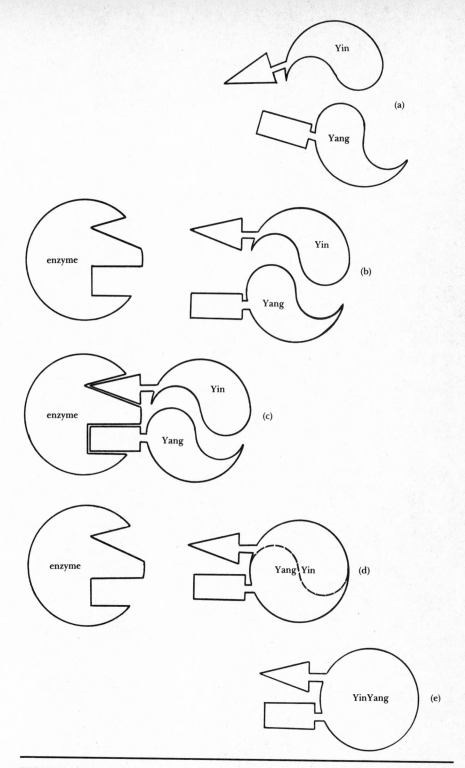

FIGURE 27–3. *The saga of Yin and Yang, two random molecules inside the cytoplasm of a cell. In (a), each drifts alone, unaware of the other. In (b), they approach an* enzyme. *In (c), the enzyme recognizes that each of them fits one of its* active sites, *and snares them. Then it performs its catalytic function, which in this case unites them into one structure (d). Finally, in (e), the new YinYang unit goes on its merry way in the cytoplasm. [Drawing by David Moser.]*

*   *   *

There is nothing more important to know about cells than how enzymes work. They are what make cells run. But there is one other thing that is equally important. That is, *which* enzymes are present, and how they got there. Not all cells have the same set of enzymes, not by a long shot; that's why not all cells have the same character. What's more, a given cell's complement of enzymes can change over time, depending on both internal and external circumstances. Where do the enzymes come from? Ultimately, of course, they come from the genes, which are like blueprints, but that answer does not help at this point. What we need to discuss first is how enzymes are actually built from raw materials, and then we will discuss where their blueprints are stored.

Remember that an enzyme is a protein, and a protein is a long chain of amino acids linked end to end and curled up into a ball. You might think that since enzymes are so good at taking things apart and putting things together, *they* would be the protein-builders. However, the job is so delicate and specialized and critical that a different kind of machine exists to do it. That little machine is the *ribosome,* of which there are thousands in any living cell. A ribosome is partly composed of protein, but it is also partly composed of nucleotides. Its exact composition does not matter to us, though. After all, we are concerned only with the *logic* of the cell.

Is there one ribosome for each kind of protein? Hardly. That would lead to a terrible infinite regress. How would twenty different types of ribosome get constructed? By twenty different types of "metaribosome"? But then how would the twenty different types of metaribosome get constructed? You get the point. In reality, a ribosome is not specific to any protein; it knows nothing about the various proteins it builds. A ribosome is simply a general-purpose amino-acid hooker-upper. But then somebody must tell it *which* amino acids to hook up, and *in what sequence.* But who? For example, suppose the desired chain were *lysine-leucine-glycine-proline-cysteine-histidine-tryptophan.* (I invented this sequence purely for its rhythmic quality. It is too short to be a real protein. Proteins are usually many dozens of amino acids long, like the seemingly endless freight trains that rumble across the midwestern plains and hoot outside your motel room in Wibaux, Montana, late at night.) Who will tell the ribosome to start with lysine and to finish with tryptophan?

At the risk of seeming to invoke a different infinite regress, I shall now reveal that there is *another* train, this one composed of the chemical units of Species I: nucleotides. This train runs right through the middle of the ribosome, like a train through a station. Its cars, taken in triplets, are what tell the ribosome which amino acid goes first, second, third, and so on. This train is called *messenger RNA,* or mRNA (where "RNA" stands for "ribonucleic acid"). A molecule of mRNA is a long chain of A's, C's, G's, and U's. An mRNA chain is much, much longer than a protein chain. It may consist

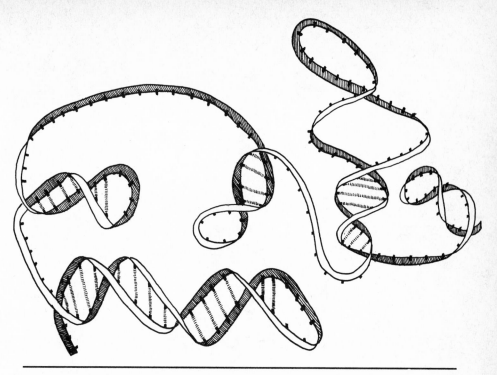

FIGURE 27–4. *A strand of messenger RNA, with some short double-helical regions where it is linked to itself by hydrogen bonds (indicated by dotted lines). [Drawing by David Moser.]*

of thousands of nucleotides strung together like beads. (See Figure 27-4.) An mRNA chain generally codes for several proteins and has special markers along it telling where the stretches representing the various proteins begin and end. That is why there are three special codons that do not stand for any amino acid. They act a little bit like semicolons, in that they convey to the ribosome: "Cut this protein off right now; don't add a single amino acid more!"

We are coming to the crux of the matter. *Where is the genetic code stored?* I have made it sound as if ribosomes "know" the code, but they do not. Although ribosomes do the translating, they know neither language involved. How can this be?

*   *   *

Imagine yourself at the United Nations (see Figure 27-5a). An important speech is about to be given by Mr. Na, the flamboyant ambassador from Nucleotidia. A simultaneous interpreter of great skill, Meri Boso, is summoned. Unfortunately, Ms. Boso has no knowledge of either the language the speech will be in or the language it must be translated into. It looks bad! But at the last moment, just before the speech begins, the members of a rescue team rush into the translating booth, where they suspend from the ceiling a huge number of tiny flash cards. Each card has on its front a word of Nucleotidian (curiously, all the words consist of three

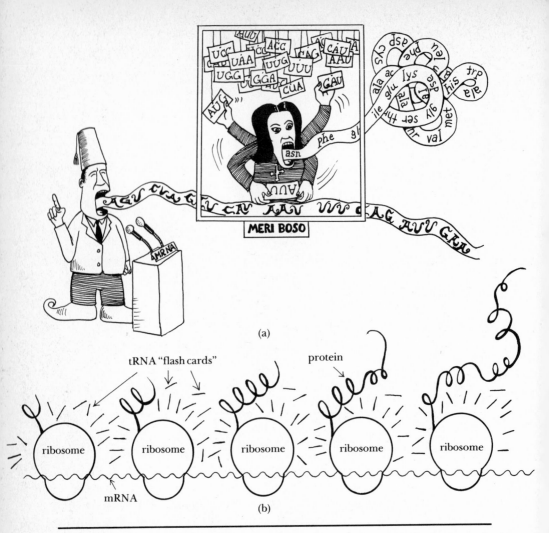

FIGURE 27–5. *Two views of the activity of translation. In (a), Meri Boso in her U.N. translating booth translates Mr. Na's speech from Nucleotidian into Aminoacidian, using the flash cards dangling all about her. In (b), a ribosome chugs down an mRNA strand, surrounded by tRNA molecules. As it reads each codon, it locates a nearby tRNA molecule with the matching anticodon; from that tRNA, it strips off the amino acid, attaching it to a growing protein that is emerging in its folded form into the cytoplasm. [Drawings by David Moser.]*

letters) and on its back the word's translation into the target language, which happens to be Aminoacidian. Meri Boso is saved! All she must do is listen carefully to Mr. Na and then, for each word she hears, find with lightning speed its flash card. Having found the card, she deftly flips it around so that she can speak its Aminoacidian translation in the nick of time into the microphone before her. Next word, please!

It's no sweat, being a ribosome. All you need to do is find the right flash card in a jiffy. But where are the flash cards in the cell? Even more to the point, *what* are they? At this juncture, it seems that the genetic code has receded from view a little; it has become more decentralized, harder to localize. Whereas at first one might have guessed that the genetic code was somehow stored inside each ribosome in the chemical equivalent of a tablet or dictionary, now it seems to lie in those flash cards. So if we want to determine how arbitrary the genetic code is, we must determine whether the flash cards could be changed and, if so, how.

The cell's flash cards are *tRNA*'s (that is, *transfer RNA*'s). The term suggests that they are made out of the same stuff as mRNA is: A's, C's, G's, and U's. This is true, except that some nucleotides are occasionally modified by enzymes, but for our purposes we can ignore that fine detail. At birth, a tRNA is just an ordinary snippet of RNA, but it is much shorter than an mRNA train. Also, quite unlike mRNA, which stays long and snaky, a young tRNA folds up just like a protein, assuming a specific tertiary structure. This is in contrast to mRNA, which merely forms rather aimless curls over short stretches. The curling-up of mRNA is nonfunctional, whereas the curling of tRNA is functional. (Or rather, we don't yet know much about mRNA's curling; probably it is functional but in a more cryptic or subtle way.) All tRNA's fold up into roughly the same shape: a chubby 'L', rather like the bent arm of Mr. America. At a more detailed level, however, the tertiary structures of tRNA's differ. In Figure 27-6, you will find a series of pictures of tRNA at various levels of abstraction.

FIGURE 27–6. *Transfer RNA, viewed at three levels of abstraction. In (a), physically the most realistic, the three-dimensional structure as it has been revealed by X-ray diffraction techniques. In (b), a more schematic "cloverleaf" representation, showing the various hydrogen-bonded loops, as well as the amino-acid attachment site and the anticodon. In (c), the most schematic representation of all, a tRNA molecule is portrayed in its barest functionality: a molecule labeled at one end by an anticodon and potentially carrying at its other end an amino acid.* [*Drawings by David Moser.*]

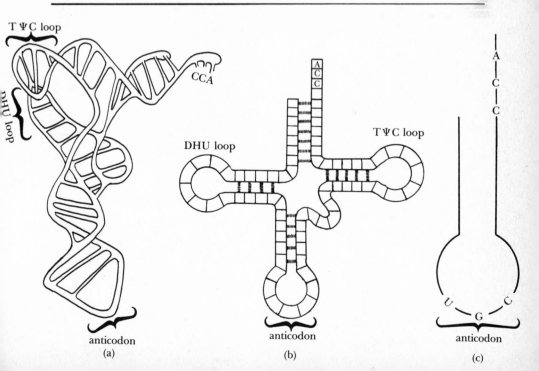

Once it is folded up, a tRNA acts like a flash card, in that it has an amino acid at one end of the 'L', and a codon at the other. Actually, it is not a codon but an *anticodon*. An anticodon is to a codon as a photographic negative is to a positive, or an engraving to a bas-relief. To make one from the other, you merely interchange A with U, and C with G. (A and U are said to be *complementary*, as are C and G.) Therefore CUC and GAG are each other's anticodons. To be more explicit about tRNA, one end of it simply *is* an anticodon. The other end is a site where an amino acid can be attached. And if you're wondering who does the attaching, you'll find out soon enough.

*   *   *

In a nutshell, a ribosome is a translating mechanism between the two intracellular languages of Nucleotidian and Aminoacidian. The words of Nucleotidian are codons; the words of Aminoacidian are amino acids. The mRNA is a long speech whose sentences are written in Nucleotidian. The ribosome is a quick but ignorant simultaneous interpreter who, guided by tRNA molecules, assembles proteins, which are the word-by-word translations of the mRNA sentences into Aminoacidian. (By "quick" I mean the following. Under normal conditions, a ribosome in a bacterial cell can translate about twenty codons per second. In a rabbit cell, things are slower: a little better than one codon per second. I have no idea why rabbits are so much slower than bacteria.)

As is shown in outline in Figure 27-5*b* and in more detail in Figure 27-7, an mRNA "speech" is constantly clicking through the ribosome, one codon at a time. On encountering a new codon, the ribosome must seek out a matching tRNA, one whose anticodon perfectly fits the codon. Of course a ribosome has no eyes, and cannot scan about as Meri Boso does. It must try one tRNA after another (again, think of Cinderella and her slipper). A mystery is how a ribosome can find a matching tRNA so quickly. In any case, having found one and clicked its anticodon into position against the mRNA codon, the ribosome snips off the tRNA's amino acid and snaps it onto the growing protein chain; then it releases the "nude" tRNA, which is free to pick up a new amino acid.

This is a salient difference between the metaphorical flash cards and tRNA molecules. Whereas flash cards can be used over and over again, each time a tRNA molecule gets used, it has to be "recharged" with the right amino acid. Just where and how does this take place? Which amino acid should it get charged with? How is this determined? Who determines it? All of a sudden, these questions loom large, because they have everything to do with the link between a codon and its amino acid. We shall return to them shortly.

It is now apparent that if the genetic code is stored anywhere, it is in a spread-about fashion, distributed among the thousands of tRNA's floating in suspension in the cell near the ribosomes. Could these tRNA's somehow be subverted? Could they falsely guide the translation process? Certainly we

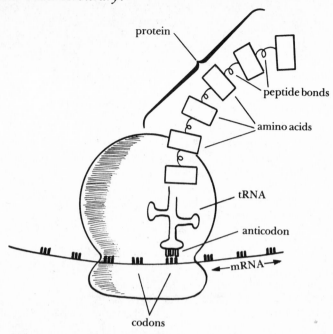

FIGURE 27-7. *The intracellular translation process, in more detail. Inside this ribosome, one can see the matching-up of one tRNA's anticodon (white) with an mRNA codon (black). The amino acid at the top of the tRNA has just been snapped onto the growing chain of amino acids with a "peptide bond", symbolized by the curly link between rectangles. [Drawing by David Moser.]*

can imagine the UN rescue team rushing in with the wrong set of flash cards, hanging them all up in Ms. Boso's booth, and her then translating Mr. Na's speech into a completely inappropriate language. Could the analogue happen in a cell? Could there conceivably be produced an entire set of "bad" tRNA's: tRNA's with wrong amino acids attached to them, tRNA's that would fool the ribosomes into manufacturing nonsensical proteins? Who could perpetrate such a nasty practical joke?

*   *   *

Well, this is the stage I was at when I started drawing pictures on the blackboard for my students. I drew a typical tRNA molecule and stated that at one end—its *AA end*—it would attract a particular amino acid. But why should it attract the *right* amino acid? Simple enough, I thought to myself. As with most chemical affinities in the cell, the AA end of the tRNA would simply have the *right shape*. Each tRNA would lure only the amino acid that (by the genetic code) corresponds to its anticodon. My supposition was that for each anticodon, the tRNA that carried it would be shaped differently at its AA end. And so that's what I drew on the board: a tRNA molecule with

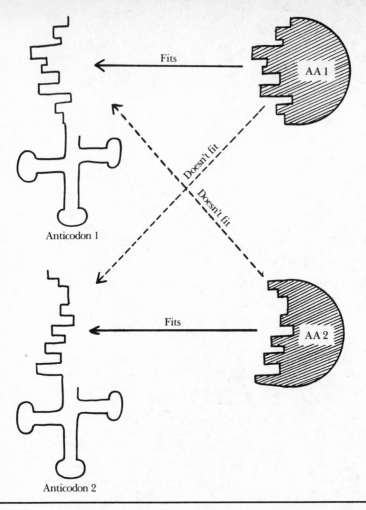

FIGURE 27–8. *My first guess at how tRNA molecules manage (nearly always) to have just the right amino acid attached at their AA ends. In this appealing but simplistic theory, which turned out to be completely wrong, the AA end and the desired amino acid are like lock and key. Thus, only the desired amino acid will fit a given tRNA's AA end. The truth of the matter is that the AA end of a tRNA is completely nonspecific: it will accept any of the twenty types of amino acid. [Drawing by David Moser.]*

a specific anticodon at one end and a specific "attractive shape" at the other end, a shape that would presumably combine with just one kind of amino acid. (See Figure 27-8.)

Here a good question arises. Why should each tRNA attract the *right* amino acid for its anticodon, "right" being the amino acid defined by the genetic code? Why couldn't some tRNA fold up in such a way as to attract some other amino acid? Or is there some intrinsic connection between the two ends of the tRNA? Does the anticodon, for instance, somehow "tell" the other end of the tRNA how to fold up? This was one thought Vahe had, and it would imply that codons and amino acids really had some *direct chemical associations*. But I didn't believe that for a moment.

The Genetic Code: Arbitrary?

I told my class that neither end of the tRNA could possibly know anything about the other. I insisted that you could surgically replace the anticodon with some other anticodon, and the AA end would not know the difference. Conversely, you could surgically lop off the specially shaped AA tip of the tRNA and graft on an alien AA tip, which would then lure the wrong amino acid, thereby making the tRNA embody a false piece of genetic code. I concluded by saying: *"Since the two ends of any tRNA are independent, the genetic code can in principle be subverted and is therefore arbitrary."* Then I blew the chalk dust off my hands and turned to another topic.

Well, it turns out that this picture I had drawn was right in spirit but wrong in detail. Contrary to my supposition, all tRNA molecules have at their AA tip precisely the *same* structure! For instance, the last three nucleotides at the AA tip are always CCA (glance back at Figure 27-6). Thus, the site where the amino acid gets attached is completely *non*specific. There is no special chemical affinity between the AA tip of a tRNA and the amino acid that goes there! When I first found this out (after class was over), I was somewhat at a loss. How, I wondered, does the tRNA always end up with the right amino acid attached to it? What lures it there? Could it be the anticodon, even though it is at the other end of the tRNA? And if so, does that mean that there is, as Vahe surmised, some special and intrinsic link between the anticodon and its amino acid partner? Is the genetic code, after all, inevitable?

*　　*　　*

By talking with biologist friends and looking in books, I found the answer. In the end, it seemed to come out supporting my side, but matters turned out to be far subtler and murkier than I had suspected. Although the AA end of a tRNA molecule is indifferent to the amino acid that docks there, so that in principle it can accept *any* amino acid, under normal circumstances only one amino acid will get attached. However, this is due not to the anticodon but to the tertiary structure of another region of the tRNA: its *DHU loop*. ("DHU" stands for "dihydrouridine", in case you were curious.) This is a loop that every tRNA molecule has, and it bends around in a characteristic shape in each different kind of tRNA. It is therefore a kind of three-dimensional signature by which the tRNA's type can be recognized from the outside. (Actually, as it turns out, probably considerably more is involved in tRNA recognition than just the DHU loop, but for simplicity's sake, I will here continue to speak as if that were the entire story.)

But who could accomplish such recognition? Why, an enzyme, of course —in fact, an *aminoacyl-tRNA synthetase*. (Sorry about that! But despite the strangeness of this name, you should try to remember it, because *these* molecules turn out to play, if not the starring roles in our saga, then certainly pivotal roles.) Such an enzyme has two active sites. One of them recognizes the tRNA's three-dimensional signature, and the other looks for an amino acid. That site, unlike the AA end of the tRNA, is *not* indifferent to the amino acid. It will bind one and only one amino acid—namely, the one coded for by the tRNA's anticodon. To be sure, the synthetase itself never looks at the anticodon. All it does is "sniff" the DHU loops of various tRNA's (and perhaps other substructures as well—as I said, this is still not entirely clear) and when it finds one it "likes", it fastens its amino acid tightly to the tRNA and bids it farewell. For each type of amino acid, there is at least one type of synthetase.

So here we have a funny thing. There are molecules floating around in the cell whose purpose it is to "instruct" the tRNA's in the genetic code. They load up each tRNA with an appropriate amino-acidic burden and then let it trudge off to encounter a ribosome somewhere. So . . . Do the tRNA's know the genetic code? No; they have to be instructed. And who instructs them? The synthetases. Well, then . . . Do the synthetases know the genetic code? No; they merely match up DHU loops of various shapes with amino acids. So in the end, we find out that *nobody* in the cell knows the genetic code!

Of course, that has to be an exaggeration. The truth, again, is simply that "knowledge" of the genetic code is extremely spread out. It is shared by the entire set of tRNA's and synthetases, and cannot be claimed by either one alone. And yet, there is *one* place where one might contend that the genetic code is stored all in one piece . . . "And where, pray tell, is *that*?" you ask. Ah—it is the DNA. You might have been wondering when we would come to DNA, usually the star in tales of molecular biology. Well, this is the moment.

*   *   *

One might regard DNA as a big, fat, aristocratic, lazy, cigar-smoking slob of a molecule. It never does anything. It is the ultimate "lump" of the cell. It merely issues orders, never condescending to do anything itself, quite like a queen bee. How did it get such a cushy position? By ensuring the production of certain enzymes, which do all the dirty work for it. And how can it make sure that these desirable enzymes will get produced? Ah, that is the trick.

DNA is a set of blueprints for all kinds of cellular constituents, lumps and doers alike. If you want to know where something in a cell comes from, the chances are the answer is: It is coded for in the DNA. *The piece of DNA that codes for some specific entity is that entity's gene.* The entity may be a protein, it may be a tRNA molecule, or it may be some RNA that will eventually become part of a ribosome. Whatever the constituent is, there is a gene for it in the long, twisty DNA molecule. Indeed, that is why DNA is so long. The length of the DNA for a mere bacterium can be a million nucleotides—and for a human being, thousands of times longer than that! A DNA strand therefore consists of a sequence of thousands, millions, or even billions of codons, constituting anywhere from a handful of genes to many thousands of them, arranged sequentially, like sentences following one another in a book, or songs in the grooves of a record.

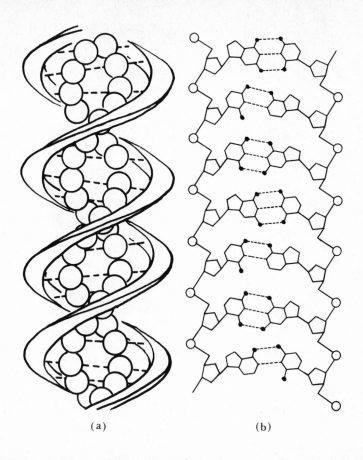

(a)   (b)

FIGURE 27–9.   *DNA at two levels of abstraction.*

In (a), an architectonic view, showing the famous double helix. The outer winding staircases are formed of non-informational matter (sugars and phosphates), while the inner core, represented here by hydrogen-bonded spheres, is where all the genes are stored, defining the entire nature of the cell or organism within which all this is occurring.

In (b), the helices are uncoiled but no bonds are broken. This "flattened" DNA is then spread out like a rug on a floor, allowing you to see exactly how the bases join up with each other in complementary pairs ("Watson-Crick bonding"). [Drawings by David Moser.]

DNA, like RNA, is made up of nucleotides, but instead of U it uses T (which stands for "thymine"). In DNA, A and T, like C and G, are complementary. For every strand of DNA there is a complementary strand that twists around it, making the entire supermolecule look like a double vine (see Figure 27-9). The reason DNA does this while RNA does not is that A and U do not fit together as tightly as A and T do, and so the twists of any would-be RNA double helix are not as stable as those in DNA. Actually, RNA *can* form a double helix for short stretches, but not for long ones. That is also why tRNA's have short double-helical hairpin turns but are not double helices all the way.

I've said several times that an entity's gene is a *coded* version of the entity. Now where there is code, there must be decoding. But attention: There are *two possible depths of decoding,* for a stretch of DNA. (See Figure 27-10.) First of all, you can decode it into RNA. This is the shallower way, and it is done merely by complementation: A codes for U, T for A, and C and G for each other. Thus, DNA stretch "TCAT" becomes RNA stretch "AGUA", which can come in handy if ever you're parched in Paraguay. The deeper way of decoding DNA involves a *second layer of decoding* (shades of the two layers of decoding of Enigma messages, described in Chapter 21)—that is, one must go further, and decode the message contained in the RNA. That, of course, is the job of all the Meri Bosos and their tRNA flash cards.

If the cell wants to make, say, a tRNA molecule, it uses only the shallow decoding, called *transcription.* It finds the tRNA's gene and in effect asks the DNA-decoding enzyme known as *RNA polymerase* to manufacture the corresponding (complementary) segment of RNA. If, however, the cell wants to make a protein, it uses *both* stages. First, as before, the cell gets an RNA polymerase to transcribe the gene for the protein. The result is a long string of messenger RNA. Then this mRNA encounters a ribosome, threads itself through it like tape through a playing head, and as the ribosome then begins clicking down the mRNA, codon after codon, out comes the desired protein. This second stage is called, naturally enough, *translation,* and truly it lies at the dead center of life. (See Figure 27-10.)

And so you see that in our UN scenario, Mr. Na (in whose name I hope you recognized "mRNA") did not write his own speech. Being merely the ambassador from Nucleotidia, he got his speech from the big boss back at home: the DNA. Mr. Na is merely a mouthpiece, a tool. He is just reading a *transcript* of the DNA's speech, and that transcript is slavishly *translated* by Meri Boso (whose name is "ribosome" rotated) into Aminoacidian, according to the genetic-code flash cards. And even those flash cards were dictated by the big boss! You see what I mean about the lazy DNA being in control of everything?

The DNA incorporates coded versions of all the tRNA molecules, of all the synthetases and polymerases, not to mention the constituents of the ribosomes. Thus *the DNA contains coded versions of its own decoders!* By decoding their own genes, the decoders manufacture more copies of themselves. You

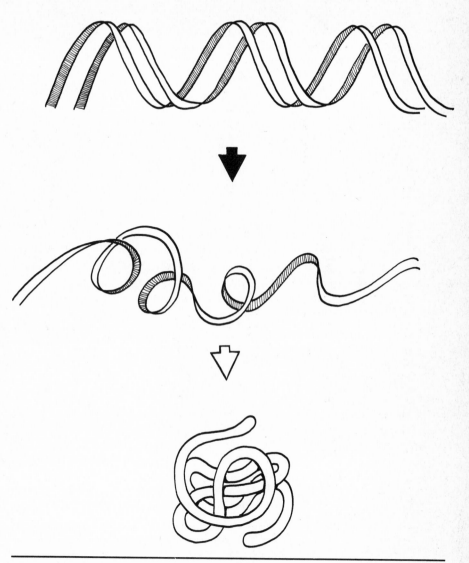

FIGURE 27–10. *The so-called Central Dogma of molecular biology:, "From DNA to RNA to protein." The first conversion is* transcription *(black arrow); the second conversion is* translation *(white arrow). It is now known that* reverse transcription *(from RNA to DNA) takes place in certain organisms and viruses, but* reverse translation *(from proteins to either DNA or RNA) has never been observed. It is safe to say that if it were observed, a tidal wave would be unleashed in biology, wreaking havoc with all sorts of fundamental tenets of the science. It would entail a full-scale return to now-discredited Lamarckian ideas about evolution. In my estimation, a comparable event in physics would be the discovery of a method of accelerating electrons to superluminal speeds, or perhaps the invention of a perpetual-motion machine.* [Drawing by David Moser.]

can see that this is a Grand Loop indeed. The genetic code is *locked in,* because the decoders cannot help but produce more copies of themselves. Not only that, they also produce enzymes that will replicate the DNA they came from, ensuring that new cells will have exactly the same DNA, and will use exactly the same code.

*   *   *

Now, a code that is locked in is not the same thing as a code that is inevitable. For instance, the French language is locked into France: not only do the adults in France speak French among themselves, but also they teach it to their children. Moreover, they publish dictionaries and grammars that stabilize the language. This does not mean, though, that French is the only possible language in the world! The French word for something is not intrinsically tied to that thing by some God-given rule. French is an arbitrary code, as arbitrary as any other human language is—*n'est-ce pas?*

Could it be that the genetic code is likewise changeable, despite being locked in? How could one conceivably subvert it? What would be the chemical equivalent of tampering with the letters in the table of the genetic code? What kind of magic wand would I have to wave over a strand of DNA if I wanted to institute my own personal genetic code?

Let us set up the following hypothetical scenario. We take an ordinary functioning cell, reach into it, and magically remove all its mRNA and tRNA. We throw those molecules into the garbage can. Then we reach back in and remove all the DNA (but we hold onto it), leaving behind a lot of random flotsam and random jetsam, including some ribosomes and some enzymes (RNA polymerases particularly). Now these enzymes and ribosomes have nothing to do, since there is nothing left for them to transcribe or translate. But if they will just be patient, we will be right back! We go off and tamper with the DNA we extracted, and then we inject it back into the unsuspecting cell. Is it possible that not only *this* cell, but also its progeny, will now and forever use our new genetic code? What kinds of changes would we have to have wrought on that piece of DNA for the cell still to *function exactly as before,* except with the new code?

What does "function exactly as before" mean, in this strange context? It means that the cell's behavior should look, to an outside observer, as it did before. What determines its overall functioning from that global point of view? The answer is: its complement of proteins. Proteins, as I said early on, are what endow a cell with its character, its personality. Given this fact, how can we ensure that our cell's *external personality* is unchanged even though its *internal language* has been subverted?

Well, the instant we insert the altered DNA into the cell, many RNA polymerases will start working on it. They will transcribe it into strands of RNA—both short tRNA snippets and long mRNA trains. The tRNA snippets will fold up into their characteristic 'L' shape. At this point, various synthetases encountering the fresh tRNA's will slap amino acids onto them. Then the ribosomes will obediently use the charged-up tRNA's to translate the mRNA's. So, if we are to produce the *same proteins* as before, we have to make sure of two things:

(1) that the new tRNA's embody the new genetic code, and
(2) that the new protein genes are written in the new code.

Goal 1 is tantamount to making sure each tRNA has the right anticodon, according to the *new* code. To achieve this goal, we just need to change three nucleotides in each tRNA gene in the DNA. When those genes are transcribed, they will make tRNA's that are "wrong" in just that one spot. Therefore, the first set of changes to be made is as follows: Find all tRNA genes. In each gene, alter that DNA codon which dictates the tRNA anticodon. To accomplish goal 2, we simply rewrite all the "literature"— that is, all the genes that code for proteins—in the new language. (We're making "wrong" flash cards, but we're also changing Mr. Na's speech, so that what Meri Boso says will in the end be exactly the same.)

*   *   *

Now will things really work as we hoped? For instance, will the synthetases really do the right thing? Well, each tRNA will fold up exactly as before. (Remember that the anticodon has no effect on the way the rest of the tRNA folds up—in particular, no effect on the DHU loop.) Now a synthetase comes along and encounters a familiar-seeming DHU loop. It sticks on the very same amino acid it would have stuck on before. The enzyme has been "fooled" in exactly the way we wished. It has even become an accomplice to our deviltry, because according to the *old* genetic code, the tRNA is carrying the *wrong* anticodon for that amino acid, but according to the *new* code, it is carrying the *right* one.

If you think this scheme through, you'll see that it really works. A piece of "alien" DNA can be inserted into a cell whose *shallow* decoding machinery (a set of RNA polymerases) is present, and the cell will proceed to manufacture new *deep* decoding machinery (ribosomes and tRNA's), and therewith to produce all the proteins coded for in the alien way in the alien DNA. Collectively, these proteins will then imbue the cell with the same external character as it had before, when it was using the ordinary genetic code.

Thus we have succeeded in our aim of showing that the genetic code is just as arbitrary as Gödel numbering is. And along the way, we have accomplished an unforeseen goal: We have spelled out, in quite some detail, just what "arbitrary" means, in the context of cellular codes.

Gratifyingly, it turns out that inside the mitochondria of many organisms, just such a code switch as I described above has taken place! (Mitochondria are semi-autonomous organelles inside cells, and their purpose is to carry out respiration and to produce energy-carrying ATP (adenosine triphosphate) for consumption by the host cell. They contain their own private stock of DNA, tRNA's, ribosomes, and so forth.) The genetic code of mitochondria is very similar to the standard genetic code: it differs in only four codons. Thus it is a *dialect* of Nucleotidian, rather than a completely new language, somewhat as Joual, spoken in parts of Québec, is a dialect of French. Joual is just as locked in to those areas as Parisian is to Paris. Mitochondria have their own tRNA's and their own genes, which would not be properly understood in the main body of the cell—and yet they get along perfectly well, thus confirming my original contention.

* * *

This excursion through the workings of the cell has provided only the barest glimpse of the complexities and subtleties of the interlocking mechanisms that add up to life. Why do there have to be so many stages and so many intermediaries? Why are things accomplished so indirectly?

I am reminded of a visit I made to the R. R. Donnelley plant in Chicago, where *Scientific American* is printed. I was astonished by the degree of indirectness—that is, the layers and layers of intermediary elements—of the complex machines. I kept asking my guide about various wheels and gears and pulley systems that I saw: "What is *this* for?" It always turned out that it gave the plant an extra degree of flexibility in some way that might not have been anticipated at first. In the development of almost any machine, the earliest model is crude. Only the most straightforward applications and circumstances of use are taken into consideration. Then refinements introduced over the years result in levels of complexity that make the evolved system hard, for someone not familiar with it, to understand all at once. In fact, it may become almost impenetrable. This certainly holds for the mechanisms of cars, airplanes, radios, televisions, computers—even of pianos! And on a more intangible plane, this of course applies to human language and culture, and computer software.

In this light, it is not surprising that the cell has so many delicately balanced mechanisms, some of which are there just to compensate for errors made by others. Sometimes biologists and biochemists write about these things in a way that makes it seem they have a wonderful view of the trees but have forgotten about the forest. The way I see the machinery of the living cell, the type of counterfactual thought experiment that comes to my mind—in short, the view presented here—is assuredly not the way a specialist sees it. My counterfactual thought experiment is perhaps not experimentally feasible, but it serves the purpose of casting in stark relief the processes that "pull" a cell's dynamic, sparking unity out of its silent, inert DNA. It highlights the vital intermediary roles that ribosomes and tRNA's play. It serves to focus attention on the logical crux of a cell, as we understand it: the mechanisms of *gene expression*.

The Genetic Code: Arbitrary?

What I get out of a lucid and thorough treatise such as Lehninger's *Biochemistry* is a silhouette: the shadow projected by cellular processes into the space of information-processing concepts. I hope this is not an invalid way of looking at things, because to me that shadow has an eerie but beautiful shape.

Post Scriptum.

Picture this: One fine Easter morn, A-ooga and Duhhh, two strong young protohumans of the year 198,016 B.C., are out looking for brightly colored bison eggs. Unbeknowst to them, a ferocious green snaarfbeest, hidden in the limbs of a nearby billaboo tree, is greedily eying them and looking forward to a couple of nice juicy raw protohuman-burgers (without the bun). The unsuspecting pair are approaching the tree, and as the snaarfbeest tenses up and prepares to leap upon them, A-ooga spies it and begins to yell, "Watch out for antidisestablishmentarianism snaarfbeest!!"—but before she has gotten all the way through that slightly awkward definite article, the beest has leapt, and . . .

Well, I shall spare you the gory details, but one thing I can tell you for sure: Our two brave proto-language-users have become mere sidelines on the evolutionary tree, and aside from the fact that their tragic tale will be retold some 200,000 years later in a postscript concerning ergonomics and the evolution of the genetic code, they are ciphers in the vast scheme of things. What's more, the snaarfbeest has unwittingly done a great favor for users of rival proto-languages: It has reduced by two the number of speakers of A-ooga and Duhhh's proto-language, and thereby strengthened the relative position of all rival proto-languages. With enough such events, it may turn out that the chief rival proto-language, which uses "the" in place of "antidisestablishmentarianism" (and vice versa), will move ahead in the "Top 40" charts for proto-languages.

Obviously this is a ridiculous tale, but I think it gets across an idea: Efficiency in communication really matters. Let us think about what this allegory is saying. A language in which "the" and "antidisestablishmentarianism" were reversed might well survive in the world, if there were no rival languages competing with it. (I can imagine robots using it very happily.) There is nothing *intrinsically* wrong with "the" being an obscure noun, and "antidisestablishmentarianism" being the most common word in the language. Words and things don't really have intrinsic affinities for each other. On the other hand, if you make too many of a language's high-frequency words be long and awkward, sooner or later you are going to cross a threshold where users of the language will be unable to keep up with the speed of events in the world, and their survival will be imperiled. In that sense, things and their names are not *totally* independent, either—

at least not if the names are to be parts of a communication system helping beings to survive. A more compact, more elegant, more *efficient* language will be more able to keep up with real-time needs. Thus the first moral is: *Efficiency matters.*

A second moral, more implicit, is: *Having variants matters.* If you have a number of variants, they can all fight it out and the best will survive while the weaker ones will be pruned. Then the best ones will sprout new variants, and the same process of selection will take place. The ratchet of evolution will advance you towards ever more efficient variants. If, however, there is no mechanism for producing variants, then the unique candidate will live or die simply on the basis of its own qualities *vis-à-vis* the rest of the world.

None of this is new to anyone who knows the first thing about evolution, but it is more general than one might think. In particular, it applies directly to the question of the uniqueness or arbitrariness of the genetic code. Various rival codes certainly would have different efficiencies. If we presume that over many millions of years, a bunch of rival codes arose and competed, and that out of that struggle there emerged a winner, then it is fair to say that the winning code was *not* arbitrary, and that the connections it established between codons and amino acids are preferable to a vast number of other possibilities.

* * *

The allegory indicates one type of pressure: Important words should be short. There is a pressure in all living languages toward short words (think how "car" has supplanted "automobile", and think also of the immense number of abbreviations and acronyms we use). There is a counterpressure, this one towards clarity and a bit of redundancy, so that not every tiny sound is crucial. A difference between classical Chinese and contemporary Chinese is that many words that formerly were just one syllable long now are two. Why this shift away from shortness? Because no language can afford to become too dense. There's not enough room in logological space—or more precisely, in phonological space. We simply can't efficiently distinguish between sounds that are too close together. You can't pack more than so many monosyllabic words into a language before communication begins to suffer. Therefore, among languages there may be fluctuations from denser to lighter, but all languages hover about a norm, which is why translated versions of a given passage printed side by side are all about equally long. A third pressure is of course towards making crucial differences very salient. How risky it would be if the words for "yes" and "no" were as close as, say, "yes" and "yef". In fact, it is rather strange that "can" and "can't" are so close, and it is quite fascinating to observe how many phonetic tricks we native speakers of English unconsciously employ to get around that problem, such as glottal stops, subtle distinctions between the 'a' sounds, and different intonation patterns in order to convey whether we are being positive or negative.

I have been a little glib in using the word "pressure" here. What does "pressure toward shortness", for example, actually mean? Does it mean that each speaker actually feels an obligation to invent shorter words? No, of course not—but if by chance a shorter way to say something comes that speaker's way, it will not be surprising if it gets picked up and adopted, perhaps even without any conscious awareness on the speaker's part. But this means that there must be some source of variety. Otherwise, to speak of "pressure" has no meaning. "Pressure toward X" really is a shorthand way of saying that variants with X will do better than ones without it.

I must slightly qualify this remark about the meaning of "pressure". In a case where beings with goals are capable of deliberately tailoring their behaviors, "pressure toward X" may mean that such a being will sense that having quality X would be more advantageous and will think up a way of getting quality X in its behavior. Thus the source of variety in the being's behavior is internal to the being, and a given variant is purposefully chosen by that being. Another way of looking at this is to say that for sufficiently intelligent beings, variant possibilities can compete in their *minds,* and the outcome of that simulation can determine their behavior. That way, instead of the beings having to gamble with their lives, they just spend a little time "programming" an internal simulation, and tailor their actual behavior according to the results. Thus pressures are actually experienced as such, by individual intelligent beings.

But in most of evolution, the beings are not bright enough to be able to sense pressures, model them internally, and respond to them consciously. Such beings simply must accept the hand the world deals them and do the best they can. In such cases, the meaning of "pressure" is the one involving differential selection rates in the real world (not in a mental one) among variants with and without X—variants whose source is external chance rather than internal reflection. One perceives the effects of such pressure not in an *individual,* but in the shifting statistical makeup of a *population.*

Such is of course the case with tiny bacteria and viruses—the most primitive life forms that presumably tried out variant versions of the genetic code way back when. Although all of these variant codes were "viable", nonetheless some turned out to be "more viable". Now, my column was really about the "although" clause of the preceding sentence. All I was trying to say is that in principle, one could associate "AGA" with *any* amino acid, not just arginine. The "nonetheless" clause of the sentence includes *ergonomic* considerations: those that have to do with efficiency and waste of effort. When you take ergonomics into account as well, then you realize that, for various reasons, certain codings simply are more efficient in terms of information theory, and those are the ones that will eventually emerge on top of the heap.

*   *   *

Along these lines, I had a very interesting letter from Robert J. Gailer of Seattle, who pointed out to me that, contrary to what I had claimed about the telephone system, there was a definite ergonomic basis for the way area codes were assigned to various cities and larger regions. The reason that New York City's area code is 212 and San Diego's is 619, rather than the reverse, is that on dial phones (which were universal when area codes were invented), it takes much longer to dial 619 than to dial 212. And from the point of view of long-term efficiency, when you consider that billions of long-distance phone calls will be made, these choices are pretty sensible. The largest cities have the shortest-dialing-time area codes. Moreover, as Gailer writes:

> Similar codes were assigned to non-contiguous areas to minimize confusion. It was theorized that if New York City had 212 while New Jersey had 213, people could easily get the two mixed up. By assigning 213 to Los Angeles, AT&T hoped to minimize this potential source of confusion.

Thus the next two metro areas in terms of phone population, Chicago and Los Angeles, got 312 and 213, respectively.

Gailer's point (similar to my point about "yes" and "yef") is that you want to make sure that easily confusable but critically different meanings have very different codes. Clearly, if mistakes are inevitable but some are fatal and some aren't, it's obviously much smarter to engineer your code so that nonfatal mistakes will tend to occur rather than fatal ones. In the case of the genetic code, this comes down to the following. Amino acids tend to fall in families (hydrophilic-*vs.*-hydrophobic being the most important class distinction, though there are others). If you make a totally random "spelling error" in a protein (one wrong amino acid), it will almost always destroy the desired function. However, if the mistaken amino acid belongs to the *same family* as the replaced one, then the chance of salvaging some of the functional behavior is much higher. Therefore any code that has *similar codons coding for same-family amino acids* will be highly favored over other codes. This favoring will have to be a result, just to say it once more, of *selection among rival codes*, just as in the allegory that opened this *P.S.*

In sum, there are two different kinds of arbitrariness we are dealing with here. In the column's sense of the term, since alternate genetic codes might survive for a while in the absence of rivals, they are all viable and hence the one that wound up inside our cells *is* arbitrary. In this *P.S.*'s sense of the term, since rivalry *is* a part of the real world, and since selection *will* necessarily take place, the winning code is *not* arbitrary, because it is the most efficient of a bunch of imaginable schemes. (You were right, Vahe!) A number of people made this point very eloquently to me, including Henriette and Miroslav Nadj, to whom I am indebted for the term "ergonomics", Rosemarie Swanson, Nelson Max, and Barry Bunow. The most complete response along these lines was provided by J. M. Labouygues of Clermont-Ferrand, France, who sent me a series of articles by himself and colleagues, in which they describe mathematical studies that have determined the properties of a maximally robust code—that is, one maximally resistant to random mutations. Their work shows that the actual genetic code has those properties, and it also shows how it could have evolved that way.

As it turns out, the idea that the genetic code might be mathematically optimized against mutations was first suggested—I learned this from Labouygues' papers—about twenty years ago by my friend, the late Tracy Sonneborn, an outstanding geneticist at Indiana University and a wonderfully alive, endlessly inquisitive, and deeply warm human being. I laugh to think what Tracy would have thought of my original proposal that the genetic code is arbitrary!

*   *   *

There is another way to look at the question of arbitrariness of the genetic code, and that is to ask whether it is conceivable that one could actually carry out the trick I suggested in the article: namely, instantaneously switch codes in some actual living organism without impairing its life functions. I argued that one could do so by rewriting the "literature" stored in its DNA (namely, the genes coding for all proteins), as well as changing small pieces of its tRNA genes (namely, those coding for the anticodon regions). To the zeroth order, this will succeed. That is, after the switch, the two-step decoding of the DNA would produce all the same proteins as before. If this were all that were needed to make the cell run exactly as it did before the code switch, then we'd be in fine shape. But things are much more complicated than that.

When a cell's DNA looks radically different from how it used to look, but all the same old enzymes are acting on it, something bad is sure to happen. A most articulate discussion of this difficulty was supplied to me by Maurice Guéron in a letter.

> A cell has to know how its DNA is organized: where a gene begins, where it ends, etc. For this, typographical signals are needed. Some determine, on the DNA, the loci which are recognized by RNA polymerases (the enzymes that create the messenger RNA). Others provide clues on the messenger for the ribosome machinery. A few such typographical signals are known—for instance, the 'Pribnow sequence' TATGTTG, which is involved in the recognition of the beginning of a gene. The existence of such signals means that there is in essence a *second* code carried by DNA: Besides the genetic code, which is used in the translation of nucleic acid into protein, there is also a *typographical* code.
>
> The typographical signals may well ruin any efforts to change the genetic code. Indeed, two things would be necessary for your scheme to work. First, the typographical signals should not be on a stretch of DNA that is also part of the genetic messages to be expressed as protein (the so-called "structural" genes); this is in order that rewriting the structural genes in the new code will not mess up the typographical signals. Second, there should be no possibility

that rewriting the structural genes will lead, by chance, to the appearance of new, spurious typographical signals.

It appears that neither of these conditions is satisfied. Regarding the first, typographical sequences, presumably active ones, have been found within structural genes. As for the second, one cannot exclude the creation of spurious typographical sequences by a new genetic code, except through direct checking of all the structural genes.

Lastly, let me make two remarks. First, the existence of two codes means that even with the existing machinery, some messenger RNA sequences must be forbidden, namely those that would have typographical significance. Second, *the locking-in of the code means that the logic of the machinery is deeply interwoven with its hardware.* One may still distinguish various logical 'levels', but the connections between them will not let themselves be forgotten.

Touché! This is precisely my own favorite point used back on me, and I am so delighted by it. It is none other than the statement that in deep translation, you cannot translate content alone—you must pay attention to the interaction between form and content. It harks back to the idea of translating self-referential sentences from one language to another—and what is more self-referential than the beautifully tangled DNA-RNA-proteins loop?

Consider "This sentence is in English." Graduates of the A-ooga-Duhhh Memorial School of Translation, if asked to translate it into Chinese, would produce a Chinese sentence asserting of itself that it is in English. But that's nonsense! Either it should refer to the original English sentence and say that *that* sentence is in *English,* or it should refer to *itself* and say that it is in *Chinese.* This wishy-washy halfway stuff won't do! And such remarks hold with a vengeance for the self-documenting sentences that Lee Sallows struggled with so (see Chapters 2 and 3).

If you substitute the word "pun" or "allusion via form" for "typographical signal" in Guéron's commentary, and think of switching one natural language for another, then you have another way of seeing what he is getting at. It is quite strange but absolutely undeniable that DNA is *full* of puns and allusions via form. One remarkable example is the overlapping-genes pun found in the DNA of the virus φX174. There, two completely different proteins were discovered to be coming from the same section of DNA. Two genetic codes? No, nothing of the sort. Two different *reading frames.* That is, by shifting the DNA over one notch, you get a new set of codons. For example, what reads as ". . .-TGC-CAA-GGT-C . . ." one way reads as ". . .T-GCC-AAG-GTC- . . ." when you shift the frame to the right. And both ways code for proteins indispensable to the virus's tiny quasi-life! Such incredible literary creativity is something nobody could ever have anticipated. DNA is a marvel of self-referential game-playing that I daresay has no rival in human literature. But then, human literature hasn't had three billion years to develop!

I couldn't agree more with Guéron's concluding point (which is why I put it in italics), and it puts me to shame, in a way. If anything, that is my own theme song, and here I wrote an entire article attacking it! To be sure, I had intuited that there was something of this sort going on in the cell, and I half-expected to be taken severely to task by many molecular biologists for blatantly ignoring such form-content (or structure-function) interactions—but in fact, only M. Guéron did so, for which I am very grateful to him.

It just goes to show that if you dare knock existing establishments (in this case, nature's chosen genetic code), thus setting yourself up as a proponent of a rebellious disestablishmentarianism, you can be sure that somebody cleverer than you will come back and blow your points out of antidisestablishmentarianism water, one by one, with arguments inspired by a conservative but highly flexible the.

28

Undercut, Flaunt, Pounce, and Mediocrity: Psychological Games with Numbers

August, 1982

IN the summer of 1962, Robert Boeninger and I, both young mathematics students at Stanford, were riding in a bus somewhere in southern Germany on the way back from a brief trip to Prague, when we got bored. Out of the blue, we invented a curious game with numbers. Though the rules of our game were very simple, it was nonetheless very tricky to play, for it involved trying to "psych each other out" in devious ways. The rules we initially made up went like this: The game would consist of ten turns. On each turn, we'd each choose a number in secret, and then we'd compare them. One of us would choose a number—an integer—in the range 1–5, the other one an integer in the range 2–6. Each of us would get to "keep" his own number, that is, to add it to his score—provided they did not differ by 1. But in the case of two successive integers, the player with the *lower* of the two numbers collected *both* of them. So if I said 2 but Robert said 3, well then, I'd get 5 points, and poor Robert, none. Very jolly! At least until I said 5 and Robert said 4. Then not so jolly.

It seemed amusing to have the ranges not quite coincide, since it's hard to sort out who really has the advantage. One's intuitive first impression might be that the 2–6, or "larger", player has an advantage, but that is nicely counterbalanced by the fact that if you name 6, you're running the risk of being undercut by your opponent's 5, whereas your 6 itself can undercut nothing! Moreover, the "small" player can always name 1 safely, without any risk of being undercut.

Although the asymmetry seemed charming, we soon decided that having equal ranges—both 1–5—was probably preferable. And that was the way we played the game, which I shall here call "Undercut". A table showing how much both of us stand to lose or gain for each possible pair of choices is shown in Figure 28-1a. Such a table is known as a *payoff matrix*.

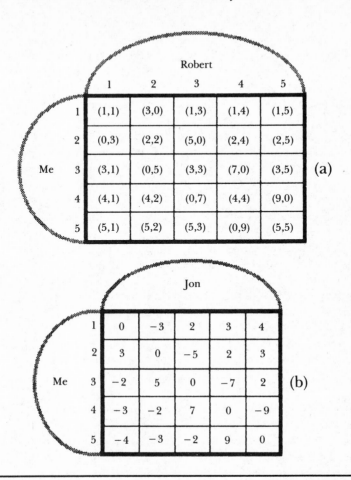

(a)

(b)

FIGURE 28–1. *In (a), the payoff matrix for the game of Undercut, as Robert Boeninger and I originally invented it. In each parenthesis pair, my payoff is on the left and Robert's is on the right. In (b), Jon Peterson's way of looking at things. This matrix exhibits the* difference *between Jon's profit and my profit—in other words, his net gain over me—for each choice of moves we might make. Looked at this way, Undercut is a zero-sum game.*

Competition was pretty fierce. The lovely thing about this game was how level upon level of "outpsyching" could pile up in our minds. For instance, I could "tease" Robert by choosing 4 a few times in a row, trying to lure him into naming 3, and just at that moment plan to switch my move on him, jumping to 2 and outfoxing him. But Robert of course would be most keenly aware of my ploy, and would have his own way of playing naïve, leading me on, making me think I could get away with such tricks, and then pulling a higher-order one on me just when I least expected it.

When I returned to Stanford from Europe that fall, I was eager to get a computer to play this game. My friend Charles Brenner had recently written a program that compiled frequencies of letters and letter groups (trigrams, to be precise) in a piece of text in English (or any other language), and then, using a random-number generator, produced pseudo-English output whose trigram frequencies reproduced those of the input text sample. I had been very impressed by the way that seemingly deep patterns of English could be so aptly captured by such an algorithm, and I saw how this idea could be adapted to a game-playing program. In particular, I was taken by the idea of getting a program to detect patterns in the move sequences of its opponent, and then using them to generate predictions—in short, having the computer itself try to "outpsych" its opponent. All the better if the opponent program were also trying to do something similar to my program! The more vicious, the better! I was in a combative mood, ready to take on all comers.

I have vivid memories of standing over the loud line-printer where the output would spill out, and watching the progress of games emerge line by line. We would have our programs play games of several hundred turns, thus giving them a serious test. My program often started out on a losing track, given that it had not yet "smelled" any patterns in the opposing program's behavior. But sooner or later, there would be a moment when it would appear to "catch the scent", and it would make a decisive undercut or two in a row, and then I would see it start to surge forward, often leaping quickly into the lead and wiping the opponent out. This was a feeling of overwhelming power, the power of insight defeating raw strength. It reminds me now of one of my favorite book titles: *Chess for Fun and Chess for Blood,* by Edward Lasker. I'm not anything as a chess player, but I love that title. It captures exactly that subtle blend of goodwill and rivalry that one feels in a highly competitive game with friends.

Since then, I have realized how universal, how primitive, such a feeling is. It is probably the most engrossing aspect of all sports, that feeling of pitting one strategy against another and watching them fight it out. Dogs certainly seem to experience this feeling with pleasure. When I play a chasing game with my friend Shandy the Airedale, I detect in him a precise sense for how well I can anticipate his moves: in his dodging tactics, he always stays one ply—one level of trickery—ahead of me. Whenever I think I have caught on to his pattern, he somehow senses it, and just at that moment shifts his strategy so that I wind up lunging for a dog that is not there.

Oh, to be sure, he lets me win sometimes—just to keep me interested. He even has the instinct of teasing, dropping his prized stick right in front of me, acting nonchalant about it and coolly tempting me to make a move for it. But he has it all calculated out. He knows how quick I am, how quick he is, and what my patterns of trying to fake him out are.

What's more, Shandy often seems to come up with new ways of shifting his strategy, so that I cannot simply catch onto the "meta-pattern" of his strategy-shifts and thereby outwit him. No, there is something extremely cunning in his dog's mind, and clearly the joyous exercise of that native intelligence reflects a deeper quality of dogs and people in general, namely the enormous evolutionary advantage that intelligence seems to confer on beings that have it, in this dog-eat-dog, people-eat-people world.

*   *   *

But back to Undercut. One day, a math graduate student named Jon Peterson who used to hang around the Stanford "comp center", as it was known back then, challenged my program with his own. He said he had used game theory in his program. I wasn't worried. But when I pitted my program against his, I soon saw that there was good cause for worry. It's not that my program got trounced by his; rather that it just never caught on to any patterns, and simply wound up more or less tying him each time. This was baffling. Jon explained that he had computed appropriate weights for each different choice, 1 through 5, weights that had nothing to do with the opponent's strategy, but merely with the amount of payoff for each set of possibilities. The payoff matrix he was talking about is shown in Figure 28-1*b*. It shows, for each combination of numbers, how much Jon stands to gain relative to me. Notice the *antisymmetry*—the fact that each number, when reflected across the diagonal of zeros, changes sign, signaling that what is good for me is bad for him. (That is the definition of a *zero-sum game*: when the two players' scores in any turn always cancel out.) And of course the zeros down the diagonal mean that when we name identical numbers, it does neither of us any good (other than carrying us one turn closer to the end).

Since the game is completely symmetric for the two players, there can be no winning strategy, for otherwise both players could use it and be guaranteed to beat each other. Nonetheless, there is an *optimal* strategy, according to game theory, which in a statistical sense will guarantee you long-term parity with your opponent. This strategy is based on assigning statistical weights to the five numbers. To find those weights, you have to solve five simultaneous homogeneous linear equations. Each equation is based on making your expectation equal to zero. If Jon's weights for playing 1, 2, 3, 4, 5 are, respectively, $a, b, c, d, e,$ then my expectation, when I choose, say, 3, will be $-2a + 5b - 7d + 2e$ (read straight off the third row of the payoff matrix). Set this expectation to zero and you have one of the five equations. The other four arise analogously. The system to solve is thus:

$$
\begin{aligned}
(1) &\quad & -3b + 2c + 3d + 4e &= 0 \\
(2) &\quad 3a & -5c + 2d + 3e &= 0 \\
(3) &\quad -2a + 5b & -7d + 2e &= 0 \\
(4) &\quad -3a - 2b + 7c & -9e &= 0 \\
(5) &\quad -4a - 3b - 2c + 9d & &= 0
\end{aligned}
$$

This amounts to inverting a 4×4 matrix. Jon had done so, and came up with the following weights: 10, 26, 13, 16, and 1, for choosing 1, 2, 3, 4, and

5 respectively. Thus, according to game theory, an optimal player should play 5 very seldom: one time out of 66. And 2 should be the most common choice. However, it would do little good to play ten 1's in a row, followed by 26 2's, 13 3's, and so on. One must choose completely randomly, given these weights.

Imagine a 66-sided die on ten of whose faces the number 1 appears, only one face having 5 written on it, and so on. Each move, you must throw such a die (or a suitable computational simulation thereof). In other words, you must avoid any and all patterned behavior when you play according to this strategy. No matter how tempting it might be, you must not yield! Even if your opponent plays 5 a dozen times in a row, you must totally ignore it and merely keep on throwing your 66-sided die obliviously. That's the way Jon's program played, and it's why my program found nothing to pick up on. Had Jon's program ever given in to temptation and tried to outguess me, my program would likely have picked up some pattern and twisted it back to work against his. But his program knew nothing of temptations or teasing. It just played blindly on, and the longer the game, the more surely it would break even. If it won, so much the better, but it had only a fifty-fifty chance of that. That's the "optimum strategy" for you!

It was a humiliating and infuriating experience for me to watch my program, with all its "intelligence", struggle in vain to overcome the blind randomness of Jon's program. But there was no way out. I was most disappointed to learn that, in some sense, the "most intelligent" strategy of all not only was dumb—it even paid no attention whatsoever to the enemy's moves! Something about this seemed directly opposite to the original aim of Undercut, which was to have players trying to psych each other out to ever deeper levels.

*   *   *

When I saw the game so completely demolished by game theory, I abandoned it. Recently, however, I have returned to thinking about such games in which patterns in one's play can be taken advantage of, even if game theory in some theoretical sense can find the optimal strategy. There is still something curiously compelling and fascinating about the teasing and flirting and other ploys that arise in these games, something that vividly recalls strategies in evolution, and even seems relevant to many political situations today.

Furthermore, there is something strikingly academic and bookish to adopting a purely game-theoretic strategy when playing against a human opponent, especially in the face of "teasing" strategies. Obviously, humans have more complex goals in life than merely winning the game, and this fact determines a lot about how they play a game. Impatience and audacity, for instance, are both important psychological elements in human game-playing, and an optimal strategy in the ordinary game-theoretic sense does not take those into account. Therefore I feel games of this sort are still important models of how· people and larger organizations tackle complex challenges and threats.

Let me describe, therefore, some more recent variations on Undercut that I have been experimenting with. They all involve extending the degree to which one can go out on a limb by "baiting" one's opponent. My purpose was to encourage teasing, which means that one player flaunts a pattern for a while, implicitly saying, "I dare you—just try an undercut!" So to encourage this kind of pattern-flaunting, it seemed reasonable to award patterns points whenever they are not picked up on by the enemy. Let's call this version "Flaunt".

Suppose that you and I are playing Flaunt. I say 4, you say 1. As in Undercut, I get 4 points, you get 1. Now on my next turn, suppose I again say 4, and you say 2. If we were playing Undercut, I would again get 4. But in Flaunt, repetitions are rewarded. Therefore, I am given the *product* of my two numbers: 4×4, or 16. Now suppose that on my next turn I again play 4, while you again play 2. My bravado now earns me $4 \times 4 \times 4$, or 64, points, while you get 2×2, or 4. So in these three turns, I have gained $4 + 16 + 64$, or 84 points, to your $1 + 2 + 4$, or 7. Of course, you have not been oblivious to my prancing-about in front of you—you have merely been biding your time. Now you make your move—a 3—hoping to undercut me. Too bad— I chose 2! I get 5 points, you get nothing. Sorry, sucker.

But what if I had been so dumb as to let you catch me at this? If I had indeed said 4 this time, I would have been hoping for 256 points. But as you successfully undercut my pattern with your 3, you get a high reward for this, namely 259 points (your 3 points plus "my" 256 points).

$$* \quad * \quad *$$

Now Flaunt, like haircuts, can come in various styles. The one I have just presented is the simplest. But more complex patterned behavior can also be rewarded, if you like. I am not sure of the best way to do this, so what follows —the game I call "Superflaunt"—is only one possible way to reward pattern-flaunting. Suppose that instead of playing 1-2-2 against my 4-4-4 moves, you had played 2, then 1, then 2. You might well have had a reason for doing so. Maybe it was the continuation of a pattern for you and was worth your while keeping up for the moment. If your previous four moves had been 2-1-2-1, then your recent three moves would have continued that pattern. Depending on how it's scored, extending your own established pattern might be more worthwhile than undermining my relatively new one. If 2-1-2-1-2-1-2 is worth the product of its elements, then that amounts to 16 points. (Actually, it's worth 16 only if it was preceded by another 2-1, but that's beside the point.) By the time you've picked up on my 4-4 pattern, maybe it seems worth it to you to let me have my third 4 while you name one more 2, thinking that that will lull me a bit and at that moment, you will suddenly strike, and undercut me.

So what constitutes a pattern in this game of Superflaunt? At the moment, I'm inclined to limit it to one fairly simple definition, although it might be possible to have more complex definitions. The main idea is that a pattern exists when, in a given situation, you do what you did last time you were in "that situation". So it all hinges on what we mean by the notion of "same situation". Let's say that you have just played x, and are about to play y. We'll say that you are making a pattern provided that the most recent time you played x you also followed it with y. If, for instance, your last seven moves had been 3-4-1-5-3-4-1, then to make your pattern continue, you must play 5, and after that, 3, 4, 1, 5, 3, 4, 1, and so on. When you first establish the sequence 3-4-1-5, you of course get no bonus points, because until the repetitions start, there is not any pattern. Thus only when the *second* 4 is played has a pattern started, and it nets you 12 (3×4) points. The next patterned move, 1, nets you another 12 points, and then saying 5 gives you 60 points (as long as it is not undercut)! But as soon as you break the pattern, your cumulative product must start out again from scratch.

If you had played 3-4-1-5-3-4, and were worried about the obviousness of playing 1 now, you might choose to play 4, which, although it breaks one pattern, establishes another pattern (*viz.*, 4-4). Now in ordinary Flaunt, this in itself would already be worth 16 points, but in Superflaunt only on your *next* 4 would you begin to reap the benefits of your patterned playing, since only then would you have made "the same choice" in "the same situation" twice in a row.

A limitation of Undercut and Flaunt is that both confine your moves to a small range. I wanted a game in which numbers of arbitrary size were permitted. It was not too hard to come up with the following game, which I call "Underwhelm". You and I both think of positive integers. Now, if they are unequal and do not differ by 1, then whoever named the *lower* one gets that number (the other player getting, of course, nothing). If they differ by 1, then the namer of the *upper* of the two is awarded both numbers. In that respect, Underwhelm is like a tipped-over version of Undercut (another name for it was "Overcut"). If our two numbers are equal, then neither of us gets anything for this turn.

The goal can be a fixed number of points—any number. For example, 1,000 seems a good choice, although 100 or even a million will do just fine. Think about what this does to the game. Clearly it is not useful for you to name huge numbers, because I am likely to name a lower number and then you will get nothing while I will get something. So there is pressure on both of us—it seems—to play fairly small numbers. But if we stick to very small numbers, then the likelihood of being "overcut" is fairly high. Furthermore, the scores will advance very, very slowly. If we are progressing toward the goal of 1,000 points at a snail's pace, someone will want to speed things up. And so someone will go out on a limb, naming a big number like 81. Of course, doing so just once is not useful, because the other player will not have known in advance that that 81 was coming.

But suppose that I say 81 several times in a row. (Pattern-flaunting is not rewarded in Underwhelm, by the way—at least not in this version.) You will soon catch on, and may well be tempted to say 82, to overcut me. Or perhaps you will want to make points more conservatively off my foolishness, by simply choosing numbers close to but below 81, such as 70. Aha! Once I've managed to lure you up into my vicinity, then of course I can start trying to jump *below you*. And maybe I can even anticipate just when you'll "bite". If so, then I can really take you to the cleaners.

The interesting thing about Underwhelm is that by using obvious patterns as bait to lure the opponent, either one of us can in essence establish one or more Undercut-like games at various positions along the number line. I can set one up in the vicinity of 81, trying to coax you into saying 82 just when I anticipate it. Meanwhile, you may be playing a baiting game down around 30, getting 30 points each time I extravagantly bait you with my 81, and you know that sooner or later I am bound to try to catch you there, either going below you or overcutting you.

What I find fascinating is how many parallel subgames of this type can arise spontaneously in a game of Underwhelm. Particularly interesting is what happens toward the end, when one player has a significant lead. At that point, the trailing player will tend to play very conservatively, naming very small numbers. This means that the possibilities for overcutting are much enhanced. Moreover, there is an entirely psychological element to this game having to do with human impatience. Nobody wants to dawdle to victory by choosing smallish numbers over and over again several hundred times. Therefore, the simple quest for some variety will inevitably lead to some quirky, daring play every once in a while, and that will of course be exploitable.

*   *   *

Much of the spontaneous and creative teasing behavior that tends to occur in these games has its parallels in evolution. The most picturesque and vivid portrayal that I know of the uncanny patterns and counterpatterns that are set up by living beings competing against each other is provided by Richard Dawkins in his book *The Selfish Gene*. The discussion centers around the notion of an *evolutionary stable strategy*, or ESS—a term due to J. Maynard Smith. An ESS is defined as: "a strategy which, if most members of a population adopt it, cannot be bettered by an alternative strategy". However, here, "adoption of a strategy by an individual" really means that that individual has *genes* for that behavioral policy. It's not a question of *choice*.

Dawkins' first example of this concept involves rival genes for two types of aggressive behavior in a given species. The two strategies are named "hawk" and "dove", and have the recent political connotations of those terms. If x positive points are assigned for winning a fight, y negative points for wasting time, and z negative points for getting injured, one can calculate, as a function of x, y, and z, the eventual optimal balance of hawks and doves in the population. This may be an average over time, involving swings back and forth between mostly having hawks and mostly having doves, or it may represent an eventual equilibrium in which the ratio is stable.

Dawkins considers a wide variety of colorful everyday examples in human life, carefully comparing them to strategies in the world of nonhuman evolution. Such things as gas wars, with their price-fixing and treacherous undercutting, fall very neatly into line with the game-theoretic analysis that he brings to bear. Some other strategies considered are: "retaliator" (an individual who, when attacked by a hawk, behaves like a hawk, and when attacked by a dove, behaves like a dove); "bully" (who goes around behaving like a hawk until somebody hits back, then immediately runs away); "prober-retaliator" (who is like a retaliator, but who occasionally tries a brief experimental escalation of the contest). These five strategies can all be activated simultaneously in a computer simulation of a large population, just as the strategies in Undercut could fight each other on a computer. From such simulations, one can learn about the optimum strategies without doing the game theory. In essence, Dawkins maintains, this is what nature has done over eons: Vast numbers of strategies have fought each other, nature's profligacy paying off in the long run in the development of species with optimal strategies, in some sense of the term.

Dawkins uses this concept to show how *group selection* can seem to be taking place in a population, when in fact mere *gene selection* can account for what is observed. He says:

> Maynard Smith's concept of the ESS will enable us, for the first time, to see clearly how a collection of independent selfish entities can come to resemble a single organized whole . . . . Selection at the low level of the single gene can give the impression of selection at some higher level.

The book contains many other provocative examples of peculiar strategies that offer sometimes frightening parallels to situations in the world of human politics, often reminding me of the dangers of the current arms race. In fact, the connection is made explicitly by Dawkins more than once. He refers to "evolutionary arms races" and the survival value of deception of one species by another.

One of the funnier parts of Dawkins' book, although it is dead serious, is concerned with the evolution of sexuality. To show how sexuality might have evolved, he invents "sneaky" versus "honest" gametes (fertilized eggs) and shows how, over many generations, the former will slowly evolve into males, the latter into females. Along the way, such amusingly named strategies are discussed as the "domestic-bliss strategy", the "he-man strategy", the "coy" and "fast" strategies (limited to females), and the "faithful" and "philanderer" strategies (limited to males). Dawkins emphasizes that these are only metaphors, and are not to be taken literally (and certainly not anthropomorphically). When one takes them with the proper grain of salt, however, they can enormously illuminate the mechanisms of evolution. And many of these strategies find their counterparts in such number games as we have discussed above.

*   *   *

As I was preparing this article, I had a long phone conversation with Robert Boeninger in which we tried out various versions of these games. One idea that intrigued me was to play Underwhelm, but with no specific target number of points, such as 1,000, in mind. Instead, a convention of another sort would terminate play. My candidate for that convention was "Stop when the two players' numbers differ by 2." Thus if I say 10 and you say 8, that marks the game's end (and neither of us gets any points on that turn).

Robert and I tried this version out, and quickly discovered that whenever somebody started losing, they would have no option but to go for a stale-mate—a nonterminating game. One way for the losing player to do this is to name huge arbitrary numbers, so that they cannot be anticipated and so the condition for termination is never met. The player who is ahead, having nothing to lose, will cooperate by naming small numbers all the time, thereby gaining even more points and building up even more of a lead. So you get a kind of vicious circle in which both players wind up cooperating in a stalemate.

Robert suggested that one way to prevent this is to add the condition that if either player wins five turns in a row (*i.e.,* gets a positive number of points five times in a row), then the game is over. This prevents the player who is trailing from going for the stalemate, because such behavior will now ensure loss. As Robert amusingly pointed out, even if you are behind, you can start to "wind things up" by trying to win five turns in a row, for by the time those five turns have passed, you may be in the lead! My name for this game is "Pounce", since it made me feel like a tiger hunting down a giraffe in the savannah, bringing down my prey in one swift, sudden move.

*   *   *

One day, several years after the Undercut episode, my sister Laura and our friend Michael Goldhaber and I were having lunch in the Peninsula Creamery and jotting down various trivia on napkins, as was our wont, and somehow it came to us to play a number game involving *three* persons. We decided that on each turn, each of us would choose a number in a certain range, and, since it seemed too boring to let the biggest number win, and equally boring if the littlest number won, it became obvious that the *middlemost* number should be rewarded. So we decided that on each turn, only the "most mediocre" player's score would be allowed to increase. It would increase, of course, by the mediocre number; the other players' scores would stay fixed. (A bit of a problem was posed when two players chose tying numbers, but we found a makeshift way of handling that case.)

Thus at the end of, say, five turns, we would all compare our scores, and the highest one . . . No, wait a minute. Why should we let the *highest* score win? To do that would be, after all, contrary to the spirit of each turn. We saw quite clearly that, if the spirit of the *whole* was to be consistent with the spirit of its *parts,* then the player whose score was in the *middle* should win! We called our game "Mediocrity", but occasionally I like to refer to it as "Hruska".

This name was inspired by a famous remark by the then senator from Nebraska, Roman Hruska. In those days (the early 1970's), President Nixon was attempting to get G. Harrold Carswell appointed to the Supreme Court, against the vehement opposition of Indiana senator Birch Bayh and others. In a radio interview defending Carswell against his critics, Senator Hruska came out with the following profundity:

> Even if he were mediocre, there are a lot of mediocre judges and people and lawyers. They are entitled to a little representation, aren't they, and a little chance? We can't have all Brandeises and Frankfurters and Cardozos and stuff like that there.

Alas for mediocrity, Carswell's nomination was defeated. But it worked out fine for Hruska, who shall forevermore be known as a champion of mediocrity—and stuff like that there.

Speaking of champs, after eating our sandwiches and drinking our thick, rich, old-fashioned milkshakes (served in metal containers, to boot!), the three of us sat in our booth and played a few rounds of this quirky game, and what came to our minds but the inspired idea of determining the World Champion of Mediocrity! So we totaled up our scores over several games, to see who had come out highest. *Highest?!* Again something seemed wrong. The pervasive spirit of mediocrity that had settled on us that day like a heavy smog urged us to deem Champion not the player who had won the *most* games, not the player who had won the *fewest* games, but the player who had won the *middlemost* number of games. Which we did, and I forget who it was. This may be appropriate.

At that point, a general principle seemed to be emerging, which created a hierarchy of levels of Mediocrity. To win at Level Two (that is, our "Championship" level), it's best to be a mediocre player at Level One (the single-game level). This means that whereas before it was desirable to be *extremely* mediocre at choosing mediocre numbers, now it's desirable to be *mediocrely* mediocre at choosing mediocre numbers. How perverse! How wonderful! How wonderfully perverse! It fits in with a general principle of perversity, a Zen-flavored principle, that applies to many aspects of life: Try too hard, and you wind up a loser.

*   *   *

After the initial session at the Peninsula Creamery in which the game of Mediocrity was born, I worked on a number of versions of it, trying to polish it and make it into an elegant game. I am not sure if I succeeded, but I would like to present the rules as they presently stand.

The major issue is how to avoid ties—not only ties at Level Zero, but at all higher levels. My current best solution is the following: Let each player have a slightly shifted range, relative to the other two. More concretely, let player A pick integers from, say, 1 to 5. Then players B and C will have staggered ranges: B picks numbers of the form $n + 1/3$, and C picks numbers of the form $n + 2/3$, where n runs from 1 to 5. Clearly, then, there can be no ties at Level Zero.

Now what happens at Level One? Recall that a Level One game consists of five Level Zero games, in each of which the middlemost number is awarded to the player who chose it, with the other two players getting zero. Well, the first part of this scoring scheme is fine, but the second part has to be modified very slightly in order to avoid ties at higher levels. Suppose that the numbers chosen are as follows: A: 3; B: $2\frac{1}{3}$; C: $4\frac{2}{3}$. Having the middle number, A receives 3 points. B and C, however, do not receive zero points each, but the closest positive approximation to zero that they can, given their staggered ranges. Thus, 1/3 of a point goes to B, and 2/3 of a point to C.

The reasoning behind this goes as follows: After five turns, each player has received five numbers of the same *form*. Player A's five pure integers will add up to a pure integer. Player B's five numbers of the form $n + 1/3$ will add up to a number of the form $n + 2/3$, and player C's five numbers of the form $n + 2/3$ will add up to a number of the form $n + 1/3$. Thus at the next level up, B and C have exchanged roles in terms of the form of their numbers. Consequently, the three total scores at the new level are all of different form and cannot tie, hence there will always be a most mediocre Level One score: a winner.

If we now go on to consider a game at Level Two, we must award points to each Level One game. The "winner" of a Level One game gets, of course, that middling number of points, while the other two players once again receive the closest positive approximations to zero possible, in their respective forms. For player A, this means exactly zero points, as before. However, for B it now means 2/3, and for C, 1/3. Five games at Level One constitute one game at Level Two. The heretofore tacit "Principle of Uniformity of Levels" compels us to sum up the five Level One numbers to produce Level Two scores. Needless to say, the same reasons as before will prevent tie scores from arising, and so there will always be a Level Two winner.

The same general principle will of course allow us to extend the game of Mediocrity to *any* number of levels. One game of Level $N + 1$ Mediocrity consists of five Level N games. The winner of each Level N game is awarded their Level N score, and the other two players get the minimum amount (*i.e.*, 0, 1/3, or 2/3) of the form of their scores at that level. These five Level N numbers are added up to yield totals for the three players, and the middlemost one wins.

Actually, there is nothing sacred about always having *five* Level N turns in a Level $N+1$ game; the "width" could as well be four, or even two. (Multiples of three must be avoided, since after three moves, all three players have scores that are perfect integers, thus allowing ties.) With a width as narrow as two, this allows a very *deep* (*i.e.*, many-leveled) game to be played much more easily. For instance, with width two, a five-level game of Mediocrity requires only 32 Level Zero turns—whereas with the standard width of five, merely three levels of depth will require 125 Level Zero turns. Moreover, there is nothing sacred about the Level Zero choices being confined to numbers bounded by 5; they could run from 1 to infinity! This is just one of the many possible variants of Mediocrity.

$$* \quad * \quad *$$

I can testify that the strategy for playing even Level Two Mediocrity gets mighty confusing very quickly. I have played Level Three Mediocrity on a couple of occasions, and found it completely beyond my reach. I find this both fascinating and frustrating. And think what it implies about world politics, if such simple games as the ones described in this article are so baffling. How much more complex are the "games" of international bargaining, bluffing, and war-making! All of the conceptual messes that we have discussed above have their counterparts (only "squared", so to speak) in international politics. As one watches these huge themes being played out on the world stage, one can hardly help feeling like a single cell in some vast organism whose strategy was set long ago, the consequences of which one can only watch, hoping all will turn out for the best.

***

Post Scriptum.

Suppose you are playing a very, very short game of Undercut: one turn long. The number of points you receive will be multiplied by 1,000 and then paid to you in dollars. What would you play? The answer must depend on what your goal is. Which interests you more: beating your opponent, or amassing as much money as possible? If the former is your priority, then a score of 9 to 0 (your favor) is no better than a score of 3 to 1: Either way, you win just as fully. But if money is your desire, the former is $6,000 more favorable than the latter. Even more striking: If you both name 5, you have a tie game—a big disappointment for someone out to win, but for someone out for money, $5,000 is a fine take.

There is thus a big difference between the original payoff matrix for Undercut and Jon Peterson's modified matrix (both shown in Figure 28-1). Jon's matrix looks at the game solely from the point of view of someone who wants to beat the other player. By taking the difference between payoffs, Jon managed to convert Undercut into a zero-sum game, which he knew to be tractable by the methods of game theory. But if he had left it as Robert and I had formulated it to begin with, it would not have been so easy.

In fact, the original (non-zero-sum) formulation of Undercut subsumes the most famous of all non-zero-sum games: the Prisoner's Dilemma (a treacherous Gordian knot with which the next few chapters deal). I have extracted, in Figure 28-2, just a small fragment of the Undercut payoff

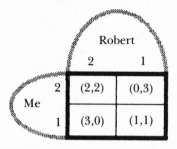

FIGURE 28–2. *A portion of the original Undercut payoff matrix, showing how Undercut contains a Prisoner's Dilemma matrix. (In fact, it contains several.)*

matrix, geometrically rearranged but otherwise intact. This miniature payoff matrix has virtually all the same mathematical qualities as does the standard Prisoner's Dilemma payoff matrix (Figure 29-1). Thus Undercut actually poses a more severe problem than Jon Peterson said. His trick of subtracting one person's payoff from the other's will turn any symmetric game into a zero-sum game, which is tractable by standard techniques of game theory. But that ignores significant aspects of the original game; in particular, for any normal person, *losing* by 3 to 5 (and getting $3,000) is precisely as good as *winning* by 3 to 1 (also getting $3,000). But in Jon's matrix, these two events are as opposite as night and day—as opposite as −2 and +2.

*   *   *

One real-life counterpart to Undercut is given by the following amusing observation. The long-distance telephone rates get much cheaper at 11:00 at night, and so as 11 approaches, the lines get less and less busy, until suddenly, when the hour strikes, the lines get *very* crowded. In some parts of the country, this "rush hour" prevents you from being able to get a line at all, which is most annoying. So you have the option of calling just before 11 and getting an expensive line, or calling just after 11 and taking your chances. Maybe you decide that it's better to call just before 11, and pay the extra amount for the security of getting through. But if *everybody* thinks of this strategy, then calling just before 11 is self-defeating! So then you have to start pushing your calls back earlier, perhaps even into a more expensive time period . . . I guess this is just a new variant on the old "Nobody ever goes there any more because it's always too crowded" joke.

Games of this sort and jokes do indeed have a lot in common. In an article in the British journal *Manifold* titled "A Pandora's Box of non-Games", Anatole Beck and David Fowler set forth a panoply of rather silly games that are halfway between true games and pure jokes. The tragedy is that so many of them resemble current global political behavior. For instance, consider the game they call *Finchley Central*:

> Two players alternate naming the stations on the London Underground. The first to say 'Finchley Central' wins. It is clear that the 'best' time to say 'Finchley Central' is exactly before your opponent does. Failing that, it is good that he should be considering it. You could, of course, say 'Finchley Central' on your second turn. In that case, your opponent puffs on his cigarette and says, 'Well, . . .' Shame on you.

Another amusing game, quite similar to the ones described in the column, is called *Penny Pot*:

> Players alternate turns. At each turn, a player either adds a penny to the pot or takes the pot. Winning player makes first move in next game. Like Finchley Central, this games defies analysis. There is, of course, the stable situation in which each player takes the pot whenever it is not empty. This is a solution?

At the end of their article, Beck and Fowler add:

> M. Henton of New Addington noted with horror that there is an isomorphism between Finchley Central and the game commonly known as 'Nuclear Deterrent'. 'It occurs to me that we should work very fast to analyse the non-games, before we are left with a non-world.'

*   *   *

Several readers wrote in to tell me that they had worked out by game theory the optimal strategy for playing my game of Underwhelm, and that they had found it involves playing only the numbers between 1 and 5, in the ratios 25:19:27:16:14. Numbers higher than 5 should never be played at all! This was a surprise to me, taking away most of the interest of the game. Oh, well . . . as they say in game theory, "You win some, you lose some."

29

The Prisoner's Dilemma Computer Tournaments and the Evolution of Cooperation

May, 1983

LIFE is filled with paradoxes and dilemmas. Sometimes it even feels as if the essence of living is the sensing—indeed, the savoring—of paradox. Although all paradoxes seem somehow related, some paradoxes seem abstract and philosophical, while others touch on life very directly. A very lifelike paradox is the so-called "Prisoner's Dilemma", discovered in 1950 by Melvin Dresher and Merrill Flood of the RAND Corporation. Albert W. Tucker wrote the first article on it, and in that article he gave it its now-famous name. I shall here present the Prisoner's Dilemma—first as a metaphor, then as a formal problem.

The original formulation in terms of prisoners is a little less clear to the uninitiated, in my experience, than the following one. Assume you possess copious quantities of some item (money, for example), and wish to obtain some amount of another item (perhaps stamps, groceries, diamonds). You arrange a mutually agreeable trade with the only dealer of that item known to you. You are both satisfied with the amounts you will be giving and getting. For some reason, though, your trade must take place in secret. Each of you agrees to leave a bag at a designated place in the forest, and to pick up the other's bag at the other's designated place. Suppose it is clear to both of you that the two of you will never meet or have further dealings with each other again.

Clearly, there is something for each of you to fear: namely, that the other one will leave an empty bag. Obviously, if you both leave full bags, you will both be satisfied; but equally obviously, getting something for nothing is even more satisfying. So you are tempted to leave an empty bag. In fact, you can even reason it through quite rigorously this way: "If the dealer brings a full bag, I'll be better off having left an empty bag, because I'll have gotten all that I wanted and given away nothing. If the dealer brings an empty bag, I'll be better off having left an empty bag, because I'll not have been cheated. I'll have gained nothing but lost nothing either. Thus it seems that *no matter what the dealer chooses to do,* I'm better off leaving an empty bag. So I'll leave an empty bag."

The dealer, meanwhile, being in more or less the same boat (though at the other end of it), thinks analogous thoughts and comes to the parallel conclusion that it is best to leave an empty bag. And so both of you, with your impeccable (or impeccable-seeming) logic, leave empty bags, and go away empty-handed. How sad, for if you had both just cooperated, you could have each gained something you wanted to have. *Does logic prevent cooperation?* This is the issue of the Prisoner's Dilemma.

* * *

In case you're wondering why it is called "Prisoner's Dilemma", here's the reason. Imagine that you and an accomplice (someone you have no feelings for one way or the other) committed a crime, and now you've both been apprehended and thrown in jail, and are fearfully awaiting trials. You are being held in separate cells with no way to communicate. The prosecutor offers each of you the following deal (and informs you both that the identical deal is being offered to each of you—and that you both know *that* as well!): "We have a lot of circumstantial evidence on you both. So if you both claim innocence, we will convict you anyway and you'll both get two years in jail. But if you will help us out by admitting your guilt and making it easier for us to convict your accomplice—oh, pardon me, your *alleged* accomplice— why, then, we'll let you out free. And don't worry about revenge—your accomplice will be in for five years! How about it?" Warily you ask, "But what if we *both* say we're guilty?" "Ah, well, my friend—I'm afraid you'll both get four-year sentences, then."

Now you're in a pickle! Clearly, you don't want to claim innocence if your partner has sung, for then you're in for five long years. Better you should both have sung—then you'll only get four. On the other hand, if your partner claims innocence, then the best possible thing for you to do is sing, since then you're out scot-free! So at first sight, it seems obvious what you should do: Sing! But what is obvious to you is equally obvious to your opposite number, so now it looks like you both ought to sing, which means —Sing Sing for four years! At least that's what *logic* tells you to do. Funny, since if both of you had just been *illogical* and maintained innocence, you'd both be in for only half as long! Ah, logic does it again.

* * *

Let us now go back to the original metaphor and slightly alter its conditions. Suppose that both you and your partner very much want to have a regular supply of what the other has to offer, and so, before conducting your first exchange, you agree to carry on a lifelong exchange, once a month. You still expect never to meet face to face. In fact, neither of you has any idea how old the other one is, so you can't be very sure of how long this lifelong agreement may go on, but it seems safe to assume it'll go on for a few months anyway, and very likely for years.

Now, what do you do on your first exchange? Taking an empty bag seems fairly nasty as the opening of a relationship—hardly an effective way to build up trust. So suppose you take a full bag, and the dealer brings one as well. Bliss—for a month. Then you both must go back. Empty, or full? Each month, you have to decide whether to *defect* (take an empty bag) or to *cooperate* (take a full one). Suppose that one month, unexpectedly, your dealer defects. Now what do you do? Will you suddenly decide that the dealer can never be trusted again, and from now on always bring empty bags, in effect totally giving up on the whole project forever? Or will you pretend you didn't notice, and continue being friendly? Or—will you try to punish the dealer by some number of defections of your own? One? Two? A random number? An increasing number, depending on how many defections you have experienced? Just how mad will you get?

This is the so-called *iterated* Prisoner's Dilemma. It is a very difficult problem. It can be, and has been, rendered more quantitative and in that form studied with the methods of game theory and computer simulation. How does one quantify it? One builds a *payoff matrix* presenting point values for the various alternatives. A typical one is shown in Figure 29-1a. In this matrix, mutual cooperation earns both parties 2 points (the subjective value of receiving a full bag of what you need while giving up a full bag of what you have). Mutual defection earns you both 0 points (the subjective value of gaining nothing and losing nothing, aside from making a vain trip out to the forest that month). Cooperating while the other defects stings: you get −1 point while the rat gets 4 points! Why so many? Because it is so pleasurable to get something for nothing. And of course, should *you* happen to be a rat some month when the dealer has cooperated, then you get 4 points and the dealer loses 1.

It is obvious that in a *collective* sense, it would be best for both of you to always cooperate. But suppose you have no regard whatsoever for the other. There is no "collective good" you are both working for. You are both supreme egoists. Then what? The meaning of this term, "egoist", can perhaps be made clear by the following. Suppose you and your dealer have developed a trusting relationship of mutual cooperation over the years, when one day you receive secret and reliable information that the dealer is quite sick and will soon die, probably within a month or two. The dealer has no reason to suspect that you have heard this. Aren't you highly tempted to defect, all of a sudden, despite all your years of cooperating? You are, after all, out for yourself and no one else in this cruel, cruel world. And since it seems that this may very well be the dealer's last month, why not profit

Dealer

	Cooperates	Defects
Cooperate	(2,2)	(−1,4)
Defect	(4,−1)	(0,0)

You ——

(a)

Your accomplice

	Stays mum	Sings
Stay mum	(−2,−2)	(−5,0)
Sing	(0,−5)	(−4,−4)

You ——

(b)

Player B

	Cooperates	Defects
Cooperates	(3,3)	(0,5)
Defects	(5,0)	(1,1)

Player A ——

(c)

FIGURE 29–1.   *The Prisoner's Dilemma.*

In (a), a Prisoner's Dilemma payoff matrix in the case of a dealer and a buyer of commodities or services, in which both participants have a choice: to cooperate *(i.e., to deliver the goods or the payment) or to* defect *(i.e., to deliver nothing). The numbers attempt to represent the degree of satisfaction of each partner in the transaction.*

In (b), the formulation of the Prisoner's Dilemma to which it owes its name: in terms of prisoners and their opportunities for double-cros-sing or collusion. The numbers are negative because they represent punishments: the length of both prisoners' prospective jail sentences, in years. This metaphor is due to Albert W. Tucker.

In (c), a Prisoner's Dilemma formulation where all payoffs are nonnegative numbers. This is my canonical version, following the usage in Robert Axelrod's book, The Evolution of Cooperation.

as much as possible from your secret knowledge? Your defection may never be punished, and at the worst, it will be punished by one last-gasp defection by the dying dealer.

The surer you are that this next turn is to be the very last one, the more you feel you *must* defect. Either of you would feel that way, of course, on learning that the other one was nearing the end of the rope. This is what is meant by "egoism". It means you have no feeling of friendliness or goodwill or compassion for the other player; you have no conscience; all you care about is amassing points, more and more and more of them.

What does the payoff matrix for the other metaphor, the one involving prisoners, look like? It is shown in Figure 29-1*b*. The equivalence of this matrix to the previous matrix is clear if you add a constant—namely, 4—to all terms in this one. Indeed, we could add any constant to either matrix and the dilemma would remain essentially unchanged. So let us add 5 to this one so as to get rid of all negative payoffs. We get the canonical Prisoner's Dilemma payoff matrix, shown in Figure 29-1*c*. The number 3 is called the *reward for mutual cooperation,* or *R* for short. The number 1 is called the *punishment,* or *P.* The number 5 is *T,* the *temptation,* and 0 is *S,* the *sucker's payoff.* The two conditions that make a matrix represent a Prisoner's Dilemma situation are these:

$$(1) \quad T > R > P > S$$

$$(2) \quad (T+S)/2 < R$$

The first one simply makes the argument go through for each of you, that "it is better for me to defect no matter what my counterpart does". The second one simply guarantees that if you two somehow get locked into out-of-phase alternations (that is, "you cooperate, I defect" one month and "you defect, I cooperate" the next), you will not do better—in fact, you will do worse—than if you were cooperating with each other each month.

Well, what would be your best strategy? It can be shown quite easily that there is no universal answer to this question. That is, there is no strategy that is better than all other strategies under all circumstances. For consider the case where the other player is playing *ALL D*—the strategy of defecting each round. In that case, the best you can possibly do is to defect each time as well, including the first. On the other hand, suppose the other player is using the *Massive Retaliatory Strike* strategy, which means "I'll cooperate until you defect and thereafter I'll defect forever." Now if you defect on the very first move, then you'll get one *T* and all *P*'s thereafter until one of you dies. But if you had waited to defect, you could have benefited from a relationship of mutual cooperation, amassing many *R*'s beforehand. Clearly that bunch of *R*'s will add up to more than the single *T* if the game goes on for more than a few moves. This means that against the *ALL D* strategy, *ALL D* is the best counterstrategy, whereas "Always cooperate unless you learn that you or the other player is just about to die, in which case defect" is the best counterstrategy against *Massive Retaliatory Strike.* This simple argument shows that *how* you should play depends on *who* you're playing.

The whole concept of the "quality" of a strategy takes on a decidedly more operational and empirical meaning if one imagines an ocean populated by dozens of little beings swimming around and playing Prisoner's Dilemma over and over with each other. Suppose that each time two such beings encounter each other, they recognize each other and remember how previous encounters have gone. This enables each one to decide what it wishes to do this time. Now if each organism is continually swimming around and bumping into the others, eventually each one will have met every other one numerous times, and thus all strategies will have been given the opportunity to interact with each other. By "interact", what is meant here is certainly not that anyone knocks anyone else out of the ocean, as in an elimination tournament. The idea is simply that each organism gains zero or more points in each meeting, and if sufficient time is allowed to elapse, everybody will have met with everybody else about the same number of times, and now the only question is: Who has amassed the most points? Amassing points is truly the name of the game.

It doesn't matter if you have "beaten" anyone, in the sense of having gained more from interacting with them than they gained from interacting with you. That kind of "victory" is totally irrelevant here. What matters is not the number of "victories" rung up by any individual, but the individual's *total point count*—a number that measures the individual's overall viability in this particular "sea" of many strategies. It sounds nearly paradoxical, but you could lose many—indeed, *all*—of your individual skirmishes, and yet still come out the overall winner.

As the image suggests very strongly, this whole situation is highly relevant to questions in evolutionary biology. Can totally selfish and unconscious organisms living in a common environment come to evolve reliable cooperative strategies? Can cooperation emerge in a world of pure egoists? In a nutshell, *can cooperation evolve out of noncooperation?* If so, this has revolutionary import for the theory of evolution, for many of its critics have claimed that this was one place that it was hopelessly snagged.

* * *

Well, as it happens, it has now been demonstrated rigorously and definitively that such cooperation can emerge, and it was done through a computer tournament conducted by political scientist Robert Axelrod of the Political Science Department and the Institute for Public Policy Studies of the University of Michigan in Ann Arbor. More accurately, Axelrod first studied the ways that cooperation evolved by means of a computer tournament, and when general trends emerged, he was able to spot the underlying principles and prove theorems that established the facts and conditions of cooperation's rise from nowhere. Axelrod has written a fascinating and remarkably thought-provoking book on his findings, called *The Evolution of Cooperation,* published in 1984 by Basic Books, Inc. (Quoted sections below are taken from an early draft of that book.) Furthermore, he and evolutionary biologist William D. Hamilton have worked out and published many of the implications of these discoveries for evolutionary theory. Their work has won much acclaim—including the 1981 Newcomb Cleveland Prize, a prize awarded annually by the American Association for the Advancement of Science for "an outstanding paper published in *Science* ".

There are really three aspects of the question "Can cooperation emerge in a world of egoists?" The first is: How can it get started at all? The second is: Can cooperative strategies survive better than their noncooperative rivals? The third one is: Which cooperative strategies will do the best, and how will they come to predominate?

*   *   *

To make these issues vivid, let me describe Axelrod's tournament and its somewhat astonishing results. In 1979, Axelrod sent out invitations to a number of professional game theorists, including people who had published articles on the Prisoner's Dilemma, telling them that he wished to pit many strategies against one another in a round-robin Prisoner's Dilemma tournament, with the overall goal being to amass as many points as possible. He asked for strategies to be encoded as computer programs that could respond to the 'C' or 'D' of another player, taking into account the remembered history of previous interactions with that same player. A program should always reply with a 'C' or a 'D', of course, but its choice need not be deterministic. That is, consultation of a random-number generator was allowed at any point in a strategy.

Fourteen entries were submitted to Axelrod, and he introduced into the field one more program called *RANDOM,* which in effect flipped a coin (computationally simulated, to be sure) each move, cooperating if heads came up, defecting otherwise. The field was a rather variegated one, consisting of programs ranging from as few as four lines to as many as 77 lines (of Basic). Every program was made to engage each other program (and a clone of itself) 200 times. No program was penalized for running slowly. The tournament was actually run five times in a row, so that pseudo-effects caused by statistical fluctuations in the random-number generator would be smoothed out by averaging.

The program that won was submitted by the old Prisoner's Dilemma hand, Anatol Rapoport, a psychologist and philosopher from the University of Toronto. His was the shortest of all submitted programs, and is called *TIT FOR TAT. TIT FOR TAT* uses a very simple tactic:

Cooperate on move 1;
thereafter, do whatever the other player did the previous move.

That is all. It sounds outrageously simple. How in the world could such a program defeat the complex stratagems devised by other experts?

Well, Axelrod claims that the game theorists in general did not go far enough in their analysis. They looked "only two levels deep", when in fact they should have looked *three* levels deep to do better. What precisely does this mean? He takes a specific case to illustrate his point. Consider the entry called *JOSS* (submitted by Johann Joss, a mathematician from Zürich, Switzerland). *JOSS*'s strategy is very similar to *TIT FOR TAT*'s, in that it begins by cooperating, always responds to defection by defecting and *nearly* always responds to cooperation by cooperating. The hitch is that *JOSS* uses a random-number generator to help it decide when to pull a "surprise defection" on the other player. *JOSS* is set up so that it has a 10 percent probability of defecting right after the other player has cooperated.

In playing *TIT FOR TAT*, *JOSS* will do fine until it tries to catch *TIT FOR TAT* off guard. When it defects, *TIT FOR TAT* retaliates with a single defection, while *JOSS* "innocently" goes back to cooperating. Thus we have a "DC" pair. On the next move, the 'C' and 'D' will switch places since each program in essence echoes the other's latest move, and so it will go: CD, then DC, CD, DC, and so on. There may ensue a long reverberation set off by *JOSS*'s D, but sooner or later, *JOSS* will randomly throw in *another* unexpected D after a C from *TIT FOR TAT*. At this point, there will be a "DD" pair, and that determines the entire rest of the match. Both will defect forever, now. The "echo" effect resulting from *JOSS*'s first attempt at exploitation and *TIT FOR TAT*'s simple punitive act lead ultimately to complete distrust and lack of cooperation.

This may seem to imply that both strategies are at fault and will suffer for it at the hands of others, but in fact the one that suffers from it most is *JOSS*, since *JOSS* tries out the same trick on partner after partner, and in many cases this leads to the same type of breakdown of trust, whereas *TIT FOR TAT*, never defecting first, will never be the initial cause of a breakdown of trust. Axelrod's technical term for a strategy that never defects before its opponent does is *nice*. *TIT FOR TAT* is a nice strategy, *JOSS* is not. Note that "nice" does not mean that a strategy *never* defects! *TIT FOR TAT* defects when provoked, but that is still considered being "nice".

Axelrod summarizes the first tournament this way:

> A major lesson of this tournament is the importance of minimizing echo effects in an environment of mutual power. A sophisticated analysis must go at least three levels deep. First is the direct effect of a choice. This is easy, since a defection always earns more than a cooperation. Second are the indirect effects, taking into account that the other side may or may not punish a defection. This much was certainly appreciated by many of the entrants. But third is the fact that in responding to the defections of the other side, one may be repeating or even amplifying one's own previous exploitative choice. Thus a single defection may be successful when analyzed for its direct effects, and perhaps even when its secondary effects are taken into account. But the real costs may be in the tertiary effects when one's own isolated defections turn into unending mutual recriminations. Without their realizing it, many of these rules actually wound up punishing themselves. With the other player serving as a mechanism to delay the self-punishment by a few moves, this aspect of self-punishment was not perceived by the decision rules . . . .
>
> The analysis of the tournament results indicates that there is a lot to be learned about coping in an environment of mutual power. Even expert strategists from political science, sociology, economics, psychology, and mathematics made the systematic errors of being too competitive for their own

good, not forgiving enough, and too pessimistic about the responsiveness of the other side.

Axelrod not only analyzed the first tournament, he even performed a number of "subjunctive replays" of it, that is, replays with different sets of entries. He found, for instance, that the strategy called *TIT FOR TWO TATS,* which tolerates two defections before getting mad (but still only strikes back once), *would* have won, had it been in the line-up. Likewise, two other strategies he discovered, one called *REVISED DOWNING* and one called *LOOK-AHEAD,* would have come in first had they been in the tournament.

In summary, the lesson of the first tournament seems to have been that it is important to be *nice* ("don't be the first to defect") and *forgiving* ("don't hold a grudge once you've vented your anger"). *TIT FOR TAT* possesses both these qualities, quite obviously.

*   *   *

After this careful analysis, Axelrod felt that significant lessons had been unearthed, and he felt convinced that more sophisticated strategies could be concocted, based on the new information. Therefore he decided to hold a second, larger computer tournament. For this tournament, he not only invited all the participants in the first round, but also advertised in computer hobbyist magazines, hoping to attract people who were addicted to programming and who would be willing to devote a good deal of time to working out and perfecting their strategies. To each person who entered, Axelrod sent a full and detailed analysis of the first tournament, along with a discussion of the "subjunctive replays" and the strategies that would have won. He described the strategic concepts of "niceness" and "forgiveness" that seemed to capture the lessons of the tournament so well, as well as strategic pitfalls to avoid. Naturally, each entrant realized that all the other entrants had received the same mailing, so that everyone knew that everyone knew that everyone knew that . . .

There was a large response to Axelrod's call for entries. Entries were received from six countries, from people of all ages, and from eight different academic disciplines. Anatol Rapoport entered again, resubmitting *TIT FOR TAT* (and was the only one to do so, even though it was explicitly stated that anyone could enter any program written by anybody). A ten-year-old entered, as did one of the world's experts on game theory and evolution, John Maynard Smith, professor of biology at the University of Sussex in England, who submitted *TIT FOR TWO TATS.* Two people separately submitted *REVISED DOWNING.*

Altogether, 62 entries were received, and generally speaking, they were of a considerably higher degree of sophistication than those in the first tournament. The shortest was again *TIT FOR TAT,* and the longest was a program from New Zealand, consisting of 152 lines of Fortran. Once again,

RANDOM was added to the field, and with a flourish and a final carriage return, the horses were off! Several hours of computer time later, the results came in.

*   *   *

The outcome was nothing short of stunning: *TIT FOR TAT*, the simplest program submitted, won again. What's more, the two programs submitted that had won the subjunctive replays of the first tournament now turned up way down in the list: *TIT FOR TWO TATS* came in 24th, and *REVISED DOWNING* ended up buried in the bottom half of the field.

This may seem horribly nonintuitive, but remember that a program's success depends entirely on the environment in which it is swimming. There is no single "best strategy" for all environments, so that winning in one tournament is no guarantee of success in another. *TIT FOR TAT* has the advantage of being able to "get along well" with a great variety of strategies, while other programs are more limited in their ability to evoke cooperation. Axelrod puts it this way:

> What seems to have happened is an interesting interaction between people who drew one lesson and people who drew another lesson from the first round. Lesson One was "Be nice and forgiving." Lesson Two was more exploitative: "If others are going to be nice and forgiving, it pays to try to take advantage of them." The people who drew Lesson One suffered in the second round from those who drew Lesson Two.

Thus the majority of participants in the second tournament really had not grasped the central lesson of the first tournament: the importance of being willing to initiate and reciprocate cooperation. Axelrod feels so strongly about this that he is reluctant to call two strategies playing against each other "opponents"; in his book he always uses neutral terms such as "strategies" or "players". He even does not like saying they are playing *against* each other, preferring "with". In this article, I have tried to follow his usage, with occasional departures. One very striking fact about the second tournament is the success of "nice" rules: of the top fifteen finishers, only one (placing eighth) was not nice. Amusingly, a sort of mirror image held: of the bottom fifteen finishers, only one was nice!

Several non-nice strategies featured rather tricky probes of the opponent (sorry!), sounding it out to see how much it "minded" being defected against. Although this kind of probing by a program might fool occasional opponents, more often than not it backfired, causing severe breakdowns of trust. Altogether, it turned out to be very costly to try to use defections to "flush out" the other player's weak spots. It turned out to be more profitable to have a policy of cooperation as often as possible, together with a willingness to retaliate swiftly against any attempted undercutting. Note, however, that strategies featuring *massive* retaliation were less successful than *TIT FOR TAT* with its more gentle policy of *restrained* retaliation.

Forgiveness is the key here, for it helps to restore the proverbial "atmosphere of mutual cooperation" (to use the phrase of international diplomacy) after a small skirmish.

"Be nice and forgiving" was in essence the overall lesson of the first tournament. Apparently, though, many people just couldn't get themselves to believe it, and were convinced that with cleverer trickery and scheming, they could win the day. It took the second tournament to prove them dead wrong. And out of the second tournament, a third key strategic concept emerged: that of *provocability*—the notion that one should "get mad" quickly at defectors, and retaliate. Thus a more general lesson is: "Be nice, provocable, and forgiving."

*   *   *

Strategies that do well in a wide variety of environments are called by Axelrod *robust,* and it seems that ones with "good personality traits"—that is, nice, provocable, and forgiving strategies—are sure to be robust. *TIT FOR TAT* is by no means the only possible strategy with these traits, but it is the canonical example of such a strategy, and it is astonishingly robust.

Perhaps the most vivid demonstrations of *TIT FOR TAT*'s robustness were provided by various subjunctive replays of the second tournament. The principle behind any replay involving a different environment is quite simple. From the actual playing of the tournament, you have a 63×63 matrix documenting how well each program did against each other program. Now, the effective "population" of a program in the environment can be manipulated mathematically by attaching a weight factor to all that program's interactions, then just retotaling all the columns. This way you can get subjunctive *instant* replays without having to rerun the tournament.

This simple observation means that the results of a huge number of potential subjunctive tournaments are concealed in, but potentially extractable from, the 63×63 matrix of program-*vs.*-program totals. For instance, Axelrod discovered, using statistical analysis, that there were essentially six classes of strategies in the second tournament. For each of these classes, he conducted a subjunctive instant replay of the tournament by quintupling the importance (the weight factor) of that class alone, thus artificially inflating a certain strategic style's population in the environment. When the scores were retotaled, *TIT FOR TAT* emerged victorious in five out of six of those hypothetical tournaments, and in the sixth it placed second.

Undoubtedly the most significant and ingenious type of subjunctive replay that Axelrod tried was the *ecological tournament.* Such a tournament consists not merely of a single subjunctive replay, but of a whole cascade of hypothetical replays, each one's environment determined by the results of the previous one. In particular, if you take a program's score in a tournament as a measure of its "fitness", and if you interpret "fitness" as meaning "number of progeny in the next generation", and finally, if you let "next generation" mean "next tournament", then what you get is that each tournament's results determine the environment of the next one—and in particular, successful programs become more copious in the next tournament. This type of iterated tournament is called "ecological" because it simulates ecological adaptation (the shifting of a *fixed* set of species' populations according to their mutually defined and dynamically developing environment), as contrasted with evolution via mutation (where *new* species can come into existence).

As one carries an ecological tournament through generation after generation, the environment gradually changes. In a paraphrase of how Axelrod puts it, here is what happens. At the very beginning, poor programs and good programs alike are equally represented. As time passes, the poorer ones begin to drop out while the good ones flourish. But the rank order of the good ones may now change, because their "goodness" is no longer being measured against the same field of competitors as initially. Thus success breeds ever more success—but only provided that the success derives from interaction with other similarly successful programs. If, by contrast, some program's success is due mostly to its ability to milk "dumber" programs for all they're worth, then as those programs are gradually squeezed out of the picture, the exploiter's base of support will be eroded and it will suffer a similar fate.

A concrete example of ecological extinction is provided by *HARRINGTON*, the only non-nice program among the top fifteen finishers in the second tournament. In the first 200 generations of the ecological tournament, while *TIT FOR TAT* and other successful nice programs were gradually increasing their percentage of the population, *HARRINGTON* too was increasing its percentage. This was a direct result of *HARRINGTON*'s exploitative strategy. However, by the 200th generation, things began to take a noticeable turn. Weaker programs were beginning to go extinct, which meant fewer and fewer dupes for *HARRINGTON* to profit from. Soon the trend became apparent: *HARRINGTON* could not keep up with its nice rivals. By the 1,000th generation, *HARRINGTON* was as extinct as the dodos it had exploited. Axelrod summarizes:

> Doing well with rules that do not score well themselves is eventually a self-defeating process. Not being nice may look promising at first, but in the long run it can destroy the very environment it needs for its own success.

Needless to say, *TIT FOR TAT* fared spectacularly well in the ecological tournament, increasing its lead ever more. After 1,000 generations, not only was *TIT FOR TAT* ahead, but its rate of growth was greater than that of any other program. This is an almost unbelievable success story, all the more so because of the absurd simplicity of the "hero". One amusing aspect of it is that *TIT FOR TAT* did not defeat a single one of its rivals in their encounters. This is not a quirk; it is in the nature of *TIT FOR TAT. TIT FOR TAT* simply *cannot* defeat anyone; the best it can achieve is a tie, and often it loses (though not by much).

Axelrod makes this point very clear:

> *TIT FOR TAT* won the tournament, not by beating the other player, but by
> eliciting behavior from the other player which allowed both to do well. *TIT FOR
> TAT* was so consistent at eliciting mutually rewarding outcomes that it attained
> a higher overall score than any other strategy in the tournament.
>
> So in a non-zero-sum world you do not have to do better than the other
> player to do well for yourself. This is especially true when you are interacting
> with many different players. Letting each of them do the same or a little better
> than you is fine, as long as you tend to do well yourself. There is no point in
> being envious of the success of the other player, since in an iterated Prisoner's
> Dilemma of long duration the other's success is virtually a prerequisite of your
> doing well for yourself.

He gives examples from everyday life in which this principle holds. Here is
one:

> A firm that buys from a supplier can expect that a successful relationship will
> earn profit for the supplier as well as the buyer. There is no point in being
> envious of the supplier's profit. Any attempt to reduce it through an
> uncooperative practice, such as by not paying your bills on time, will only
> encourage the supplier to take retaliatory action. Retaliatory action could take
> many forms, often without being explicitly labeled as punishment. It could be
> less prompt deliveries, lower quality control, less forthcoming attitudes on
> volume discounts, or less timely news of anticipated market conditions. The
> retaliation could make the envy quite expensive. Instead of worrying about the
> relative profits of the seller, the buyer should worry about whether another
> buying strategy would be better.

Like a business partner who never cheats anyone, *TIT FOR TAT* never beats
anyone—yet both do very well for themselves.

*   *   *

One idea that is amazingly counterintuitive at first in the Prisoner's
Dilemma is that the best possible strategy to follow is *ALL D* if the other
player is unresponsive. It might seem that some form of random strategy
might do better, but that is completely wrong. If I have laid out all my moves
in advance, then playing *TIT FOR TAT* will do you no good, nor will flipping
a coin. You should simply defect every move. It matters not what pattern
I have chosen. Only if I can be influenced by your play will it ever do you
any good to cooperate.

Fortunately, in an environment where there are programs that cooperate
(and whose cooperation is based on reciprocity), being unresponsive is a
very poor strategy, which in turn means that *ALL D* is a very poor strategy.
The single unresponsive competitor in the second tournament was
RANDOM, and it finished next to last. The last-place finisher's strategy was
responsive, but its behavior was so inscrutable that it *looked* unresponsive.

And in a more recent computer tournament conducted by Marek Lugowski and myself in the Computer Science Department at Indiana University, three *ALL-D*'s came in at the very bottom (out of 53), with a couple of *RANDOM*'s giving them a tough fight for the honor.

One way to explain *TIT FOR TAT*'s success is simply to say that it *elicits cooperation*, via friendly persuasion. Axelrod spells this out as follows:

> Part of its success might be that other rules anticipate its presence and are designed to do well with it. Doing well with *TIT FOR TAT* requires cooperating with it, and this in turn helps *TIT FOR TAT*. Even rules that were designed to see what they could get away with quickly apologize to *TIT FOR TAT*. Any rule that tries to take advantage of *TIT FOR TAT* will simply hurt itself. *TIT FOR TAT* benefits from its own nonexploitability because three conditions are satisfied:
>
> 1. The possibility of encountering *TIT FOR TAT* is salient;
> 2. Once encountered, *TIT FOR TAT* is easy to recognize; and
> 3. Once recognized, *TIT FOR TAT*'s nonexploitability is easy to appreciate.

This brings out a fourth "personality trait" (in addition to niceness, provocability, and forgiveness) that may play an important role in success: recognizability, or straightforwardness. Axelrod chooses to call this trait *clarity,* and argues for it with clarity:

> Too much complexity can appear to be total chaos. If you are using a strategy that appears random, then you also appear unresponsive to the other player. If you are unresponsive, then the other player has no incentive to cooperate with you. So being so complex as to be incomprehensible is very dangerous.

How rife this is with morals for social and political behavior! It is rich food for thought.

*   *   *

Anatol Rapoport cautions against overstating the advantages of *TIT FOR TAT*; in particular, he believes that *TIT FOR TAT* is too harshly retaliatory on occasion. It can also be persuasively argued that *TIT FOR TAT* is too lenient on other occasions. Certainly there is no evidence that *TIT FOR TAT* is the ultimate or best possible strategy. Indeed, as has been emphasized repeatedly, the very concept of "best possible" is incoherent, since all depends on environment. In the tournament at Indiana University mentioned earlier, several *TIT-FOR-TAT*-like strategies did better than pure *TIT FOR TAT* did. They all shared, however, the three critical "character traits" whose desirability had been so clearly delineated by Axelrod's prior analysis of the important properties of *TIT FOR TAT*. They were simply a little better than *TIT FOR TAT* at detecting nonresponsiveness, and when they were convinced the other player was unresponsive, they switched over to an *ALL-D* mode.

In his book, Axelrod takes pains to spell out the answers to three fundamental questions concerning the temporal evolution of cooperation in a world of raw egoism. The first concerns *initial viability*: How can cooperation get started in a world of unconditional defection—a "primordial sea" swarming with unresponsive *ALL-D* creatures? The answer (whose proof I omit here) is that invasion by small clusters of conditionally cooperating organisms, even if they form a tiny minority, is enough to give cooperation a toehold. One cooperator alone will die, but small clusters of cooperators can arrive (via mutation or migration, say) and propagate even in a hostile environment, provided they are defensive like *TIT FOR TAT*. Complete pacifists—Quaker-like programs—will *not* survive, however, in this harsh environment.

The second fundamental question concerns *robustness*: What type of strategy does well in unpredictable and shifting environments? We have already seen that the answer to this question is: Any strategy possessing the four fundamental "personality traits" of niceness, provocability, forgiveness, and clarity. This means that such strategies, once established, will tend to flourish, especially in an ecologically evolving world.

The final question concerns *stability*: Can cooperation protect itself from invasion? Axelrod proved that it can indeed. In fact, there is a gratifying asymmetry to his findings: Although a world of "meanies" (beings using the inflexible *ALL-D* strategy) is penetrable by cooperators in clusters, a world of cooperators is *not* penetrable by meanies, even if they arrive in clusters of any size. Once cooperation has established itself, it is permanent. As Axelrod puts it, "The gear wheels of social evolution have a ratchet."

The term "social" here does not mean that these results necessarily apply only to higher animals that can think. Clearly, four-line computer programs do not think—and yet, it is in a world of just such "organisms" that cooperation has been shown to evolve. The only "cognitive abilities" needed by *TIT FOR TAT* are: (1) recognition of previous partners, and (2) memory of what happened last time with this partner. Even bacteria can do this, by interacting with only one other organism (so that recognition is automatic) and by responding only to the most recent action of their "partner" (so that memory requirements are minimal). The point is that the entities involved can be on the scale of bacteria, small animals, large animals, or nations. There is no need for "reflective rationality"; indeed, *TIT FOR TAT* could be called "reflexive" (in the sense of being as simple as a knee-jerk reflex) rather than "reflective".

*   *   *

For people who think that moral behavior toward others can emerge only when there is imposed some totally external and horrendous threat (say, of the fire-and-brimstone sort) or soothing promise of heavenly reward (such as eternal salvation), the results of this research must give pause for thought. In one sentence, Axelrod captures the whole idea: *Mutual cooperation can emerge in a world of egoists without central control, by starting with a cluster of individuals who rely on reciprocity.*

There are so many situations in the world today where these ideas seem of extreme relevance—indeed, urgency—that it is very tempting to draw morals all over the place. In the later chapters of his book, Axelrod offers advice about how to promote cooperation in human affairs, and at the very end the political scientist in him cautiously ventures some broad conclusions concerning global issues, which are a fitting way for me to conclude as well:

> Today, the most important problems facing humanity are in the arena of international relations where independent, egoistic nations face each other in a state of near anarchy. Many of these problems take the form of an iterated Prisoner's Dilemma. Examples can include arms races, nuclear proliferation, crisis bargaining, and military escalation. Of course, a realistic understanding of these problems would have to take into account many factors not incorporated into the simple Prisoner's Dilemma formulation, such as ideology, bureaucratic politics, commitments, coalitions, mediation, and leadership. Nevertheless, we can use all the insights we can get.
>
> Robert Gilpin [in his book *War and Change in World Politics*] points out that from the ancient Greeks to contemporary scholarship all political theory addresses one fundamental question: "How can the human race, whether for selfish or more cosmopolitan ends, understand and control the seemingly blind forces of history?" In the contemporary world this question has become especially acute because of the development of nuclear weapons.
>
> The advice given in this book to players of the Prisoner's Dilemma might also serve as good advice to national leaders as well: Don't be envious, don't be the first to defect, reciprocate both cooperation and defection, and don't be too clever. Likewise, the techniques discussed in this book for promoting cooperation in the Prisoner's Dilemma might also be useful in promoting cooperation in international politics.
>
> The core of the problem is that trial-and-error learning is slow and painful. The conditions may all be favorable for long-run developments, but we may not have the time to wait for blind processes to move us slowly towards mutually rewarding strategies based upon reciprocity. Perhaps if we understand the process better, we can use our foresight to speed up the evolution of cooperation.

Post Scriptum.

In the course of writing this column and thinking the ideas through, I was forced to confront over and over again the paradox that the Prisoner's Dilemma presents. I found that I simply could not accept the seemingly flawless logical conclusion that says that a rational player in a *noniterated* situation will always defect. In turning this over in my mind and trying to articulate my objections clearly, I found myself inventing variation after variation after variation on the basic situation. I would like to describe just a few here.

A version of the dealer-and-buyer scenario involving bags exchanged in a forest actually occurs in a more familiar context. Suppose I take my car in to get the oil changed. I know little about auto mechanics, so when I come in to pick it up, I really have no way to verify if they've done the job. For all I know, it's been sitting untouched in their parking lot all day, and as I drive off they may be snickering behind my back. On the other hand, maybe *I've* got the last laugh, for how do *they* know if that check I gave them will bounce?

This is a perfect example of how either of us *could* defect, but because the situation is iterated, neither of us is likely to do so. On the other hand, suppose I'm on my way across the country and have some radiator trouble near Gillette, Wyoming, and stop in town to get my radiator repaired there. There is a decent chance now that one party or the other will attempt to defect, because this kind of situation is not an iterated one. I'll probably never again need the services of this garage, and they'll never get another check from me. In the most crude sense, then, it's not in my interest to give them a good check, nor is it in theirs to fix my car. But do I really defect? Do I give out bad checks? No. Why not?

Consider this related situation. Late at night, I bang into someone's car in a deserted parking lot. It's apparent to me that nobody witnessed the incident. I have the choice of leaving a note, telling the owner who's to blame, or scurrying off scot-free. Which do I do? Similarly, suppose I have given a lecture in a classroom in a university I am visiting for one day, and have covered the board with chalk. Do I take the trouble of erasing the board so that whoever comes in the next morning won't have to go to that trouble? Or do I just leave it?

*　　*　　*

I was recently waiting to board an airplane when a voice announced: "Passengers holding seats in rows 24 to 36 may now board." Well, my seat was in row 4, so I waited. A few minutes later, the voice said that passengers in rows 18 to 36 were free to board. A group of people got up and went in. Then after a couple of minutes, rows 10 to 36 were told they could board. A dozen people or so remained in the waiting area. For a while, we were all patient, waiting for the final announcement allowing us to board, but after about five minutes, people started fidgeting a bit and edging up toward the gate. Then, after another two or three minutes, a couple of people just went right on. And then the rest of us wondered, "Should we get on, too? Will we be left behind?" For most of the people, the answer was obvious: they rushed to board. And once *they* had boarded, then the rest of us felt kind of like suckers, and we just got on too. In effect, there was a stampede that converted cooperators into defectors. Even the people who triggered the stampede had originally been cooperating, but after a while, the temptation got to be too great, and they broke down. At that point, some sort of phase transition, or collective shift, took place, and the stable state of patient cooperation collapsed into a chaotic scrambling for places. Actually, it wasn't that bad, and there was a good reason for the relatively polite way we did board, defectors though we were: we all had seat assignments, so it didn't matter who got on first. But imagine if the earliest defectors were sure to get the best remaining seats! The contemporary aphorist Ashleigh Brilliant has found just the right *bons mots* to describe this sort of dilemma:

> Should I abide by the rules until they're changed, or help speed the change by breaking them?

> Better start rushing before the rush begins!

In pondering the Prisoner's Dilemma, I could not help but be reminded of horrible scenarios in Nazi concentration camps, where large herds of unarmed people would be led to their deaths by small herds of armed people. It seems that a stampede by the masses could quickly have overcome a small number of guards, at least in certain critical narrow passageways here and there. The trouble is, it would require certain death on the part of a few ultra-cooperators, in exchange for the liberation of a large number of other people. Generally speaking, individuals are not willing to perform such an exchange. Nobody wants to be in the front lines of a protest demonstration facing troops with machine guns. Everyone wants to be in the rear. But not everyone can be in the rear! If nobody is willing to be in the front lines, then there will be no front lines, and consequently no demonstration at all.

* * *

Driving a car has a certain primitive quality to it that brings out the animal in us all, and probably that's why it confronts us with Prisoner's-Dilemma-like situations so often—more often than any other activity I can think of. How about those annoying drivers who, when there's a long line at a freeway exit, zoom by all the politely lined-up cars and then butt in at the very last moment, getting off 50 cars ahead of you? Are you angry at such people, or do you do it too? Or, worse—do you do it and yet resent others who have such gall?

I have been struck by the relative savagery of the driving environment in the Boston area. I know of no other city in which people are so willing to take the law into their own hands, and to create complete anarchy. There seems to be less respect for such things as red lights, stop signs, lines in the street, speed limits, other people's cars, and so forth, than in any other city, state, or country that I have ever driven in. This incessant "me-first" attitude seems to be a vicious, self-reinforcing circle. Since there *are* so many people who do whatever they want, nobody can afford to be polite and let other people in ahead of them (say), for then they will be taken advantage of repeatedly and will wind up losing totally. You simply *must* assert yourself in many situations, and that means you must *defect*. Of course, just one defection does not an *ALL-D* player make. In fact, a retaliatory defection is just good old *TIT-FOR-TAT* playing. However, very often in Boston driving, there is no way you can get back at a nasty driver who cuts in front of you and then takes off screeching around the corner. That person is gone forever. You can take out your frustrations only on the rest of the people near you, who are not to blame for *that* driver. You can cut in ahead of *them*. Does this do any good? That is, does it teach anybody a lesson? Obviously it will teach them only that it pays to defect. And thus the spiral starts.

Is there any way to put a halt to the descending spiral, the vortex towards oblivion? Is there any point at which the people of Boston will collectively come to realize that it has gotten *so* bad that they will all suddenly "flip" and begin to cooperate in situations where they formerly would have defected? Can there be a stampede toward cooperation, just as there can be a stampede toward defection?

Clearly, if large numbers of people were to start driving much less aggressively and nastily, everybody would benefit. Huge snarls would unsnarl—in fact would never form. Traffic would flow smoothly and regularly. The shoulders—those favorite illegal passing lanes for defectors—would be completely clear. So clear, in fact, that—just think—you and I could make sensational progress by swerving onto an empty shoulder and passing everybody. Wheee! Isn't this fun? Aren't those other people *suckers*, staying in the slow lane and glaring at us? Say, how come other people are barging in on us? This is *our* lane. Oh, so that person in the yellow car wants to play dirty, eh? Okay, I'll show them what playing dirty's *really* like!

Sound familiar? Is there any solution to such terrible spirals? Sometimes I am very pessimistic on that subject. Anatol Rapoport and I exchanged letters concerning this matter, and he related a frightening anecdote. I quote from his letter:

> Do you know of the experiment performed by Martin Shubik, in which a dollar bill was auctioned off for $3.40? This was a consequence of a rule (the implications of which dawned on the subjects only when they were already hooked) specifying that while the highest bidder got the dollar, the *second*-highest bidder would also have to pay what he last bid. It thus became imperative to keep going, since the second-highest bidder (whoever he was at each stage) had progressively more to lose as the bids went up. Are Reagan and Andropov too stupid to see the point? . . . .
>
> I believe the "technological imperative" is driving our species to extinction. Ever more horrendous weapons must be produced, simply because it is *possible* to produce them. Eventually they must be used, to justify the insane waste. It thus becomes imperative to seal off the "logic" of the paradigm based on "deterrence", "balance of power", and similar metaphors—to make it unassailable.

I don't think intelligence plays any part in the vicious cycle of the arms race. The rulers only think they make the decisions. If they were C players, they would not be where they are. If they started to play C while in office, they would be impeached, overthrown, or assassinated. Does this mean that D players are selected for? Possibly in the short run, but not on the time scale of evolution. *H. sapiens* is apparently not the last word, but for me, a homocentric, this is no consolation.

Pretty sobering words from one of the leading rational thinkers of our era.

Section VII:
Sanity and Survival

Section VII:
Sanity and Survival

In the four chapters of this concluding section, themes of the previous section are carried further and brought into contact with common social dilemmas and, eventually, the current world situation. On a small scale, we are constantly faced with dilemmas like the Prisoner's Dilemma, where personal greed conflicts with social gain. For any two persons, the dilemma is virtually identical. What would be sane behavior in such situations? For true sanity, the key element is that each individual must be able to recognize both that the dilemma is symmetric and that the other individuals facing it are equally able. Such individuals—individuals who will cooperate with one another despite all temptations toward crude egoism—are more than just *rational*; they are *superrational*, or for short, *sane*. But there are dilemmas and "egos" on a suprahuman level as well. We live in a world filled with opposing belief systems so similar as to be nearly interchangeable, yet whose adherents are blind to that symmetry. This description applies not only to myriad small conflicts in the world but also to the colossally blockheaded opposition of the United States and the Soviet Union. Yet the recognition of symmetry—in short, the sanity—has not yet come. In fact, the insanity seems only to grow, rather than be supplanted by sanity. What has an intelligent species like our own done to get itself into this horrible dilemma? What can it do to get itself out? Are we all helpless as we watch this spectacle unfold, or does the answer lie, for each one of us, in recognition of our own typicality, and in small steps taken on an individual level toward sanity?

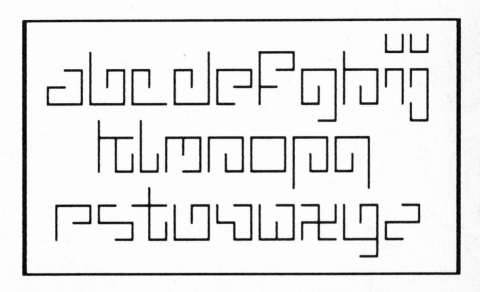

30

Dilemmas for Superrational Thinkers, Leading Up to a Luring Lottery

June,1983

AND then one fine day, out of the blue, you get a letter from S. N. Platonia, well-known Oklahoma oil trillionaire, mentioning that twenty leading rational thinkers have been selected to participate in a little game. "You are one of them!" it says. "Each of you has a chance at winning one billion dollars, put up by the Platonia Institute for the Study of Human Irrationality. Here's how. If you wish, you may send a telegram with just your name on it to the Platonia Institute in downtown Frogville, Oklahoma (pop. 2). You may reverse the charges. If you reply within 48 hours, the billion is yours—unless there are two or more replies, in which case the prize is awarded to no one. And if no one replies, nothing will be awarded to anyone."

You have no way of knowing who the other nineteen participants are; indeed, in its letter, the Platonia Institute states that the entire offer will be rescinded if it is detected that any attempt whatsoever has been made by any participant to discover the identity of, or to establish contact with, any other participant. Moreover, it is a condition that the winner (if there is one) must agree in writing not to share the prize money with any other participant at any time in the future. This is to squelch any thoughts of cooperation, either before or after the prize is given out.

The brutal fact is that no one will know what anyone else is doing. Clearly, everyone will want that billion. Clearly, everyone will realize that if their name is *not* submitted, they have no chance at all. Does this mean that twenty telegrams will arrive in Frogville, showing that even possessing transcendent levels of rationality—as you of course do—is of no help in such an excruciating situation?

This is the "Platonia Dilemma", a little scenario I thought up recently in trying to get a better handle on the Prisoner's Dilemma, of which I wrote last month. The Prisoner's Dilemma can be formulated in terms resembling this dilemma, as follows. Imagine that you receive a letter from the Platonia Institute telling you that you and just *one* other anonymous leading rational thinker have been selected for a modest cash giveaway. As before, both of you are requested to reply by telegram within 48 hours to the Platonia Institute, charges reversed. Your telegram is to contain, aside from your name, just the word "cooperate" or the word "defect". If two "cooperate"s are received, both of you will get $3. If two "defect"s are received, you both will get $1. If one of each is received, then the cooperator gets nothing and the defector gets $5.

What choice would you make? It would be nice if you both cooperated, so you'd each get $3, but doesn't it seem a little unlikely? After all, who wants to get suckered by a nasty, low-down, rotten defector who gets $5 for being sneaky? Certainly not *you*! So you'd probably decide not to cooperate. It seems a regrettable but necessary choice. Of course, both of you, reasoning alike, come to the same conclusion. So you'll both defect, and that way get a mere dollar apiece. And yet—if you'd just both been willing to risk a bit, you could have gotten $3 apiece. What a pity!

*   *   *

It was my discomfort with this seemingly logical analysis of the "one-round Prisoner's Dilemma" that led me to formulate the following letter, which I sent out to twenty friends after having cleared it with *Scientific American*:

Dear X:

I am sending this letter out via Special Delivery to twenty of 'you' (namely, various friends of mine around the country). I am proposing to all of you a one-round Prisoner's Dilemma game, the payoffs to be monetary (provided by *Scientific American*). It's very simple. Here is how it goes.

Each of you is to give me a single letter: 'C' or 'D', standing for 'cooperate' or 'defect'. This will be used as your move in a Prisoner's Dilemma with *each* of the nineteen other players. The payoff matrix I am using for the Prisoner's Dilemma is given in the diagram [see Figure 29-1c].

Thus if everyone sends in 'C', everyone will get $57, while if everyone sends in 'D', everyone will get $19. You can't lose! And of course, anyone who sends in 'D' will get at least as much as everyone else will. If, for example, 11 people send in 'C' and 9 send in 'D', then the 11 C-ers will get $3 apiece from each of the other C-ers (making $30), and zero from the D-ers. So C-ers will get $30 each. The D-ers, by contrast, will pick up $5 apiece from each of the C-ers, making $55, and $1 from each of the other D-ers, making $8, for a grand total of $63. No matter what the distribution is, D-ers always do better than C-ers. Of course, the more C-ers there are, the better *everyone* will do!

By the way, I should make it clear that in making your choice, you should not aim to be the *winner*, but simply to get as much *money* for yourself as possible. Thus you should be happier to get $30 (say, as a result of saying 'C' along with 10 others, even though the 9 D-sayers get more than you) than to get $19 (by saying 'D' along with everybody else, so nobody 'beats' you). Furthermore, you are not supposed to think that at some subsequent time you will meet with and be able to share the goods with your co-participants. You are not aiming at maximizing the total number of dollars *Scientific American* shells out, only at maximizing the number that come to *you!*

Of course, your hope is to be the *unique* defector, thus really cleaning up: with 19 C-ers, you'll get $95 and they'll each get 18 times $3, namely $54! But why am I doing the multiplication or any of this figuring for you? You're very bright. So are all of you! All about equally bright, I'd say, in fact. So all you need to do is tell me your choice. I want all answers by telephone (call collect, please) the day you receive this letter.

It is to be understood (it *almost* goes without saying, but not quite) that you are not to try to get in touch with and consult with others who you guess have been asked to participate. In fact, please consult with no one at all. The purpose is to see what people will do on their own, in isolation. Finally, I would very much appreciate a short statement to go along with your choice, telling me *why* you made this particular choice.

Yours, . . .

P.S.—By the way, it may be helpful for you to imagine a related situation, the same as the present one except that you are told that all the other players have *already* submitted their choice (say, a week ago), and so you are the last. Now what do you do? Do you submit 'D', knowing full well that *their* answers are already committed to paper? Now suppose that, immediately after having submitted your 'D' (or your 'C') in that circumstance, you are informed that, in fact, the others really *haven't* submitted their answers yet, but that they are all doing it today. Would you retract your answer? Or what if you knew (or at least were told) that you were the *first* person being asked for an answer?

And—one last thing to ponder—what would you do if the payoff matrix looked as shown in Figure 30-1*a* ?

FIGURE 30–1. *In (a), a modification of Figure 29-1(c). Here, the incentive to defect seems considerably stronger. In (b), the payoff matrix for a Wolf's-Dilemma situation involving just two participants. Compare it to that in Figure 29-1(c).*

Player B

		Cooperates	Defects
Player A	Cooperates	(3,3)	(0,50)
	Defects	(50,0)	(.01, .01)

(a)

Player B

		Refrains	Pushes button
Player A	Refrains	(1000,1000)	(0,100)
	Pushes button	(100,0)	(100,100)

(b)

*　　*　　*

I wish to stress that this situation is not an *iterated* Prisoner's Dilemma (discussed in last month's column). It is a one-shot, multi-person Prisoner's Dilemma. There is no possibility of learning, over time, anything about how the others are inclined to play. Therefore all lessons described last month are inapplicable here, since they depend on the situation's being iterated. All that each recipient of my letter could go on was the thought, "There are nineteen people out there, somewhat like me, all in the same boat, all grappling with the same issues as I am." In other words, there was nothing to rely on except pure reason.

I had much fun preparing this letter, deciding who to send it out to, anticipating the responses, and then receiving them. It was amusing to me, for instance, to send Special Delivery letters to two friends I was seeing every day, without forewarning them. It was also amusing to send identical letters to a wife and husband at the same address.

Before I reveal the results, I invite you to think how you would play in such a contest. I would particularly like you to take seriously the assertion "everyone is very bright". In fact, let me expand on that idea, since I felt that people perhaps did not really understand what I meant by it. Thus please consider the letter to contain the following clarifying paragraph:

> All of you are very rational people. Therefore, I hardly need to tell you that you are to make what you consider to be your maximally rational choice. In particular, feelings of morality, guilt, vague malaise, and so on, are to be disregarded. Reasoning alone (of course including reasoning about the others' reasoning) should be the basis of your decision. And please always remember that everyone is being told this (including *this*!)!

I was hoping for—and expecting—a particular outcome to this experiment. As I received the replies by phone over the next several days, I jotted down notes so that I had a record of what impelled various people to choose as they did. The result was not what I had expected—in fact, my friends "faked me out" considerably. We got into heated arguments about the "rational" thing to do, and everyone expressed much interest in the whole question.

I would like to quote to you some of the feelings expressed by my friends caught in this deliciously tricky situation. David Policansky opened his call tersely by saying, "Okay, Hofstadter, give me the $19!" Then he presented this argument for defecting: "What you're asking us to do, in effect, is to press one of two buttons, knowing nothing except that if we press button D, we'll get more than if we press button C. Therefore D is better. That is the essence of my argument. I defect."

Martin Gardner (yes, I asked Martin to participate) vividly expressed the emotional turmoil he and many others went through. "Horrible dilemma", he said. "I really don't know what to do about it. If I wanted to maximize my money, I would choose D and expect that others would also; to maximize my satisfactions, I'd choose C, and hope other people would do the same (by the Kantian imperative). I don't know, though, how one should behave *rationally*. You get into endless regresses: 'If they all do X, then I should do Y, but then they'll anticipate that and do Z, and so . . .' You get trapped in an endless whirlpool. It's like Newcomb's paradox." So saying, Martin defected, with a sigh of regret.

In a way echoing Martin's feelings of confusion, Chris Morgan said, "More by intuition than by anything else, I'm coming to the conclusion that there's no way to deal with the paradoxes inherent in this situation. So I've decided to flip a coin, because I can't anticipate what the others are going to do. I think—but can't know—that they're all going to negate each other." So, while on the phone, Chris flipped a coin and "chose" to cooperate.

Sidney Nagel was very displeased with his conclusion. He expressed great regret: "I actually couldn't sleep last night because I was thinking about it. I *wanted* to be a cooperator, but I couldn't find any way of justifying it. The way I figured it, what I do isn't going to affect what anybody else does. I might as well consider that everything else is already fixed, in which case the best I can do for myself is to play a D."

Bob Axelrod, whose work proves the superiority of cooperative strategies in the *iterated* Prisoner's Dilemma, saw no reason whatsoever to cooperate in a one-shot game, and defected without any compunctions.

Dorothy Denning was brief: "I figure, if I defect, then I always do at least as well as I would have if I had cooperated. So I defect." She was one of the people who faked me out. Her husband, Peter, cooperated. I had predicted the reverse.

*   *   *

By now, you have probably been counting. So far, I've mentioned five D's and two C's. Suppose you had been me, and you'd gotten roughly a third of the calls, and they were 5–2 in favor of defection. Would you dare to extrapolate these statistics to roughly 14–6? How in the world can seven individuals' choices have anything to do with thirteen *other* individuals' choices? As Sidney Nagel said, certainly one choice can't influence another (unless you believe in some kind of telepathic transmission, a possibility we shall discount here). So what justification might there be for extrapolating these results?

Clearly, any such justification would rely on the idea that people are "like" each other in some sense. It would rely on the idea that in complex and tricky decisions like this, people will resort to a cluster of reasons, images, prejudices, and vague notions, some of which will tend to push them one way, others the other way, but whose overall impact will be to push a certain percentage of people toward one alternative, and another percentage of people toward the other. In advance, you can't hope to predict what those percentages will be, but given a sample of people in the situation, you can hope that their decisions will be "typical". Thus the notion that early returns running 5–2 in favor of defection can be extrapolated to a final result of 14–6 (or so) would be based on assuming that the seven people are acting "typically" for people confronted with these conflicting mental pressures.

The snag is that the mental pressures are not completely explicit; they are evoked by, but not totally spelled out by, the wording of the letter. Each person brings a unique set of images and associations to each word and concept, and it is the set of those images and associations that will collectively create, in that person's mind, a set of mental pressures like the set of pressures inside the earth in an earthquake zone. When people decide, you find out how all those pressures pushing in different directions add up, like a set of force vectors pushing in various directions and with strengths influenced by private or unmeasurable factors. The assumption that it is valid to extrapolate has to be based on the idea that everybody is alike inside, only with somewhat different weights attached to certain notions.

This way, each person's decision can be likened to a "geophysics experiment" whose goal is to predict where an earthquake will appear. You set up a model of the earth's crust and you put in data representing your best understanding of the internal pressures. You know that there unfortunately are large uncertainties in your knowledge, so you just have to choose what seem to be "reasonable" values for various variables. Therefore no single run of your simulation will have strong predictive power, but that's all right. You run it and you get a fault line telling you where the simulated earth shifts. Then you go back and choose other values in the ranges of those variables, and rerun the whole thing. If you do this repeatedly, eventually a pattern will emerge revealing where and how the earth is likely to shift and where it is rock-solid.

This kind of simulation depends on an essential principle of statistics: the idea that when you let variables take on a few sample random values in their ranges, the overall outcome determined by a cluster of such variables will start to emerge after a few trials and soon will give you an accurate model. You don't need to run your simulation millions of times to see valid trends emerging.

This is clearly the kind of assumption that TV networks make when they predict national election results on the basis of early returns from a few select towns in the East. Certainly they don't think that free will is any "freer" in the East than in the West—that whatever the East chooses to do, the West will follow suit. It is just that the cluster of emotional and intellectual pressures on voters is much the same all over the nation. Obviously, no individual can be taken as representing the whole nation, but a well-selected group of residents of the East Coast can be assumed to be representative of the whole nation in terms of how much they are "pushed" by the various pressures of the election, so that their choices are likely to show general trends of the larger electorate.

Suppose it turned out that New Hampshire's Belknap County and

California's Modoc County had produced, over many national elections, very similar results. Would it follow that one of the two counties had been exerting some sort of causal influence on the other? Would they have had to be in some sort of eerie cosmic resonance mediated by "sympathetic magic" for this to happen? Certainly not. All it takes is for the electorates of the two counties to be similar; then the pressures that determine how people vote will take over and automatically make the results come out similar. It is no more mysterious than the observation that a Belknap County schoolgirl and a Modoc County schoolboy will get the same answer when asked to divide 507 by 13: the laws of arithmetic are the same the world over, and they operate the same in remote minds without any need for "sympathetic magic".

This is all elementary common sense; it should be the kind of thing that any well-educated person should understand clearly. And yet emotionally it cannot help but feel a little peculiar since it flies in the face of free will and regards people's decisions as caused simply by combinations of pressures with unknown values. On the other hand, perhaps that is a better way to look at decisions than to attribute them to "free will", a philosophically murky notion at best.

*   *   *

This may have seemed like a digression about statistics and the question of individual actions versus group predictability, but as a matter of fact it has plenty to do with the "correct action" to take in the dilemma of my letter. The question we were considering is: To what extent can what a *few* people do be taken as an indication of what *all* the people will do? We can sharpen it: To what extent can what *one* person does be taken as an indication of what *all* the people will do? The ultimate version of this question, stated in the first person, has a funny twist to it: To what extent does *my* choice inform me about the choices of the other participants?

You might feel that each person is completely unique and therefore that no one can be relied on as a predictor of how other people will act, especially in an intensely dilemmatic situation. There is more to the story, however. I tried to engineer the situation so that everyone would have the same image of the situation. In the dead center of that image was supposed to be the notion that everyone in the situation was using *reasoning alone*—including reasoning about the reasoning—to come to an answer.

Now, if reasoning dictates an answer, then everyone should independently come to that answer (just as the Belknap County schoolgirl and the Modoc County schoolboy would independently get 39 as their answer to the division problem). Seeing this fact is itself the critical step in the reasoning toward the correct answer, but unfortunately it eluded nearly everyone to whom I sent the letter. (That is why I came to wish I had included in the letter a paragraph stressing the rationality of the players.) Once you realize this fact, then it dawns on you that *either* all rational players will choose D *or* all rational players will choose C. This is the crux.

Any number of ideal rational thinkers faced with the same situation and undergoing similar throes of reasoning agony will necessarily come up with the identical answer eventually, so long as reasoning alone is the ultimate justification for their conclusion. Otherwise reasoning would be subjective, not objective as arithmetic is. A conclusion reached by reasoning would be a matter of preference, not of necessity. Now *some* people may believe this of reasoning, but rational thinkers understand that a valid argument must be *universally* compelling, otherwise it is simply not a valid argument.

If you'll grant this, then you are 90 percent of the way. All you need ask now is, "Since we are all going to submit the same letter, which one would be more logical? That is, which world is better for the *individual* rational thinker: one with all C's or one with all D's?" The answer is immediate: "I get $57 if we all cooperate, $19 if we all defect. Clearly I prefer $57, hence cooperating is preferred by this particular rational thinker. Since I am typical, cooperating must be preferred by *all* rational thinkers. So I'll cooperate." Another way of stating it, making it sound weirder, is this: "If I choose C, then everyone will choose C, so I'll get $57. If I choose D, then everyone will choose D, so I'll get $19. I'd rather have $57 than $19, so I'll choose C. Then everyone will, and I'll get $57."

*   *   *

To many people, this sounds like a belief in voodoo or sympathetic magic, a vision of a universe permeated by tenuous threads of synchronicity, conveying thoughts from mind to mind like pneumatic tubes carrying messages across Paris, and making people resonate to a secret harmony. Nothing could be further from the truth. This solution depends in no way on telepathy or bizarre forms of causality. It's just that the statement "I'll choose C and then everyone will", though entirely correct, is somewhat misleadingly phrased. It involves the word "choice", which is incompatible with the compelling quality of logic. Schoolchildren do not *choose* what 507 divided by 13 is; they figure it out. Analogously, my letter really did not allow choice; it demanded reasoning. Thus, a better way to phrase the "voodoo" statement would be this: "If reasoning guides *me* to say C, then, as I am no different from anyone else as far as rational thinking is concerned, it will guide everyone to say C."

The corresponding foray into the opposite world ("If I choose D, then everyone will choose D") can be understood more clearly by likening it to a musing done by the Belknap County schoolgirl before she divides: "Hmm, I'd guess that 13 into 507 is about 49—maybe 39. I see I'll have to calculate it out. But I know in advance that *if* I find out that it's 49, then sure as shootin', that Modoc County kid will write down 49 on his paper as well; and if I get 39 as my answer, then so will he." No secret transmissions are involved; all that is needed is the universality and uniformity of arithmetic.

Likewise, the argument "Whatever I do, so will everyone else do" is simply a statement of faith that reasoning is universal, at least among rational thinkers, not an endorsement of any mystical kind of causality.

This analysis shows why you should cooperate even when the opaque envelopes containing the other players' answers are right there on the table in front of you. Faced so concretely with this unalterable set of C's and D's, you might think, "Whatever they have done, I am better off playing D than playing C—for certainly what I *now* choose can have no retroactive effect on what they chose. So I defect." Such a thought, however, assumes that the logic that now drives you to playing D has no connection or relation to the logic that earlier drove them to their decisions. But if you accept what was stated in the letter, then you must conclude that the decision you now make will be mirrored by the plays in the envelopes before you. If logic now coerces *you* to play D, it has *already* coerced the others to do the same, and for the same reasons; and conversely, if logic coerces you to play C, it has also already coerced the others to do *that*.

Imagine a pile of envelopes on your desk, all containing other people's answers to the arithmetic problem, "What is 507 divided by 13?" Having hurriedly calculated *your* answer, you are about to seal a sheet saying "49" inside your envelope, when at the last moment you decide to check it. You discover your error, and change the '4' to a '3'. Do you at that moment envision all the answers inside the other envelopes suddenly pivoting on their heels and switching from "49" to "39"? Of course not! You simply recognize that what is changing is your *image* of the contents of those envelopes, not the contents themselves. You used to think there were many "49"s. You now think there are many "39"s. However, it doesn't follow that there was a moment in between, at which you thought, "They're all switching from '49' to '39'!" In fact, you'd be crazy to think that.

It's similar with D's and C's. If at first you're inclined to play one way but on careful consideration you switch to the other way, the other players obviously won't retroactively or synchronistically follow you—but if you give them credit for being able to see the logic you've seen, you have to assume that their answers are what yours is. In short, you aren't going to be able to undercut them; you are simply "in cahoots" with them, like it or not! Either all D's, or all C's. Take your pick.

Actually, saying "Take your pick" is 100 percent misleading. It's *not* as if you could merely "pick", and then other people—even in the past—would magically follow suit! The point is that since you are going to be "choosing" by using what you believe to be compelling *logic,* if you truly respect your logic's compelling quality, you would have to believe that others would buy it as well, which means that you are certainly *not* "just picking". In fact, the more convinced you are of what you are playing, the more certain you should be that others will also play (or have already played) the same way, and for the same reasons. This holds whether you play C or D, and it is the real core of the solution. Instead of being a paradox, it's a self-reinforcing solution: a benign circle of logic.

*   *   *

If this still sounds like retrograde causality to you, consider this little tale, which may help make it all make more sense. Suppose you and Jane are classical music lovers. Over the years, you have discovered that you have incredibly similar tastes in music—a remarkable coincidence! Now one day you find out that two concerts are being given simultaneously in the town where you live. Both of them sound excellent to you, but Concert A simply cannot be missed, whereas Concert B is a strong temptation that you'll have to resist. Still, you're extremely curious about Concert B, because it features Zilenko Buznani, a violinist you've always heard amazing things about.

At first, you're disappointed, but then a flash crosses your mind: "Maybe I can at least get a first-hand report about Zilenko Buznani's playing from Jane. Since she and I hear everything through virtually the same ears, it would be almost as good as my going if *she* would go." This is comforting for a moment, until it occurs to you that something is wrong here. For the same reasons as you do, Jane will insist on hearing Concert A. After all, she loves music in the same way as you do—that's precisely why you wish she would tell you about Concert B! The more you feel Jane's taste is the same as yours, the more you wish she would go to the other concert, so that you could know what it was like to have gone to it. But the more her taste is the same is yours, the less she will want to go to it!

The two of you are tied together by a bond of common taste. And if it turns out that you are different enough in taste to disagree about which concert is better, then that will tend to make you lose interest in what she might report, since you no longer can trust her opinion as that of someone who hears music "through your ears". In other words, hoping she'll choose Concert B is pointless, since it undermines your reasons for *caring* which concert she chooses!

The analogy is clear, I hope. Choosing D undermines your reasons for doing so. To the extent that all of you really *are* rational thinkers, you really will think in the same tracks. And my letter was supposed to establish beyond doubt the notion that you are all "in synch"; that is, to ensure that you can depend on the others' thoughts to be rational, which is all you need.

Well, not quite. You need to depend not just on their being rational, but on their depending on everyone else to be rational, *and* on their depending on everyone to depend on everyone to be rational—and so on. A group of reasoners in this relationship to each other I call *superrational.* Superrational thinkers, by recursive definition, include in their calculations the fact that they are in a group of superrational thinkers. In this way, they resemble elementary particles that are *renormalized.*

A renormalized electron's style of interacting with, say, a renormalized photon takes into account that the photon's quantum-mechanical structure includes "virtual electrons" and that the electron's quantum-mechanical structure includes "virtual photons"; moreover it takes into account that all these virtual particles (themselves renormalized) also interact with one another. An infinite cascade of possibilities ensues but is taken into account in one fell swoop by nature. Similarly, superrationality, or renormalized reasoning, involves seeing all the consequences of the fact that other renormalized reasoners are involved in the same situation—and doing so in a finite swoop rather than succumbing to an infinite regress of reasoning about reasoning about reasoning . . .

*   *   *

'C' is the answer I was hoping to receive from everyone. I was not so optimistic as to believe that literally everyone would arrive at this conclusion, but I expected a majority would—thus my dismay when the early returns strongly favored defecting. As more phone calls came in, I did receive some C's, but for the wrong reasons. Dan Dennett cooperated, saying, "I would rather be the person who bought the Brooklyn Bridge than the person who sold it. Similarly, I'd feel better spending $3 gained by cooperating than $10 gained by defecting."

Charles Brenner, who I'd figured to be a sure-fire D, took me by surprise and C'd. When I asked him why, he candidly replied, "Because I don't want to go on record in an international journal as a defector." Very well. Know, World, that Charles Brenner is a cooperator!

Many people flirted with the idea that everybody would think "about the same", but did not take it seriously enough. Scott Buresh confided to me: "It was not an easy choice. I found myself in an oscillation mode: back and forth. I made an assumption: that everybody went through the same mental processes I went through. Now I personally found myself wanting to cooperate roughly one third of the time. Based on that figure and the assumption that I was typical, I figured about one third of the people would cooperate. So I computed how much I stood to make in a field where six or seven people cooperate. It came out that if I were a D, I'd get about three times as much as if I were a C. So I'd have to defect. Water seeks out its own level, and I sank to the lower righthand corner of the matrix." At this point, I told Scott that so far, a substantial majority had defected. He reacted swiftly: "Those rats—how can they all defect? It makes me so mad! I'm really disappointed in your friends, Doug."

So was I, when the final results were in: Fourteen people had defected and six had cooperated—exactly what the networks would have predicted! Defectors thus received $43 while cooperators got $15. I wonder what Dorothy's saying to Peter about now? I bet she's chuckling and saying, "I told you I'd do better this way, didn't I?" Ah, me . . . What can you do with people like that?

A striking aspect of Scott Buresh's answer is that, in effect, he treated his own brain as a simulation of other people's brains and ran the simulation enough to get a sense of what a "typical person" would do. This is very much in the spirit of my letter. Having assessed what the statistics are likely to be, Scott then did a cool-headed calculation to maximize his profit, based on the assumption of six or seven cooperators. Of course, it came out in favor of defecting. In fact, it would have, no matter what the number of cooperators was! Any such calculation will always come out in favor of defecting. As long as you feel your decision is *independent* of others' decisions, you should defect. What Scott failed to take into account was that cool-headed calculating people should take into account that cool-headed calculating people should take into account that cool-headed calculating people should take into account that . . .

This sounds awfully hard to take into account in a finite way, but actually it's the easiest thing in the world. All it means is that all these heavy-duty rational thinkers are going to see that they are in a symmetric situation, so that *whatever reason dictates to one, it will dictate to all.* From that point on, the process is very simple. Which is better for an individual if it is a universal choice: C or D? That's all.

*   *   *

Actually, it's not quite all, for I've swept one possibility under the rug: maybe throwing a die could be better than making a deterministic choice. Like Chris Morgan, you might think the best thing to do is to choose C with probability p and D with probability $1-p$. Chris arbitrarily let p be 1/2, but it could be any number between 0 and 1, where the two extremes represent D'ing and C'ing respectively. What value of p would be chosen by superrational players? It is easy to figure out in a two-person Prisoner's Dilemma, where you assume that both players use the same value of p. The expected earnings for each, as a function of p, come out to be $1+3p-p^2$, which grows monotonically as p increases from 0 to 1. Therefore, the optimum value of p is 1, meaning certain cooperation. In the case of more players, the computations get more complex but the answer doesn't change: the expectation is always maximal when p equals 1. Thus this approach confirms the earlier one, which didn't entertain probabilistic strategies.

Rolling a die to determine what you'll do didn't add anything new to the standard Prisoner's Dilemma, but what about the modified-matrix version I gave in the *P.S.* to my letter? I'll let you figure that one out for yourself. And what about the Platonia Dilemma? There, two things are very clear: (1) if you decide not to send a telegram, your chances of winning are zero; (2) if everyone sends a telegram, your chances of winning are zero. If you believe that what you choose will be the same as what everyone else chooses because you are all superrational, then neither of these alternatives is very appealing. With dice, however, a new option presents itself: to roll a die with probability p of coming up "good" and then to send in your name if and only if "good" comes up.

Now imagine twenty people all doing this, and figure out what value of

p maximizes the likelihood of exactly *one* person getting the go-ahead. It turns out that it is $p = 1/20$, or more generally, $p = 1/N$ where N is the number of participants. In the limit where N approaches infinity, the chance that exactly one person will get the go-ahead is $1/e$, which is just under 37 percent. With twenty superrational players all throwing icosahedral dice, the chance that *you* will come up the big winner is very close to $1/(20e)$, which is a little below two percent. That's not at all bad! Certainly it's a *lot* better than zero percent.

The objection many people raise is: "What if my roll comes up bad? Then why shouldn't I send in my name anyway? After all, if I fail to, I'll have no chance whatsoever of winning. I'm no better off than if I had never rolled my die and had just voluntarily withdrawn!" This objection seems overwhelming at first, but actually it is fallacious, being based on a misrepresentation of the meaning of "making a decision". A *genuine* decision to abide by the throw of a die means that you really *must* abide by the throw of the die; if under certain circumstances you ignore the die and do something else, then you never *made* the decision you claimed to have made. Your decision is revealed by your actions, not by your words before acting!

If you like the idea of rolling a die but fear that your will power may not be up to resisting the temptation to defect, imagine a third "Policansky button": this one says 'R' for "Roll", and if you press it, it rolls a die (perhaps simulated) and then instantly and irrevocably either sends your name or does not, depending on which way the die came up. This way you are never allowed to go back on your decision after the die is cast. Pushing *that* button is making a *genuine* decision to abide by the roll of a die. It would be easier on any ordinary human to be thus shielded from the temptation, but any superrational player would have no trouble holding back after a bad roll.

*   *   *

This talk of holding back in the face of strong temptation brings me to the climax of this column: the announcement of a Luring Lottery open to all readers and nonreaders of *Scientific American*. The prize of this lottery is $1,000,000/N$, where N is the number of entries submitted. Just think: If you are the only entrant (and if you submit only one entry), a cool million is yours! Perhaps, though, you doubt this will come about. It does seem a trifle iffy. If you'd like to increase your chances of winning, you are encouraged to send in multiple entries—no limit! Just send in one postcard per entry. If you send in 100 entries, you'll have 100 times the chance of some poor slob who sends in just one. Come to think of it, why should you have to send in multiple entries separately? Just send *one* postcard with your name and address and a positive integer (telling how many entries you're making) to:

Luring Lottery
c/o Scientific American
415 Madison Avenue
New York, N.Y. 10017

You will be given the same chance of winning as if you had sent in that number of postcards with '1' written on them. Illegible, incoherent, ill-specified, or incomprehensible entries will be disqualified. Only entries received by midnight June 30, 1983 will be considered. Good luck to you (but certainly not to any *other* reader of this column)!

Post Scriptum.

The emotions churned up by the Prisoner's Dilemma are among the strongest I have ever encountered, and for good reason. Not only is it a wonderful intellectual puzzle, akin to some of the most famous paradoxes of all time, but also it captures in a powerful and pithy way the essence of a myriad deep and disturbing situations that we are familiar with from life. Some are choices we make every day; others are the kind of agonizing choices that we all occasionally muse about but hope the world will never make us face.

My friend Bob Wolf, a mathematician whose specialty is logic, adamantly advocated choosing D in the case of the letters I sent out. To defend his choice, he began by saying that it was clearly "a paradox with no rational solution", and thus there was no way to know what people would do. Then he said, "Therefore, I will choose D. I do better that way than any other way." I protested strenuously: "How *dare* you say 'therefore' when you've just gotten through describing this situation as a paradox and claiming there is no rational answer? How dare you say *logic* is forcing an answer down your throat, when the premise of your 'logic' is that *there is no logical answer?*" I never got what I considered a satisfactory answer from Bob, although neither of us could budge the other. However, I did finally get some insight into Bob's vision when he, pushed hard by my probing, invented a situation with a new twist to it, which I call "Wolf's Dilemma".

Imagine that twenty people are selected from your high school graduation class, you among them. You don't know which others have been selected, and you are told they are scattered all over the country. All you know is that they are all connected to a central computer. Each of you is in a little cubicle, seated on a chair and facing one button on an otherwise blank wall. You are given ten minutes to decide whether or not to push your button. At the end of that time, a light will go on for ten seconds, and while it is on, you may either push or refrain from pushing. All the responses will then go to the central computer, and one minute later, they will result in consequences. Fortunately, the consequences can only be good. If you pushed your button, you will get $100, no strings attached, emerging from a small slot below the button. If *nobody* pushed their button, then *everybody* will get $1,000. But if there was even a single button-pusher, the refrainers will get nothing at all.

Bob asked me what I would do. Unhesitatingly, I said, "Of course I would not push the button. It's obvious!" To my amazement, though, Bob said he'd push the button with no qualms. I said, "What if you knew your co-players were all logicians?" He said that would make no difference to him. Whereas I gave credit to everybody for being able to see that it was to everyone's advantage to refrain, Bob did not. Or at least he expected that there is enough "flakiness" in people that he would prefer not to rely on the rationality of nineteen other people. But of course in assuming the flakiness of others, he would be his own best example—ruining everyone else's chances of getting $1,000.

What bothered me about Wolf's Dilemma was what I have come to call *reverberant doubt.* Suppose you are wondering what to do. At first it's obvious that everybody should avoid pushing their button. But you do realize that among twenty people, there might be one who is slightly hesitant and who might waver a bit. This fact is enough to worry you a tiny bit, and thus to make you waver, ever so slightly. But suddenly you realize that if *you* are wavering, even just a tiny bit, then most likely *everyone* is wavering a tiny bit. And that's considerably worse than what you'd thought at first—namely, that just *one* person might be wavering. Uh-oh! Now that you can imagine that everybody is at least *contemplating* pushing their button, the situation seems a lot more serious. In fact, now it seems quite probable that at least one person will push their button. But if that's the case, then pushing your own button seems the only sensible thing to do. As you catch yourself thinking this thought, you realize it must be the same as everyone else's thought. At this point, it becomes plausible that the *majority* of participants —possibly even all—will push their button! This clinches it for you, and so you decide to push yours.

Isn't this an amazing and disturbing slide from certain restraint to certain pushing? It is a cascade, a stampede, in which the tiniest flicker of a doubt has become amplified into the gravest avalanche of doubt. That's what I mean by "reverberant doubt". And one of the annoying things about it is that the brighter you are, the more quickly and clearly you see what there is to fear. A bunch of amiable slowpokes might well be more likely to unanimously refrain and get the big payoff than a bunch of razor-sharp logicians who all think perversely recursively reverberantly. It's that "smartness" to see that initial flicker of a doubt that triggers the whole avalanche and sends rationality a-tumblin' into the abyss. So, dear reader . . . if you push that button in front of you, do you thereby lose $900— or do you thereby gain $100?

*　　*　　*

Wolf's Dilemma is not the same as the Prisoner's Dilemma. In the Prisoner's Dilemma, pressure towards defection springs from *hope for asymmetry* (*i.e.,* hope that the other player might be dumber than you and thus make the opposite choice) whereas in Wolf's Dilemma, pressure towards button-pushing springs from *fear of asymmetry* (*i.e.,* fear that the other player might be dumber than you and thus make the opposite choice). This difference shows up clearly in the games' payoff matrices for the two-person case (compare Figure 30-1*b* with Figure 29-1*c*). In the Prisoner's Dilemma, the temptation T is greater than the reward R (5 > 3), whereas in Wolf's Dilemma, R is greater than T (1,000 > 100).

Bob Wolf's choice in his own dilemma revealed to me something about his basic assessment of people and their reliability (or lack thereof). Since his adamant decision to be a button-pusher even in this case stunned me, I decided to explore that cynicism a bit more, and came up with this modified Wolf's Dilemma.

Imagine, as before, that twenty people have been selected from your high school graduation class, and are escorted to small cubicles with one button on the wall. This time, however, each of you is strapped into a chair, and a device containing a revolver is attached to your head. Like it or not, you are now going to play Russian roulette, the odds of your death to be determined by your choice. For anybody who pushes their button, the odds of survival will be set at 90 percent—only one chance in ten of dying. Not too bad, but given that there are twenty of you, it means that almost certainly one or two of you will die, possibly more. And what happens to the refrainers? It all depends on how many of them there are. Let's say there are N refrainers. For each one of them, their chance of being shot will be one in N^2. For instance, if five people don't push, each of them will have only a 1/25 chance of dying. If ten people refrain, they will each get a 99 percent chance of survival. The bad cases are, of course, when nearly everybody pushes their button ("playing it safe", so to speak), leaving the refrainers in a tiny minority of three, two, or even one. If you're the *sole* refrainer, it's curtains for you—one chance in one of your death. Bye-bye! For two refrainers, it's one chance in four for each one. That means there's nearly a 50 percent chance that at least one of the two will perish.

Clearly the crossover line is between three and four refrainers. If you have a reasonable degree of confidence that at least three other people will hold back, you should definitely do so yourself. The only problem is, *they're* all making *their* decisions on the basis of trying to guess how many people will refrain, too! It's terribly circular, and you hardly know where to start. Many people, sensing this, just give up, and decide to push their button. (Actually, of course, how do I know? I've never seen people in such a situation—but it seems that way from evidence of real-life situations resembling this, and of course from how people respond to a mere description of this situation, where they aren't really faced with any dire consequences at all. Still, I tend to believe them, by and large.) Calling such a decision "playing it safe" is quite ironic, because if only everybody "played it dangerous", they'd have a chance of only one in 400 of dying! So I ask you: Which way is safe, and which way dangerous? It seems to me that this Wolf Trap epitomizes the phrase "We have nothing to fear but fear itself."

Variations on Wolf's Dilemma include some even more frightening and unstable scenarios. For instance, suppose the conditions are that each button-pusher has a 50 percent chance of survival, but if there is unanimous refraining from pushing the button, everyone's life will be spared—and as before, if anyone pushes their button, all refrainers will die. You can play around with the number of participants, the survival chance, and so on. Each such variation reveals a new facet of grimness. These visions are truly horrific, yet all are just allegorical renditions of ordinary life's decisions, day in, day out.

*   *   *

I had originally intended to close the column with the following paragraph, but was dissuaded from it by friends and editors:

> I am sorry to say that I am simply inundated with letters from well-meaning readers, and I have discovered, to my regret, that I can barely find time to *read* all those letters, let alone *answer* them. I have been racking my brains for months trying to come up with some strategy for dealing with all this correspondence, but frankly I have not found a good solution yet. Therefore, I thought I would appeal to the collective genius of you-all out there. If you can think of some way for me to ease the burden of my correspondence, please send your idea to me. I shall be most grateful.

31

Irrationality Is the
Square Root of All Evil

September, 1983

T HE Luring Lottery, proposed in my June column, created quite a stir. Let me remind you that it was open to anyone; all you had to do was submit a postcard with a clearly specified positive integer on it telling how many entries you wished to make. This integer was to be, in effect, your "weight" in the final drawing, so that if you wrote "100", your name would be 100 times more likely to be drawn than that of someone who wrote '1'. The only catch was that the cash value of the prize was inversely proportional to the sum of all the weights received by June 30. Specifically, the prize to be awarded was $1,000,000/W,$ where W is the sum of all the weights sent in.

The Luring Lottery was set up as an exercise in *cooperation* versus *defection*. The basic question for each potential entrant was: "Should I restrain myself and submit a small number of entries, or should I 'go for it' and submit a large number? That is, should I cooperate, or should I defect?" Whereas in previous examples of cooperation versus defection there was a clear-cut dividing line between cooperators and defectors, here it seems there is a continuum of possible answers, hence of "degree of cooperation". Clearly one can be an extreme cooperator and voluntarily submit nothing, thus in effect cutting off one's nose to spite one's face. Equally clearly, one can be an extreme defector and submit a giant number of entries, hoping to swamp everyone else out but destroying the prize in so doing. However, there remains a lot of middle ground between these two extremes. What about someone who submits two entries, or one? What about someone who throws a six-sided die to decide whether or not to send in a single entry? Or a million-sided die?

Before I go further, it would be good for me to present my generalized and nonmathematical sense of these terms "cooperation" and "defection". As a child, you undoubtedly often encountered adults who admonished you for walking on the grass or for making noise, saying "Tut, tut, tut—just think if *everyone* did that!" This is the quintessential argument used against the defector, and serves to define the concept:

> A defection is an action such that, if everyone did it, things would clearly be worse (for everyone) than if everyone refrained from doing it, and yet which tempts everyone, since if only one individual (or a sufficiently small number) did it while others refrained, life would be sweeter for that individual (or select group).

Cooperation, of course, is the other side of the coin: the act of resisting temptation. However, it need not be the case that cooperation is passive while defection is active; often it is the exact opposite: The cooperative option may be to participate industriously in some activity, while defection is to lay back and accept the sweet things that result for everybody from the cooperators' hard work. Typical examples of defection are:

* loudly wafting your music through the entire neighborhood on a fine summer's day;
* not worrying about speeding through a four-way stop sign, figuring that the people going in the crosswise direction will stop anyway;
* not being concerned about driving a car everywhere, figuring that there's no point in making a sacrifice when other people will just continue to guzzle gas anyway;
* not worrying about conserving water in a drought, figuring "Everyone else will";
* not voting in a crucial election and excusing yourself by saying "One vote can't make any difference";
* not worrying about having ten children in a period of population explosion, leaving it to other people to curb their reproduction;
* not devoting any time or energy to pressing global issues such as the arms race, famine, pollution, diminishing resources, and so on, saying "Oh, of course I'm very concerned—but there's nothing one person can do."

When there are large numbers of people involved, people don't realize that their own seemingly highly idiosyncratic decisions are likely to be quite typical and are likely to be recreated many times over, on a grand scale; thus, what each couple feels to be their own isolated and private decision (conscious or unconscious) about how many children to have turns into a population explosion. Similarly, "individual" decisions about the futility of working actively toward the good of humanity amount to a giant trend of apathy, and this multiplied apathy translates into insanity at the group level. In a word, *apathy at the individual level translates into insanity at the mass level.*

*   *   *

Garrett Hardin, an evolutionary biologist, wrote a famous article about this type of phenomenon, called "The Tragedy of the Commons". His view was that there are two types of rationality: one (I'll call it the "local" type) that strives for the good of the individual, the other (the "global" type) that strives for the good of the group; and that these two types of rationality are in an inevitable and eternal conflict. I would agree with his assessment, provided the individuals are unaware of their joint plight but are simply blindly carrying out their actions as if in isolation.

However, if they are fully aware of their joint situation, and yet in the face of it they blithely continue to act as if their situation were not a communal one, then I maintain that they are acting totally irrationally. In other words, with an enlightened citizenry, "local" rationality is *not* rational, period. It is damaging not just to the group, but to the individual. For example, people who defected in the One-Shot Prisoner's Dilemma situation I described in June did worse than if all had cooperated.

This was the central point of my June column, in which I wrote about *renormalized rationality,* or *superrationality.* Once you know you are a typical member of a class of individuals, you must act as if your own individual actions were to be multiplied manyfold, because they inevitably will be. In effect, to sample yourself is to sample the field, and if *you* fail to do what you wish the rest would do, you will be very disappointed by the rest as well. Thus it pays a lot to reflect carefully about one's situation in the world before defecting, that is, jumping to do the naïvely selfish act. You had better be prepared for a lot of *other* people copping out as well, and offering the same flimsy excuse.

People strongly resist seeing themselves as parts of statistical phenomena, and understandably so, because it seems to undermine their sense of free will and individuality. Yet how true it is that each of our "unique" thoughts is mirrored a million times over in the minds of strangers! Nowhere was this better illustrated than in the response to the Luring Lottery. It is hard to know precisely what constitutes the "field", in this case. It was declared universally open, to readers and nonreaders alike. However, we would be safe in assuming that few nonreaders ever became aware of it, so let's start with the circulation of *Scientific American,* which is about a million. Most of them, however, probably did no more than glance over my June column, if that; and of the ones who did more than that (let's say 100,000), still only a fraction—maybe one in ten—read it carefully from start to finish. I would thus estimate that there were perhaps 10,000 people motivated enough to read it carefully and to ponder the issues seriously. In any case, I'll take this figure as the population of the "field".

In my June column, I spelled out plainly, for all to see, the superrational argument that applies to the Platonia Dilemma, for rolling an N-sided die and entering only if it came up on the proper side. Here, a similar argument goes through. In the Platonia Dilemma, where more than one entry is fatal to all, the ideal die turned out to have N faces, where N is the number of players—hence, with 10,000 players, a 10,000-sided die. In the Luring Lottery, the consequences aren't so drastic if more than one entry is submitted. Thus, the ideal number of faces on the die turns out to be about 2/3 as many—in the case of 10,000 players, a 6,667-sided die would do admirably. Giving the die fewer than 10,000 sides of course slightly increases each player's chance of sending in one entry. This is to make it quite likely that at least one entry will arrive!

With 6,667 faces on the die, each superrational player's chance of winning is not quite 1 in 10,000, but more like 1 in 13,000; this is because there is about a 22 percent chance that no one's die will land right, so no one will send in any entry at all, and no one will win. But if you give the die still fewer faces—say 3,000—the expected size of the pot gets considerably smaller, since the expected number of entrants grows. And if you give it more faces —say 20,000—then you run a considerable risk of having no entries at all. So there's a trade-off whose ideal solution can be calculated without too much trouble, and 6,667 faces turns out to be about optimal. With that many faces, the expected value of the pot is maximal: nearly $520,000—not to be sneered at.

Now this means that had everyone followed my example in the June column, I would probably have received a total of one or two postcards with '1' written on them, and one of those lucky people would have gotten a huge sum of money! But do you think that is what happened? Of course not! Instead, I was inundated with postcards and letters from all over the world —over 2,000 of them. What was the breakdown of entries? I have exhibited part of it in a table, below:

1: 1,133
2: 31
3: 16
4: 8
5: 16
6: 0
7: 9
8: 1
9: 1
10: 49
100: 61
1,000: 46
1,000,000: 33
1,000,000,000: 11
602,300,000,000,000,000,000,000 (Avogadro's number): 1
10^{100} (a googol): 9
$10^{10^{100}}$ (a googolplex): 14

Curiously, many if not most of the people who submitted just one entry patted themselves on the back for being "cooperators". Hogwash! The *real* cooperators were those among the 10,000 or so avid readers who calculated the proper number of faces of the die, used a random-number table or something equivalent, and then—most likely—rolled themselves out. A few people wrote to tell me they had rolled themselves out in this way. I appreciated hearing from them. It is conceivable, just barely, that among the thousand-plus entries of '1' there was one that came from a lucky super-rational cooperator—but I doubt it. The people who simply withdrew *without* throwing a die I would characterize as well-meaning but a bit lazy, not true cooperators—something like people who simply contribute money to a political cause but then don't want to be bothered any longer about it. It's the lazy way of claiming cooperation.

By the way, I haven't by any means finished with my score chart. However, it is a bit disheartening to try to relate what happened. Basically, it is this. Dozens and dozens of readers strained their hardest to come up with inconceivably large numbers. Some filled their whole postcard with tiny '9's, others filled their card with rows of exclamation points, thus creating iterated factorials of gigantic sizes, and so on. A handful of people carried this game much further, recognizing that the optimal solution avoids all pattern (to see why, read Gregory Chaitin's article "Randomness and Mathematical Proof"), and consists simply of a "dense pack" of definitions built on definitions, followed by one final line in which the "fanciest" of the definitions is applied to a relatively small number such as 2, or better yet, 9.

I received, as I say, a few such entries. Some of them exploited such powerful concepts of mathematical logic and set theory that to evaluate which one was the largest became a very serious problem, and in fact it is not even clear that I, or for that matter anyone else, would be able to determine which is the largest integer submitted. I was strongly reminded of the lunacy and pointlessness of the current arms race, in which two sides vie against each other to produce arsenals so huge that not even teams of experts can meaningfully say which one is larger—and meanwhile, all this monumental effort is to the detriment of *everyone*.

*   *   *

Did I find this amusing? Somewhat, of course. But at the same time, I found it disturbing and disappointing. Not that I hadn't expected it. Indeed, it was precisely what I had expected, and it was one reason I was so sure the Luring Lottery would be no risk for the magazine.

This short-sighted race for "first place" reveals the way in which people in a huge crowd erroneously consider their own fancies to be totally unique. I suspect that nearly everyone who submitted a number above 1,000,000 actually believed they were going to be the *only* one to do so. Many of those who submitted numbers such as a googolplex, or a '9' followed by thousands of factorial signs, explicitly indicated that they were pretty sure that they were going to "win". And then those people who pulled out all the stops and sent in definitions that would boggle most mathematicians were *very* sure they were going to win. As it turns out, I don't know who won, and it doesn't matter, since the prize is zero to such a good approximation that even God wouldn't know the difference.

Well, what conclusion do I draw from all this? None too serious, but I do hope that it will give my readers pause for thought next time they face a "cooperate-or-defect" decision, which will likely happen within minutes for each of you, since we face such decisions many times each day. Some of them are small, but some will have monumental repercussions. The globe's future is in *your* hands—and yes, I mean *you* (as well as every other reader of this column).

*   *   *

And with this perhaps sobering conclusion, I would like to draw my term as a columnist for *Scientific American* to a close. It has been a valuable and beneficial opportunity for me. I have enjoyed having a platform from which to express my ideas and concerns, I have—at least sometimes—enjoyed receiving the huge shipments of mail forwarded to me from New York several times a month, and I have certainly been happy to make new friends through this channel. I won't miss the monthly deadline, but I will undoubtedly come across ideas, from time to time, that would have made perfect "Metamagical Themas". I will be keeping them in mind, and maybe at some future time will write a similar set of essays.

But for now, it is time for me to move on to other territory: I look forward to a return to my professional work, and to a more private life. Good-bye, and best wishes to you and to all other readers of this magazine, this issue, this copy, this piece, this page, this column, this paragraph, this sentence, and, last but not least, this "this".

Post Scriptum.

What do you do when in a crushingly cold winter, you hear over the radio that there is a severe natural gas shortage in your part of the country, and everyone is requested to turn their thermostat down to 60 degrees? There's no way anyone will know if you've complied or not. Why shouldn't you toast in your house and let all the rest of the people cut down their consumption? After all, what *you* do surely can't affect what anyone *else* does.

This is a typical "tragedy of the commons" situation. A common resource has reached the point of saturation or exhaustion, and the questions for each individual now are: "How shall I behave? Am I typical? How does a lone person's action affect the big picture?" Garrett Hardin's article "The Tragedy of the Commons" frames the scene in terms of grazing land shared by a number of herders. Each one is tempted to increase their own number of animals even when the land is being used beyond its optimum capacity, because the individual gain outweighs the individual loss, even though in the long run, that decision, multiplied throughout the population of herders, will destroy the land totally.

The real reason behind Hardin's article was to talk about the population explosion and to stress the need for rational global planning—in fact, for coercive techniques similar to parking tickets and jail sentences. His idea is that families should be allowed to have many children (and thus to use a large share of the common resources) but that they should be penalized by society in the same way as society "allows" someone to rob a bank and then applies sanctions to those who have made that choice. In an era when resources are running out in a way humanity has never had to face heretofore, new kinds of social arrangements and expectations must be imposed, Hardin feels, by society as a whole. He is a dire pessimist about any kind of superrational cooperation, emphasizing that cooperators in the birth-control game will breed themselves right out of the population. A perfect illustration of why this is so is the man I heard about recently: he secretly had ten wives and by them had sired something like 35 children by the time he was 30. With genes of that sort proliferating wildly, there is little hope for the more modest breeders among us to gain the upper hand. Hardin puts it bluntly: "Conscience is self-eliminating." He goes even further and says:

> The argument has here been stated in the context of the population problem, but it applies equally well to any instance in which society appeals to an individual exploiting a commons to restrain himself for the general good—by means of his conscience. To make such an appeal is to set up a selective system that works toward the elimination of conscience from the race.

An even more pessimistic vision of the future is proffered us by one Walter Bradford Ellis, a hypothetical speaker representing the views of his inventor, Louis Pascal, in a hypothetical speech:

> The United States—indeed the whole earth—is fast running out of the resources it depends on for its existence. Well before the last of the world's supplies of oil and natural gas are exhausted early in the next century, shortages of these and other substances will have brought about the collapse of our whole economy and, indeed, of our whole technology. And without the wonders of modern technology, America will be left a grossly overpopulated, utterly impoverished, helpless, dying land. Thus I foresee a whole world full of wretched, starving people with no hope of escape, for the only countries which could have aided them will soon be no better off than the rest. And thus unless we are saved from this future by the blessing of a nuclear war or a truly lethal

pestilence, I see stretching off into eternity a world of indescribable suffering and hopelessness. It is a vision of truly unspeakable horror mitigated only by the fact that try as I might I could not possibly concoct a creature more deserving of such a fate.

Whew! The circularity of the final thought reminds me of an idea I once had: that it will be just as well if humanity destroys itself in a nuclear holocaust, because civilizations that destroy themselves are barbaric and stupid, and who would want to have one of *them* around, polluting the universe?

Pascal's thoughts, expressed in his article "Human Tragedy and Natural Selection" and in his rejoinder to an article by two critics called "The Loving Parent Meets the Selfish Gene" (which is where Ellis' speech is printed), are strikingly reminiscent of the thoughts of his earlier namesake Blaise, who in an unexpected use of his own calculus of probabilities managed to convince himself that the best possible way to spend his life was in devotion to a God who he wasn't sure (and couldn't be sure) existed. In fact, Pascal felt, even if the chances of God's existence were one in a million, faith in that God would pay off in the end, because the potential rewards (or punishments) if Heaven and Hell exist are infinite, and all earthly rewards and punishments, no matter how great, are still finite. The favored behavior is to *be* a believer, Pascal "calculated"—regardless of what you *do* believe. Thus Blaise Pascal devoted his brilliant mind to theology.

Louis Pascal, following in his forebear's mindsteps, has opted to devote his life to the world's population problem. And he can produce mathematical arguments to show why you should, too. To my mind, there is no question that such arguments have considerable force. There are always points to nitpick over, but in essence, thinkers like Hardin and Pascal and Anne and Paul Ehrlich and many others have recognized and internalized the novelty of the human situation at this moment in history: the moment when humanity has to grapple with dwindling resources and overwhelmingly huge weapons systems. Not many people are willing to wrestle with this beast, and consequently the burden falls all the more heavily on those few who are.

*   *   *

It has disturbed me how vehemently and staunchly my clear-headed friends have been able to defend their decisions to defect. They seem to be able to digest my argument about superrationality, to mull it over, to begrudge some curious kind of validity to it, but ultimately to feel on a gut level that it is wrong, and to reject it. This has led me to consider the notion that my faith in the superrational argument might be similar to a self-fulfilling prophecy or self-supporting claim, something like being absolutely convinced beyond a shadow of a doubt that the Henkin sentence "This sentence is true" actually *must* be true—when, of course, it is equally defensible to believe it to be false. The sentence is undecidable; its truth value is stable, whichever way you wish it to go (in this way, it is the diametric opposite of the Epimenides sentence "This sentence is false", whose truth value flips faster than the tip of a happy pup's tail). One difference, though, between the Prisoner's Dilemma and oddball self-referential sentences is that whereas your beliefs about such sentences' truth values usually have inconsequential consequences, with the Prisoner's Dilemma, it's quite another matter.

I sometimes wonder whether there haven't been many civilizations Out There, in our galaxy and beyond, that have already dealt with just these types of gigantic social problems—Prisoner's Dilemmas, Tragedies of the Commons, and so forth. Most likely some would have survived, some would have perished. And it occurs to me that perhaps the ultimate difference in those societies may have been the survival of the meme that, in effect, asserts the logical, rational validity of cooperation in a one-shot Prisoner's Dilemma. In a way, this would be the opposite thesis to Hardin's. It would say that *lack* of conscience is self-eliminating—provided you wait long enough that natural selection can act at the level of entire societies.

Perhaps on some planets, Type I societies have evolved, while on others, Type II societies have evolved. By definition, members of Type I societies believe in the rationality of lone, uncoerced, one-shot cooperation (when faced with members of Type I societies), whereas members of Type II societies reject the rationality of lone, uncoerced, one-shot cooperation, irrespective of who they are facing. (Notice the tricky circularity of the definition of Type I societies. Yet it is not a vacuous definition!) Both types of society find their respective answer to be obvious—they just happen to find opposite answers. Who knows—we might even happen to have some Type I societies here on earth. I cannot help but wonder how things would turn out if my little one-shot Prisoner's Dilemma experiment were carried out in Japan instead of the U.S. In any case, the vital question is: Which type of society survives, in the long run?

It could be that the one-shot Prisoner's Dilemma situations that I have described are undecidable propositions within the logic that we humans have developed so far, and that new axioms can be added, like the parallel postulate in geometry, or Gödel sentences (and related ones) in mathematical logic. (Take a look at Figure 31-1, and see what kind of logic will extract those two poor devils from their one-shot dilemma.) Those civilizations to which cooperation appears axiomatic—Type I societies— wind up surviving, I would venture to guess, whereas those to which defection appears axiomatic—Type II societies—wind up perishing. This suggestion may seem all wet to you, but watch those superpowers building those bombs, more and more of them every day, helplessly trapped in a rising spiral, and think about it. Evolution is a merciless pruner of ill logic.

Most philosophers and logicians are convinced that truths of logic are "analytic" and *a priori*; they do not like to think that such basic ideas are grounded in mundane, arbitrary things like survival. They might admit that

"The problem is how to turn loose without letting go."

FIGURE 31-1. *One powerful metaphor for the absurdity we have collectively dug ourselves into. The symmetry of the situation is acutely portrayed in this cartoon drawn by Bill Mauldin in 1960. Note that if either person releases his rope, thus chopping off his counterpart's head, that person's hand will go limp, thus releasing* his *rope and causing the other blade to fall and chop off the head of the instigator. That idea is a centerpiece of our current nuclear deterrence strategy: Even if we are wiped off the globe, our trUSty missiles will still wreak divine revenge on the evil empire of Satanic Uglies who dared do harm to US.*

natural selection tends to *favor* good logic—but they would certainly hate the suggestion that natural selection *defines* good logic! Yet truth and survival value *are* all tangled together, and civilizations that survive certainly *have* glimpsed higher truths than those that perish. When you argue with someone whose ideas you are sure are wrong but who dances an infuriatingly inconsistent yet self-consistent verbal dance in front of you, your one solace is that something in *life* may yet change this person's mind, even though your own best logic is helpless to do so. Ultimately, beliefs have to be grounded in experience, whether that experience is the organism's or its ancestors' or its peer group's. (That's what Chapter 5, particularly its

P.S., was all about.) My feeling is that the concept of superrationality is one whose truth will come to dominate among intelligent beings in the universe simply because its adherents will survive certain kinds of situations where its opponents will perish. Let's wait a few spins of the galaxy and see. After all, healthy logic is whatever remains after evolution's merciless pruning.

*   *   *

I was describing the Copycat project (Chapter 24) to physicist Victor Weisskopf, and I gave him our canonical example: "If *abc* goes to *abd,* what does *xyz* go to?" After we had discussed various possible answers and settled on *wyz* as the most compelling for reasons of symmetry, he surprised me by saying this: "You know, the root of the world's deepest problems is the tragic inability on the part of the world's leaders to see such basic symmetries. For instance, that the U.S. is to the S.U. what the S.U. is to the U.S.—that is too much for them to accept." Oh, but how could Weisskopf be so silly? After all, *we're* not trying to export communism to the entire world!

Logician Raymond Smullyan, who first heard about the Prisoner's Dilemma from me and who was absolutely delighted by it, also surprised me, but in a different way: He vehemently insisted on the correctness of defection in a one-shot situation no matter who might be on the other side, including his twin or his clone! (He did waver about his mirror image.) But just as I was giving up on him as a lost cause, he conceded this much to me: "I suspect, Doug, that this problem is a lot knottier than you or I suspect." Indeed, I suspect so, Raymond.

32

The Tale of Happiton

June, 1983

HAPPITON was a happy little town. It had 20,000 inhabitants, give or take 7, and they were productive citizens who mowed their lawns quite regularly. Folks in Happiton were pretty healthy. They had a life expectancy of 75 years or so, and lots of them lived to ripe old ages. Down at the town square, there was a nice big courthouse with all sorts of relics from WW II and monuments to various heroes and whatnot. People were proud, and had the right to be proud, of Happiton.

On the top of the courthouse, there was a big bell that boomed every hour on the hour, and you could hear it far and wide—even as far out as Shady Oaks Drive, way out nearly in the countryside.

One day at noon, a few people standing near the courthouse noticed that right after the noon bell rang, there was a funny little sound coming from up in the belfry. And for the next few days, folks noticed that this scratching sound was occurring after every hour. So on Wednesday, Curt Dempster climbed up into the belfry and took a look. To his surprise, he found a crazy kind of contraption rigged up to the bell. There was this mechanical hand, sort of a robot arm, and next to it were five weird-looking dice that it could throw into a little pan. They all had twenty sides on them, but instead of being numbered 1 through 20, they were just numbered 0 through 9, but with each digit appearing on two opposite sides. There was also a TV camera that pointed at the pan and it seemed to be attached to a microcomputer or something. That's all Curt could figure out. But then he noticed that on top of the computer, there was a neat little envelope marked "To the friendly folks of Happiton". Curt decided that he'd take it downstairs and open it in the presence of his friend the mayor, Janice Fleener. He found Janice easily enough, told her about what he'd found, and then they opened the envelope. How neatly it was written! It said this:

Grotto 19, Hades
June 20, 1983

Dear folks of Happiton,

I've got some bad news and some good news for you. The bad first. You know your bell that rings every hour on the hour? Well, I've set it up so that each time it rings, there is exactly one chance in a hundred thousand—that is, 1/100,000—that a Very Bad Thing will occur. The way I determine if that Bad Thing will occur is, I have this robot arm fling its five dice and see if they all land with '7' on top. Most of the time, they won't. But if they do—and the odds are exactly 1 in 100,000—then great clouds of an unimaginably revolting-smelling yellow-green gas called "Retchgoo" will come oozing up from a dense network of underground pipes that I've recently installed underneath Happiton, and everyone will die an awful, writhing, agonizing death. Well, that's the bad news.

Now the good news! You all can prevent the Bad Thing from happening, if you send me a bunch of postcards. You see, I happen to like postcards a whole lot (especially postcards of Happiton), but to tell the truth, it doesn't really much matter what they're of. I just *love* postcards! Thing is, they have to be written personally—not typed, and especially not computer-printed or anything phony like that. The more cards, the better. So how about sending me some postcards—batches, bunches, boxes of them?

Here's the deal. I reckon a typical postcard takes you about 4 minutes to write. Now suppose just one person in all of Happiton spends 4 minutes one day writing me, so the next day, I get one postcard. Well, then, I'll do you all a favor: I'll slow the courthouse clock down a bit, for a day. (I realize this is an inconvenience, since a lot of you tell time by the clock, but believe me, it's a lot more inconvenient to die an agonizing, writhing death from the evil-smelling, yellow-green Retchgoo.) As I was saying, I'll slow the clock down for *one day,* and by how much? By a factor of 1.00001. Okay, I know that doesn't sound too exciting, but just think if all 20,000 of you send me a card! For *each* card I get that day, I'll toss in a slow-up factor of 1.00001, the next day. That means that by sending me 20,000 postcards a day, you all, working together, can get the clock to slow down by a factor of *1.00001 to the 20,000th power,* which is just a shade over 1.2, meaning it will ring every 72 minutes.

All right, I hear you saying, "72 minutes is just barely over an hour!" So I offer you more! Say that one day I get 160,000 postcards (heavenly!). Well then, the very next day I'll show my gratitude by slowing your clock down, all day long, midnight to midnight, by 1.00001 to the 160,000th power, and that ain't chickenfeed. In fact, it's about 5, and that means the clock will ring only every 5 hours, meaning those sinister dice will only get rolled about 5 times (instead of the usual 24). Obviously, it's better for both of us that way. You have to bear in mind that I don't have any personal interest in seeing that awful Retchgoo come rushing and gushing up out of those pipes and causing every last one of you to perish in grotesque, mouth-foaming, twitching convulsions. All I care about is getting postcards! And to send me 160,000 a day wouldn't cost you folks that much effort, being that it's just 8 postcards a day—just about a half hour a day for each of you, the way I reckon it.

So my deal is pretty simple. On any given day, I'll make the clock go off once every X hours, where X is given by this simple formula:

$$X = 1.00001^N$$

Here, N is the number of postcards I received the previous day. If N is 20,000, then X will be 1.2, so the bell would ring 20 times per day, instead of 24. If N is 160,000, then X jumps way up to about 5, so the clock would slow way down—just under 5 rings per day. If I get *no* postcards, then the clock will ring once an hour, just as it does now. The formula reflects that, since if N is 0, X will be 1. You can work out other figures yourself. Just think how much safer and securer you'd all feel knowing that your courthouse clock was ticking away so slowly!

I'm looking forward with great enthusiasm to hearing from you all.

Sincerely yours,
Demon #3127

The letter was signed with beautiful medieval-looking flourishes, in an unusual shade of deep red . . . ink?

"Bunch of hogwash!" spluttered Curt. "Let's go up there and chuck the whole mess down onto the street and see how far it bounces." While he was saying this, Janice noticed that there was a smaller note clipped onto the back of the last sheet, and turned it over to read it. It said this:

P.S. —It's really not advisable to try to dismantle my little set-up up there in the belfry: I've got a hair trigger linked to the gas pipes, and if anyone tries to dismantle it, pssssst! Sorry.

Janice Fleener and Curt Dempster could hardly believe their eyes. What gall! They got straight on the phone to the Police Department, and talked to Officer Curran. He sounded poppin' mad when they told him what they'd found, and said he'd do something about it right quick. So he hightailed it over to the courthouse and ran up those stairs two at a time, and when he reached the top, a-huffin' and a-puffin', he swung open the belfry door and took a look. To tell the truth, he was a bit ginger in his inspection, because one thing Officer Curran had learned in his many years of police experience is that an ounce of prevention is worth a pound of cure. So he cautiously looked over the strange contraption, and then he turned around and quite carefully shut the door behind him and went down. He called up the town sewer department and asked them if they could check out whether there was anything funny going on with the pipes underground.

Well, the long and the short of it is that they verified everything in the Demon's letter, and by the time they had done so, the clock had struck five more times and those five dice had rolled five more times. Janice Fleener had in fact had her thirteen-year-old daughter Samantha go up and sit in a wicker chair right next to the microcomputer and watch the robot arm throw those dice. According to Samantha, an occasional 7 had turned up now and then, but never had two 7's shown up together, let alone 7's on all five of the weird-looking dice!

* * *

The next day, the Happiton *Eagle-Telephone* came out with a front-page story telling all about the peculiar goings-on. This caused quite a commotion. People *everywhere* were talking about it, from Lidden's Burger Stop to Bixbee's Druggery. It was truly the talk of the town.

When Doc Hazelthorn, the best pediatrician this side of the Cornyawl River, walked into Ernie's Barbershop, corner of Cherry and Second, the atmosphere was more somber than usual. "Whatcha gonna do, Doc?" said big Ernie, the jovial barber, as he was clipping the few remaining hairs on old Doc's pate. Doc (who was also head of the Happiton City Council) said the news had come as quite a shock to him and his family. Red Dulkins, sitting in the next chair over from Doc, said he felt the same way. And then the two gentlemen waiting to get their hair cut both added their words of agreement. Ernie, summing it up, said the whole town seemed quite upset. As Ernie removed the white smock from Doc's lap and shook the hairs off it, Doc said that he had just decided to bring the matter up first thing at the next City Council meeting, Tuesday evening. "Sounds like a good idea, Doc!" said Ernie. Then Doc told Ernie he couldn't make the usual golf date this weekend, because some friends of his had invited him to go fishing out at Lazy Lake, and Doc just couldn't resist.

Two days after the Demon's note, the *Eagle-Telephone* ran a feature article in which many residents of Happiton, some prominent, some not so prominent, voiced their opinions. For instance, eleven-year-old Wally Thurston said he'd gone out and bought up the whole supply of picture postcards at the 88-Cent Store, $14.22 worth of postcards, and he'd already started writing a few. Andrea McKenzie, sophomore at Happiton High, said she was really worried and had had nightmares about the gas, but her parents told her not to worry, things had a way of working out. Andrea said maybe her parents weren't taking it so seriously because they were a generation older and didn't have as long to look forward to anyway. She said she was spending an hour each day writing postcards. That came to 15 or 16 cards each day. Hank Hoople, a janitor at Happiton High, sounded rather glum: "It's all fate. If the bullet has your name on it, it's going to happen, whether you like it or not." Many other citizens voiced concern and even alarm about the recent developments.

But some voiced rather different feelings. Ned Furdy, who as far as anyone could tell didn't do much other than hang around Simpson's bar all day (and most of the night) and buttonhole anyone he could, said, "Yeah, it's a problem, all right, but I don't know nothin' about gas and statistics and such.

It should all be left to the mayor and the Town Council, to take care of. They know what they're doin'. Meanwhile, eat, drink, and be merry!" And Lulu Smyth, 77-year-old proprietor of Lulu's Thread 'N Needles Shop, said "I think it's all a ruckus in a teapot, in my opinion. Far as I'm concerned, I'm gonna keep on sellin' thread 'n needles, and playin' gin rummy every third Wednesday."

*   *   *

When Doc Hazelthorn came back from his fishing weekend at Lazy Lake, he had some surprising news to report. "Seems there's a demon left a similar set-up in the church steeple down in Dwaynesville", he said. (Dwaynesville was the next town down the road, and the arch-rival of Happiton High in football.) "The Dwaynesville demon isn't threatening them with gas, but with radioactive water. Takes a little longer to die, but it's just as bad. And I hear tell there's a demon with a subterranean volcano up at New Athens." (New Athens was the larger town twenty miles up the Cornyawl from Dwaynesville, and the regional center of commerce.)

A lot of people were clearly quite alarmed by all this, and there was plenty of arguing on the streets about how it had all happened without anyone knowing. One thing that was pretty universally agreed on was that a commission should be set up as soon as possible, charged from here on out with keeping close tabs on all subterranean activity within the city limits, so that this sort of outrage could never happen again. It appeared probable that Curt Dempster, who was the moving force behind this idea, would be appointed its first head.

Ed Thurston (Wally's father) proposed to the Jaycees (of which he was a member in good standing) that they donate $1,000 to support a postcard-writing campaign by town kids. But Enoch Swale, owner of Swale's Pharmacy and the Sleepgood Motel, protested. He had never liked Ed much, and said Ed was proposing it simply because his son would gain status that way. (It was true that Wally had recruited a few kids and that they spent an hour each afternoon after school writing cards. There had been a small article in the paper about it once.) After considerable debate, Ed's motion was narrowly defeated. Enoch had a lot of friends on the City Council.

Nellie Doobar, the math teacher at High, was about the only one who checked out the Demon's math. "Seems right to me", she said to the reporter who called her about it. But this set her to thinking about a few things. In an hour or two, she called back the paper and said, "I figured something out. Right now, the clock is still ringing very close to once every hour. Now there are about 720 hours per month, and so that means there are 720 chances each month for the gas to get out. Since each chance is 1 in 100,000, it turns out that each month, there's a bit less than a 1-in-100 chance that Happiton will get gassed. At that rate, there's about 11 chances in 12 that Happiton will make it through each year. That may sound pretty good, but the chances we'll make it through any 8-year period are almost exactly 50–50, exactly the same as tossing a coin. So we can't really count on very many years . . ."

This made big headlines in the next afternoon's *Eagle-Telephone*—in fact, even bigger than the plans for the County Fair! Some folks started calling up Mrs. Doobar anonymously and telling her she'd better watch out what she was saying if she didn't want to wind up with a puffy face or a fat lip. Seems like they couldn't quite keep it straight that Mrs. Doobar wasn't the one who'd set the thing up in the first place.

After a few days, though, the nasty calls died down pretty much. Then Mrs. Doobar called up the paper again and told the reporter, "I've been calculating a bit more here, and I've come up with the following, and they're facts every last one of them. If all 20,000 of us were to spend half an hour a day writing postcards to the Demon, that would amount to 160,000 postcards a day, and just as the Demon said, the bell would ring pretty near every *five* hours instead of every hour, and that would mean that the chances of us getting wiped out each month would go down considerable. In fact, there would only be about 1 chance in 700 that we'd go down the tubes in any given month, and only about a chance in 60 that we'd get zapped each year. Now I'd say that's a darn sight better than 1 chance in 12 per year, which is what it is if we don't write any postcards (as is more or less the case now, except for Wally Thurston and Andrea McKenzie and a few other kids I heard of). And for every 8-year period, we'd only be running a 13 percent risk instead of a 50 percent risk."

"That sounds pretty good", said the reporter cheerfully.

"Well," replied Mrs. Doobar, "it's not too bad, but we can get a whole lot better by doublin' the number of postcards."

"How's that, Mrs. Doobar?" asked the reporter. "Wouldn't it just get twice as good?"

"No, you see, it's an exponential curve," said Mrs. Doobar, "which means that if you *double N,* you *square X.* "

"That's Greek to me", quipped the reporter.

"*N* is the number of postcards and *X* is the time between rings", she replied quite patiently. "If we all write a half hour a day, *X* is 5 hours. But that means that if we all write a whole *hour* a day, like Andrea McKenzie in my algebra class, *X* jumps up to 25 hours, meaning that the clock would ring only about once a day, and obviously, that would reduce the danger a *lot.* Chances are, hundreds of years would pass before five 7's would turn up together on those infernal dice. Seems to me that under those circumstances, we could pretty much live our lives without worrying about the gas at all. And that's for writing about an hour a day, each one of us."

The reporter wanted some more figures detailing how much different amounts of postcard-writing by the populace would pay off, so Mrs. Doobar obliged by going back and doing some more figuring. She figured out that if 10,000 people—half the population of Happiton—did 2 hours a day for the year, they could get the same result—one ring every 25 hours. If only 5,000 people spent 2 hours a day, or if 10,000 people spent one hour a day, then it would go back to one ring every 5 hours (still a lot safer than one every hour). Or, still another way of looking at it, if just 1250 of them worked *full-time* (8 hours a day), they could achieve the same thing.

"What about if we all pitch in and do 4 minutes a day, Mrs. Doobar?" asked the reporter.

"Fact is, 'twouldn't be worth a damn thing! (Pardon my French.)" she replied. "N is 20,000 that way, and even though that sounds pretty big, X works out to be just 1.2, meaning one ring every 1.2 hours, or 72 minutes. That way, we still have about a chance of 1 in 166 every month of getting wiped out, and 1 in 14 every year of getting it. Now that's real scary, in my book. Writing cards only starts making a noticeable difference at about 15 minutes a day per person."

* * *

By this time, several weeks had passed, and summer was getting into full swing. The County Fair was buzzing with activity, and each evening after folks came home, they could see loads of fireflies flickering around the trees in their yards. Evenings were peaceful and relaxed. Doc Hazelthorn was playing golf every weekend, and his scores were getting down into the low 90's. He was feeling pretty good. Once in a while he remembered the Demon, especially when he walked downtown and passed the courthouse tower, and every so often he would shudder. But he wasn't sure what he and the City Council could do about it.

The Demon and the gas still made for interesting talk, but were no longer such big news. Mrs. Doobar's latest revelations made the paper, but were relegated this time to the second section, two pages before the comics, right next to the daily horoscope column. Andrea McKenzie read the article avidly, and showed it to a lot of her school friends, but to her surprise, it didn't seem to stir up much interest in them. At first, her best friend, Kathi Hamilton, a very bright girl who had plans to go to State and major in history, enthusiastically joined Andrea and wrote quite a few cards each day. But after a few days, Kathi's enthusiasm began to wane.

"What's the point, Andrea?" Kathi asked. "A handful of postcards from me isn't going to make the slightest bit of difference. Didn't you read Mrs. Doobar's article? There have got to be 160,000 a day to make a big difference."

"That's just the point, Kath!" replied Andrea exasperatedly. "If you and everyone else will just do your part, we'll *reach* that number—but you can't cop out!" Kathi didn't see the logic, and spent most of her time doing her homework for the summer school course in World History she was taking. After all, how could she get into State if she flunked World History?

Andrea just couldn't figure out how come Kathi, of all people, so interested in history and the flow of time and world events, could not see her *own* life being touched by such factors, so she asked Kathi, "How do you know there will be any *you* left to *go* to State, if you don't write postcards? Each year, there's a 1-in-12 chance of you and me and all of us being wiped out! Don't you even want to work against that? If people would just *care,* they could *change* things! An hour a day! Half an hour a day! Fifteen minutes a day!"

"Oh, come *on,* Andrea!" said Kathi annoyedly, "Be realistic."

"Darn it all, *I'm* the one who's *being* realistic", said Andrea. "If you don't help out, you're adding to the burden of someone else."

"For Pete's sake, Andrea", Kathi protested angrily, "I'm *not* adding to anyone else's burden. Everyone can help out as much as they want, and no one's obliged to do anything at all. Sure, I'd like it if everyone were helping, but you can see for yourself, practically nobody is. So I'm not going to waste my time. I need to pass World History."

And sure enough, Andrea had to do no more than listen each hour, right on the hour, to hear that bell ring to realize that nobody was doing much. It once had sounded so pleasant and reassuring, and now it sounded creepy and ominous to her, just like the fireflies and the barbecues. Those fireflies and barbecues really bugged Andrea, because they seemed so *normal,* so much like any *other* summer—only *this* summer was *not* like any other summer. Yet nobody seemed to realize that. Or rather, there was an undercurrent that things were not quite as they should be, but nothing was being done . . .

One Saturday, Mr. Hobbs, the electrician, came around to fix a broken refrigerator at the McKenzies' house. Andrea talked to him about writing postcards to the Demon. Mr. Hobbs said to her, "No time, no time! Too busy fixin' air conditioners! In this heat wave, they been breakin' down all over town. I work a 10-hour day as it is, and now it's up to 11, 12 hours a day, includin' weekends. I got no time for postcards, Andrea." And Andrea saw that for Mr. Hobbs, it was true. He had a big family and his children went to parochial school, and he had to pay for them all, and . . .

Andrea's older sister's boyfriend, Wayne, was a star halfback at Happiton High. One evening he was over and teased Andrea about her postcards. She asked him, "Why don't you write any, Wayne?"

"I'm out lifeguardin' every day, and the rest of the time I got scrimmages for the fall season."

"But you could take some time out—just 15 minutes a day—and write a few postcards!" she argued. He just laughed and looked a little fidgety. "I don't know, Andrea", he said. "Anyway, me 'n Ellen have got better things to do—huh, Ellen?" Ellen giggled and blushed a little. Then they ran out of the house and jumped into Wayne's sports car to go bowling at the Happi-Bowl.

*   *   *

Andrea was puzzled by all her friends' attitudes. She couldn't understand why everyone had started out so concerned but then their concern had fizzled, as if the problem had gone away. One day when she was walking home from school, she saw old Granny Sparks out watering her garden. Granny, as everyone called her, lived kitty-corner from the McKenzies and was always chatty, so Andrea stopped and asked Granny Sparks what she thought of all this. "Pshaw! Fiddlesticks!" said Granny indignantly. "Now Andrea, don't you go around believin' all that malarkey they print in the newspapers! Things are the same here as they always been. I oughta know —I've been livin' here nigh on 85 years!"

Indeed, that was what bothered Andrea. *Every*thing seemed so annoyingly *normal.* The teenagers with their cruising cars and loud motorcycles. The usual boring horror movies at the Key Theater down on the square across from the courthouse. The band in the park. The parades. And especially, the damn fireflies! Practically nobody seemed moved or affected by what to her seemed the most overwhelming news she'd ever heard. The only other truly sane person she could think of was little Wally Thurston, that eleven-year-old from across town. What a ridiculous irony, that an eleven-year-old was saner than all the adults!

Long about August 1, there was an editorial in the paper that gave Andrea a real lift. It came from out of the blue. It was written by the paper's chief editor, "Buttons" Brown. He was an old-time journalist from St. Jo, Missouri. His editorial was real short. It went like this:

The Disobedi-Ant

The story of the Disobedi-Ant is very short. It refused to believe that its powerful impulses to play instead of work were anything but unique expressions of its very unique self, and it went its merry way, singing, "What I choose to do has nothing to do with what any-ant else chooses to do! What could be more self-evident?"

Coincidentally enough, so went the reasoning of all its colony-mates. In fact, the same refrain was independently invented by every last ant in the colony, and each ant thought it original. It echoed throughout the colony, even with the same melody.

The colony perished.

Andrea thought this was a terrific allegory, and showed it to all her friends. They mostly liked it, but to her surprise, not one of them started writing postcards.

All in all, folks were pretty much back to daily life. After all, nothing much seemed really to have changed. The weather had turned real hot, and folks congregated around the various swimming pools in town. There were lots of barbecues in the evenings, and every once in a while somebody'd make a joke or two about the Demon and the postcards. Folks would chuckle and then change the topic. Mostly, people spent their time doing what they'd always done, and enjoying the blue skies. And mowing their lawns regularly, since they wanted the town to look nice.

Post Scriptum.

> *The atomic bomb has changed everything*
> *except our way of thinking. And so we*
> *drift helplessly towards unparalleled disaster.*
> —Albert Einstein

People of every era always feel that their era has the severest problems that people have ever faced. At first this sounds silly. How can *every* era be the toughest? But it's not silly. Things can be getting constantly more dangerous and frightful, and that would mean that each new generation truly *is* facing unprecedentedly serious problems. As for us, we have the problem of extinction on our hands.

Someone once said that our current situation *vis-à-vis* the Soviet Union is like two people standing knee-deep in a room filled with gasoline. Both hold open matchbooks in their hands. One person is jeering at the other: "Ha ha ha! *My* matchbook is *full,* and yours is only *half* full! Ha ha ha!"

The reality of our situation is about that simple. The vast majority of people, however, refuse to let this reality seep into their systems and change their day-to-day behaviors. And thus the validity of Einstein's gloomy utterance.

* * *

I remember many years ago reading an estimate that the famous geneticist George Wald had made about nuclear war. He said he figured there was a two percent chance per year of a nuclear war taking place. This amounts to throwing one 50-sided die (or a couple of seven-sided dice) once a year, and hoping that it doesn't come up on the bad side. How Wald arrived at his figure of two percent per year, I don't know. But it was vivid. The figure has stuck with me for a couple of decades. I tend to think that the chances are greater nowadays than they were back then: maybe about five percent per year. But who can say?

The *Bulletin of the Atomic Scientists* features a clock on its cover. This clock doesn't tick, it just hovers. It hovers near midnight, sometimes getting closer, sometimes receding a bit. Right now, it's at three minutes to midnight. Back at the signing of SALT I, it was at twelve minutes before midnight. The closest it ever came was two minutes before midnight, and I think that was at the time of the Cuban missile crisis.

The purpose of the clock is to symbolize the current danger of a nuclear holocaust. It's a little like those "Danger of Fire Today" signs that Smokey the Bear holds up for you as you enter a national forest in the summer. It is a subjective estimate, made by the magazine's board of directors. Now what is the meaning of "danger", if not *probability of disaster per unit time*? Surely, the more dangerous a place or situation, the faster you want to get out of it, for just that reason. Therefore, it seemed to me that the *Bulletin*'s number of minutes before midnight, *B*, was really a coded way of expressing a Wald number, *W*—a probability of nuclear war per year. And so I decided to make a subjective table, matching up the values of *B* that I knew about with my own best estimates of *W*. After a bit of experimentation, I came up with the following table:

Bulletin Clock (minutes before midnight)	Wald's percentage (probability per year)
1 min	20 percent
2 mins	10 percent
3 mins	7 percent
4 mins	5 percent
5 mins	4 percent
7 mins	3 percent
10 mins	2 percent
12 mins	1.5 percent
20 mins	1 percent

A fairly accurate summary of this subjective correspondence is given by the following simple equation:

$$W = 20/B$$

This estimates for you the holocaust danger per orbit of the earth, as a function of the current setting of the *Bulletin*'s clock.

W and *B* may not be estimable in any truly scientific way, but there is a definite reality behind them, even if not so simple as that of *N* and *X* in Happiton. Obviously it is not a "random", dicelike process that will determine whether nuclear war erupts in any given year. Nonetheless, it makes good sense to think of it in terms of a probability per year, since what actually *does* determine history is a lot of things that are *in effect* random, from the point of view of any less-than-omniscient being. What other people (or countries) do is unpredictable and uncontrollable: it might as well be random.

If tensions get unbearably high in the Middle East or in Central America, that is not something that we could have predicted or forestalled. If some terrorist group manufactures and uses or threatens to use a nuclear bomb, that is essentially a "random" event. If overpopulation in Asia or starvation in Africa or crop failures in the Soviet Union or oil gluts or shortages create huge tensions between nations, that is like a random variable, like a throw of dice. Who could have predicted the crazy flareup between Britain and Argentina over the silly Falkland Islands? Who knows where the next hot spot will turn out to be? The global temperature can change as swiftly and capriciously as a bright summer day can turn sultry and menacing—even in Happiton.

*　　*　　*

It is the vivid imagery behind the Wald number and the *Bulletin* clock that first got me thinking in terms of the Happiton metaphor. The story was pretty easy to write, once the metaphor had been concocted. I had to work out the mathematics as I went along, but otherwise it flowed easily. It was crucial to me that the numbers in the allegory seem realistic. The most important numbers were: (1) the chance of devastation per year, which came out about right, as I see it; and (2) the amount of time per day that I think would begin to make a significant difference if devoted by a typical person to some sort of activity geared toward the right ends. In Happiton, that threshold turned out to be about fifteen minutes per day per person. Fifteen minutes a day is just about the amount of time that I think would begin to make a real difference in the real world, but there are two ways that one might draw a distinction between the situation in Happiton and the actual case.

Firstly, some people say that the situation in Happiton is much simpler than that of global competition and potential nuclear war. In Happiton, it's obvious that writing postcards will do some good, whereas it's not so obvious (they claim) what kind of action will do any good in the real world. Working hard for a freeze or for a reduction of US-SU tensions might even be harmful, they claim! The situation is *so* complex that nothing corresponds to the simplistic and sure-fire recipe of writing postcards.

Ah, but there is a big fallacy here. Writing postcards in Happiton is *not* sure-fire. The gas could *still* come oozing up at any time. All that changes is the *odds.* Now in the real world, we must follow our own best estimates, in the absence of perfect information, as to what actions are likely to be positive and what ones to be negative. You can only follow your nose. You can never be *sure* that any action, no matter how well intended, is going to improve the situation. That's just the way life is.

I happen to believe that the odds of a holocaust will be reduced (perhaps by a factor of 1.0000001) by writing to my representatives and senators fairly regularly, by attending local freeze meetings, by contributing to various organizations, by giving lectures here and there on the topic, and by writing articles like this. How can I *know* that it will do any good? I can't, of course. And it's no different in Happiton. The best of intentions can backfire for totally unforeseeable reasons. It might turn out that little Wally Thurston, by moving his pencil in a certain graceful curlicue motion one afternoon while writing his 1,000th postcard to the Demon, stirs up certain air molecules which, by bouncing and jouncing against other ones helter-skelter, wind up giving that tiny last push to the caroming icosahedral dice atop the belfry, and bang! They all come up '7'! Wally, oh Wally, why such folly? Why did you ever write those postcards?

Those who would caution people that it *might* be counter-productive to work against the arms race—unless they believe one should work *for* the arms race—are in effect counseling paralysis. But would they do so in other areas of life? You never know if that car trip to the grocery store won't be the last thing you do in your life. All life is a gamble.

The second distinction between Happiton and reality is this. In Happiton, for fifteen minutes a day to make a noticeable dent, it would have had to be donated by all 20,000 citizens, adults and children. Obviously I do not think that is realistic in our country. The fifteen minutes a day per person that I would like to see spent by real people in this country is limited to adults (or at least people of high-school age), and I don't even include most adults in this. I cannot realistically hope that everyone will be motivated to become politically active. Perhaps a highly active minority of five percent would be enough. It is amazing how visible and influential an articulate and vocal minority of that size can be! So, being realistic, I limit my desires to an average of fifteen minutes of activity per day for five percent of the adult American population. I sincerely believe that with about this much work, a kind of turning point would be reached—and that at 30 minutes or 60 minutes per day (exactly as in Happiton), truly significant changes in the national mood (and hence in the global danger level) could be effected.

* * *

I think I have explained what Happiton was written for. Trigger activity it may not. I'm growing a little more realistic, and I don't expect much of anything. But I would like to understand human nature better, to understand what it is that makes us so much like stupid gnats dully buzzing above a freeway, unable to see the onrushing truck, 100 yards down the road, against whose windshield we are about to be smashed.

One last thought: Although to me it seems that nuclear war is the gravest threat before us, I would grant that to other people it might appear otherwise. I don't care so much what kinds of efforts people invest their time in, as long as they do something. The exact thing that corresponds to the threat to Happiton doesn't much matter. It could be nuclear weapons, chemical or biological weapons, the population explosion, the U.S.'s ever-deepening involvement in Central America, or even something more contained, like the environmental devastation inside the U.S. What it seems to me is needed is a healthy dose of indignation: a spark, a flame, a fire inside. Until that happens, that courthouse clock'll be tickin' away, once every hour, on the hour, until . . .

Post Post Scriptum.

Two magazines are devoted to the prevention of nuclear war. They are: the *Bulletin of the Atomic Scientists* and *Nuclear Times*. The *Bulletin*, founded in 1945, aims to forestall nuclear holocaust by promoting awareness and understanding of the issues involved. It describes itself as "a magazine of science and world affairs". Its address is: 6042 South Kimbark Avenue, Chicago, Illinois 60637.

Nuclear Times is a more recent arrival, and calls itself "the news magazine of the antinuclear weapons movement". Its articles are shorter and lighter than those of the *Bulletin*, but it keeps you up to date on what's happening all over the country and the world. Its address is: Room 512, 298 Fifth Avenue, New York, New York 10001. (As of the current edition of this book, *Nuclear Times* is no longer in existence.)

The following organizations are effective and important forces in the attempt to slow down the arms race and to reduce global tensions. Most of them put out excellent literature, which is available in large quantities at low prices (sometimes free) for distribution. Needless to say, they can always use more members and more funding. Many have local chapters.

Council for a Livable World
110 Maryland Avenue, N.E.
Washington, D.C. 20002

Peace Action (formerly SANE)
1819 H Street N.W., Suite 640
Washington, D.C. 20006-3603

Center for Defense Information
1500 Massachusetts Avenue, N.W.
Washington, D.C. 20077-1724

Physicians for Social Responsibility
1101 14th Street N.W., Suite 700
Washington, D.C. 20005

Union of Concerned Scientists
2 Brattle Square
Cambridge, Massachusetts
02138-9105

33

The Tumult of Inner Voices, or, What Is the Meaning of the Word "I"?

Grace Adams Tanner Lecture in Human Values
Southern Utah State College
Cedar City, Utah
May, 1982

I pushed a button, and in an instant, Utah ceased to exist. Totally obliterated, beyond recall. There was nothing I could do now, no matter how much I wished I hadn't pushed that button, no matter how recent the action was. A mere second after it had been pushed, there was no way to undo my action. A miscalculation with dire consequences.

Utah, a sandy state, full of deserts and strange, barren scenery. A beautiful state, a place I had passed through many times, always with a sense of wonder. Eerie, resonant names like "Uintah", "Wasatch", "Moab", "Koosharem", "Shivwits", "Tavaputs", "Panguitch" . . .

Now all those names had been destroyed, leaving not a trace. All those names would have to be retyped. But that was not the bad part. The bad part was that all my ideas, the inspired ones I'd had a few days ago, had gone down the drain as well. Of course, the first thing I'd checked was if there were any backup copies. There had been two. But I'd destroyed both of them as well. Just moments ago there had been three, yes, *three,* copies of Utah on my directory, and now they were all gone. The disk space had been released, and perhaps for a few seconds my file's bit-patterns had continued to spin around, no longer protected, no longer shielded, yet still intact. But then, inevitably, somebody had wanted to write a file, and mercilessly, the operating system handed my space over to them. Now somebody else's bit-patterns had overwritten mine.

I was desperate. I hoped against hope that there was still some way to get

Utah back. They checked for me if there were any versions on tape, and sure enough, it appeared there was one—a day old. Whew! At least *something* remained of my work. But within minutes we discovered that it was only four lines long. It had been taped before I'd had those good ideas. So every last shred of hope was destroyed.

I knew I would be unable to reconstruct my ideas. I would just have to start again from scratch. It was a horrible feeling, to have deleted my file called "Utah". And just one single bad keystroke had done it. And yet, here I am again, typing away, creating a new file by the same name, trying to construct new ideas to replace the old ones. Or rather, here I am, standing before you, reading these week-old words to you, words that came off that same spinning disk where my others had spun before them.

Which is right? Am I really here at my terminal, typing, or am I really *here,* standing before you? I can't decide.

And that is one aspect of the question that I wish to confront today, in this Grace Adams Tanner Lecture on Human Values. The question is an intellectual, philosophical one: What is the meaning of the word "I"? Yet the question is also a pragmatic, real-life, soul-ripping issue: Which one of the many people who I am, the many inner voices inside me, will dominate? Who, or how, will I be? Which part of me decides? And can that part in turn have inner conflicts about how to decide which version of me it wants to let dominate?

* * *

Such were the questions I was confronting when I was typing that earlier version of Utah that I spoke about at the beginning. But I have to confide that I made a slight distortion. I spoke about those place names like "Uintah" and "Wasatch" being destroyed when I pushed the button. Actually, those place names were not in the original version of my talk. Only after the irreversible calamity of deleting that original version of Utah the *file* had taken place did the metaphor of destroying Utah the *state* by pushing a button lead me to imagine the places in it that would be obliterated. Thus, only in my *new* file called "Utah" did I actually type those place names one by one. And that is *this* file, upon which I am now typing. Thus, only if I make the same mistake once again—and I certainly hope I won't be *that* careless —will those names be lost.

However, it is no accident that the analogy between destroying Utah the state and Utah the computer file arose in my mind. Far from an accident. In fact, what was that old file of inspired ideas about? It was about an inner fight between two voices inside of me, two competing selves, two major facets of the person Doug Hofstadter, vying against each other. It was, in fact, a hypothetical dialogue taking place before you, a dialogue between two persons both of whom are inside me, both of whom are genuinely myself, but who are at odds, in some sense, with each other.

Those of you who are familiar with my book *Gödel, Escher, Bach* will remember the characters of Achilles and the Tortoise (as well as a few others), whom I used in my Dialogues. I wrote another dialogue more recently, with Achilles and the Tortoise once again, in which they discuss the soul-searching question "Who shoves whom around inside the careenium?" In that dialogue [Chapter 25 of this book], I deal—or should I say, *they* deal—with the question of what governs the soul. How can we understand the nature of our selves when we are composed of so many myriads of parts, none of which we understand? How are those parts put together? How does the total add up to a self, a soul, a you or a me?

That is the metaphor of the "careenium"—a kind of enormous arena in which billions of marbles careen around, bashing into each other unconsciously, and yet which gives rise, when one stands back and looks at the *whole,* to a vast consciousness. The title question, "Who shoves whom around inside the careenium?" concerns how to look at such a system, one that has various levels of description. The central issue is whether the marbles shove the total system around, or whether the desires of the system as a whole shove the marbles around. It's a slippery issue, and pinning down the nature of free will—referred to also as *free won't* in the dialogue!—is the name of the game.

Part of me was intending to read that dialogue to you tonight. One inner voice spoke for it. But another, more urgent inner voice spoke eloquently against it. And the transcript of the debate between *those* inner selves was what was deleted by my careless finger. That was the original file called "Utah".

Who was this other Doug Hofstadter that was so rudely intruding? And what did he want to say? Why was he fighting for control of my top level? What in him insisted that the story of the careenium was not appropriate? Well, I know of no better way to explain this than to let him speak for himself. So without further ado, here he is.

*   *   *

Thank you, but I feel a little awkward about this. After all, I don't really feel as if I am a different person from you. That is, from the person who graciously consented to let me have a few words. I am really the same person, am I not? Sometimes I am not sure. After all, who is it that was invited to give the Grace Adams Tanner Lecture on Human Values? Was it Douglas Hofstadter the author, or Douglas Hofstadter the person? The former won a Pulitzer Prize, writes a monthly column, is publicly visible, and hence seems a likely candidate for being invited to give a lecture. And the latter is unknown, except to friends. Yet all the energy of the former comes from an invisible, inner person, a hot fiery core of combatting ideas and hopes and goals. And so in essence, all that was just done is to symbolically yield the floor to the *real* person in whose mind all those ideas seethe and tangle with each other.

And since I am now the *real* person, not the image, I can feel free to talk about what it is that is gripping me these days, and splitting me into several people (of which *I* am one). The thing that is gripping me, the thing that is splitting me into subselves, that thing is an ever-increasing sense of the reality of that other ludicrously simple act of button-pushing, the act that will destroy not a computer file but a real state, in fact all our states, all our towns. It is the "like-it-or-not-it's-real" issue of nuclear war. It's what has occasionally been called "the unthinkable".

I have always liked that name for it. "The unthinkable"—it carries with it the notion that it truly will never happen. That it is so awful that no one could ever conceivably start a nuclear war. That nuclear war is synonymous with the end of the world, with Armageddon. But it seems that such a view has faded, over the years. It seems that "the unthinkable" is being contemplated more and more by the governments of the nuclear powers.

*   *   *

I live in Bloomington, Indiana, a town of somewhere around 60,000 inhabitants. Somewhere in the Soviet Union, there is a missile silo with a missile inside it destined specially for Bloomington, Indiana. After all, there are several thousands of nuclear-tipped missiles (I love the delicacy of that word "nuclear-tipped"!—one can almost visualize a cute little hood ornament on the tip of a gracefully streamlined rocket, an artistic flourish added merely as an extravagant but stylish afterthought)—several thousands of these that will form a first-strike force. There are workers who have never heard of Bloomington, Indiana who daily do routine checks to make sure this missile will hit its mark and do its duty. Some of them may know where it is going. They pronounce it "Bluminktoan, Eendianna", and maybe then they chuckle.

Somewhere else, perhaps in Utah, perhaps not, there are American workers, people very similar to those in the Soviet Union, who are taking loving care of very similar missiles aimed at Gorky, Novosibirsk, Omsk, towns in the Soviet Union. They pronounce those names with accents as atrocious as those of their Soviet counterparts.

These people on opposite continents are very similar to each other, and bear each other no ill will. Yet each missile is there for no other purpose than to put out the soul-flames of hundreds of thousands of human beings in a distant town, instantaneously or over several weeks. The horror is not imagined.

*   *   *

The American people are known the world over for their generosity and warmth. America has given the world many wonderful things. There is the notion of freedom, the image of the western frontier, the Hollywood movies, jazz of many types, a vision of how technology and science can make life increasingly pleasurable, a looseness of language and dressing style that have inspired people the world over, an informality and palsiness that are deep products of the American character. I am proud to write for the magazine *Scientific American,* to be part of that wonderfully characteristic American institution, and to have my writings represent my country. I feel my mentality is a uniquely American product, and I am proud of it. This country has contributed greatly to humanity's self-image; it has become a special symbol in the minds of people all over the world.

Similarly, the Russian people have contributed monumentally to world culture, through their novelists like Dostoyevsky and Tolstoy, through the noble music of such composers as Rachmaninoff and Prokofiev and Scriabin and Shostakovich (some of my favorites), through their scientists and their mathematicians and philosophers, through their rich and sad culture, so full of torment and resignation. In the town of Gorky lives the towering spirit of Andrei Sakharov, often known as the father of the Russian H-bomb, now a dissident struggling to turn around his government. In ordinary towns all over the Soviet Union live ordinary individuals like you and me, who desire nothing more than simple lives without a Sword of Damocles hanging over them.

But we all have this sword hanging over us, and the thread by which it is hanging is getting thinner every day. The "unthinkable" has not only been thought about, it has been planned in infinite detail, on both sides. Dozens of nuclear war scenarios have been considered and weighed in the balance, on both sides. Dozens of versions of Armageddon have been played out either on computers or in the minds of war planners. These people plant colored pins on maps and think only about numbers, not about lives.

Not only have plans—software—been drawn up, but of course, all the hardware, the materiel, is there. It is in place. But as if that were not enough, each day, on the surface of the earth, somewhere between three and five *new* atomic warheads are brought into being. We all know that only two atomic bombs have ever been dropped on people, each one killing somewhere around sixty thousand people. And those two were just tiny, "cute" bombs compared to the ones we are making every day.

* * *

There are some people who, ostrich-like, wish to think that nothing has changed radically since the days of World War II. A friend of mine told me recently of a lunchtime conversation he had with his boss. My friend had brought up the horrors of nuclear war, and his boss—a kindly man with good intentions—protested that in reality, a nuclear war wouldn't be much different from any previous war. "How do you see that?" my friend asked. "What's the smallest nuclear weapon?" said his boss. "Oh, about 500 tons of TNT", replied my friend, up on his statistics. "And the largest conventional bomb?" asked his boss. "Maybe around 20 tons of TNT", said my friend. "There—you see?" said the boss. "Only a factor of 25."

FIGURE 33–1. *One common posture to take, in light of the distressing news constantly bombarding us.* [*Drawing by David Moser.*]

When my friend repeated this story to me, I couldn't believe my ears. My friend's boss felt *comforted* by the thought that the smallest nuclear bombs are "only" 25 times bigger than the largest conventional bombs. He was willing to completely neglect a factor of 25—in itself preposterous. But 25 is not the right number at all. A *typical* nuclear bomb is more like one megaton, which is to say more like 50,000 times bigger than the largest of conventional bombs. Not 25 times, but 50,000 times larger. If you compare it to a *typical,* rather than the *largest,* conventional bomb, the ratio becomes something on the order of a million. One million. The figure is staggering, literally incomprehensible. To make it more graphic, though certainly no more conceivable, I can quote the statistic that a single nuclear warhead of moderate size—2.2 megatons—carries more destructive power than *all* the bombs dropped on Germany during all of World War II. The bomb destined for Bloomington, Indiana, is in all likelihood about half that big. Maybe knowing that it's only half that big would comfort my friend's boss. Yet somehow I doubt it. (See Figure 33-1.) ʼ

*　　*　　*

But why am I speaking to you about these ghoulish issues? Did I come here to speak about nuclear war, or about the nature of the soul? Why am

I talking so passionately about issues that terrify rather than issues that fascinate and enchant? Because I am more than one person. Because I cannot keep *this* inner voice quiet. Because something in me has woken up after being dormant for many years.

Now if I turn my attention to this strange notion of "inner voices" and "dormant ideas", I will have come right back to the other topic—the nature of the soul. So my own internal split forces me to oscillate back and forth between a topic I hate to contemplate and one I love to contemplate. I am the victim of my own mind, a mind that has woken up to the nightmare of today's realities. Or seen on another level, my mind is "shoved around" by internal agents of my brain that have become activated after a long dormancy. And where my brain shoves, so must my mind follow. How can I be saying these strange things? What is the meaning of "mind as slave to brain"?

To confront this, let me shift gears, and talk about brains. A brain is a collection of many, many parts—some 100 billion neurons. These parts are linked up to each other in fantastically complex ways. Most neurons, for instance, are connected up with several thousand other neurons, often quite far removed ones. Neurons come in a few types, but basically, they are all quite similar to one another. Systems with large numbers of identical parts have been studied for years in that branch of physics known as statistical mechanics. For example, think of a lattice—a three-dimensional lattice, like an enormous three-dimensional checkerboard—in which a particle can sit with its "spin" pointing either up or down. Each particle directly affects only its immediate neighbors through the magnetic field created by its spinning. So now each particle must decide if it "wishes" to be pointing up or pointing down. Its "decision" is deterministic, being governed by the states of its neighbors. But the strange thing is that *their* states are in turn governed by *their* neighbors, and so the whole thing turns out to be interconnected in a vast interlocked way, despite the fact that any given particle "feels" only its immediate neighbors.

As a consequence, the behavior of such a system has some striking properties that go under the general heading of "collective phenomena". This is an elusive notion. An example is the common phenomenon of the traffic jam. You won't locate a traffic jam if you restrict your search to the insides of a single taxi. After failing to find it inside the glove compartment, you move on to the trunk, and then you look inside the hood, opening up the battery, the radiator, then moving on to the gas tank . . . . A ludicrous image! A traffic jam is just not on the level of an individual car. It is a pattern composed of cars, a pattern that moreover has deep repercussions on the cars it is composed of.

The nature of collective phenomena is that they are patterns composed of parts, and they in turn exert powerful influences on their parts, acting to keep them in line. Think of hurricanes, life, intelligence. You don't see a hurricane by looking inside an atom of oxygen. You don't see life by looking at an amino acid. Nor do you see intelligence by looking at one isolated teeny part of its substrate, whether that part is a synapse, as in a brain, or a binary digit, as in a machine.

*　　*　　*

What is the dynamics of a collective phenomenon? How does a thought get generated and propagate inside a brain? Nobody really knows. All we can do is resort to simpler analogies. For instance, imagine a school of fish swimming through the sea, when the lead fish encounter a danger: a strange shadow passing overhead, or a sudden movement ahead. In a split second the whole school has wheeled about in unison, and is hightailing it away at top speed. Such collective actions, such cooperative effects, take place when all the components are in phase with each other. They act as a unit. In a similar way, the spins in a magnetic substance act in concert. If certain pivotal spins flip, a whole "school" of spins—a magnetic domain—flips in unison.

These so-called magnetic domains are a little like countries, and their individual spins are like people. Of course, people are much more complicated than spins or particles are. They do not just have two states, up or down! However, in times of crisis, people are often placed in just such positions, of having to see a situation in very black-and-white terms. I remember vividly how Iran seemed to switch overnight from a country favorably inclined toward the United States—and I mean at Iran's grass-roots level, not just at its governmental level—to a country full of bitter hatred toward this country. Just all at once it became *polarized*—a term, incidentally, applying to magnetic substances as well, in which all the spins (or a substantial majority of them) point in one direction.

And external events can swing these polarizations the other way in dizzyingly short times. We all know how countries can seemingly "flip" in very short periods of time. (Think of Argentina's amazingly sudden near-total unification behind a junta that, up until a few weeks ago, was bitterly hated!) But such a thing can happen only if the country is polarized or otherwise coalesces so as to act as one large collective entity. During much of any country's existence, it does not act as a single, polarized, black-and-white entity, but rather as a much more complex entity composed of dozens and dozens of smaller entities—domains—all of which are pursuing their own goals and coexisting in one way or other with each other.

A brain is much like this. A brain, with its billions of neurons, resembles a community made up of smaller communities, each in turn made up of smaller ones, and so on. The highest-level communities just below the level of the whole are what I like to call "subselves" or "inner voices". I mean the latter not in the sense of a schizophrenic who hallucinates inner voices, but rather, in the sense of competing facets that try to commandeer the whole, something like hijackers, although often benevolent hijackers. Perhaps it is more like a vote of the passengers on a plane where they want to go, *after* it is in the air!

To give an example of such commandeering, one that I pursue in some depth in that Achilles-Tortoise dialogue on the careenium, consider what happens when I pick up a Rubik's Cube. What makes me pick one up, in the first place? I see it sitting there on the shelf, glittering just as always—but once in a while its glittering catches my eye a little more than usual. Somehow, I am "primed" for a cubing session. So I pick it up. I may have a million other more pressing things to do than solve the cube, but somehow I cannot resist picking up that cube and scrambling it, and then twisting its colorful faces one way and another, very enjoyably, back to order. And having done it once, I do it over again. Then I may say to myself, "Really! You have better things to do! Just *once* more!" And so I do it once more, but somehow that is not satisfying enough, and with a slightly guilty feeling, I do it again. And again.

As the old potato-chip ad said, "Betcha can't eat just one!" I am also reminded of the Académie Française, which tries to keep a tight hold on the French language, to prevent the people from following certain natural trends that are apparent to everyone. But the Académie tries to exert what I call "top-down pressure" to prevent things that obviously cannot be prevented. There is too much "bottom-up" momentum to put the lid on. You cannot keep the lid on a boiling pressure cooker. You cannot keep the lid on an angry populace, such as the Iranian people or the Salvadoran people or the Chilean people. No matter how tightly you press down, the pressure from inside will in the end overwhelm you.

*   *   *

And thus it is within the brain. Competing subselves cannot be held in check indefinitely. They cannot be clamped down, forbidden to act. For each "inner voice" is in actuality composed of millions of smaller parts, each of which is active, and under the proper circumstances, those small activities will someday all "point in the same direction", and at that moment the inner voice will crystallize, will undergo what is called a *phase transition,* will emerge from obscurity and proclaim itself an active member of the community of selves. And if it is powerful enough, it will try to exert pressure and to be recognized. It will attempt to seize power. It will not want to relinquish power, once it has it. That's what I mean by "commandeering the soul".

I attempted to illustrate this with the Rubik's Cube example, but I see it happening in me all the time. I have a "piano-playing subself", who, once he is given the floor, refuses to relinquish it for hours on end—until, say, my back—*his* back?—grows achy, or until he gets sleepy. Or until the phone rings or my watch beeps at me, telling me that some other facet of life must be attended to.

And somehow, in such circumstances, there is a governing personality who can grab control away from the "hijacker". In fact, it is not at all hard to dislodge the piano-playing hijacker, or the cubing hijacker, or any other subself, when a phone call comes. Isn't that a sign of the times? This kind of selective interruptibility is one of the most critical characteristics of a hierarchically organized system such as a brain, which has evolved to deal with a world in which there are events of various priorities that have to be dealt with sequentially. Choices must be made, so there must be a highest-level body whose purpose is to make choices rapidly and reliably, one that sorts out the priority of subselves and allows only the one deemed most important to take charge.

An interesting problem arises when even this *deciding* agent must be preempted by some sort of extreme emergency situation that arises. The decider can be in the midst of trying to sort out some ordinary conflict of subselves when—

*   *   *

Believe it or not, I was just interrupted by a phone call from a friend. Just as well, I think. For it gives me the opportunity to review where I am going and to resume some of the ideas that I had left off in the middle. In particular, it allows me to make the analogy that I wished to make between a country that is dormant and a subself that is dormant.

In each of us right now—and I am quite confident of this—there are competing inner voices, perhaps one of them dormant, but still present, in some implicit sense—that say opposite things about the prospect of nuclear war. One of them says something like "Nuclear war certainly would be bad, in fact, unthinkable—and everyone knows that, including all the military people and all—so there will never really *be* a nuclear war, it's all just a way of maneuvering and bluffing and so on. So I'll just go on with everyday life." This is the voice in me that has been dominant for years and years! And then there is the other voice, the one that tries more seriously to envision the true nature not only of the devastation that such a war would cause, but also that tries more seriously to evaluate what is going on in the proliferation of nuclear arms.

Certainly it is hard to keep shutting one's eyes to the fact that all over the world, people are thinking about "the unthinkable". Some people are talking about how unthinkable it is, but others are talking about how it might be thinkable, after all. There are those who are telling you how to build a shelter, how to stockpile food, how to keep a gun to ward off neighbors, friends, and strangers when they try to burst into your safe little haven. There are those who are talking about evacuating whole cities into rural areas—as if we'd have the time to do such an incredible thing, or as if people would stomach the idea. These people—people involved in civil defense— have a vested interest in reassuring us all that nuclear war is indeed conceivable, survivable—not all that bad, in fact! What an incredible kind of job to have.

But worst of all, there are those who seem blind to the idea that nuclear war would truly spell the end of the world as we know it. There are millions of ordinary citizens who are somehow relieved when they see a map of their city with the various circles drawn around the downtown square, because they notice that where *they* live is outside the "90 percent killed" circle, even outside the "50 percent killed" circle. So no need to worry. They'll survive. And that's as far as they choose to think about it! Some vague apprehension, maybe, about fallout, or difficulty of getting gas for the car, but that's about all.

Now I shouldn't really be accusing *some people* of thinking this way. The strange thing is that we *all* tend to think this way—or at least parts of us do. (At least I can speak for myself, and I think I am a very typical person.) For we are dealing with something that not only is very vague and unknowable, but also something that is unimaginably catastrophic, something the likes of which has never happened on this planet. So we are not equipped to imagine it (but see Figure 33-2 for some help). And so we turn off. And this turning-off happens to some extent in each and every one of us. Certainly it has been the dominant mode in me for many years. One develops and encourages a sense of security in the ridiculousness of nuclear war.

But the stockpiles are increasing every day. The dangers are increasing every day. The warmongering talk is increasing every day. The number of flashpoints around the world is increasing every day. The mistrust and suspicion and polarization of peoples is increasing every day. The only thing we have on our side is the hope that apathy is *not* increasing. The hope that a country as large as our own can itself undergo a "phase transition", an awakening, a realization of the insanity of the course on which we are embarked.

*   *   *

Phase transitions take place in simple physical systems, schools of fish, individual brains, and in countries as well, when there are sufficiently strong and numerous interactions between the components of the system, and when those interactions add up in such a way as to make for large-scale correlations, or, put another way, long-distance effects despite the short-range nature of the direct interaction. When such long-distance effects occur, then a new kind of entity springs up, an entity on a higher level of organization than its constituents, and that entity obeys certain laws of its own.

Performers are highly aware of this collective aspect of crowds, for instance. A singer will speak of the interaction between herself and the crowd, of how she senses the mood of the crowd *as a whole.* Yet how can this be? Isn't a crowd composed of individuals who are totally unknown to each other, individuals with nothing in common? Yes; however, they do have *one* thing in common: they are all there, physically, listening to the same performer, and so they are influencing each other whenever they laugh at her jokes or applaud her, or encourage her or seem impatient in any way. Such collective modes tend to lock in very quickly, to create self-reinforcing loops of interaction between performer and audience.

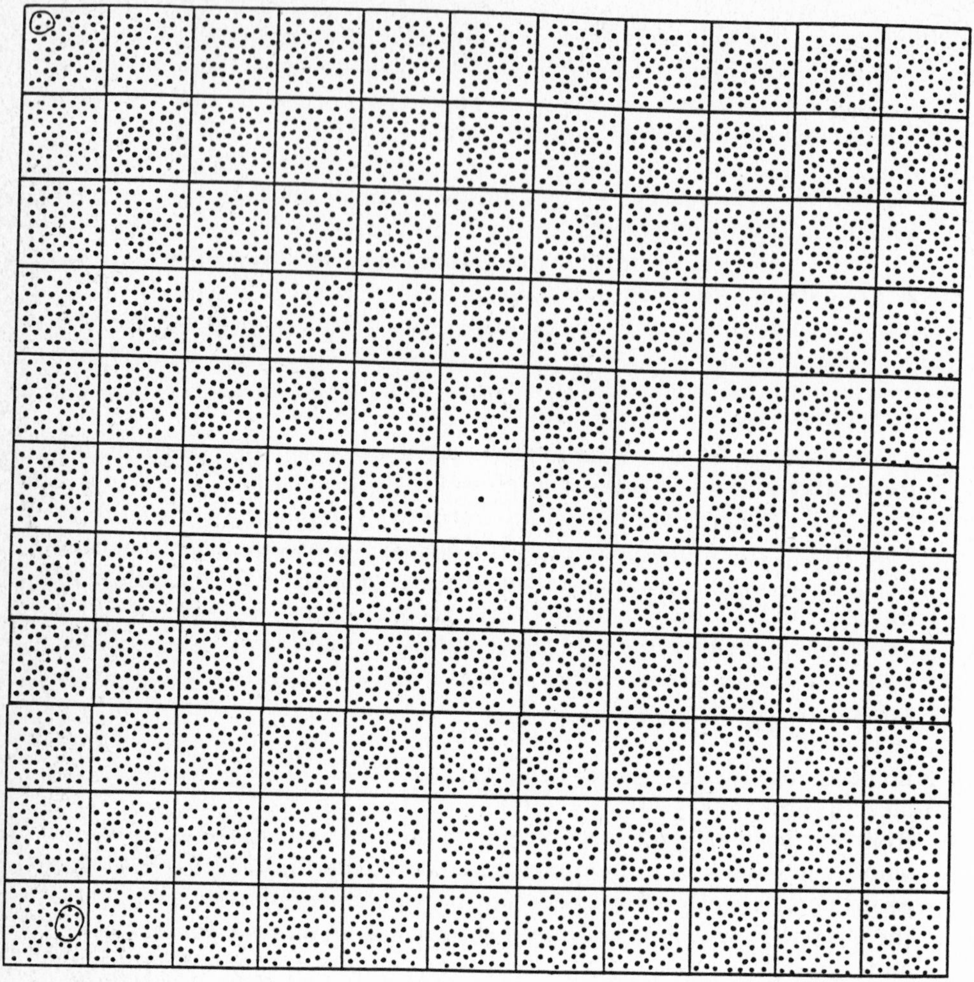

FIGURE 33–2. *A realistic view of the world armaments situation. The chart shows the world's current firepower in terms of the firepower of World War II. The dot in the center square represents all the firepower of World War II (including the atomic bombs dropped on Hiroshima and Nagasaki): three megatons. The other dots represent the world's present nuclear weaponry. This comes to 18,000 megatons, which equals 6,000 World War II's. The U.S. (and allies) and the S.U. (and allies) share this firepower approximately equally.*

The top lefthand circle enclosing nine megatons represents the weapons in just one Poseidon submarine. This is equal to the firepower of three World War II's and is enough to destroy over 200 of the Soviet Union's largest cities. We have 31 such submarines and ten similar Polaris submarines. The bottom lefthand circle enclosing 24 megatons represents one new Trident submarine with the power of eight World War II's: enough to destroy every major city in the Northern hemisphere. The Soviets have similar levels of destructive power.

Just two squares on this chart (300 megatons) represent enough firepower to destroy all the large and medium-size cities in the entire world. [Designed by Jim Geier and Sharyl Green in 1981.]

Such self-reinforcing loops are of the essence in phase transitions and collective modes, for they are what tend to keep the whole thing going. And thus it is with the collective mode of my neurons, the one that has somehow gotten triggered into activity after many years of dormancy. This new inner voice is one that I am not yet entirely comfortable living with. But it is one that haunts me and will not leave me alone. It has seized some power inside me and it will not let go. And the "government" that it has to some extent usurped is not entirely displeased with the state of affairs.

Let me now try to return control to the more dispassionate and objective "top-level" self who began this talk.

*   *   *

Thank you. It has been interesting to me to observe the flipping back and forth that has taken place as the previous subself tried to express himself. One thing that it clearly shows is that there are no clear boundary lines to be drawn between "that subself", "this subself" and any other subselves of Doug Hofstadter the person. All of them are fictions, because the only real thing is the sum total, the integrated person. And that integrated person is clearly not the same person he was a few months ago, when he was blithely ignoring the notion of nuclear horror, somehow unwilling to face the possibility squarely.

This phase transition has not been an entirely pleasant thing to undergo, no more than any *coup d'état* would be. Not that it was so revolutionary. It all arose peacefully, nonviolently, from within. There were no provocateurs from without. Or perhaps, I should say, there was *one*—a 92-year-old lady who was my neighbor last year, and whom I befriended. I would visit her on occasion, and we would have wonderful conversations that rambled from the music of Chopin to the pangs of sad romances to the secrets of the mind and—once in a while—to politics.

One day as I was leaving after one of these discussions, Hildegarde said to me, in a very gentle way, "One thing I'd like to ask you someday is how it is that with your very alert mind, you don't seem to feel the need to do something—or to *try* to do something—about nuclear war." It was a very gentle nudge, really only a passing remark indicating her puzzlement about me. But it did set *me* to wondering how it was that I could systematically ignore the biggest thing in all our lives, day after day after day.

Partially, the reason that I gave to myself was that it was just *too* big. There was no use in worrying about it. But that rang phony to me. It didn't sound like *me*! So actually, I had no answer, and that realization began to eat at me. It began to feel like either pure "ostrichism", or pure egotism. Either way, I didn't like it. But a sense of shame or guilt is never the way to bring about a phase transition. It's got to come from somewhere far deeper than that. And fortunately, there were seething, churning forces down deep inside me that slowly aligned, slowly started to bring about that collective mode that crystallized in the inner voice that you heard.

This "waking up" of an individual has its parallel in the collective waking up of a nation. It will happen when enough citizens band together, seeing some common interest, sensing some common goal. There is a sort of "critical point" when that number reaches a threshold and suddenly, there is a turnaround at a national level. But just how or when that will happen is very tricky to say.

*   *   *

In a recent book entitled *Science: Good, Bad, and Bogus,* Martin Gardner quoted a beautiful passage, written by psychologist William James roughly a century ago, about the act of waking up and rising in the morning. This passage captures for me something very deep about the way the soul of a person arises from a myriad smaller actions that are completely unknowable and yet that are somehow coordinated. I would like to quote that passage to you:

> We know what it is to get out of bed on a freezing morning in a room without a fire, and how the very vital principle within us protests against the ordeal. Probably most persons have lain on certain mornings for an hour at a time unable to brace themselves to the resolve. We think how late we shall be, how the duties of the day will suffer; we say, "I *must* get up, this is ignominious," etc.; but still the warm couch feels too delicious, the cold outside too cruel, and resolution faints away and postpones itself again and again just as it seemed on the verge of bursting the resistance and passing over into the decisive act. Now how do we *ever* get up under such circumstances? If I may generalize from my own experience, we more often than not get up without any struggle or decision at all. We suddenly find that we *have* got up. A fortunate lapse of consciousness occurs; we forget both the warmth and the cold; we fall into some revery connected with the day's life, in the course of which the idea flashes across us, "Hollo! I must lie here no longer!"—an idea which at that lucky instant awakens no contradictory or paralyzing suggestions, and consequently produces immediately its appropriate motor effects. It was our acute consciousness of both the warmth and the cold during the period of struggle, which paralyzed our activity then and kept our idea of rising in the condition of *wish* and not of *will.* The moment these inhibitory ideas ceased, the original idea exerted its effects.

I find this to be a remarkably perceptive passage, so accurate in its understanding of the way people really work. In the "Careenium" dialogue, I expressed some similar ideas of my own, with which perhaps it would be appropriate to conclude my talk tonight.

Achilles says to the Tortoise, "In emotionally wrenching cases, you can hardly *decide* what you will feel. Something just *happens* inside you. Subtle forces shift deep inside you, hidden, subterranean. It's quite scary, in a way, because in real crises like that, instead of being able to *decide* how you'll act, you *find out* what sort of stuff you're made of. It's more passive than active —or more accurately put, the action is on levels of yourself that are far lower —far more microscopic—than you have direct control over."

The Tortoise replies, "Correct. You and your neurons are not on speaking terms, any more than a country could be on speaking terms with its citizens. There is, in both cases, a kind of collective action of a myriad tiny elements on low levels that tips the balance—exactly as in a country that 'decides' to go to war or not. It will flip or not, depending on the polarization of its citizens. And they seem to align in larger and larger groups, aided by communication channels and rumors and so on. All of a sudden, a country that seemed undecided will just 'swing' in a way that surprises everyone."

Achilles continues, "Or, to shift imagery again, it's like an avalanche caused by the collective outcome of the way that billions upon billions of snow crystals are poised. One tiny event can get amplified into stupendous proportions—a chain reaction. But the crystals have to be poised in the right way, otherwise nothing will happen."

And Mr. Tortoise takes over: "In cases of judgment, whether it be of one musical composer over another, one potential title or subtitle for a book over another, or whatever, the top level pretty much has to wait for decisions to percolate up from the bottom level. The masses down below are where the decision *really* gets made, in a time of brooding and rumination. Then the top level may struggle to articulate the seething activity down below, but those verbalized reasons it comes up with are always *a posteriori*. Words alone are never rich enough to explain the subtlety of a difficult choice. Reasons may sound plausible but they are never the essence of a decision. The verbalized reason is just the tip of an iceberg. Or, to change images, conflicts of ideas are like wars, in which *every reason has its army*. When reasons collide, the real battleground is not at the verbal level (although some people would love to believe so); it's really a battle between opposing armies of neural firings, bringing in their heavy artillery of connotations, imagery, analogies, memories, residual atavistic fears, and ancient biological realities."

Finally, Achilles exclaims, "My goodness, it sounds terrifying! You make the battlefield of the mind sound like a vast mined battlefield! Or a treacherous ice field on a steep mountain face. I never realized that a mechanistic explanation of thinking could sound so organic and living. It's sort of awful and yet it's sort of awe-inspiring as well."

* * *

Achilles' remarks hit the nail on the head, for me. Life, when you contemplate its basis in biology, is in many ways terrifying; yet there is a kind of majesty to the depth and complexity of it all. The same holds for humanity as a whole. In many ways, we are a shocking bunch, doing the most terrible things to each other and to other living beings; yet there is also an element of the sacred in humanity, something sacred in spite of the profane in each one of us.

The pile of contradictions that each one of us is still often adds up to something beautiful and cherishable. To preserve that sacred and beautiful facet from the menace created by the profane and awful facet is worth every effort that we can muster, drawing on the power of the many subselves and inner voices that resonate within us and make us what we are.

* * *
* *
*

Epilogue

After writing such a long book, I have a long list of people to whom I owe genuine thanks for many different reasons. I find it very hard to draw the line between people who have contributed directly to this book, and people whose contributions, though real, are indirect. Yet I must attempt to do so, for the sphere of indebtedness extends out hazily to encompass practically everyone I know. In what follows, I shall do my best.

To begin with, I should like to thank Dennis Flanagan and Gerard Piel for offering me the opportunity to write for their distinguished magazine. Each month, I worked with Dennis on the microscopic level of the columns, and I thank him for his good judgment. Though we had our share of disagreements, we developed a warm friendship that I value.

Martin Gardner suggested that I might be his successor. To be recommended by someone of Martin's honesty, wit, and insight is a very high compliment. Thank you, Martin, for that and for all the wonderful things you have written and continue to write.

This book was written in many places. The first few columns were written when I was a John Simon Guggenheim Fellow visiting Stanford University's Computer Science Department, and I would like to thank the Guggenheim Foundation for its support. The majority of the columns were written in Bloomington, Indiana, where for seven years I have been on the faculty of the Computer Science Department of Indiana University. Some new material was written while I visited the Institute for Cognitive Science at the University of California at San Diego in early 1984, and the rest at MIT's Artificial Intelligence Laboratory in Cambridge, Massachusetts, where I spent most of my sabbatical year.

I would particularly like to thank two of my hosts. Donald Norman made me feel most welcome at UCSD's Institute for Cognitive Science. I enjoyed not only all the facilities there, but also a couple of runs along the beautiful Del Mar ocean front with Don. At MIT, I was truly lucky to have the interest and support of Marvin Minsky, who went out of his way to make my stay especially comfortable. He even supported two people working with me, something I will never forget.

To Indiana University, however, I owe the most. IU offered me a job in

1977 when I had little to show by way of achievements in cognitive science. Since then, my department has been an extremely supportive and friendly environment. I would like to thank several close colleagues, whose friendship I cherish: Dan Friedman, John O'Donnell, Frank Prosser, Cindy Brown, Mitch Wand, Dave Wise, Paul Purdom, Ed Robertson, Stan Kwasny, Bob Filman, Will Clinger, George Epstein, and the three JB's: Jim Burns, John Buck, and John Barnden. I have exchanged ideas with all of them, and together, they have markedly influenced this book. The Computer Science staff has also been a joy to work with over the years. I would like to single out Kathy Thompson, whose spunk and wacky humor have brightened many a dismal day.

In Bloomington, I have made friends too numerous to mention. As happens in any university town, many of them have left, but they have all made Bloomington a special and wonderful place to live.

The tremendous interchange of ideas I've had with Don Byrd over these past seven years is reflected on all scales of this book, and the generous companionship he has offered is reflected on all scales of its author's life.

Two friends whose intellectual influence on me has been profound are Gray Clossman and Marsha Meredith. But even if their intellectual influence had been nil, they have been friends in need, friends in deed. For that I have to thank them deeply.

Ann Trail, with her sparkling sense of humor, her optimism and generosity, and especially her sense of mortality, has deeply and permanently enriched my life. This book reflects her style in so many ways.

Other Bloomington friends have made such a difference as well. John and Joanie Woodcock have long been close friends, and have always been warm and lively conversation partners. With Scott and Ruth Sanders I have shared political hopes and disappointments, and many exuberant discussions. The Leake family—Roy and Alice, David and Patsy—have been true friends, full of interest and empathy. Ruth Sonneborn and her late husband Tracy were among my very first Bloomington friends, and I will never forget evenings spent at their house engaged in delightfully passionate philosophical arguments. I have relished many consonances and dissonances over musical matters with Al and Helga Winold. Over stimulating lunches with Mike Dunn and others, I gained a new kind of respect for philosophers. It has been my privilege to know University Chancellor Herman B Wells, who, if it could be said of anyone, is the soul of Indiana University.

I have shared enthusiasms for all sorts of ideas—often over meals or coffee—with other Bloomington friends of then and now: Ann McMillan, Sue Wintsch, Vahe Sarkissian, Adrienne Gnidec, Al and Linda David, Tulle Hazelrigg, Jimmy and Gilan Tocco, Judy Mahy and Rich Shiffrin, Marlene Mannella and Evan Smith, Tom Ernst, John Goldsmith, Enrico Predazzi, Marion O'Connor, Sujan Yang, Vicky Grossack, and Anneke Campbell— and the list goes on. I wish I could say something about each one of them, but that would take a whole book—and then I would have to write the acknowledgments to *that* book, which could get to be a problem.

When I was at Stanford in 1980–81, I benefited from contact with many people. Scott Kim, as usual, was full of inspiration and creative energy. I shared Mexican food and long talks with Pentti and Dianne Kanerva. Conversations with Scott Buresh, Marcia Bianchi, Debbie Schweninger, Louis Mendelowitz, Louella Kates, Liz Powers, Allen Wheelis, Larry Breed, Margie and Sia Khosrovi, Eric Hamburg, Debbie Starbuck, Fanya Montalvo, and Stan Isaacs made my Guggenheim year at Stanford a lively and memorable one.

My upstairs neighbor that year was a most remarkable woman in her 90's, named Hildegarde Kneeland. An economist by profession, Hildegarde taught many of today's most influential economists. Even today, she is passionately concerned about the fate of humanity. Hildegarde touched off the fire in me concerning nuclear madness. I wish I could do for others what Hildegarde did for me.

Several of my family's oldest and dearest friends have died in these past few years: Dan Mendelowitz, George Feigen, and Felix Bloch. I had known them all since I was very small, and each has left indelible tracks in my soul. Traces of Dan, George, and Felix lurk throughout this book.

My love for writing and alphabets was heightened by my grandmother, Mary Givan, who shared her love for letterforms with her grandson by showing him wondrous and eye-opening alphabetic books. A little later, in 1960, my uncle and aunt, Albert and Manya Hofstadter, introduced their 15-year-old nephew to abstract art at New York's Guggenheim Museum and Museum of Modern Art, and though I protested how silly it all was, they taught me things that changed my way of looking at visual forms. These experiences started me down many of the artistic and scientific pathways described herein.

I have known Ernest and Edith Nagel for almost 25 years, and they remain a beacon of sanity in this crazy world. Many echoes of fondly remembered conversations are found in this book.

During my enjoyable stay in San Diego, I fruitfully exchanged memes with Paul Smolensky, Don Norman, Dave Rumelhart, Larry West, Karen Pickens, Wendie Maurer, Liam Bannon, Larry McGilvery, and many others. I would also like to thank the ICS staff for making things work smoothly while I was there. It's a superlative place to do cognitive science.

While in the Boston-Cambridge area, I was overwhelmed by the number of good friends I have there. Gloria Minsky is one of the world's warmest people, and it is such a pleasure to enjoy a "Min Chin Din" with her and Marvin and whoever happens by. Betty Dexter is much more than a fantastic secretary—she is a great friend. Her laughing presence on the seventh floor of the AI Lab was a delight. Dan Dennett, as always, was bubbling over with ideas and enthusiasm. I just wish I had gotten to see more of Dan and his wife Susan.

Working with me at the Lab were Marek Lugowski and Melanie Mitchell, graduate students, and David Rogers (also known affectionately as "Dr. Ogers"), post-doc. It is so much fun to bat about research ideas with people who are as intrigued with them as I am!

Some other people who made the Boston atmosphere so exciting are Henry Lieberman, Bernie Greenberg, Chris Morgan, Greg Huber, David Levitt, John Amuedo, Marek Hołyński, Margaret Minsky, Joe Shipman, Russell Brand, Randy Davis, Fanya Montalvo, Carl Hewitt, Jay McClelland, and Dedre Gentner.

Scattered around the globe are numerous other friends who have contributed to my thoughts and moods over the past few years. I'd like to mention Charles Brenner, David Policansky, Mary Adele and Norman Mather, Elwyn and Darlene Wolcott, Pete Rimbey, Francisco Claro, Inga Karliner, Marek Demiański, Maria Nosowska, Zamir Bavel, Peter Suber, Phil and Sarah Taylor, Bob Wolf, Len Shar, Dorothy and Peter Denning, Betty and David Hamburg, and Piet Hoenderdos.

Of special interest are three translators of *Gödel, Escher, Bach* into other languages: Bob French and Jacqueline Henry (into French, of course), and Ronald Jonkers (into Dutch). In the summer of 1983 in Paris, we thrashed out many tricky translation problems, and those stimulating sessions contributed to my understanding of the connection between translation and analogy.

For lending their expertise in specific columns or articles, I would like to thank Bob Axelrod, Bill Huff, Dave Martin, Merald Wrolsted, David Singmaster, Mitch Feigenbaum, Dan Mauldin, and Paul Stein. At *Scientific American,* I was helped numerous times by Adele Premice, Brian Hayes, Sally Jenks, Mary Knight, and Sam Howard.

I am most pleased to be doing this, my third book, with Basic Books once again. Martin Kessler, president, was interested in such a book from the moment he heard about the column, and has been very supportive. My day-to-day contact has been with Maureen Bischoff, just as with my earlier books, and I must say, I get a great kick out of our lengthy telephone calls. Maureen's rapier wit takes the tension out of many difficult situations. Other people who have helped a great deal at Basic include Vincent Torre, Ellen Prior, Elizabeth Werter, Linda Carbone, Ann Rudick, Thalia Doukas, Ruth Elwell, and Jeremy Orgel. Thanks also to Debra Manette, Sandra Dohls, Sabrina Soares, Michael Wilde, David Graf, John Masur, Kathi Lee, Donna Singer, Lisa Adams, and John McAusland.

As always, my family looms large in my life. My parents, Robert and Nancy Hofstadter, have constantly served as critics and supporters. My love for them truly cannot be expressed. My sister Laura is a wonderful person, full of ideas and humor. Her company is cherished. My sister Molly, for unknown reasons unable to talk or understand, has all my love. Her sad plight was a deep mystery that made me start to wonder, many years ago, about mind, brain, and soul.

Finally I come to two special people without whom I would be a very different person today. David Moser has not only been my closest consultant on this book; he has also been like a brother: empathetic, generous, and caring. Carol Brush has been the light in my life over the past two years. We have gained so much from each other, and deeply changed each other's lives.

Taking leave is so hard, and I am going to leave Bloomington soon, for other pastures. A unique constellation of colleagues at the University of Michigan hoped that they could find a way to have me join them. Largely through their efforts, I was offered the Walgreen Chair in Human Understanding—a rare opportunity. After brooding on it for a while, I concluded that this was just too good to pass up, and so, with feelings of excitement tempered by sadness, I accepted the offer.

As any reader can tell, what I have written here is much more than simply acknowledgments for this book. I wanted to thank the many people who have "been there". I also wanted to express special feelings of affection for Bloomington, Indiana, a town where I have flourished in every aspect of life over the past few years. These acknowledgments are a sentimental "thank you" and "farewell" to Bloomington. I will certainly miss Howard's, the Spoon, the Grind, the Horn, and the Harmonica, but it is time for me to move on to Ann Arbor, and, as Hildegard Kneeland once said so memorably to me, to "welcome the future".

—D.R.H.
Bloomington, November 1984.

<pre>
      *
   *   *
*   *   *
</pre>

Bibliography

Anderson, Alan Ross, ed. *Minds and Machines.* Englewood Cliffs, N.J.: Prentice-Hall, 1964. A classic collection of stimulating articles on the mind-body problem, including Alan Turing's important paper "Computing Machinery and Intelligence", and J.R. Lucas' provocative "Minds, Machines, and Gödel".

Applewhite, Philip. *Molecular Gods: How Molecules Determine Our Behavior.* Englewood Cliffs, N.J.: Prentice-Hall, 1981. A book whose subtitle helped spark my "Careenium" dialogue (Chapter 25).

Atlan, Henri. *Entre le cristal et la fumée: Essai sur l'organisation du vivant.* Paris: Editions du Seuil, 1979. A biologist who views life as the emergence of ordered complexity out of randomness here explains his philosophy.

Axelrod, Robert. *The Evolution of Cooperation.* New York: Basic Books, 1984. A beautifully written account of how mutually beneficial behavior—that is, cooperation—can emerge among purely egoistic organisms that share an environment over time.

Ayala, Francisco José and Theodosius Dobzhansky, eds. *Studies in the Philosophy of Biology: Reduction and Related Problems.* Berkeley: University of California Press, 1974. The proceedings of one of the most fascinating conferences I have ever heard about. Many great biological thinkers were present, and discussed the most fundamental questions about how life and minds can be reconciled with physical law. I wish I had been there!

Bandelow, Christoph. *Inside Rubik's Cube and Beyond.* Boston: Birkhäuser, 1982. A clear and mathematically oriented book about the Cube and its successors. Perhaps the best of all the Cube books in English.

Barr, Avron. "Artificial Intelligence: Cognition as Computation". In *The Study of Information: Interdisciplinary Messages,* edited by Fritz Machlup and Una Mansfield. New York: Wiley-Interscience, 1983. A paper putting forward the orthodox AI dogma: that mental activity is "information processing"—more specifically, that the manipulation of "symbols" (representational data structures) by suitable computer programs is no more and no less than what minds do.

Beck, Anatole and David Fowler. "A Pandora's Box of Non-Games". In *Seven Years of Manifold,* edited by Ian Stewart and John Jaworski. Nantwich, England: Shiva Publications, 1981. An amusing collection of silly pseudo-games with more of a moral than might first meet the eye.

Beckett, Samuel. *Waiting for Godot.* New York: Evergreen, 1954. The classic existential drama in which meaninglessness pervades—and at whose core is the famous nonsensical verbal vomit emitted from the mouth of the sad sack named "Lucky".

Benton, William. *Normal Meanings.* Paducah, Ky.: Deer Crossing Press, 1978. A book of strangely evocative poems, a bit comprehensible in parts, totally incomprehensible in others.

Bernstein, Leonard. *The Joy of Music.* New York: Simon & Schuster, 1959. An exciting medley of ideas by this exuberantly articulate thinker and musician. His dialogues are especially enjoyable.

Biggs, John R. *Letterforms and Lettering.* Poole, England: Blandford Press, 1977. One of the best books on the fluidity of letterforms I have run across. Includes short sections on other languages, such as Hebrew, Chinese, and Arabic.

Blesser, Barry et al. "Character Recognition based on Phenomenological Attributes". *Visible Language* 7, no. 3 (Summer 1973). An early article by researchers who clearly had come to appreciate the depth of the letter-recognition problem.

Bloch, Arthur. *Murphy's Law*; *Murphy's Law, Book Two*; *Murphy's Law, Book Three.* Los Angeles: Price/Stern/Sloan, 1977, 1980, 1982. Humorous and cynical observations about the human condition, featuring many self-referential or self-undermining aphorisms.

Boeke, Kees. *Cosmic View: The Universe in 40 Jumps.* New York: John Day, 1957. A book to instill humility and awe in anyone, as well as a vivid sense of the meaning of the term "astronomical number".

Bibliography

Bombaugh, Charles Carroll. *Oddities and Curiosities of Words and Literature,* edited and annotated by Martin Gardner. New York: Dover, 1961. An antique collection of palindromes, acrostics, pangrams, and so on, for wordmongers and people who love the bizarre fringes of language. (Gardner's footnotes at the back are the best part of the book.)

Boole, George. *The Laws of Thought.* New York: Dover, 1961. A reprint of the old classic from the mid-1850's. The hubris of the title is quite remarkable, especially in light of what 130 years' progress has revealed!

Brams, Steven, Morton D. Davis, and Philip Strafin, Jr. "The Geometry of the Arms Race". *International Studies Quarterly* 23, no. 4 (December 1979): 567–88. Looking at international behavior in terms of the iterated Prisoner's Dilemma and similar payoff matrices.

Brilliant, Ashleigh. *I May Not Be Perfect, but Parts of Me Are Excellent*; *I Have Abandoned My Search for Truth, and Am Now Looking for a Good Fantasy*; *Appreciate Me Now and Avoid the Rush*; *I Feel Much Better, Now That I've Given Up Hope.* Santa Barbara, Calif.: Woodbridge Press, 1979–1984. Four books containing many incisive epigrams about life, death, love, relationships, greed, egotism, loneliness, fear, and so on. None is longer than seventeen words.

Bush, Donald J. *The Streamlined Decade.* New York: George Braziller, 1975. Showing how the style of an era permeates its creations in all media. Full of elegant photos. Compare with the books by Loeb and McCall (see below).

Byrd, Donald. "Music Notation by Computer". Ph.D. thesis, Indiana University Computer Science Department. Bloomington, 1984. About the problems of developing a computer program that will have some "understanding" of the subtleties of music notation. The program, SMUT, produced the musical examples in this book.

Chaitin, Gregory. "Randomness and Mathematical Proof". *Scientific American* 232, no. 5 (May 1975): 47–52. An enlightening way of defining the meaning of "random pattern", and the unexpected and deep resonances with Gödel's incompleteness theorem and other metamathematical results.

Charniak, Eugene, C. K. Riesbeck, and Drew V. McDermott. *Artificial Intelligence Programming.* Hillsdale, N.J.: Lawrence Erlbaum Associates, 1980. Sophisticated ways of using Lisp and Lisp-like languages in artificial-intelligence research.

Collet, Pierre and Jean-Pierre Eckmann. *Iterated Maps on the Interval as Dynamical Systems.* Boston: Birkhäuser, 1980. An in-depth study of the iteration of simple smooth functions on the interval [0,1], and the resulting roads to turbulent behavior as parameters are varied.

Compugraphic Corporation. *A Portfolio of Text and Display Type.* Wilmington, Mass.: Compugraphic Corporation, 1982. A collection of many typefaces (mostly book faces), including several extensions to other alphabets of faces originally conceived only for our alphabet.

Conway, John Horton, Elwyn Berlekamp, and Richard K. Guy. *Winning Ways (for your mathematical plays).* New York: Academic Press, 1982. A two-volume set of remarkable games, some analyzed, some unanalyzed, filled with humorous pictures and an oddball type of creative wordplay that is Conway's hallmark. Included in Volume 2 are discussions of the Cube and Conway's game of Life.

Coueignoux, Philippe. "La reconnaissance des caractères". *La Recherche* 12, no. 126 (October, 1981): 1094–1103. An interesting article about the workings of some computer systems that can read text in a variety of typefaces. The ideas derive from work by Blesser and colleagues (see above). This work is rather practically oriented, and does not come close to giving computers an understanding of letterforms in their full generality.

Csányi, Vilmos. *General Theory of Evolution.* Budapest: Akadémiai Kiadó, 1982. A thorough investigation of the process of simultaneous evolution on both genetic and "memetic" fronts.

Davies, Paul. *God and the New Physics.* New York: Simon & Schuster, 1983. In this book, Davies tackles the biggies of metaphysics: creation, free will, religion, souls, and more. Since he is a good scientist as well as a good writer, his musings are articulate and penetrating.

————. *Other Worlds*. New York: Simon & Schuster, 1980. A popular account, by a highly reliable professional, of the mysteries at the base of quantum mechanics.

Davis, Morton D. *Game Theory: A Nontechnical Introduction*. New York: Basic Books, 1983. An excellent overview of all the main ideas of game theory, including many unresolved issues such as the Prisoner's Dilemma.

Dawkins, Richard. *The Selfish Gene*. New York: Oxford University Press, 1976. A book that views organisms as by-products of a purely molecular-level competition for efficient self-replication. A topsy-turvy, disorienting, yet powerfully revealing viewpoint.

DeLong, Howard. *A Profile of Mathematical Logic*. Reading, Mass.: Addison-Wesley, 1970. A wonderful book on the philosophical and technical issues of logic by someone who knows how to achieve an artistic balance between formalisms and ideas that appeal to the intuition.

Dennett, Daniel C. *Brainstorms: Philosophic Essays on Mind and Psychology*. Cambridge, Mass.: Bradford Books, MIT Press, 1978. A collection of penetrating analyses of problems of mind, brain, and computer models of thought, perception, and sensation. Excellent rebuttals of such figures as B. F. Skinner and J. R. Lucas, and a wonderful "dessert" at the end. Dennett's graceful style and comparative lack of in-references and jargon make this book much more engaging than the average philosophy book.

————. "Can Machines Think?" Unpublished manuscript. A fresh look at the power of the Turing Test, completely in sympathy with my view that the test as originally posed is as valid as ever, despite the doubts of many.

————. "Cognitive Wheels: The Frame Problem of Artificial Intelligence". In *Minds, Machines, and Evolution*. Edited by C. Hookway. New York: Cambridge University Press, 1985. An article about why artificial intelligence is so far from achieving common sense.

————. *Elbow Room: The Varieties of Free Will Worth Wanting*. Cambridge, Mass.: Bradford Books, MIT Press, 1984. One provocative metaphor after another, all building up towards an image of "self-made selves" that enjoy as much free will as it is reasonable to hanker after. Although I resist some of this book's conclusions, I find it the best writing on the subject that I know.

————. "The Logical Geography of Computational Approaches (A View from the East Pole)". Unpublished manuscript. Dennett wittily divides the world of AI approaches into two camps: the orthodox one ("High Church Computationalism"), centered on the "East Pole" (located at MIT), and the unorthodox one ("New Connectionism"), scattered hither and yon.

————. "The Myth of the Computer: An Exchange". *New York Review of Books*. (June 24, 1982): 56. A civil reply to John Searle's biting review of *The Mind's I*.

————. "The Self as a Center of Narrative Gravity". In *Self and Consciousness*, edited by P. M. Cole et al. New York: Praeger, 1985. A wonderful new metaphor for thinking about abstractions such as "I".

Dewdney, A. K. "Computer Recreations: A computational garden sprouting anagrams, pangrams, and few weeds". *Scientific American* 251, no. 4 (October 1984): 20–27. Includes a description of Lee Sallows' pangram machine, concluding with a public challenge to discover a computer-generated pangram.

DeWitt, Bryce S. and Neill Graham, eds. *The Many-Worlds Interpretation of Quantum Mechanics*. Princeton, N.J.: Princeton University Press, 1973. A well-rounded presentation of one of the most disorienting yet irrefutable ways of thinking about reality yet invented.

Dyer, Michael. *In-Depth Understanding*. Cambridge, Mass.: MIT Press, 1983. A book summarizing a Ph.D. project in language understanding that puts together many recent AI ideas about how memory is organized and how flexible control structures fit into the picture.

Dylan, Bob. *Tarantula*. New York: Macmillan, 1971. The poet-singer lets loose some stream-of-consciousness musings that are sometimes reasonably intelligible and sometimes totally wacko.

Edson, Russell. *The Clam Theater*. Middletown, Conn.: Wesleyan University Press, 1973. Strange and surrealistically written fantasies permeated by a tragic vision of life. To me, the most haunting passage is a description of a human head: "this teetering bulb of dread and dream . . .".

Eidswick, Jack. "How to Solve the $n \times n \times n$ Cube". Mathematics and Statistics Department, University of Nebraska, Lincoln, 1982. A useful manual for those beset by generalized cubic frustration.

Endl, Kurt. *Rubik's Cube Made Simple*; *The Pyramid*; *Pyraminx Cube; Impossiball*; *Megaminx*; *Rubik's Master Cube*. Giessen, Germany: Würfel-Verlag GmbH, 1982. These booklets, together with the previous entry, will give you a good reference shelf on Cubology.

Erman, Lee D. et al. "The Hearsay-II Speech-Understanding System: Integrating Knowledge to Resolve Uncertainty". *ACM Computing Surveys* 2, no. 2 (June 1980): 213–53. A review article describing, with a good deal of hindsight, the architecture and performance of what I consider to be the most inspiring work yet done in artificial intelligence.

Evans, Thomas G. "A Program for the Solution of a Class of Geometric-Analogy Intelligence Test Questions". In *Semantic Information Processing*, edited by Marvin Minsky. Cambridge, Mass.: MIT Press (1968): 271–353. Perhaps the first truly large system ever written in Lisp, this impressive project is often pointed to as having "solved" the problem of analogies. How far from the truth that is!

Ewing, John and Czes Kośniowski. *Puzzle It Out: Cubes, Groups, and Puzzles*. New York: Cambridge University Press, 1982. Another lively and mathematically interesting book about the Cube and its variants.

Fahlman, Scott E., Geoffrey Hinton, and Terrence J. Sejnowski. "Massively Parallel Architectures for AI: NETL, Thistle, and Boltzmann Machines". In "Proceedings of the AAAI-83 Conference", Washington, D. C., August, 1983. A comparison of some recent ideas for AI architectures, including ones inspired by physics, in which statistical emergence plays a central role.

Falletta, Nicholas. *The Paradoxicon*. New York: Doubleday, 1983. An excellent collection of paradoxes of all sorts, ranging from Zeno and Epimenides to modern-day voting paradoxes, optical illusions, and antinomies in the philosophies of science and mathematics.

Fauconnier, Gilles. *Espaces Mentaux: Aspects de la construction du sens dans les langues naturelles*. Paris: Les Editions de Minuit, 1984. A probing inquiry into how we make sense of frame-crossing and counterfactual statements of this sort: "If Clark Gable had been a woman, Scarlett O'Hara would have been a man". Full of ideas about reference, identity, slippability, and essence.

Feigenbaum, Mitchell. "Universal Behavior in Nonlinear Systems". *Los Alamos Science* 1, no. 1 (Summer 1981): 4–27. An excellent introduction to the studies of chaos that come from iterating simple functions. Excellent graphics, some of which are reproduced in Chapter 16.

Feldman, Jerome and Dana Ballard. "Connectionist Models and Their Properties". *Cognitive Science* 6, no. 3 (July–September 1982): 205–54. One of several intriguing approaches to computational collectivism currently being investigated in cognitive science. Like most such models, this one is strongly influenced by ideas about human perception.

Feynman, Richard P. *The Character of Physical Law*. Cambridge, Mass.: MIT Press, 1967. The transcripts of five marvelous lectures delivered at Cornell University in 1964 by a physicist who is not only sharp as a nail but also sparklingly witty.

Feynman, Richard P., Robert B. Leighton, and Matthew Sands. *The Feynman Lectures in Physics*. Reading, Mass.: Addison-Wesley, 1965. Lectures given to beginning students, often more suitable for advanced students, filled with the excitement of science and the peppery observations of Feynman.

Franke, H. W. *Computer Graphics—Computer Art*. New York: Phaidon, 1971. By far the best collection of examples and discussion of computers and art I have yet seen, even though it is quite old by now.

Frey, Alexander H., Jr. and David Singmaster. *Handbook of Cubik Math*. Hillside, N.J.: Enslow, 1982. The group-theoretical ideas behind the cube are developed systematically, for possible use in an advanced high-school or college-level course.

Friedman, Daniel P. *The Little LISPer.* Chicago: Science Research Associates, 1974. An engaging and elegant introduction to the recursive ideas of Lisp, by one of its most ardent and accomplished practitioners.

Fromkin, Victoria A, ed. *Errors in Linguistic Performance: Slips of the Tongue, Ear, Pen, and Hand.* New York: Academic Press, 1980. A compendium of articles on how we can infer general mechanisms of thought from observing everyday surface-level quirks that show up ubiquitously in speech, typing, handwriting, listening, and so on.

Frutiger, Adrian. *Type Sign Symbol.* Zurich: Editions ABC, 1980. This book by the creator of Frutiger and Univers, two of today's most popular and elegant sans-serif typefaces, deals with the interaction of technical constraints and artistic creation, and features many beautiful letterforms and styles.

Gablik, Suzi. *Progress in Art.* New York: Rizzoli International, 1976. A heretical theory suggesting that there is a reason why art is where it is today, and that art follows a comprehensible trajectory in some abstract space, even if it is explicable only after the fact.

Gardner, Martin. *Fads and Fallacies.* New York: Dover, 1952. A book that comes about as close as one could hope to being a course on common sense. Many chapters of this classic bunk-puncturer should be part of the cultural education of everyone.

Gardner, Martin. "Mathematical Games: White and brown music, fractal curves, and one-over-f fluctuations". *Scientific American* 238, no. 4 (April 1978): 16–32. A solid introduction to the concepts of fractals and multi-level statistical structures.

Gardner, Martin. *Science: Good, Bad, and Bogus.* Buffalo: Prometheus Press, 1981. Following in the footsteps of *Fads and Fallacies,* this book sharp-swordedly decapitates one dragon of hokum after another—and yet sadly, like a hydra's many heads, they always seem to pop back up again.

Gardner, Martin. *Wheels, Life, and Other Mathematical Amusements.* New York: W. H. Freeman, 1983. Gardner's final collection of his sparkling *Scientific American* columns, showing why mathematics is the ultimate metamagic.

Gebstadter, Egbert B. *Thetamagical Memas: Seeking the Whence of Letter and Spirit.* Perth: Acidic Books, 1985. A curious pot-pourri, bloated and muddled—yet remarkably similar to the present work. This is a collection of Gebstadter's monthly rows in *Literary Australian* together with a few other articles, all with prescripts. Gebstadter is well known for his love of twisty analogies, such as this one (unfortunately not found in his book): "Egbert Gebstadter is the Egbert Gebstadter of indirect self-reference."

Geisel, T. and J. Nierwetberg. "Universal Fine Structure of the Chaotic Region in Period-Doubling Systems". *Physical Review Letters* 47, no. 14 (October 5, 1981): 975–78. An investigation turning up remarkable order deep in the heart of chaos.

Gentner, Dedre. "Structure-Mapping: A Theoretical Framework for Analogy". *Cognitive Science* 7 (1983): 155–70. One in a series of papers attempting to draw guidelines that will help distinguish good analogies from bad.

Gödel, Kurt. *On Formally Undecidable Propositions in "Principia Mathematica" and Related Systems.* Translated by Bernard Meltzer, edited and with an introduction by R. B. Braithwaite. New York: Basic Books, 1962. The fundamental work that opened up worlds to logicians, philosophers, computer scientists, and others.

Golden, Michael. "Don't Rewrite the Bible". *Newsweek* (November 7, 1983): 47. This piece sounds almost like my "Person Paper" in the way it exploits crude mockery of progressive new usages in order to make oppressive old usages sound good. The difference is that Golden is not being sarcastic.

Golomb, Solomon. "Rubik's Cube and Quarks". *American Scientist* 70, no. 3 (May–June 1982): 257–59. In which Golomb puts forth his analogy between twisted corners on the Cube and fractionally charged subatomic particles.

Gonick, Larry and Mark Wheelis. *A Cartoon Guide to Genetics.* New York: Barnes & Noble, 1983. A short and snappy—but most informative and entertaining—overview of molecular biology, genetics, and their human import.

Gould, Stephen Jay. *Hen's Teeth and Horse's Toes.* New York: W. W. Norton, 1983. Written in a lively and engaging way, this book (like its predecessors, *Ever Since Darwin* and

The Panda's Thumb) relates with great clarity the issues of evolution as illustrated in nature's unbelievable variety of quirks.

Gray, J. Patrick and Linda Wolfe. "The Loving Parent Meets the Selfish Gene". *Inquiry* 23: 233–42. A cogent set of arguments against the pessimistic theses of Louis Pascal, to which he then replies (see below).

Grossman, I. and W. Magnus. *Groups and Their Graphs*. New York: Random House New Mathematical Library, 1975. An excellent visual introduction to group theory, featuring so-called "Cayley diagrams", which lay bare the structure of a group in a marvelously clear way.

Ground Zero. *Nuclear War: What's in It for You?* New York: Pocket Books, 1982. A dispassionate, calmly reasoned discussion for "just-plain folk" of what it means for countries to make and threaten each other with nuclear weapons. The best primer that I have come across on the subject, and one that I wish everyone would read and take to heart.

Guillemin, Victor. *The Story of Quantum Mechanics*. New York: Scribner's, 1968. A thoughtful survey of where quantum mechanics came from and what it has revealed about the physical world, concluding with a long discussion of the philosophical implications of quantum mechanics for such things as free will and causality.

Haab, A., A. Stocker, and W. Hättenschweiler. *Lettera 1* and *Lettera 2*. Arthur Niggli, 1954 and 1961. Some deliciously fanciful alphabets are featured in these books. Books like these ought to make anyone marvel at the fluidity of letters and human perception.

Hachtman, Tom. *Double Takes*. New York: Harmony Books, 1984. Two-in-one caricatures, done in a very satisfying way. It makes one wonder: could a computer ever do this kind of thing?

Hansel, C. E. M. *ESP and Parapsychology: A Critical Re-Evaluation*. Buffalo: Prometheus Books, 1980. A scholar looks at the claims of parapsychologists and finds them wanting.

Hanson, Norwood Russell. *Patterns of Discovery*. New York: Cambridge University Press, 1969. A philosophically rich discussion of how physical concepts progressed by fits and starts through the centuries, culminating in the strange world of elementary particles that we are now finding are not so elementary.

Hardin, Garrett. "The Tragedy of the Commons". *Science* 162 , no. 3859 (December 13, 1968): 1243–48. Using the metaphor of shared grazing land that gets ruined by overuse without any specific person being responsible, Hardin argues with great force that people's narrow-minded selfishness will drive us to extinction unless we band together and form strong organizations dedicated to global ecological goals, particularly population control.

Harel, David. "Response to Scherlis and Wolper". *Communications of the ACM* 23, no. 12 (December 1980): 736–37. An amusing list of tricky self-references known to Harel, sporting a footnote claiming to be the first published example of "a reference to the very self-reference being discussed in the very letter being read!"

Hart, H. L. A. "Self-Referring Laws". In *Festskrift Tillägnad Karl Olivecrona*. Stockholm: Kungliga Boktryckeriet, P. A. Norstedt och Söner, 1964. A reply to Alf Ross (see below), in which Hart, perhaps today's leading philosopher of law, claims that self-amending laws are legally possible.

Harth, Erich. *Windows on the Mind: Reflections on the Physical Basis of Consciousness*. New York: William Morrow, 1982. Lively discussions of brain, mind, consciousness, and how they all tie together, by a physicist with a philosophical bent.

Haugeland, John, ed. *Mind Design: Philosophy, Psychology, and Artificial Intelligence*. Cambridge, Mass.: Bradford Books, MIT Press, 1981. A self-proclaimed sequel to Alan Ross Anderson's *Minds and Machines*, now that artificial intelligence has had a couple of decades to develop. This volume is of uneven quality, but it contains many articles sure to provoke thought.

Heiser, Jon F. et al. "Can Psychiatrists Distinguish a Computer Program Simulation of Paranoia from the Real Thing? The Limitations of Turing-Like Tests as Measures of the Adequacy of Simulations". *Journal of Psychiatric Research* 15, no. 3 (1979): 149–62. Researchers who developed the program called "Parry" claim that the Turing Test's validity is cast in doubt because Parry passed a pseudo-Turing Test.

Hinton, Geoffrey and James Anderson, eds. *Parallel Models of Associative Memory.* Hillsdale, N.J.: Lawrence Erlbaum Associates, 1981. A collection of articles exploring different models, all of which are based on the thesis that human perception and memory are statistically emergent phenomena and are best modeled as such.

Hintze, Wolfgang. *Der Ungarische Zauberwürfel.* Berlin: VEB Deutscher Verlag der Wissenschaften, 1982. This is an excellent book about the Cube, including new mathematical results on subgroups. Serious cubists who can read German should obtain this book.

Hobby, John, and Gu Guoan. "A Chinese Meta-Font". Stanford, Calif.: Stanford University Computer Science Department Technical Report STAN-CS-83-974, 1983. A description of a system for creating Chinese characters whose style is "tunable" to some extent. This system is very similar in some ways to Hàn Zì, the variable-style Chinese-character-generation system developed by David Leake and myself at Indiana, at about the same time.

Hodges, Andrew. *Alan Turing: The Enigma.* New York: Simon & Schuster, 1983. The definitive biography of this extremely important figure in mathematics and computer science, written with empathy, insight, and charm.

Hofstadter, Douglas R. *Ambigrams.* Forthcoming in 1985 from the Centre d'Art Contemporain (Geneva, Switzerland). A collection of ambigrams (or "inversions", as Scott Kim prefers to call them) aptly subtitled "A Panoply of Palindromic Pinwheels". The foreword is by Scott (to whose book I symmetrically wrote the foreword). Also included are a lecture on ambigrams (in French) and an interview (in English).

———. "Analogies and Metaphors to Explain Gödel's Theorem". *Two-Year College Mathematics Journal* 13, no. 2 (March, 1982): 98–114. A collection of simple images and ideas that help to build up one's intuitions to the point where one can fully digest the intricacies of Gödel's clever construction.

———. "The Architecture of Jumbo". In "Proceedings of the Second Machine Learning Workshop", Monticello, Illinois, 1983. A description of the architecture of a biologically-inspired AI system that exploits parallelism and randomness in order to build up "well-chunked wholes" out of isolated parts and then to allow them to seek maximal "happiness" by internally reconfiguring themselves in a manner governed by a computational temperature.

———. "The Copycat Project: An Experiment in Nondeterminism and Creative Analogies". Cambridge, Mass.: MIT Artificial Intelligence Laboratory AI Memo 755, April 1984. A description of ongoing research in my group, first at Indiana, then at MIT, and now at Michigan. The goal is to make a system capable of doing analogies in a tiny domain, but with the insight (and short sight) of humans. Describing how the Jumbo architecture can be adapted to this task is the burden of this paper.

———. "512 Words on Recursion". *Math Bulletin,* Bronx High School of Science (1984): 18–19. A recursively structured article, which quotes a compressed version of itself (in which, of course, a more compressed version is quoted, etc.).

———. *Gödel, Escher, Bach: an Eternal Golden Braid.* New York: Basic Books, 1979. Originally titled "Gödel's Theorem and the Human Brain", this book tries to build up the concepts necessary to show how ideas that arise in Gödel's proof will one day be at the core of explanations of consciousness and the meaning of the word "I". Analogy, humor, and contrapuntal dialogues are essential features of *GEB.*

———. "Gridfonts". Unpublished manuscript, 1984. A still-growing collection of (currently) about 300 skeletal typefaces—that is, distinctive and artistically consistent ways of rendering all 26 letters in a highly confined grid. The purpose is to provide material for discussion of the roots of style and the fluid nature of categories.

———. "On Seeking Whence". Unpublished manuscript, 1982. A discussion of the immense difficulty of extrapolating linear patterns, and its relationship to such deeply human experiences as scientific induction, analogy-making, understanding music, and creativity.

———. "Poland: A Quest for Personal Meaning". *Poland* (May, 1981): 42–47. A compression of a longer piece called "Poland: A Mythical Quest", describing my powerful emotional reactions to Poland on my first visit there, in 1975.

————. "In Search of Essence". Unpublished manuscript, 1983. A discussion of what the "essence" of a written passage is, by way of a description of problems encountered in translating the dialogue *Contracrostipunctus* (from *Gödel, Escher, Bach*) into French.

————. "Simple and Not-So-Simple Analogies in the Copycat Domain". Unpublished manuscript. A collection of 84 analogy problems in the Copycat domain that show how diverse the tiny alphabetic universe can be.

Hofstadter, Douglas R., Gray Clossman, and Marsha Meredith. "Shakespeare's Plays Weren't Written by Him, but by Someone Else of the Same Name". Bloomington: Indiana University Computer Science Department Technical Report 96, 1980. "A study of intensionality in frame-based representation systems" is the subtitle. The paper's main purpose is to discuss the relationship between slots that an entity fills, and that entity's identity. The notion of "Core ID" is presented and explored in an analysis of the title sentence. Closely related to the problems discussed by Fauconnier (see above).

Hofstadter, Douglas R. and Daniel C. Dennett, eds. *The Mind's I: Fantasies and Reflections on Self and Soul.* New York: Basic Books, 1981. Our purpose was to jolt people on all sides of the fence in this anthology of fictional and evocative pieces on the curious fact (or illusion) that something we call an "I" is somehow connected to some hunk of matter floating somewhere and somewhen in some universe.

Holland, John et al. *Induction: Processes of Inference, Learning, and Discovery.* Forthcoming. A wide-ranging study of how learning takes place in genuine human minds, and how aspects of it might be modeled in computer simulations. This pioneering work is by a computer scientist, two psychologists, and a philosopher, and its breadth reflects their diversity.

Hollis, Martin and Steven Lukes, eds. *Rationality and Relativism.* Cambridge, Mass.: MIT Press, 1982. Showing how scholars can get mired in the endlessly circular reasoning that arises when any system of belief tries to provide its own justification. Reminiscent of "Münchhausens Zopf": the quandary of Baron Münchhausen trying to pull himself out of a quicksand quagmire simply by tugging on his own braid.

Hopfield, J. J. "Neural networks and physical systems with emergent computational abilities". In *Proceedings of the National Academy of Sciences* 79, Washington, D.C., 1982: 2554–58. A system that makes use of statistics and parallelism to model familiar properties of the mind: reconstructing a full memory from a partial one, generalizing, and so on.

How to Learn Lettering. Hong Kong: Nạm San Publisher, n.d. A collection of Chinese characters in modern artistic styles, mostly for the use of advertisers, but also a treasure trove for those interested in the elusive "sameness" we see in wildly different renderings of a single Platonic essence.

Huff, William. "A Catalogue of Parquet Deformations". School of Architecture, State University of New York at Buffalo. Approximately 35 parquet deformations created by students in William Huff's studio, assembled and commented on by Huff.

Hughes, Patrick and George Brecht. *Vicious Circles and Infinity: An Anthology of Paradoxes.* New York: Penguin, 1975. A delightful and provocative collection of paradoxical material and choice epigrams embodying paradoxes of every conceivable sort. Includes lengthy discussions of the paradox of the Unexpected Examination, also known as the "Hangman's Paradox".

Huneker, James Gibbons. *Chopin: The Man and His Music.* New York: Dover, 1966. Originally published around the turn of the century, this romantic biography features florid prose so rich that it verges on meaninglessness, and yet it is wonderfully evocative.

Jaspert, W. Pincus, W. Turner Berry, and A. F. Johnson. *The Encyclopaedia of Type Faces.* Poole, England: Blandford Press, 1983. An engrossing catalogue of over 1,000 typefaces. One of its unique features is its thorough crediting of designers: the people whose exquisitely developed sense of line and space is far too often completely taken for granted. An index to designers and an index to typefaces are included. This way you can get a sense for the stylistic range of various designers.

Johnson-Laird, P. N. and P. C. Wason, eds. *Thinking.* New York: Cambridge University Press, 1975. A diverse collection of pieces by top-notch authors about nearly all aspects of human thought, including imagery, perception, inference, categories, memory, language, and so on.

Kadanoff, Leo. "Roads to Chaos". *Physics Today* 36, no. 12 (December 1983): 46–53. A good summary of the connection between mathematical approaches to chaos and the physical phenomena to be explained.

Kamack, H. J. and T. R. Keane. "The Rubik Tesseract". Unpublished manuscript, 1982. A companion piece to the one by Eidswick (see above), this discussion of a $3 \times 3 \times 3 \times 3$ "Rubik's hypercube" is a *tour de force* in visualization—for its readers almost as much as for its authors!

Kahneman, Daniel and Amos Tversky. "The Simulation Heuristic". In *Judgment Under Uncertainty: Heuristics and Biases*, edited by Daniel Kahneman, P. Slovic, and A. Tversky. New York: Cambridge University Press, 1982: 202–8. In what ways are people inclined to let situations mentally glide into alternate versions of themselves, and how do subtle cognitive pressures modify those tendencies? Two outstanding cognitive psychologists here describe their findings about this sort of slippability or "alternity" in their subjects' minds.

Kanerva, Pentti. *Self-Propagating Search: A Unified Theory of Memory*. Stanford, Calif.: Center for the Study of Language and Information, Technical Report, Stanford University, 1984. A neurocomputational theory of memory: How to make similarity-sensitive, reconstructive software out of statistically robust addressing hardware. This may sound a bit abstruse, but in my opinion Kanerva's elegant work constitutes one of the most important steps toward reconciliation of fluidity and mechanism—mind and brain —that I have seen taken.

Kennedy, Paul E. *Modern Display Alphabets*. New York: Dover, 1974. An excellent collection of visually attractive letterforms in styles of all sorts, and exhibiting all degrees of wildness.

Kim, Scott E. "The Impossible Skew Quadrilateral: A Four-Dimensional Optical Illusion". In *Proceedings of the 1978 AAAS Symposium on Hypergraphics: Visualizing Complex Relationships in Art and Science*. Boulder, Colo.: Westview Press, 1978. Like the work by Kamack and Keane (see above), this article takes a powerful visual imagination to write or to read. The familiar "impossible triangle" is here carried one step further, by a bold process of analogy. The writing style is full of typical Kimian recursive trickery and parallel passages.

———. *Inversions*. Peterborough, N.H.: Byte Books, 1981. A collection of inversions (or "ambigrams", as I prefer to call them) aptly subtitled "A Catalogue of Calligraphic Cartwheels". The foreword is by myself (to whose book Scott symmetrically wrote the foreword). The prose is as full of illusions and tricks of parallelism as the drawings are, although it is not immediately apparent.

———. "Noneuclidean Harmony". In *The Mathematical Gardner*, edited by David A. Klarner. Belmont, Calif.: Wadsworth International/Prindle, Weber, and Schmidt, 1981. A serious treatise on atonal geometry masquerading as a piece of humor. Describes Georg Cantor's important results on supersonic pitches, 2 against 3 correspondence, and uncountable rhythms.

———. "Visual Art: The Creative Cycle". *Response*, no. 1 (December, 1981). Minnesota Artists Exhibition Program. A short article putting forth a thesis about creative activity, at the heart of which is a loop similar to the one I describe in the *P.S.* to Chapter 12 of this book.

Kirkpatrick, S., C. D. Gelatt, Jr., and M. P. Vecchi. "Optimization by Simulated Annealing". *Science* 220, no. 4598 (May 13, 1983): 671–80. An article describing a new statistical technique for seeking the optimal state of a system with a huge number of degrees of freedom, based on local improvement modulated by occasional local degradation, the amount of which is determined by a "cooling schedule", analogous to the annealing of a metal to strengthen it.

Kleppner, Daniel, Michael G. Littman, and Myron L. Zimmerman. "Highly Excited Atoms". *Scientific American* 244, no. 5 (May 1981): 130–49. A bridge between microscopic quantum-mechanical phenomena and macroscopic classical phenomena is provided by the study of "Rydberg atoms", with which this paper is concerned.

Bibliography

Knuth, Donald. "The Concept of a Meta-Font". *Visible Language* 16, no. 1 (Winter 1982): 3–27. Knuth describes his program, METAFONT, for creating an entire family of typefaces at one fell swoop, by parametrizing all letters of the alphabet with a consistent set of parameters. This is the paper that triggered my response printed as Chapter 13.

————. *TEX and Metafont: New Directions in Typesetting.* Bedford, Mass.: Digital Press, 1979. Knuth describes in detail how to use his programs that facilitate typesetting and letter design.

Kobylańska, Krystyna. *Chopin in His Own Land.* Cracow: Polish Music Publishers, 1956. A collection of hard-to-find Chopiniana, beautifully reproduced in large format.

Koestler, Arthur. *The Act of Creation.* New York: Dell, 1964. The great novelist and amateur psychologist delves into his own mind to come up with interesting but often wrong-headed speculations on humor, creativity, and insight.

————. *The Roots of Coincidence: An Excursion into Parapsychology.* New York: Vintage, 1972. Koestler reveals his belief in the occult, somewhat reminiscent of Arthur Conan Doyle's belief in fairies.

Kolata, Gina. "Does Gödel's Theorem Matter to Mathematics?" *Science* 218 (November 19, 1982): 779–80. An excellent description of how the specification of what today would be considered "large integers" depends on abstruse concepts from mathematical logic.

Koning, H. and J. Eizenberg. "The language of the prairie: Frank Lloyd Wright's prairie houses". *Environment and Planning B,* 8 (1981): 295–323. A description of how pseudo-Frank-Lloyd-Wright houses can be generated from a "shape grammar".

Kripke, Saul. *Naming and Necessity.* Cambridge, Mass.: Harvard University Press, 1972. This theory of extensions, intensions, "rigid designators", and identity is in direct opposition to my views (put forward in the "Shakespeare" paper cited above) that the roots of identity come from patterns of embeddedness in a network.

Kuwayama, Yasaburo. *Trademarks and Symbols, Volume 1: Alphabetical Designs.* New York: Van Nostrand Reinhold, 1973. A celebration of the craziness of letterforms, this collection contains letters to boggle anyone's mind.

Larcher, Jean. *Fantastic Alphabets.* New York: Dover, 1976. The letterforms in Kuwayama's book are highly imaginative, but by comparison, the letterforms in Larcher's book are downright zany. It just goes to show that a computer program that could deal with letters in their full complexity would be able to deal with the entire world.

Lehninger, Albert. *Biochemistry,* 2d. ed. New York: Worth, 1975. A well-written treatise covering the huge field of biochemistry in a lucid manner.

Lennon, John. *In His Own Write* and *A Spaniard in the Works.* New York: Simon & Schuster, 1964 and 1965. This Beatle had a wonderful sense of silliness, all his own. Stories, pictures, poems—even a little drama or two. Perhaps better than the Beatles' music, if I may venture a heretical opinion.

Letraset, Inc. *Graphic Art Materials Reference Manual.* Paramus, N.J.: Letraset, Inc., 1981. This is the best inexpensive typeface manual available, and is well worth the price. It contains nearly all the common typefaces, as well as many unusual ones. This is a great way to get into typefaces.

Letraset, Inc. *Letraset Greek Series.* Athens: A. Pallis, 1984. This to me ia an amazing catalogue of typefaces, because each and every one of them represents a "transalphabetic leap". Here we see such faces as Blanchard, Futura Black, Helvetica, Korinna, Optima, Souvenir, University Roman, Zipper, and a number of others—*all in Greek!* All I can say is, "Wow!"

Levin, Michael. "Mathematical Logic for Computer Scientists". Cambridge, Mass.: MIT Laboratory for Computer Science Technical Report LCS TR 131, June, 1974. An unorthodox treatment of logic for those more comfortable with Lisp than with number theory.

Lipman, Jean and Richard Marshall. *Art about Art.* New York: E. P. Dutton, 1978. An amusing annotated collection of self-conscious art, including pieces by Roy Lichtenstein, Andy Warhol, Robert Rauschenberg, Jasper Johns, Robert Arneson, Tom Wesselmann, Larry Rivers, Mel Ramos, Peter Saul, and many others.

Loeb, Marcia. *New Art Deco Alphabets.* New York: Dover, 1975. Showing how somebody can perfectly recreate the spirit of an era in many different ways. These original alphabets all look like they came straight out of the 1930's. Compare with the books by Bush (see above) and McCall (see below).

Lucas, J. R. "Minds, Machines, and Gödel". Reprinted in *Minds and Machines,* edited by Alan Ross Anderson. Englewood Cliffs, N.J.: Prentice-Hall, 1964. The article that launched a thousand rebuttals. Although I find its conclusions totally unjustified, the issues are well raised and deserve to be thought about far more. Closely related to the issues discussed in the book by Webb (see below).

Machlup, Fritz and Una Mansfield, eds. *The Study of Information: Interdisciplinary Messages.* New York: Wiley-Interscience, 1983. A multi-faceted collection of articles by high-level scholars about information theory, cybernetics, artificial intelligence, cognitive science, librarianship, and more. Many of the pieces are statements of opinion, and conflict with one another. This volume includes the exchange between Allen Newell and myself, which is continued in the *P.S.* to my Chapter 26.

Mandelbrot, Benoît. *The Fractal Geometry of Nature.* New York: W. H. Freeman, 1982. This latest presentation of Mandelbrot's visions is rich in imagery and ideas. Fractal shapes constitute one of the twentieth century's most fertile mathematical playgrounds.

Marek, George R. and Maria Gordon-Smith. *Chopin.* New York: Harper & Row, 1978. One of the many biographies of Chopin in English. I find it to be balanced and well written.

Marx, George, Eva Gajzágó, and Peter Gnädig. "The universe of Rubik's cube". *European Journal of Physics* 3 (1982): 34–43. The implications of extending the concept of entropy to the surface of the Cube are reported on, with the results of several statistical studies done on computer.

Max, Nelson. *Space-Filling Curves.* Chicago: International Film Bureau, Inc., 1974. A marvelous movie done with computer graphics showing the fascination of some of the earliest-discovered fractal curves. The computer-produced music adds extra appeal and eerieness.

May, Robert. "Simple Mathematical Models with Very Complicated Dynamics". *Nature* 261, no. 5560 (June 10, 1976): 459–67. An early review article about the chaos one is led to when one iterates simple functions on the interval [0,1]. Very informative, except for the reversal of two figures!

Maynard-Smith, John. *Evolution and the Theory of Games.* New York: Cambridge University Press, 1982. Showing how new depths of understanding of the mechanisms of evolution can come from modeling the interactions of organisms in a common environment in terms of mathematical game theory.

McCall, Bruce. *Zany Afternoons.* New York: Alfred A. Knopf, 1982. A genius for capturing the flavor of things turns all his talents to recreating the twenties, thirties, forties, and fifties—not only in America but also abroad. This very funny book is an amazing achievement. Compare with the books by Bush and Loeb (see above).

McCarthy, John. "Attributing Mental Qualities to Machines". In *Philosophical Perspectives on Artificial Intelligence,* edited by Martin Ringle, 161–95. Atlantic Highlands, N. J.: Humanities Press, 1979. Under what circumstances does it make sense to speak of machines having desires or beliefs? When to adopt this "intentional stance" is here discussed by one of the pioneers of artificial intelligence.

————. "History of Lisp". In *History of Programming Languages,* edited by Richard L. Wexelblatt, 217–23. New York: Academic Press, 1980. Lisp's genesis as recalled by its inventor.

McCarty, L. T. and N. S. Sridharan. "A Computational Theory of Legal Argument". New Brunswick, N.J.: Rutgers University Computer Science Department Technical Report LPR-TR-13, January 1982. A professor of law and a computer scientist work together to implement "prototype deformation", which they consider the key to computer modeling of argumentation by precedent.

McClelland, James L., David E. Rumelhart, and Geoffrey E. Hinton, eds. *Interactive Activation: A Framework for Information Processing.* Forthcoming. A collection of articles on statistically emergent parallel computation featuring "temperature" as a regulator of randomness.

McKay, Michael D. and Michael S. Waterman. "Self-Descriptive Strings". *Mathematical Gazette* 66, no. 435 (1982): 1–4. A mathematical study of a simple class of sentences that attempt to inventory themselves.

Meehan, James R. "Tale-Spin, an Interactive Program that Writes Stories". In *Proceedings of the 5th International Conference on Artificial Intelligence–1977,* Vol. 1: 91–98. Pittsburgh: Department of Computer Science, Carnegie-Mellon University. One of the funniest (and therefore most sensible) articles ever written on computers and their "understanding" of language. Meehan explains how, in one "mis-spun tale", his program made gravity fall into a river and drown. Meehan's discussions of mis-spun tales in general are among the most illuminating passages I know on that perennial vexation: the fact that AI programs stubbornly resist acquiring common sense, no matter how hard or how often they are spanked.

Miller, Casey and Kate Swift. *The Handbook of Nonsexist Writing for Writers, Editors, and Speakers.* New York: Barnes and Noble, 1980. A first-class book. It is a scandal that this book is not found on every newsman's desk. Thinking about these issues is not only socially important, but challenging and fascinating.

———. *Words and Women.* New York: Doubleday, Anchor Press, 1977. A magnificent and devastating job of showing the absurdity of thinking that women have been dealt a fair hand by our society. If language is the thermometer of society, then this book reveals that collectively, we are gravely ill.

Mills, George. "Gödel's Theorem and the Existence of Large Numbers". Northfield, Minnesota: Mathematics Department, Carleton College, October 1981. Showing how the descriptions of some stupendously large numbers can be obtained via recursive definitions of functions and the process of diagonalization. Repeated jootsing leads to some remarkable destinations!

Minsky, Marvin. "A Framework for Representing Knowledge". In *The Psychology of Computer Vision,* edited by P. H. Winston. New York: McGraw-Hill, 1975. The paper that defined such now-central notions to AI as frames, slots, and default assumptions.

———. "Matter, Mind, and Models". In *Semantic Information Processing,* edited by Marvin Minsky. Cambridge, Mass.: MIT Press, 1968. Some cogent speculations about the meaning of the word "I" in computational terms.

———. "Why People Think Computers Can't". *AI Magazine* (Fall 1982): 3–15. With his usual unusual insight, Minsky tears people apart for not understanding how to think about thinking (or about machines, for that matter).

Mondrian, Piet. *Tout l'Œuvre Peint de Piet Mondrian.* Paris: Flammarion, 1976. For a grand overview of the evolution of the style of one painter, I know of no better book than this, which traces Mondrian from his earliest representational paintings to his most abstract and geometrical ones, revealing the sweep to be continuous and logical, but no less dramatic for that.

Morrison, Philip and Phylis, and the Office of Charles and Ray Eames. *Powers of Ten.* New York: Scientific American Books, 1983. Kees Boeke's inspirational *Cosmic View* (see above) revisited some thirty years later, with considerably more commentary. A charming and worthy successor.

Myhill, John. "Some Philosophical Implications of Mathematical Logic: Three Classes of Ideas". *Review of Metaphysics* 6, no. 2 (December 1952): 165–98. In 1982 I met Myhill and asked him about this paper. He told me he considered it to be a piece of junk. That astonished me, since I consider it to be a thoughtful and important piece of philosophizing making use of mathematical metaphors, something that hardly anyone dares to do. You never know how someone will evaluate their own work 30 years later!

Nagel, Ernest, and J. R. Newman. *Gödel's Proof.* New York: New York University Press, 1958. A gracious and highly accessible introduction to the twists in Gödel's reasoning, as well as to the philosophical issues surrounding his work.

Nakanishi, Akira. *Writing Systems of the World.* Rutland, Vt.: Tuttle, 1980. For a sampling of the many different spirits residing in letterforms, see this book. It features reproductions of newspaper pages showing each writing system in several different styles.

Newell, Allen. "Physical Symbol Systems". *Cognitive Science* 4, no. 2 (April–June 1980): 135–83. A lengthy article putting forth the orthodox dogma on which artificial intelligence has traditionally considered itself founded: the idea that a universal computer embodies all the prerequisites for intelligent behavior.

Norman, Donald. "Categorization of Action Slips". *Psychology Review* 88, no. 1 (January 1981): 1–15. A delightful compendium of types of error that people commit in performing everyday actions such as putting water on to boil, answering the telephone, unbuckling one's seatbelt, driving home from work, and so on. There is remarkable regularity behind the seeming chaos, and Norman's purpose is to chart and exploit that regularity in order to reveal hidden mechanisms of thought.

Nozick, Robert. *Philosophical Explanations.* Cambridge, Mass.: Harvard University, Belknap Press, 1981. A very wide-ranging book on philosophy, intended for lay readers as well as for professionals. It covers matters from personal identity and the meaning of reference to free will and morality, and only occasionally lapses into brief spasms of absolutely inscrutable jargon.

Parfit, Derek. *Reasons and Persons.* Oxford: Clarendon Press, 1984. A very thoughtful treatise on moral dilemmas and ethical behavior, rooted in close consideration of the deepest roots of caring: why do I care about my present and future self? Parfit knows that to make any serious attempt to answer this riddle, one must look long and hard at the meaning of the word "I" in the real world and in many counterfactual ones, which he does with skill and insight.

Pascal, Louis. "Human Tragedy and Natural Selection". *Inquiry* 21: 443–60. Taking up where Garrett Hardin left off (see above), Pascal paints a gloomy picture of a population explosion as the natural outcome of selection itself, something built into the nature of societies just as deeply as the sexual drive is built into individuals.

———. "Rejoinder to Gray and Wolfe". *Inquiry* 23: 242–51. A bitter indictment of the human race's apathy before visibly onrushing catastrophe.

Peattie, Lisa. "Normalizing the Unthinkable". *Bulletin of the Atomic Scientists* 40, no. 3 (March 1984): 32–36. Like Pascal, Peattie is concerned with people's apparent ability to turn off their sensitivities and to focus on the very local to such an extent that great tragedies are allowed to ensue. In a striking analogy, she likens the current public apathy about the arms madness to the inhumanity of collaborators in Hitler's concentration camps.

Pérec, Georges. *La Disparition.* Paris: Editions Denoël, 1969. If anything, writing without 'e's is harder in French than in English, yet here is an entire novel in that bizarre dialect. Naturally enough, its subject is the mysterious disappearance of item number five in a collection of twenty-six objects. It was probably inspired by the 'e'-less novel *Gadsby,* written in English by Ernest Vincent Wright in the late 1930's.

Perfect, Christopher and Gordon Rookledge. *Rookledge's International Typefinder.* New York: Frederick C. Beil, 1983. A wonderful (though expensive) compendium of typefaces, indexed in such a way that you can look a typeface up by its features, thus allowing you to home in quickly on an unknown specimen instead of spending hours leafing through catalogues. Lovers of my "horizontal and vertical problems" will delight in this book.

Phillips, Tom. *A Humument: A Treated Victorian Novel.* New York: Thames & Hudson, 1980. Like a child who has covered a wall with colorful crayon drawings, Tom Phillips has completely obliterated the pages of an old novel (*A Human Document* by W. H. Mallock) with his colorful scribblings, and only here and there do traces of the original show through. A droll stunt!

Poincaré, Henri. "On Mathematical Creation". In *The World of Mathematics,* vol. 4, edited by James R. Newman, 2041–50. New York: Simon & Schuster, 1956. A lecture presented to the Psychological Society in Paris in the early part of this century. Anticipating developments in cognitive science some eighty years later, Poincaré speculates on the nature of the events taking place inside his skull as he makes mathematical discoveries.

Pólya, George. *How to Solve It.* New York: Doubleday, 1957. Pólya, like Poincaré a mathematician fascinated by thought processes, attempts here to give recipes for how to attack mathematical problems. The problem with this is that there is—and can be—no failsafe recipe. Even trying to give guidelines is probably futile. The nose that smells the right route is simply rare, and there are no two ways about it.

Post, Emil. "Absolutely Unsolvable Problems and Relatively Undecidable Propositions: Account of an Anticipation". In *The Undecidable,* edited by Martin Davis, 338–443. New York: Raven, 1965. In this paper, Post concludes that mathematicians' thought processes are essentially creative and non-mechanizable. Related to the article by Myhill (see above).

Poundstone, William. *The Recursive Universe: Cosmic Complexity and the Limits of Scientific Knowledge.* New York: William Morrow, 1985. A superlative account of the reductionist miracle: fantastically complex entities—living, self-reproducing organisms—turn out to be vast arrays of very simply interacting parts. Von Neumann's trick of self-reproduction without infinite regress (adapted from Gödel) is one of the main topics explored here, and with the help of Conway's absorbing game of Life, which plays the starring role in his book, Poundstone does a masterful job of explaining how it comes about.

Racter. *The Policeman's Beard Is Half Constructed.* New York: Warner Books, 1984. "Racter" is a program written by Bill Chamberlain and Thomas Etter. *The Policeman's Beard* is a book written by Racter. It is all somewhat tongue-in-cheek, because Racter does not really know much about half-constructed beards, but what is lovely about Racter's prose is the way it skirts the fringes of meaning, weaving drunkenly across the boundary between sense and senselessness.

Rapoport, Anatol. *Two-Person Game Theory.* Ann Arbor: University of Michigan Press, Ann Arbor Science Library, 1966. A sound treatment of game theory, featuring a most interesting personal discussion on opinions about the meaning of "rationality" in Prisoner's-Dilemma-like situations.

Reps, Paul. *Zen Flesh, Zen Bones.* New York: Doubleday, Anchor Press, n.d. An easily available collection of Zen koans, highly amusing and, perhaps, even enlightening—providing you take it all with sufficiently many grains of salt.

Rogers, Hartley. *Theory of Recursive Functions and Effective Computability.* New York: McGraw-Hill, 1967. A standard reference work on many advanced concepts in metamathematics, including such concepts as "productive" and "creative" sets, referred to in my Chapter 13.

Ross, Alf. "On Self-Reference and a Puzzle in Constitutional Law". *Mind* 78, no. 309 (January 1969): 1–24. A conundrum in the philosophy of law: Can laws modify themselves, or is that paradoxical? Ross' view is that logical inconsistency is unacceptable in law, and therefore that self-amendment is impossible. For a response, see the paper by Hart (above).

Rucker, Rudy. *Infinity and the Mind.* Boston: Birkhäuser, 1982. A book that had to be written. Presenting the most abstruse concoctions of the mind in language that is not abstruse, and connecting it with thoughts about consciousness and the mystery of existence—this is what Rucker excels in.

Ruelle, David. "Les Attracteurs Etranges". *La Recherche* 11, no. 108 (February 1980): 131–44. A good article relating these wispy mathematical clouds to the physics they are supposed to explain, featuring a number of excellent illustrations.

Rumelhart, David E. and Donald A. Norman. "Simulating a Skilled Typist: A Study of Skilled Cognitive-Motor Performance". *Cognitive Science* 6, no. 3 (July–September 1982): 1–36. Anticipating the current "parallel distributed processing" project at the University of California at San Diego, the research described in this article is among the most interesting work on modeling human performance that I have encountered.

Russett, Bruce. *The Prisoners of Insecurity.* New York: W.H. Freeman, 1983. The arms race as an iterated Prisoner's Dilemma, and how we might be able to break out of the deadlock. The last section—"Responsibility"—concludes with this admonition (and how I wish people would take it to heart!): "In a democracy, silence about nuclear issues carries an implication not just of indifference but of acceptance. If we stand silent in the face of an arms race—and the war to which it may lead us—we must share responsibility for the outcome. 'Silence gives consent.' "

Ryder, Frederick, and Company. *Ryder Types,* 2 vols. (with periodic supplements). Chicago: Frederick Ryder and Company. The best catalogue of typefaces I have run across—but it is expensive. Some of the oddest faces I have ever seen are found in the four supplements I own.

Sagan, Carl, ed. *Communication with Extraterrestrial Intelligence.* Cambridge, Mass.: MIT Press, 1973. An entertaining transcript of an international meeting in the days when Russians and Americans spoke to each other. Sagan puts forth his notion of various types of earth-life-based "chauvinisms". A hopeful and intellectually refreshing book—the kind one wishes there were hundreds of.

Sampson, Geoffrey. "Is Roman Type an Open-Ended System? A Response to Douglas Hofstadter". *Visible Language* 17, no. 4 (Autumn 1983): 410–12. Sampson disputes my claim that it is impossible to capture the fluid spirit of letters of the alphabet in parametrized computer subroutines; he suggests that it *is* possible, if you limit your goals to capturing the spirit of letters suitable for printing serious books in.

Schank, Roger. *Dynamic Memory: A Theory of Reminding and Learning in Computers and People.* New York: Cambridge University Press, 1982. Although I disagree with some of his theorizing, I agree fully with Schank's focus on the types of problems that cognitive science ought to be most concerned with, and I like the examples he uses.

Scherlis, William L. and Pierre L. Wolper. "Self-Referenced Referenced, and Self-Referenced". *Communications of the ACM* 23, no. 12 (December 1980): 736. A humorous short note on papers that cite themselves.

Schrödinger, Ernst. *What Is Life?* and *Mind and Matter.* New York: Cambridge University Press, 1967 (reprint of 1944 edition). This philosophically-minded physicist prophetically speculates on the nature of the hereditary message, before the days when DNA's structure or function were known. Also venturing into the deep waters of consciousness, he comes up with this mystical conclusion: "The over-all number of minds is just one."

Schwenk, Theodor. *Sensitive Chaos.* New York: Schocken, 1976. A book about fluids in the wild and in the laboratory, filled with striking patterns and fantastic photographs strongly suggesting this mystical conclusion: "The over-all number of fluids is just one."

Searle, John. "Minds, Brains, and Programs". *The Behavioral and Brain Sciences* 3 (September 1980): 417–57. Like the Lucas article (see above), this one launched a thousand rebuttals (including two by me). In this, its original setting, it was followed by nearly thirty rebuttals and counter-rebuttals. It is amusing and educational to read them all. I view this article as a litmus test, in the sense that someone convinced by Searle's imagery is almost sure to have a very negative opinion of AI.

———. "The Myth of the Computer: An Exchange". *New York Review of Books* (June 24, 1982): 56–57. Searle portrays Dennett and me as blind advocates of "strong AI"—the notion that "the appropriately programmed computer literally has a mind". He resents the fact that such views are "well financed and backed by prestigious teams of research workers", and tells how he is constantly working toward "the relentless exposure of its preposterousness".

———. "The Myth of the Computer". *New York Review of Books* (April 29, 1982): 3–6. A rather negative review of *The Mind's I* by someone roundly criticized in the book. The one good thing Searle does in this review is to stress the central epistemological problem facing AI: to explain the nature of the fluid reference, or *semanticity,* that mental activity exhibits.

Serafini, Luigi. *Codex Seraphinianus.* Milan: Franco Maria Ricci, 1981. A complete pseudo-encyclopedia, in two volumes. The writing system, the page numbering, the weird diagrams, and especially the wonderful color illustrations are all products of the cryptic mind of Serafini, an Italian architect. Also available in one volume, and more inexpensively, from Abbeville Press in New York, N.Y.

Seuss, Dr. *On Beyond Zebra.* New York: Random House, 1955. Humorous ways of extending the alphabet beyond 'z': a metaphor for jumping out of the system ("jootsing").

Simon, Herbert A. "Cognitive Science: The Newest Science of the Artificial". *Cognitive Science* 4, no. 2 (April–June 1980): 33–46. Simon's confident assertion that computers already have what it takes to possess full intelligence: symbol manipulation. In his conclusion, he claims: "Wherever the boundary is drawn, there exists today a science of intelligent systems that extends beyond the limits of any single species."

———. "Studying Human Intelligence by Creating Artificial Intelligence". *American Scientist* 69, no. 3 (May–June 1981): 300–309. This is the lecture in which Simon spoke of the all-important 100-millisecond barrier, below which it is of no interest to cognitive scientists to know what happens (see my Chapter 26). In print he changed "100 milliseconds" to "ten milliseconds", although the point remains the same either way.

Singmaster, David. *Notes on Rubik's "Magic Cube".* Hillside, N.J.: Enslow, 1981. On its cover, this book proudly boasts: " 'The definitive treatise.'—*Scientific American.* " Well, if *Scientific American* says so, who am I to dispute it?

Sloman, Aaron. *The Computer Revolution in Philosophy: Philosophy, Science, and Models of Mind.* Atlantic Highlands, N.J.: Humanities Press, 1978. A book that warns philosophers that they will miss the boat if they don't jump on the AI bandwagon. This a good though somewhat tendentious discussion of the philosophical import of AI.

Smith, Brian C. "Reflection and Semantics in Lisp". Palo Alto, Calif.: Xerox Palo Alto Research Center Report, 1983. A boiling-down of a 500-page Ph.D. thesis into a dozen pages or so, concerning a system capable of reasoning about itself (and reasoning about such reasoning, etc.). This work exemplifies the "meta-meta" style of AI research (see my Chapter 23—especially its *P.S.*).

Smith, Stephen B. *The Great Mental Calculators: The Psychology, Methods, and Lives of Calculating Prodigies.* New York: Columbia University Press, 1983. A set of portraits of very strange yet very human beings whose aberrant minds let them do calculating tasks that we ordinary people—no matter how number-loving—could not conceivably do.

Smolensky, Paul. "Harmony Theory: A Mathematical Framework for Stochastic Parallel Processing". University of California at San Diego Institute for Cognitive Science Technical Report ICS No. 8306. Based on the ideas of statistical mechanics, this project utilizes stochastic parallelism, regulated by a "temperature" that gradually drops to zero, to search most efficiently for the optimal global state of a system.

Smullyan, Raymond. *This Book Needs No Title: A Budget of Living Paradoxes.* Englewood Cliffs, N.J.: Prentice-Hall, 1980. A very humorous collection of observations about the constant intermingling of life and paradox, by a logician whose awareness of paradox is especially keen.

Solo, Dan X. *Sans Serif Display Alphabets.* New York: Dover, 1979. A collection of elegant letterforms whose subdued flair resides exclusively in their gentle curves, stroke taperings, line endings, and the interplay between positive and negative space.

————. *Special-Effects and Topical Alphabets.* New York: Dover, 1978. After you've looked at a book like this, you know why the problem of letterforms is synonymous with the problem of full human intelligence.

Sonneborn, Tracy M. "Degeneracy of the Genetic Code: Extent, Nature, and Genetic Implications". In *Evolving Genes and Proteins,* edited by Vernon Bryson and Henry J. Vogel. New York: Academic Press, 1965. Perhaps the first paper to suggest that there is evolutionary rhyme and reason to the particular match-up between codons and amino acids that exists in the arbitrary-seeming genetic code.

Soppeland, Mark. *Words.* Los Altos, Calif.: William Kaufmann, 1980. A witty collection of words drawn as self-referential pictures: stacks of books whose shapes spell out "books", and so on.

Sorrels, Bobbye. *The Nonsexist Communicator.* Englewood Cliffs, N.J.: Prentice-Hall, 1983. A sizable collection of sexist usages and nonsexist remedies; some of the remedies, however, are needlessly awkward.

Sperry, Roger. "Mind, Brain, and Humanist Values". In *New Views on the Nature of Man,* edited by John R. Platt. Chicago: University of Chicago Press, 1965. The article that asks, and tries to answer, the question "Who pushes whom around in the population of causal forces that occupy the cranium?" My answer is in Chapter 25.

Spinelli, Aldo. *Loopings.* Amsterdam: Multi-Art Points Edition, 1976. A somewhat clumsy realization of a good idea: a book whose pages are all self-inventorying sentences, or close to that. At the end, Spinelli discusses the kinds of locked-in loops and entryways into looping that can arise.

Stanley, H. Eugene et al. "Interpretation of the Unusual Behavior of H_2O and D_2O at Low Temperature: Are Concepts of Percolation Relevant to the 'Puzzle of Liquid Water'?" *Physica.* 106a (1981): 260–77. An article that gives a fairly recent picture of the "flickering-cluster" view of water.

Stein, Gertrude. *How to Write.* Craftsbury Common, Vt.: Sherry Urie, 1977. (Originally published in 1931.) Classic volume of absurdities. It would be a nightmare to try to read the whole thing; each page is like a unique grain of sand, but all the pages taken together add up to a rather bland beach. Better to savor it in small bits.

Steiner, George. *After Babel: Aspects of Language and Translation.* New York: Oxford University Press, 1975. This fascinating exploration of the depths of personal meaning is full of obscurities, but it is still one of the best statements on the subtlety of language that I know.

Strich, Christian. *Fellini's Faces: 418 Photographs from the Archives of Federico Fellini.* New York: Holt, Rinehart, and Winston, 1981. A celebration of the remarkable diversity of human physiognomies, as well as their amazing expressivity.

Stryer, Lubert. *Biochemistry.* New York: W. H. Freeman, 1975. An elegantly produced text of biochemistry, on a slightly lower level than Lehninger. The figures—many in color —are especially appealing.

Suber, Peter. "A Bibliography of Works on Reflexivity". Unpublished manuscript, 1984. An extensive collection of pointers to the vast universe of writings on self-modifying laws, self-referring literature, self-fulfilling prophecies, self-replicating machines, self-monitoring computer programs, and on and on.

———. *The Paradox of Self-Amendment: A Study of Logic, Law, Omnipotence, and Change.* Forthcoming. This is the first (and only) book-length study of logical paradoxes in law. It focuses on one family of paradoxes (rules that authorize change being applied to themselves), but eventually tries to address the general problem of how legal reasoning copes with logical paradox. The game of Nomic is featured in an appendix.

Thomas, Dylan. *Collected Poems.* New York: New Directions, 1953. I am the first to admit I don't understand poetry well, but this beguiling collection, with its combination of beautiful language and bewildering opacity, leaves me especially disturbed.

Thomas, Lewis. *The Medusa and the Snail.* New York: Viking, 1979. Short essays on science and life, some of which are very insightful, others of which I find puzzling or just plain silly.

Turing, Alan. "Computing Machinery and Intelligence". Reprinted in *Minds and Machines,* edited by Alan Ross Anderson. Englewood Cliffs, N.J.: Prentice-Hall, 1964. In which the now-infamous "Turing Test" for machine thought is proposed. Lucid and straightforward, this article contains plenty of provocative ideas, even today.

Turner-Smith, Ronald. *The Amazing Pyraminx.* Hong Kong: Mèffert Novelties, 1981. A simple guide to the notation, group theory, and solution of the Pyraminx puzzle.

Ulam, Stanislaw. *Adventures of a Mathematician.* New York: Scribners, 1976. A fascinating account of the intellectual life of a highly innovative and fun-loving mathematician, including some speculations on mind and consciousness.

Vetterling-Braggin, Mary, ed. *Sexist Language: A Modern Philosophical Analysis.* Totowa, N.J.: Littlefield, Adams & Co., 1981. A valuable collection of articles from many perspectives on sexism. Includes one section on the comparison between sexism and racism.

von Neumann, John. *Theory of Self-Reproducing Automata,* edited and completed by Arthur W. Burks. Urbana, Illinois: University of Illinois Press, 1966. Here von Neumann describes how a machine could construct replicas of itself, or even machines more complex than itself, out of raw materials. At the core of this work by von Neumann, however, is Gödel's method of achieving mathematical self-reference—a fact that many people seem to overlook or downplay.

Walker, Alan, ed. *The Chopin Companion.* New York: W. W. Norton, 1966. An excellent collection of articles by composers, performers, and musicologists on Chopin, his music, and its influence.

Watzlawick, Paul. *Change.* New York: W. W. Norton, 1974. A theory of psychological disorder as caused by encounters with paradox in daily living. The suggested therapy is to fight paradox with paradox (if that isn't paradoxical!).

Webb, Judson. *Mechanism, Mentalism, and Metamathematics.* Hingham, Mass.: D. Reidel, 1980. A penetrating scholarly analysis of the import of the most important metamathematical work by Gödel, Church, Turing, Kleene, Tarski, Post, and others. In particular, it seeks to elucidate the relationship of the famous "limitative theorems" applicable to formal systems with attempts to mechanize human mental activity. In essence, its conclusion is that those results strengthen, rather than diminish, the notion that mental activity can in principle be mechanized.

Wells, Carolyn. *A Nonsense Anthology.* New York: Dover Press, 1958. A wonderful collection of strange poems and silly snippets of pseudo-sense.

Wheelis, Allen. *The Scheme of Things.* New York: Harcourt Brace Jovanovich, 1980. A moving story of human and animal attachments, and the emotional resonances of places and scenes. The story hints at many self-references, most notably to another novel called *The Way Things Are,* written by Wheelis' protagonist Oliver Thompson who, like Wheelis himself, is an unorthodox psychiatrist living in San Francisco. Wheelis within Wheelis!

Williams, J. D. *The Compleat Strategyst.* New York: McGraw-Hill, 1954. One of the earliest and clearest books ever written on game theory. It is a very elementary book, with humorous drawings and droll stories to illustrate its points.

Wilson, E. O. *The Insect Societies.* Cambridge, Mass.: Harvard University, Belknap Press, 1971. It is heavy going to read this book, but parts of it are engrossing, as they tell how order at a high level comes from the independent activity of low-level organisms acting without knowledge of one another.

Winston, Patrick Henry. *The Psychology of Computer Vision.* New York: McGraw-Hill, 1975. Although fairly old by AI standards, this collection of six articles is still of considerable interest. One is Minsky's article on frames; another is an article by Winston on learning and recognition; and another is an article by Waltz on economical strategies for vision in the "blocks world".

Winston, Patrick Henry and Berthold Horn. *Lisp.* Reading, Mass.: Addison-Wesley, 1981. A solid textbook on Lisp, starting right at the beginning and going up to AI programming techniques.

Yaguello, Marina. *Alice au pays du langage.* Paris: Editions du Seuil, 1981. A French linguist examines the phonetics, syntax, semantics, and gradual shifts of her own language to illuminate the strangeness of human thought. Many of the examples in the book are taken from popular language, slang, or humor, which makes it particularly unstuffy.

————. *Les mots et les femmes.* Paris: Petite Bibliothèque Payot, 1978. On fighting the battle against sexist language in French—but unfortunately, because of the two-gender system of romance languages, eradication of sexism will prove far more difficult in those languages than in English (where it is already ferociously hard).

Zapf, Hermann. *About Alphabets.* Cambridge, Mass.: MIT Press, 1960. A famous contemporary type designer describes his adventures among the alphabets of the world.

Acknowledgments

Grateful acknowledgment is made to the following individuals and publishers for permission to quote from the following sources. Every effort has been made to locate the copyright owners of material reproduced in this book. Omissions brought to our attention will be corrected in subsequent editions.

Letters in Chapter 1 reprinted by permission of A. J. Dale and George Brabner.

In Chapter 2, "This Is the Title of This Story, Which Is Also Found Several Times in the Story Itself." Copyright 1982, David Moser.

In Chapter 3: excerpt from Jacques Monod, *Chance and Necessity,* translated by Austryn Wainhouse (New York: Alfred A. Knopf, Inc., 1972) Reprinted by permission. Excerpt from Richard Dawkins, *The Selfish Gene,* (Oxford, UK: Oxford University Press). Copyright 1976. Used by permission. Extract from letter used by permission of Stephen Walton. Excerpt from *The Scheme of Things,* by Allen Wheelis, Harcourt, Brace Jovanovich, 1980. Reprinted with the author's permission. Excerpt from personal communication to Douglas Hofstadter used by permission of Lee Sallows.

Letter in Chapter 4 reprinted by permission of William Popkin.

In Chapter 5: excerpt from Ray Hyman, "Cold Reading: How to Convince Strangers that You Know All About Them," *Skeptical Enquirer,* Spring/Summer 1977. Reprinted by permission. Excerpt from Victor A. Benassi and Barry Singer, "Fooling Some of the People All of the Time", from *Skeptical Enquirer,* Winter 1980/81. Reprinted by permission. Excerpt from personal communication to Douglas Hofstadter used by permission of Marcello Truzzi.

In chapter 7: excerpt from address given by Donald Kennedy. Used with permission.

In Chapter 10: excerpt from personal communication included with the permission of William Huff.

In Chapter 11: excerpt from Gertrude Stein, *How to Write,* copyright © 1973 by the Something Else Press. Excerpt from Samuel Beckett, *Waiting for Godot,* translated from the French by Samuel Beckett. Copyright 1954. Reprinted with permission from Grove Press, Inc., New York. Excerpt from Dylan Thomas, *The Poems of Dylan Thomas.* Copyright 1953 by Dylan Thomas. Reprinted by permission of New Directions Publishing Corporation. Excerpt from Bob Dylan, *Tarantula,* copyright 1971. Reprinted by permission of Macmillan Publishing Company, Inc. "I Sat Belonely," from *In His Own Write,* copyright © 1964 by John Lennon. Reprinted by permission of Simon & Schuster, Inc. "The Faulty Bagnose," from *A Spaniard in the Works,* copyright © 1965 by John Lennon. Reprinted by permission of Simon & Schuster, Inc. Excerpt from William Benton, *Normal Meanings,* copyright © 1979. Paducah, Kentucky, Deer Crossing Press. Used by permission. "When Science Is in the Country," copyright © 1973 by Russell Edson. Reprinted from *The Clam Theater* by permission of Wesleyan University Press. Excerpt from Paul Reps, *Zen Flesh, Zen Bones,* Tokyo, Japan, Charles E. Tuttle Co., Inc., 1957. "Twirl, Twirl", from *Songs of the Pogo.* Copyright © 1956 by Walt Kelly. Reprinted by permission of Simon & Schuster, Inc. Excerpt from *A Humument* by Tom Phillips. Copyright © 1980 by Tom Phillips. Published by Thames and Hudson.

In Chapter 13: pp. 260–287 are from Douglas Hofstadter, "Metafont, Metamathematics, and Metaphysics . . . ," *Visible Language,* XVI, no. 4, pp. 309–338. Excerpt from Geoffrey Sampson, "Is Roman Type an Open-Ended System? A Reply to Hofstadter," on p. 288 originally appeared in *Visible Language,* XVII, no. 4, pp. 410–12. Douglas Hofstadter, "A Reply to Sampson," on pp. 288–92, was first published in *Visible Language,* XVII, no. 4, pp. 413–16. Reprinted by permission from *Visible Language.* Copyright © 1982 & 1983 by *Visible Language,* Box 1972, CMA, Cleveland, OH 44106.

In Chapter 15: excerpt from "Four-Axis Puzzles," copyright © 1982 by Anthony E. Durham.

In Chapter 16: lyrics from "It's All Right with Me," copyright © 1953 by Cole Porter. Copyright renewed, assigned to Robert H. Montgomery, Jr., trustee of the Cole Porter Musical and Literary Property Trust. Chappell & Co., Inc., owner of publication and allied rights. International copyright secured. All Rights Reserved. Used by permission.

In Chapter 21: Douglas Hofstadter, "Review of *Alan Turing: The Enigma* by Andrew Hodges." Copyright © 1983 by The New York Times Company. Reprinted by permission.

In Chapter 23: excerpt from J. R. Lucas, "Minds, Machines, and Godel," in *Minds and Machines,* edited by A. R. Anderson, Englewood Cliffs, New Jersey, Prentice-Hall, 1964. *See also* J. R. Lucas, *The Freedom of the Will,* Oxford University Press, 1970, δ 28, pp. 143–45.

In Chapter 24: lyrics from "Something's Gotta Give," by Johnny Mercer. Copyright © 1954 (Renewed), WB Music Corp. All Rights Reserved. Used with permission.

In Chapter 26: excerpt from *The Medusa and the Snail* by Lewis Thomas. Copyright © 1979 by Lewis Thomas. Reprinted by permission of Viking Penguin Inc.

In Chapter 28: excerpt from Richard Dawkins, *The Selfish Gene,* Oxford, UK, Oxford University Press. Copyright 1976. Used by permission.

In Chapter 29: excerpt from letter reprinted with the permission of Anatol Rapoport. Excerpts from *The Evolution of Cooperation* used with the permission of Robert Axelrod.

In Chapter 33: "The Tumult of Inner Voices" was presented as the Grace Adams Tanner Lecture in Human Values at Southern Utah State College in May, 1982, and was published in that year by the Grace A. Tanner Center for Human Values. Reprinted by permission of the Tanner Center.

Illustrations:

Figure 1–1 originally appeared in *Scientific American,* copyright © January 1981, all rights reserved. Figure 2–2 first appeared in *Scientific American,* copyright © January

Acknowledgments

1982, all rights reserved. Figure 6–1, photo by Ray Atkeson, originally published in *Scientific American*, copyright © May 1982, all rights reserved. Characters in figures 7–1 and 13–1 printed by the Hàn Zì program, developed by David B. Leake and the author at Indiana University. Figures 9–1, 9–4, 9–5, 9–6, 9–7, and 9–8 were printed by the SMUT music-printing program, courtesy of Donald Byrd. Figures 9–1 through 9–8 were first published in *Scientific American*, copyright © April 1982, all rights reserved. Figure 9–2 used by arrangement with G. Schirmer, Inc., publisher and copyright owner. Figures 10–1 through 10–13 and 10–15 reprinted by permission of William Huff. Figures 10–1 through 10–13 originally appeared in *Scientific American*, copyright © July 1983, all rights reserved. Figure 12–1 was first published in Donald B. Knuth, "The Concept of a Meta-Font," in *Visible Language*, XVI, no. 1, p. 15. Reprinted by permission from *Visible Language*. Copyright © 1982 by *Visible Language*, Box 1972, CMA, Cleveland, OH 44106. Figure 12–2 is from Federico Fellini, *Fellini's Faces*, Copyright © 1981 by Diogenes Verlag, AG, Zurich, all rights reserved. Figures 12–3, 13–1, 13–2, 13–4, 13–9, 13–11, and 13–12: copyright © Letraset USA. Reprinted by permission of Letraset USA. Figure 12–1 and 12–3 were originally published in *Scientific American*, copyright © October 1982, all rights reserved. Figures 14–2 and 14–3b originally appeared in *Scientific American*, copyright © March 1981, all rights reserved. Figures 15–1 through 15–13 first appeared in *Scientific American*, copyright © July 1982, all rights reserved. Figures 16–1, 16–5, 16–8, 16–9 and 16–10 were originally published in *Scientific American*, copyright © November 1981, all rights reserved. Figures 16–2, 16–3, and 16–4 first appeared in Mitchell J. Feigenbaum, "Universal Behavior in Nonlinear Systems," *Los Alamos Science*, Volume 1, number 1, Summer 1980, pp. 4-26. University of California, Los Alamos Scientific Laboratory. Figure 16–6 is reprinted from Leo P. Kadanoff, "Roads to Chaos", *Physics Today*, December 1983, p. 51, American Institute of Physics, New York. Originally published in J. Crutchfield, J. Farmer, and B.A. Huberman, *Physics Reports*, Volume 92, pp. 45-82, December 1982. Figure 16–7 reprinted by permission of Schocken Books, Inc. from *Sensitive Chaos* by Theodor Schwenk. Copyright © 1965 by Rudolf Steiner Press. Figure 16–9 originally appeared in *Scientific American*, copyright © November 1981, all rights reserved. Figure 18–1 was first published in *Scientific American*, copyright © March 1983, all rights reserved. Figure 20–2 is from Bryce S. DeWitt and Neill Graham, eds., *The Many-Worlds Interpretation of Quantum Mechanics*, p. 156, copyright © 1973 by Princeton University Press. Reprinted by permission of Princeton University Press. In figure 24–12, the computer graphics programming is courtesy of Phillip G. Apley of Bitstream, Inc. and Richard L. Bryan, Symbolics, Inc. Letter design by David A. Berlow of Bitstream, Inc. Figure 27–6 originally appeared in *Scientific American*, copyright © March 1982, all rights reserved. Figure 29–1 was first published in *Scientific American*, copyright © May 1983, all rights reserved. Figure 30–1 originally appeared in *Scientific American*, copyright © June 1983, all rights reserved. Figure 31–1 used by permission of Bill Mauldin and Wil-Jo Associates, Inc.

Index

Collet, Pierre, 383
Collins, John, 25
color vector on Cube, 361
commandeering the soul, 789–90
Committee for the Scientific Investigation of Claims of the Paranormal (CSICOP), 95–97, 105, 111
common sense, 640; and artificial intelligence, 96; Department of, 94; nonalgorithmic nature of, 96; perception and categorization as roots of, 107–8; as root of science, 93–94; self-applied, 94
commutators, 326, 342, 348, 349
compactness, 578
completeness, *see* consistency-completeness tradeoff
complexity: computational, 361; *see also* compounding complexity
composers: most melodically gifted, 541–42; *see also specific composers*
composite identity of knight's move, 593
compounding complexity: in Lisp, 404; in programming languages, 452
compression: of letterforms in coarse grids, 601; of text, 596
computability, 449
computation, cognition as, 648, 654
computational complexity, 361
computational glasses, 496
computational reality and perceptual shifts, 443
computational rules and epiphenomena, 647
computational vertigo, 486, 487
computer languages: compounding complexity in, 452; grain size and, 449–54; interactive *vs.* noninteractive, 397–98; *see also* Lisp
computer mail, 522
computers: and boredom, 532–34; and creativity, 205–9; don't make mistakes, 508; evolution of, 507; and nonsense, 231; self-modifying, 83; self-watching, 212; speed of, 127–28; and Turing, 489; *see also* programs
computer simulations of evolution, 708
"Concept of Meta-Font, The" (Knuth), 240, 260
"concept" as proto-scientific notion, 234, 254
concepts, 209; chemistry of, 250; and connectivity, 528–29; interdependent, 553; knobbed, 234; nature of, 528; and orbits, 254; physics of, 250; Platonic, 528
conceptual categories, 261–62
conceptual skeletons, 249; of letterforms, *xix*
Concert-goer's Dilemma, 748
cond, 407
cond clauses, 407
conflicting views, balance of, 254, 568
Confrey, Zez, 199
conjugate elements, 316–18
conjugates on Cube, 320

"connectionist" models of mind, 658
connectivity and concepts, 528–29
cons, 402, 403
conscience: elimination of, 762; lack of, elimination of, 764
consciousness, *xxvi;* of animals, 505, 525; flame as prerequisite to, 623; in math-space, 661; in melody-space, 661; mystery of, 487; self-model as prerequisite for, 503; and self-watching, 614–15; splitting, 469–73; synonyms for, 631; and transcendence of Gödel's theorem, 536
conscious *vs.* unconscious slippage, 237–38, 254
consistency-completeness tradeoff, 8, 58, 263–65, 475
"Consternation" (Grady), 199
constitutive rules *vs.* rules of skill, 77
constraints: and creativity, 598; desires as, 616, 622
constructive categories, 539–40
content: as fancy form, 22, 445; reading form from, 109
Conway, John, 314, 352
Conway-Berlekamp-Guy nomenclature, 352
Coombs, Stephen, 34
Cooper, Harold, 36
cooperation: *vs.* defection, 756–57; defined, 717; emergence of, 729; evolution of, 715–30; general definition of, 757; logic of, 746–47; stampede toward, 733; survival of, 729; voluntary, extinction of, 762
Copernicus, Nicholas, 108
Coppélia (ballet), 520
Copycat project, 285, 563–85, 658, 766; alphabet in, 570–72; basic theme of, 574; crux of, 563–67; group-types in, 573; insight into insight via, 585; richness of, 602
copy-group, 573
Corps a ses raisons, Le, 587
CORTEX, 562
Cortot, Alfred, 186
Cosby, Bing, 273
Cosmic View: The Universe in Forty Jumps (Boeke), 130–31
counterfactual conditionals, 13–14, 36, 139, 189, 197, 239, 258, 448; plausibility of, 258
counterfactuality and slop, 629
counterfactual self-referential questions, 36
counterfactual worlds, 232
counterparts, 567; and counterroles, 568
counterpoint, *xxviii;* in writing, *xxvi–xxvii*
counterroles, 567; and counterparts, 568
crackpotism, 224; detection of, 108
"Crazy Cogs" (Larson), 201
creative mechanisms, universality of, *xxix–xx*
creative processes, modeling of, *xxix–xx*
creative sets, 540
creative spark, 526, 527

Index

The text of this book is printed in Baskerville, and the display face is Raleigh.